Novel Hybrid Intelligence Techniques in Engineering

Novel Hybrid Intelligence Techniques in Engineering

Editors

Danial Jahed Armaghani
Yixia Zhang
Pijush Samui
Ahmed Hussein Kamel Ahmed Elshafie
Aydin Azizi

MDPI • Basel • Beijing • Wuhan • Barcelona • Belgrade • Manchester • Tokyo • Cluj • Tianjin

MDPI

Editors
Danial Jahed Armaghani
University of Technology Sydney
Australia

Yixia Zhang
Western Sydney University
Australia

Pijush Samui
NIT Patna
India

Ahmed Hussein Kamel
Ahmed Elshafie
University of Malaya
Malaysia

Aydin Azizi
Oxford Brookes University
UK

Editorial Office
MDPI
St. Alban-Anlage 66
4052 Basel, Switzerland

This is a reprint of articles from the Special Issue published online in the open access journal *Applied Sciences* (ISSN 2076-3417) (available at: https://www.mdpi.com/journal/applsci/special_issues/Intelligence_Techniques_Engineering).

For citation purposes, cite each article independently as indicated on the article page online and as indicated below:

LastName, A.A.; LastName, B.B.; LastName, C.C. Article Title. *Journal Name* **Year**, *Volume Number*, Page Range.

ISBN 978-3-0365-7106-5 (Hbk)
ISBN 978-3-0365-7107-2 (PDF)

Cover image courtesy of Petai Jantrapoon

Contents

About the Editors

Danial Jahed Armaghani

Dr. Danial Jahed Armaghani is a lecturer and researcher in Civil Engineering. He received his postdoctorate degree from Amirkabir University of Technology, Iran, in 2018, and his Ph.D. and master's degrees in civil geotechnics from Universiti Teknologi Malaysia, Malaysia, in 2015 and 2012, respectively. His areas of research are mechanised tunnelling, rock mechanics, piling technology, concrete and green concrete technology, transportation geotechnics, blasting environmental issues, reliability and risk assessment, applying artificial intelligence, and optimisation algorithms in different areas of civil engineering. Dr. Armaghani is internationally renowned for his significant number of publications and citations. He has published over 250 papers in top-ranked journals and conferences in the field of civil engineering. Based on his excellent academic achievement during his PhD and the years following, Dr. Armaghani has been recognised with several prestigious international awards. He was among the World's Top 2% of Scientists in 2020, 2021, and 2022, according to Stanford University. Dr. Armaghani has served or currently serves as an Associate Editor, Editorial Board Member and Guest Editor for several well-established journals.

Yixia Zhang

Yixia Zhang is a Professor of Engineering at the School of Engineering, Design and the Built Environment at Western Sydney University. She is the Deputy Director of the Urban Transformation Research Centre, and Discipline Lead of Civil and Environmental Engineering. Before moving to WSU in 2019, she worked at the University of New South Wales for 15 years, 12 of which were spent at UNSW, Canberra. She received her PhD in structural engineering from the University of Hong Kong in 2001. Prof. Zhang is a research leader in construction sustainability and structural resilience, and has strong expertise in computational mechanics and advanced composite materials and structures, including construction and building materials in civil engineering and composites in mechanical and aeronautical engineering. She has published one book, edited four books, and has 24 book chapters and over 360 peer-reviewed scholarly research papers (including more than 170 research papers in top international journals) to her name. She has been awarded research grants of over AUD 14 million from various funding schemes, including the ARC, defence, industries and government.

Pijush Samui

Pijush Samui is a working professor in the Civil Engineering Department of NIT Patna, India. He received his PhD in Geotechnical Engineering from the Indian Institute of Science Bangalore in 2008. His research interests include geohazards; earthquake geotechnical engineering; concrete technology, pile foundation and slope stability and application of AI in civil engineering. He has published more than 21 books, 32 book chapters and over 200 research papers in high impact factor journals as well as 30 conference proceedings.

Ahmed Hussein Kamel Ahmed Elshafie

A. El-Shafie received his B.Sc. and M.Sc. from the Department of Civil Engineering from Cairo University, Giza, Egypt in 1993 and 1998, respectively. In 2003, he received his Ph.D. in water resources management and planning from the Department of Civil Engineering, Cairo University under a collaborative academic channel program with the Civil Engineering Department, University of Calgary; Calgary, Alberta, Canada. Between 2004 and 2007, he was a postdoctoral fellow at the Department of Electrical and Computer Engineering at both Royal Military College of Canada and Queen's University, Kingston, ON, Canada. Between 2007–2015, he is an Associate Professor with the Smart Engineering System, Department of Civil & Structural Engineering, University Kebangsaan Malaysia; Malaysia. Presently he is a Professor with the Department of Civil Engineering, University of Malaya. His research interest is related to artificial intelligence techniques with their applications to several engineering applications giving emphasis to hydrological process, environmental and water resources, dam and reservoir operation, and multi-sensor system integration. He has more than 350 research manuscripts that have been published in highly prestigious scientific journals and published 5 books as well.

Aydin Azizi

Dr. Aydin Azizi holds a PhD degree in Mechanical Engineering-Mechatronics, an MSc in Mechatronics and a BSc in Mechanical Engineering-Heat & Fluids. Certified as the Fellow of the Higher Education Academy (FHEA), official instructor for the Siemens Mechatronic Certification Program (SMSCP) and the editor of the book series Emerging Trends in Mechatronics publishing by Springer Nature Group, he currently serves as a Senior Lecturer at the Oxford Brookes University. His current research focuses on investigating and developing novel techniques to model, control and optimize complex systems. Dr. Azizi's areas of expertise include Control & Automation, Artificial Intelligence and Simulation Techniques. Dr. Azizi is the recipient of the National Research Award of Oman for his AI-focused research, DELL EMC's "Envision the Future" competition award in IoT for "Automated Irrigation System", and 'Exceptional Talent' recognition by the British Royal Academy of Engineering.

Preface to "Novel Hybrid Intelligence Techniques in Engineering"

The focus of this book is the development of novel intelligence techniques for solving various problems in engineering. These techniques, due to their ability to create complex relationships between dependent and independent variables, can be implemented in a faster and more reliable way. Such techniques utilise algorithms/approaches such as artificial neural networks, fuzzy logic, evolutionary theory, learning theory, and probabilistic theory, making them a suitable and useful fit for real-life complex problems. This book introduces the process of selecting, applying, and developing such techniques in different engineering designs and applications. In addition, the validation process of intelligence systems as an alternative is discussed in this book. Overall, this book forms an excellent introduction to these systems for engineers who are not familiar with them.

Danial Jahed Armaghani, Yixia Zhang, Pijush Samui, Ahmed Hussein Kamel Ahmed Elshafie, and Aydin Azizi
Editors

applied sciences

MDPI

Article

Modeling the Impact of Liquid Polymers on Concrete Stability in Terms of a Slump and Compressive Strength

Ahmed Salih Mohammed [1,2,*], Wael Emad [3], Warzer Sarwar Qadir [1], Rawaz Kurda [4,5,6,*], Kawan Ghafor [1] and Raed Kadhim Faris [7]

1 Civil Engineering Department, College of Engineering, University of Sulaimani, Kurdistan Region, Sulaymaniyah 46001, Iraq
2 Engineering Department, American University of Iraq, Sulaimani, Sulaymaniyah 46001, Iraq
3 Department of Civil Engineering, Komar University, Sulaymaniyah 46001, Iraq
4 Department of Highway and Bridge Engineering, Technical Engineering College, Erbil Polytechnic University, Erbil 44001, Iraq
5 Department of Civil Engineering, College of Engineering, Nawroz University, Duhok 42001, Iraq
6 CERIS, Civil Engineering, Architecture and Georresources Department, Instituto Superior Técnico, Universidade de Lisboa, Av. Rovisco Pais, 1049-001 Lisbon, Portugal
7 Gasin & Tasluja Cement Company, Sulaymaniyah 46001, Iraq
* Correspondence: ahmed.mohammed@univsul.edu.iq (A.S.M.); rawaz.kurda@epu.edu.iq or rawaz.kurda@tecnico.ulisboa.pt (R.K.); Tel.: +964-770-158-8695 (A.S.M.)

Abstract: It is generally known that the two most crucial elements of concrete that depend on the slump value of the mixture are workability and compressive strength. In addition, slump retention is more delicate than the commonly used slump value since it reflects the concrete mixture's durability for usage in civil engineering applications. In this study, the effect of three water-reducer additives was tested on concrete's workability and compressive strength from 1 day to 28 days of curing. The slump of the concrete was measured at the time of adding water to the mix and after 30 min of adding water. This study employed 0–1.5% (%wt) water-reducer additives. The original ratio between water and cement (wc) was 0.65, 0.6, and 0.56 for mixtures incorporating 300, 350, and 400 kg of cement. It was lowered to 0.3 by adding water-reducer additives based on the additives type and cement content. Depending on the kind and amount of water-reducer additives, w/c, gravel content, sand content, crushed content, and curing age, adding water-reducer additives to the concrete increased its compressive strength by 8% to 186%. When polymers were added to the concrete, they formed a fiber net (netting) that reduced the space between the cement particles. As a result, joining the cement particles quickly enhanced the fresh concrete's viscosity and the hardened concrete's compressive strength. The study aims to establish mathematical models (nonlinear and M5P models) to predict the concrete compressive strength when containing water-reducer additives for construction projects without theoretical restrictions and investigate the impact of mix proportion on concrete compressive strength. A total of 483 concrete samples modified with 3 water-reducer additives were examined, evaluated, and modeled for this study.

Keywords: concrete; water-reducer contents; workability; compressive strength; slump retention

Citation: Mohammed, A.S.; Emad, W.; Sarwar Qadir, W.; Kurda, R.; Ghafor, K.; Kadhim Faris, R. Modeling the Impact of Liquid Polymers on Concrete Stability in Terms of a Slump and Compressive Strength. *Appl. Sci.* **2023**, *13*, 1208. https://doi.org/10.3390/app13021208

Academic Editor: Asterios Bakolas

Received: 27 December 2022
Revised: 10 January 2023
Accepted: 13 January 2023
Published: 16 January 2023

1. Introduction

Cement, fine, and coarse aggregates are combined with water to make a composite material called concrete [1]. Concrete is a flexible material in a fresh condition that can be quickly blended to fit a range of particular demands and molded into almost any shape. Ordinary portland cement is the most often used cement for manufacturing concrete [2]. The study of concrete characteristics and their practical applications are covered by concrete technology [3]. Floors, columns, beams, slabs, and other load-bearing components are made of concrete in building construction [4–6].

Chemical admixtures known as water-reducer additives are added to concrete mixtures to lower the water content or slow the concrete setting rate while preserving the mixture flowability. Several liquid and powdered water reduction additives are available [7,8]. Chemically, water-reducer additives fall into three groups. First, Sulfonate Naphthalene Formaldehyde (SNF); then, Formaldehyde Sulfonate (SMF); and finally, sulfonate and carboxylic copolymers [9–11]. Polycarboxylate ether (PCE) (high-scale water reduction) is one of the most common water-reducer additive types [12]. Through the adsorption and dispersion of cement components, water-reducer additives are active in the cement waterways network [13–15]. Water-reducer additives improve concrete flowability by dispersing agglomerated cement particles [16].

Concrete compressive strength, a significant mechanical characteristic, is often measured using concrete specimens after a standard curing period of 28 days. Various factors affect the strength of concrete, including cement strength, water content, w/c, and aggregate quality. The conventional method for modeling the impact of these factors on the concrete compressive strength begins with an assumed form of an analytical equation and is followed by a regression analysis utilizing experimental data to identify the equation's parameters [17]. Polymers are one of the chemical admixtures used to improve the properties of fresh and hardened concrete [15–18]. Polymers affect cement setting times, hydration, flowability, and strength. Many types of polymers are present in liquid and powder forms. Polycarboxylate (PCE) (high-scale water reduction) is one of the most common polymer types [19]. The currently available superplasticizers can be divided into three categories according to the chemical compound. The first is condensed with Sulfonate Naphthalene Formaldehyde (SNF), the second Formaldehyde Sulfonate (SMF), and the last is made up of sulfonate and carboxylic copolymers, for example, Polycarboxylate Superplasticizers (PC) in the Sulphonate group. Concrete quality and durability can be significantly enhanced with PC superplasticizers [20]. Superplasticizers are activated in the cement waterways network by adsorption and dispersion of cement parts. The main way in which polymers increase the flowability of concrete is to disperse agglomerated cement particles. The fluidity of superplasticizers depends mainly on their adsorption on concrete surfaces [21–27]. The effects of polymers (Polycarboxylate–Superplasticizer) in liquid form have been studied to enhance concrete's mechanical properties, such as compressive strength [12]. There are several methods for modeling the properties of materials, including computational modeling, statistical techniques, and recently developed tools such as regression analyses and Artificial Neural Networks (ANN) [33]. Multilinear regression analysis, M5P-tree, and ANN are techniques widely used to solve problems in construction project applications [18–22].

Nonlinear regression, multilinear regression analysis, and M5P-tree are construction problem-solving methodologies [28–30]. M5P-tree was initially introduced by [31]. This tree technique adapts to each sub-location by classifying or dividing data into various spaces. Error is estimated using each node's M5P-tree tree division criterion. Variance measures class mistakes. Any node function uses the attribute that minimizes errors. The M5P-tree tree division criterion is the error computations per node. Node-class standard deviation calculates M5P error. Node division reduces errors by evaluating each node's characteristics. Parent nodes have more StDev than child nodes (more significant nodes). Choose the structure with the best error-reduction potential. This split is tree-like. Second, linear regression functions replace the clipped sub-trees. Thus, the effect of numerous parameters such as water-reducer content, w/c, and curing duration of 1 day to 28 days was quantified using nonlinear regressions, multi-regression, and M5P-tree-based approaches to forecast concrete compressive strength, utilizing 483 tested samples for each model.

Research Significance

The main objective of this study is to propose two systematic multiscale equations to estimate the maximum stress of concrete modified with polymers. Thus, experimental data of 483 tested samples using three different types of liquid polymer with polymer contents, mix proportion, curing period, and the water-to-cement ratio was considered with different analysis approaches. (i) The effect of polymers on the slump retention and compression strength of concrete is investigated and quantified in the early curing period (ii) to guarantee the construction industry to use the proposed models without any experimental work, and (iii) to quantify and propose a systematic multiscale model to predict the compression strength of concrete containing small amounts of polymers (up to 1.5%) with various water-to-cement ratios and curing time up to 28 days.

2. Materials and Methods

2.1. Ordinary Portland Cement

This investigation used ordinary portland cement (OPC) from the Gasin Cement Company in Sulaimani, Iraq. Table 1 summarizes the chemical and mineralogical constitution of the OPC.

Table 1. Composition of the ordinary portland cement.

Chemical composition	CaO	63.9%
	SiO_2	20.1%
	Al_2O_3	4.08%
	Fe_2O_3	5.10%
	MgO	1.48%
	SO_3	2.20%
	LOI	3.41%
Mineralogical composition	Ca_3SiO_5	66.3%
	Ca_2SiO_4	7.67%
	$Ca_3Al_2O_6$	2.19%
	$Ca_4Al_2Fe_2O_{10}$	15.5%

2.2. Aggregate

In this study, natural sand was used. Crushed stone was used as fine aggregate, and gravel passing a sieve of 20 mm was used as coarse aggregate.

2.3. Additives

In this study, three additives were used to enhance the compressive strength of concrete. SP62 is a liquid brown Polycarboxylic ether. It is a highly concentrated fluidizing admixture. An admixture can obtain a homogeneous mixture with minimized frictional forces between the mixed components. RC897 is a superplasticizer that produces high-quality ready-mix and precast concrete with reduced water needs and high workability retention. This water-reducer extends processing time and meets industry requirements. PC180 is a high-performance superplasticizer that was purposefully designed for concretes having high consistencies and low w/c ratios in precast applications. In this study, up to 1.5% of the additives were used. The properties of the three types of additives are summarized in Table 2.

Table 2. Properties of the additives.

Additives	SP62	RC897	PC180
Color	Brown	Light yellow	Amber
State	Liquid	Liquid	Liquid
Density, (gm/cm^3)	1.1	1.08 ± 0.02	1.07 ± 0.02
pH	-	4.5 ± 1.0	5 ± 1
Chloride content	0.1%	<0.10 mass-%	<0.10 mass.-%
Alkali content (Na_2O equivalent)	<8.5 mass-%	<8.5 mass-%	<8.5 mass-%

2.4. Slump

In this study, the additives' consistency and effectiveness on the concrete mixes' flowability according to ASTM C143 and EN-12350 were assessed using a concrete slump test (Figure 1a,b). The slump values of the modified concrete with additives and the control sample ranged from 200 to 220 mm. In order to assess the effectiveness of the additives on the workability of the concrete to the control sample, slump retention was also carried out.

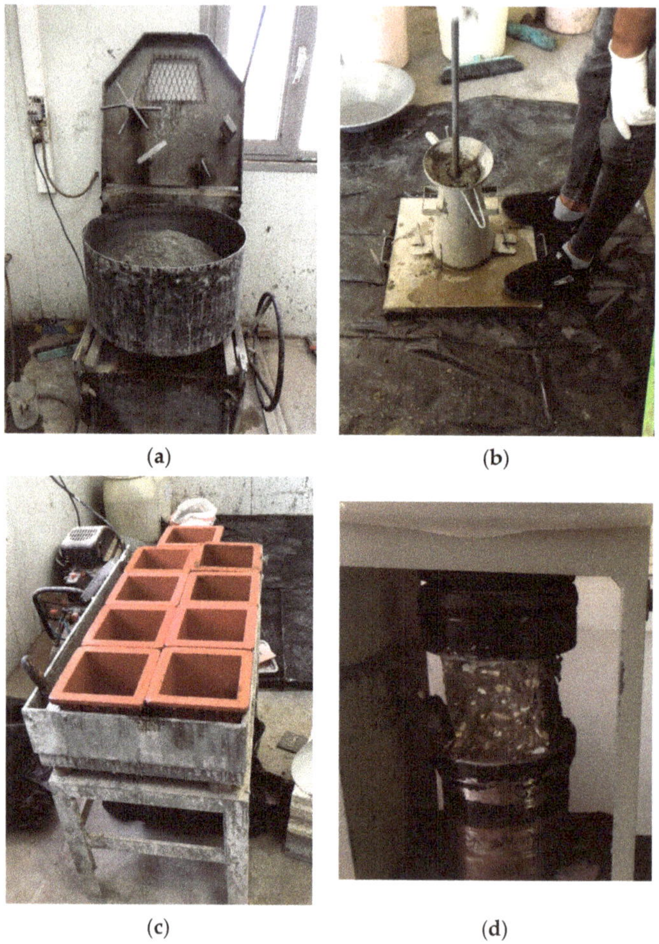

Figure 1. Experimental work (**a**) concrete mixer, (**b**) slump test, (**c**) cubic molds, (**d**) compressive strength test.

2.5. Compressive Strength

For this investigation, a cube sample (150 × 150 × 150 mm) was employed (Figure 1c). There was a (0.5 MPa)/sec loading speed. Based on EN-12390-3 [8], the three-sample average was chosen as the concrete strength for the analysis during a specific curing period (Figure 1d).

2.6. Concrete Mix

The range of additive content was 0% to 1.5%. Due to the addition of the additives, less water was used to make the mixture, and the w/c ratio was gradually lowered so that the slump value remained between 200 and 220 mm. The specimens were kept in water with a humidity level of 95 percent and a temperature of 25 °C for the appropriate curing age. Table 3 provides a summary of the concrete mixtures. The slumps were controlled between 200 to 220 mm, and 0, 0.5, 0.75, 1, 1.25, and 1.5% of the three additives, such as SP62, PC180, and RC897, were used (Table 4).

Table 3. Concrete mix design.

Materials	Mix 1	Mix 2	Mix 3
Cement, kg	300	350	400
Coarse aggregate, kg	788	669	557
Crushed stone, kg	98	96	186
Sand, kg	1083	1145	1115
Water, kg	195.73	221	225.3

Table 4. Impact of the additives on the workability of concrete.

		Slump Retention, mm					
Cement, kg	Additive, %	SP62		PC180		RC897	
		10 min.	30 min.	10 min.	30 min.	10 min.	30 min.
300	0	210	208	200	190	210	210
	0.5	200	80	200	80	200	60
	0.75	210	100	210	90	210	80
	1	200	85	200	80	210	80
	1.25	200	120	215	100	210	90
	1.5	200	90	210	95	210	90
350	0	210	215	210	215	210	215
	0.5	200	90	200	90	200	90
	0.75	200	0	200	100	210	100
	1	220	90	210	110	200	90
	1.25	210	130	210	80	200	100
	1.5	200	100	220	130	210	110
400	0	200	100	200	100	200	100
	0.5	210	80	200	80	210	70
	0.75	205	90	210	90	220	90
	1	220	100	210	120	210	100
	1.25	210	110	210	50	210	115
	1.5	200	90	215	70	220	80

2.7. Modelling

A total of 483 datasets (161 samples for each polymer) containing tested results for each modification were examined. The water–cement ratio (w/c), curing age (t, days), cement content (C, kg), gravel content (G, kg), sand content (S, kg), crushed stone content (CRS, kg), curing time (t, days), and the additives' content (Add.,%) are all included in the set of input data, with the tested compressive strength (MPa) of the concrete provided as the target value.

2.7.1. Nonlinear Regression Model

To develop a nonlinear regression model, the following formula (Equation (1)) can be considered a general form [2,8,12]. Equation (1) represents the interrelation between the variables to estimate the compressive strength of the conventional and concrete components.

$$\sigma_c = \beta_1 \times w/c^{\beta_2} + \beta_3 \times C^{\beta_4} + \beta_5 \times S^{\beta_6} + \beta_7 \times CRS^{\beta_8} + \beta_9 \times G^{\beta_{10}} + \beta_{11} \times t^{\beta_{12}} + \beta_{13} P^{\beta_{14}} \quad (1)$$

2.7.2. M5P Model

One of the most significant advantages of model trees is their ability to efficiently solve problems, dealing with many data sets with a substantial number of attributes and dimensions. They are also noted for being powerful while dealing with missing data [31]. The M5P-tree approach establishes a linear regression at the terminal node by classifying or partitioning diverse data areas into numerous separate spaces. It fits on each sub-location in a multivariate linear regression model. The error is estimated based on the default variance value inserted into the node. The general formula for the M5P-tree model is shown in Equation (2).

$$\sigma_c = \beta_1 \times \left(\frac{w}{c}\right) + \beta_2 \times (C) + \beta_3 \times (S) + \beta_4 \times (CRS) + \beta_5 \times (G) + \beta_6 \times (C.T) + \beta_7 \times (P) + \beta_8 \quad (2)$$

w/c: ratio of water-to-cement content
C: cement content
S: sand content
CRS: crushed stone content
G: gravel content
t: curing time
P.: additive (SP62 or PC180 or RC897) ranged from 0% to 1.5 and β_1 to β_{14} are modeli paramters (Tables 5 and 6).

Table 5. NLR model paramters.

Model Parameter	Additive		
	SP62	RC897	PC180
β_1	52.60	282.2	303.8
β_2	−0.491	−0.13	−0.116
β_3	652.5	298	273
β_4	0.006	0.029	0.051
β_5	2.018	2.017	2.01
β_6	−1.33	−1.33	−1.36
β_7	−33.1	69.07	124.8
β_8	−0.125	0.008	−0.248
β_9	1.297	1.467	1.467
β_{10}	0.303	−0.37	−0.377
β_{11}	−720	−712	−715
β_{12}	−0.008	−0.009	−0.01
β_{13}	0.209	2.229	2.120
β_{14}	2.00	0.574	0.634
R^2	0.89	0.92	0.94
RMSE (MPa)	4.220	3.867	3.556

Table 6. M5P-tree model paramters.

Additive	$\sigma_c=\beta_1\times(\frac{w}{c})+\beta_2\times(C)+\beta_3\times(S)+\beta_4\times(CRS)+\beta_5\times(G)+\beta_6\times(C.T)+\beta_7\times(P)+\beta_8$										
	LM Number	β_1	β_2	β_3	β_4	β_5	β_6	β_7	β_8	R^2	RMSE (MPa)
SP62	1	78.4	0.0051	0	0	0	0.3114	5.491	−15.22	0.91	3.784
	2	74.03	0.0051	0	0	0	0.3114	5.491	−13.45		
	3	51.98	0.0051	0	0	0	0.3114	5.967	−9.737		
	4	38.08	0.0127	0	0	0	0.888	5.499	0.2689		
	5	−7.795	−0.0006	0	0	0	0.331	3.6161	36.01		
	6	−7.795	−0.006	0	0	0	0.29	3.174	36.02		
	7	−58.81	0.0122	−0.0153	0	0	0.2877	0.4122	65.65		
	8	−50.1	0.0122	−0.0153	0	0	0.2877	0.4122	61.94		
	9	−61.61	0.0122	0	0	0	0.2877	0.4122	49.89		
	10	−77.31	0.0227	0	0	0	0.4121	0.4122	61.9		

Table 6. *Cont.*

Additive	LM Number	β_1	β_2	β_3	β_4	β_5	β_6	β_7	β_8	R^2	RMSE (MPa)
		$\sigma_c=\beta_1\times(\frac{w}{c})+\beta_2\times(C)+\beta_3\times(S)+\beta_4\times(CRS)+\beta_5\times(G)+\beta_6\times(C.T)+\beta_7\times(P)+\beta_8$									
PC180	1	−77.311	0.0227	0	0	0	0.4121	0.4122	61.9	0.92	4.00
	2	−98.45	0.052	0	0	0	0.278	0	61.51		
	3	−112.6	0.0261	0	0	0	0.3861	0	84.71		
	4	−93.21	0.0326	0	0	0	0.3681	0	64.14		
	5	−106.8	0.0335	0	0	0	0.4932	0	76.37		
RC897	1	−60.77	0	0.0374	0	0	2.654	1.353	9.7022	0.96	2.846
	2	−50.87	0	0.041	0	0	2.654	1.125	2.481		
	3	−49.95	0	0.0257	0	0	2.654	1.584	21.82		
	4	−60.34	0	0.0257	0	0	2.654	1.743	26.2		
	5	−73.07	0	0.0228	0	0	4.13	0.8277	31.23		
	6	−37.81	0	0.0629	0	0	0.4122	3.87	−20.7		

2.8. Performance Evaluation and Model Criteria

To assess the accuracy and efficacy of the model predictions, the coefficient of determination (R^2), root mean squared error (RMSE), and mean absolute error (MAE) were used. The reliability of the suggested models and the effect of mix proportions on the concrete compressive strength were investigated using the nonlinear and M5P models, which were evaluated using several common assessment criteria. Their equations are as follows:

$$R^2 = \left[\frac{\sum_{p=1}^{p}(y_i - \overline{y})(x_i - \overline{x})}{\sqrt{\left[\sum_{n=1}^{n}(y_i - \overline{y})^2\right]\left[\sum_{p=1}^{p}(x_i - \overline{x})^2\right]}} \right]^2 \quad (3)$$

$$RMSE = \sqrt{\frac{\sum_{n=1}^{n}(y_i - x_i)^2}{n}} \quad (4)$$

$$MAE = \frac{\sum_{p=1}^{p}|(y_i - x_i)|}{n} \quad (5)$$

yi = laboratory-tested values; xi = estimated value; $\overline{y}$ = average of yi; $\overline{x}$ = average of xi, and n is the number of datasets.

3. Results and Analysis

3.1. Water-Reducing Additives

In this research paper, three types of additives (SP62, PC180, and RC897) were used to enhance the performance of the concrete. The additives content ranged from 0 to 1.5%. Adding the additives reduced the water in the mixture, and the w/c ratio gradually decreased, thus keeping the slump value in the range of 200–220 mm. Regarding the concrete mixture, which contains 300 kg of cement, an addition of 0.5% of SP62 reduced the mixture's water content by 12.6%, while it was reduced by 17.6% and 9.1% when modified with 0.5% of PC180 and RC 897, respectively. Compared with 300 kg and 400 kg cement content in the mixture, the percentage of water-content reduction was higher for the mixture containing 350 kg of cement for the three types of additives, as shown in Figure 2. By increasing the content of the additive, the water-content reduction gradually increased (Figure 2). Modified the concrete with the SP62, PC180, and RC897 decreased the water content required to achieve the desired workability by 9.1% to 46.7%, based on the types and content of additives and based on the cement content, as shown in Figure 2.

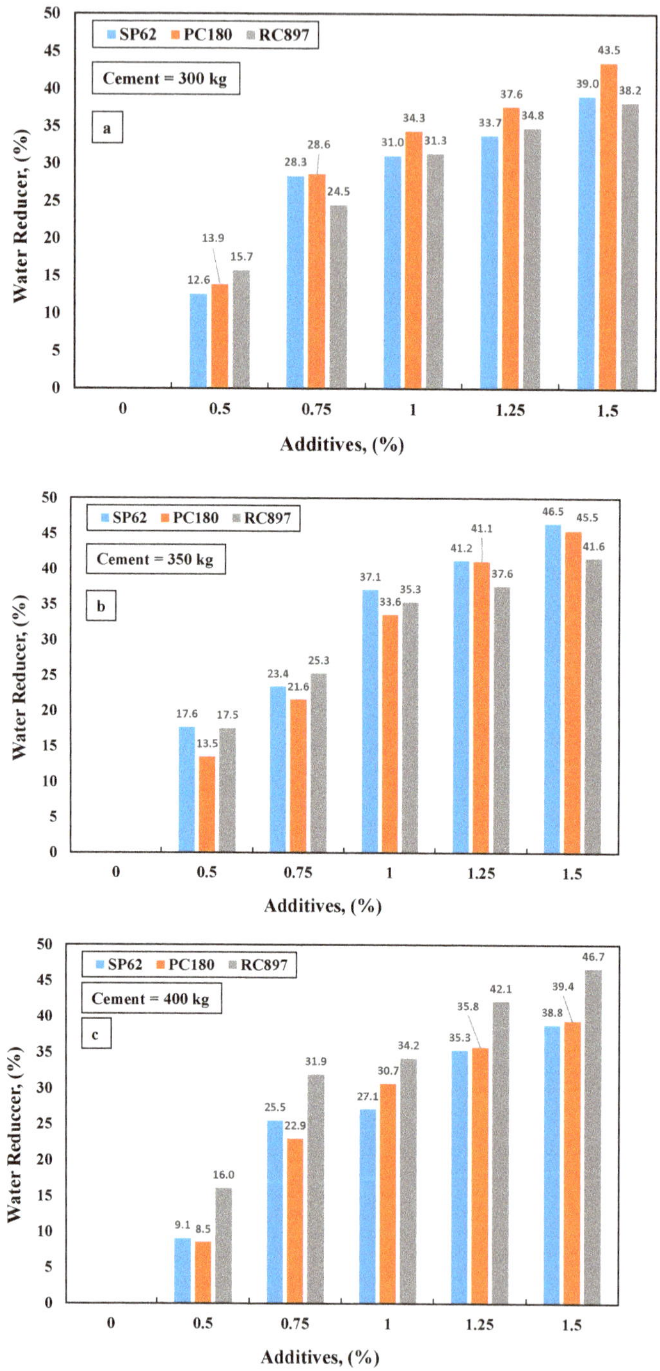

Figure 2. Percentage of water reduction caused by the addition of 3 types of additives in concrete mixes with (**a**) 300 kg, (**b**) 350 kg, and (**c**) 400 kg of cement.

3.2. Slump Retention (ASTM C 143-90)

Fresh concrete loses its workability due to stiffening with time—a well-known phenomenon called "slump loss". The consistency changes because chemical and physical factors brought about by early hydration gradually reduce the system's free water and build up the inner skeleton structure. It is well known that the workability of concrete in the concrete industry faces slump loss, which is different for various grades of concrete. Slump loss also varies with time. A study must determine the factors affecting slump loss in the concrete mix. Factors such as cement content, water content, admixtures, weather, and concrete volume influence the workability loss rate. The main objectives of this project are to study the variation of a slump with the time of transportation, which is dependent on the slump value of the concrete mixture.

Moreover, slump retention is the most sensitive compared to a well-known slump value because it represents the durability of the concrete mixture for its applications in civil engineering. Slump loss is the rapid stiffening of fresh concrete. Slump loss becomes significant when polymers are used with cement. The stiffening of concrete becomes accelerated under hot climates. This is due to the evaporation of mixing water, hydration of cement, and even water absorption by the aggregates. Retarders lower the rate of hydration of cement. The concrete compressive strength linearly increases with a mixing time of up to 180 min. This increase was 10% after mixing for 180 min [3]. The dispersant remaining in the aqueous phase can influence slump retention. Rapidly adsorbed dispersant from the aqueous phase has a higher rate of slump loss than that was absorbed more slowly from the aqueous phase [7]. The slump loss in the field can be regained by redosing the polymer in the concrete. Besides enhancing the concrete compressive strength, monitoring the slump retention of the fresh concrete modified with water-reducer additives is necessary. In this study, slump retention of the fresh concrete modified with SP62, PC180, and RC897 was monitored when adding water to the mixture and after 30 min of adding water, as summarized in Table 4. The slump of the fresh concrete with and without water-reducer additives was controlled between 200 mm and 220 mm. After 30 min, the concrete modified with water-reducer additives lost its workability (Table 4). Workability loss is affected by cement, water, admixtures, weather, concrete volume, and other factors. The rapid stiffening of fresh concrete is known as slump loss. A hot environment accelerates concrete stiffening due to the evaporation of mixing water, cement hydration, and aggregate water absorption [32]. There were many ways to control the slump loss of fresh concrete. One of the methods was by adding retarder admixture to the mix. By slowing the cement's rate of hydration, retarding admixtures delay the setting. As a result, the water combined with cement decreases due to the decreased hydration rate throughout a particular period. The slump loss in such a mix for a specific period will be significantly lower than that without a retarder [32,33].

Modifying the concrete with water-reducer additives enhances the concrete compressive strength from 1 day up to 28 days of curing for 3 different contents of cement (300, 350, and 400 kg), as shown in Figures 3–5. For the mixture containing 300 kg of cement at 1 day of curing, the compressive strength was 11.41 MPa, while it was 16.52 MPa and 20.17 MPa for 350 and 400 kg of cement, respectively. Regarding the mixture containing 300 kg of cement, adding 1% of SP62, PC180, and RC897 enhanced the concrete compressive strength by 104%, 150%, and 129%, respectively, as shown in Figure 2. While it was 97%, 141%, and 150%, the mixture contained 350 kg of cement (Figure 3). The growth percentage decreased when the mixture contained 400 kg of cement modified with 1% water-reducer additives (Figure 4).

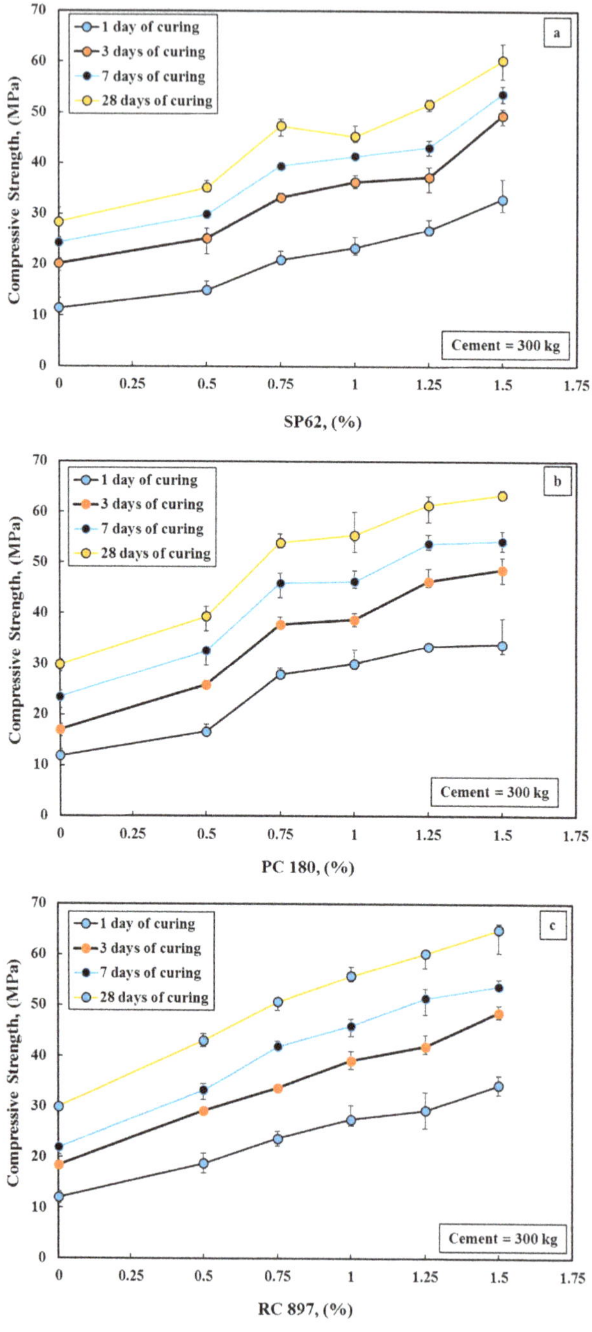

Figure 3. Additives' impact on the concrete compressive strength using 300 kg cement (**a**) SP62, (**b**) PC 180, and (**c**) RC 897.

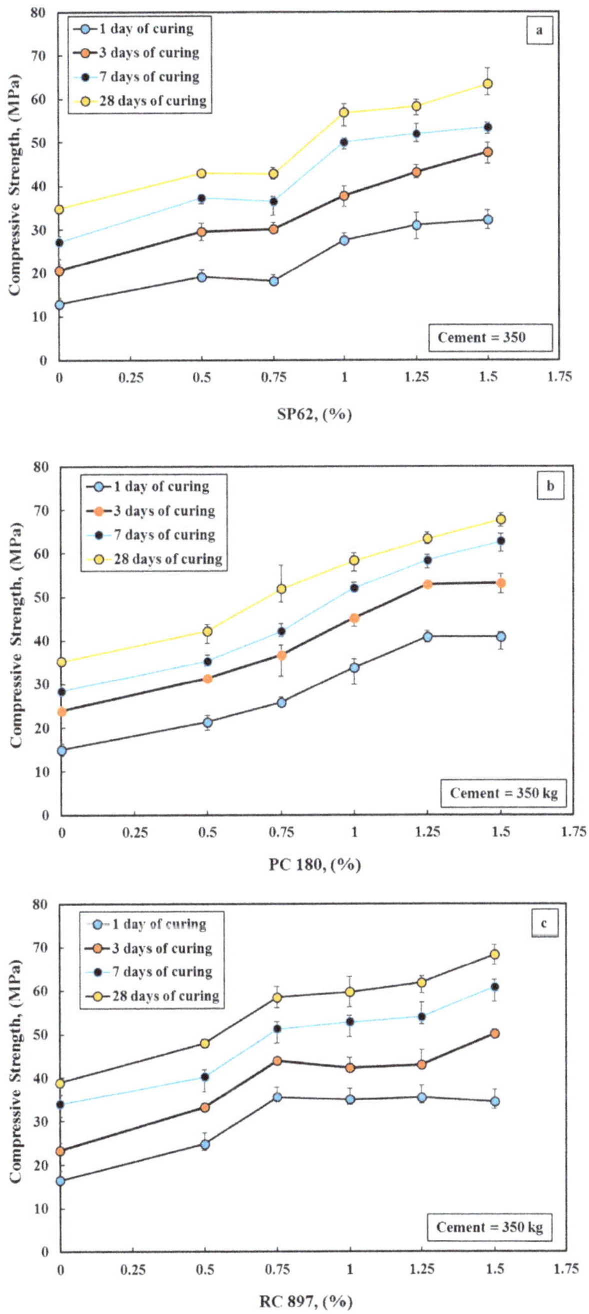

Figure 4. Additives' impact the concrete compressive strength using 350 kg of cement (**a**) SP62, (**b**) PC 180, and (**c**) RC 897.

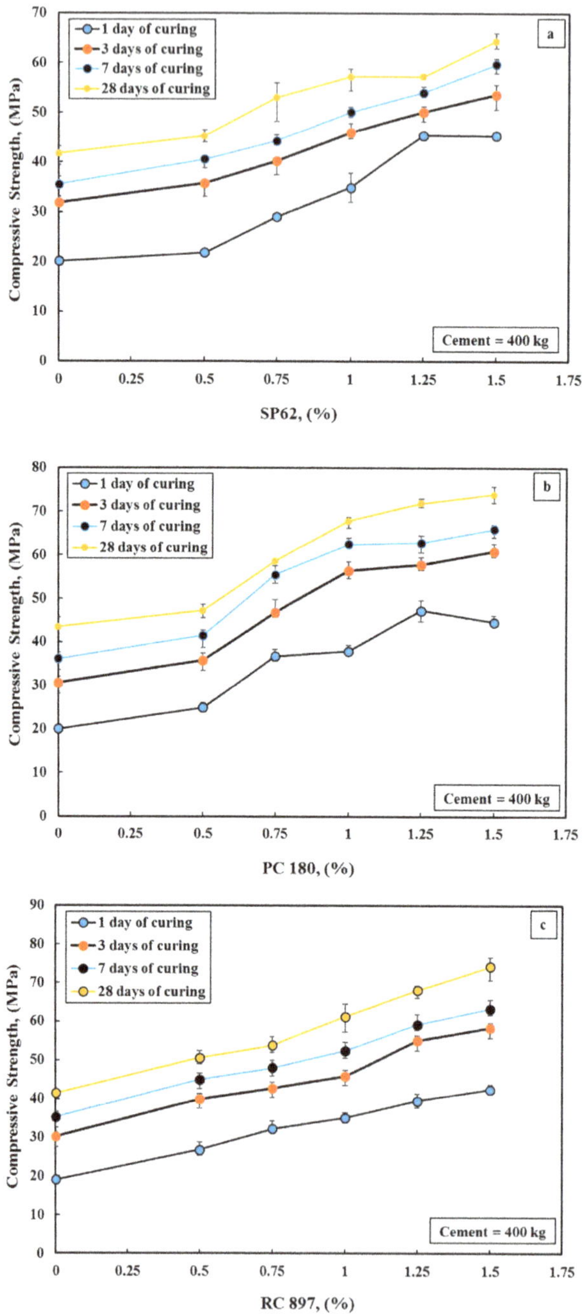

Figure 5. Additives' impact on the concrete compressive strength using 400 kg cement (**a**) SP62, (**b**) PC 180, and (**c**) RC 897.

3.3. Compressive Strength

After 28 days of curing, the concrete compressive strength was enhanced up to 74 MPa, depending on the content of cement and the types and content of water-reducer additives. In the case of polycarboxylate-based superplasticizers and naphthalene- or melamine-based superplasticizers, electrostatic and steric repulsion mechanisms work together to weaken the cohesiveness of the cement particles.

The compressive strength of concrete was predicted using nonlinear and M5P models based on data from 483 tests using three distinct mixtures and three different water-reducer additives, as shown in Figure 6. Additionally, it explores how mixed proportions affect concrete compressive strength.

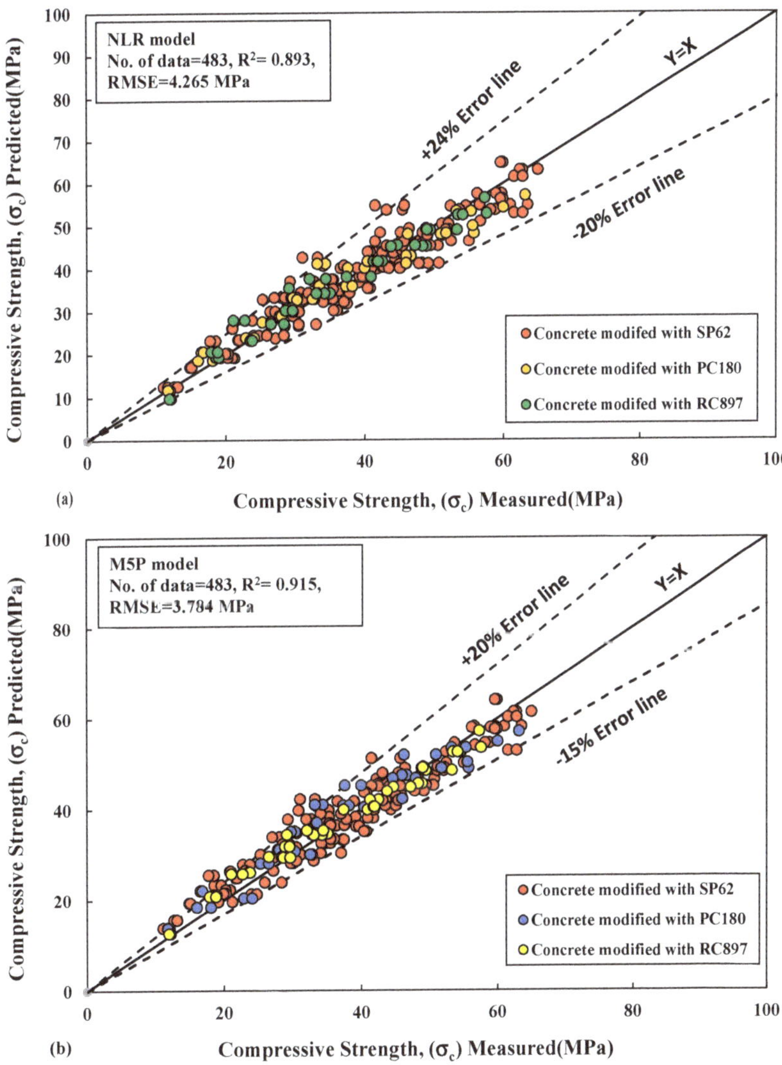

Figure 6. The mix proportions' impact on the concrete compressive strength using (**a**) nonlinear and (**b**) M5P models.

3.4. Predicted and Measured Compressive Strength Relationships

3.4.1. The Nonlinear Regression Model (NLR)

The connection between the anticipated and actual compressive strengths of normal concrete, including polymer content, is shown in Figure 6a. The most significant parameters affecting compressive strength were curing time and cement content. The following formula was derived for the NLR model with numerous changeable parameters (Table 5).

3.4.2. M5P-Tree Model

In this study, the M5P-tree model tree is utilized to forecast the compressive strength of conventional concrete using 483 mix-design data. The coefficient of determination (R^2), root mean square error (RMSE), and goal were all employed to assess the suggested performance of the model in this research. The M5P-tree technique, as seen in Figure 6b, divides the input space (independent variables) into linear tree regression functions (marked LM1 through LM8). $Y = b_o + b_1 \times X_1 + b_2 \times X_2$, where b_o, b_1, and b_2 are linear regression constants representing the model parameters. The model parameters are listed in Table 6. The study dataset has a 25% and 35% error line, indicating that all measured values fall within the 20% and −15% error line. The coefficient of determination R^2 for this model indicates that the model performance is better than the NLR.

Therefore, from the result of slump retention and compressive strength, SP62 (FM) can be used to produce a precast concrete member. The admixture should maintain a liquid consistency and good workability when used with concrete that has a low w/c and a high quantity of mineral additives. High early strength developments are made possible by the PCE-based superplasticizer even at low ambient temperatures and without additional external heat. This might make it possible to shorten the stripping periods, which could lead to a more effective production process. The compaction energy used to compact concrete may be lessened with concrete admixture. Therefore, concrete producers, builders, and installers may profit economically and technically. The three types of water-reducer additives can be used to produce the precast concrete member.

A similar study was also conducted on the effect of two water-reducer polymers with smooth and rough surfaces on the workability and the compression strength of concrete from an early age (1 day) up to 28 days of curing. The polymer contents used in this study varied from 0 to 0.25% (%wt.). The initial ratio between water and cement was 60%, and it slowly reduced to 0.46 by increasing the polymer contents. The compression strength of concrete was increased significantly by increasing the polymer contents by 24% to 95% depending on the polymer type, polymer content, w/c, and curing age. Because of a fiber net (netting) in the concrete when the polymers were added, which led to a decreased void between the particles, binding the cement particles increased the viscosity of the fresh concrete and the compression strength of the hardened concrete rapidly. This study also aims to establish systematic multiscale models to predict the compression strength of concrete containing polymers and to be used by construction projects with no theoretical restrictions. For that purpose, 88 concrete samples modified with two types of polymer (44 samples for each modification) have been tested, analyzed, and modeled. Linear and nonlinear regression, M5P-tree, and Artificial Neural Network (ANN) approaches were used for the qualifications. In the modeling process, the most relevant parameters affecting the strength of concrete were polymer incorporation ratio (0–0.25% of cement's mass), water-to-cement ratio (0.46–0.6), and curing ages (1 to 28 days). Among the used approaches and based on the training data set, the model made based on the nonlinear regression, ANN, and M5P-tree models seem to be the most reliable. The sensitivity investigation concludes that the curing time is the most dominating parameter for predicting concrete's maximum stress (compression strength) with this dataset [12].

4. Conclusions

The following conclusions are drawn based on the tested data and the simulation of the compression strength of concrete at 483 different ratios between the water and the cement, polymer content, and curing ages.

1. The compression strength of cement increased from 84% to 250%, depending on the mix proportion. Based on NLR parameters, polymer RC897 had the highest impact on increasing the compression strength of concrete as compared to polymer SP62 and PC180. This improvement in compression strength was due to the dispersion of cement particles and increasing the friction between the particles, reducing the void ratio and increasing the density of concrete.
2. With a cement content of 300 kg, the polymer PC180 had the highest effect on reducing the water content of the other two types of the polymer by 43.5%, while, at a cement content of 400 kg, the polymer RC 897 had the highest effect on reduction in water content compared with the other two polymers, by 46.7%.
3. The compressive strength of the concrete mixes was calculated using NLR and M5P-tree models. The correlation of the coefficient (R^2) and the root mean square error (RMSE) are used as assessment criteria. The order of the models was M5P-tree and NLR; the M5P-tree was the best model offered in this study, based on data obtained from the experimental work, and provided a higher R^2 and a lower MAE and RMSE.

Author Contributions: Investigation, Writing an original draft, A.S.M. and W.E., Methodology, Supervision, R.K. and K.G., Conceptualization, W.S.Q. and R.K.F., Conceptualization, Methodology, A.S.M. and R.K., Methodology and Supervision, W.E., Validation and Funding acquisition, K.G., Editing, Supervision, R.K.F. All authors have read and agreed to the published version of the manuscript.

Funding: This research received no external funding.

Conflicts of Interest: The authors declare no conflict of interest.

References

1. Fu, T.C.; Yeih, W.; Chang, J.J.; Huang, R. The influence of aggregate size and binder material on the properties of pervious concrete. *Adv. Mater. Sci. Eng.* **2014**, *2014*, 963971. [CrossRef]
2. Burhan, L.; Ghafor, K.; Mohammed, A. Modeling the effect of silica fume on the compressive, tensile strengths and durability of NSC and HSC in various strength ranges. *J. Build. Rehabil.* **2019**, *4*, 19. [CrossRef]
3. Mohammed, A.; Burhan, L.; Ghafor, K.; Sarwar, W.; Mahmood, W. Artificial neural network (ANN), M5P-tree, and regression analyses to predict the early age compression strength of concrete modified with DBC-21 and VK-98 polymers. *Neural Comput. Appl.* **2021**, *33*, 7851–7873. [CrossRef]
4. Sarir, P.; Jiang, H.; Asteris, P.G.; Formisano, A.; Armaghani, D.J. Iterative Finite Element Analysis of Concrete-Filled Steel Tube Columns Subjected to Axial Compression. *Buildings* **2022**, *12*, 2071. [CrossRef]
5. Asteris, P.G.; Armaghani, D.J.; Hatzigeorgiou, G.D.; Karayannis, C.G.; Pilakoutas, K. Predicting the shear strength of reinforced concrete beams using Artificial Neural Networks. *Comput. Concr. Int. J.* **2019**, *24*, 469–488.
6. Kamal, S.M.; Saeed, J.A.; Mohammed, A. The Characterization and Modeling the Mechanical Properties of High Strength Concrete (HSC) Modified with Fly Ash (FA). *Eng. Technol. J.* **2020**, *38*, 173–184. [CrossRef]
7. Abdalla, L.B.; Ghafor, K.; Mohammed, A. Testing and modeling the young age compressive strength for high workability concrete modified with PCE polymers. *Results Mater.* **2019**, *1*, 100004. [CrossRef]
8. Liu, X.; Wang, Z.; Zheng, Y.; Cui, S.; Lan, M.; Li, H.; Zhu, J.; Liang, X. Preparation, characterization and performances of powdered polycarboxylate superplasticizer with bulk polymerization. *Materials* **2014**, *7*, 6169–6183. [CrossRef]
9. Lu, S.; Koopialipoor, M.; Asteris, P.G.; Bahri, M.; Armaghani, D.J. A novel feature selection approach based on tree models for evaluating the punching shear capacity of steel fiber-reinforced concrete flat slabs. *Materials* **2020**, *13*, 3902. [CrossRef]
10. Ferrari, L.; Kaufmann, J.; Winnefeld, F.; Plank, J. Multi-method approach to study influence of superplasticizers on cement suspensions. *Cem. Concr. Res.* **2011**, *41*, 1058–1066. [CrossRef]
11. Burhan, L.; Ghafor, K.; Mohammed, A. Enhancing the Fresh and Hardened Properties of the Early Age Concrete Modified with Powder Polymers and Characterized Using Different Models. *Adv. Civ. Eng. Mater.* **2020**, *9*, 227–249. [CrossRef]
12. Kar, A. *Characterizations of Concretes with Alkali-Activated Binder and Correlating Their Properties from Micro-to Specimen Level*; West Virginia University: Morgantown, WV, USA, 2013.
13. Sarwar, W.; Ghafor, K.; Mohammed, A. Regression analysis and Vipulanandan model to quantify the effect of polymers on the plastic and hardened properties with the tensile bonding strength of the cement mortar. *Results Mater.* **2019**, *1*, 100011. [CrossRef]

14. Sarwar, W.; Ghafor, K.; Mohammed, A. Modeling the rheological properties with shear stress limit and compressive strength of ordinary Portland cement modified with polymers. *J. Build. Pathol. Rehabil.* **2019**, *4*, 25. [CrossRef]
15. Mohammed, A.; Mahmood, W.; Ghafor, K. TGA, rheological properties with maximum shear stress and compression strength of cement-based grout modified with polycarboxylate polymers. *Constr. Build. Mater.* **2020**, *235*, 117534. [CrossRef]
16. Noorzaei, J.; Hakim, S.J.S.; Jaafar, M.S.; Thanoon, W.A.M. Development of artificial neural networks for predicting concrete compressive strength. *Int. J. Eng. Technol.* **2007**, *4*, 141–153.
17. Psyllaki, P.; Stamatiou, K.; Iliadis, I.; Mourlas, A.; Asteris, P.; Vaxevanidis, N. Surface treatment of tool steels against galling failure. *MATEC Web Conf.* **2018**, *188*, 04024. [CrossRef]
18. Armaghani, D.J.; Asteris, P.G.; Fatemi, S.A.; Hasanipanah, M.; Tarinejad, R.; Rashid, A.S.A.; Huynh, V.V. On the Use of Neuro-Swarm System to Forecast the Pile Settlement. *Appl. Sci.* **2020**, *10*, 1904. [CrossRef]
19. He, B.; Armaghani, D.J.; Lai, S.H. Assessment of tunnel blasting-induced overbreak: A novel metaheuristic-based random forest approach. *Tunn. Undergr. Space Technol.* **2023**, *133*, 104979. [CrossRef]
20. Huang, J.; Asteris, P.G.; Manafi Khajeh Pasha, S.; Mohammed, A.S.; Hasanipanah, M. A new auto-tuning model for predicting the rock fragmentation: A cat swarm optimization algorithm. *Eng. Comput.* **2022**, *38*, 2209–2220. [CrossRef]
21. Asteris, P.G.; Gavriilaki, E.; Touloumenidou, T.; Koravou, E.E.; Koutra, M.; Papayanni, P.G.; Pouleres, A.; Karali, V.; Lemonis, M.E.; Mamou, A.; et al. Genetic prediction of ICU hospitalization and mortality in COVID-19 patients using artificial neural networks. *J. Cell. Mol. Med.* **2022**, *26*, 1445–1455. [CrossRef]
22. Mohammed, A.; Rafiq, S.; Mahmood, W.; Noaman, R.; Ghafor, K.; Qadir, W.; Kadhum, Q. Characterization and modeling the flow behavior and compression strength of the cement paste modified with silica nano-size at different temperature conditions. *Constr. Build. Mater.* **2020**, *257*, 119590. [CrossRef]
23. Mohammed, A.; Mahmood, W. Estimating the efficiency of the sandy soils-cement based grout interactions from particle size distribution (PSD). *Geomech. Geoeng.* **2021**, *16*. [CrossRef]
24. Asteris, P.G.; Douvika, M.G.; Karamani, C.A.; Skentou, A.D.; Chlichlia, K.; Cavaleri, L.; Daras, T.; Armaghani, D.J.; Zaoutis, T.E. A Novel Heuristic Algorithm for the Modeling and Risk Assessment of the COVID-19 Pandemic Phenomenon. *Comput. Model. Eng. Sci.* **2020**, *125*, 815–828. [CrossRef]
25. Asteris, P.G.; Mamou, A.; Hajihassani, M.; Hasanipanah, M.; Koopialipoor, M.; Le, T.-T.; Kardani, N.; Armaghani, D.J. Soft computing based closed form equations correlating L and N-type Schmidt hammer reboundnumbers of rocks. *Transp. Geotech.* **2021**, *29*, 100588. [CrossRef]
26. Saridemir, M. Prediction of Compression Strength of Concretes Containing Metakaolin and Silica Fume by Artificial Neural Networks. *Adv. Eng. Softw.* **2009**, *40*, 350–355. [CrossRef]
27. Cavaleri, L.; Chatzarakis, G.E.; Di Trapani, F.; Douvika, M.G.; Roinos, K.; Vaxevanidis, N.M.; Asteris, P.G. Modeling of Surface Roughness in Electro-Discharge Machining Using Artificial Neural Networks. *Adv. Mater. Res.* **2017**, *6*, 169–184.
28. Gavriilaki, E.; Asteris, P.G.; Touloumenidou, T.; Koravou, E.E.; Koutra, M.; Papayanni, P.G.; Karali, V.; Papalexandri, A.; Varelas, C.; Chatzopoulou, F.; et al. Genetic justification of severe COVID-19 using a rigorous algorithm. *Clin. Immunol.* **2021**, *226*, 108726. [CrossRef]
29. Asteris, P.G.; Lemonis, M.E.; Le, T.-T.; Tsavdaridis, K.D. Evaluation of the ultimate eccentric load of rectangular CFSTs using advanced neural network modeling. *Eng. Struct.* **2021**, *248*, 113297. [CrossRef]
30. Sihag, P.; Jain, P.; Kumar, M. Modelling of impact of water quality on recharging rate of storm water filter system using various kernel function-based regression. *Model. Earth Syst. Environ.* **2018**, *4*, 61–68. [CrossRef]
31. Khayat, K.H.; Hester, W.T. Evaluation of concrete mixtures for underwater pile repairs. *Cem. Concr. (New Delhi) Aggreg.* **1991**, *13*, 32–41.
32. Jayasree, C.; Gettu, R. Experimental study of the flow behaviour of superplasticized cement paste. *Mater. Struct.* **2008**, *41*, 1581–1593. [CrossRef]
33. Ali, R.; Muayad, M.; Mohammed, A.S.; Asteris, P.G. Analysis and prediction of the effect of Nanosilica on the compressive strength of concrete with different mix proportions and specimen sizes using various numerical approaches. *Struct. Concr.* **2022**. [CrossRef]

applied sciences

MDPI

Article

Three Steps towards Better Forecasting for Streamflow Deep Learning

Woon Yang Tan [1], Sai Hin Lai [1,*], Fang Yenn Teo [2], Danial Jahed Armaghani [3], Kumar Pavitra [4] and Ahmed El-Shafie [1,*]

[1] Department of Civil Engineering, Faculty of Engineering, University of Malaya, Kuala Lumpur 50603, Malaysia
[2] Faculty of Science and Engineering, University of Nottingham Malaysia, Semenyih 43500, Malaysia
[3] Department of Urban Planning, Engineering Networks and Systems, Institute of Architecture and Construction, South Ural State University, 454080 Chelyabinsk, Russia
[4] Department of Geography and Planning, University of Liverpool, Liverpool L69 3BX, UK
* Correspondence: laish@um.edu.my (S.H.L.); elshafie@um.edu.my (A.E.-S.)

Abstract: Elevating the accuracy of streamflow forecasting has always been a challenge. This paper proposes a three-step artificial intelligence model improvement for streamflow forecasting. Step 1 uses long short-term memory (LSTM), an improvement on the conventional artificial neural network (ANN). Step 2 performs multi-step ahead forecasting while establishing the rates of change as a new approach. Step 3 further improves the accuracy through three different kinds of optimization algorithms. The Stormwater and Road Tunnel project in Kuala Lumpur is the study area. Historical rainfall data of 14 years at 11 telemetry stations are obtained to forecast the flow at the confluence located next to the control center. Step 1 reveals that LSTM is a better model than ANN with R 0.9055, MSE 17,8532, MAE 1.4365, NSE 0.8190 and RMSE 5.3695. Step 2 unveils the rates of change model that outperforms the rest with R = 0.9545, MSE = 8.9746, MAE = 0.5434, NSE = 0.9090 and RMSE = 2.9958. Finally, Stage 3 is a further improvement with R = 0.9757, MSE = 4.7187, MAE = 0.4672, NSE = 0.9514 and RMSE = 2.1723 for the bat-LSTM hybrid algorithm. This study shows that the δQ model has consistently yielded promising results while the metaheuristic algorithms are able to yield additional improvement to the model's results.

Keywords: optimization; metaheuristic algorithms; streamflow forecasting

Citation: Tan, W.Y.; Lai, S.H.; Teo, F.Y.; Armaghani, D.J.; Pavitra, K.; El-Shafie, A. Three Steps towards Better Forecasting for Streamflow Deep Learning. *Appl. Sci.* **2022**, *12*, 12567. https://doi.org/10.3390/app122412567

Academic Editor: Giancarlo Mauri

Received: 31 October 2022
Accepted: 5 December 2022
Published: 8 December 2022

1. Introduction

The natural water movement on our planet is known as the hydrological cycle. Streamflow is one of the main components of this cycle. The streamflow characteristic is often associated with climate and land use conditions [1]. Under-capacity rivers can trigger frequent flooding in the surrounding catchment due to excess runoff. On the other hand, water scarcity can also happen during dry weather. Therefore, the state of streamflow can transpire in future events. Streamflow forecasting can optimize water resource allocation [2].

For this reason, researchers have been developing various methods to forecast streamflow [3]. The conventional approach relies on preserving mass, momentum and energy [4] to retrieve broad basin information. However, data collection is time-consuming and costly as the conventional method requires a wide range of parameters. As more and more flooding occurs due to climate change, a more accurate forecasting model is required to pursue better flood management and disaster preparedness [5]. Artificial intelligence is seen as a better alternative to the conventional method. A study has shown that the adaptation of artificial intelligence allows better river and drought management [1]. It can establish the association of predictors and predictand variables without considering hydrological complexity.

Although many studies have shown promising results, standalone models (e.g., artificial neural network) display specific drawbacks of overfitting due to large datasets. In addition, past states of network retrieved from time-series data are not kept for the benefit of information related to data sequence [6]. These drawbacks can be tackled through the implementation of deep learning that can generate higher accuracy through better extraction of obscure data with higher computing power and complex mapping ability. This ability has contributed to significant developments in many fields, such as speech recognition, language processing and hydrological studies, such as river flood forecasting, runoff forecasting, streamflow forecasting and groundwater level forecasting [7].

Xiang and Demir (2020) proposed a study applying a deep recurrent neural network, specifically the neural runoff model, to predict streamflow in the state of Iowa. The model successfully incorporated multiple measurements and model results to produce long-term rainfall–runoff modeling [7]. Ahmed et al. (2021) applied a deep-learning hybrid model to forecast the monthly streamflow water level in the Murray Darling Basin that yielded improved results when optimized with Boruta [1]. Lin et al. (2021) developed three components of the hybrid DIFF-FFNN-LSTM model to forecast hourly streamflow, which accomplished better results than statistical methods [6]. Granata et al. (2022) performed a comparison study between the stacked model of random forest and the multilayer perceptron algorithm with bidirectional LSTM. The bidirectional LSTM model significantly outperformed the stacked model for low-flow prediction [8]. Elbeltagi et al. (2022) developed a study comparing four machine learning algorithms, namely random subspace, M5P, random forest and bagging, to predict streamflow in the Des Moines watershed. The M5P algorithm yielded the best prediction [9].

Increasing accessibility to the latest research has triggered tremendous advancement in science and technology. A modern measuring device can quickly secure physical hydrological data with standard intervals. As more significant obscured knowledge is extracted, more demands for complex engineering optimization start to the surface [10]. This requirement comes with multiple purposes, multi-level conditions and numerous restrictions.

In response, more recent research has been integrating machine learning methods with a metaheuristic algorithm to solve the optimization complexity [11]. This integration leads to a more efficient, effective and robust search, resulting in faster convergence.

Khosravi et al. (2022) introduced an optimized deep learning model integrating a convolutional neural network (CNN) with the BAT metaheuristic algorithm to predict daily streamflow in the Korkorsar catchment in northern Iran. This model outperformed the other algorithms [12].

Machine learning is a subset of artificial intelligence that exploits algorithms and statistical methods to provide computers with learning ability [13]. It aims to optimize experimental arrangements for a data structure [14]. A continuous source of data from actual observation is fed into the system, improving the learning over time. Artificial intelligence closely resembles how human brains capture internal data relationship patterns [15]. The acquired knowledge enriches the machine's ability to generalize a real-world position [16].

Metaheuristics denote high-level computational intelligence algorithm frameworks that are problem-independent and are employed to solve complex optimization demands [17]. A robust, iterative search process is involved in the metaheuristics algorithm to generate an approximation that does not guarantee an optimum solution [18] but instead an adequately good global solution within a reasonable computational time. The algorithm can self-tune the global exploration and local exploitation to reach greater search abilities [19].

Metaheuristics can be categorized into nature-inspired and non-nature-inspired. The nature-inspired category can be further classified into evolutionary algorithms [20] and swarm intelligence. Evolutionary algorithms include genetic algorithms, genetic programming, evolution strategy and differential evolution based on biological transformation. Swarm intelligence includes artificial bee colony algorithm, ant colony optimization, crow search algorithm, jellyfish search optimizer, firefly optimization and bat algorithm. The non-nature-inspired category consists of the Jaya algorithm, imperialist competitive algorithm, simulated annealing, harmony search and forensic-based investigation algorithm.

All evolutionary and swarm intelligence algorithms involve proper tuning of standard controlling parameters such as population size and generation boundary. In addition, each algorithm has its algorithm-specific control parameters such as mutation probability, crossover probability and selection operator for the genetic algorithm. Failure to properly tune can decrease computational speed and entrap in local optimal. Swarm intelligence algorithms are also subjected to slow convergence and are challenging to integrate with a particular artificial intelligence model [21]. In order to avoid algorithm-specific non-performance, the teaching learning-based optimization algorithm and the Jaya algorithm can be implemented [22].

The bat algorithm is used in tuning residential HVAC controller parameters to optimize energy consumption and obtain thermal comfort. It is also used for controlling illumination and air quality [23]. Other applications are wind power forecasting [24] and transportation [25].

The firefly algorithm has been used in numerous fields to solve complex applications such as breast cancer recognition, vehicle communication problem, path planning, privacy protection and forecast power consumption. It can also be used in structural optimization and image processing [26].

The Jaya algorithm has been developed for many engineering works such as structural damage identification [27], welding optimization, heat exchangers optimization, path selection for a wireless network, waterjet machines, dam monitoring [28], wind power systems and cart position control [29].

From the authors' observation, there is a lack of research in the area of optimization for deep learning using hybrid models.

In order to fill this gap, this study aims to improve the deep learning model for better streamflow simulation and forecasting using optimization algorithm hybrid models, which will lead to a better early warning system.

The contributions of this paper can be simplified as follows:

1. Application of the LSTM model as a deep learning model for simulation and multi-step ahead streamflow forecasting;
2. A new approach to using rates of change in the artificial learning model to minimize input errors;
3. To improve the performance of LSTM models by introducing a novel method in deep learning through metaheuristic algorithms to form hybrid models.

2. Methodology

This study involves numerous deep learning models and metaheuristic algorithms such as the bat, firefly and Jaya algorithms. The study area and model development are also discussed.

2.1. Long Short-Term Memory (LSTM)

LSTM is an improved version of a recurrent neural network (RNN) [30]. It is a deep learning algorithm that has been set up to perform forecasting in the field of hydrology and water resources [31]. It eliminates the issue of overfitting and can yield better generalization than standalone models. The network captures long-term dependencies and deals with vanishing gradient limitations that exist in the original RNN [32]. The LSTM network (see Figure 1) comprises blocks of memory cells, an input gate, an output gate and a forget gate. The network operates like a chain [33] and can deal with delays such as seasonal and trend patterns [34]. The input gate manages the extra information added to the cell state.

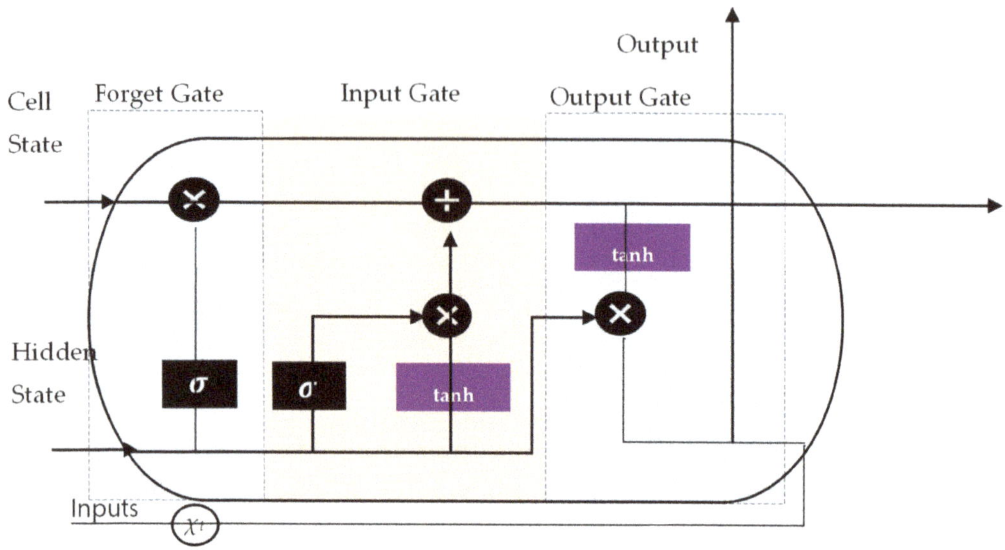

Figure 1. Architecture of LSTM blocks.

The forget gate eliminates information from the cell state. The ability of LSTM to store or remove information outperforms other neural networks [35]. Information can be carried over multiple time steps and provide for learning of sequential dependency in the input data, making it relevant even for long time series [36]. This gives an advantage to the LSTM when it comes to modeling time series, particularly hydrologic variables, which employ common hyperparameters, such as precipitation, flow or water level, for streamflow prediction, water quality modeling and flood forecasting [37]. Although the training process is longer than other data-driven models, LSTM can yield higher accuracy [38].

2.2. Bat Algorithm

The bat algorithm (see Figure 2) is a swarm intelligence algorithm inspired by the echolocation produced by bats when interacting with their surroundings [23]. The echolocation starts with the emission of short and loud sound waves released by bats to identify their prey, obstacles or resting cracks in the dark. The time-lapse for the emitted sound to bounce back reveals the prey's distance, direction and speed. All bats use echolocation to measure distance and distinguish between targets and obstacles [39]. The algorithm keeps a record of the bat's velocity, position, frequency, varying wavelength, loudness and pulse emission. The loudness is measured in the range between A_{min} and A_0, while the pulse emission is logged between 0 and 1, where 0 represents no pulse, and 1 refers to the highest rate of the bat's emission. The bat algorithm is suitable to handle both continuous and discrete optimization matters. One of the advantages of this algorithm is the ability to reach quick convergence at the initial stage and shift from exploration to exploitation when optimality is near [40].

The mathematical equations that relate to the velocity and location can be defined as:

$$f_i = fmin + (fmax - fmin)\beta \tag{1}$$

$$v_i^t = v_i^{t-1} + \left(x_i^{t-1} - x_*\right) f_i \tag{2}$$

$$x_i^t = x_i^{t-1} + v_i^t \tag{3}$$

where:

$\beta \in [0,1]$ is the random vector from a uniform distribution;
f_i is the initial frequency;
v_i^t is the velocity at t iteration;
x_i^t is location at t iteration in a d-dimensional search or solution space.
The loudness and pulse emission rates are represented below:

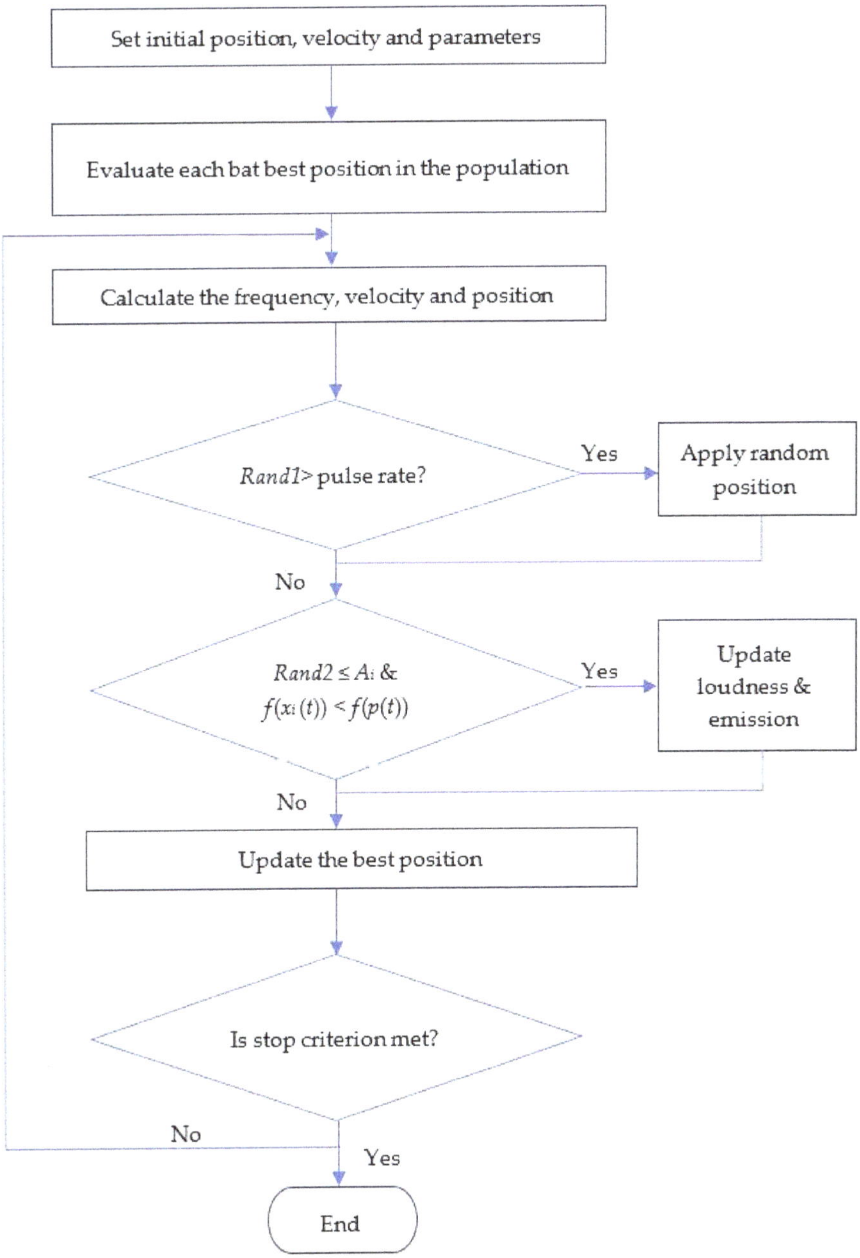

Figure 2. Bat algorithm flowchart.

$$A_i^{t+1} = \alpha A_i^t \quad (4)$$

$$r_i^{t+1} = r_i^0[1 - \exp(-\gamma t)] \quad (5)$$

where:

$0 < \alpha < 1$ and $\gamma > 0$ are constants;

α is the constant reducing loudness, and γ is the constant increasing pulse rate.

2.3. *Firefly Algorithm*

Bioluminescence refers to the biochemical process that provides the insects' ability to flicker. The flashing light is visible, particularly at night, to court potential mates and gives a warning signal for potential predators nearby. The emission of the rays can be controlled towards brighter or dimmer light [41].

The firefly algorithm (see Figure 3) is considered a swarm intelligence algorithm that originated from the flickering behaviors of insects. It is a popular algorithm in the swarm intelligence domain [42]. Flashlight without gender distinction is simulated to entice fireflies with less brightness to draw toward the individual. Under this algorithm (see Figure 3), two significant features are considered, mainly brightness and attractiveness. The brightness echoes the firefly's position and establishes the path of movement. At the same time, the attraction indicates the distance the firefly travels. The algorithm's goal is to continuously update the brightness and attractiveness status [15].

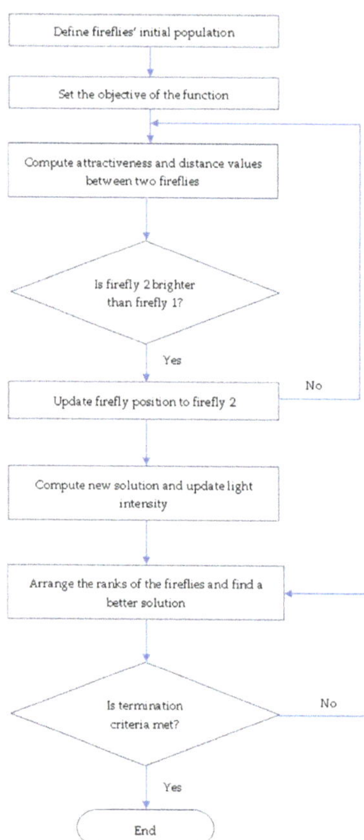

Figure 3. Firefly algorithm flowchart.

The light brightness will decrease as distance increases. Since the brighter fireflies attract the dimmer ones, the latter will move toward the former position. The brightness indicates the fitness value of the algorithm. The greater the brightness, the better will be the fitness value. If two adjacent fireflies transmit similar brightness, the fireflies will move randomly.

The algorithm is set to adhere to the three following rules [43]:

(a) All fireflies are considered unisex, and therefore, they are attracted to others regardless of their sex;
(b) Attractiveness is based on the brightness of the light. The dimmer one will move towards the brighter one. If brightness is equal, movement will be random;
(c) The brightness is associated with the objective of the function.

When firefly i is attracted to j, then the new position of the firefly i will be computed as follows:

$$x_i^{t+1} = x_i^t + \beta_0 e^{-\gamma r^2 ij}\left(x_j^t - x_i^t\right) + \alpha_t \epsilon_i^t \tag{6}$$

where:
x_i^{t+1} is the new position of the firefly i;
x_i^t is the original position of the firefly i;
β_0 is the attractiveness parameter;
γ is the absorption coefficient;
α_t is randomization parameter (0 to 1);
r is the distance between two fireflies;
ϵ_i^t is random number.

2.4. Jaya Algorithm

Jaya algorithm (see Figure 4) is a population-based algorithm that constantly searches for the best solutions and avoids bad ones [29,44]. Two main parameters, the population size and the maximum number of iterations, are used to define the framework of the algorithm [45]. The iteration process will continue to be executed to find a better solution [46] than the current state with the following equation:

$$X'_{i,j,k} = X_{i,j,k} + r_{1,j,k}\left(X_{i,best,k} - \left|X_{i,j,k}\right|\right) - r_{2,j,k}\left(\dot{X}_{i,worst,k} - \left|X_{j,j,k}\right|\right) \tag{7}$$

where:
$X_{i,j,k}$ is the current state;
$r_{1,j,k}\left(X_{i,best,k} - \left|\dot{X}_{i,j,k}\right|\right)$ is the best solution;
$r_{2,j,k}\left(\dot{X}_{i,worst,k} - \left|X_{j,j,k}\right|\right)$ is the worst solution.

The process will remain until the stopping criteria are met. Jaya algorithm is suitable for controlled and unrestricted optimization [22].

2.5. Rates of Change

Rates of change (δQ) is introduced as a new model development method to replace the conventional method of utilizing flow or water level as the prediction model output. The current research on streamflow forecasting concentrates mainly on the prediction of the flow or water level as the output variables of the forecasted value (Q_f). The mathematical expression of a forecast flowrate is as follows:

$$Q_f = Q_i + \delta Q \tag{8}$$

where:
Q_f is the forecast flowrate;
Q_i is the initial flowrate at the time, t;
δQ is the rate of change.

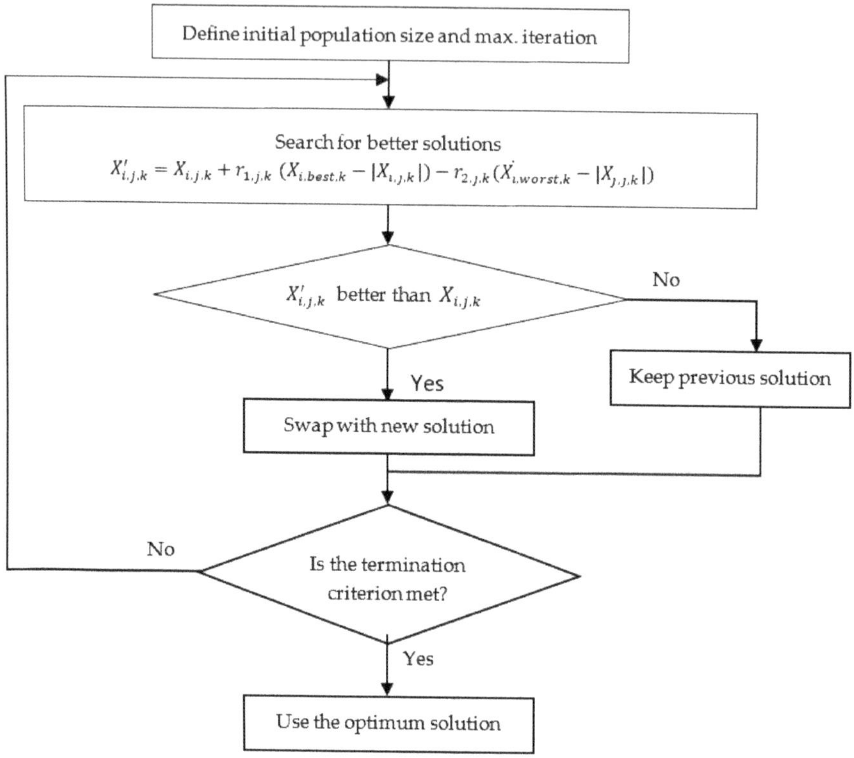

Figure 4. Jaya algorithm.

A rate of change is proposed in this study based on the mathematical relationship as follows:

$$\delta Q = \frac{Q_t - Q_i}{t - t_i} \quad (9)$$

where:

δQ is the rate of change;
Q_t is the flowrate at current time, t;
Q_i is the initial flowrate at a previous time interval;
t is the current time;
t_i is the last time interval.

By applying the rates of change (δQ), the fluctuation can be controlled to improve the model's accuracy. For this research, the δQ will be based on 30 min.

2.6. Model Performance Evaluation

In this study, the performance of each model is evaluated based on four types of performance indices. The evaluation includes both the absolute and relative aspects of the errors, such as the root mean square error (*RMSE*), mean absolute error (*MAE*), correlation coefficient (*R*), Nash–Sutcliffe efficiency (*NSE*) and mean absolute percentage error (MAPE).

2.6.1. Root Mean Square Error, RMSE

RMSE measures the deviations between predicted values and observed values. The variations, also known as the prediction errors, are developed from computation performed over out-of-sample data. *RMSE* is sensitive to maximum and minimum errors and can better reflect the predicted results. However, it is not sensitive to linear offsets between the observed and simulated values resulting in a low *RMSE* value [47]. *RMSE* with a value close to 0 indicates a higher level of prediction accuracy.

$$RMSE = \sqrt{\frac{\sum_{i=1}^{n}(y_i - x_i)^2}{n}}, 0 \le RMSE < +\infty \tag{10}$$

where:

x_i are the observed values of the criterion;
y_i are the simulated values of the criterion;
n = sample size.

2.6.2. Mean Absolute Error, MAE

MAE measures the significance of average error in a model with the same criteria [48]. The mathematical representation of *MAE* is as follows:

$$MAE = \frac{\sum_{i=1}^{n}|y_i - x_i|}{n}, 0 \le MAE < +\infty \tag{11}$$

where:

x_i are the observed values of the criterion;
y_i are the simulated values of the criterion;
n = sample size.

2.6.3. Nash-Sutcliffe Efficiency (NSE)

NSE measures the relative differences between the observed and predicted values. A higher value of *NSE* indicates the model's superiority. When *NSE* is 1, it means a perfect match of the observed and predicted. Otherwise, if *NSE* is 0, the predicted values are similar to the average of the observed values [49]. The model accuracy can be categorized as very good for $0.75 < NSE \le 1$, good for $0.65 < NSE \le 0.75$, satisfactory for $0.50 < NSE \le 0.65$ or unsatisfactory for $NSE \le 0.50$ [50].

The mathematical representation of NSE is as follows:

$$NSE = 1 - \frac{\sum_{i=0}^{n}(Y_i - Y_t)^2}{\sum_{i=0}^{n}(Y_i - \overline{Y})^2}, -\infty < NSE \le 1 \tag{12}$$

where:

Y_i is the predicted values of the criterion;
Y_t is the measured value of the criterion variable (dependent) variable Y;
$\overline{Y}$ is the mean of the measured values of Y;
n = sample size.

2.6.4. Mean Absolute Percentage Error (MAPE)

MAPE is an error metric used to measure the accuracy of forecasting values. It denotes the average absolute percentage deviation of each dataset entry between actual and forecast values [51]. As absolute values are applied, the possibility of negative and positive errors canceling each other out can be avoided. The lower the value of MAPE, the better the model will forecast.

$$\text{MAPE} = \frac{100\%}{n}\sum_{t=1}^{n}\left|\frac{A_t - F_t}{A_t}\right| \tag{13}$$

where:

A_t is the actual value;
F_t is the forecast value;
n = sample size.

2.7. *Study Area and Data Description*

Malaysia's climate is hot and with high humidity all year round. The country is exposed to two major monsoon seasons, mainly the north-east monsoon from November to February and the south-west monsoon from May to August. During the north-east monsoon, a significant increase in rainfall occurrence can be detected in the eastern and southern regions of the country. Moreover, the south-west monsoon and inter-monsoon seasons of March to April and September to October can cause intense convective rainfall on the country's west coast.

Kuala Lumpur is Malaysia's capital city, as shown in Figure 5. The city is highly urbanized and covers an area of 243 km^2 with an estimated population density of 6696 residents per square kilometer [52]. Changes in land use and land cover have been intense since the 1980s due to the economic boom. The city receives an average annual rainfall of 2600 mm and is subjected to flash floods. It is situated in the middle of the Klang River basin with a watershed area of 1288 km^2. The Klang River flows through a 120 km distance [53], with 11 major tributaries flowing across Selangor state and Kuala Lumpur [54]. Batu, Gombak, Ampang and upper Klang River at the upper catchment of Kuala Lumpur are the main tributaries of Klang River that contribute significantly to the flow at the downstream point of Masjid Jamek, which is a famous historical site and a tourist attraction.

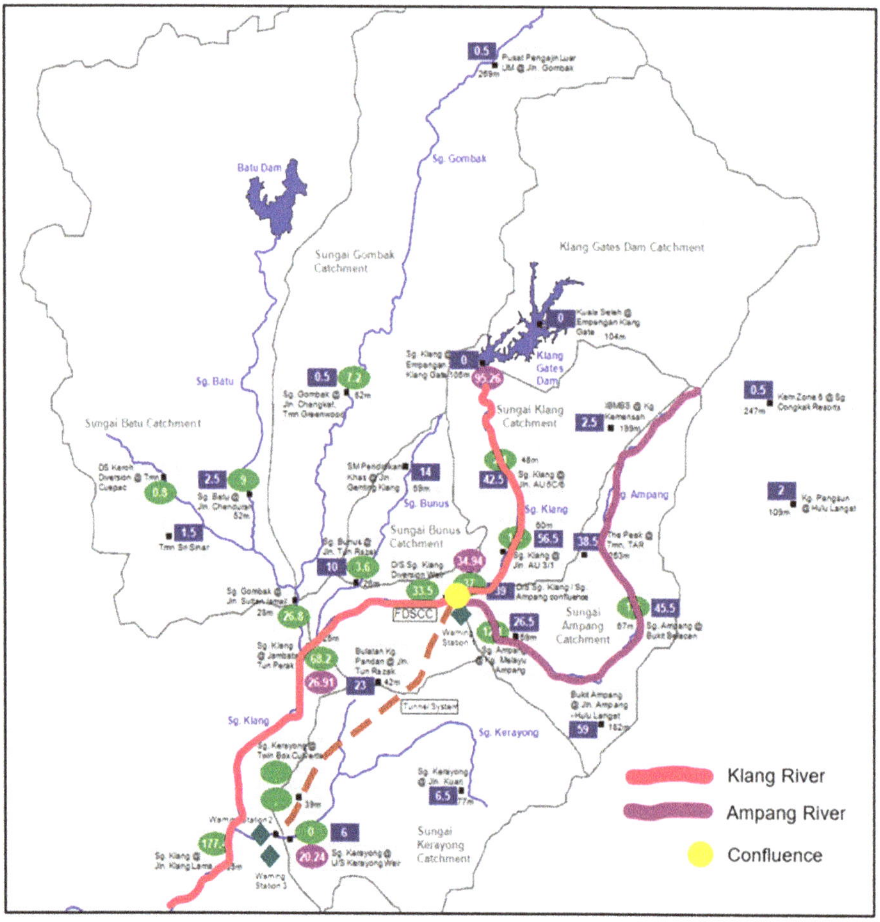

Figure 5. Map of study location at the Klang River Catchment.

The flash flood occurrence in 1971 that lasted for five days with massive damage of RM36 million prompted the government to develop a comprehensive Kuala Lumpur Flood Mitigation Plan (KLFM) [55]. The Stormwater Management and Road Tunnel (SMART) project built in 2007 is part of the early plan to divert flow from the upper catchment of the Klang River and Ampang River to the Kerayong River downstream [56].

SMART is a mega project to construct a 9.7 km tunnel that combines wet and dry systems [52]. During a major storm, mode 2 is activated when the flow reaches more than 70 m^3/s at the confluence of the Klang and Ampang rivers [57]. Moreover, Mode 3 is activated when the flow at the confluence reaches 150 m^3/s. Total storage of 3 $million^3$ infrastructure is available to cater for the excess stormwater. During regular days, a total length of 3 km is available for dual-deck motorway use [57].

Study Data

The SMART catchment has an area of 160 km^2 equipped with a rain gauge and doppler current meter at 28 hydrological stations. The sensors collect rainfall and flow data and transmit the data to the control center using telemetry. Within the 28 hydrological stations, data from 11 telemetry stations are used for modeling. The rest of the stations are meant for observation only. This study collects historical data of 30 min interval rainfall at the 11 telemetry stations and the flow at the confluence of the Klang River and Ampang River from January 2008 to August 2021. Seventy percent of the historical data from January 2008 to August 2019 are used for training, while the rest are used for testing. Normal flow at the confluence of Ampang and Klang Rivers is generally within the range of 5 to 10 m^3/s. However, this flow can increase tremendously above 150 m^3/s depending on the intensity of the precipitation.

2.8. Model Development

As shown in Figure 6, the proposed artificial intelligence model is intended to seek the best fit that yields the best results for deployment purposes. Input data for the model consist of historical rainfall data from 11 telemetry stations at the upper catchment of the Klang River basin taken from 1 January 2008 to August 2021 with an interval of 30 min. Moreover, the target data consist of flow data at the confluence between the Ampang–Klang rivers with equal intervals and similar time ranges. The confluence is considered the point of interest in this study as the current flow will determine the mode of operation, as mentioned earlier. Three steps of model development are introduced to pursue the best relationship between historical data and predictors.

Step 1 employs the LSTM model as the deep learning framework for streamflow prediction, and ANN is the benchmark model. Several performance indices are performed to compare the models.

Step 2 introduces the novel rates of change method and implements multi-step ahead forecasting to analyze the results better. The models' performance on fitness and errors are checked.

Step 3 develops the novel optimization method for deep learning using metaheuristics to find the near-optimum weights and biases. Three optimization algorithms were picked for this study: bat algorithm, firefly algorithm and Jaya algorithm. After going through the optimization algorithm, the data are fed into the LSTM model. Performances on fitness and errors are checked. The best model is deployed after the three steps.

3. Results and Discussion

This section unveils the results acquired from the training and testing of various LSTM models. There are three steps involved (refer to Figure 6). For Step 1, numerous LSTM and ANN models are employed to perform streamflow prediction. The performance is evaluated for the goodness of fit by executing several measures listed in Section 2.6. Table 1 lists the best model results of the LSTM and ANN. Figures 7 and 8 show the graphs of observed flow vs. forecast flow for the ANN model and LSTM model, respectively.

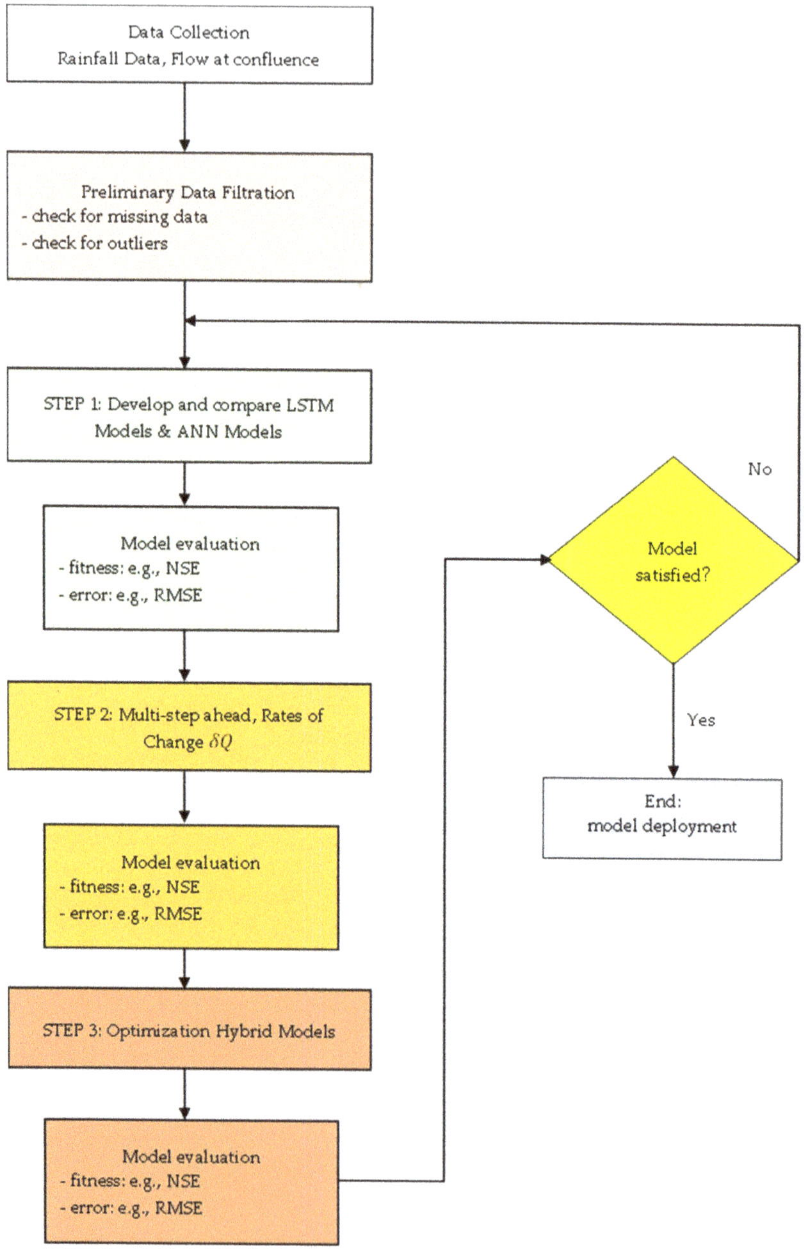

Figure 6. Model development flowchart.

Table 1. Best LSTM and ANN models for prediction.

Best ANN Model								
# of Neurons in Layer 1	**R.Train**	**R.Test**	**MSE**		**MAE**		**NSE**	**RMSE**
10	0.4520	0.4254	78.4215		3.7135		0.1994	8.8556
Best LSTM Model								
Model	**R.Train**	**R.Test**	**MSE.Train**	**MSE.Test**	**MAE.Train**	**MAE.Test**	**NSE**	**RMSE**
Simul.	0.9055	0.8586	17.8532	28.8315	1.4365	2.4208	0.8190	5.3695

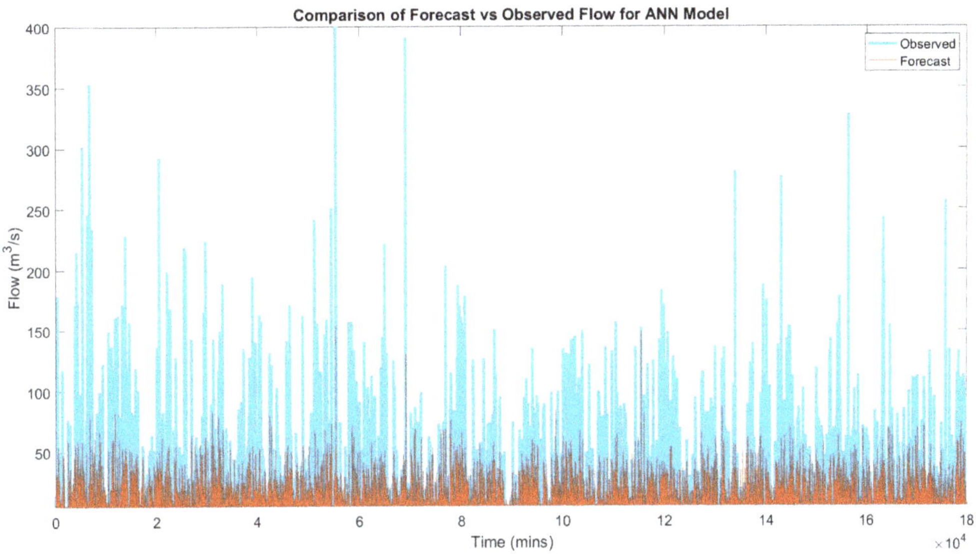

Figure 7. Graph of ANN model observed flow vs. simulated flow.

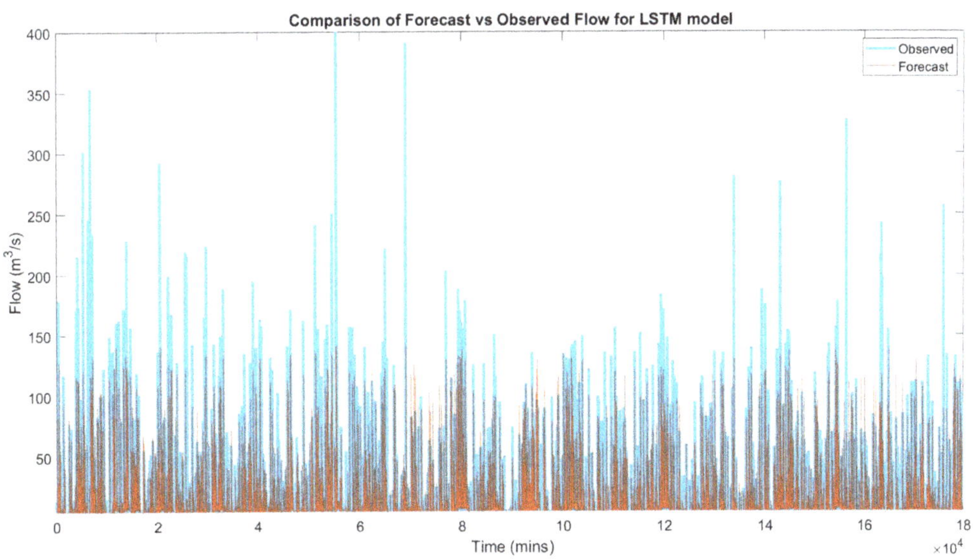

Figure 8. Graph of LSTM observed flow vs. simulated flow.

Step 2 introduces rates of change and executes multi-step ahead forecasting to facilitate the flood mitigation operation better. LSTM models are performed on multiple conditions, mainly simulation, 30 min ahead forecasting, 1 h ahead forecasting and rates of change δQ. Table 2 lists the performance of this exercise.

Table 2. LSTM forecasting models.

Model	R.Train	R.Test	MSE.Train	MSE.Test	MAE.Train	MAE.Test	NSE	RMSE
0 min.	0.9055	0.8586	17.8532	28.8315	1.4365	2.4208	0.8190	5.3695
30 min.	0.9470	0.9476	10.2326	10.0042	0.5640	0.6935	0.8963	3.1629
1 h	0.8849	0.8677	21.4296	25.0978	0.9397	1.2829	0.7828	5.0098
δQ	0.9545	0.9214	8.9746	15.6981	0.5434	0.8108	0.9090	2.9958

0 min refers to simulation of the streamflow in real time.

Step 3 develops several new hybrids of artificial intelligence. Three metaheuristic frameworks are selected for execution with the deep learning LSTM models: the bat algorithm, firefly algorithm and Jaya algorithm. Table 3 shows the streamflow prediction performance models for the hybrid model of the bat algorithm and LSTM. The parameters set for these models consist of maximum iteration = 40, alpha = 0.95, gamma = 0.95, bat numbers = 4, bat minimum frequency = 0, bat maximum frequency = 1 and maximum epochs = 500.

Table 3. Performance of LSTM with bat algorithm models for streamflow.

Model	R.Train	R.Test	MSE.Train	MSE.Test	MAE.Train	MAE.Test	NSE	RMSE
0 min.	0.9237	0.8525	14.2629	29.9912	1.4419	2.5535	0.8529	3.7766
30 min.	0.9512	0.9355	9.5954	11.4159	0.6157	0.7815	0.9043	3.0976
1 h	0.8932	0.8473	19.5178	28.9255	0.9876	1.3884	0.7976	4.4179
δQ	0.9757	0.9046	4.7187	19.8966	0.4672	0.8565	0.9514	2.1723

Table 4 displays the performance of the LSTM model after integration with the firefly algorithm. The parameters set for these models consist of maximum iteration = 40, alpha = 0.95, betta = 1, gamma = 0.95, firefly numbers = 4 and maximum epochs = 500.

Table 4. Performance of LSTM with firefly algorithm models for streamflow.

Model	R.Train	R.Test	MSE.Train	MSE.Test	MAE.Train	MAE.Test	NSE	RMSE
0 min.	0.9491	0.8214	9.6326	43.8420	1.3235	2.5607	0.9006	3.1036
30 min.	0.9743	0.9291	5.0785	12.5037	0.5178	0.7324	0.9493	2.2536
1 h	0.9146	0.8447	15.8306	30.6191	0.8815	1.2835	0.8365	3.9788
δQ	0.9733	0.8990	5.1913	20.4748	0.4910	0.8525	0.9465	2.2784

Table 5 shows the performance of the LSTM model with the Jaya algorithm. The parameters set for these models consist of maximum iteration = 30, population = 5 and maximum epochs = 500.

Table 5. Performance of LSTM with Jaya algorithm models for streamflow simulation.

Model	R.Train	R.Test	MSE.Train	MSE.Test	MAE.Train	MAE.Test	NSE	RMSE
0 min.	0.9420	0.7773	10.9191	46.4891	1.4316	2.9865	0.8873	3.3044
30 min.	0.9741	0.9102	5.1395	15.7782	0.5536	0.8486	0.9487	2.2670
1 h	0.9401	0.7928	11.2678	38.2200	0.8736	1.5270	0.8836	3.3568
δQ	0.9738	0.9010	5.1252	20.5746	0.4766	0.8563	0.9475	2.2639

Figure 9 displays graphs of observed vs. forecast flow based on simulation, 30 min ahead forecasting, 1 h ahead forecasting and rates of change model.

A further check is performed on the hybrid optimization models to determine the MAPE, MAE and maximum error values for the flows equal to or greater than 150 m^3/s. This ensures the accuracy of forecasting high flow values, which is important in a flood mitigation operation.

3.1. Performance of Step 1

In Step 1, the LSTM and ANN algorithms were developed and compared. It was found that LSTM performed much better than ANN. Several literature reviews also supported this by identifying the LSTM as the best deep learning model for time series data due to its ability to keep selective memory. LSTM algorithm could also filter the hydrological noise and retrieve the intrinsic characteristics of the hydrological series for simulation and future forecasting purposes.

Table 1 indicated that the ANN model had a regression of 0.4520, MSE 78.4215, MAE 3.7135 m^3/s, NSE 0.1994 and RMSE 8.8556 m^3/s. Furthermore, the best LSTM had regression 0.9055, MSE 17,8532, MAE 1.4365 m^3/s, NSE 0.8190 and RMSE 5.3695 m^3/s. Generally, it had shown a double improvement in overall results.

When comparing the graphs between Figures 7 and 8 on peak-to-peak values between the observed and forecast flows, it was evident that LSTM was much better than ANN models. Therefore, LSTM was chosen as the primary research model for this study.

3.2. Performance of Step 2

Step 2 introduced rates of change as an innovative approach to the model development. In addition, multi-step ahead forecasting was performed as a requirement for flood mitigation operations. Table 2 revealed that the worst result was acquired for the 1 h ahead forecasting, where the regression value for training was the lowest at 0.8849. However, it had a better regression value for testing when compared to simulation. This trend was applicable to MSE and MAE for having the worst values. The NSE value also turned out to be the worst. Considering the longer forecasting time, the results of this study were still regarded as logical and satisfactory. The longer the forecasting time, the more uncertainties and missing information would appear.

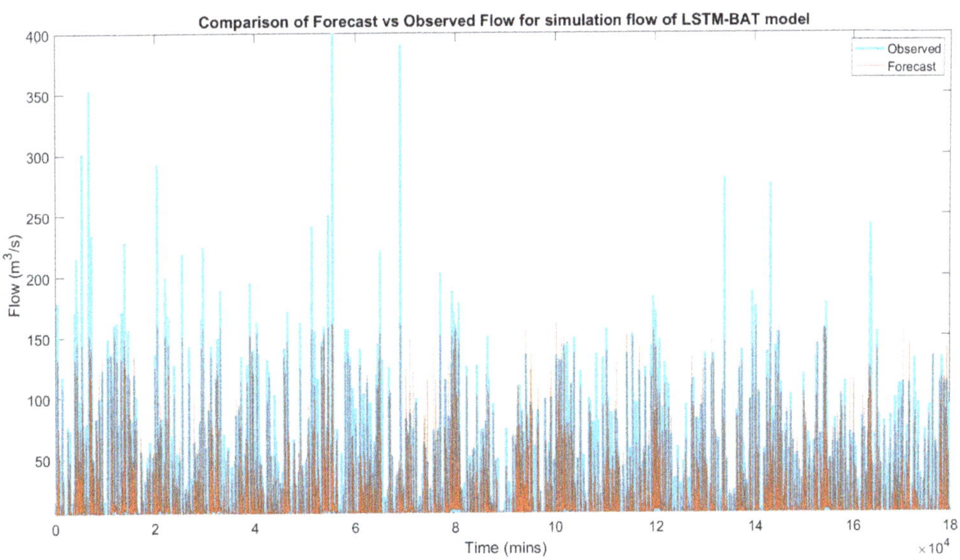

Bat algorithm: observed vs. forecast—flow simulation

Figure 9. *Cont.*

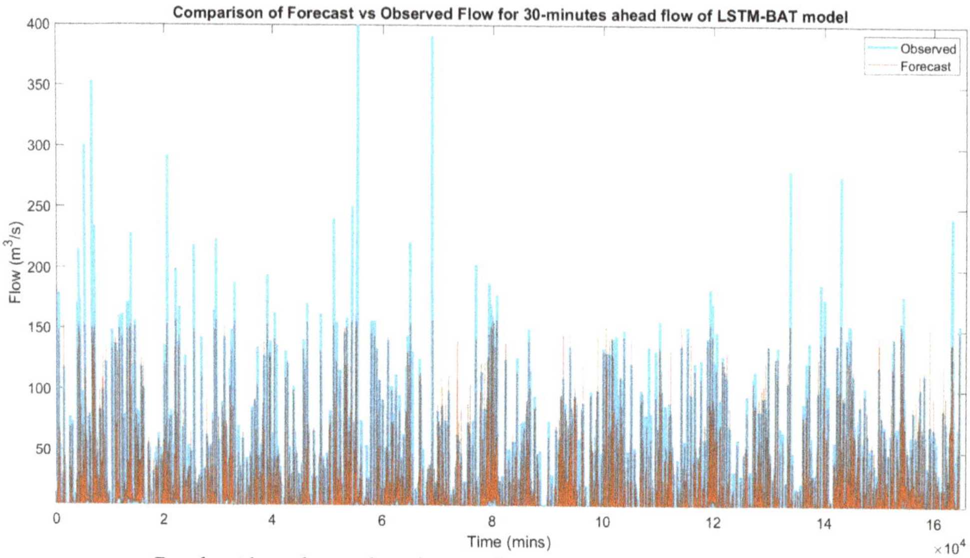

Bat algorithm: observed vs. forecast flow—30 min ahead of forecasting

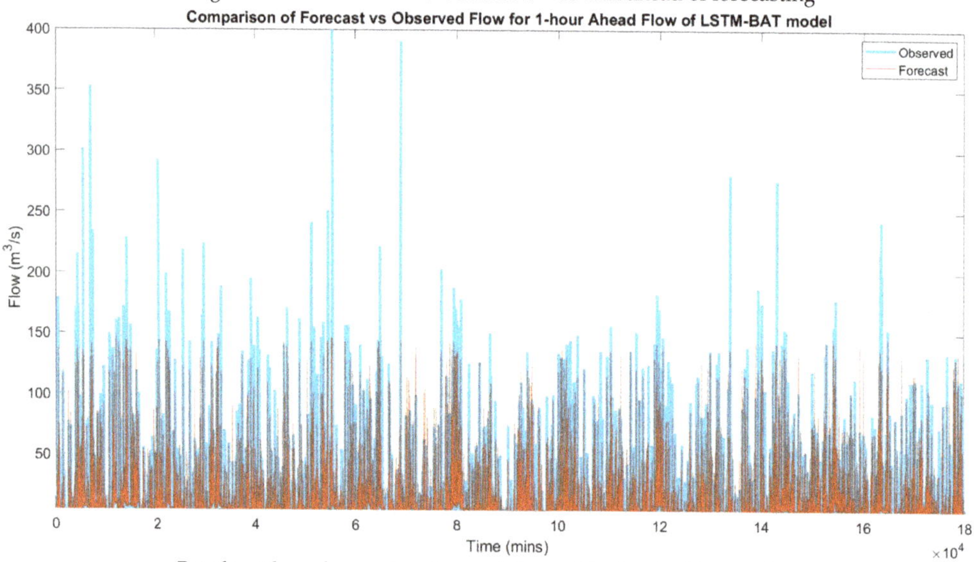

Bat algorithm: observed vs. forecast flow—1 h ahead of forecasting

Figure 9. *Cont.*

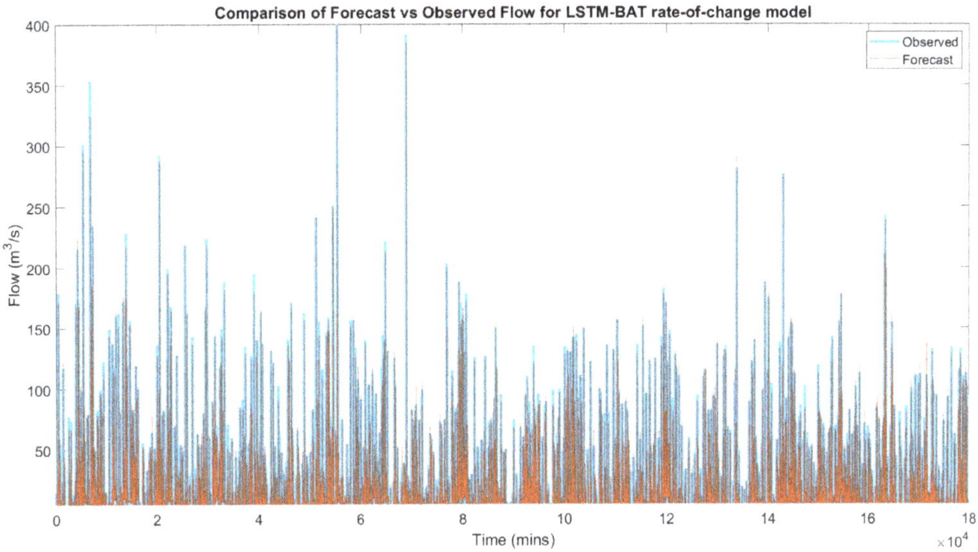

Bat algorithm: observed vs. forecast flow—rates of change model

Figure 9. Graphs of optimization-LSTM hybrid models.

The output from 30 min ahead forecasting turned out to be quite good as it had a regression training value of 0.9470, and the regression value for testing was not far off, which was 0.9476. The MSE and MAE values were low, which was good, with acceptable values for NSE and RMSE.

However, the best performance of the model was discovered with the novel method when applying rates of change as the target values. The regression value for training was the highest, 0.9545, while the error values were the lowest. The NSE value was the highest, 0.9090, while the RMSE was the lowest at 2.9958 m^3/s. The δQ model was the most superior among the four models tested.

The study did not seek an experiment of more than 1 h forecasting as the lag time determined was 30 min for this catchment. The results would deteriorate further as the time of forecasting increased.

3.3. *Performance of Step 3*

Step 3 was one of the main contributions of this study. Current metaheuristics studies mainly concentrate on developing a hybrid model with ANN or primary neural networks. Therefore, this study initiated the hybrid models for the deep learning algorithm, mainly the LSTM. Three metaheuristic frameworks, the bat algorithm, firefly algorithm and Jaya algorithm, were selected for this study. The bat algorithm and firefly algorithm belonged to swarm intelligence algorithms. They required trials on nature-based characteristics to find the optimum yield. Jaya algorithm, on the other hand, was designed based on searching for the best solutions. The effort to introduce numerous hybrid optimization algorithms was intended to further enhance the model performance results from steps 1 and 2.

Tables 3–5 represent each of the selected optimization algorithms. From the three tables, it was determined that all the hybrid models produced better results. However, the best model identified was the bat-LSTM hybrid algorithm where the δQ model yielded R.train 0.9757, R.test 0.9046, MSE.train 4.7187, MSE.test 19.8966, MAE.train 0.4672 m^3/s, MAE.test 0.8565 m^3/s, NSE 0.9514 and RMSE 2.1723 m^3/s.

The results also proved that the choice of metaheuristic algorithms did not significantly impact the performance. The performance inclination is still the same as the LSTM-only model in step 2, where 30 min ahead of forecasting yielded the best results. As the time of forecasting increased, the results deteriorated accordingly. δQ models consistently yielded the best results by keeping the error values to a minimum.

This process was then followed by the plotting of a peak-to-peak flow graph between the observed and the forecast values. Figure 9 indicated that the best graph with the highest accuracy was the δQ models.

A further experiment was performed to seek high flow performance for each hybrid model in terms of MAPE, MAE and maximum error. This study concentrates on the flows equal to or greater than 150 m^3/s, which was the high flow indicator to initiate modes 3 and 4 in the SMART control center's standard operating procedure. The results are tabulated in Table 6. From the results, it could be seen that δQ models again outperformed the rest with the smallest error values, where the bat and Jaya algorithms yielded the best with MAPE 6.33%, MAE 12.2865 m^3/s and maximum error 97.70% for bat-LSTM algorithm while MAPE 6.22%, MAE 12.6687 m^3/s and maximum error 97.70% for Jaya-LSTM algorithm. The maximum error values could be ignored in this case as they could not represent the overall performance of the models.

Table 6. Performance of LSTM-optimization algorithm models for streamflow forecasting.

	MAPE	MAE m^3/s	Max. Error
Bat-LSTM algorithm			
0 min	26.41%	59.6951	64.71%
30 min	24.03%	55.7999	64.02%
1 h	37.63%	80.4564	83.81%
δQ	6.33%	12.2865	97.70%
Firefly-LSTM algorithm			
0 min	15.11%	31.7240	56.92%
30 min	11.33%	22.8895	56.52%
1 h	27.55%	55.1539	86.05%
δQ	7.40%	14.6589	97.71%
Jaya-LSTM algorithm			
0 min	17.29%	37.1065	61.75%
30 min	11.08%	24.7873	53.09%
1 h	16.98%	35.3379	78.41%
δQ	6.22%	12.6687	97.70%

4. Conclusions

The effectiveness of flood management and disaster preparedness is in tandem with the ability to accurately forecast the immediate condition of streamflow in the catchment area. This study intended to develop the best deep learning model for the SMART control center in managing the river flow through streamflow forecasting. The aim was to create a novel approach in using rates of change for model development and introduce new metaheuristic algorithms with LSTM hybrid models to enhance the performance results.

This study employed LSTM models to develop and train historical data at the Ampang River and Klang River. The task is to forecast river streamflow with simulation, 30 min ahead, 1 h ahead and rates-of-change models. In order to ascertain the best performance that can be achieved, three steps of the improvement process were introduced.

Step 1 is where the comparison of ANN and LSTM models is performed. The best results come from the LSTM model with regression 0.9055, MSE 17,8532, MAE 1.4365 m^3/s, NSE 0.8190 and RMSE 5.3695 m^3/s. ANN yielded weaker results, and therefore LSTM model is the center of this research.

Step 2 introduces rates of change and performs multi-step ahead streamflow forecasting. The best result comes from the δQ model with performance values of R (training) = 0.9545, R (testing) = 0.9214, MSE (training) = 8.9746, MSE (testing) = 15.6981 , MAE (training) = 0.5434 m^3/s, MAE (testing) = 0.8108 m^3/s, NSE = 0.9090 and RMSE = 2.9958 m^3/s. The finding reveals that a shorter forecasting time yields better performance results. The second finding shows that applying new rate changes in model development has significantly improved the model results.

The last step of the experiment is to introduce new hybrid models between optimization and LSTM algorithms. The bat algorithm, firefly algorithm and Jaya algorithm were selected for this study. From the results, all hybrid models demonstrate better outcomes. Therefore, the third finding shows that metaheuristic algorithms play a role in model improvement. Under this study, it is also noticeable that the selection of an optimization algorithm does not significantly affect performance.

δQ model for the bat algorithm with LSTM hybrid model yielded the best results with R (training) = 0.9757, R (testing) = 0.9046, MSE (training) = 4.7187, MSE (testing) = 19.8966, MAE (training) = 0.4672 m^3/s, MAE (testing) = 0.8565 m^3/s, NSE = 0.9514 and RMSE = 2.1723 m^3/s.

Findings from this study are beneficial to improving the deep learning process so that the performance can yield better results with higher precision. This knowledge also helps elevate a new approach to flood mitigation operations. This study is significant as it has presented several new steps to improve the learning process leading to a better relationship between the input and output data. The current study is limited to a small catchment area and several optimization models. The results may differ for bigger catchments and with more optimization models. In order to further improve the experiment, it is suggested to try reinforcement learning for future studies.

Author Contributions: All authors contributed to the study and design. Material preparation, data collection, investigation, resources, data curation and analysis were performed by W.Y.T.; S.H.L. contributed to conceptualization, methodology, software teaching and supervision; F.Y.T. contributed to supervision; D.J.A. contributed to resources and funding acquisition; K.P. contributed to the software development; and A.E.-S. contributed to project management, visualization and supervision. The first draft of the manuscript was written by W.Y.T. and all authors commented on previous versions of the manuscript. All authors have read and agreed to the published version of the manuscript.

Funding: This research did not receive any specific grant from funding agencies in the public, commercial, or not-for-profit sectors.

Institutional Review Board Statement: Not applicable.

Informed Consent Statement: Not applicable.

Data Availability Statement: The data is available but subject to approval from the original data owner first.

Conflicts of Interest: The authors declare no conflict of interest.

References

1. Ahmed, A.M.; Deo, R.C.; Feng, Q.; Ghahramani, A.; Raj, N.; Yin, Z.; Yang, L. Deep learning hybrid model with Boruta-Random forest optimiser algorithm for streamflow forecasting with climate mode indices, rainfall, and periodicity. *J. Hydrol.* **2021**, *599*, 126350. [CrossRef]
2. Reis, G.B.; da Silva, D.D.; Filho, E.I.F.; Moreira, M.C.; Veloso, G.V.; Fraga, M.D.S.; Pinheiro, S.A.R. Effect of environmental covariable selection in the hydrological modeling using machine learning models to predict daily streamflow. *J. Environ. Manag.* **2021**, *290*, 112625. [CrossRef] [PubMed]
3. Ni, L.; Wang, D.; Wu, J.; Wang, Y.; Tao, Y.; Zhang, J.; Liu, J. Streamflow forecasting using extreme gradient boosting model coupled with Gaussian mixture model. *J. Hydrol.* **2020**, *586*, 124901. [CrossRef]
4. Marechal, D. A Soil-Based Approach to Rainfall-Runoff Modelling in Ungauged Catchments for England and Wales. Ph.D. Thesis, Cranfield University, Cranfield, UK, 2004; p. 145.
5. Kao, I.-F.; Liou, J.-Y.; Lee, M.-H.; Chang, F.-J. Fusing stacked autoencoder and long short-term memory for regional multistep-ahead flood inundation forecasts. *J. Hydrol.* **2021**, *598*, 126371. [CrossRef]
6. Lin, Y.; Wang, D.; Wang, G.; Qiu, J.; Long, K.; Du, Y.; Xie, H.; Wei, Z.; Shangguan, W.; Dai, Y. A hybrid deep learning algorithm and its application to streamflow prediction. *J. Hydrol.* **2021**, *601*, 126636. [CrossRef]
7. Xiang, Z.; Demir, I. Distributed long-term hourly streamflow predictions using deep learning—A case study for State of Iowa. *Environ. Model. Softw.* **2020**, *131*, 104761. [CrossRef]

8. Granata, F.; Di Nunno, F.; de Marinis, G. Stacked machine learning algorithms and bidirectional long short-term memory networks for multi-step ahead streamflow forecasting: A comparative study. *J. Hydrol.* **2022**, *613*, 128431. [CrossRef]
9. Elbeltagi, A.; Di Nunno, F.; Kushwaha, N.L.; de Marinis, G.; Granata, F. River flow rate prediction in the Des Moines watershed (Iowa, USA): A machine learning approach. *Stoch. Environ. Res. Risk Assess.* **2022**, *36*, 3835–3855. [CrossRef]
10. Peng, H.; Xiao, W.; Han, Y.; Jiang, A.; Xu, Z.; Li, M.; Wu, Z. Multi-strategy firefly algorithm with selective ensemble for complex engineering optimization problems. *Appl. Soft Comput.* **2022**, *120*, 108634. [CrossRef]
11. Karimi-Mamaghan, M.; Mohammadi, M.; Meyer, P.; Karimi-Mamaghan, A.M.; Talbi, E.-G. Machine learning at the service of meta-heuristics for solving combinatorial optimization problems: A state-of-the-art. *Eur. J. Oper. Res.* **2022**, *296*, 393–422. [CrossRef]
12. Khosravi, K.; Golkarian, A.; Tiefenbacher, J.P. Using Optimized Deep Learning to Predict Daily Streamflow: A Comparison to Common Machine Learning Algorithms. *Water Resour. Manag.* **2022**, *36*, 699–716. [CrossRef]
13. Gambella, C.; Ghaddar, B.; Naoum-Sawaya, J. Optimization problems for machine learning: A survey. *Eur. J. Oper. Res.* **2021**, *290*, 807–828. [CrossRef]
14. Bayındır, Y.; Yolcu, O.C.; Temel, F.A.; Turan, N.G. Evaluation of a cascade artificial neural network for modeling and optimization of process parameters in co-composting of cattle manure and municipal solid waste. *J. Environ. Manag.* **2022**, *318*, 115496. [CrossRef] [PubMed]
15. Zhang, W.; Gu, X.; Tang, L.; Yin, Y.; Liu, D.; Zhang, Y. Application of machine learning, deep learning and optimization algorithms in geoengineering and geoscience: Comprehensive review and future challenge. *Gondwana Res.* **2022**, *109*, 1–17. [CrossRef]
16. Ibrahim, K.S.M.H.; Huang, Y.F.; Ahmed, A.N.; Koo, C.H.; El-Shafie, A. A review of the hybrid artificial intelligence and optimization modelling of hydrological streamflow forecasting. *Alex. Eng. J.* **2021**, *61*, 279–303. [CrossRef]
17. Horne, A.; Szemis, J.M.; Kaur, S.; Webb, J.A.; Stewardson, M.J.; Costa, A.; Boland, N. Optimization tools for environmental water decisions: A review of strengths, weaknesses, and opportunities to improve adoption. *Environ. Model. Softw.* **2016**, *84*, 326–338. [CrossRef]
18. Song, H.; Triguero, I.; Özcan, E. A review on the self and dual interactions between machine learning and optimisation. *Prog. Artif. Intell.* **2019**, *8*, 143–165. [CrossRef]
19. Li, H.; Song, B.; Tang, X.; Xie, Y.; Zhou, X. A multi-objective bat algorithm with a novel competitive mechanism and its application in controller tuning. *Eng. Appl. Artif. Intell.* **2021**, *106*, 104453. [CrossRef]
20. Ben, U.C.; Akpan, A.E.; Urang, J.G.; Akaerue, E.I.; Obianwu, V.I. Novel methodology for the geophysical interpretation of magnetic anomalies due to simple geometrical bodies using social spider optimization (SSO) algorithm. *Heliyon* **2022**, *8*, e09027. [CrossRef]
21. Ahmed, A.N.; Van Lam, T.; Hung, N.D.; Van Thieu, N.; Kisi, O.; El-Shafie, A. A comprehensive comparison of recent developed meta-heuristic algorithms for streamflow time series forecasting problem. *Appl. Soft Comput.* **2021**, *105*, 107282. [CrossRef]
22. Rao, R.V. Jaya: A simple and new optimization algorithm for solving constrained and unconstrained optimization problems. *Int. J. Ind. Eng. Comput.* **2016**, *7*, 19–34. [CrossRef]
23. Malek, M.R.A.; Aziz, N.A.A.; Alelyani, S.; Mohana, M.; Baharudin, F.N.A.; Ibrahim, Z. Comfort and energy consumption optimization in smart homes using bat algorithm with inertia weight. *J. Build. Eng.* **2022**, *47*, 103848. [CrossRef]
24. Lu, P.; Ye, L.; Zhao, Y.; Dai, B.; Pei, M.; Tang, Y. Review of meta-heuristic algorithms for wind power prediction: Methodologies, applications and challenges. *Appl. Energy* **2021**, *301*, 117446. [CrossRef]
25. Calvet, L.; de Armas, J.; Masip, D.; Juan, A.A. Learnheuristics: Hybridizing metaheuristics with machine learning for optimization with dynamic inputs. *Open Math.* **2017**, *15*, 261–280. [CrossRef]
26. Yang, X.-S. Multiobjective firefly algorithm for continuous optimization. *Eng. Comput.* **2012**, *29*, 175–184. [CrossRef]
27. Ding, Z.; Hou, R.; Xia, Y. Structural damage identification considering uncertainties based on a Jaya algorithm with a local pattern search strategy and L0.5 sparse regularization. *Eng. Struct.* **2022**, *261*, 114312. [CrossRef]
28. Kang, F.; Wu, Y.; Li, J.; Li, H. Dynamic parameter inverse analysis of concrete dams based on Jaya algorithm with Gaussian processes surrogate model. *Adv. Eng. Inform.* **2021**, *49*, 101348. [CrossRef]
29. Degertekin, S.; Bayar, G.Y.; Lamberti, L. Parameter free Jaya algorithm for truss sizing-layout optimization under natural frequency constraints. *Comput. Struct.* **2020**, *245*, 106461. [CrossRef]
30. Xu, Y.; Hu, C.; Wu, Q.; Jian, S.; Li, Z.; Chen, Y.; Zhang, G.; Zhang, Z.; Wang, S. Research on particle swarm optimization in LSTM neural networks for rainfall-runoff simulation. *J. Hydrol.* **2022**, *608*, 127553. [CrossRef]
31. Alizadeh, B.; Bafti, A.G.; Kamangir, H.; Zhang, Y.; Wright, D.B.; Franz, K.J. A novel attention-based LSTM cell post-processor coupled with bayesian optimization for streamflow prediction. *J. Hydrol.* **2021**, *601*, 126526. [CrossRef]
32. Johny, K.; Pai, M.L.; Adarsh, S. A multivariate EMD-LSTM model aided with Time Dependent Intrinsic Cross-Correlation for monthly rainfall prediction. *Appl. Soft Comput.* **2022**, *123*, 108941. [CrossRef]
33. Dikshit, A.; Pradhan, B.; Huete, A. An improved SPEI drought forecasting approach using the long short-term memory neural network. *J. Environ. Manag.* **2021**, *283*, 111979. [CrossRef] [PubMed]
34. Ishii, K.; Sato, M.; Ochiai, S. Prediction of leachate quantity and quality from a landfill site by the long short-term memory model. *J. Environ. Manag.* **2022**, *310*, 114733. [CrossRef] [PubMed]
35. Ni, L.; Wang, D.; Singh, V.P.; Wu, J.; Wang, Y.; Tao, Y.; Zhang, J. Streamflow and rainfall forecasting by two long short-term memory-based models. *J. Hydrol.* **2020**, *583*, 124296. [CrossRef]

36. Anshuman, A.; Eldho, T. Entity aware sequence to sequence learning using LSTMs for estimation of groundwater contamination release history and transport parameters. *J. Hydrol.* **2022**, *608*, 127662. [CrossRef]
37. Sadler, J.M.; Appling, A.P.; Read, J.S.; Oliver, S.K.; Jia, X.; Zwart, J.A.; Kumar, V. Multi-Task Deep Learning of Daily Streamflow and Water Temperature. *Water Resour. Res.* **2022**, *58*, e2021WR030138. [CrossRef]
38. Han, H.; Morrison, R.R. Improved runoff forecasting performance through error predictions using a deep-learning approach. *J. Hydrol.* **2022**, *608*, 127653. [CrossRef]
39. Yang, X.-S.; He, X. Bat algorithm: Literature review and applications. *Int. J. Bio-Inspired Comput.* **2013**, *5*, 141–149. [CrossRef]
40. Fister, I.; Yang, X.-S.; Fong, S.; Zhuang, Y. Bat algorithm: Recent advances. In Proceedings of the 2014 IEEE 15th International symposium on computational intelligence and informatics (CINTI), Budapest, Hungary, 19–21 November 2014; pp. 163–167. [CrossRef]
41. Kumar, V.; Kumar, D. A Systematic Review on Firefly Algorithm: Past, Present, and Future. *Arch. Comput. Methods Eng.* **2020**, *28*, 3269–3291. [CrossRef]
42. Li, J.; Wei, X.; Li, B.; Zeng, Z. A survey on firefly algorithms. *Neurocomputing* **2022**, *500*, 662–678. [CrossRef]
43. Yang, X. Firefly Algorithms for Multimodal Optimization. In *International Symposium on Stochastic Algorithms*; Springer: Berlin/Heidelberg, Germany, 2009; Volume 5792, pp. 169–178.
44. Abu Zitar, R.; Al-Betar, M.A.; Awadallah, M.A.; Abu Doush, I.; Assaleh, K. An Intensive and Comprehensive Overview of JAYA Algorithm, Its Versions and Applications. *Arch. Comput. Methods Eng.* **2021**, *29*, 763–792. [CrossRef] [PubMed]
45. Aslay, S.E.; Dede, T. 3D cost optimization of 3 story RC constructional building using Jaya algorithm. *Structures* **2022**, *40*, 803–811. [CrossRef]
46. Zhao, F.; Zhang, H.; Wang, L.; Ma, R.; Xu, T.; Zhu, N. A surrogate-assisted Jaya algorithm based on optimal directional guidance and historical learning mechanism. *Eng. Appl. Artif. Intell.* **2022**, *111*, 104775. [CrossRef]
47. Jackson, E.K.; Roberts, W.; Nelsen, B.; Williams, G.P.; Nelson, E.J.; Ames, D.P. Introductory overview: Error metrics for hydrologic modelling—A review of common practices and an open source library to facilitate use and adoption. *Environ. Model. Softw.* **2019**, *119*, 32–48. [CrossRef]
48. Althoff, D.; Rodrigues, L.N. Goodness-of-fit criteria for hydrological models: Model calibration and performance assessment. *J. Hydrol.* **2021**, *600*, 126674. [CrossRef]
49. Feng, Z.-K.; Shi, P.-F.; Yang, T.; Niu, W.-J.; Zhou, J.-Z.; Cheng, C.-T. Parallel cooperation search algorithm and artificial intelligence method for streamflow time series forecasting. *J. Hydrol.* **2022**, *606*, 127434. [CrossRef]
50. Kim, T.; Yang, T.; Gao, S.; Zhang, L.; Ding, Z.; Wen, X.; Gourley, J.J.; Hong, Y. Can artificial intelligence and data-driven machine learning models match or even replace process-driven hydrologic models for streamflow simulation? A case study of four watersheds with different hydro-climatic regions across the CONUS. *J. Hydrol.* **2021**, *598*, 126423. [CrossRef]
51. Ding, Y.; Zhu, Y.; Feng, J.; Zhang, P.; Cheng, Z. Interpretable spatio-temporal attention LSTM model for flood forecasting. *Neurocomputing* **2020**, *403*, 348–359. [CrossRef]
52. Mohtar, W.H.M.W.; Abdullah, J.; Maulud, K.N.A.; Muhammad, N.S. Urban flash flood index based on historical rainfall events. *Sustain. Cities Soc.* **2020**, *56*, 102088. [CrossRef]
53. Zabidi, H.; De Freitas, M.H. Re-evaluation of rock core logging for the prediction of preferred orientations of karst in the Kuala Lumpur Limestone Formation. *Eng. Geol.* **2011**, *117*, 159–169. [CrossRef]
54. Othman, F.; Alaaeldin, M.; Seyam, M.; Ahmed, A.N.; Teo, F.Y.; Fai, C.M.; Afan, H.A.; Sherif, M.; Sefelnasr, A.; El-Shafie, A. Efficient river water quality index prediction considering minimal number of inputs variables. *Eng. Appl. Comput. Fluid Mech.* **2020**, *14*, 751–763. [CrossRef]
55. Kim-Soon, N.; Isah, N.; Ali, M.B.; Bin Ahmad, A.R. Relationships Between Stormwater Management and Road Tunnel Maintenance Works, Flooding and Traffic Flow. *Adv. Sci. Lett.* **2016**, *22*, 1845–1848. [CrossRef]
56. Bell, V.; Rehan, B.; Hasan-Basri, B.; Houghton-Carr, H.; Miller, J.; Reynard, N.; Sayers, P.; Stewart, E.; Toriman, M.E.; Yusuf, B.; et al. Flood Impacts across Scales: Towards an integrated multi-scale approach for Malaysia. In Proceedings of the 4th European Conference on Flood Risk Management (FLOODrisk2020), Online, 22–24 June 2021. [CrossRef]
57. Alrabie, N.A.; Mohamat-Yusuff, F.; Rohasliney, H.; Zulkeflee, Z.; Amal, M.N.A.; Arshad, A.; Zulkifli, S.Z.; Wijaya, A.R.; Masood, N.; Sani, M.S.A. Preliminary Evaluation of Heavy Metal Contamination and Source Identification in Kuala Lumpur SMART Stormwater Pond Sediments Using Pb Isotopic Signature. *Sustainability* **2021**, *13*, 9020. [CrossRef]

applied sciences

MDPI

Article

Technical Energy Assessment and Sizing of a Second Life Battery Energy Storage System for a Residential Building Equipped with EV Charging Station

Farhad Salek, Shahaboddin Resalati *, Denise Morrey, Paul Henshall and Aydin Azizi

Faculty of Technology, Design and Environment, Oxford Brookes University, Oxford OX3 0BP, UK
* Correspondence: sresalati@brookes.ac.uk

Abstract: This study investigates the design and sizing of the second life battery energy storage system applied to a residential building with an EV charging station. Lithium-ion batteries have an approximate remaining capacity of 75–80% when disposed from Electric Vehicles (EV). Given the increasing demand of EVs, aligned with global net zero targets, and their associated environmental impacts, the service life of these batteries, could be prolonged with their adoption in less demanding second life applications. In this study, a technical assessment of an electric storage system based on second life batteries from electric vehicles (EVs) is conducted for a residential building in the UK, including an EV charging station. The technical and energy performance of the system is evaluated, considering different scenarios and assuming that the EV charging load demand is added to the off-grid photovoltaic (PV) system equipped with energy storage. Furthermore, the Nissan Leaf second life batteries are used as the energy storage system in this study. The proposed off-grid solar driven energy system is modelled and simulated using MATLAB Simulink. The system is simulated on a mid-winter day with minimum solar irradiance and maximum energy demand, as the worst case scenario. A switch for the PV system has been introduced to control the overcharging of the second life battery pack. The results demonstrate that adding the EV charging load to the off-grid system increased the instability of the system. This, however, could be rectified by connecting additional battery packs (with a capacity of 5.850 kWh for each pack) to the system, assuming that increasing the PV installation area is not possible due to physical limitations on site.

Keywords: second life batteries; off-grid PV system; residential building; EV charging station

Citation: Salek, F.; Resalati, S.; Morrey, D.; Henshall, P.; Azizi, A. Technical Energy Assessment and Sizing of a Second Life Battery Energy Storage System for a Residential Building Equipped with EV Charging Station. *Appl. Sci.* **2022**, *12*, 11103. https://doi.org/10.3390/app122111103

Academic Editor: Luisa F. Cabeza

Received: 29 September 2022
Accepted: 31 October 2022
Published: 2 November 2022

1. Introduction

Global concerns surrounding the decarbonization of energy systems have notably increased over the past years [1]. Distributed energy generation systems such as PV panels are one of the most promising technologies primarily contributing to the building service industry [2]. However, the main improvement to the technology has been in connection with the electrochemical efficiency of the PV cells [3]. Despite notable technological advancements, there are various technical challenges associated with their adoption in the building sector including the mismatch in the supply and demand timing. One of the possible solutions to address this challenge is to install electric storage systems (ESS) [4]. The ESS, integrated with the renewable energy systems equipped with PV panels, especially in the stand-alone (off-grid) systems, is used for peak shaving and power shifting from day time to peak load hours (mostly evenings) [5]. In stand-alone renewable energy systems in buildings, the total energy demand is supplied by solar or other renewable energy sources [6], making the energy supply and demand management an integral part of the system [7,8].

The energy storage systems although contributing positively to the energy management solutions, have considerable environmental impacts [9]. This is mainly associated with the extraction of raw materials such as Cobalt, Nickel, and Lithium, and energy intensive processes when manufacturing Lithium-ion batteries [10]. This impact, however, could be reduced by prolonging the service life of the batteries retired from their first application in EVs, to less demanding applications such as residential buildings [11]. The initial state of health (SoH) of the second life batteries in such applications is generally around 75–80% of their nominal capacity [12]. Such second life applications are also expected to provide financial benefits making renewable energy more affordable and desirable for the end-users [13].

Lithium-ion batteries used in electric vehicles are considered second life when their capacity reaches 80% of their initial value. The lithium-ion batteries can be used in less-stressed applications such as buildings until their end of life. In order to achieve highest life span of the SLBSs, the load stress applied to them should be minimized. For grid-connected systems, it will be managed by the battery management system (BMS) which controls the energy flow through the SLBs, and mostly the extra demand will be applied to the grid. However, in stand-alone systems, the stress level and variations of the load applied to the SLBs are higher than grid-connected systems. In addition, the size of the PVs and SLBs plays a key role in the stand-alone system to find the optimum energy performance of the system as well as achieving the highest life span for the SLBs. On the other hand, as the number of electric vehicles increases, more buildings are equipped with EV charging stations applying a significant extra load to the building energy storage system which may directly affect the SLBs service life. This is the case especially when these systems are designed to cover the building demands excluding EVCS.

Numerous studies have investigated the application of second life batteries for ESS in residential buildings. Hart et al. [14] studied second life batteries in a micro-grid using an equivalent circuit model (ECM) and validated the model against the experimental data. Furthermore, the performance of the microgrid with different architectures was assessed. The results demonstrated that the second life batteries could be successfully installed in grid-connected or islanded microgrid applications uninterrupting the normal operation of the system. Sun et al. [13] have introduced the integration of a 3 MW second life battery ESS with the grid for peak shaving in China. The mathematical modelling of the system as well as a cost-effective model for the BSS is developed. It has been demonstrated that employment of second life batteries in the grid for peak shaving in China is cost beneficial, especially for the grid companies. The impacts of the second life battery packs with a different state of health (SoH) on the performance of the system was investigated by Mathews et al. [15]. The semi-empirical degradation model was used for modelling demonstrating that second life batteries are comparatively more profitable than first life batteries in PV systems. Cusenza et al. [9] developed a mathematical model for the second life battery sizing and optimization of a stand-alone PV system for a net zero energy residential building. The second life battery sizing was performed to achieve the best load match of the building and the results confirmed the optimum ratio of battery size to PVs total power to achieve the best load match in the residential buildings.

Further, Uddin et al. [16], modelled a grid-connected residential building equipped with PV and second life ESS considering building demand in various times during the year. The ECM was used to predict the battery parameters at different times and estimated the battery degradation parameters. The results of their work demonstrated that by considering degradation effects on financial parameters, the second life batteries are no longer cost effective for the customers. The technical assessment of integration of second life batteries with grid-connected PV systems for a residential building is demonstrated in Assuncao et al. [17], by considering a typical European residential building load demand. MATLAB Simulink was used to model the proposed system for three scenarios: without storage, large (Nissan Leaf), and small (Citroen C0) second life battery energy storage system. In the first year, the employment of second life BSS resulted in a reduction of 82.1% and 78.8% in energy exchange between the building and the grid for large and small BSS, respectively. Tong et al. [18] has investigated the integration of second life batteries with an off-grid EV charging station in the United States, where MATLAB SIMULINK has been applied for mathematical modelling of the proposed system. The charging station cost was significantly reduced in some locations, along with the similar performance compared to new batteries in other places. It was evident from the reviewed literature that the integration of the second life battery ESS for a residential building with EV charging station has not been investigated. The main contribution of this study is to reveal the impacts of load increase on the sizing of the second life battery energy storage system. The load applied to the second life battery storage system in this study is the residential building electricity load plus EV charging station.The EV charging stations apply an extra load to the residential building load demand [19].

Accordingly, in this study, the design and sizing of the second life battery ESS applied to a residential building with an EV charging station is investigated. The proposed system is modelled using MATLAB SIMULINK. The performance and stability of the system is assessed in a day in the middle of the winter, with the lowest solar irradiance and highest demand. The assessment considers the second life battery ESS with a different number of packs. It is assumed that the roof area is fully covered with PVs, therefore, the energy supply demand mismatch and the system stability maintenance is accomplished by adjusting the ESS size. The energy assessment and SoH analysis are performed to compare the system energy exchange, degradation, and energy supply demand mismatch in various scenarios.

2. System Description

The study is based on an off-grid PV system designed for the energy consumption of a typical house located in Oxford, UK. The study assesses the impacts of adding EVCS demand on the ESS technical parameters, energy exchange, and degradation. The proposed off-grid renewable energy system with an EVCS component could be listed as PV panels, DC-DC converter, second life battery packs, DC-AC inverter, residential building' load, and EVCS. The block diagram and components of the proposed energy system is shown in Figure 1. According to the figure, the solar energy is converted to electrical energy by PV panels and some of the generated electrical energy will be stored in the second life battery packs, while the rest of the energy would be consumed directly by the AC consumers such as the residential building electric consumers and EVCS. The load demand profiles are presented in Figure 2 [20,21]. The demand profile represents the average UK household load according to the CREST demand model for 15,000 households in the UK [21]. The red line in Figure 2 indicates the building's daily electricity load. The EV charging station daily load applied to the system is also shown by the black line in Figure 2. The aggregate hourly load is also calculated based on the building and EV charging station loads as presented in Figure 2.

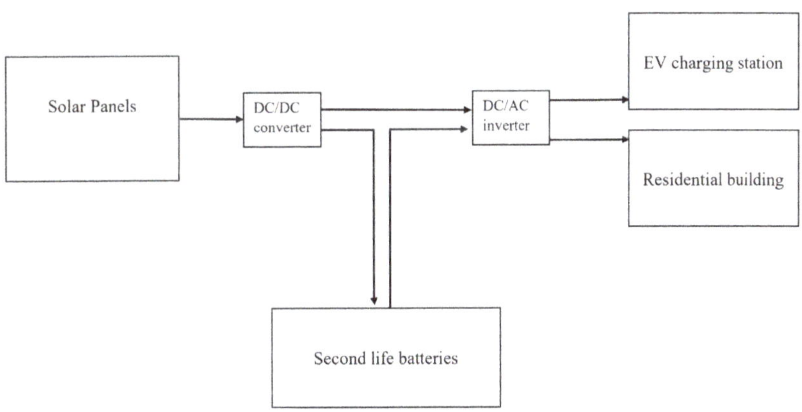

Figure 1. Schematic block diagram of the proposed off-grid energy system.

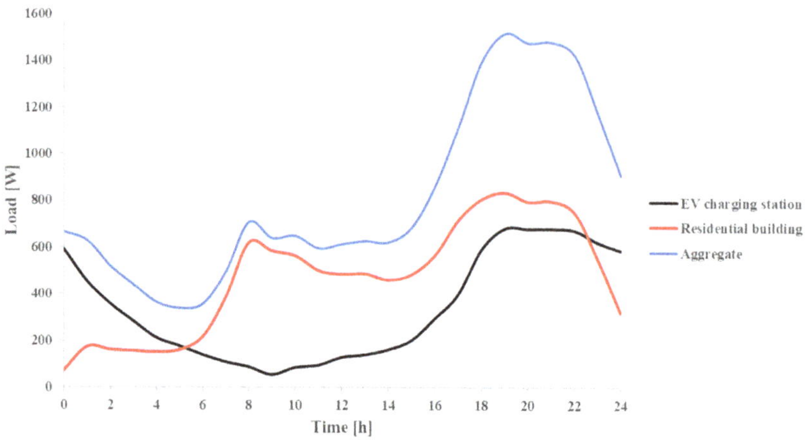

Figure 2. Different load demands applied to system [20,21].

Figures 3 and 4 demonstrate the block diagrams of the solar system and the second life battery pack. The SoC is monitored frequently during the solution of the model and is used for controlling the switches in the PV system to prevent battery packs from over charging. A MATLAB function is used to calculate the solar irradiance in different times during the day, the details of which will be presented in the next section (see Figure 4).

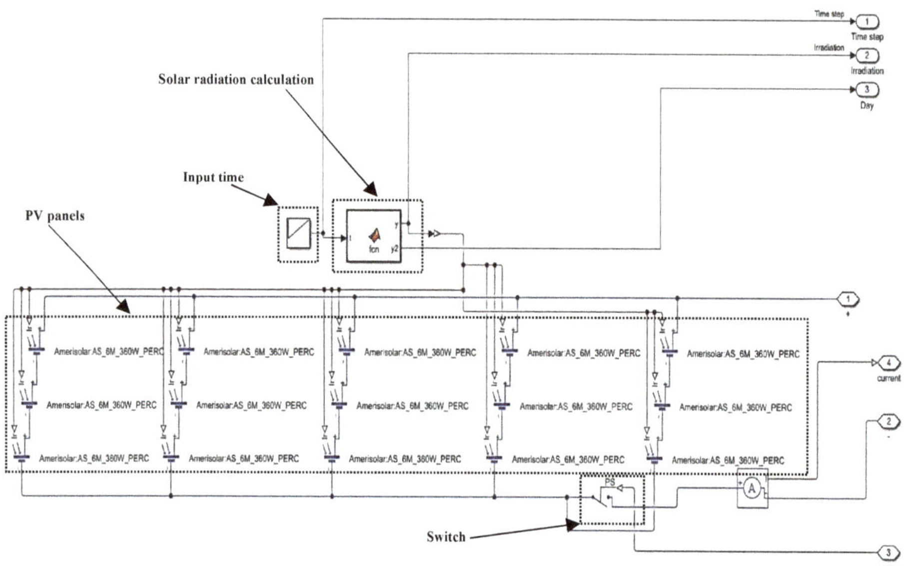

Figure 3. Block diagram of the solar PV system in MATLAB SIMULINK.

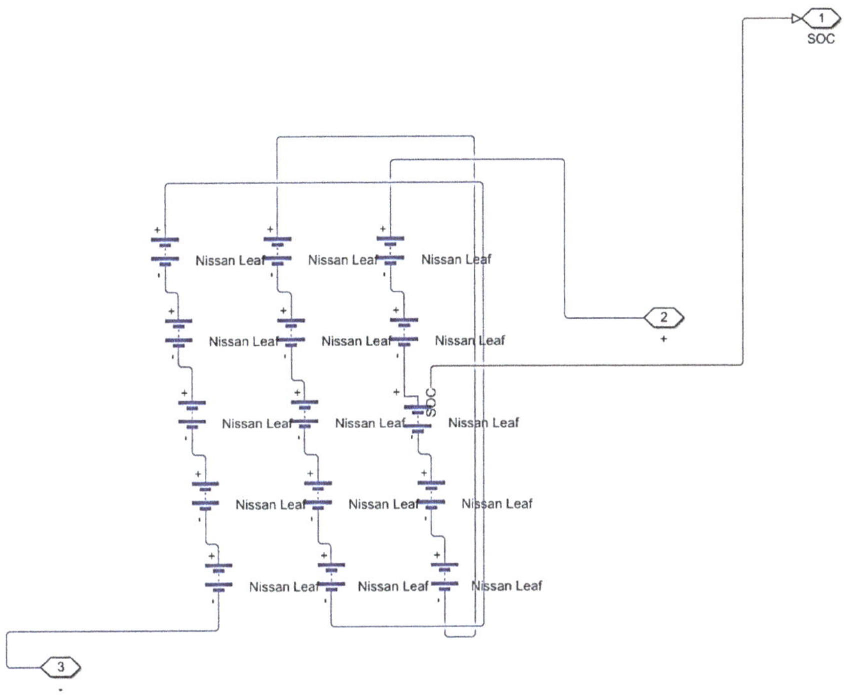

Figure 4. Block diagram of the 2nd life Nissan Leaf battery pack in MATLAB SIMULINK.

In this study, three scenarios for the off-grid PV system are defined and assessed as presented in Table 1. For the base scenario, two 2nd life battery packs connected in parallel are used, and only residential building demand is applied to the system. The second life batteries and the solar PVs specifications are provided in Tables 2 and 3. There are 15 s life modules in each battery pack (Figure 4) and the PV panels are connected with a 5 parallel and 3 series configuration (Figure 3).

Table 1. The defined scenarios in this study.

Scenarios	Number of Battery Packs	Number of PVs	Load Demand
Base	2	15	RB
EV-2P	2	15	RB + EVCS
EV-3P	3	15	RB + EVCS
EV-4P	4	15	RB + EVCS

Table 2. Second life battery pack specifications [22].

Parameter	Value
Model	Nissan Leaf
Number of modules in the pack	15
Modules configuration in the pack	series
Module nominal voltage [V]	7.5
Module maximum voltage [V]	8.3
Module minimum voltage [V]	5
Initial state of charge [%]	60
Second life module initial capacity [Ah]	47.026

Table 3. Solar PV panel specifications [23].

Parameter	Value
Model	Amerisolar-6 M 360 W
Voltage at maximum power [V]	38.7
Current at maximum power [A]	9.31
Open circuit voltage [V]	47.3
Panel efficiency [%]	18.55
Maximum power [W]	360
Cell number	72

3. Mathematical Modelling

As mentioned in the previous section, the mathematical modelling of the proposed system is performed in MATLAB SIMULINK software using the Simscape toolbox. The PV panel and batteries with other components are added to the SIMULINK environment and connected to each other with the desired architecture.

3.1. Solar PV Panels

For calculation of the solar irradiance based on the geographic location (Oxford, UK) and other technical parameters such as the tilt angle of the panel, a model has been designed in MATLAB function in SIMULINK, which calculates the solar irradiance in various simulation steps. For the calculation of beam radiation incidence angle on a surface (θ), Equation (1) is employed [24]:

$$\begin{aligned}\cos\theta = \sin\delta\sin\varphi\cos\beta - \sin\delta\cos\varphi\sin\beta\cos\gamma + \cos\delta\cos\varphi\cos\beta\cos\omega \\ + \cos\delta\sin\varphi\sin\beta\cos\gamma\cos\omega + \cos\delta\sin\beta\sin\gamma\sin\omega\end{aligned} \tag{1}$$

where δ, φ, β, ω and γ are declination, latitude, slope, hour angle and surface azimuth angle, respectively [24]. γ and β are assumed to be $0°$ and $30°$, respectively, since most of the houses in the UK has $30°$ slope on their ceilings, where PV panels arebe installed. The equation of Cooper is used for calculation of declination [24]:

$$\delta = 23.45\sin\left(360\frac{284+n}{365}\right) \tag{2}$$

where n is the number of days during the year. Further, the radiation on the tilted plane (G_o) could be calculated by Equation (3) [24,25]:

$$G_o = G_{sc}\left(1 + 0.0033\cos\frac{360n}{365}\right) \tag{3}$$

where G_{sc} is extraterrestrial radiation and assumed as 1367 W/m^2 in this study [1]. To calculate the beam and diffuse radiations transmitted through a clear atmosphere, the following equations are applied based on Hottel's method [1,24]:

$$\tau_b = a_0 + a_1 exp\left(\frac{-k}{\cos\theta_z}\right) \tag{4}$$

$$\tau_d = 0.271 - 0.294\tau_b \tag{5}$$

τ_b and τ_d are the atmospheric transmittance for beam and diffuse radiations, consecutively. Further information about the parameters used in Equations (4) and (5) (such as k, a_0 and a_1) can be found in this reference [24]. Finally, the clear-sky radiation (G_c) is obtained using Equation (6) [24]:

$$G_c = G_o(\tau_b + \tau_d) \tag{6}$$

The output current of the PV panel is given by [26]:

$$I_{pv} = I_{ph} - I_s\left(e^{\frac{V+I*R_S}{N*V_t}} - 1\right) - I_{s2}\left(e^{\frac{V+I*R_S}{N_2*V_t}} - 1\right) - \frac{V + I * R_s}{R_p} \quad (7)$$

where, I_s and I_{s2} are diode saturation currents for diodes 1 and 2 shown in Figure 5, respectively. V_t is the thermal voltage, N and N_2 are diode emission coefficients and I_{ph} is solar-generated current, respectively. The mentioned PV parameters are obtained from MATLAB SIMULINK Simscape library for Amerisolar PV panel the specifications of which are provided in Table 3 [26].

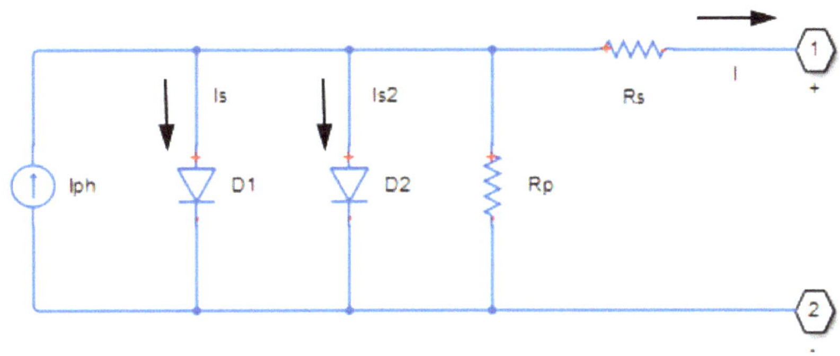

Figure 5. Block diagram of the equivalent circuit used for modeling PV cell in MATLAB.

3.2. Second Life Battery Pack

Rint ECM [27] was used for modelling the second life batteries in this study. Besides, the ECM model contains an ideal voltage source representative of OCV as the function of SoC with resistors to calculate the internal ohmic losses [28]. The output of this systems is calculated by the following Equation (8):

$$V_k = V_{OC,k} - I_k R_s \quad (8)$$

The experimental data for the parameters in this model is obtained from the literature [22,29–31]. In Nissan Leaf prismatic module, there are two cells integrated in a series configuration. In the reference [22], the Nissan Leaf battery is aged using an accelerated ageing profile (Figure 6) [22] in which the second life modules are put under constant current-voltage charging (=1C) and constant current discharging (=1C) at 25 °C environmental temperature. The reference performance test (RTP) is done every 25 cycles to measure the module capacity fade and HPPC test [22]. The HPPC test results are used to fit ECM Rint model parameters the results of which are shown in Figures 7–9.

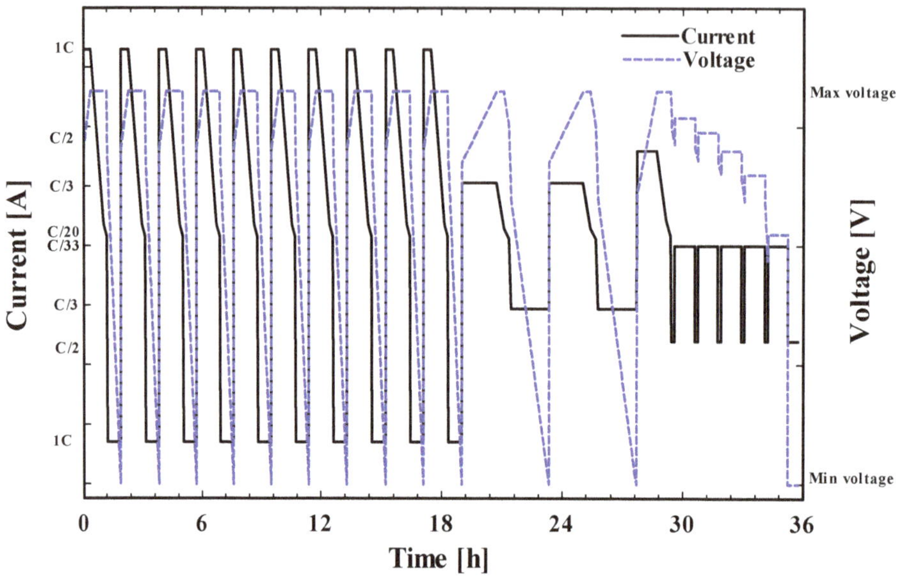

Figure 6. The accelerated ageing profile and RTP test for degradation analysis of the Nissan Leaf second life battery.

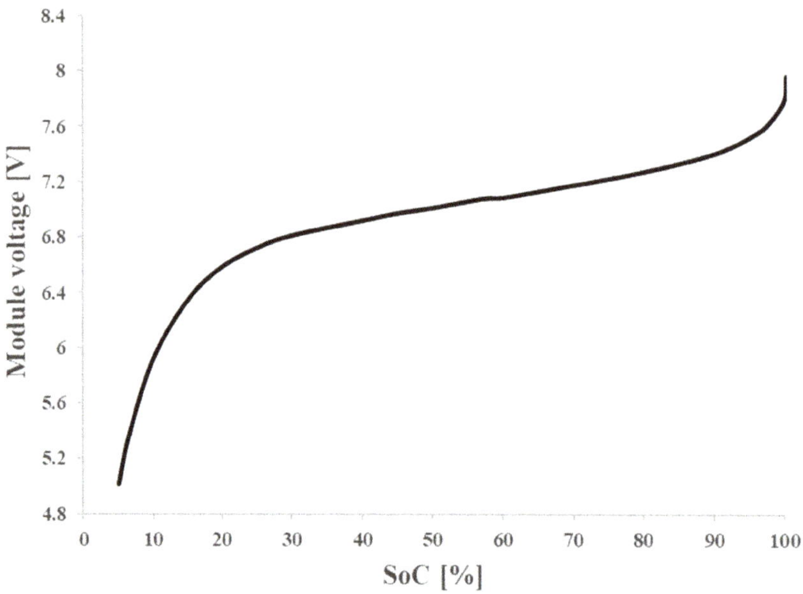

Figure 7. Nissan leaf second life battery voltage variations in various SoCs [22,29–31].

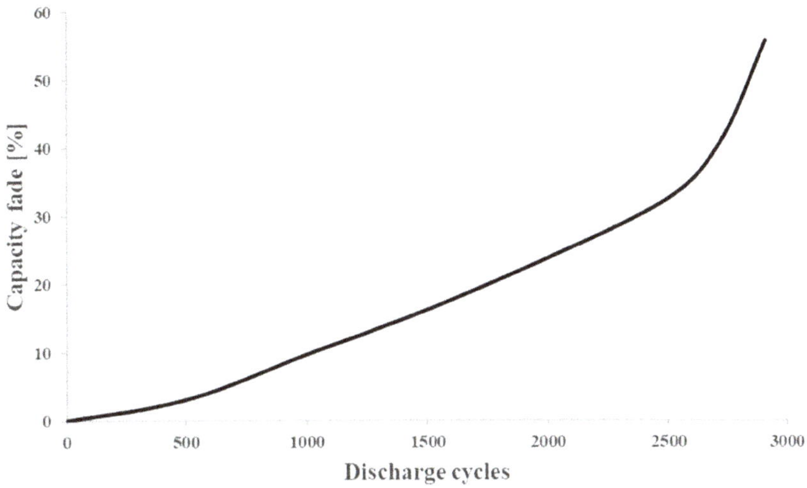

Figure 8. Nissan leaf second life battery capacity fade in various discharge cycles [22,29–31].

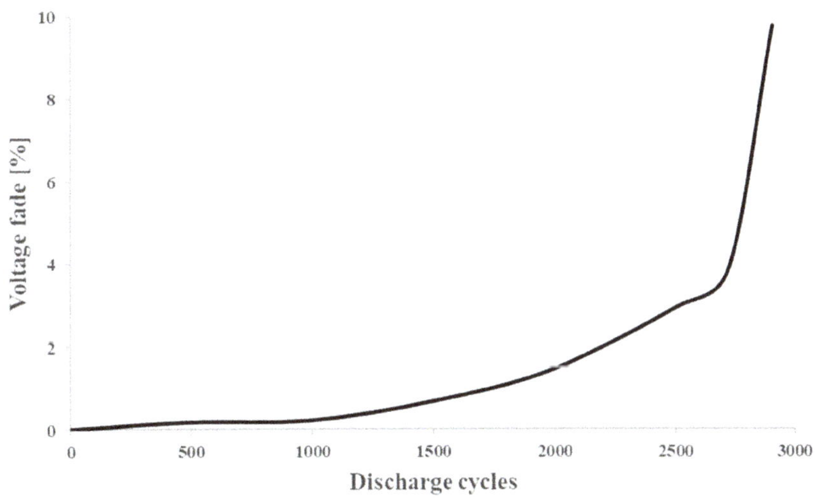

Figure 9. Nissan leaf second life battery voltage fade in various discharge cycles [22,29–31].

4. Results and Discussions

As the main aim of this study is to assess the impacts of second life battery sizing and demand variations on the energy performance of the system, a two-way coupling between the developed second life batteries and solar cells is performed. The operating voltage of the solar cells will be affected by the variation of second life batteries voltage leading to their efficiency variations which are considered in this paper. Accordingly, the solar system modelling is also performed and the impacts of extra load addition and SLB size increase on solar power generation are presented in Section 4.1. The energy exchange between the components and battery operational parameters is demonstrated during the day in Sections 4.2 and 4.3.

4.1. Impacts on Solar Power Generation

The developed mathematical model has been solved for 24 h in a day (17 January) in the middle of winter in Oxford, UK. In particular, the main reason for choosing a day at the middle of winter is to assess the performance of the system and ESS when the energy input (solar irradiance) is at its minimum values [4]. The solar irradiance is calculated using the model presented in Section 3.1. In addition, the simulation has considered 3600s as the time step. Figure 10 shows the aggregation of solar beam and diffuse radiations transmitted through the atmosphere installed on the tilted plane (PV panels).

Figure 11 demonstrates the current output of the PVs to the system during the day for different scenarios. As mentioned before, a switch, controlled by the batteries' SoC parameter is adopted to prevent the batteries from overcharging. Therefore, the switch will break the connection between the PVs and the battery storage system when the batteries are overcharged. The activation time of the switch can be figured out in Figure 11. For the base scenario (black line), in which only residential building demand is applied to the system, the switch is turned to active mode right after reaching the peak current value at 1 PM. The extra power generated by PVs would not be directed to the system afterward resulting in a sharp increase in PV current flow through the system after 1 PM. An increase of second life battery pack size by an increment of the number of modules from 2 to 4, when extra EVCS load is applied, leads to an increased PV system current flow to the system after reaching peak hours. This is due to an increase in the overcharging limit of the energy storage system and higher amounts of stored energy. Accordingly, the switch activation time is delayed by an increase in the number of packs.

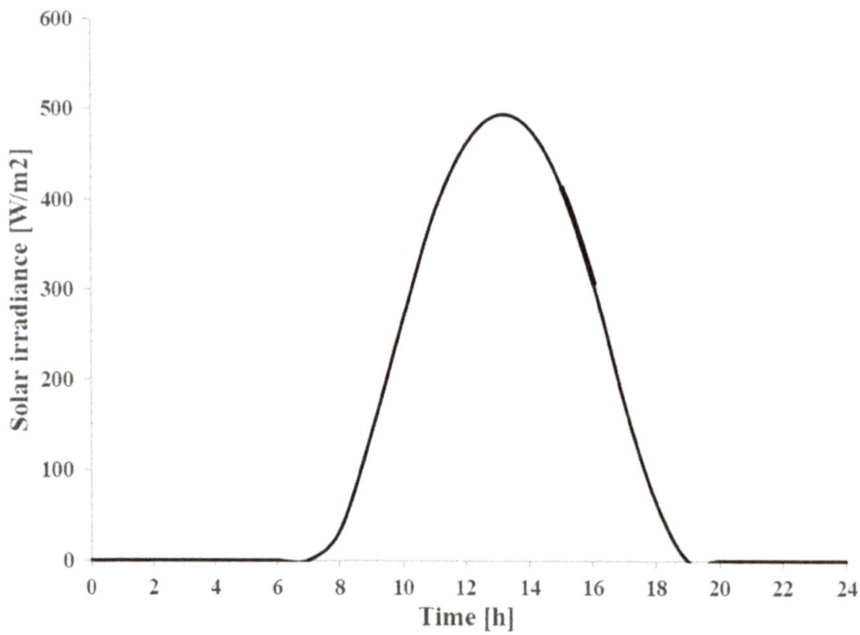

Figure 10. Solar irradiance (clear-sky radiation) at various times during the day on 17 January.

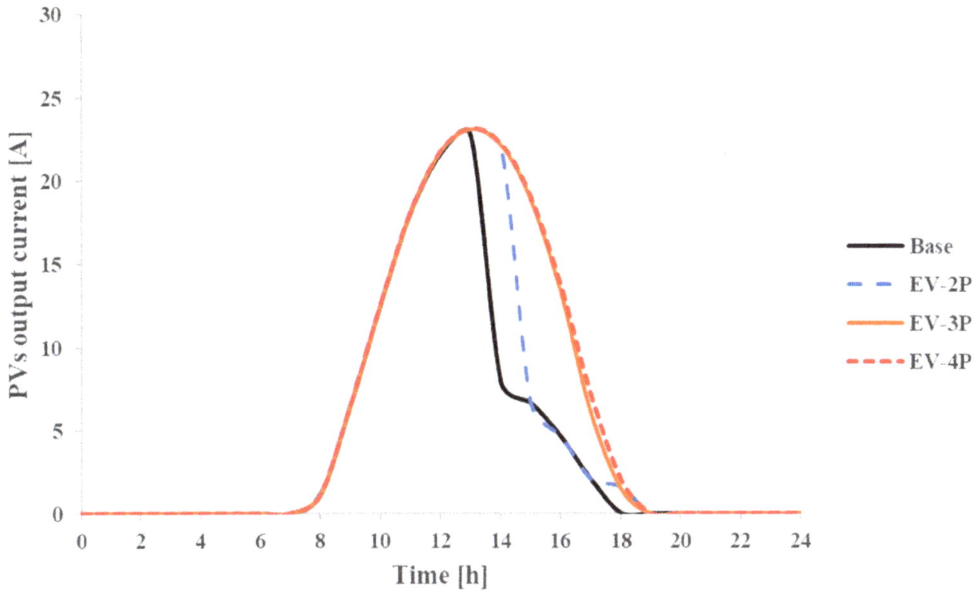

Figure 11. Second life battery SoC for different scenarios at various times during the day.

4.2. Impacts on Second Life Battery

Figures 12 and 13 show the current and the voltage of the second life battery packs during a day under different scenarios. The addition of the EVCS load demand leads to a decrease of the SLBs charging current peak up to approximately 10% as shown in Figure 12 due to the increment of the system demand. Furthermore, the peak discharge current is also increased and when the EVCS load is applied to the system. Figure 12 also reveals that the pack size increment will extend the charging capacity of the ESS as the integral of the charging current curve for the orange and red lines (3 and 4 packs) are higher than the curve representing the EV-2P scenario. This is due to the activation of the switch reaching the maximum charge capacity of the batteries. The peak voltage of the second life battery packs decreased with the increase in the number of packs as shown in Figure 13, primarily due to the increase in the total capacity of the system and gaining a more stable operational voltage. In Figure 13, reaching the maximum voltage of the ESS is delayed by an increase in the number of SLB packs. The maximum voltage of the ESS on a full charge. By increment of SLB size to 17.55 kWh and 23.4 kWh, in EV-3P and EV-4P scenarios, the peak voltages dropped to 116 V and 111 V, respectively. This is due to an increase in capacity resulting in a decrement in voltage variations of ESS and an extension of the SLB life span.

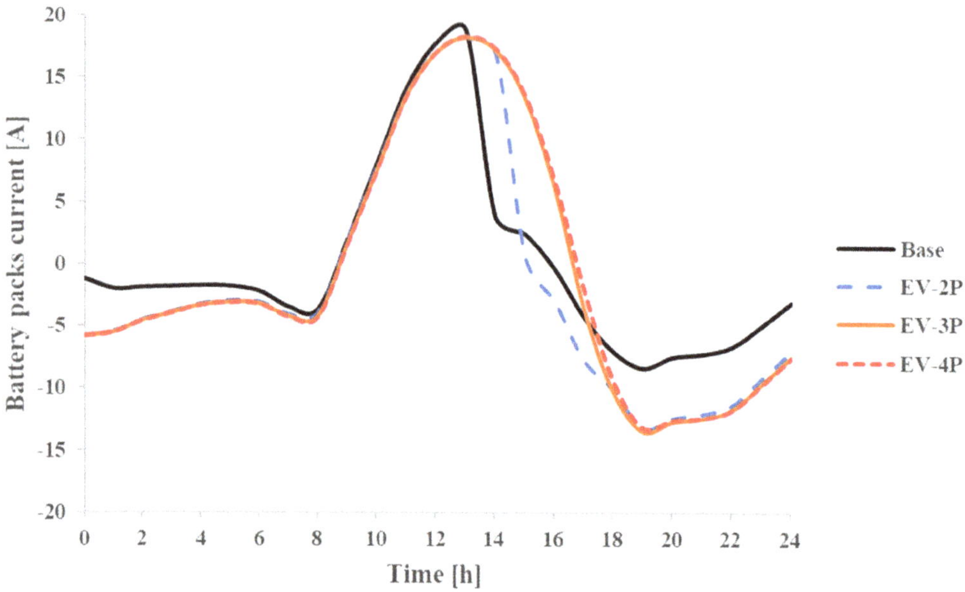

Figure 12. Second life battery current for different scenarios during the day.

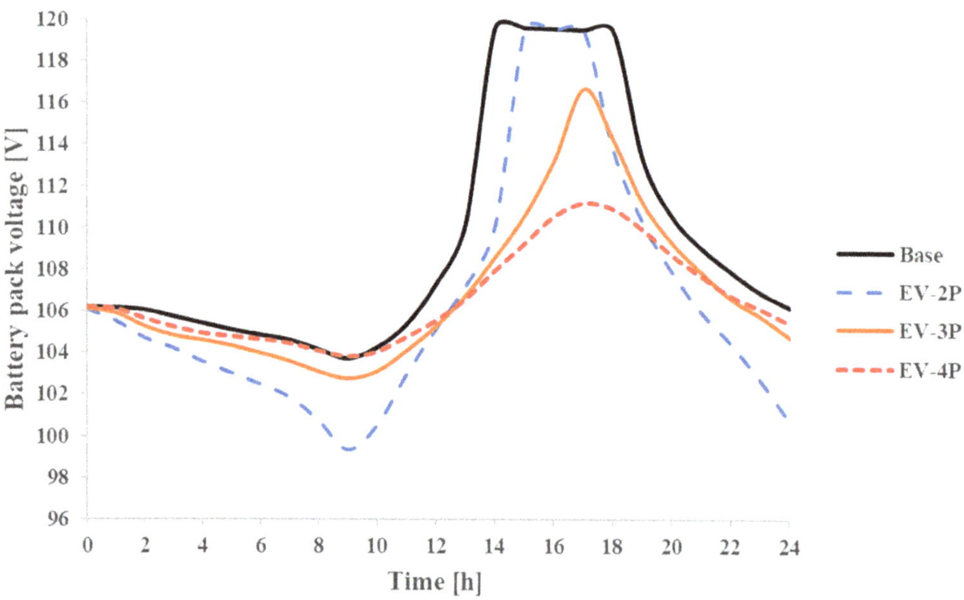

Figure 13. Second life battery voltage for different scenarios in various times during the day.

Figure 14 presents the SoC variations of the second life battery packs in transient conditions during the day for different scenarios. In particular, the comparison of the base and EV-2P scenarios (both with two battery packs) indicates that adding extra load demand to the system (EV charging station load) results in a rapid discharge of the battery packs from 00.00 to 01.00. Furthermore, the second life battery packs in the EV-2P scenario would be fully charged by approximately 1 h delay compared to the base scenario (which is fully charged at around 14.00). The final SoC of the battery packs also decreased dramatically from 60% (base scenario) to 26% (EV-2P scenario) by adding EVCS load to the system which could result in a significant mismatch between load and energy generation for the next day. To solve this issue the number of battery packs are increased in scenarios EV-3P and EV-4P to 3 and 4 packs, by assuming that the number of PVs are constant. According to Figure 10, by increasing the number of battery packs, the discharging curve between 10.00 to 12.00 is shifted upward by nearly 10%, and the SoC peak has decreased steeply due to the increased capacity of the ESS. Furthermore, the final SoC has increased by escalating the number of battery packs to nearly 53% for the EV-4P scenario. This suggests that increasing the SLB EES size up to 23.4 kWh would be beneficial in gaining a stable energy exchange between the components and reducing the energy generation-consumption mismatch in the proposed system. The main drawback of the size increment of energy storage system would be the increase in its cost, which might be solved by the employment of SLBs given their relatively lower price when compared with brand new batteries.

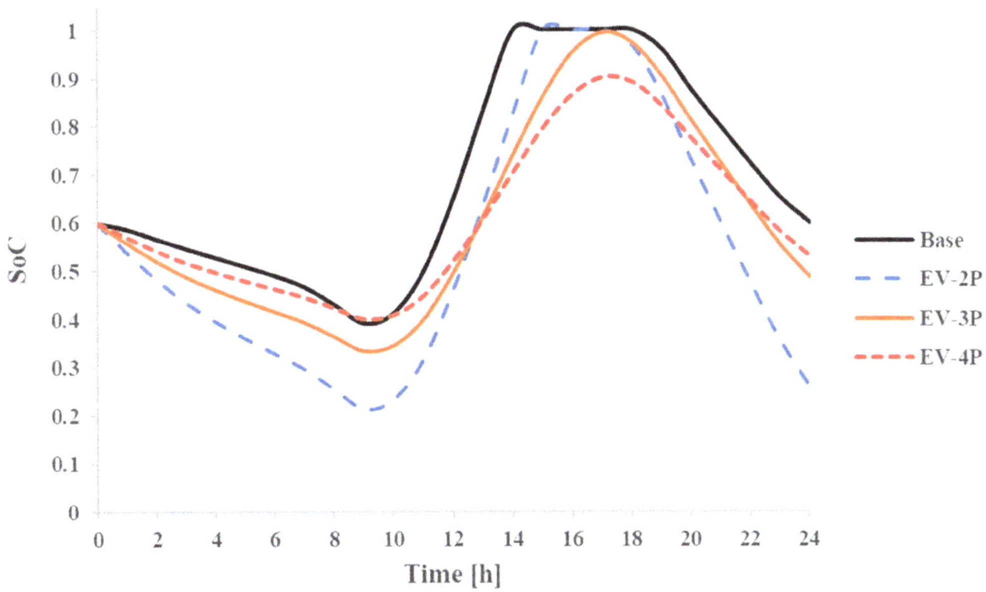

Figure 14. Second life battery SoC for different scenarios in various times during the day.

4.3. Energy Exchange Analysis

In this section, the energy exchange between the main components of the system; ESS, PVs, and the demand side is demonstrated. Table 4 presents the energy generated by the solar system against the demand side's consumption. For the base scenario, in which the solar system is designed to cover only the residential building demand, a significant difference is not observed between energy supply and demand. However, the aggregated demand has overcome the supply energy value by adding an extra load to the solar system. The difference between the energy supply and demand needs to be compensated by the amount of energy available in the batteries (the initial SoC of the batteries equals 60% in the simulations) in this scenario (EV-2P). This must be noted that the simulations are performed by considering the worst-case scenario; the solar irradiance and residential building energy demand are at their minimum and maximum rates during the year, respectively. The mismatch between energy supply and demand is expected to be minimum in the warmer months of the year. Table 4 indicates that increasing the number of second life batteries results in minimizing the difference between the energy supply and demand, which could increase the stability of the proposed energy system. The hourly energy exchange rate between the system components is shown in Figure 15a–d for different scenarios.

Table 4. Energy generation and demand for different scenarios for the proposed day.

Parameter	Availability of Input Energy by PV Panels [kWh/day]	Demand Energy [kWh/day]
Base	11.34	10.6
EV-2P	12.88	17.2
EV-3P	15.58	17.2
EV-4P	15.62	17.2

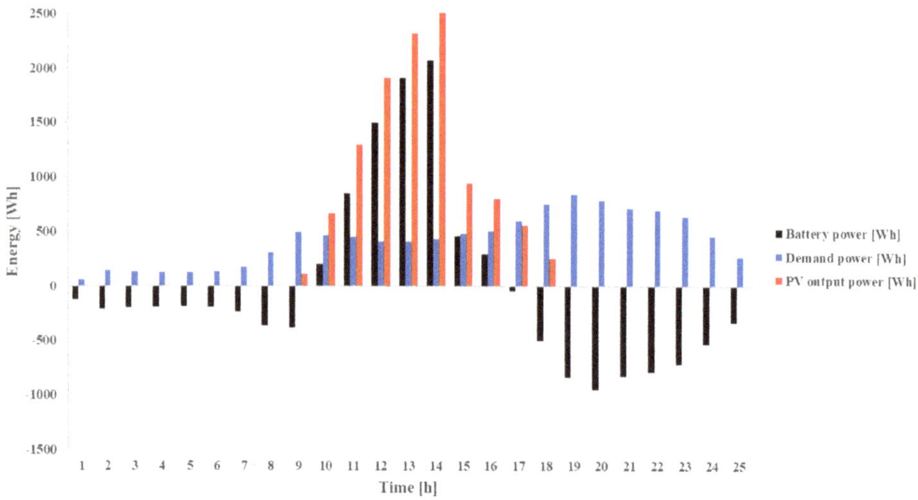

(a)

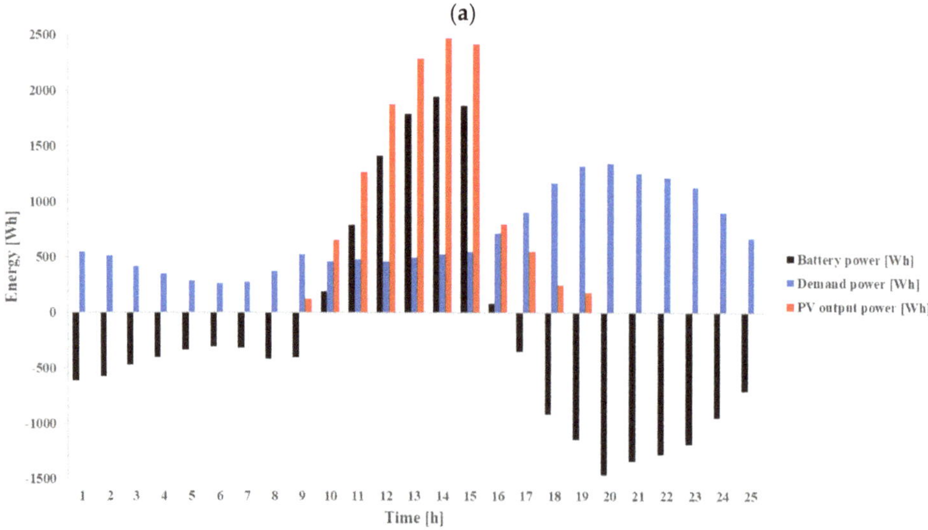

(b)

Figure 15. *Cont.*

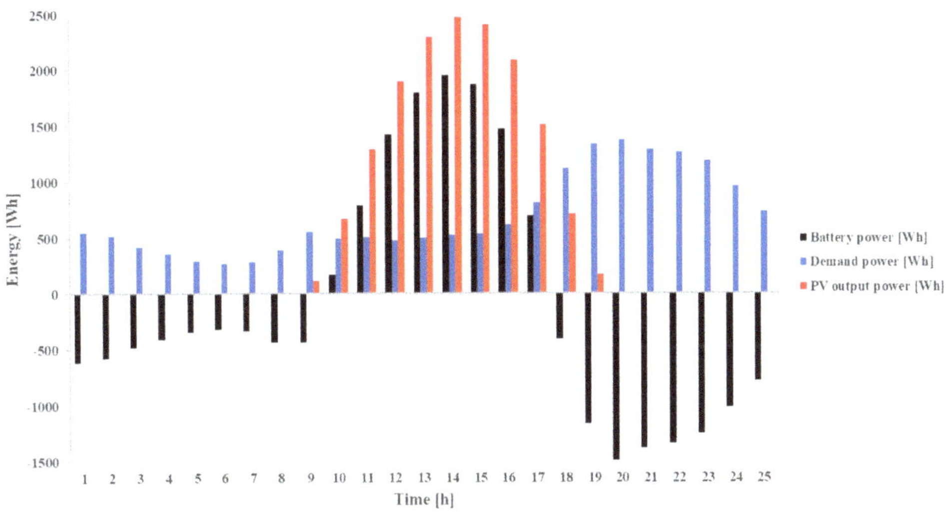

(c)

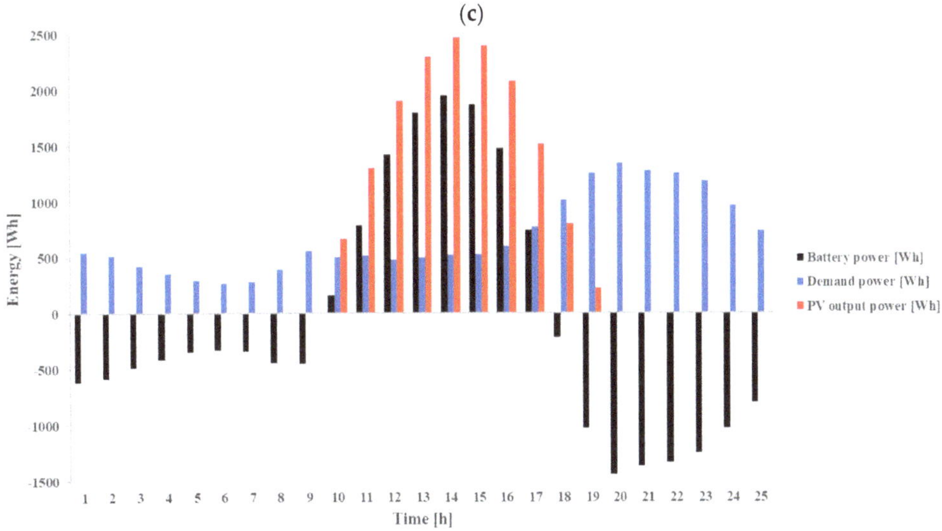

(d)

Figure 15. Proposed system hourly energy exchange for different components in (**a**) Base, (**b**) EV-2P, (**c**) EV-3P and (**d**) EV-4P scenarios.

Table 4 presents the energy generation and demand for different scenarios in this study. By adding extra EVCS load to the system in the EV-2P scenario, while the number of SLB packs is kept constant (compared to the base scenario), the minimal increase in PV panel power generation is seen due to variations of their operating voltage affected by battery packs voltage variations shown in Figure 13. An increase in SLB packs' size to 17.55 kWh (EV-3P) and 23.4 kWh (EV-4P) leads to an increase in PV panel energy generation by 21% and 21.27%, respectively, resulting from an increase in electricity storage capacity which allows higher rates of energy storage and lowering variations of PV panels operating voltage.

4.4. Impacts of ESS' SoH on Stability of the System

The SoH of the second life batteries would be decreased over time and reduce the total capacity of the ESS system. This could directly affect the off-grid system stability in terms of energy exchange between ESS and the other components. Additionally, the difference between the initial charge of the SLBs and their state of charge at the end of the day can be used as the indicator of energy storage system stability. If the SoC at the end of the day would be much lower than SoC at the beginning of the day, it suggests that the discharging rate of the SLBs is higher than their charging rate. Therefore, the ESS charging and discharging stability depends on two parameters: storage size and generation rate. The storage size is increased by the increment of the number of SLB packs; however, it will be also affected by cycle ageing of the batteries leading to decrement in size withbthe passing of time (also charging and discharging cycles). In this section, the impacts of SLBs sizing and cycle ageing on ESS stability are investigated. The difference between the initial and the final SoC is defined as the primary indicator of system stability in this study, which could be expressed as:

$$SoC_D = SoC_i - SoC_f \tag{9}$$

where SoC_D, SoC_i, and SoC_f are second life battery SoC difference, initial SoC and final SoC, respectively.

The effects of second life batteries' ageing on the system stability is illustrated in Figures 16 and 17 for the scenarios with extra EVCS loads. When EVCS load is applied, the stability of the system could be maintained by increasing the number of battery packs, due to the reduction of SoC difference. This translates to 4 battery packs in this study to hold the system stability at an acceptable level before reaching 2000 discharge cycles and approximately 60% SoH, since the SoC difference remains almost constant until reaching these points, as shown in Figures 16 and 17.

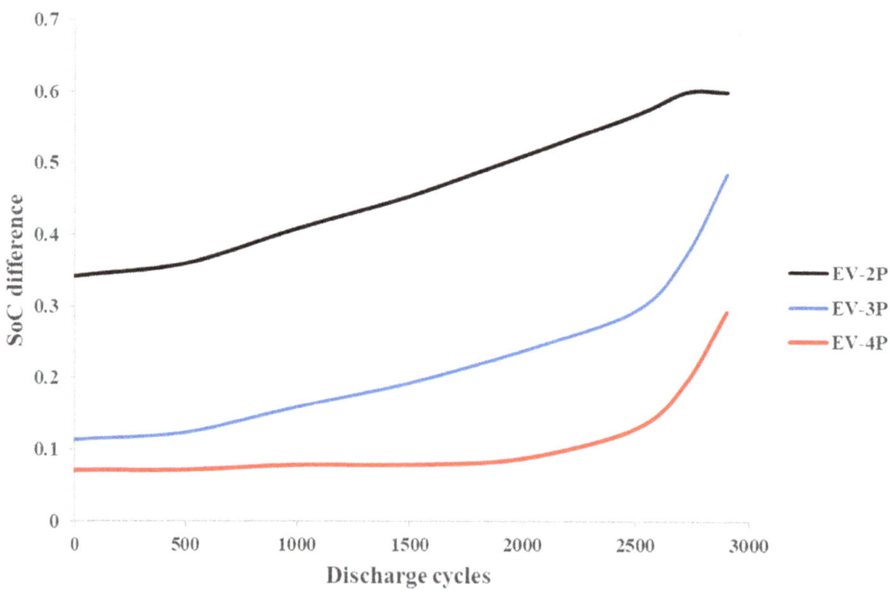

Figure 16. The SoC difference in different second life batteries discharge cycles for various scenarios.

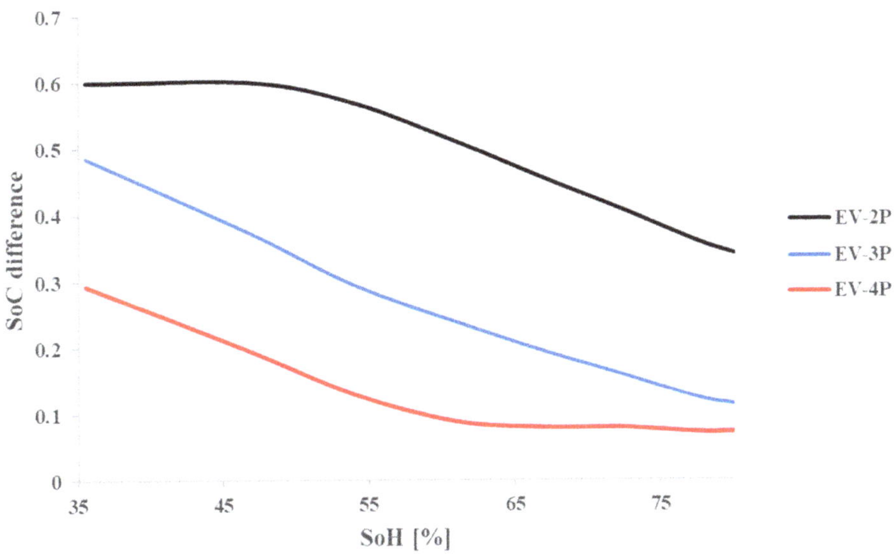

Figure 17. The SoC difference in different second life batteries SoHs for various scenarios.

The impacts of battery sizing on the energy flow in a stand-alone PV system equipped with second life ESS is investigated while the system is designed based on the worst-case scenario. The results of this study revealed the relation between SLB SoC and SoH which can be used for programming battery management systems. Since most of the PVs in residential buildings are connected to the grid, the effects of employment of different SLB ESS sizes on the energy performance of such a system is not considered which can be counted as the disadvantage of this study.

The experimental data for Nissan Leaf, collected from the literature, are obtained by degradation analysis of the SLBs using accelerated ageing profile (charging and discharging c-rates of 1C) and consequently the ESS model is not validated against the battery empirical tests at various C-rates. Its impacts, therefore, on the degradation of the batteries is neglected in this study.

5. Conclusions

In this paper, the battery sizing and technical assessment of an energy system with a second life energy storage system and an off-grid PV energy system is performed. The main aim of this paper is to investigate the effects of adding extra EV charging station load on the ESS performance applied to a residential building. In addition, a parametric study is performed to assess the SLBs' size variations in the ESS when an extra load is applied. The proposed case study residential building is located in Oxford. This paper has developed a novel methodology for assessing the off-grid PV system stability and minimizing the energy supply–demand mismatch. The proposed off-grid system with second life ESS has been mathematically modelled in MATLAB SIMULINK. The system is simulated considering the worst-case scenario on a day in the middle of winter, when the solar irradiation and demand are at their minimum and maximum levels, respectively. The configuration of the second life ESS is accomplished by utilizing Nissan Leaf retired battery modules with an 80% SoH. The main conclusions drawn from the analysis can be listed as:

- An increase in SLBs size, when an extra EV charging load is applied, leads to a voltage peak drop in the second life battery. The increase in the number of SLB packs to 4 resulted in a 7.5% voltage peak drop of ESS.

- Adding EV charging station demand to the off-grid PV driven system, which has been designed to cover residential building demand (with two second life battery packs), expectedly resulted in instabilities in energy exchange between different components of the system during the year. Assuming that there is no extra space left on the building roof to add PV panels, increasing the number of second life battery packs was explored with the findings suggesting that an installed capacity equivalent to 4 battery packs for the studied residential building would minimize the energy mismatch between the energy supply and demand. This occurs before reaching 2000 discharge cycles and approximately 60% SoH (the final SoC of the ESS increased to nearly 53% for the case with 4 battery packs).
- When EVCS load has been applied to the residential load demand, the stability of the system could be improved by increasing the number of second life batteries due to the minimal differences in the initial and final SoC of the second life ESS. This is also beneficial in terms of cost, given that second life batteries have a lower price than brand new batteries.

Author Contributions: Conceptualization, S.R., D.M., P.H. and A.A.; Formal analysis, F.S.; Funding acquisition, A.A.; Methodology, S.R., D.M., P.H. and A.A.; Software, F.S.; Supervision, S.R., D.M., P.H. and A.A.; Validation, F.S.; Writing—original draft, F.S., S.R., D.M. and P.H.; Writing—review and editing, F.S. All authors have read and agreed to the published version of the manuscript.

Funding: This research received no external funding.

Institutional Review Board Statement: Not applicable.

Informed Consent Statement: Not applicable.

Data Availability Statement: The related data are presented within the manuscript.

Conflicts of Interest: The authors declare no conflict of interest.

Nomenclature

EV	Electric vehicle
PV	Photovoltaic panel
ESS	Energy storage system
SoC	State of charge
SLB	Second life battery
SoH	State of health
EVCS	Electric vehicle charging station
CS	Charging station
ECM	equivalent circuit model

References

1. Salek, F.; Eshghi, H.; Zamen, M.; Ahmadi, M.H. Energy and exergy analysis of an atmospheric water generator integrated with the compound parabolic collector with storage tank in various climates. *Energy Rep.* **2022**, *8*, 2401–2412. [CrossRef]
2. Cui, C.; Zou, Y.; Wei, L.; Wang, Y. Evaluating combination models of solar irradiance on inclined surfaces and forecasting photovoltaic power generation. *IET Smart Grid* **2019**, *2*, 123–130. [CrossRef]
3. Obeidat, F. A comprehensive review of future photovoltaic systems. *Sol. Energy* **2018**, *163*, 545–551. [CrossRef]
4. Mohamed, A.A.R.; Morrow, D.J.; Best, R.J.; Cupples, A.; Bailie, I.; Pollock, J. Distributed battery energy storage systems operation framework for grid power levelling in the distribution networks. *IET Smart Grid* **2021**, *4*, 582–598. [CrossRef]
5. Campana, P.E.; Cioccolanti, L.; François, B.; Jurasz, J.; Zhang, Y.; Varini, M.; Stridh, B.; Yan, J. Li-ion batteries for peak shaving, price arbitrage, and photovoltaic self-consumption in commercial buildings: A Monte Carlo Analysis. *Energy Convers. Manag.* **2021**, *234*, 113889. [CrossRef]
6. Uzair, M.; Rehman, N.U.; Yousuf, M.U. Sensitivity analysis of capital and energy production cost for off-grid building integrated photovoltaic systems. *Renew. Energy* **2022**, *186*, 195–206. [CrossRef]
7. Zand, M.; Nasab, M.A.; Sanjeevikumar, P.; Maroti, P.K.; Holm-Nielsen, J.B. Energy management strategy for solid-state transformer-based solar charging station for electric vehicles in smart grids. *IET Renew. Power Gener.* **2020**, *14*, 3843–3852. [CrossRef]

8. Li, C.; Zheng, Y.; Li, Z.; Zhang, L.; Zhang, L.; Shan, Y.; Tang, Q. Techno-economic and environmental evaluation of grid-connected and off-grid hybrid intermittent power generation systems: A case study of a mild humid subtropical climate zone in China. *Energy* **2021**, *230*, 120728. [CrossRef]
9. Cusenza, M.A.; Guarino, F.; Longo, S.; Mistretta, M.; Cellura, M. Reuse of electric vehicle batteries in buildings: An integrated load match analysis and life cycle assessment approach. *Energy Build.* **2019**, *186*, 339–354. [CrossRef]
10. Costa, C.; Barbosa, J.; Gonçalves, R.; Castro, H.; Del Campo, F.; Lanceros-Méndez, S. Recycling and environmental issues of lithium-ion batteries: Advances, challenges and opportunities. *Energy Storage Mater.* **2021**, *37*, 433–465. [CrossRef]
11. Cusenza, M.A.; Guarino, F.; Longo, S.; Ferraro, M.; Cellura, M. Energy and environmental benefits of circular economy strategies: The case study of reusing used batteries from electric vehicles. *J. Energy Storage* **2019**, *25*, 100845. [CrossRef]
12. Wang, T.; Jiang, Y.; Kang, L.; Liu, Y. Determination of retirement points by using a multi-objective optimization to compromise the first and second life of electric vehicle batteries. *J. Clean. Prod.* **2020**, *275*, 123128. [CrossRef]
13. Sun, B.; Su, X.; Wang, D.; Zhang, L.; Liu, Y.; Yang, Y.; Liang, H.; Gong, M.; Zhang, W.; Jiang, J. Economic analysis of lithium-ion batteries recycled from electric vehicles for secondary use in power load peak shaving in China. *J. Clean. Prod.* **2020**, *276*, 123327. [CrossRef]
14. Hart, P.; Kollmeyer, P.; Juang, L.; Lasseter, R.; Jahns, T. (Eds.) Modeling of second-life batteries for use in a CERTS microgrid. In Proceedings of the 2014 Power and Energy Conference at Illinois (PECI), Champaign, IL, USA, 28 February–1 March 2014.
15. Mathews, I.; Xu, B.; He, W.; Barreto, V.; Buonassisi, T.; Peters, I.M. Technoeconomic model of second-life batteries for utility-scale solar considering calendar and cycle aging. *Appl. Energy* **2020**, *269*, 115127. [CrossRef]
16. Uddin, K.; Gough, R.; Radcliffe, J.; Marco, J.; Jennings, P. Techno-economic analysis of the viability of residential photovoltaic systems using lithium-ion batteries for energy storage in the United Kingdom. *Appl. Energy* **2017**, *206*, 12–21. [CrossRef]
17. Assunção, A.; Moura, P.S.; de Almeida, A.T. Technical and economic assessment of the secondary use of repurposed electric vehicle batteries in the residential sector to support solar energy. *Appl. Energy* **2016**, *181*, 120–131. [CrossRef]
18. Tong, S.J.; Same, A.; Kootstra, M.A.; Park, J.W. Off-grid photovoltaic vehicle charge using second life lithium batteries: An experimental and numerical investigation. *Appl. Energy* **2013**, *104*, 740–750. [CrossRef]
19. Torres-Sanz, V.; Sanguesa, J.A.; Martinez, F.J.; Garrido, P.; Marquez-Barja, J.M. Enhancing the Charging Process of Electric Vehicles at Residential Homes. *IEEE Access* **2018**, *6*, 22875–22888. [CrossRef]
20. Sprake, D.; Vagapov, Y.; Lupin, S.; Anuchin, A. (Eds.) Housing estate energy storage feasibility for a 2050 scenario. In Proceedings of the 2017 Internet Technologies and Applications (ITA), Guangzhou, China, 26–28 May 2017.
21. Pimm, A.J.; Cockerill, T.T.; Taylor, P.G. The potential for peak shaving on low voltage distribution networks using electricity storage. *J. Energy Storage* **2018**, *16*, 231–242. [CrossRef]
22. Braco, E.; Martín, I.S.; Berrueta, A.; Sanchis, P.; Ursúa, A. Experimental assessment of cycling ageing of lithium-ion second-life batteries from electric vehicles. *J. Energy Storage* **2020**, *32*, 101695. [CrossRef]
23. Available online: https://www.enfsolar.com/pv/panel-datasheet/crystalline/44143 (accessed on 1 September 2022).
24. Duffie, J.A.; Beckman, W.A. *Solar Engineering of Thermal Processes*; Wiley: Hoboken, NJ, USA, 2013.
25. Salek, F.; Rahnama, M.; Eshghi, H.; Babaie, M.; Naserian, M.M. Investigation of Solar-Driven Hydroxy gas production system performance integrated with photovoltaic panels with single-axis tracking system. Renew. *Energy Res. Appl.* **2022**, *3*, 31–40.
26. Solar Cell. Available online: https://uk.mathworks.com/help/physmod/sps/ref/solarcell.html;jsessionid=bdf3b2865e01ab9c5896c9e033dc (accessed on 1 August 2022).
27. He, H.; Xiong, R.; Fan, J. Evaluation of Lithium Ion Battery Equivalent Circuit Models for State of Charge Estimation by an Experimental Approach. *Energies* **2011**, *4*, 582–598. [CrossRef]
28. Nejad, S.; Gladwin, D.; Stone, D. A systematic review of lumped-parameter equivalent circuit models for real-time estimation of lithium-ion battery states. *J. Power Sources* **2016**, *316*, 183–196. [CrossRef]
29. Braco, E.; San Martín, I.; Sanchis, P.; Ursúa, A. (Eds.) Characterization and capacity dispersion of lithium-ion second-life batteries from electric vehicles. In Proceedings of the 2019 IEEE International Conference on Environment and Electrical Engineering and 2019 IEEE Industrial and Commercial Power Systems Europe (EEEIC/I&CPS Europe), Genova, Italy, 11–14 July 2019.
30. Braco, E.; San Martin, I.; Ursúa, A.; Sanchis, P. (Eds.) Incremental capacity analysis of lithium-ion second-life batteries from electric vehicles under cycling ageing. In Proceedings of the 2021 IEEE International Conference on Environment and Electrical Engineering and 2021 IEEE Industrial and Commercial Power Systems Europe (EEEIC/I&CPS Europe), Bari, Italy, 7–10 September 2021.
31. Braco, E.; San Martin, I.; Berrueta, A.; Sanchis, P.; Ursua, A. Experimental Assessment of First-and Second-Life Electric Vehicle Batteries: Performance, Capacity Dispersion and Aging. *IEEE Trans. Ind. Appl.* **2021**, *57*, 4107–4117. [CrossRef]

applied sciences

MDPI

Article

Electrification of a Class 8 Heavy-Duty Truck Considering Battery Pack Sizing and Cargo Capacity

Aiden Thomas Leonard [1], Farhad Salek [1], Aydin Azizi [2] and Shahaboddin Resalati [1,*]

[1] Faculty of Technology, Design and Environment, Oxford Brookes University, Oxford OX3 0BP, UK
[2] School of Engineering, Computing and Mathematics, Wheatley Campus, Oxford Brookes University, Oxford OX33 1HX, UK
* Correspondence: sresalati@brookes.ac.uk

Abstract: The design and performance optimization of fully electric trucks constitute an integral goal of the transport sector to meet climate emergency measures and local air quality requirements. Most studies in the literature have determined the optimum pack size based on economic factors, without accounting for the details of pack behavior when varying the size. In this paper, the effect of battery pack sizing and cargo capacity of a class 8, 41-ton truck on its overall energy performance and technical parameters of its powertrain is investigated. For this purpose, the proposed electric truck is designed and mathematically modelled using AVL CRUISE M software. The second-order equivalent circuit model is developed to predict the battery packs' parameters. The proposed battery pack model is extracted from experimental analysis on SONY VTC6 lithium-ion batteries performed in the lab. The weight changes due to adding the battery packs to the truck are also estimated and have been taken into account. The mathematical model of the powertrain is simulated in the long-haul driving cycle considering different cargo capacities and battery pack sizes. The results of this study revealed that the battery pack voltage reached its minimum value when the maximum cargo capacity was applied for the 399 kWh battery pack. In addition, increasing the occupied cargo capacity from 10% to 100% resulted in an increase in the regenerative brake energy of up to 9.87 kWh, while changing the battery size imposed minimal impacts on regenerative brake energy recovery as well as energy consumption.

Keywords: battery electric; battery pack; energy performance; simulation

Citation: Leonard, A.T.; Salek, F.; Azizi, A.; Resalati, S. Electrification of a Class 8 Heavy-Duty Truck Considering Battery Pack Sizing and Cargo Capacity. *Appl. Sci.* **2022**, *12*, 9683. https://doi.org/10.3390/app12199683

Academic Editor: Giuseppe Lacidogna

Received: 6 September 2022
Accepted: 23 September 2022
Published: 27 September 2022

1. Introduction

As climate change becomes an increasingly pressing issue, the automotive industry is shifting towards electrification of vehicles to reduce carbon emissions. EV sales rose by 68% globally between 2017–2018 [1]. However, the heavy-duty vehicle industry has not seen the same volume of electric vehicle sales [2–4]. In 2020, global sales of electric heavy-duty trucks accounted for less than 1% of the total sales [5,6]. The trucking industry contributed 23% of transport GHG emissions in the US [7]. The low level of electrification of this space presents a large opportunity for a reduction in carbon emissions.

A primary reason for the dependence of the truck manufacturing industry on fossil fuels is the high energy demands compared to passenger cars [8]. Heavy-duty EVs require larger energy storage due to their greater range, weight, and aerodynamic load demands compared to passenger EVs [9–11]. Most heavy-duty trucks must travel long distances between each stop. This requires large batteries to be installed [12] or electric roads, which require large investments for infrastructure [13]. Since battery energy density is lower than fuel tanks used in Internal Combustion Engine (ICE) trucks, the energy storage system in heavy-duty vehicles becomes expensive and heavy due to electrification [9,14]. Accordingly, the extra mass of the battery packs reduces the cargo capacity since the total weight of the truck needs to remain constant during electrification [9,15].

Various studies have analyzed the electrification of heavy-duty vehicles addressing aforementioned challenges. Mauler et al. [9] conducted an economic analysis between diesel, fuel cell, and battery electric long-haul trucks in the United States. Current battery electric trucks were seen to be nearly competitive to diesel trucks for trips below 500 km. For weight constrained cargo exceeding 500 km, fuel cells were nearly competitive to diesel. The more volume constrained the cargo, the greater the advantage to the battery electric truck, since the added mass of the battery pack does not limit cargo capacity. Energy and fuel price had a large effect on all of the comparisons, and such comparisons do not consider charging nor hydrogen refueling availability. Tanco et al. [16] also found that energy and fuel price influence electric truck feasibility in Latin America, with upfront costs being the largest barrier. Besides the cost of electricity, the cost of a battery pack influences the economic feasibility of EV heavy-duty vehicles. Vijayagopal and Rousseau [14] determined that the cost of battery pack depends on desired range and influences electric truck upfront costs. Nykvist and Olson [17] proposed using smaller battery packs for heavy-duty trucks and improving the fast charging network in order to reduce both the cost of battery electric trucks and their weight, improving their weight carrying capacity. Optimizing heavy-duty truck battery pack size would maximize the economic viability of such vehicles [18]. Baek et al. [18] were able to find an ideal battery pack size by factoring in charging costs, battery costs, depreciation costs, and revenue. Large battery packs maximize range and payload size, but unnecessarily increase the weight and require a larger initial investment. On the contrary, a small battery pack limits the range and carrying capacity and relies more on the charging network. The previously mentioned studies rely on economic factors to determine the pack size, but details of pack behavior are not accounted for when varying the size. The subject of this research is to perform the optimal pack sizing considering the distance, power, weight, size of truck, and the volume of transported goods in order to maximize efficiency and energy usage of a heavy-duty truck. Combining cell parametrization through lab testing for a pack model to test various pack sizes of electric truck provides detailed insight into pack behavior variation. Cell testing was conducted using pulse discharges to obtain transient and steady-state behavior at various SOC increments. This data was then parametrized using MATLAB Simulink R2022a with a nonlinear least-squares optimization and a Trust-Region-Reflective algorithm. Error was minimized using sum squared error. This parametrization yielded ECM parameters which were used for the pack model. An AVL pack model was used with the cell parameters for the sizing optimization. The pack model was used to evaluate various pack configurations and capacities to optimize power, weight, range, and cell health for the 41-ton truck.

This article brings a new viewpoint to the existing literature in the battery electric truck modelling and optimal pack sizing of batteries based on experimental data. Results of the study inform an ideal pack configuration based on the truck weight, and how varying truck parameters influences pack behavior.

This paper is organized as follows; Section 2.1 (battery electric powertrain modelling) discusses the methodology used to mathematically model the proposed powertrain, Section 2.2 (parametric analysis) contains the variables assigned in the model to determine the impacts of cargo capacity and battery pack sizing on the model output parameters, and Section 2.3 (battery cell characterization) describes the methodology used to parametrize second-order ECM used for modelling of batteries using lab data. Sections 3.1 and 3.2, located in the results section, discuss battery parametrization and parametric study results, respectively.

2. Methodology

2.1. Battery Electric Powertrain Modelling

The battery electric powertrain of a heavy-duty class 8 truck is modelled using AVL CRUISE M R2022.1 software [19]. The modelling process is indicated in Figure 1. The Mercedes Benz Actros 41-ton truck, featuring 8 wheels and 4-wheel drive (8×4) is considered as the case study with its technical parameters presented in Table 1. The battery electric powertrain components in AVL CRUISE M R2022.1 software are presented in Figure 2 with the electric motor efficiency presented in Figure 3. The electrical network of the proposed powertrain includes two electric motors and six packs of batteries. The battery packs are sized and modelled based on specifications of Sony VTC6 lithium-ion battery. The battery pack specifications are provided in Table 2.

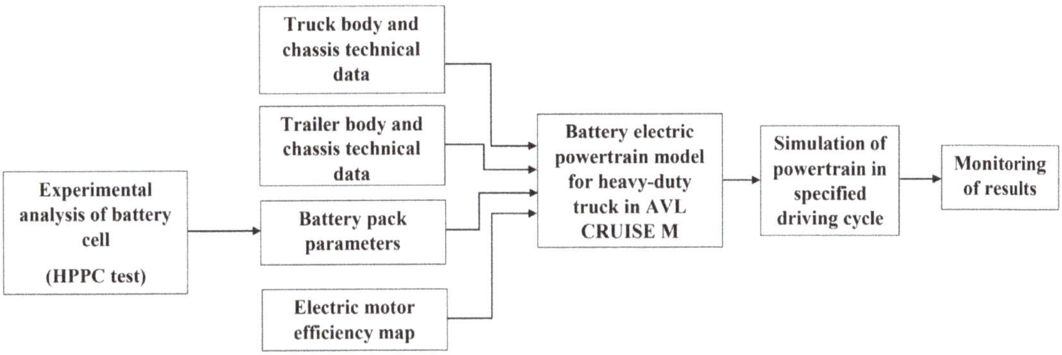

Figure 1. Flow diagram of modelling process.

Table 1. Truck and trailer bodies technical parameters.

Component	Parameter	Value
Truck body dimensions	Distance from hitch to front axle [mm]	3450
	support point height, bench test [mm]	1050
	Wheel base [mm]	3900
Trailer body dimensions	Distance from hitch to axle [m]	7.7
	Pitching moment coefficient	1
Truck nominal weight	Curb weight [kg]	8000
	Gross weight [kg]	8500
Trailer nominal weight	Curb weight [kg]	19,000
	Gross weight [kg]	32,500
Total carrying capacity	Truck [kg]	500
	Trailer [kg]	13,500
Aerodynamic Properties	Frontal Area [m^2]	8.48
	Drag coefficient	0.6

Table 2. Battery pack parameters.

Component	Parameter	Value
Cell	Nominal voltage [V]	3.6
	Current capacity [Ah]	3
	Energy capacity [Wh]	10.8
	Average weight [g]	46.6
	Estimated energy density [Wh/kg]	232
Pack	Cells in series	176
	Cells in parallel	35
	Module energy capacity [kWh]	66.5
	Number of Packs	6
	Packs total energy capacity [kWh]	399

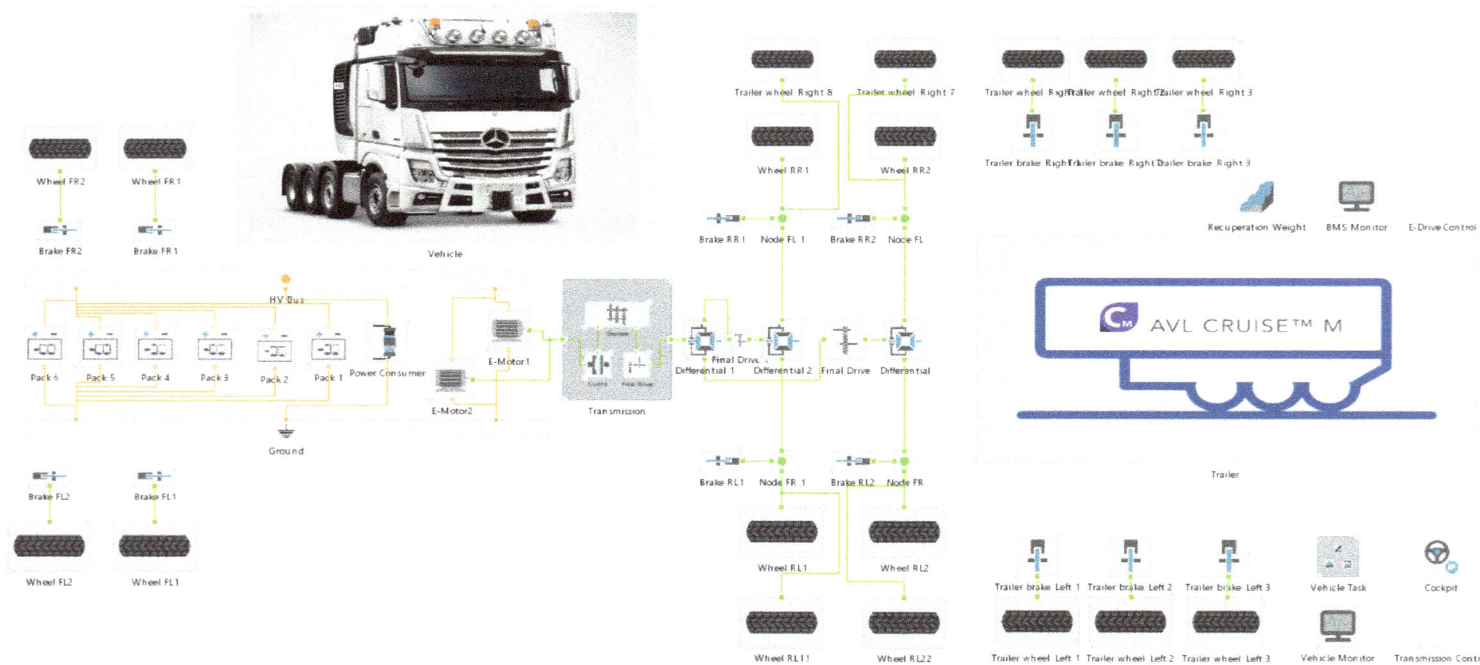

Figure 2. Block diagram for AVL vehicle model.

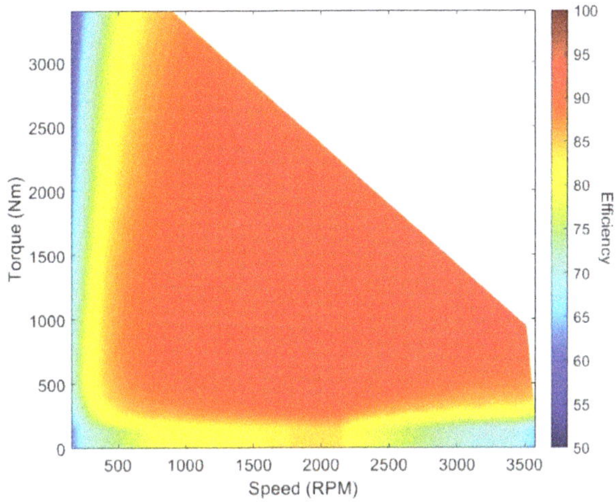

Figure 3. The electric motor efficiency map.

The originally diesel Mercedes Actros truck is considered the case study in this paper. Therefore, the weight change associated with the electrification of this truck is obtained based on the weight of components exchanged. These values are presented in Table 3. The transmission and drivetrain weight were assumed to be equal for the diesel and electric powertrains. The mass loss from removing the fuel and fuel tank was estimated by using 50% of the full capacity tank weight for the Actros as a conservative estimate for the average amount of fuel in the truck during long-haul driving. The electric vehicle weight addition includes weight from the two motors, battery pack, inverter, and DC/DC converter. Dual motors were used to obtain high efficiency compared to a single larger motor [20], and permanent magnet motors were selected for their high specific torque [21]. In order to determine the effect of pack size on weight, the required structural mass for a given energy capacity was determined based on previous studies. The cell energy density was used to calculate pack energy density, based on a 60% percentage weight of cells to total pack weight. This value was chosen based on typical gravimetric energy densities achieved with current battery packing methods [22–25]. The resulting pack gravimetric energy density is 140 kg/kWh. The pack weight was then calculated for each pack size by dividing the power capacity by 140 Wh/kg.

A long-haul driving cycle (Figure 4) was used in order to evaluate the transient performance of the 41-ton truck at various pack capacities. The driving cycle represents a typical trip of a heavy-duty truck, which undergoes brief accelerations and decelerations, but generally cruises at highway speeds of approximately 85 kph. Inclination also varies during this cycle to accurate modelling of real-life driving conditions [26,27].

Table 3. Weight of Diesel and Electric powertrain components.

Component	Parameter	Value
Diesel	Engine Weight [kg]	1240
	Max engine power [kW]	460
	Fuel tank weight at 50% capacity [kg]	382
Electric	Weight of two motors [kg]	680
	Combined motor power [kW]	500
	Total battery pack weight [tons]	1.73–2.54
	Inverter weight [kg]	72
	DC/DC converter weight [kg]	35
	Total electric powertrain weight [tons]	3.64–6.49

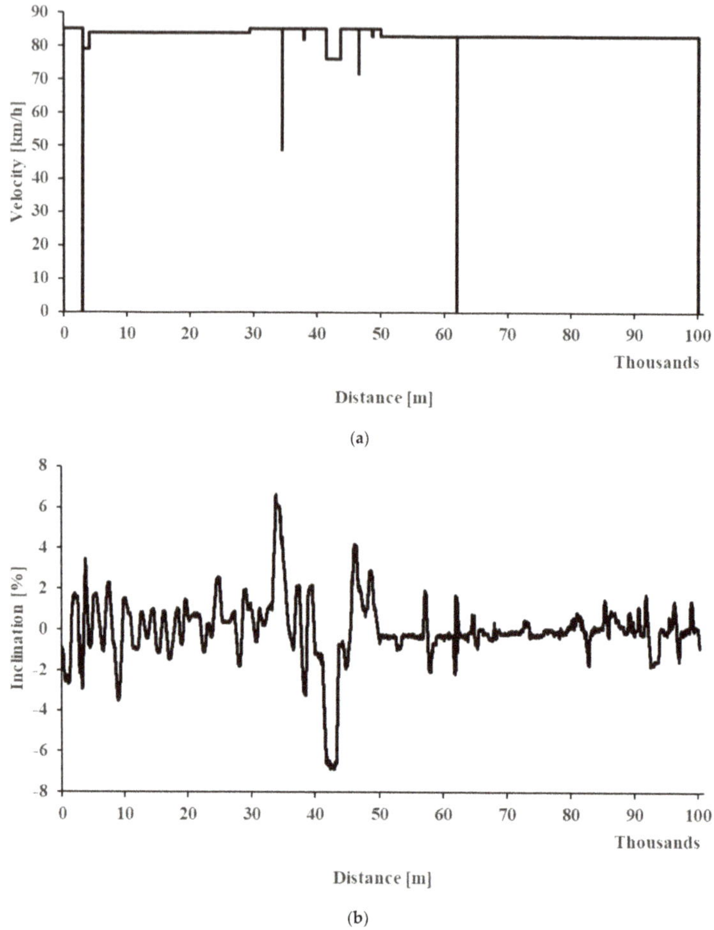

Figure 4. The (**a**) velocity profiles and (**b**) road inclination variations in the Long-haul driving cycle [26,27].

2.2. Parametric Analysis

A parametric study is performed to size the battery packs and investigate their impacts on technical parameters of the powertrain. The battery pack size is determined by changing the number of cell rows in each pack and considering the variations in cargo capacity. Cells in parallel were varied from 35 to 45, increasing the total pack energy from 399 to 513 kWh. This was the primary variable of the study since pack size variation is a crucial design choice for a heavy-duty vehicle. The occupied cargo capacity is also varied between 10% and 100%. Different case scenarios considered in this paper are shown in Table 4.

2.3. Battery Cell Characterization

The second-order equivalent circuit model is used for the mathematical modelling of each battery pack shown in Figure 2. For this purpose, the SONY VTC6 cylindrical battery cells were tested in a High Voltage Energy Storage (HVES) lab at Oxford Brookes University. The HPPC tests were performed for characterization of the battery cells in the lab at a temperature of 25 °C. In HPPC tests, the battery cell underwent a discharge test at 1.5 C, which provided a picture of voltage at various SOC, as well as the transient response to current pulses. The cell has a maximum voltage of 4.2 V, and a capacity of

3000 mAh. As an approximation, 1.5 C was chosen as a typical discharge rate seen for commercial automobiles [17,18,28,29]. The cell was placed in a testing chamber to hold the temperature constant (Figure 5). A clamping apparatus which utilizes spring force was used to provide firm contact between the cell terminals and the power supply. Voltage across the cell was measured by clipping leads to metal at each plate, while current was recorded using an Arbin battery tester (ARBIN LBT21084). Thermocouples were used to obtain the temperature of the cell, and the experiment was conducted at 25 °C to gain an understanding of behavior at average ambient temperature.

Table 4. Case scenarios for parametric study performed.

Case	Number of Cell Rows	Cargo Capacity [%]
R35L10	35	10
R40L10	40	10
R45L10	45	10
R35L50	35	50
R40L50	40	50
R45L50	45	50
R35L100	35	100
R40L100	40	100
R45L100	45	100

After data was obtained from the HPPC tests, a second-order ECM was developed in MATLAB Simulink to characterize the cell behavior for use in pack modelling with AVL CRUISE M. The model uses a battery block, which is controlled by defining parameters such as internal resistance, dynamic time constants and resistances, capacity, and cell voltage. Current is controlled through the "Current Source" block, which received the same input current as the experimental tests. The "Voltage Sensor" block outputs cell voltage, while the "SOC" block outputs the state of charge (Figure 6). The model received parameter values from a separate MATLAB workspace, and it sends outputs and input current to the parameter estimate module as well as the workspace.

Figure 5. Sony VTC6 cell experimental setup.

Each parameter can be expressed as an array of values, each value being defined at a specific SOC. In this particular case, 16 discharge pulses were conducted experimentally to take the cell from 4.2 V to 3.4 V, so all parameter arrays had 16 items.

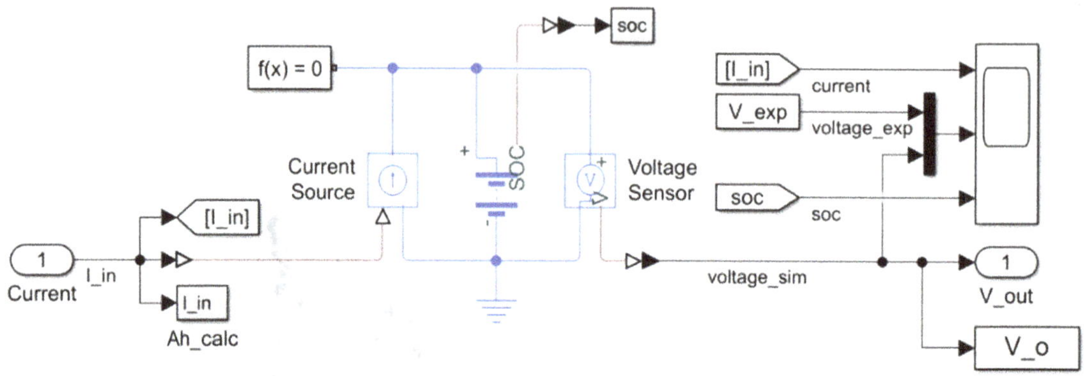

Figure 6. MATLAB model used for cell parameter estimation.

Simulink's built-in parameter estimator was used to determine parameter values which matched experimental results. The varied parameters were voltage, first and second dynamic resistances, first and second-time constants, and static resistance. The Levenberg–Marquardt algorithm was used to iteratively find parameters that resulted in minimizing the least-squares difference from the experimental results. The parametrization resulted in closely matching voltage vs SOC graphs, with an R^2 value of 99% (Figure 7).

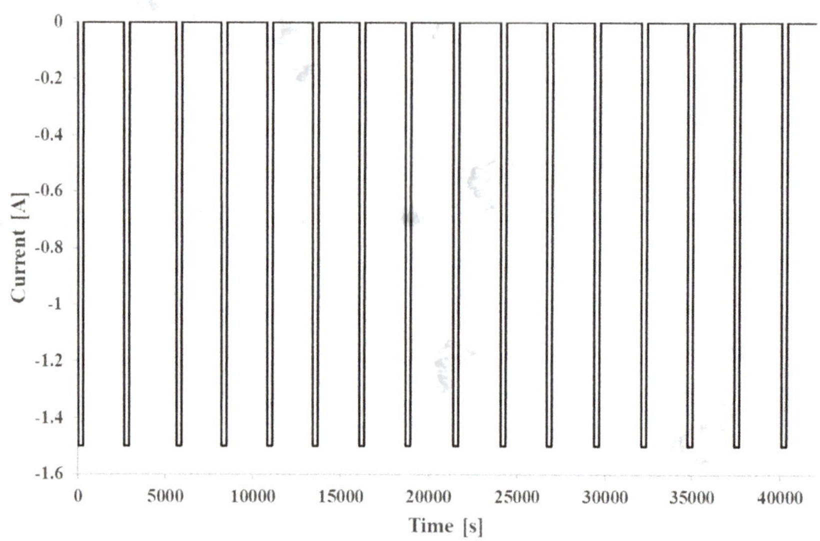

(**a**)

Figure 7. *Cont.*

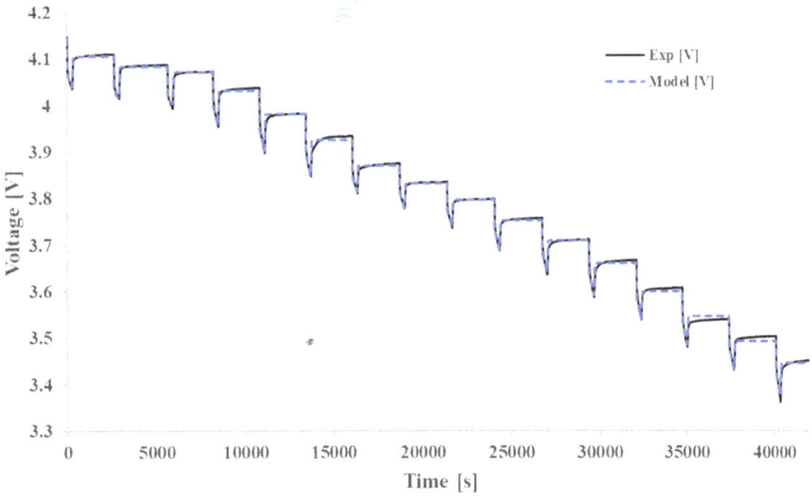

(b)

Figure 7. Parametrization results for the second-order ECM compared to experimental results, (a) applied current profile in HPPC test versus time and (b) cell voltage response in different times calculated by model and obtained in HPPC tests.

3. Result and Discussion

3.1. Experimental Tests Results

The estimated parameters based on HPPC lab data for second-order ECM are shown in Figure 8a–e. These parameters are used as the input of the battery pack model shown in Figure 2 for simulation of lithium-ion battery behavior under the transient conditions of the driving cycle.

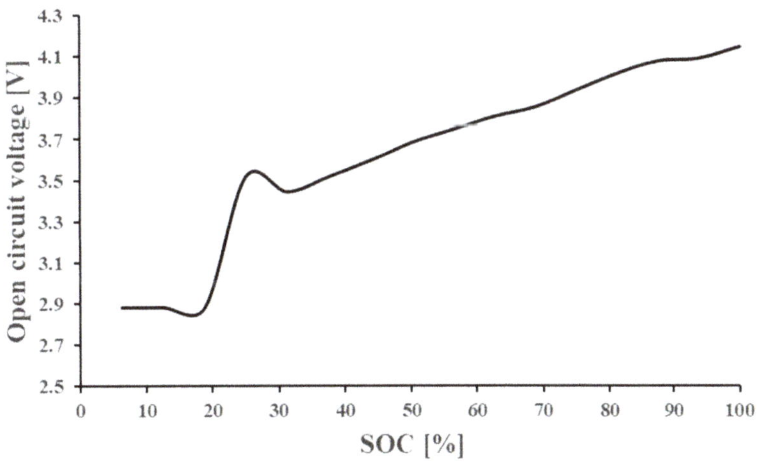

(a)

Figure 8. *Cont.*

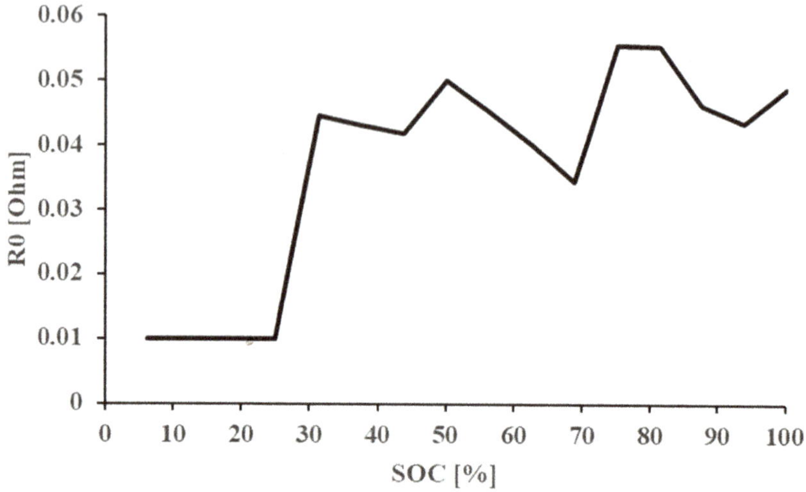

(**b**)

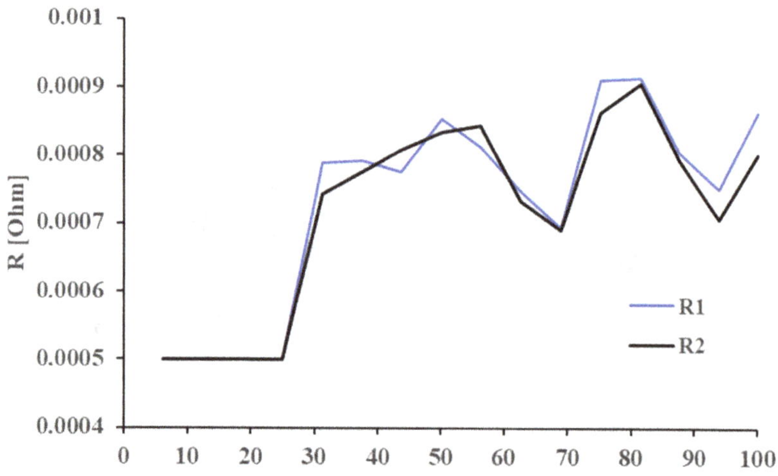

(**c**)

Figure 8. *Cont.*

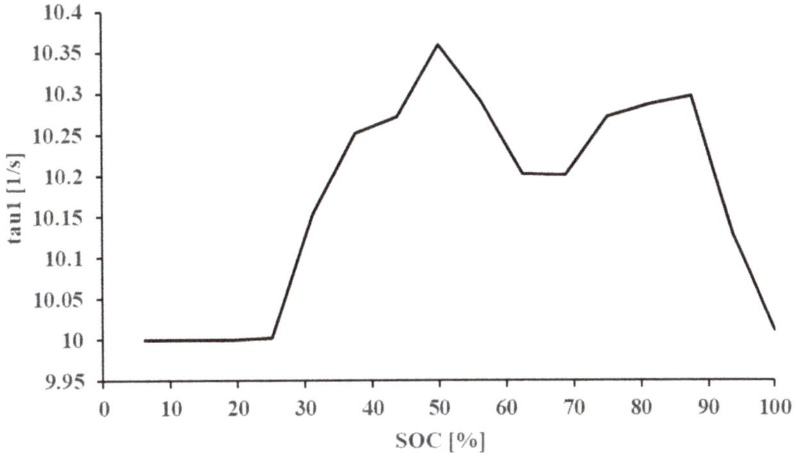

(**d**)

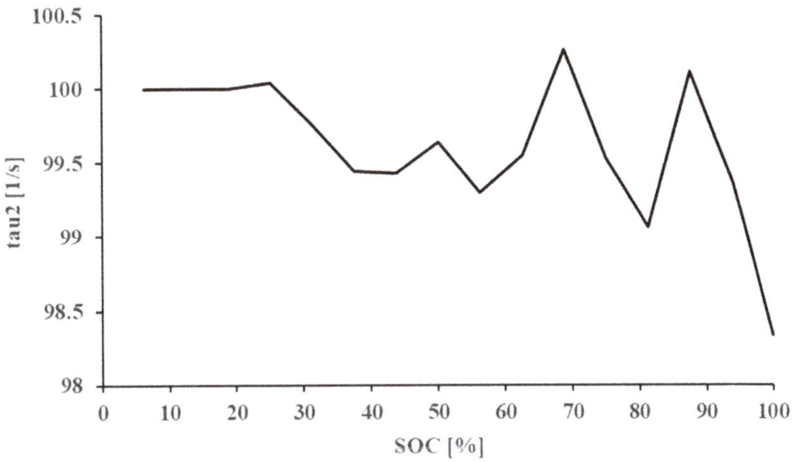

(**e**)

Figure 8. The estimated parameters for second order ECM using HPPC data: (**a**) open circuit voltage, (**b**) static ohmic resistance, (**c**) dynamic constants resistances, (**d**) first time constant and (**e**) second time constant in various state of charges.

3.2. Parametric Analysis

The effects of battery pack size variations on the cargo capacity of the proposed battery electric truck are shown in Figure 9. Electrification of the truck resulted in a decrease in its cargo capacity from 13.5 tons to 11.77 tons, 11.36 tons, and 10.96 tons by increasing the battery pack size to 399 kWh (35 rows), 456 kWh (40 rows), and 513 kWh (45 rows), respectively.

The difference between the desired and actual velocity of the proposed battery electric truck for different battery pack sizes at 100% cargo load is shown in Figure 10. The fluctuations in actual velocity in cruising velocities (desired profile) are due to the existence of road inclination, resulting in activation and deactivation of the brake pedal. The desired velocity is caught in most of the points, and minor deviation is observed in sudden acceleration point following from electric motor moment of inertia. Furthermore, Figure 10 also indicates that desired torque at various speeds is delivered by dual motors installed in the battery electric powertrain.

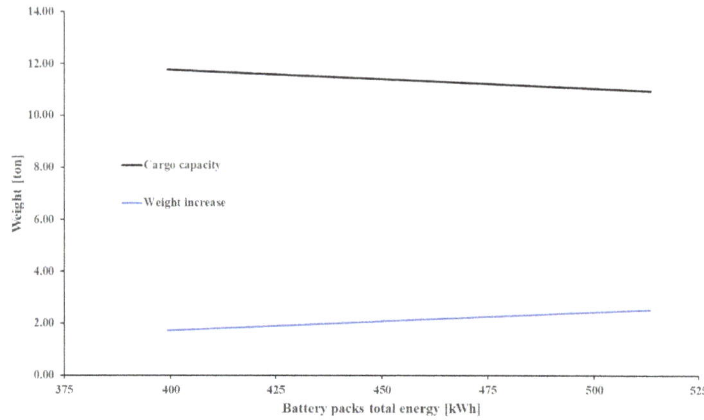

Figure 9. Effect of battery pack size on cargo capacity.

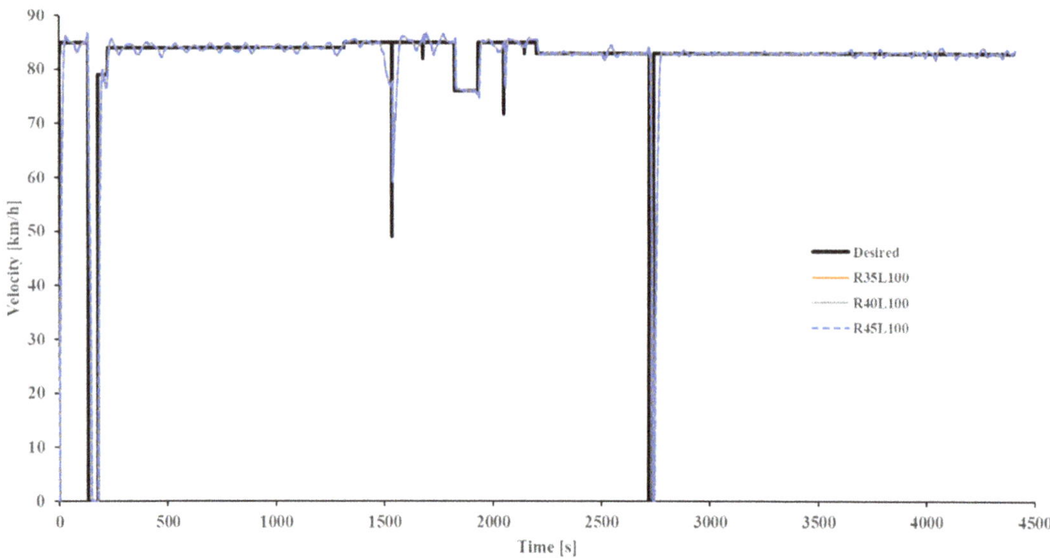

Figure 10. Truck desired and actual velocity for different case studies in various times in proposed driving cycle.

The battery pack voltage change during travel for different case scenarios is shown in Figure 11. By increasing the battery size, the overall voltage of the pack increased, while increment of occupied cargo capacity resulted in decrease of battery pack voltage due to faster discharging of the pack leading to sharp fade of voltage. There are four sharp increases in voltage during travel due to sharp changes in velocity as can be seen in Figure 11. When the velocity decreases sharply, a drastic increase in voltage is also observed. These sudden voltage jumps can be controlled by employing a programmed battery management system. In the last period of travel, steep increase of voltage is seen for R35L50, R35L100 and R40L100 due to reaching below 30% in SOC (Figure 12). As indicated in Figure 8a, there is a fluctuation in open circuit voltage for SOCs between 20% and 30%. The same trend is reflected in battery pack voltage, as is demonstrated in Figure 11. The most voltage fade belongs to R35L100 case scenario with lowest size of the battery and highest cargo capacity. This led to a sharper decrease in the battery pack voltage due to high power demand, resulting in a lower SOC being reached at the end of travel compared to other scenarios.

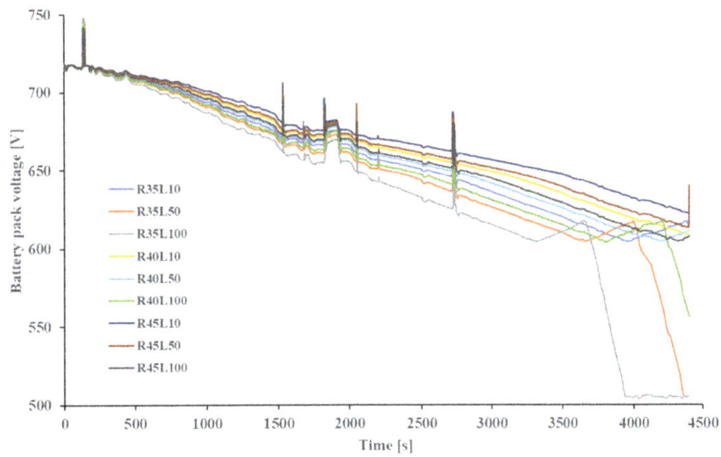

Figure 11. The battery pack voltage variations during travel for different scenarios.

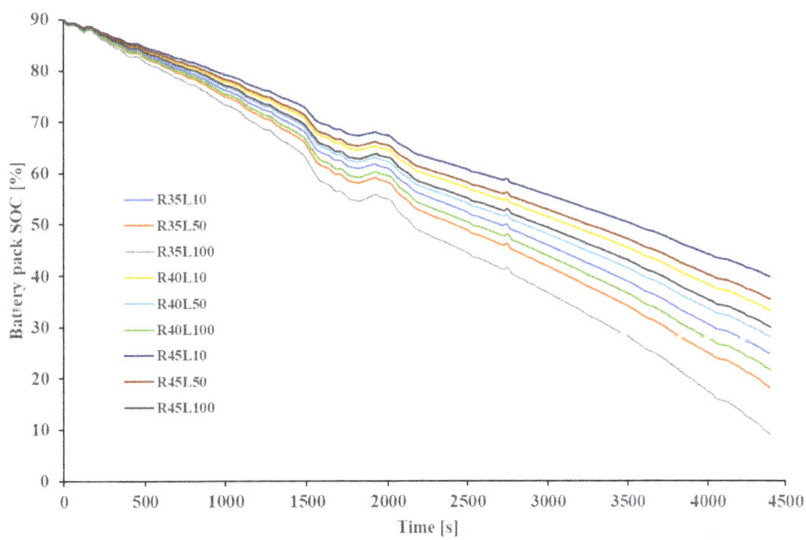

Figure 12. The battery pack SOC variations during travel for different scenarios.

The battery power change is shown in Figure 13 in various time steps of the driving cycle for different scenarios. The discharging power increased during truck cruising speeds due to the increase in the cargo load. The highest charging peaks are achieved in the R45L100 scenario due to the existence of a larger battery and a large load (cargo) resulting in higher values of regenerative brake energy. However, the charging sudden peaks should be capped as well as discharging sharp peaks to prevent the battery from being damaged due to high temperature.

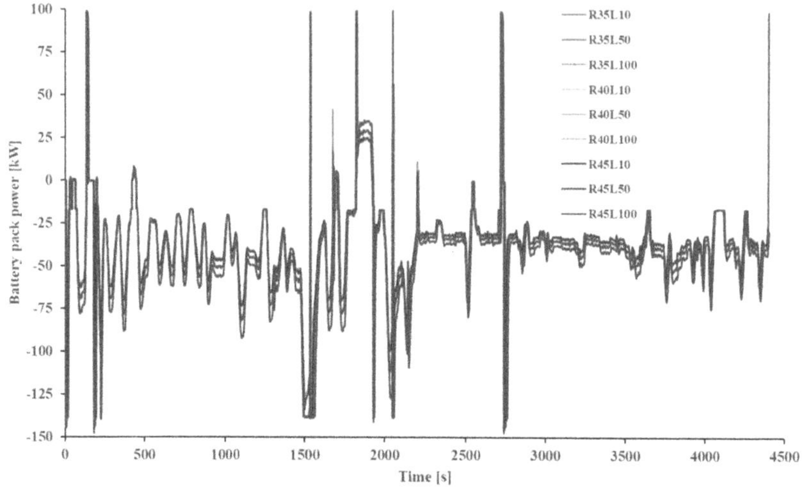

Figure 13. The battery pack power variations during travel for different scenarios.

The cumulated energy recovered by regenerative brakes and consumed energy during travel for different scenarios is shown in Figure 14a,b. Regenerative brake energy increased from 6.94 kWh to approximately 9.87 kWh due to the increment in the truck cargo capacity between 10% and 100%. In addition, an increase in the battery cell rows in the packs from 35 to 40 and 45 resulted in energy recovery increases of 1.73% and 3.46% for 10% load, 0.98% and 2.2% for 50% load and 0.41% and 0.74% for 100% load, respectively. In addition, there were negligible changes in the energy consumption due to the increase in battery size since power demand remains constant.

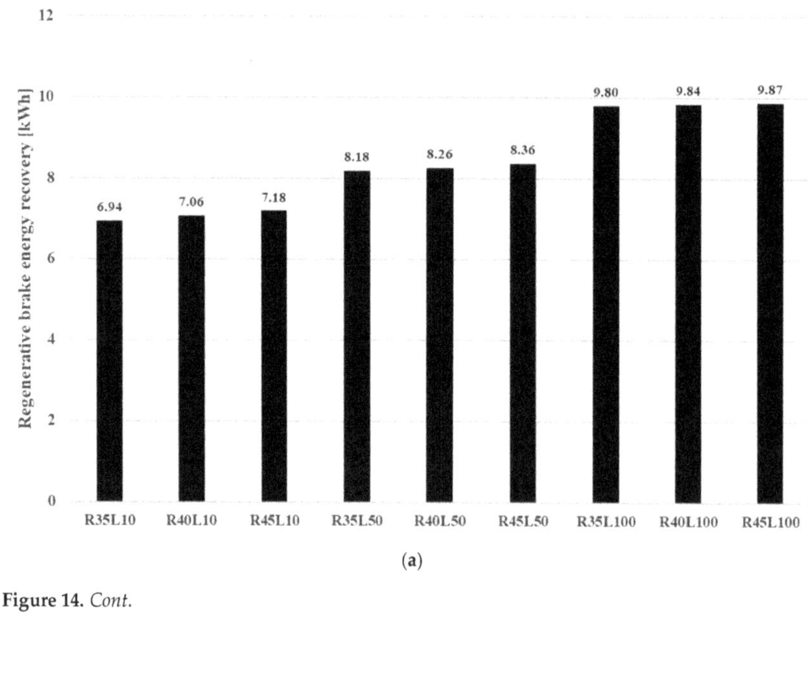

(a)

Figure 14. *Cont.*

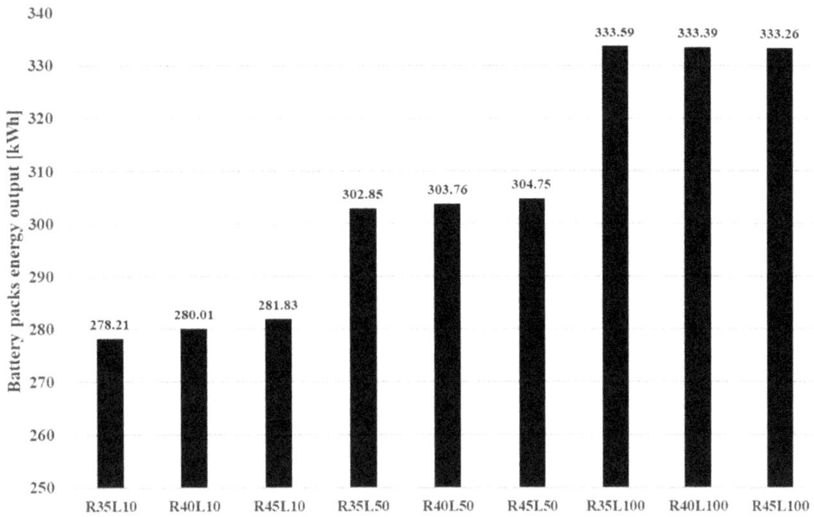

(b)

Figure 14. (**a**)The regenerative brake energy recovery and (**b**) battery pack energy output during travel for different scenarios.

4. Conclusions

In this paper, a parametric analysis is performed to assess the effect of battery pack sizing and cargo capacity of a truck on its energy performance. As the case study, the electrification of a class 8, 41 tons' truck is performed by modelling and simulation of its battery electric powertrain in AVL CRUISE M software. The experimental lab battery testing data is used to parametrize the battery pack model in this study. Issues such as weight changes due to the addition of battery packs to the truck are also addressed. A parametric study is employed consisting of various case scenarios to evaluate the impacts of the battery pack size and cargo capacity on truck performance and energy consumption. The main conclusions drawn from this study can be listed as:

- Electrification of the Mercedes–Benz Actros 41-ton truck resulted in decreasing of its cargo capacity from 13.5 tons to 12.27 tons, 11.86 tons, and 11.46 tons by increasing battery pack size from 399 kWh to 456 kWh and 513 kWh, respectively.
- Employment of dual electric motors in the powertrain resulted in gaining desired torque at various speeds during travel in long haul driving cycle.
- Increasing the battery size boosted the overall voltage of the pack, while an increment in occupied cargo capacity resulted in a decreased battery pack voltage due to faster discharging of the pack, leading to a sharp decrease in voltage.
- The greatest decrease in voltage was achieved in battery packs when the cargo capacity was equal to 100% and the lowest size of battery pack was used. This leads to a sharper decrease in the battery pack voltage due to high power demand, resulting in a lower SOC at the end of travel compared to other scenarios.
- The highest charging peaks were achieved for the scenario with the biggest battery pack size and full cargo capacity due to recovering higher values of regenerative brake energy.
- Regenerative brake energy increased from 6.94 kWh to approximately 9.87 kWh by incrementing the truck occupied cargo capacity up to 100%.
- Increase of total battery packs size from 399 kWh to 456 kWh and 513 kWh resulted in energy recovery increases of 1.73% and 3.46% for 10% load, 0.98% and 2.2% for 50% load and 0.41% and 0.74% for 100% load, respectively.

As a limitation of this research, it can be mentioned that the employed second-order ECM used to model the battery packs is not capable of predicting their capacity fade after driving in various number of cycles, which should be addressed and solved in future steps of this research. The developed methodology coupled with the experimental data presented in this paper would assist the future research and development activities aiming to truly optimize the performance of heavy-duty trucks to tackle the global net-energy targets as well as the local air quality requirements.

Author Contributions: Conceptualization, methodology, software, validation, A.T.L. and F.S.; formal analysis, S.R., A.A. and F.S.; resources, S.R. and A.A.; writing—original draft preparation, A.T.L., F.S. and S.R.; writing—review and editing, F.S., A.T.L. and S.R.; visualization, A.T.L. and F.S.; supervision, S.R., A.A. and F.S. All authors have read and agreed to the published version of the manuscript.

Funding: This research received no external funding.

Institutional Review Board Statement: Not applicable.

Informed Consent Statement: Not applicable.

Data Availability Statement: The related data are presented within the manuscript.

Acknowledgments: Special thanks to AVL List GmbH for providing the simulation tools through their University Partnership Program with Oxford Brookes University.

Conflicts of Interest: The authors declare no conflict of interest.

Nomenclature

EV	Electric Vehicle
ECM	Equivalent Circuit Model
SOC	State of Charge
HPPC	Hybrid Pulse Power Characterization
ICE	Internal Combustion Engine

References

1. Ding, S.; Li, R. Forecasting the sales and stock of electric vehicles using a novel self-adaptive optimized grey model. *Eng. Appl. Artif. Intell.* **2021**, *100*, 104148. [CrossRef]
2. Borlaug, B.; Muratori, M.; Gilleran, M.; Woody, D.; Muston, W.; Canada, T.; Ingram, A.; Gresham, H.; McQueen, C. Heavy-duty truck electrification and the impacts of depot charging on electricity distribution systems. *Nat. Energy* **2021**, *6*, 673–682. [CrossRef]
3. Machado, P.G.; Teixeira, A.C.; Collaço, F.M.; Mouette, D. Review of life cycle greenhouse gases, air pollutant emissions and costs of road medium and heavy-duty trucks. *Wiley Interdiscip. Rev. Energy Environ.* **2021**, *10*, e395. [CrossRef]
4. Fenton, D.; Kailas, A. Redefining Goods Movement: Building an Ecosystem for the Introduction of Heavy-Duty Battery-Electric Vehicles. *World Electr. Veh. J.* **2021**, *12*, 147. [CrossRef]
5. IEA. *Trucks and Buses*; IEA: Paris, France, 2022. Available online: https://www.iea.org/reports/trucks-and-buses (accessed on 1 August 2022).
6. Borlaug, B.; Moniot, M.; Birky, A.; Alexander, M.; Muratori, M. Charging Needs for Electric Semi-Trailer Trucks. *SSRN Electron. J.* **2022**. [CrossRef]
7. Giuliano, G.; Dessouky, M.; Dexter, S.; Fang, J.; Hu, S.; Miller, M. Heavy-duty trucks: The challenge of getting to zero. *Transp. Res. Part D Transp. Environ.* **2021**, *93*, 102742. [CrossRef]
8. Kyle, P.; Kim, S.H. Long-term implications of alternative light-duty vehicle technologies for global greenhouse gas emissions and primary energy demands. *Energy Policy* **2011**, *39*, 3012–3024. [CrossRef]
9. Mauler, L.; Dahrendorf, L.; Duffner, F.; Winter, M.; Leker, J. Cost-effective technology choice in a decarbonized and diversified long-haul truck transportation sector: A US case study. *J. Energy Storage* **2022**, *46*, 103891. [CrossRef]
10. Xie, S.; Lang, K.; Qi, S. Aerodynamic-aware coordinated control of following speed and power distribution for hybrid electric trucks. *Energy* **2020**, *209*, 118496. [CrossRef]
11. Mareev, I.; Becker, J.; Sauer, D.U. Battery dimensioning and life cycle costs analysis for a heavy-duty truck considering the requirements of long-haul transportation. *Energies* **2018**, *11*, 55. [CrossRef]
12. Cunanan, C.; Tran, M.-K.; Lee, Y.; Kwok, S.; Leung, V.; Fowler, M. A Review of Heavy-Duty Vehicle Powertrain Technologies: Diesel Engine Vehicles, Battery Electric Vehicles, and Hydrogen Fuel Cell Electric Vehicles. *Clean Technol.* **2021**, *3*, 474–489. [CrossRef]

13. Coban, H.H.; Rehman, A.; Mohamed, A. Analyzing the Societal Cost of Electric Roads Compared to Batteries and Oil for All Forms of Road Transport. *Energies* **2022**, *15*, 1925. [CrossRef]
14. Vijayagopal, R.; Rousseau, A. Electric truck economic feasibility analysis. *World Electr. Veh. J.* **2021**, *12*, 75. [CrossRef]
15. Mareev, I.; Sauer, D.U. Energy Consumption and Life Cycle Costs of Overhead Catenary Heavy-Duty Trucks for Long-Haul Transportation. *Energies* **2018**, *11*, 3446. [CrossRef]
16. Tanco, M.; Cat, L.; Garat, S. A break-even analysis for battery electric trucks in Latin America. *J. Clean. Prod.* **2019**, *228*, 1354–1367. [CrossRef]
17. Nykvist, B.; Olsson, O. The feasibility of heavy battery electric trucks. *Joule* **2021**, *5*, 901–913. [CrossRef]
18. Baek, D.; Chen, Y.; Chang, N.; Macii, E.; Poncino, M. Optimal Battery Sizing for Electric Truck Delivery. *Energies* **2020**, *13*, 709. [CrossRef]
19. Available online: https://www.avl.com/cruise-m (accessed on 1 August 2022).
20. Mantriota, G.; Reina, G. Dual-Motor Planetary Transmission to Improve Efficiency in Electric Vehicles. *Machines* **2021**, *9*, 58. [CrossRef]
21. Agamloh, E.; von Jouanne, A.; Yokochi, A. An Overview of Electric Machine Trends in Modern Electric Vehicles. *Machines* **2020**, *8*, 20. [CrossRef]
22. Dixit, M.; Parejiya, A.; Essehli, R.; Muralidharan, N.; Haq, S.U.; Amin, R.; Belharouak, I. SolidPAC is an interactive battery-on-demand energy density estimator for solid-state batteries. *Cell Rep. Phys. Sci.* **2022**, *3*, 100756. [CrossRef]
23. Sankaran, G.; Venkatesan, S. Standardization of electric vehicle battery pack geometry form factors for passenger car segments in India. *J. Power Sources* **2021**, *502*, 230008. [CrossRef]
24. Yang, C. Running battery electric vehicles with extended range: Coupling cost and energy analysis. *Appl. Energy* **2022**, *306*, 118116. [CrossRef]
25. Li, J.; Du, Z.; Ruther, R.E.; An, S.J.; David, L.A.; Hays, K.; Wood, M.; Phillip, N.D.; Sheng, Y.; Mao, C. Toward low-cost, high-energy density, and high-power density lithium-ion batteries. *JOM* **2017**, *69*, 1484–1496. [CrossRef]
26. Pell, J.; Schörghuber, C.; Schubert, T.; Ozli, S. Energy and Lifetime Management for Fuel Cell Powered Trucks. *ATZheavy Duty Worldw.* **2020**, *13*, 42–47. [CrossRef]
27. Basma, H.; Beys, Y.; Rodríguez, F. Battery Electric Tractor-Trailers in the European Union: A Vehicle Technology Analysis. Retrieved from the International Council on Clean Transportation. 2021. Available online: https://theicct.org/publications/eu-tractor-trailers-analysis-aug21 (accessed on 1 August 2022).
28. Hales, A.; Brouillet, E.; Wang, Z.; Edwards, B.; Samieian, M.A.; Kay, J.; Mores, S.; Auger, D.; Patel, Y.; Offer, G. Isothermal Temperature Control for Battery Testing and Battery Model Parameterization. *SAE Int. J. Electrified Veh.* **2021**, *10*, 105–122. [CrossRef]
29. Ramesh Babu, A.; Andric, J.; Minovski, B.; Sebben, S. System-Level Modeling and Thermal Simulations of Large Battery Packs for Electric Trucks. *Energies* **2021**, *14*, 4796. [CrossRef]

applied sciences

MDPI

Article

Understanding and Predicting the Usage of Shared Electric Scooter Services on University Campuses

Seyed Mohammad Hossein Moosavi [1,*], Zhenliang Ma [2], Danial Jahed Armaghani [3,*], Mahdi Aghaabbasi [4], Mogana Darshini Ganggayah [5], Yuen Choon Wah [1] and Dmitrii Vladimirovich Ulrikh [3]

1 Department of Civil Engineering, Faculty of Engineering, Universiti of Malaya, Kuala Lumpur 50603, Malaysia
2 Department of Civil and Architectural Engineering, KTH Royal Institute of Technology, 114 28 Stockholm, Sweden
3 Department of Urban Planning, Engineering Networks and Systems, Institute of Architecture and Construction, South Ural State University, 76 Lenin Prospect, 454080 Chelyabinsk, Russia
4 Transportation Institute, Chulalongkorn University, Bangkok 10330, Thailand
5 Malaysia School of Business, Monash University, Subang Jaya 47500, Selangor, Malaysia
* Correspondence: mh.moosavi65@gmail.com (S.M.H.M.); danialarmaghani@susu.ru (D.J.A.)

Abstract: Electric vehicles (EVs) have been progressing rapidly in urban transport systems given their potential in reducing emissions and energy consumptions. The Shared Free-Floating Electric Scooter (SFFES) is an emerging EV publicized to address the first-/last-mile problem in travel. It also offers alternatives for short-distance journeys using cars or ride-hailing services. However, very few SFFES studies have been carried out in developing countries and for university populations. Currently, many universities are facing an increased number of short-distance private car travels on campus. The study is designed to explore the attitudes and perceptions of students and staff towards SFFES usage on campus and the corresponding influencing factors. Three machine learning models were used to predict SFFES usage. Eleven important factors for using SFFESs on campus were identified via the supervised and unsupervised feature selection techniques, with the top three factors being daily travel mode, road features (e.g., green spaces) and age. The random forest model showed the highest accuracy in predicting the usage frequency of SFFESs (93.5%) using the selected 11 variables. A simulation-based optimization analysis was further conducted to discover the characterization of SFFES users, barriers/benefits of using SFFESs and safety concerns.

Keywords: green campus; shared free-floating electric scooter; usage frequency prediction; decision tree; random forest

Citation: Moosavi, S.M.H.; Ma, Z.; Armaghani, D.J.; Aghaabbasi, M.; Ganggayah, M.D.; Wah, Y.C.; Ulrikh, D.V. Understanding and Predicting the Usage of Shared Electric Scooter Services on University Campuses. *Appl. Sci.* **2022**, *12*, 9392. https://doi.org/10.3390/app12189392

Academic Editor: Mirco Peron

Received: 2 August 2022
Accepted: 13 September 2022
Published: 19 September 2022

Publisher's Note: MDPI stays neutral with regard to jurisdictional claims in published maps and institutional affiliations.

1. Introduction

The advancement of information technology and sharing economy business models is changing traditional models of ownership and transport services. New modes of travel are emerging in urban areas, such as transport network company services, bike-sharing and scooter-sharing, etc. Shared micro-mobility (SMM, the shared utilization of an e-/bicycle, e-/scooter, or other low-speed modes) is a newly developed transportation mode [1]. SMM provides users with a short-term access to a transportation service over an as-needed basis [2].

Early documented impacts of SMM include increased mobility [3], decreased greenhouse gas emissions [4], and decreased automobile use [5,6]. Since 2017, over USD 5.7 billion have been devoted to SMM start-up companies, mostly in China. A steady customer pool has been established in the SMM market, which is two to three times faster than ride-hailing or car-sharing services. The combined value of SMM start-ups is estimated to exceed USD 1 billion [7].

Shared Free-Floating Electric Scooters (SFFESs) have been altering travel in cities and on university campuses. Though SFFESs have swiftly obtained popularity and approval over the past couple of years, limited studies have been reported on their use. The analysis of Berg Insight shows that the COVID-19 crisis lead to a lower shared-scooter ridership in 2021. However, in the long term, ridership is projected to reach over 4.6 million people in 2024 worldwide, with a base of 774,000 people in 2019 [8].

New mobility services, such as Uber/Lyft, have been changing the landscape of urban mobility. SFFESs have become increasingly popular and utilized by communities given their acceptable cost, zero-emission power and minimal environmental footprint. In addition, given the present pandemic, city planners are looking for new methods, such as SFFESs, to reconcile urban mobility need and social distance requirements. While SFFESs offer promising opportunities, they also bring negative externalities, including safety and equity issues for pedestrians, cyclists and disabled/elder citizens [6,9]. Many cities proscribed SFFES services, particularly in the initial boom of SMM, due to vital vandalism and street clutter, including Austin, Nantes, Amsterdam, Bordeaux, and recently Kuala Lumpur. These cities revisited their decisions afterwards and devised novel regulation provisions to optimize SFFES benefits while limiting their drawbacks. Some cities banned the usage of SFFESs due to regulation requirements; for example, the New York State Department of Motor Vehicles legislation requires the registration of any electric vehicle, which makes SFFES service impossible.

Effective regulation faces two major setbacks emanating from the organizational culture/climate mismatch between local authorities and service operators. The operators need a high vehicle density to guarantee a high service quality and ultimately foster their market [10,11]. However, local authorities are wary of street clutter and intend to limit the fleet size. While technology and investments are essential for service implementation, equally important is to signify the impact of shared micro-mobility on the urban mobility ecosystem and its evolution trend over time, in order to better design and integrate it into sustainable mobility as a whole [12]. However, very few studies examined SFFESs, and in addition the existing studies were limited in the analysis approaches used, which may fail to capture the complex nonlinear relationship between variables. In addition, most studies on SFFES services were conducted in the United States, China and, most recently, European cities, but are yet very limited in developing countries. It is presumed that this study will be the first step taken to assess the adoption of SFFESs and usage behavior with respect to a Malaysian context. The paper identifies public concerns, SFFES benefits and barriers, and the choice and usage behavior of the university population (students and non-/academic staff).

Choice behavior in new mobility services is usually assessed and modeled using traditional statistical models, such as regression, mixed logit, multinomial and binary logit models [13,14]. Recently, [15] used the Chi^2 and Kruskal–Wallis tests to analyze the frequency of e-scooter use. Given the strict assumptions of statistical models, they have limited capabilities to capture the complex relationships between factors and choices, nonlinear correlations among factors, and to deal with factors with various categories [16]. Machine learning (ML) methods have been widely utilized in civil engineering [17–23] and transportation studies [24,25]. They can model the nonlinear associations between independent and target variables as well as among independent variables [26,27]. Therefore, it can be argued that the current study is one of the first attempts to predict SFFES usage frequency and identify significant factors impacting its use by adopting ML techniques.

Malaysian universities are currently adopting new sustainable strategies in moving their campuses towards becoming green campuses. Specifically, the management of the University of Malaya is planning to launch SFFES service in the near future. This paper aims to predict the usage frequency of SFFESs among the students and staff on the campus. As a summary of above discussion, the main contributions are:

1. This research study is one of the first efforts made to scrutinize the usage of SFFESs on a large university campus. In addition, this is perhaps the first study on SFFES services in "developing countries" such as Malaysia.

2. This research is one of the first studies which aims to predict the usage frequency of SFFESs and pinpoint significant attributes affecting the use of SFFESs by adopting various supervised and unsupervised machine learning techniques.

The remaining of the article is organized as follows: Section 2 concerns a literature review on related works, followed by the survey design and data collection in Section 3. Section 4 proposes the analysis methodology, including feature selection and model development. Sections 5 and 6 present the model output, analysis results and simulation-based optimization and discussion. The final part offers the obtained findings of the study and suggests future directions.

2. Related Works

It is believed that, in terms of urban features and population, higher education organizations mirror smaller cities [28]. Moreover, there exist many activities occurring on university campuses that exert both direct and indirect effects on the natural milieu [29]. Therefore, practitioners in these academic contexts need to apply green practices and provide support in offering multidisciplinary green technical solutions to achieve sustainable development on campuses [30]. The United States Green Building Council [31] revealed that a green campus is a higher education community seeking to enhance its resource conservation, energy efficiency, and ecological quality via training on healthy living, sustainability, and convenience learning environments for all.

In the context of higher education, green practices are rising rapidly. However, achieving sustainability in Malaysian universities in this regard has yet remained an issue [32]. Malaysia has committed itself to buttress sustainability on university campuses after signing the Talloires Declaration. Thereafter, enthusiasm for focusing more on sustainable development has increased in Malaysia. Nevertheless, many universities yet lag behind in green practices in order to attain sustainability as an institutional policy. This runs counter to the outline of higher education institutions since the 1992 Earth Summit in Rio. Universities are facing pressure from non-/governmental organizations to incorporate green practices in their activities following several sustainability declarations.

Shared micro-mobility (SMM)—the short-term rent of micro-mobility vehicles such as (e-)scooters and (e-)bicycles—is regarded as a mobility (sub)system that can alter the present transport system in terms of cars [2,33]. This technology was first presented in 2017 and has now become an important mode of transport emerging in more than 1000 cities and college campuses worldwide. Such web-based SFFES services are managed by rental networks and operated using smartphones.

Academic studies on SFFESs have been emerging. For example, ref. [34] examined anonymized SFFES trip data and concluded that users ride SFFESs for about 8 min for 0.7 miles, with an average speed of 5.23 miles per hour. The SFFES service could be used as an appropriate travel model for last-mile transport or short-distance trips. Ref. [35] found considerable differences in temporal and spatial usage patterns between SFFESs and docked bike-sharing ridership [36]. Ref. [37] assessed the behavioral determinants of travelers' purpose for using SFFESs and found that the perceived compatibility of SFFESs significantly affected usage intention. Ref. [15] performed Kruskal–Wallis and Chi^2 test with e-scooter-associated survey data and pointed to the importance of sociodemographic characteristics in affecting SFFES usage. Ref. [38] evaluated the API data of SFFES vendors and found significant SFFES ridership variations between weekends and weekdays, but not between morning and afternoon trips.

As mentioned before, most of the academic studies in this field were conducted in the US. Surveys were conducted by a few cities to complement assessing the e-scooter pilot programs. It was found that e-scooters were popular or generally considered to present a respected service, even among non-users [15]. For example, the Portland report stated that over 30% of people had tried e-scooters. Over 70% of Portlanders riding an e-scooter stated that they utilized e-scooters most commonly for transportation, but not recreation. The reasons for use included reliability, speed, cost, convenience and fun [39]. Unequal adoptions between population groups were suggested by surveys. The gender (female/male) splits were 64/34 and 70/30 for Portland and Denver. In total, 69% of e-scooter users were aged 20–39 in Portland, while the figure was over 50% in Denver [39,40].

An online survey was performed by the [41] on 1250 individuals in the five largest cities of Germany (Hamburg, Berlin, Frankfurt, Cologne, and Munich) in September 2019. It explored their overall mobility behavior and utilization of SFFES systems. It revealed that 42.7% of e-scooter users were aged between 18–25, and 28.8% between 26–35. The SFFES service substituted 49.1% of walking trips and 64.5% of public transport trips. A quantitative study was performed by the French [42]. They gathered 4382 user responses after various semi-structured and exploratory interviews. It reported that e-scooter renters were young (52% younger than 34), male (66%), highly educated (19% students, 53% work executives), and with a significant share of non-locals (42%). For the modal shift, users substituted walking (44%), public transport (30%), and bike trips (3% owned a bike; 9% shared a bike).

Supervised learning algorithms learn correlation patterns from data (independent and target variables) and make decisions/predictions based on a specific objective. Decision trees (DT) are widely used in data-driven prediction analysis [43–46]. Decision trees have been used for model evaluation and identifying important variables. Random forests (RF), a derivation of decision trees, can work in both supervised and unsupervised modes. It can handle continuous as well as categorical data in classification or regression tasks [47,48]. Random forests are prioritized over other techniques, as it can manage highly non-linear data, and demonstrates many features, such as agility in locating noise in data and adjustability to parameters [49]. It has three main features: (i) estimating missing values automatically, (ii) Weighted Random Forest (WRF) for balancing errors in imbalanced data, and (iii) estimation of the significance of variables utilized for categorization [50]. Naïve Bayes (NB) classifiers are also able to handle continuous and categorical variables and quickly make real-time predictions [51].

Unsupervised learning is designed to analyze unlabeled data [52]. As the amount of unlabeled data is exponentially rising, it is essential to explore unsupervised learning to perform feature selection. Data clustering (feature selection) is an important problem in knowledge discovery to improve the understandability, scalability and accuracy of resulting models. The clusters correspond to hidden models and the resulting outcomes represent data notions. In the context of supervised learning, feature selection refers to predictions based on provided outputs, while in unsupervised learning the features are clustered without any prior knowledge of the expected output. The importance of feature clustering is to improve prediction performance and provide a deeper understanding of the underlying process that produces the data. Examples of clustering algorithms are k-means, partitioning around medoids (PAM) and hierarchical clustering. This paper uses both supervised and unsupervised learning techniques for feature selections and predicting the usage frequency of SFFESs on campus.

3. Methodology

3.1. *Survey Design and Data Collection*

The survey was designed to understand the adoption, choice behaviour and usage of SFFES services on the university campus. The questionnaire consisted of 55 mandatory questions covering the following aspects:

- Sociodemographic information, including information about age, gender, marital status, residential area, highest level of education, employment status, race, household monthly income, private vehicle ownership, shared mobility and membership and frequency of usage of e-hailing services.

- Commuting characteristics, including commuting mode to and from the campus, and the travel mode, frequency, distance, time and cost on campus.
- Perceptions and choices regarding the SFFES service, including (1) perceptions regarding using SFFESs and concerns of safety, equity, costs, comfort, and social distancing due to COVID-19; (2) service attributes, such as accessibility, payment methods, and the advantages and disadvantages of shared e-scooters compared to other transport modes; and (3) infrastructure and built environment, such as separated lanes for scooters, green spaces, quality of road surfaces and connectivity.
- Usage frequency of the SFFES service, including four levels of response: (1) not using an e-scooter at all; (2) using an e-scooter as a mode of transport occasionally (sometimes but infrequently); (3) using an e-scooter frequently; and (4) using an e-scooter regularly as a main mode of transport. (Table 1 presents the information on the data and attributes used in this study).

Table 1. Variables used in this study for analysis.

Attribute	Description	Values
Sociodemographic		
Age	Age	(1) 18 to 29; (2) 30 to 44; (3) 45 to 60; (4) Over 60
Gender	Gender	(1) Male; (2) Female
Education	Highest education level	(1) Secondary; (2) Diploma; (3) Bachelor's degree; (4) Master's degree; (5) Doctorate degree
Position	Job position	(1) Undergraduate student; (2) Postgraduate student; (3) Academic staff; (4) Non-academic staff
Status	Employment/education status	(1) Full-time; (2) Part-time
Race	Race	(1) Chinese; (2) Malay; (3) Indian; (4) Other
Monthly Income	Monthly household income	(1) Less than RM 2000; (2) Between RM 2000 RM 4000; (3) Between RM 4000 and RM 6000; (4) Between RM 6000 and RM 12,000; (5) More than RM 12,000
Private vehicle	Private vehicle ownership	(1) Yes; (2) No
E-hailing	Usage of e-hailing services per week	(1) Not using at all; (2) Less than 3 times; (3) 3 to 6 times; (4) More than 6 times
SMS Membership	Membership of shared mobility services	(1) Yes; (2) No
Travel characterization		
Travel mode	Usual travel mode for going to campus	(1) E-hailing taxi; (2) Private car; (3) Private motorcycle; (4) Public transportation; (5) Walking/cycling
Camp.Hrs/d	Hours usually spent on the campus per day	(1) 1 to 3 h; (2) 3 to 5 h; (3) 5 to 8 h; (4) More than 8 h
Camp.Tra/d	Number of journeys onto or to outside of the campus per day	(1) Less than 2 journeys; (2) 2 to 4 journeys; (3) 4 to 6 journeys; (4) More than 6 journeys
Camp.mod/d	Travel mode on the campus	(1) E-hailing taxi; (2) Private car; (3) Private motorcycle; (4) Public transportation; (5) Walking/cycling
Camp.tra.time/d	Duration of daily travel on the campus	(1) Less than 10 min; (2) 10 to 20 min; (3) 20 to 30 min; (4) More than 30 min
Camp.tra.cost/d	Daily travel cost on the campus	(1) Less than RM 5; (2) Between RM 5 and RM15; (3) Between RM15 and RM 25; (4) More than RM25
Attitudinal factors: impact of infrastructure		
Sep.lane	Bike/scooter lane separate from road traffic	(1) Strongly discourage; (2) Discourage; (3) Encourage; (4) Strongly encourage
On-road.Lane	Bike/scooter lane on the road with traffic	(1) Strongly discourage; (2) Discourage; (3) Encourage; (4) Strongly encourage
No-Lane	Road with no bike/scooter lane	(1) Strongly discourage; (2) Discourage; (3) Encourage; (4) Strongly encourage
Greenery	Green Space (e.g., road-side trees, greenery, water)	(1) Strongly discourage; (2) Discourage; (3) Encourage; (4) Strongly encourage
Smooth.Surf	A smooth road surface	(1) Strongly discourage; (2) Discourage; (3) Encourage; (4) Strongly encourage
Connectivity	Pathways/roads connectivity	(1) Strongly discourage; (2) Discourage; (3) Encourage; (4) Strongly encourage
e-scooter Usage *(Target variable)*	Shared e-scooter frequency of usage	(1) Not using at all; (2) Sometimes/infrequently; (3) Frequently; (4) Regularly as the main mode of transport.

The survey was carried out on students and staff of the University of Malaya (UM). The UM is situated in the southwest of Kuala Lumpur. It has a 373.12-hectare campus and houses around 20,000 students and 6000 staff. In addition to these numbers, many daily operations, activities, and events require continuous mobility access to different transportation modes. Consequently, integrated transportation system management on the university campus is pivotal. The current transportation services on the UM campus include bus services (campus and traditional buses), a bicycling facility, and car and pedestrian accessibility. Figure 1 shows the University Campus Map and the road line map.

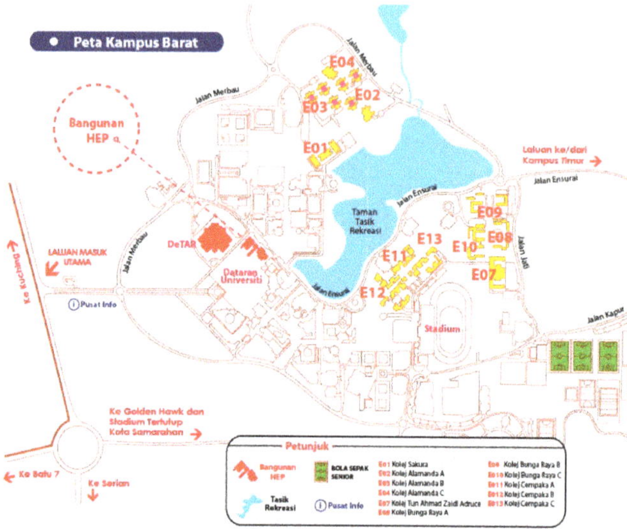

Figure 1. University of Malaya campus map.

The online Google questionnaire survey was disseminated to over 30,000 faculty, non-academic staff and university students in December 2020. The survey was estimated to take 10 min to complete. The survey link was active for a period of three weeks. We received 1023 responses and 1000 surveys were valid for further analysis (response rate: 1.7%).

Table 2 captures the sociodemographic characteristics of the sample, the UM population and the overall university populations in Malaysia. For the university population, we used the data statistics of 2020. The gender distribution in the sample is overall representative, with the female population slightly overrepresented. Shares for occupation composition are comparable. Given the similarities of gender and occupation, we believe that the sample sufficiently reflects the socioeconomic features of the targeted population.

Table 2. Comparison between the survey sample and the university population in percentage.

Socio-Demographics	Total Sample (n = 1000)	UM University	All Universities in Malaysia
Gender			
Male	45.6	49.0	47.0
Female	54.4	51.0	53.0
Occupation			
Undergraduate students	51.5	51.7	48.5
Graduate students	36.5	27.6	33.5
Part-time graduate students	2.1	6.3	6.3
Faculty and staff	9.9	16.3	11.7

3.2. Feature Selection

Feature selection is an option in statistics to detect significant factors that use measures of confidence intervals as well as hypothesis testing. After conducting model evaluation, the elements (independent variables) must be examined further to see how they lead to measurement accuracy. Hence, machine learning algorithms are built-in with the feature selection technique to analyze the variables or features in the input data. The distribution of these features contributes to the prediction of the final outcome using machine learning models. Feature selection helps to understand the model better by focusing only on the important variables. This statistical technique eliminates variables which are insignificant or highly correlated with any other variable. Based on significance score, the order of variables can be illustrated to realize the accuracy of prediction. The reliability of important variables depends on the accuracy of a specific algorithm. The objectives of feature selection in machine learning are to reduce the complexity of the mode and to promote the performance of the model. Feature selection evaluates the relationship between the input variables and target variable.

The supervised and unsupervised feature selection methods vary considering the target variables. While the supervised learning model requires a target variable to specify the important variables, the unsupervised learning model disregards the target variable and chooses important variables using correlation. Figure 2 shows the study methodology workflow.

Clustering

The unsupervised learning model clusters the input variables based on correlation between each other, and without considering the target variable. The important variables obtained from the random forest feature selection are used to perform clustering. There are two steps in clustering: (a) determination of the optimal number of clusters, and (b) hierarchical clustering.

To determine the optimal number of clusters:

The optimal number of clusters is specified using the Gap Statistics method. The *fviz_nbclust()* function in *factoextra* R package is employed to compute the optimal number of clusters. The Gap Statistics algorithm works as follows [53]:

- The observed data of 1000 samples with n variables is analyzed by changing the number of clusters from $k = 1, \ldots, k_{max}$, and the total within intra-cluster variation W_k is computed.
- B reference datasets with a random uniform distribution is generated. Each reference dataset is clustered with varied number of clusters $k = 1, \ldots, k_{max}$, and the corresponding total within intra-cluster variation W_{kb} is computed.
- The estimated gap statistic is computed as the deviation of the observed W_k value from its expected value, W_{kb} under the null hypothesis: $Gap(k) = 1B\sum b = 1Blog(W * kb) - log(Wk)$. The standard deviation of the statistics is also computed.
- The number of clusters is chosen as the smallest value of k such that the gap statistic is within one standard deviation of the gap at $k + 1$: $Gap(k) \geq Gap(k + 1) - sk + 1$.

The optimal number of clusters is used to perform hierarchical clustering using the *hclust* R package. Hierarchical clustering is an agglomerative clustering algorithm, which can be cut at a chosen height to produce the desired number of clusters [54,55]. The clusters produced in a dendrogram are joined together in order of their closeness measured by dissimilarity. The steps of hierarchical clustering are as follows:

- Divide n variables into k groups by cutting at a desired similarity level.
- Calculate the dissimilarity matrix between variables using function *dist ()* in *hclust* package.
- Plot the dendrogram using *fviz_dend ()* function in *factoextra* package with dissimilarity matrix as the input.

Correlation analysis is performed using R *corrplot* function to assess the relationship between the variables.

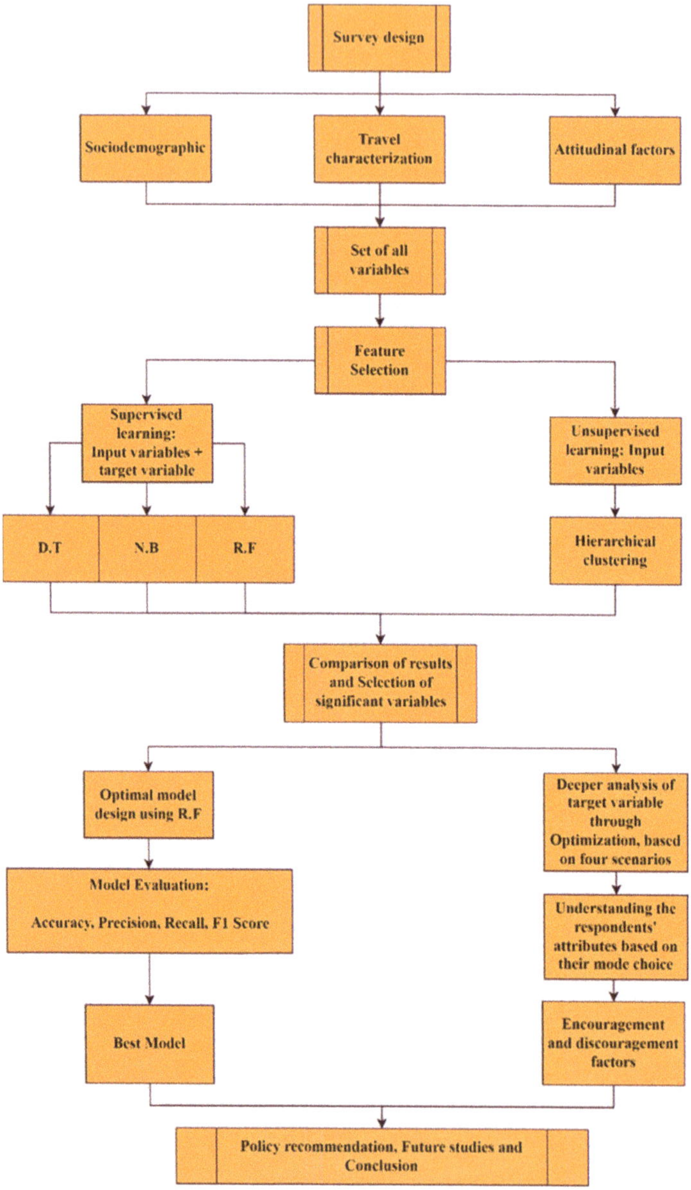

Figure 2. Methodology workflow.

3.3. The Optimal Model Design

The model assessment is performed using the important variables selected through supervised (random forest, decision tree and Naïve Bayes) and unsupervised learning methods. After selection of significant variables, the random forest classifier is used to assess the model performance using the test and out-of-bag errors by changing the total number of trees (*ntree*) and predictors at each split (*mtry*). The best *ntree* and *mtry* are obtained using the measures of the mean squared error and variance, calculated using the out-of-bag errors. A total of 2/3 of the data is used for training and 1/3 for validating the trees. The final model is developed using the best *ntree* and *mtry*. Random forest algorithm is a joint and collaborative learning algorithm that is derived from decision trees. It follows the rules of decision trees but constructs numerous decision trees during training time and outputs the class with maximum vote. For example, the random forest algorithm constructs trees of different classes using the similar input data. The tree structures can be explained using subset matrices as shown in Figure 3. Three random subsets are created during the training process. Three different trees are explained using three subsets (S1, S2 and S3). Different samples are grouped into different subsets based on the correlation between input features (independent variables). Decision trees are built based on the subset values. The decisions or the final predicted output from each decision tree is considered a class. The class, which receives maximum votes from the total number of trees, will be chosen as the final output. Class 1 has two votes whereas class 2 has one vote in Figure 3, therefore class 1 is the final predicted output. This class 1 will be used to rank the variables based on importance score.

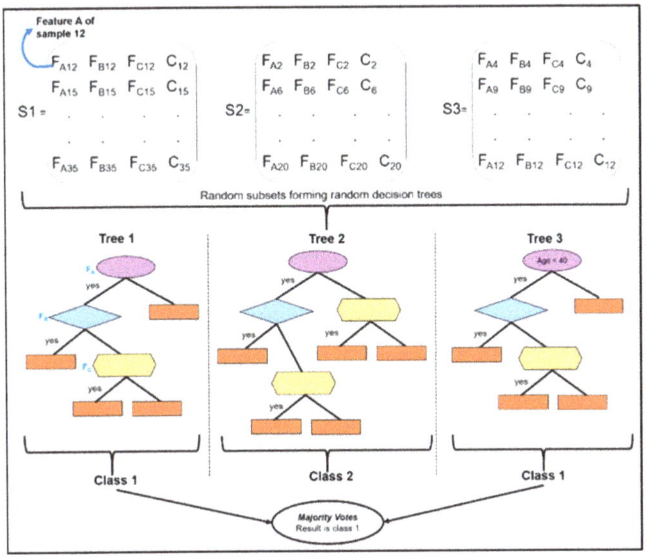

Figure 3. Random forest algorithm workflow.

The R package *random-forest* is used to perform feature selection for n variables, and the number of important variables is determined in three stages: initial, threshold and prediction. The most important variables are selected from the final prediction stage. Random forest considers a random subset of predictors, p, each time when splitting the training set. The trees find all the predictors while performing a split and select the best amongst them. The total number of predictors at each split is calculated using the formula $mtry = \sqrt{n}$. The default number of trees used in random forest feature selection is $ntree = 500$ and the total number of predictors used to construct the trees is $\sqrt{n}$.

3.4. Model Evaluation

Model evaluation in machine learning is an alternative to the assessment of effect size in conventional statistics [56]. It is a key step in machine learning, as the ability of the model to make predictions on unseen or future samples will enhance the trust on the model to be used in a particular dataset. The measurement for model evaluation is accuracy in percentage (estimate of generalization of a model on future data). The most popular model evaluation technique is cross-validation. Cross-validation divides the data into test (independent dataset) and training (subset of data used to train the model for future predictions) sets; 5-fold cross-validation was performed. The accuracy is assessed based on the overall error estimation comparing the test and training sets. An interchange of test and training sets reduces bias and variance in the method. Cross-validation can be used to compare the performance of different machine-learning algorithms on the same data, as this will make it easier to select the best algorithm to perform further analyses. A confusion matrix is the most common interpretation of model performance in supervised learning. A confusion matrix can produce model accuracy, precision, recall and F1 score. In this study, the total number of samples (n = 1000) was divided into 80% of the training set and 20% of the testing set. The model evaluation was performed using three different algorithms: decision tree, random forest and Naïve Bayes, and the accuracy measures based on the confusion matrix were recorded.

4. Results

4.1. Descriptive Analysis (Encouragement and Discouragement Factors)

This section presents the results of the last part of the survey, which measured encouragement and discouragement factors for using SFFESs. In other words, after predictions of SFFES usage, important factors and levels of acceptance between different groups of respondents, this section was designed to answer the following questions: 1—Why will certain respondents never use SFFESs (8% of total respondents according to Figure 4)? 2—What are their main concerns? 3—What are the benefits of the SFFES service from our respondents' point of view?

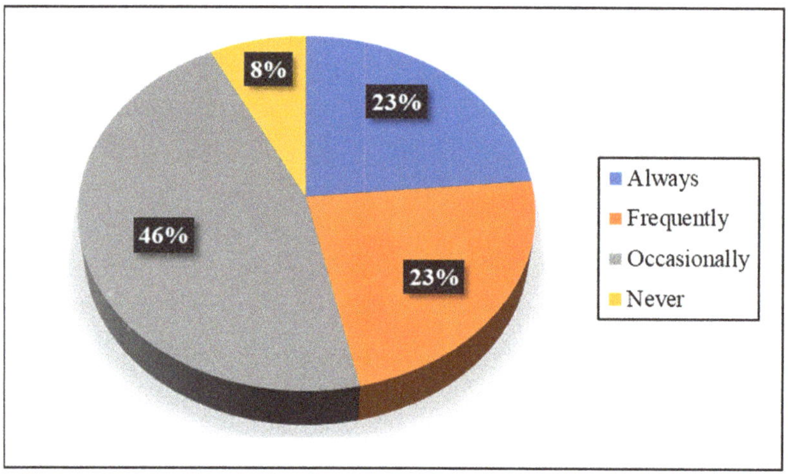

Figure 4. Percentage of SFFES use frequency based on four categories.

Survey participants were asked to express their perceptions and feelings regarding the encouragement and discouragement factors of using SFFES services. In the first part, we asked the participants about the benefits and advantages of SFFESs. Figure 5 presents the overall responses to questions about the benefits/advantages of using SFFESs.

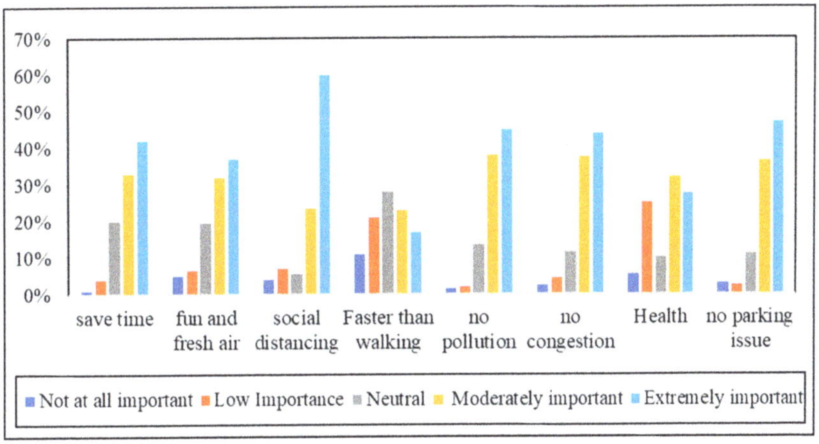

Figure 5. Participants' perceptions about the advantages/benefits of using SFFESs.

According to Figure 5, most of the respondents were almost agreed on all mentioned benefits of SFFESs, except for "speed" and "physical/mental health". A considerable number of participants used a private vehicle as their main mode of transport on the campus. This could be one of the reasons why most of the participants selected a neutral option about the speed. The second, less-important advantage of the SFFES was health benefits, based on participants' responses. Indeed, the physical and mental benefits of e-scooters are not well investigated. However, using an e-scooter obviously requires less physical activity comparing to walking and cycling. Interestingly, social distancing during the pandemic was selected as the most important benefit of SFFESs. Recently, the COVID-19 virus hit Malaysia badly, and the number of new positive cases reached 4500 per day. This was the main concern of the current situation and people were seeking safe ways to go about their daily activities. 47% of survey participants indicated that they would not have car park issues by using SFFESs, and 45% believed that the environmental benefits (no pollution) of SFFESs were extremely important. In addition, "saving time" and "no traffic congestion" were indicated as extremely important benefits of SFFESs by 42% and 44% of participants, respectively.

The next series of questions were designed to ask respondents about their concerns about SFFESs, and what factors would prevent them from using this service, as presented in Figure 6. Safety was indicated as an extremely important concern of using SFFESs by 59% of respondents, moderately important by 26% of respondents, and not at all an important concern by only 4% of respondents. Surprisingly, the cost of riding SFFESs was selected as the second most important concern by 75% of respondents. In total, 53% and 22% of respondents indicated the "cost" as an extremely important and moderately important preventative factor, respectively. Due to the hot and humid tropical weather of Malaysia throughout the year, which is also interspersed with tropical rain showers, "adverse weather" is always a significant concern. Accordingly, almost 55% of respondents indicated the weather as an important preventative factor.

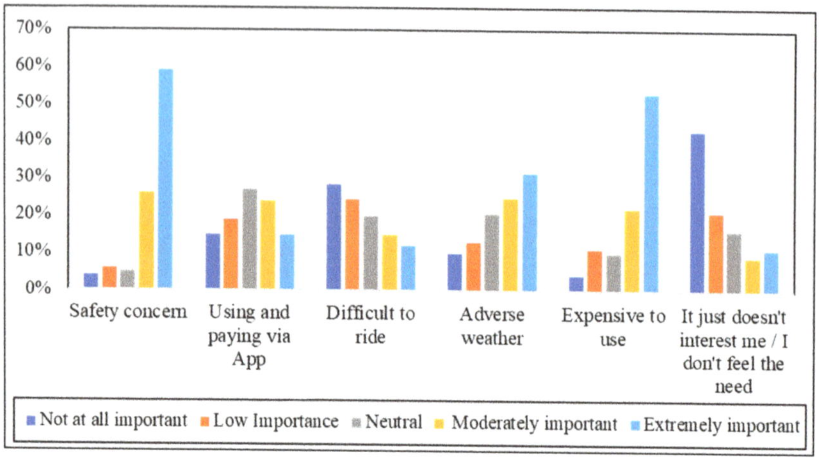

Figure 6. Participants' opinions about what reasons would prevent them from using the SFFESs.

As explained above, safety was indicated as the most important concern by almost 85% of the survey participants. Therefore, we decided to further explore this concern to gain better insights for policy making discussions and recommendations. Figure 7 illustrates SFFES users' perception of safety concerns based on their willingness to use the service in future. Respondents who would never ride e-scooters had the highest level of safety concern. Almost 40% of participants who belonged to this category specified that safety was an extremely important preventative factor to riding an e-scooter on campus, and 30% stated that it was moderately important.

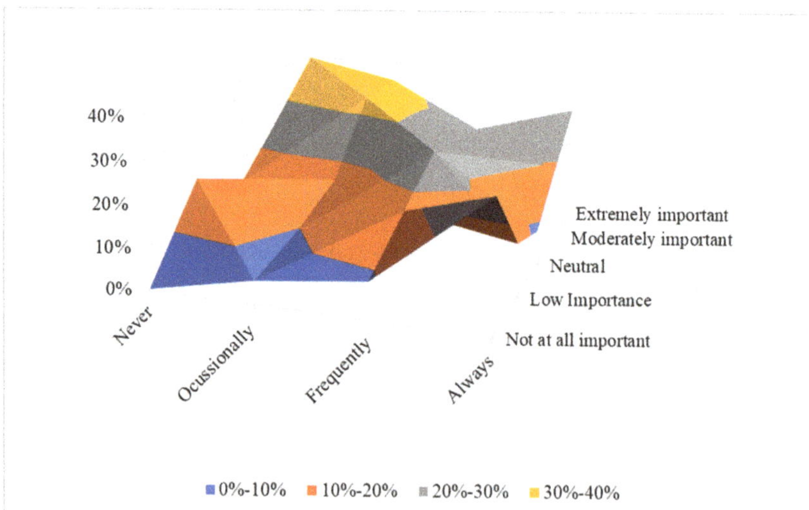

Figure 7. Safety concerns based on SFFES usage categories.

In addition, over 80% of respondents who indicated safety as an extremely preventative concern also stated that they were extremely afraid of hitting somebody or being hit while riding an e-scooter. One of the chief causes of worry about accidents was the road features. Almost 60% of respondents who were extremely worried about safety indicated that separated scooter/bicycle lanes would strongly encourage them to ride an e-scooter.

In addition, almost 67% of them specified that no separated lanes for e-scooters would strongly discourage them from riding an e-scooter. The impact of other road features such as road connectivity, the quality of the surface and the availability of water and green spaces on their willingness to ride an e-scooter is shown in Figure 8.

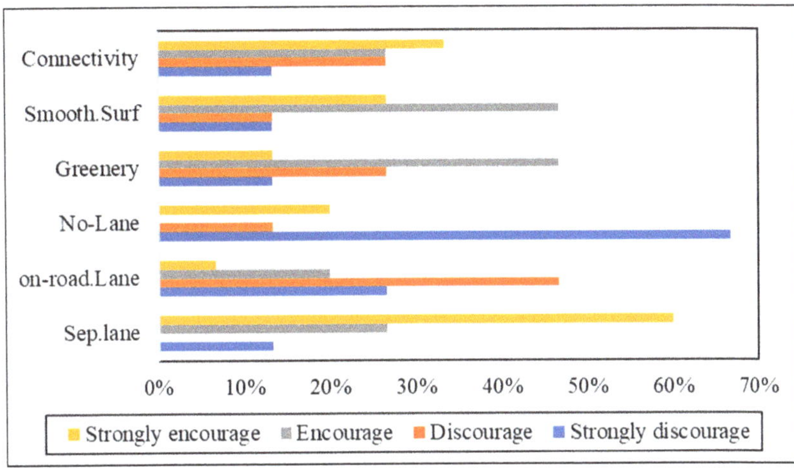

Figure 8. Impact of road features on the perceptions of respondents who believed safety was an extremely preventative factor for riding an e-scooter.

4.2. Policy Recommendation

In line with the intentions of the Malaysian government to develop green university campuses in the country, a number of universities in Malaysia have begun carrying out different green practices in an effort to improve sustainability. Accordingly, Malaysian academic centers, especially those at the higher education level, are dedicated to supporting the 40% reduction of carbon dioxide (CO_2) emissions vowed by the government at the 1992 Earth Summit in Rio [57]. Nevertheless, scholars such as [58] argue that, in Malaysian university management, practitioners and stakeholders are oblivious to green campus paradigms, which has caused most universities to ignore green practices. Currently, research on sustainability is initiated and socially certified by experts in higher education institutions [59]. However, there is still a lack of a proper method for interdisciplinary communication and cooperation among these sustainability practitioners to compile integrated data gleaned based on green indicators, which should be considered when achieving sustainability within Malaysian university campuses [60,61].

Nowadays, various sustainability practitioners in different areas of expertise work collaboratively to reach sustainability in the context of universities. However, interdisciplinary communication and collaboration is still absent among sustainability practitioners at higher education levels [62,63]. As [64] put, there is an urgent need for an interdisciplinary approach that is able to provide higher education institutions with a green campus paradigm toward accomplishing socio-economic and environmental sustainability. This is echoed by [65], who declared that there was insufficient harmonization and cooperation among practitioners from dissimilar domains that work jointly to obtain sustainability. The green campus concept aims to introduce engineering features including waste treatment, water treatment, and air pollution control, alongside personal aspects, such as promoting a laissez-faire outlook.

To develop a green campus, it is essential to assess the present data, information, and reports while focusing on enhancement. Generally, the aspects assessed in terms of green campus valuation instruments for higher education covers site and planning management, waste management, energy efficiency, sustainable transportation, water efficiency and conservation, indoor environmental quality, material and resource management, green education, and green innovation. In this regard, electricity, waste generation, and transportation were chosen as targets considering their higher influences on CO_2 emission. Promoting active and novel modes of transportation can be an effective approach to reduce carbon emissions, as future transport will probably be dominated by electric vehicles (EVs). These vehicles offer several environmental benefits, which can lead to sustainability in urban transportation. More specifically, battery electric vehicles (BEVs) are gaining worldwide popularity. With their light weight, they could be well integrated into urban transport systems.

Electric scooters are emergent vehicles that could be used as an alternate transportation mode in campus and urban areas. These scooters have the potential to improve mobility and can be used in place of short car and ride-hail journeys. On the other hand, scooters have introduced some new challenges, which include safety, negative effects on disabled people, walkway clutter, etc. It is important for cities to evaluate the benefits that may be gained by using Shared Free-Floating Electric Scooter (SFFES) systems. SFFES services have the potential to introduce a number of environmental/social benefits, e.g., saving expenses and time (since they are generally faster than walking and even driving on crowded roads), lessening traffic blocking, enhancing multimodal transport connections, and decreasing the emissions of greenhouse gas (GHG). However, all these benefits are deeply dependent on adopted policies. For instance, based on our study results, most of the respondents specified the SFFES as an expensive transportation mode for campus usage. Making reliable decisions on this issue can be of great support to the expansion of e-scooter share programs in both campuses and cities.

4.3. Selection of Significant Variables through Unsupervised Clustering

Hierarchical clustering produced a dendrogram, which divided the 22 variables into 2 different clusters—13 variables in cluster one and 9 variables in cluster two. The variables in each cluster are shown in Figure 9.

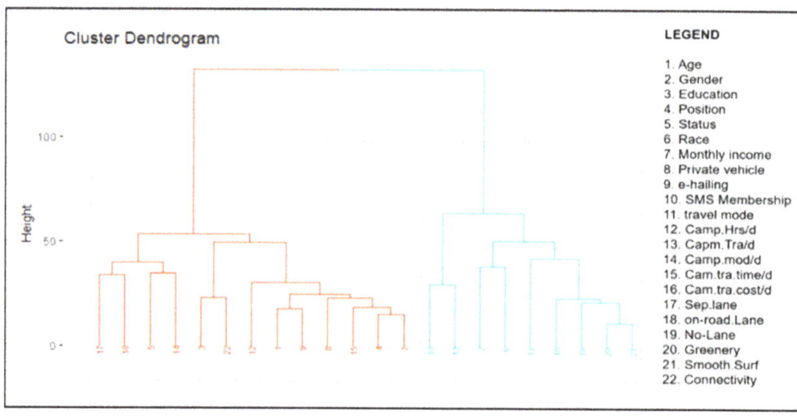

Figure 9. Cluster dendrogram of 22 variables.

The correlation between the variables was assessed using the dissimilarity matrix. The y-axis in the dendrogram in Figure 9 can be explained using the terms clade and leaves. The clusters were formed at a particular cluster cutoff value based on the number of clusters specified. As the analysis on determining optimal number of clusters regarding the dataset used in this study produced the result k = 2, the number of clusters was set as two. The specified number of clusters returned vectors containing features in each cluster. The lines showing the variables (number 1–22) are the leaves, whereas clusters 1 and 2 are clades 1 and 2 respectively. Leaves 17, 18, 5 and 14 are more similar to each other than they are to other leaves in clade 1. Leaves 3 and 22 are more similar to each other than they are to other leaves in clade 1. Leaves 12, 1, 9, 8, 15, 4 and 7 are more similar to each other than they are to other leaves in clade 1. The x-axis in the dendrogram represents the clusters. The y-axis in the dendrogram represents the closeness of the leaves/variables. For example, leaves 4 and 7 were correlated before they joined 15, 8, and the following leaves together in one clade.

The distance between two clusters was measured using the linkage method. The complete linkage method used in this study displayed the distance between clusters 1 and 2 using the longest distance between two points in each cluster. The point refers to the line height in the dendrogram (Figure 9). The similarity between the features were assessed using the dissimilarity matrix index, whereas the important variables were determined using the line height. The heights of the lines in each leaf represent the importance score of the variables. In cluster 1, the most important features were Sep.lane, On-road.Lane, Status and Camp.mod/d with similar line heights. In cluster 2, the most important features were Gender, Race and Travel mode. To further assess the correlation between the independent variables, correlation analysis was performed. Figure 10 shows the correlation between the 22 independent variables.

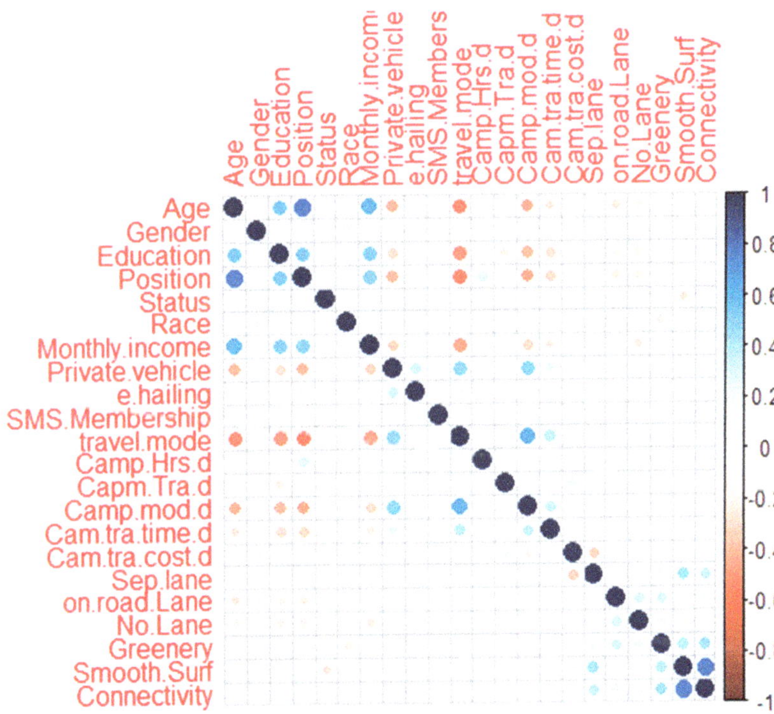

Figure 10. Correlation between the 22 independent variables.

The blue color represents positive correlation and the red color displays negative correlation. Based on the correlation analysis, two combinations are highly positively correlated, which are Position and Age and Connectivity and Smooth Surf. Moderately positively correlated combinations are Monthly Income and Age, and Camp.mod.d and travel.mode. Next, the lowly positively correlated combinations are Education and Age, Position and Education, Position and Monthly income, Monthly Income and Education, Private Vehicle and travel.mode, and Private Vehicle and Camp.mod.d.

4.4. Selection of Significant Variables Using Supervised Learning Models

The variables' importance rank of the 22 independent variables based on RF, DT and NB techniques are shown and compared in Figure 11. The present study takes advantage of various feature selection methods to pick only the important variables and design the prediction model according to selected variables. The core motive behind decreasing the number of variables (based on their level of importance and correlations) is to diminish the complexity and promote the applicability of our final model. Therefore, after implementing unsupervised clustering and identifying the correlation of the variables, we also compared the variables' importance based on three different tree-based supervised machine learning techniques. Table 3 presents the variable weights using outputs of RF, DT and NB. The mutually important variables were detected. For example, monthly income, age and private vehicle ownership were variables with high weights in all three methods.

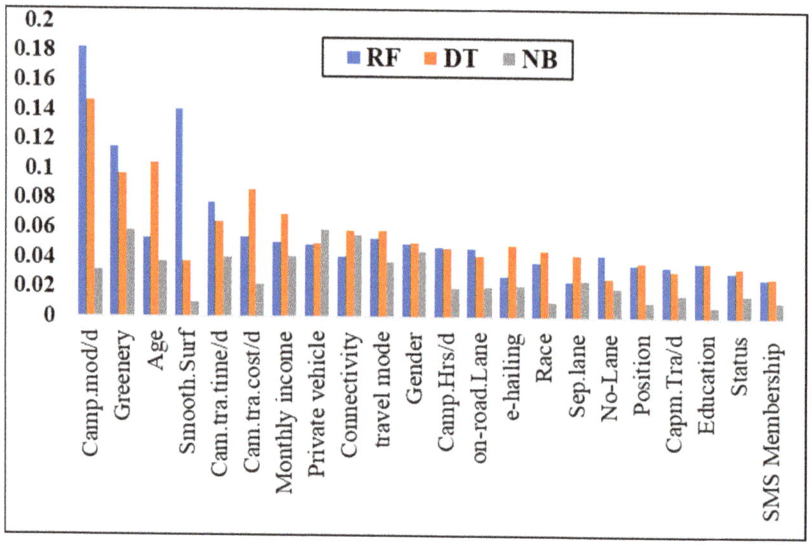

Figure 11. Importance score (weight) of variables based on three ML methods.

Furthermore, to make a more vivid conclusion from the three feature selection methods, the weight values of each variable were summarized and compared, as shown in Figure 12. Next, the amassed weight values were ranked from highest to lowest. According to Figure 12, there was a significant drop in weight values after the "Gender" variable. Therefore, we drew a line and deselected variables whose weights were below the line. The results of the selected most important variables based on three different ML techniques is summarized in Table 4.

Further random forest modelling was performed using these 11 variables. Moreover, all these variables have a threshold of MeanDecreaseGini higher than 30.

Table 3. Importance score (weight) of variables based on three ML methods.

	RF		DT		NB	
No.	**Attribute**	**Weight**	**Attribute**	**Weight**	**Attribute**	**Weight**
1	Camp.mod/d	0.1825	Camp.mod/d	0.14634	Private vehicle	0.059752
2	Smooth.Surf	0.1409	Age	0.10431	Greenery	0.058748
3	Greenery	0.1151	Greenery	0.09712	Connectivity	0.056134
4	Cam.tra.time/d	0.0777	Cam.tra.cost/d	0.08648	Gender	0.04504
5	Cam.tra.cost/d	0.0547	Monthly income	0.06964	Monthly income	0.041161
6	Travel mode	0.0538	Cam.tra.time/d	0.06434	Cam.tra.time/d	0.040235
7	Age	0.0534	Travel mode	0.0588	Travel mode	0.037877
8	Monthly income	0.0509	Connectivity	0.05861	Age	0.037555
9	Gender	0.0498	Gender	0.05055	Camp.mod/d	0.032029
10	Private vehicle	0.0490	Private vehicle	0.04992	Sep.lane	0.025078
11	Camp.Hrs/d	0.0477	e-hailing	0.04938	Cam.tra.cost/d	0.022726
12	on-road.Lane	0.0469	Camp.Hrs/d	0.04726	e-hailing	0.021732
13	No-Lane	0.0429	Race	0.0454	on-road.Lane	0.021263
14	Connectivity	0.0415	Sep.lane	0.04247	No-Lane	0.020533
15	Race	0.0376	on-road.Lane	0.04189	Camp.Hrs/d	0.019974
16	Education	0.0374	Position	0.03798	Capm.Tra/d	0.016223
17	Position	0.0366	Smooth.Surf	0.03751	Status	0.015691
18	Capm.Tra/d	0.0351	Education	0.0374	SMS Membership	0.011859
19	Status	0.0308	Status	0.03386	Position	0.010624
20	e-hailing	0.0280	Capm.Tra/d	0.03175	Race	0.010495
21	SMS membership	0.0268	SMS Membership	0.02767	Smooth.Surf	0.010309
22	Sep.lane	0.0248	No-Lane	0.02676	Education	0.0082

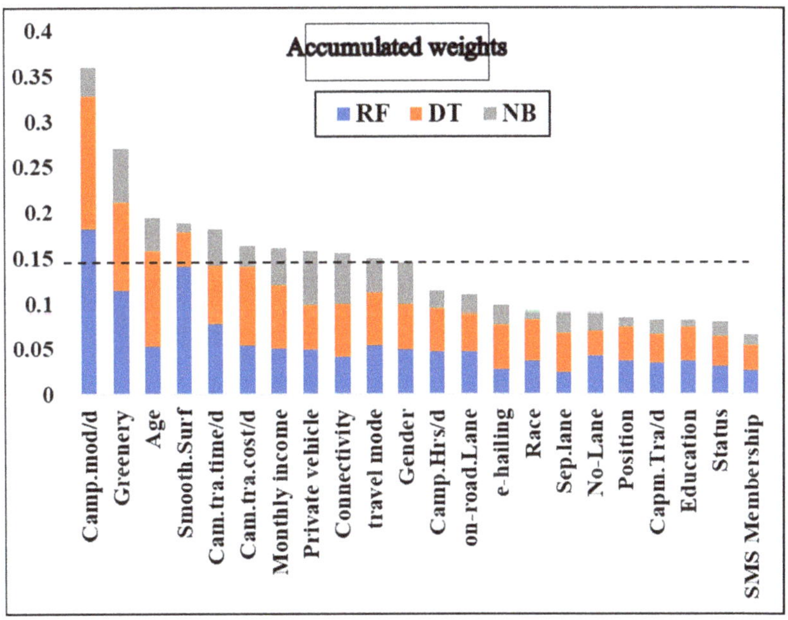

Figure 12. Accumulated weights of variables.

Table 4. Importance of 11 selected variables based on feature selection criteria.

No.	Attribute	Accumulated Weight	Mean Decrease Gini
1	Camp.mod/d	0.360867959	72.26206
2	Greenery	0.270941347	62.26634
3	Age	0.195234017	61.92460
4	Smooth.Surf	0.188729931	60.28623
5	Cam.tra.time/d	0.182241979	59.64285
6	Cam.tra.cost/d	0.16393153	57.96135
7	Monthly income	0.161725573	57.71634
8	Private vehicle	0.158708056	53.55493
9	Connectivity	0.156257276	51.93130
10	Travel mode	0.150511383	44.97282
11	Gender	0.145347998	44.94371

4.5. Model Assessment and Evaluation

Having reduced the number of variables by a comprehensive feature selection method (through both unsupervised clustering and supervised models), the random forest algorithms were conducted using eleven selected variables. The model performance of random forest is reported as below:

Call:
Number of trees: 500
No. of variables tried at each split: 3
Mean of squared residuals: 0.07049505
% Var explained: 93.02

The default *ntree* used was 500 and mtry was 3. The accuracy was 93.02% and the Mean of squared residuals was 0.07049505. The error vs number of tree graphs in Figure 13 shows that the error rate remained constant from 390 to 470. Model assessment was repeated nine times using a different number of trees from 390 to 470, and the results are presented in Table 5.

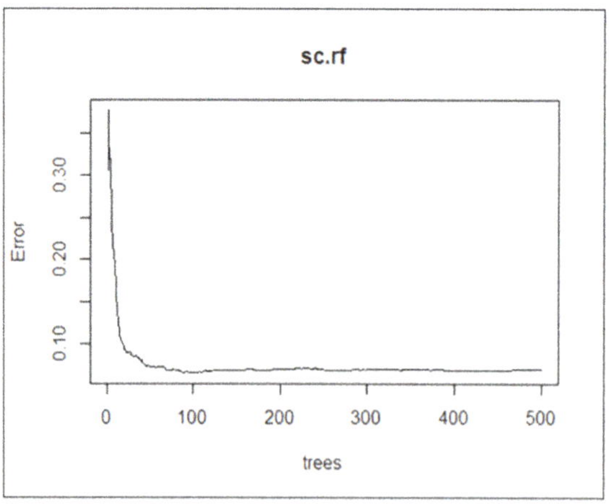

Figure 13. The error vs number of tree graphs for 11 important features.

Table 5. Random forest model.

No	Number of Trees	Accuracy (%)
1	390	93.42
2	400	93.26
3	410	93.28
4	420	93.19
5	430	93.29
6	440	93.51
7	450	93.14
8	460	93.15
9	470	93.28

The best *ntree* was 440 as shown in Table 6, since it produced the highest accuracy compared to other values. The ntree = 440 was used to assess the test error and OOB error, as shown in Figure 14.

Table 6. Model assessment for decision tree, random forest and Naïve Bayes.

Model	Algorithm	Accuracy (%)		Precision		Recall		F1 Score	
		11 Variable	22 Variable	11 Variable	22 Variable	11 Variable	22 Variable	11 Variable	22 Variable
Decision tree	*rpart* from "caret"	54.13	57.130	0.29	0.318	0.38	0.4000	0.32	0.325
Random Forest	*rf* from "caret"	93.51	99.49	0.85	0.890	0.82	0.850	0.72	0.760
Naïve Bayes	*nb* from "e1071" package	61.00	64.50	0.51	0.530	0.45	0.480	0.52	0.540

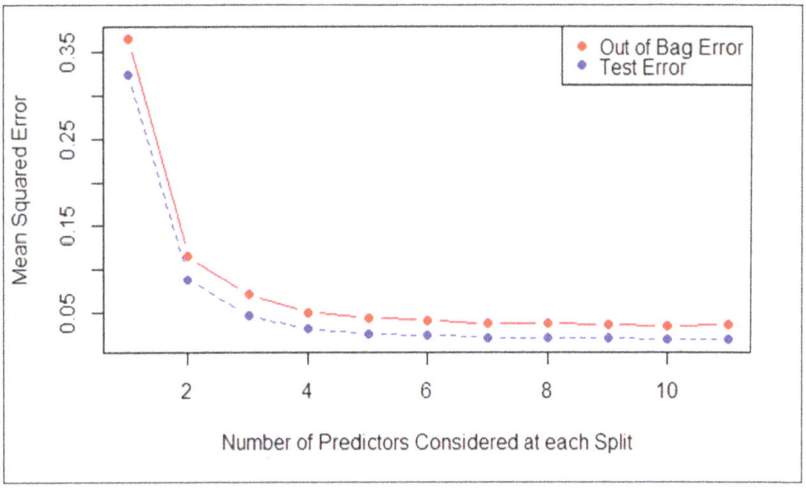

Figure 14. Test error and out-of-bag (OOB) error rate of the predicted model.

The red line represents the out-of-bag error estimates, and the blue line represents the error calculated on test set. Both curves are relatively smooth, and the error estimates are also correlated. The error inclines are reduced at around mtry = 3. Hence, the final model with the 11 most important variables produced an accuracy of 93.51%, with ntree = 440 and mtry = 3. The model performance comparison among the random forest, decision tree and Naïve Bayes methods are shown in Table 6 for both models with 22 variables and 11 variables.

4.6. Simulation-Based Optimization Analysis

To obtain deeper insights into the factors influencing SFFES usage, optimization analysis was carried out based on four different scenarios: a group of respondents who are: (1) most likely to "always" use SFFESs, (2) most likely to "frequently" use SFFESs, (3) most likely to "occasionally" use SFFESs, and (4) less likely to, or "never", use SFFESs. The scenarios were based on the target variables' response categories, as described in Table 1. The simulation-based optimization analysis was conducted on 11 significant variables (as described in the feature selection section) using RapidMiner Studio Educational Software version 9.8.001. All the figures in this section are outputs of the RapidMiner Software. The optimization was carried out and determined the best input factors to fit with our targets under the specified constraints. Additionally, the simulation-based sensitivity analysis was considered appropriate for evaluating and responding "What if" questions. For example, what if our target group is male youngsters who are between 18 to 23 years old and who use public transportation as their mode of transport on the campus (how frequently they will use SFFESs)? Table 7 presents the optimized value of attributes based on four scenarios.

Table 7. Optimized value of attributes based on four scenarios.

Attribute	Always	Frequently	Occasionally	Never
Gender	Female	Female	Male	Male
Age	18 to 29	30 to 44	45 to 60	45 to 60
Monthly income	Between RM 4000 and RM 6000	Between RM 6000 and RM 12,000	Between RM 2000 RM 4000	Between RM 6000 and RM 12,000
Travel mode	Walking/cycling	Public transportation	Private car	Private car
Private vehicle	No	Yes	Yes	Yes
Camp.mod/d	Walking/cycling	E-hailing	Public Transport	Private car
Cam.tra.cost/d	Between RM 5 and RM15	Between RM 15 and RM 25	Less than RM 5	Less than RM 5
Cam.tra.time/d	20 to 30 min	Less than 10 min	10 to 20 min	Less than 10 min
Greenery	Encourage	Strongly encourage	Strongly discourage	Encourage
Smooth.Surf	Encourage	Discourage	Encourage	Encourage
Connectivity	Encourage	Encourage	Discourage	Encourage

In the first scenario, the simulation model was adjusted to optimize the target variables on respondents who are most likely to always use SFFESs. According to the results, females between 18 and 29 years old with a monthly income between RM 4000 and RM 6000 (which is a higher-than-average income in Malaysia), whose primary mode of transport is walking or cycling, are the most likely to change their mode of transport to SFFESs. This group of respondents does not own a private vehicle and they spend RM 5 to RM 15 for their travels around the campus per day.

According to Figure 15, 95% of respondents described above will always use SFFESs as their main mode of transport on the campus, 3% will use SFFESs occasionally, 1.5% will never use it, and less than 1% will use it frequently. In addition, gender, age, and cost of travel per day are the most important factors affecting SFFESs choice and usage. The simulation model was adjusted to optimize the attributes based on the second scenario and determine the characterization of the SFFES service's frequent users. Frequent usage of the SFFES service has been defined as usage between two and five times per week, or replacing at least half of the user's current mode of transport with the SFFES service. According to Table 7, most of the frequent users of SFFESs will be women, as in the previous scenario. However, frequent users are most likely to be older users (30 to 40 years old) with a higher monthly income. While they most likely own private vehicles, they mostly use public transportation for arriving on campus and use e-hailing services to travel around the campus. According to Figure 16, 77% of described students/staff are willing to use the SFFES service frequently. In addition, road features such as connectivity and quality of road surface can strongly impact their usage. Travel mode and travel costs are other important factors for this group, according to Figure 16.

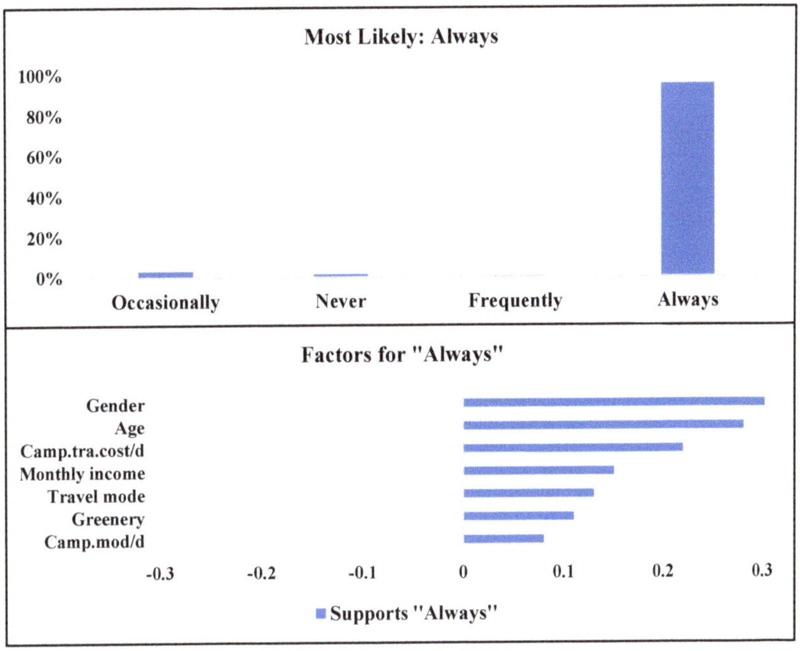

Figure 15. Optimization results and importance of variables based on the first scenario: Always use SFFESs.

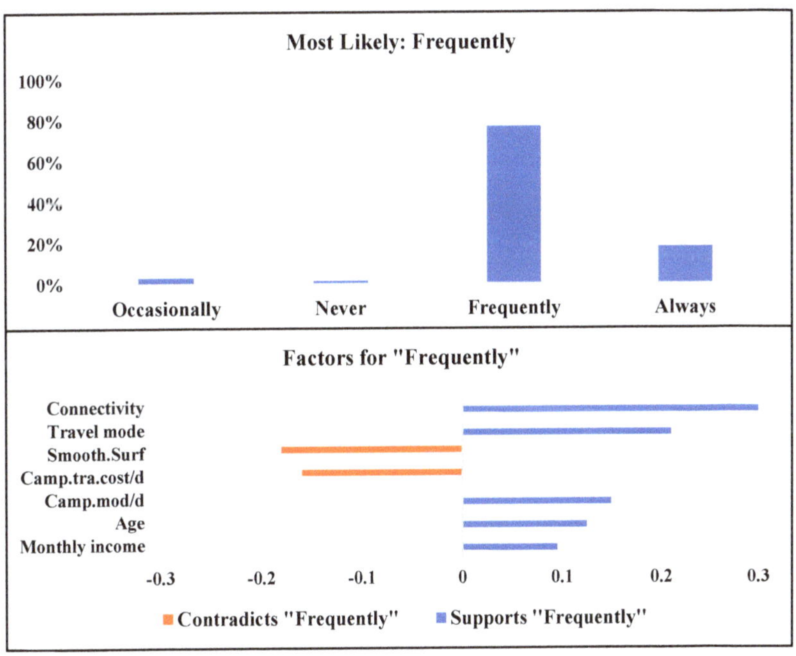

Figure 16. Optimization results and importance of variables based on the second scenario: Frequently use SFFESs.

In the third scenario, the simulation model optimized target variables on the group of users who will most likely use SFFESs occasionally (less than three times per week). Interestingly, men between 45 and 60 years old with an average monthly salary (RM 2000 to RM 4000 is considered an average monthly income in Malaysia) are most likely to use SFFESs occasionally. In addition, they own private vehicles and mostly use public transportation for their daily travels around the campus. According to Figure 17, 82% of users who are described in the third scenario will use SFFES services occasionally or less than three times per week. Moreover, travel mode, age and daily travel time are important factors which support their SFFES mode choice.

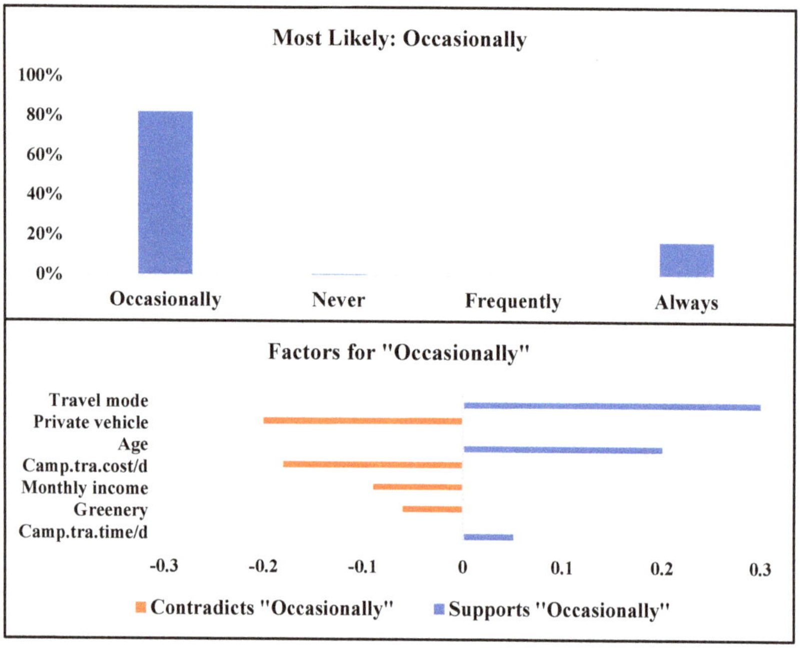

Figure 17. Optimization results and importance of variables based on the second scenario: Occasionally use SFFESs.

Respondents who are not interested in SFFESs and will never use the service were our target in the fourth scenario. According to the last column of Table 7, the sociodemographic characterization of respondents in this scenario is almost the same as the third scenario (users who will use SFFESs occasionally), with the difference being that their monthly income is much higher. In addition, their average daily travel time is significantly shorter, and they prefer to use their own car. As shown in Figure 18, 89% of users described in the fourth scenario are most likely to never use SFFESs. Moreover, road features such as green roads and smooth surfaces are the most important factors which are in contrast with the "Never" usage scenario. In other words, road features are significantly important factors that may encourage them to consider SFFES services for their future travels around the campus (as shown in Figure 18).

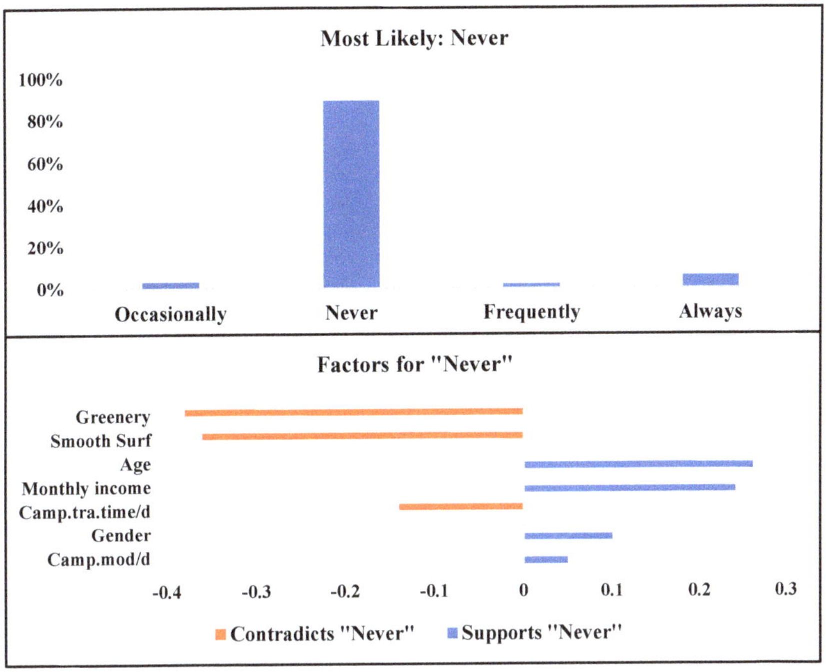

Figure 18. Optimization results and importance of variables based on the second scenario: Never use SFFESs.

5. Discussion

This study is designed to deeply explore the attitudes and perceptions of students and staff towards SFFES usage on campus. Various attributes were considered for this propose, such as the sociodemographic information of respondents, characterization of trips, road features, concerns/barriers, and benefits of riding SFFESs. In addition, this study is one of the first to predict the likelihood of usage frequency of SFFESs by employing various machine learning techniques and the first study on SFFESs in Malaysia. Previous studies have been mostly conducted in the US, China, and recently European cities. Moreover, for the first time we have employed different feature selection methods and machine learning algorithms to deeply evaluate the weight of important factors that affect the mode choice and usage of SFFESs between university students and staff. The campus of University of Malaya (UM) was selected for conducting this study because:

- Shared micromobility is new in Malaysia, and most people have limited knowledge about it. The university community is a natural laboratory to test new mobility services.
- The shared e-scooter companies such as BEAM, TRYKE and Myscooter are very interested in providing their services to university campuses in this initial stage.
- UM is the biggest university in Malaysia, with more than 30,000 students and staff. In addition, more than 5000 international students and staff are on UM campus of different races, ethics, nations and generations. The diversity of the population fits the study requirements well.

A comprehensive feature selection was conducted before developing machine learning predictive models. The main propose of this step was to accurately recognize the significant factors and importance by adopting supervised and unsupervised machine learning techniques. In addition, it decreased the complication of the final model by decreasing the number of variables based on their significance. Although decreasing the number of variables may reduce the accuracy of the final model, this reduction can be minimized by adopting proper and accurate feature selection techniques. In other words, adopting accurate feature selection methods will promote the complexity and practicality of the final model, while the accuracy remains adequately high. In this study, the initial models were developed using all 22 variables. After implementing feature selection methods, the number of variables reduced to 11.

According to the feature selection results, daily travel mode inside the campus (Camp.mod/d) was the most effective factor in determining SFFES usage frequency. Other travel characterizations, such as daily travel cost and time/duration, were among the most influential factors. Sociodemographic attributes such as age, gender, monthly income and private vehicle ownership, played significantly important roles in mode choice and SFFES usage, as also proved by previous studies [33,66]. In addition, based on the outputs of all three supervised feature selection models, road features such as greenery and the connectivity of roads influenced the mode choice significantly. The initial RF model (with 22 variables) outperformed DT and NB models with 99.45% accuracy. Therefore, we selected the RF model for further analysis and developing the final model using the 11 most important variables. As expected, reducing the number of variables caused a reduction in accuracy by 6%. However, the authors believe that the final model is a much more valuable model with acceptable accuracy and less complexity.

To unpack and shed light on the attitudes of the survey participants towards SFFES usage, a simulation-based optimization was developed. Interesting results have been gained which could be useful for future works, recommendations and policy-making. Four optimization scenarios were defined based on the four categories of possible SFFES users: always, frequently, occasionally and never. According to the optimization results, there was a strong relationship between gender and the frequency of usage of SFFESs. Surprisingly, respondents who were more likely to ride e-scooters always and frequently were mostly young to middle-aged females. This result is in contrast with previous studies [15,67] and further exploration is needed to discover the reasons for this gender gap in SFFES usage. However, this result may be biased by our survey participants' characterization, who were mostly young and highly educated.

Indeed, many interesting facts can be unveiled by adopting simulation-based optimization analysis. According to Table 6, respondents who would change their travel mode to SFFESs were mostly daily cyclists or pedestrians. The same result was observed by previous studies [15,33,68]. On the other hand, respondents who used their own private vehicle for daily trips were not interested in riding an e-scooter. These two facts can be considered as significant disadvantages of SFFES services. Undoubtedly, walking and cycling are more desirable and sustainable modes of transportation in several different aspects. Walking and cycling are healthier modes, since they require much more physical activity [69]. Moreover, while walking and cycling are the greenest possible modes of transport, the environmental impact of e-scooters is still not well-investigated [70].

Strength, Limitations and Next Steps

Before indicating the limitations, the authors would like to mention the significant strengths of this study. To the authors' knowledge, this is the first study of SFFESs on a university campus. A large number of students and staff with various sociodemographic backgrounds and undertaking different types of activities on the campus helped to shed some light on the future of SFFESs launches on other university campuses and even urban areas. Furthermore, this was the first study on SFFESs undertaken in Malaysia and one the first to employ various machine learning algorithms to predict the use frequency of SFFESs. There are also a number of limitations. One of the key limitations of this study was sample size. We forwarded the Google Form (the survey) to more than 30,000 university students and staff, and only 1.7% responded completely. The number of respondents was limited, and there must also be principal differences between respondents and non-respondents. In addition, the method of survey distribution and focus group was limited to academic and highly educated people. Undoubtedly, further studies should consider larger sample sizes which are more random and representative of potential SFFES riders. Moreover, we have not provided some specific scenarios for using SFFESs on the campus, such as estimated travel time, costs, proposals and external factors like weather. Therefore, the answers to some questions were based on the experience of respondents, which would influence the results.

Future studies should consider larger sample sizes to predict a better model with higher accuracy, which also represents all SFFES users in Malaysia. Moreover, future studies should incorporate the available information from SFFES companies, such as travel distance, travel time and proposals for travel. In this study, we have only focused on tree-based machine learning algorithms (RF, DT and NB) for predicting SFFES usage frequency. We propose that future studies should consider other types of machine learning techniques, such as neural networks and support vector machines, to clarify which technique has the best performance. Finally, off-campus and on-campus students have different requirements, and in turn, travel behaviors. Future studies can consider these differences.

6. Conclusions

This study predicts SFFES use on a university campus using supervised and unsupervised machine learning techniques. A comprehensive feature selection analysis was conducted using k-means and hierarchical clustering, decision tree, random forest and Naïve Bayes techniques. The 11 most important attributes were identified, including daily travel modes around the campus, the presence of green spaces and water, age, quality of the road surface, daily travel time and cost around the campus, monthly income, private vehicle ownership, connectivity between roads, modes of transport to/from campus, and gender.

The random forest algorithm was developed to predict the usage frequency of SFFES using the identified important attributes. Simulation-based sensitivity analysis was conducted to gain deeper insights into the characterization and specification of SFFES users. Young females between 18 and 29 years old with an average monthly income were the most likely to always use SFFESs for their travels on campus. Males between 45 and 60 years with a high monthly salary were less likely to use SFFESs. The safety concerns and cost of renting e-scooters were the most important discouragement factors, while road features and suitable infrastructure, such as green spaces and separated lanes for scooters, were the most important encouragement factors. In addition, social distancing during the pandemic and no parking issues were the most considerable benefits of riding e-scooters from the respondents' perspective.

The responsibility of the service providers and authorities is to provide all residents (especially people with limited transportation access) with accessible, equitable, safe, affordable, and sustainable transportation options. SFFES services are capable of helping to fill transportation gaps through providing an efficient, affordable alternative to cars for urban journeys. Scooters can have several benefits such as health, safety, and congestion opening, as well as some social/environmental equity benefits. To make an effective decision regarding whether and how SFFESs should be implemented in transportation systems of future cities, decision makers must first determine the definite role of these vehicles in the city. This can be determined through finding out the involved actors' visions of the future urban transport. As a result, to guarantee sustainable mobility, there is a need for not only technology and investment, but also fundamental research into related issues.

Author Contributions: S.M.H.M., conceptualization, methodology, software, formal analysis, investigation, resources, writing—original draft, supervision; Z.M., investigation, writing—review and editing, supervision; D.J.A., conceptualization, software, writing—review and editing, supervision; M.A., formal analysis, writing—review and editing; M.D.G., conceptualization, software, investigation; Y.C.W., conceptualization, resources, supervision; D.V.U., writing—review and editing. All authors have read and agreed to the published version of the manuscript.

Funding: This research received no external funding.

Informed Consent Statement: Not applicable.

Data Availability Statement: The data are available from the corresponding author upon reasonable request.

Acknowledgments: We would like to acknowledge all the experts and staff in the BEAM scooter company and the University of Malaya for providing data and information. In particular, the authors would like to acknowledge the Centre for Transportation Research (CTR), Faculty of Engineering, and University of Malaya for providing research facilities.

Conflicts of Interest: The authors declare no conflict of interest.

References

1. Kalda, K.; Pizzagalli, S.-L.; Soe, R.-M.; Sell, R.; Bellone, M. Language of Driving for Autonomous Vehicles. *Appl. Sci.* **2022**, *12*, 5406. [CrossRef]
2. Shaheen, S.; Cohen, A.; Chan, N.; Bansal, A. Sharing strategies: Carsharing, shared micromobility (bikesharing and scooter sharing), transportation network companies, microtransit, and other innovative mobility modes. In *Transportation, Land Use, and Environmental Planning*; Elsevier: Amsterdam, The Netherlands, 2019; pp. 237–262.
3. Fitt, H.; Curl, A. The early days of shared micromobility: A social practices approach. *J. Transp. Geogr.* **2020**, *86*, 102779. [CrossRef]
4. Kou, Z.; Wang, X.; Chiu, S.F.A.; Cai, H. Quantifying greenhouse gas emissions reduction from bike share systems: A model considering real-world trips and transportation mode choice patterns. *Resour. Conserv. Recycl.* **2020**, *153*, 104534. [CrossRef]
5. Li, W.; Kamargianni, M. Providing quantified evidence to policy makers for promoting bike-sharing in heavily air-polluted cities: A mode choice model and policy simulation for Taiyuan-China. *Transp. Res. Part A Policy Pract.* **2018**, *111*, 277–291. [CrossRef]
6. Lazarus, J.; Pourquier, J.C.; Feng, F.; Hammel, H.; Shaheen, S. Micromobility evolution and expansion: Understanding how docked and dockless bikesharing models complement and compete–A case study of San Francisco. *J. Transp. Geogr.* **2020**, *84*, 102620. [CrossRef]
7. McKinsey & Co. Sizing the Micro Mobility Market | McKinsey. McKinsey & Co. 2021. Available online: https://www.mckinsey.com/industries/automotive-and-assembly/our-insights/micromobilitys-15000-mile-checkup (accessed on 7 February 2021).
8. Berg Insight. The Bike and Scootersharing Telematics Market. 2020; pp. 2–5. Available online: http://www.berginsight.com/ReportPDF/ProductSheet/bi-micromobilitytelematics2-ps.pdf (accessed on 1 March 2020).
9. Tuncer, S.; Laurier, E.; Brown, B.; Licoppe, C. Notes on the practices and appearances of e-scooter users in public space. *J. Transp. Geogr.* **2020**, *85*, 102702. [CrossRef]
10. Sgarbossa, F.; Peron, M.; Fragapane, G. Cloud material handling systems: Conceptual model and cloud-based scheduling of handling activities. *Int. Ser. Oper. Res. Manag. Sci.* **2020**, *289*, 87–101. [CrossRef]
11. Lolli, F.; Coruzzolo, A.M.; Peron, M.; Sgarbossa, F. Age-based preventive maintenance with multiple printing options. *Int. J. Prod. Econ.* **2022**, *243*, 108339. [CrossRef]
12. Mont, O.; Palgan, Y.V.; Bradley, K.; Zvolska, L. A decade of the sharing economy: Concepts, users, business and governance perspectives. *J. Clean. Prod.* **2020**, *269*, 122215. [CrossRef]
13. Nguyen-Phuoc, D.Q.; Amoh-Gyimah, R.; Tran, A.T.P.; Phan, C.T. Mode choice among university students to school in Danang, Vietnam. *Travel Behav. Soc.* **2018**, *13*, 1–10. [CrossRef]
14. Rotaris, L.; Danielis, R.; Maltese, I. Carsharing use by college students: The case of Milan and Rome. *Transp. Res. Part A Policy Pract.* **2019**, *120*, 239–251. [CrossRef]
15. Sanders, R.L.; Branion-Calles, M.; Nelson, T.A. To scoot or not to scoot: Findings from a recent survey about the benefits and barriers of using E-scooters for riders and non-riders. *Transp. Res. Part A Policy Pract.* **2020**, *139*, 217–227. [CrossRef]
16. Stylianou, K.; Dimitriou, L.; Abdel-Aty, M. Big data and road safety: A comprehensive review. In *Mobility Patterns, Big Data and Transport Analytics*; Elsevier: Amsterdam, The Netherlands, 2019; pp. 297–343.
17. Yang, H.; Song, K.; Zhou, J. Automated Recognition Model of Geomechanical Information Based on Operational Data of Tunneling Boring Machines. *Rock Mech. Rock Eng.* **2022**, *55*, 1499–1516. [CrossRef]
18. Yang, H.; Wang, Z.; Song, K. A new hybrid grey wolf optimizer-feature weighted-multiple kernel-support vector regression technique to predict TBM performance. *Eng. Comput.* **2020**, *38*, 2469–2485. [CrossRef]
19. Du, K.; Liu, M.; Zhou, J.; Khandelwal, M. Investigating the slurry fluidity and strength characteristics of cemented backfill and strength prediction models by developing hybrid GA-SVR and PSO-SVR. *Min. Metall. Explor.* **2022**, *39*, 433–452. [CrossRef]
20. Zhou, J.; Qiu, Y.; Khandelwal, M.; Zhu, S.; Zhang, X. Developing a hybrid model of Jaya algorithm-based extreme gradient boosting machine to estimate blast-induced ground vibrations. *Int. J. Rock Mech. Min. Sci.* **2021**, *145*, 104856. [CrossRef]
21. Zhou, J.; Li, X.; Mitri, H.S. Classification of rockburst in underground projects: Comparison of ten supervised learning methods. *J. Comput. Civ. Eng.* **2016**, *30*, 4016003. [CrossRef]
22. Parsajoo, M.; Armaghani, D.J.; Mohammed, A.S.; Khari, M.; Jahandari, S. Tensile strength prediction of rock material using non-destructive tests: A comparative intelligent study. *Transp. Geotech.* **2021**, *31*, 100652. [CrossRef]
23. Hasanipanah, M.; Monjezi, M.; Shahnazar, A.; Armaghani, D.J.; Farazmand, A. Feasibility of indirect determination of blast induced ground vibration based on support vector machine. *Measurement* **2015**, *75*, 289–297. [CrossRef]

24. Toch, E.; Lerner, B.; Ben-Zion, E.; Ben-Gal, I. Analyzing large-scale human mobility data: A survey of machine learning methods and applications. *Knowl. Inf. Syst.* **2019**, *58*, 501–523. [CrossRef]
25. Xu, C.; Ji, J.; Liu, P. The station-free sharing bike demand forecasting with a deep learning approach and large-scale datasets. *Transp. Res. Part C Emerg. Technol.* **2018**, *95*, 47–60. [CrossRef]
26. Gao, X.; Lee, G.M. Moment-based rental prediction for bicycle-sharing transportation systems using a hybrid genetic algorithm and machine learning. *Comput. Ind. Eng.* **2019**, *128*, 60–69. [CrossRef]
27. Aghaabbasi, M.; Shekari, Z.A.; Shah, M.Z.; Olakunle, O.; Armaghani, D.J.; Moeinaddini, M. Predicting the use frequency of ride-sourcing by off-campus university students through random forest and Bayesian network techniques. *Transp. Res. Part A Policy Pract.* **2020**, *136*, 262–281. [CrossRef]
28. Čuš-Babič, N.; de Oliveira, S.F.G.; Tibaut, A. Interoperability of Infrastructure and Transportation Information Models: A Public Transport Case Study. *Appl. Sci.* **2022**, *12*, 6234. [CrossRef]
29. Bokolo, A.J. Green campus paradigms for sustainability attainment in higher education institutions—A comparative study. *J. Sci. Technol. Policy Manag.* **2020**, *12*, 117–148. [CrossRef]
30. Zakaria, R.; Alqaifi, G.; Rahim, A.; Hamid, A.R.A.; Mansur, S.A.; Resang, A.; Zen, I.S.; Bandi, M.; Khalid, M.S. UTM sustainable living laboratory campus; Are the implementations effective? In Proceedings of the Regional Conference in Engineering Education, Kuala Lumpur, Malaysia, 9–10 August 2016; pp. 1–6.
31. Humblet, E.M.; Owens, R.; Roy, L.P.; McIntyre, D.; Meehan, P.; Sharp, L. *Roadmap to a Green Campus*; U.S. Green Building Council: Washington, DC, USA, 2010.
32. Anthony, J.; Majid, M.A.; Romli, A. Emerging case oriented agents for sustaining educational institutions going green towards environmental responsibility. *J. Syst. Inf. Technol.* **2019**, *21*, 186–214. [CrossRef]
33. Baek, K.; Lee, H.; Chung, J.H.; Kim, J. Electric scooter sharing: How do people value it as a last-mile transportation mode? *Transp. Res. Part D Transp. Environ.* **2021**, *90*, 102642. [CrossRef]
34. Liu, M.; Seeder, S.; Li, H. Analysis of E-scooter trips and their temporal usage patterns. *Inst. Transp. Eng. ITE J.* **2019**, *89*, 44–49.
35. McKenzie, G. Spatiotemporal comparative analysis of scooter-share and bike-share usage patterns in Washington, D.C. *J. Transp. Geogr.* **2019**, *78*, 19–28. [CrossRef]
36. Kowald, M.; Gutjar, M.; Röth, K.; Schiller, C.; Dannewald, T. Mode Choice Effects on Bike Sharing Systems. *Appl. Sci.* **2022**, *12*, 4391. [CrossRef]
37. Eccarius, T.; Lu, C.-C. Adoption intentions for micro-mobility–Insights from electric scooter sharing in Taiwan. *Transp. Res. Part D Transp. Environ.* **2020**, *84*, 102327. [CrossRef]
38. Younes, H.; Zou, Z.; Wu, J.; Baiocchi, G. Comparing the temporal determinants of dockless scooter-share and station-based bike-share in Washington, DC. *Transp. Res. Part A Policy Pract.* **2020**, *134*, 308–320. [CrossRef]
39. Portland Bureau of Transportation. E-Scooter Findings Report. 2018. Available online: https://www.portlandoregon.gov/transportation/article/709719 (accessed on 15 June 2018).
40. Denver Dockless Mobility Program. Pilot Interim Report—February 2019. Available online: https://www.denvergov.org/wp-content/uploads/2019/05/Denver-Dockless-Mobility-Update-Feb-2019.pdf (accessed on 1 February 2019).
41. The Nunatak Group. New Urban Mobility. 2019. Available online: https://www.nunatak.com/en/topics/new-urban-mobility (accessed on 20 July 2019).
42. 6t-Bureau de Recherche. Usages et Usagers des Trottinettes Electriques en Free-Floating en France. 2019. Available online: https://6-t.co/etudes/usages-usagers-trottinettes-ff/ (accessed on 1 February 2019).
43. Sarker, I.H.; Colman, A.; Han, J.; Khan, A.I.; Abushark, Y.B.; Salah, K. BehavDT: A behavioral decision tree learning to build user-centric context-aware predictive model. *Mob. Netw. Appl.* **2019**, *25*, 1151–1161. [CrossRef]
44. Toraih, E.A.; Elshazli, R.M.; Hussein, M.H.; Elgaml, A.; Amin, M.; El-Mowafy, M.; El-Mesery, M.; Ellythy, A.; Duchesne, J.; Killackey, M.T.; et al. Association of cardiac biomarkers and comorbidities with increased mortality, severity, and cardiac injury in COVID-19 patients: A meta-regression and decision tree analysis. *J. Med. Virol.* **2020**, *92*, 2473–2488. [CrossRef]
45. Ganggayah, M.D.; Taib, N.A.; Har, Y.C.; Lio, P.; Dhillon, S.K. Predicting factors for survival of breast cancer patients using machine learning techniques. *BMC Med. Inform. Decis. Mak.* **2019**, *4*, 48. [CrossRef]
46. Lu, H.; Ma, X. Hybrid decision tree-based machine learning models for short-term water quality prediction. *Chemosphere* **2020**, *249*, 126169. [CrossRef] [PubMed]
47. Mosca, E.; Alfieri, R.; Merelli, I. A multilevel data integration resource for breast cancer study. *BMC Syst. Biol.* **2010**, *4*, 76. [CrossRef]
48. Genuer, R.; Poggi, J.-M.; Tuleau-Malot, C. VSURF: An R Package for Variable Selection Using Random Forests. *R J.* **2015**, *7*, 19–33. [CrossRef]
49. Lebedev, A.V.; Westman, E.; Van Westen, G.J.P.; Kramberger, M.G.; Lundervold, A.; Aarsland, D.; Soininen, H.; Kloszewska, I.; Mecocci, P.; Tsolaki, M.; et al. Random Forest ensembles for detection and prediction of Alzheimer's disease with a good between-cohort robustness. *NeuroImage Clin.* **2014**, *6*, 115–125. [CrossRef]
50. Khalilia, M.; Chakraborty, S.; Popescu, M. Predicting disease risks from highly imbalanced data using random forest. *BMC Med. Inform. Decis. Mak.* **2011**, *11*, 51. [CrossRef]
51. Chen, S.; Webb, G.I.; Liu, L.; Ma, X. A novel selective naïve Bayes algorithm. *Knowl.-Based Syst.* **2020**, *192*, 105361. [CrossRef]

52. Ding, S.; Zhao, H.; Zhang, Y.; Xu, X.; Nie, R. Extreme learning machine: Algorithm, theory and applications. *Artif. Intell. Rev.* **2015**, *44*, 103–115. [CrossRef]
53. Tibshirani, R.; Walther, G.; Hastie, T. Estimating the number of clusters in a data set via the gap statistic. *J. R. Stat. Soc. Ser. B Stat. Methodol.* **2001**, *63*, 411–423. [CrossRef]
54. Nidheesh, N.; Nazeer, K.A.A.; Ameer, P.M. A Hierarchical Clustering algorithm based on Silhouette Index for cancer subtype discovery from genomic data. *Neural Comput. Appl.* **2020**, *32*, 11459–11476. [CrossRef]
55. Rai, P. Data clustering: K-means and hierarchical clustering. *CS5350 6350 Mach. Learn. Oct.* **2011**, *4*, 24.
56. Raschka, S. Model Evaluation, Model Selection, and Algorithm Selection in Machine Learning. *arXiv* **2018**, arXiv:1811.12808.
57. Ramli, N.A.; Zen, I.S.; Bandi, M.; Tajuddin, H.A. Reduction in carbon dioxide emissions and global climate in campus: From policy into action. In Proceedings of the 2nd International Conference on Emerging Trends in Scientific Research, Kuala Lumpur, Malaysia, 1–2 November 2014; pp. 1–23.
58. Nejati, M.; Nejati, M. Assessment of sustainable university factors from the perspective of university students. *J. Clean. Prod.* **2013**, *48*, 101–107. [CrossRef]
59. Taghavi, M.; Bakhtiyari, K.; Taghavi, H.; Attar, V.O.; Hussain, A. Planning for sustainable development in the emerging information societies. *J. Sci. Technol. Policy Manag.* **2014**, *5*, 178–211. [CrossRef]
60. Foo, K.Y. A vision on the role of environmental higher education contributing to the sustainable development in Malaysia. *J. Clean. Prod.* **2013**, *61*, 6–12. [CrossRef]
61. Junior, B.A.; Majid, M.A.; Romli, A. Green information technology for sustainability elicitation in government-based organisations: An exploratory case study. *Int. J. Sustain. Soc.* **2018**, *10*, 20–41. [CrossRef]
62. Abdul-Azeez, I.A.; Ho, C.S. Realizing low carbon emission in the university campus towards energy sustainability. *Open J. Energy Effic.* **2015**, *4*, 15. [CrossRef]
63. Azlin, A.Z.B.; Er, A.C.; Rahman, N.B.A.; Alam, A.S.A. Consumers' roles and practices towards sustainable UKM campus. *Int. J. Adv. Appl. Sci.* **2016**, *3*, 30–34.
64. Peter, C.J.; Libunao, W.H.; Latif, A.A. Extent of education for sustainable development (ESD) integration in Malaysian community colleges. *J. Tech. Educ. Train.* **2016**, *8*, 1–13.
65. Junior, B.A. A retrospective study on green ICT deployment for ecological protection pedagogy: Insights from field survey. *World Rev. Sci. Technol. Sustain. Dev.* **2019**, *15*, 17–45. [CrossRef]
66. Hardt, C.; Bogenberger, K. Usage of e-Scooters in Urban Environments. *Transp. Res. Procedia* **2019**, *37*, 155–162. [CrossRef]
67. Laa, B.; Leth, U. Survey of E-scooter users in Vienna: Who they are and how they ride. *J. Transp. Geogr.* **2020**, *89*, 102874. [CrossRef]
68. Gössling, S. Integrating e-scooters in urban transportation: Problems, policies, and the prospect of system change. *Transp. Res. Part D Transp. Environ.* **2020**, *79*, 102230. [CrossRef]
69. Willmott, A.G.B.; Maxwell, N.S. The metabolic and physiological responses to scootering exercise in a field-setting. *J. Transp. Health* **2019**, *13*, 26–32. [CrossRef]
70. De Bortoli, A.; Christoforou, Z. Consequential LCA for territorial and multimodal transportation policies: Method and application to the free-floating e-scooter disruption in Paris. *J. Clean. Prod.* **2020**, *273*, 122898. [CrossRef]

applied sciences

MDPI

Article

Prediction of Blast-Induced Ground Vibration at a Limestone Quarry: An Artificial Intelligence Approach

Clement Kweku Arthur [1], Ramesh Murlidhar Bhatawdekar [2,*], Edy Tonnizam Mohamad [2], Mohanad Muayad Sabri Sabri [3], Manish Bohra [4], Manoj Khandelwal [5] and Sangki Kwon [6,*]

1 Department of Mining Engineering, Faculty of Mining and Minerals Technology, University of Mines and Technology, Tarkwa P.O. Box 237, Ghana
2 Centre of Tropical Geoengineering (GEOTROPIK), School of Civil Engineering, Faculty of Engineering, University Teknologi Malaysia, Johor Bahru 81310, Malaysia
3 Centre of Peter the Great St. Petersburg Polytechnic University, 195251 St. Petersburg, Russia
4 Shree Cement, Beawar 305 901, India
5 Institute of Innovation, Science and Sustainability, Federation University Australia, Ballarat, VIC 3350, Australia
6 Department of Energy Resources Engineering, Inha University, Incheon 22212, Korea
* Correspondence: rmbhatawdekar2@graduate.utm.my (R.M.B.); kwonsk@inha.ac.kr (S.K.)

Abstract: Ground vibration is one of the most unfavourable environmental effects of blasting activities, which can cause serious damage to neighboring homes and structures. As a result, effective forecasting of their severity is critical to controlling and reducing their recurrence. There are several conventional vibration predictor equations available proposed by different researchers but most of them are based on only two parameters, i.e., explosive charge used per delay and distance between blast face to the monitoring point. It is a well-known fact that blasting results are influenced by a number of blast design parameters, such as burden, spacing, powder factor, etc. but these are not being considered in any of the available conventional predictors and due to that they show a high error in predicting blast vibrations. Nowadays, artificial intelligence has been widely used in blast engineering. Thus, three artificial intelligence approaches, namely Gaussian process regression (GPR), extreme learning machine (ELM) and backpropagation neural network (BPNN) were used in this study to estimate ground vibration caused by blasting in Shree Cement Ras Limestone Mine in India. To achieve that aim, 101 blasting datasets with powder factor, average depth, distance, spacing, burden, charge weight, and stemming length as input parameters were collected from the mine site. For comparison purposes, a simple multivariate regression analysis (MVRA) model as well as, a nonparametric regression-based technique known as multivariate adaptive regression splines (MARS) was also constructed using the same datasets. This study serves as a foundational study for the comparison of GPR, BPNN, ELM, MARS and MVRA to ascertain their respective predictive performances. Eighty-one (81) datasets representing 80% of the total blasting datasets were used to construct and train the various predictive models while 20 data samples (20%) were utilized for evaluating the predictive capabilities of the developed predictive models. Using the testing datasets, major indicators of performance, namely mean squared error (MSE), variance accounted for (VAF), correlation coefficient (R) and coefficient of determination (R^2) were compared as statistical evaluators of model performance. This study revealed that the GPR model exhibited superior predictive capability in comparison to the MARS, BPNN, ELM and MVRA. The GPR model showed the highest VAF, R and R^2 values of 99.1728%, 0.9985 and 0.9971 respectively and the lowest MSE of 0.0903. As a result, the blast engineer can employ GPR as an effective and appropriate method for forecasting blast-induced ground vibration.

Keywords: artificial intelligence; backpropagation neural network; blast-induced ground vibration; Gaussian process regression

Citation: Arthur, C.K.; Bhatawdekar, R.M.; Mohamad, E.T.; Sabri, M.M.S.; Bohra, M.; Khandelwal, M.; Kwon, S. Prediction of Blast-Induced Ground Vibration at a Limestone Quarry: An Artificial Intelligence Approach. *Appl. Sci.* **2022**, *12*, 9189. https://doi.org/10.3390/app12189189

Academic Editor: José António Correia

Received: 28 July 2022
Accepted: 26 August 2022
Published: 14 September 2022

1. Introduction

Ground vibration is one of the main adverse blasting outcomes that has received significant attention in the mining and civil industries [1,2]. Ground vibration is known to have a lot of adverse impacts on the environment (cracks on building structures) and the stability of pit walls. It is worth mentioning that several factors contribute to the occurrence of these blast-induced ground vibrations. These factors can be categorized into controllable factors and uncontrollable factors [3,4]. The controllable factors are those that the blast engineer has control over and can change. These include the blast design parameters of stemming length, hole depth, spacing, burden, hole inclination and explosive parameters of delay timings, a maximum charge per delay, and total charge. The uncontrollable factors are those the blast engineer has no control over, and they include both geotechnical and geomechanical parameters such as rock strength, faults, and folds [5–9]. The peak particle velocity (PPV) is the index for assessing ground vibration induced by blasting [10]. When detonation of explosives takes place, high energy is released in the blast hole which fractures the rock surrounding the blasthole [11]. Some of the energy released is used to fragment and displace the rock mass. The rest of the energymove through the ground as ground vibration andimpacts surrounding structures.

Due to the adverse impact of blast-induced ground vibration, it has always been in the interest of the blast engineer to model and predicts its occurrence to minimize vibration level as much as possible. In that regard, a lot of research has been conducted since the 1950s [12] to develop models for predicting ground vibration arising out of blasting operations. These models have been developed using empirical techniques through to the use of artificial intelligence (AI) techniques [13]. These AI techniques have been found to produce more accurate results than the empirical techniques and hence have received worldwide attention due to their unique capabilities [14]. AI techniques that have been developed and used in the prediction of blasting outcomes (ground vibration, air overpressure, and flyrock) are outlined in Table 1. It is worth noting that all abbreviations used in this work are presented in the Abbreviations Section.

Table 1. AI Models developed and applied to predict ground vibration, air overpressure and flyrock.

References	Methods	Application
[3,15–28]	FL, SVR, ANFIS, ANN, CART, GPR, ICA, SVM, ELM, GEP, PSO, BN	Ground Vibration Prediction
[29–33]	PSO-ANN, FIS, ANN, ICA_ANN, BIENN, GP, M5DT, SVM, KNN, CHAID	Air Overpressure Prediction
[34–39]	PSO-ANN, RF, BN, BBO-ELM, ORELM, ELM, WOA-SVM, GP	Flyrock

More recently in ground vibration studies, other researchers have applied evolutionary and metaheuristic optimization algorithms to optimize simple AI techniques. Some of these works are presented in Table 2.

Table 2. Hybrid Models developed and applied to predict ground vibrations.

References	Hybrid Models
[40–54]	PSO-ANN, ICA-ANN, ABC-ANN, PSO-ANFIS, ICA-FIS, FFA-ANN, GA-ANFIS, PSO-ANFSI, PSO-XGBoost, GA-SVR, PSO-SVR, FFA-SVR, GA-ANN, GWO-RVR, BAT-RVR, HHOA-RF, ICA-XGBoost, ICA-M5DT, HHOA-ELM, GOA-ELM

Table 3 provides a detailed summary of some research on ground vibration prediction.

Table 3. Input parameters, size of data and AI techniques for prediction of ground vibration.

References	Technique	Input Parameters				No. of Datasets	R^2
		Rock Mass	Blast Design	Explosives	Other		
[55]	ANN	ν, BI, E, Pv	HD, B, S	VOD, Q	H	154	0.9864
[15]	FIS	-	-	Q	H	33	0.92
[56]	ANN		HD	Q	H	162	0.9493
[57]	SVM, ANN	-	-	Q	H	37	SVM = 0.89, ANN = 0.85
[16]	FIS	-	B, S, ST	Q	H	120	0.95
[58]	ANN	-	-	Q	H	20	0.93
[40]	ANN-PSO	RD	B, S, N, HD, SD	Q	H	44	0.94
[59]	ANN		ST, HD	Q	H	69	0.957
[60]	ANN	-	HD, ST	Q	H	115	0.98
[28]	ANN-PSO	RQD	ST, BS, SD	PF, Q	H	88	0.89
[61]	GA-ANN, ANFIS	-	-	Q	H, RD	70	GA-ANN = 0.988, ANFIS = 0.92
[62]	WNN, GMDH, ANN	-	HD, NH	PF, Q	H	210	WNN = 0.712, GMDH = 0.684, ANN = 0.729
[63]	GP, RSM, MARS			Q	H	200	GP = 0.7864, RSM = 0.7832, MARS = 0.8056
[64]	ANFIS	-	-	Q	H	90	0.983
[65]	ANN	-	-	Q	H	68	0.955
[66]	ANN			PF, Q	H	88	1
[22]	GPR, ANN	-	HD, NH	PF, Q	H	210	GPR = 0.695, ANN = 0.688
[49]	SaDE-ELM, ELM, ANN	-	HD, NH	PF, Q	H	210	SaDE-ELM = 0.759, ELM = 0.728, ANN = 0.729
[67]	MARS, ANN	-	HD, NH	PF, Q	H	210	MARS = 0.7074, ANN = 0.6879
[68]	LSSVM, ANN	-	HD, NH	PF, Q	H	210	LSSVM = 0.73, ANN = 0.729

Nevertheless, the application of single AI techniques is still of interest in this ever-growing technological world. ANN has been developed by [69] to predict the earth surface deformation. Thus, the predictive capacities of three artificial intelligence algorithms, backpropagation neural network (BPNN), ELM, and GPR, are investigated in this study using blasting data from a quarry (Ras Limestone Mine of Shree Cement) in India to estimate PPV values. A multivariate adaptive regression spline (MARS) approach, as well as a multivariate regression analysis (MVRA) model, was developed and used for comparison purposes. Studies have been made to compare the GPR and BPNN [22], MARS and BPNN [67], ELM and BPNN [70], GP and MARS [63], GPR and MVRA [71] and BPNN and MVRA [55]. However, little has been done in the literature to compare the predictive performance of GPR, MARS, BPNN, ELM and MVRA in ground vibration prediction studies. In that regard, this study is exploratory. It is worth mentioning that the empirical models developed for predicting blast-induced ground vibration were not considered in this study. The reason is that studies done by [17,40,53,57,59,72,73] have proved that these empirical models do not produce accurate results. The models used in this study consider seven effective parameters, namely the average depth, a maximum charge per delay, powder factor, spacing, burden, distance and stemming length, because, as shown in [5–7], they significantly affect the intensity of ground vibration.

2. Study Site and Data Description

The Ras Limestone Mine of Shree Cement is located 30 km from Beawar City, Ajmer District, Rajasthan, India. The mining concession of 750.0 ha lies between longitude E 74°10′5.96″ to E 74°11′9.62″ and latitudes N 26°16′57.13″ to N 26°15′36.23″, on toposheet No. 45 J/3 & 45 J/4 of the survey of India.

The projected production capacity of the mine is 25.3 million tons of limestone per year. The mining area is generally rocky with no overburden. A general strike of limestone at Ras Mine is North-South direction and dips in the eastern direction. Limestone has four major folds and one reverse fault. Limestone strata are massive, blocky and fractured in different portions of the deposit. HRB 150 (INDUS Make) drills are used for drilling hole diameter of 165 mm. ANFO with cast booster/slurry explosives and nonel detonators are used as explosives for blasting limestone. Figure 1 shows a blasting round view with Figure 2 showing the close-up view of blasted limestone at Shree Cement Ras Limestone Mine in India.

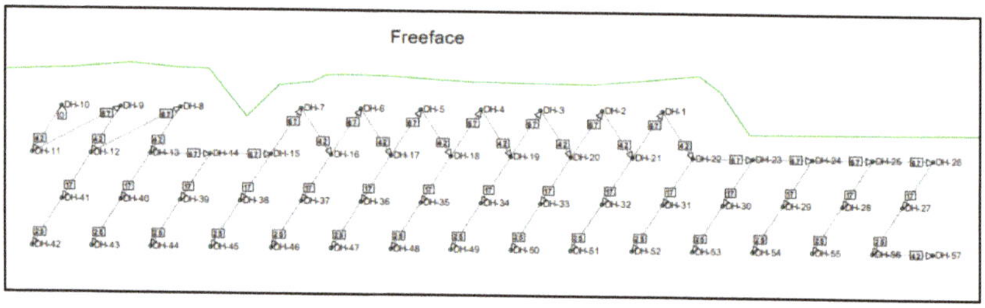

Figure 1. Blasting Round View.

Figure 2. Close-up View of Blasted Limestone at Shree Cement Ras Limestone Mine in India.

As a part of this study, for the establishment of the various models described therein, a total of 101 sets of data were collected from the Ras limestone mine. The data collected consisted of parameters such as average depth (m), spacing (m), burden (m), powder factor (t/kg), the distance between the blasting point and the monitoring station (m), stemming length (m), a maximum charge per delay (kg) and PPV (mm/s). In the creation of the various models, the input parameters were average depth (m), spacing (m), burden (m), powder factor (t/kg), the distance between the blasting site and the monitoring station (m), stemming length (m), and maximum charge per delay (kg), while the output parameter was PPV. Table 4 shows the statistical description of the dataset collected.

Table 4. Description of dataset parameters.

Parameter	Category	Symbol	Units	Minimum	Average	Maximum	Standard Deviation
Average depth	Inputs	AD	m	7.76	11.88	14.46	1.64
Burden		B	m	4.5	4.54	5.5	0.15
Spacing		S	m	5.5	6.02	7	0.31
Distance		D	m	250	1356.44	4150	906.21
Powder factor		PF	t/kg	5.38	6.30	7.83	0.44
Stemming length		SL	m	3	3.48	4	0.29
Maximum charge per Delay		MC	kg	73	129.49	180	22.94
Peak Particle Velocity	Output	PPV	mm/s	0.7	4.08	15.19	3.16

The values for the maximum charge per delay, stemming length, powder factor, spacing, burden, and average depth as statistically described in Table 2 were obtained from the daily blast plans of the mine. The distance values were calculated using the coordinates of the blasting face and monitoring locations obtained using a Global Positioning System (GPS). As shown in Figure 3, the PPV values were monitored using an Instantel Micromate ISEE Std/XM seismograph [74].

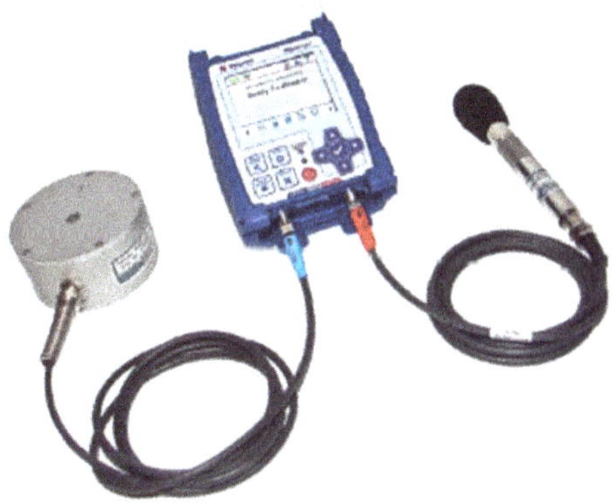

Figure 3. Instantel Micromate ISEE Std/XM seismograph.

It is worth mentioning that the mine has no permanent monitoring location due to different blasting positions. Thus, in monitoring the ground vibration due to blasting, the seismograph is positioned using pegs with an arrow on the geophone pointing towards the blast site. Figure 4 shows the portable monitoring station used by the mine. It is worth noting that the terrain of the Ras Limestone Mine is generally hilly.

The correlation coefficient matrix shows how strong the interaction between the input parameters (average depth, burden, spacing, distance, powder factor, stemming length, and maximum charge per delay) and the measured PPV is, as shown in Table 5.

Figure 4. Portable ground vibration monitoring station in realistic conditions (at Shree Cement Ras Limestone Mine in India).

Table 5. Matrix of Correlation Coefficients Between Input Parameters and PPV Measured.

	AD	B	S	D	PF	SL	MC	PPV
AD	1							
B	0.1697	1						
S	0.3347	0.8330	1					
D	0.1407	−0.0467	−0.0201	1				
PF	−0.1046	0.5294	0.4905	−0.1069	1			
SL	0.7702	−0.1507	−0.0235	0.0595	−0.1329	1		
MC	0.9301	0.3514	0.4996	0.1617	−0.2019	0.6145	1	
PPV	−0.0016	0.1492	0.2160	−0.7503	0.0789	−0.0837	0.0293	1

3. Methodology

In this section, the mathematical description of the different methods applied in this study will be briefly outlined. Furthermore, the procedure followed to develop the various models as well as the models' performance indicators will be outlined.

3.1. Study Steps

A systematic methodology was utilized in this study. First, the data collected were prepared by removing all outliers and then were partitioned into two sets (training set and testing set) and normalized into the interval [−1,1]. Then the various models were built by selecting the model's hyperparameter. The models were then trained using the training dataset. Finally, the model's results were assessed based on the test dataset by some performance indicators. The performance results were then analyzed to either finetune the model's hyperparameter or select the model as optimum. Figure 5 shows the flowchart applied in this study.

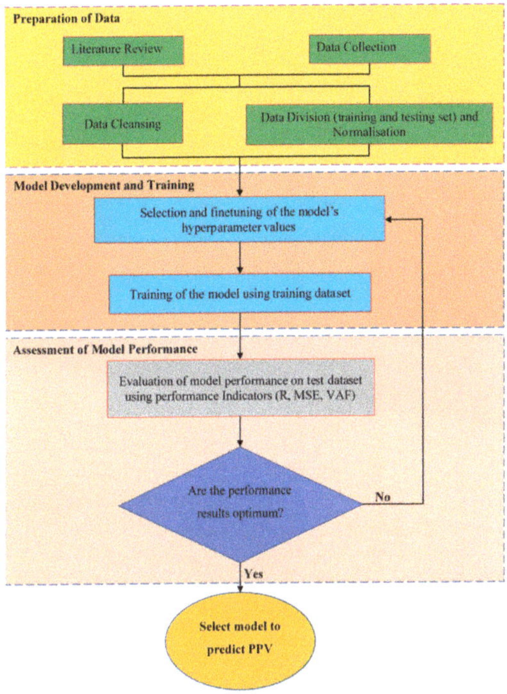

Figure 5. A Systematic Flowchart for Prediction of Blast-Induced Ground Vibration.

3.2. *Mathematical Description of the Different Methods*

3.2.1. GPR

Gaussian Process (GP)

GP is a nonparametric Bayesian technique that is used in regression modelling [75]. This GP process can be described as a finite assemblage of a set of arbitrary parameters that follow a multivariate Gaussian (normal) distribution [76]. That is, for every given input point from a set of input vectors $r = (r_1, r_2, r_3, \ldots, r_m)$, the probability distribution over its function $h(r)$ follows a Gaussian distribution. Thus, a GP $h(r)$ is precisely shown in Equation (1) as:

$$h(r) \sim \mathrm{GP}(b(r), g(r, r')) \tag{1}$$

From Equation (1) it can be deduced that a GP is fully characterized by a covariance function $g(r, r')$ and a mean function (MF) $b(r)$ as expressed in Equation (2).

$$\begin{cases} b(r) = E[h(r)] \\ g(r, r') = E[(b(r) - h(r))(b(r') - h(r'))] \end{cases} \tag{2}$$

For the basic GPR, the MF is normally set as 0, however, there many other MFs which can be applied in building the GPR model [77]. The noted MFs in literature have been categorized into two kinds, namely: simple and composite. The simple MFs include zero, one, constant, linear, polynomial, nearest neighbor MFs etc. whereas the composite ones include: the scaled version, sum, product, power and warped MFs [77]. It is worth noting that this study adopted an MF with a constant, b.

The covariance function on the other hand is the main component in the development of the GPR model. The best covariance function is dependent on the data being modelled. Literature is replete with a number of these covariance functions [70]. However, the notable ones include: the rational quadratic, matérn class, squared exponential and the exponential covariance functions. The most often used covariance function is the squared exponential covariance function [77,78].

Prediction Using GP

In the case of a regression modelling problem, an output variable q can be approximated, given function $h(r)$ with an additive noise ε_i component inherent in the dataset as shown in Equation (3).

$$q_i = h(r_i) + \varepsilon_i \tag{3}$$

Assuming this noise component ε_i has a zero mean and variance σ_n^2, the prior on the noisy data is expressed in Equation (4) as:

$$\mathrm{cov}(q) = g(r, r') + \sigma_n^2 I_n \tag{4}$$

where I_n is a matrix of the n-dimensional unit.

The GP $h(r)$ (see Equation (1)) is then precisely considered in Equation (5) as:

$$h(r) \sim \mathrm{GP}\left(b(r), g(r, r') + \sigma_n^2 I\right) \tag{5}$$

It should be emphasized that the GP model training, seeks to ascertain the best possible hyperparameter set $\Theta = \left[\beta, \chi, v_s^2, \sigma_n^2\right]$ that best fits the data sets. This can be done by the use of a maximum possible method [69] in which the log-likelihood function is maximized (Equation (6)).

$$\log(p(q|r, \Theta)) = \frac{1}{2}\log\left(\det\left(g(r, r') + \sigma_n^2 I\right)\right) - \frac{1}{2}q^T\left(g(r, r') + \sigma_n^2 I\right)^{-1} q - \frac{n}{2}\log 2\pi \tag{6}$$

Of all the maximum likelihood functions available, the conjugate gradient method is the most widely used [79] and hence was used in this study. It finds the optimal hyperparameter sets by using the partial differential of the log-likelihood function (Equation (6)) in relation to the hyperparameter set, Θ as shown in Equation (7).

$$\begin{aligned} \frac{\partial}{\partial \Theta_i}\log(p(q|r, \Theta)) &= \frac{1}{2}q^T G^{-1}\frac{\partial G}{\partial \Theta_i}G^{-1}q - \frac{1}{2}tr\left(G^{-1}\frac{\partial G}{\partial \Theta_i}\right) \\ &= \frac{1}{2}tr\left(\left(\beta\beta^T - G^{-1}\right)\frac{\partial G}{\partial \Theta_i}\right) \end{aligned} \tag{7}$$

where $\beta = G^{-1}q$ and $G = g(r, r')$.

Given the joint prior distribution of the training output variable, q at point a and the value q_* to be predicted at the test point r_* expressed in Equation (8), the GPR model is able to predict q_* by calculating the posterior distribution $p(q_*|r, q, r_*)$ (Equation (9)).

$$\begin{bmatrix} q \\ q_* \end{bmatrix} \sim GP\left(\begin{bmatrix} b(r) \\ b(r_*) \end{bmatrix}, \begin{bmatrix} g(r, r) + \sigma_n^2 I & g(r, r_*) \\ g(r_*, r) & g(r_*, r_*) \end{bmatrix}\right), \tag{8}$$

$$p(q_*|r, q, r_*) \sim GP(\overline{q}_*, cov(q_*)), \tag{9}$$

Here $\overline{q}_*$ (Equation (10)) is the mean value which is the estimation of q_* and $cov(q_*)$ (Equation (11)) is the predictive variance matrix of the test data, which reveals the credibility of the prediction values [79].

$$\overline{q}_* = b(r_*) + g(r_*, r)\left[g(r, r) + \sigma_n^2 I\right]^{-1}(q - b(r)) \tag{10}$$

$$cov(q_*) = g(r_*, r_*)\left[g(r, r) + \sigma_n^2 I\right]^{-1} g(r, r_*) \tag{11}$$

3.2.2. BPNN

BPNN is a widely used AI technique that was developed to mimic the human brain. In this, there is an input layer that takes impulses from the outside environment as inputs to the network. These inputs x_k are weighted by connecting weights w_k and relayed to the hidden layer. The hidden layer contains processing units called neurons which transform the weighted input by a transfer function, t. It is noteworthy that biases b are added to the transfer function before the transformation process. The hidden layer's output is subsequently conveyed to the output layer, which is transformed by a transfer function operating inside the hidden layer. The network's predicted values are then derived from the output, $\hat{y}$ from the output layer as shown in Equation (12).

$$\hat{y} = t\left(\sum_{k=1}^{m} w_k x_k + b\right) \tag{12}$$

In training the BPNN, a training algorithm is used in updating weights and biases based on the backpropagation error, e (divergence in true and predicted value) as shown in Equation (13) so as to produce a network with a minimum propagation error.

$$e = y - \hat{y} \tag{13}$$

Several training algorithms have been developed for such purposes. However, the Levenberg–Marquardt algorithm [80] is the widely used training function due to its high convergence speed and accuracy and thus was used in this study.

3.2.3. MARS

The MARS algorithm is a non-parametric algorithm developed by [81] to estimate the complex nonlinear correlation between model inputs and output. This estimating process is achieved by automatically building a series of linear piecewise regression models through the use of basis functions, to fit the given data pair.

In the general, the MARS model is of the form precisely considered in Equation (14):

$$\hat{f}(z) = \beta_0 + \sum_{k=1}^{N} \beta_k \lambda_k(z) \tag{14}$$

where $\hat{f}(z)$ signifies the estimated output parameter value, β_0 is constant, $\lambda_k(z)$ is the kth basis function, β_k signifies the kth basis function's coefficient and z signifies the input variable. The basis function act as a hinge function to split the data into separate sections, which can be modelled individually. Each basis function can be precisely considered in Equation (15) as:

$$\lambda_k(z) = \prod_{i=1}^{I_k} \left[s_{ik} \cdot \left(z_{v(i,k)} - h_{ik} \right) \right]_+ \tag{15}$$

where I_k is the quantity of splits that formed $\lambda_k(z)$, s_{ik} is the selected sign with value $\pm$, $v(i,k)$ labels the predictor variable and h_{ik} is the knot value on the corresponding input variables.

The MARS algorithm adopts two main steps namely: the forward selection process and the backward deletion process; to develop its model. In the forward selection process, the model is initially constructed with a constant basis function. New pairs of basis functions are thereafter iteratively included in the model to reduce the training residual sum-of-squares error; to improve the model. However, as many basis functions are added in the forward process; the model built becomes overfit and cannot generalize well with unseen data. The backward deletion process is then introduced to remove all redundant basis functions. It employs the generalized cross-validation (GCV) Equation (16) to evaluate the performance of individually created models as it eliminates the unwanted basis functions.

The individually created model with the least value of GCV is then chosen as the optimal MARS model.

$$\mathrm{GCV}(Q) = \frac{\frac{1}{H}\sum_{j=1}^{H}\left(y_j - \hat{f}_Q(z_j)\right)^2}{\left(1 - \frac{C(Q)}{H}\right)^2} \tag{16}$$

where y_j and $\hat{f}_Q(z_j)$ denotes the actual output and predicted values of the training samples, and H represents the total number of training samples. As shown in Equation (17), $C(Q)$ is a penalty for model complexity that is proportional to the model's number of basis functions.

$$C(Q) = (Q+1) + pQ \tag{17}$$

where p is the penalty cost for the optimization of every single basis function which works as a smoothing variable. The details of MARS as well as the selection of the p are in [76].

3.2.4. MVRA

MVRA is a statistical tool applied to fit a model to establish a linear relation between a set of input parameters (independent variables) and an output parameter (dependent variable) [82]. This fitted model can then be used to make predictions on new data. MVRA works by studying the correlation between the various input parameters and output parameters to construct simultaneous equations so as to acquire the best-fit equation. It uses an ordinary least squares fit on the dataset to find the best-fit equation. It forms a regression matrix in the process of solving simultaneous equations. The regression matrix is then solved using the backslash operator to obtain the regression coefficient as well as the intercept [83]. Generally, the MVRA is mathematically expressed in Equation (18) as:

$$Y = \beta_0 + \beta_2 X_2 + \beta_3 X_3 + \ldots + \beta_k X_k \tag{18}$$

where $\beta_1, \ldots, \beta_k$ are the regression coefficients, β_0 is the intercept $X_1, X_2, \ldots, X_k$ is the independent variable and Y is the dependent variable.

3.2.5. ELM

In 2004 Huang introduces the mathematical model of ELM. The ELM's basic principle is based on a single hidden layer feed-forward neural network (SLFN) (Figure 6). Because of its improved generality, simplicity, and efficient forecasting nature, the ELM has been employed in a variety of application areas [84].

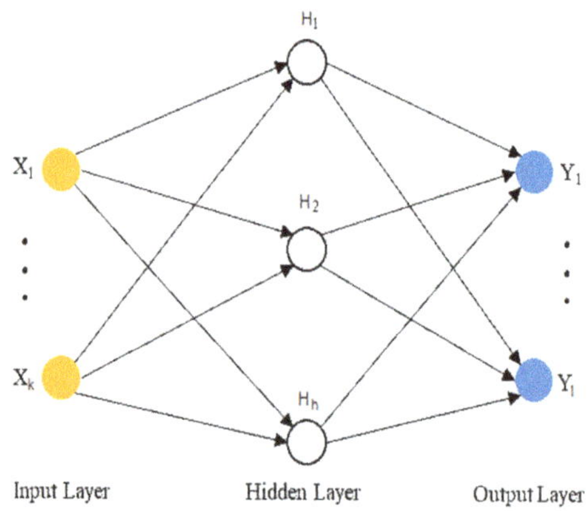

Figure 6. ELM Architecture.

The basic premise of ELM is as follows: Given N as the number of hidden units, K as the number of training samples, and the activation function $f(\,)$ in the hidden units, the output of the ELM o_m for the mth training sample is depicted in Equation (19) as:

$$o_m = \sum_{i=1}^{N} \beta_i f(w_k, b_i, x_m) \;\; m = 1, \ldots, K \tag{19}$$

where b_i is the hidden neurons' bias factor, x_m denotes the number of inputs, β_i denotes the output weight vectors and w_k denotes input weight vectors. The sigmoid function is used as an activation function. The sigmoid function's output is essentially in a range of 1 to 0. To determine the output weights, the linear equation (Equation (20)) is employed.

$$\beta = H^{\dagger} Y \tag{20}$$

where H denotes the output matrix of the hidden layer, $H^{\dagger}$ the Moore–Penrose generalized inverse [85] of H, and Y denotes the ELM output targets. In Equation (21), Equation (20) is written as:

$$H\beta = Y \tag{21}$$

Equation (22) can be used to define H, β, Y as follows:

$$H = \begin{bmatrix} p(x_1) \\ \vdots \\ p(x_K) \end{bmatrix} = \begin{bmatrix} f(w_1, b_1, x_1) & \cdots & f(w_N, b_N, x_1) \\ \vdots & \cdots & \vdots \\ f(w_1, b_1, x_j) & \cdots & f(w_N, b_N, x_j) \end{bmatrix}_{N \times K}, \beta = \begin{bmatrix} \beta_1^Y \\ \vdots \\ \beta_N^Y \end{bmatrix} \text{ and } Y = \begin{bmatrix} y_1^Y \\ \vdots \\ y_K^Y \end{bmatrix} \tag{22}$$

In this case, the hidden layer's feature mapping is $p(x)$. H is the ELM's output.

3.3. *Procedures for Model Construction*

3.3.1. Data Selection and Division

In modelling the various approaches presented in this study, the hold-out cross-validation technique was employed to partition the entire 101 datasets. The datasets were split into the 80:20 ratio. The first 80% of the total datasets were used as the training set (representing 81 training datasets). The remaining 20% (representing 20 datasets) were used as the test set. This strategy was adopted because [86,87] have proved that a ratio of 80:20 or 70:30 will produce accurate prediction results and will not cause overfitting.

3.3.2. Data Normalization

In the data preparation phase, it is expedient that the input parameters be normalized. This is because the input parameters have different input ranges order and those with the higher values have the potential to skew the prediction results to themselves. Thus, to avoid this predicament and give equal chances to each input parameter to influence the prediction outcome, the input parameters defined in Table 1 were normalized into the interval [–1,1] [88,89] utilizing Equation (23).

$$F_i = F_{\min} + \frac{(E_i - E_{\min}) \times (F_{max} - F_{\min})}{(E_{max} - E_{\min})} \tag{23}$$

where E_i signifies the actual data, E_{max} and $E_{\min}$ refer to the maximum values and minimum of the actual data, F_i are the normalized data and $F_{\min}$ and F_{max} being the min-max values of -1 and 1 in that order.

3.3.3. Model Development

For the development of the GPR model, five different models based on the squared exponential (Equation (24)), exponential (Equation (25)), rational quadratic (Equation (26)), matérn 3/2 (Equation (27)), and matérn 5/2 (Equation (28)), covariance function as well as the functions were developed. Each model had a constant MF.

$$g(r,r') = v_s^2 exp\left[\frac{-\|r-r'\|}{2\chi^2}\right] \tag{24}$$

$$g(r,r') = v_s^2 exp\left[\frac{-\|r-r'\|}{\chi}\right] \tag{25}$$

$$g(r,r') = v_s^2 exp\left[\frac{-\|r-r'\|}{2\beta\chi^2}\right]^{-\beta} \tag{26}$$

$$g(r,r') = v_s^2 exp\left[1+\frac{\sqrt{3}\|r-r'\|}{\chi}\right] exp\left[-\frac{\sqrt{3}\|r-r'\|}{\chi}\right] \tag{27}$$

$$g(r,r') = v_s^2 exp\left[1+\frac{\sqrt{5}\|r-r'\|}{\chi}+\frac{5\|r-r'\|^2}{3\chi^2}\right] exp\left[-\frac{\sqrt{5}\|r-r'\|}{\chi}\right] \tag{28}$$

where β is the rational quadratic covariance's shape parameter, χ is the length scale, and v_s^2 is the covariance function's signal variance.

The model with the lowest mean squared error and highest correlation coefficient on the test dataset was chosen as the optimum GPR model. For the BPNN model, a three-layered architecture was chosen—the first with the input layer, the second with a hidden layer and the thirdly with an output layer. A single hidden layer was used because it has been established to be a reliable predictor for any prediction problem [90]. Furthermore, in the case of hidden and output layers, hyperbolic and linear transfer functions were selected and used. The Levenberg–Marquardt algorithm was used to train this BPNN model. According to the suggested values by the previous researchers, a range of 1 to 40 for neurons was tried and the optimum number was the one that gives the lowest MSE on the test dataset [91,92]. The optimum number of neurons in the hidden layer that resulted in the lowest MSE on the test dataset was determined using a sequential experimental procedure in the construction of the ELM model. In that regard, 1 to 20 neurons were tried. It is worth stating that, the building of the MARS model, entails the choice of the highest number of basis functions to be used in the forward selection stage as well as the maximum degree of interaction. These serve as constraints in the development process. Based on their levels of interaction, three independent MARS models were built in this study–zero-degree, first degree and second-degree. Furthermore, a maximum of 20 basis functions were selected for the forward selection stage. The model with the highest correlation coefficient and lowest mean squared error (MSE) was chosen as the optimum MARS model. The MVRA model was developed using the same dataset for the development and testing of the GPR, BPNN, ELM and MARS models. The MVRA solves the multilinear regression equations established for the various input parameters and PPV using the least square technique in order to find the regression coefficient (Equation (18)) for each input parameter as well as the intercept.

3.3.4. Performance Indicators

The performance of the various models constructed in this study was assessed using performance measures such as variance accounted for (VAF), correlation coefficient (R), coefficient of determination (R^2) and mean squared error (MSE). These indicators are precisely shown in Equations (29)–(32) as:

$$\text{MSE} = \frac{1}{p}\left[\sum_{i=1}^{p}(s_i - q_i)^2\right] \tag{29}$$

$$R = \frac{\sum_{i=1}^{p} (s_i - \bar{s})(q_i - \bar{q})}{\sqrt{\sum_{i=1}^{p} (s_i - \bar{s})^2} \times \sqrt{\sum_{i=1}^{p} (q_i - \bar{q})^2}} \quad (30)$$

$$R^2 = \left[\frac{\sum_{i=1}^{p} (s_i - \bar{s})(q_i - \bar{q})}{\sqrt{\sum_{i=1}^{p} (s_i - \bar{s})^2} \times \sqrt{\sum_{i=1}^{p} (q_i - \bar{q})^2}} \right]^2 \quad (31)$$

$$VAF = \left(1 - \frac{var(s_i - q_i)}{var(s_i)} \right) \quad (32)$$

where $\bar{q}$ represents the mean of the estimated values, q_i represents the estimated values, s_i represents the measured values, p is the number of observations, while $\bar{s}$ denotes the average of the measured values.

4. Results and Discussion

4.1. Developed Models

4.1.1. Gaussian Process Regression

As shown in Table 6, the optimum GPR model that produced the MSE of 0.0903 and the highest *R*-value of 0.9986 for the testing dataset, had a matérn 3/2 covariance function with a noise variance of 0.06434, a length scale of 3.6019, and a signal variance of 7.0339. This indicates that the GPR-matérn 3/2 can generalize well with unseen datasets relative to the other GPR models. Hence, GPR-matérn 3/2 model was selected as the best GPR model in this study.

Table 6. Results of the Five different GPR Models.

Covariance Functions	Training		Testing	
	R	MSE	*R*	MSE
Matérn 3/2	0.9961	0.0798	0.9986	0.0903
Matérn 5/2	0.9978	0.0452	0.9956	0.1546
Squared exponential	0.9978	0.0458	0.9942	0.1812
Rational quadratic	0.9978	0.0458	0.9942	0.1812
Exponential	1.0000	0.0000	0.9850	0.3008

4.1.2. BPNN

As shown in Table 7, the optimal BPNN model has one neuron in the hidden layered network. Thus, having an architecture [7-1-1] which means seven input parameters and one neuron in the hidden layer, and an output layer. This is because it has the lowest MSE value on test datasets.

Table 7. Results of BPNN for Different Architectures.

Architecture	Number of Neurons in Hidden Layer	Training		Testing	
		R	MSE	*R*	MSE
7-1-1	1	0.9929	0.1453	0.9924	0.1714
7-2-1	2	0.9956	0.0902	0.9909	0.2085
7-4-1	4	0.9680	0.6452	0.8312	3.8234
7-5-1	5	0.9247	1.4831	0.9699	0.5622
7-6-1	6	0.9995	0.0105	0.4489	156.8569
7-7-1	7	1.0000	0.0007	0.9830	0.3294
7-8-1	8	1.0000	0.0002	0.9536	0.9092

Table 7. Cont.

Architecture	Number of Neurons in Hidden Layer	Training		Testing	
		R	MSE	R	MSE
7-10-1	10	1.0000	1.0244×10^{-21}	0.2794	83.2970
7-15-1	15	1.0000	1.97866×10^{-22}	0.9293	2.2797
7-20-1	20	1.0000	5.7607×10^{-24}	0.9008	2.4538
7-24-1	24	1.0000	1.5760×10^{-23}	0.8714	5.0448
7-28-1	28	1.0000	4.9485×10^{-25}	0.7352	6.7391
7-30-1	30	1.0000	3.6328×10^{-26}	0.7831	5.6888
7-34-1	34	1.0000	5.7972×10^{-20}	0.8136	8.5761
7-38-1	38	1.0000	9.7338×10^{-26}	0.5893	7.7243
7-40-1	40	1.0000	1.6407×10^{-25}	0.6707	16.7280

4.1.3. MARS

As shown in Table 8, the developed MARS model with the first order of interaction had the highest R values as well as the lowest MSE values on both the training and test datasets. Hence it was chosen as the optimum MARS model in this study.

Table 8. Results of Different MARS Models.

Interaction Order	Training		Testing	
	R	MSE	R	MSE
Zero Order	0.9924	0.1548	0.9895	0.2605
First Order	0.9944	0.1145	0.9953	0.1038
Second Order	0.9940	0.1220	0.9923	0.1506

In the developmental process of the selected first order of interaction MARS model, only eight basis functions after the backward elimination stage were used out of the 20 basis functions employed in the forward selection stage. The eight basis functions of the selected MARS model and their respective equations are shown in Table 9.

Table 9. The Relationship Between Basis Functions and their Related Equations.

Basis Function	Equation
BF1	max (0, D – 850)
BF2	max (0, 850 – D)
BF3	max (0, D – 550);
BF5	max (0, MC – 96.764);
BF6	max (0, 96.764 – MC)
BF7	max (0, D – 1750) × BF5;
BF10	max (0, MC – 119) × BF3;
BF11	max (0, 119– MC) × BF3;

The developed optimum MARS model for predicting ground vibration as a result of blasting is provided in Equation (33).

$$\begin{aligned} PPV = -2.85717 - (0.0211305 \times BF1) + (0.0270673 \times BF2) + (0.0190881 \times BF3) \\ +(0.033926 \times BF5) - (0.0570272 \times BF6) + \left(5.46015 \times 10^{-5} \times BF7\right) \\ +\left(3.56504 \times 10^{-5} \times BF10\right) + \left(2.79304 \times 10^{-5} \times BF10\right) \end{aligned} \quad (33)$$

4.1.4. ELM

With respect to the experimental results shown in Table 10, the optimum ELM model developed had 12 neurons in the hidden layer with a sigmoid activation function. Thus, having a structure [7-12-1] that represents seven inputs with 12 neurons in the hidden layer and one output.

Table 10. Training and Testing R and MSE Results for ELM.

Architecture	Number of Hidden Neurons	Training		Testing	
		R	MSE	R	MSE
7-1-1	1	0.6989	5.2395	0.8328	2.9447
7-2-1	2	0.7562	4.3848	0.8797	2.3808
7-5-1	5	0.9371	1.2477	0.9877	0.4775
7-8-1	8	0.9441	1.1130	0.9624	0.7056
7-10-1	10	0.9910	0.1832	0.9948	0.1832
7-12-1	12	0.9958	0.0870	0.9957	0.1384
7-15-1	15	0.9950	0.2181	0.9930	0.1521
7-18-1	18	0.9836	0.3341	0.9914	0.1989
7-20-1	20	0.9848	0.3080	0.9862	0.2530
7-25-1	25	0.9919	0.1656	0.9738	0.7639

4.1.5. MVRA

The developed MVRA model hsa an R-value of 0.7909 for the training dataset and 0.8310 for the test dataset. With respect to the MSE, the developed MVRA model had a value of 3.8341 for the training dataset and 3.2456 for the test dataset. Thus, the developed MVRA model using the training datasets for this study is shown in Equation (34).

$$PPV = 7.237178 + 0.714419AD - 2.80436B + 3.443905S - 0.02705MC \\ -2.33861SL - 0.67419PF - 0.00284D \quad (34)$$

4.2. Assessment of Models Performance

In evaluating the prediction capabilities of the five predictive models presented in the study, the statistical performance outcomes of the testing samples are outlined in Table 11.

Table 11. PPV Prediction Results of Various Models.

Model	MSE	R	R^2	VAF (%)
GPR	0.0903	0.9985	0.9971	99.1728
MARS	0.1038	0.9953	0.9906	98.8692
ELM	0.1381	0.9957	0.9915	98.5469
BPNN	0.1714	0.9924	0.9848	98.2273
MVRA	3.2456	0.8310	0.6906	66.0603

Notionally, a predictive model is said to be accurate if R and R^2 are 1, MSE is 0 and VAF is 100%. In that regard, it can the seen that the GPR with the MSE value of 0.0903 closest to 0, R values of 0.9985 closest to 1, R^2 values of 0.9971 closest to 1 and VAF value of 99.1728% closest to 100% outperformed all the techniques applied in this study. This shows the reliability of the GPR in predicting ground vibration. The MARS performed better than the ELM by having had MSE value of 0.1038 and a VAF value of 98.5469% with the ELM having an MSE value of 0.1381 and a VAF value of 98.2273%. The ELM also performed better than the BPNN with MSE and VAF values of 0.2178 and 98.1919%. It is worth mentioning that the GPR, MARS, ELM and BPNN were superior in predicting ground vibration to the simple MVRA model which had an MSE of 3.2456, R-value of 0.8310, the R^2 value of 0.6906 and VAF value of 66.0603%. Figure 7 depicts the interpretation of the obtained results.

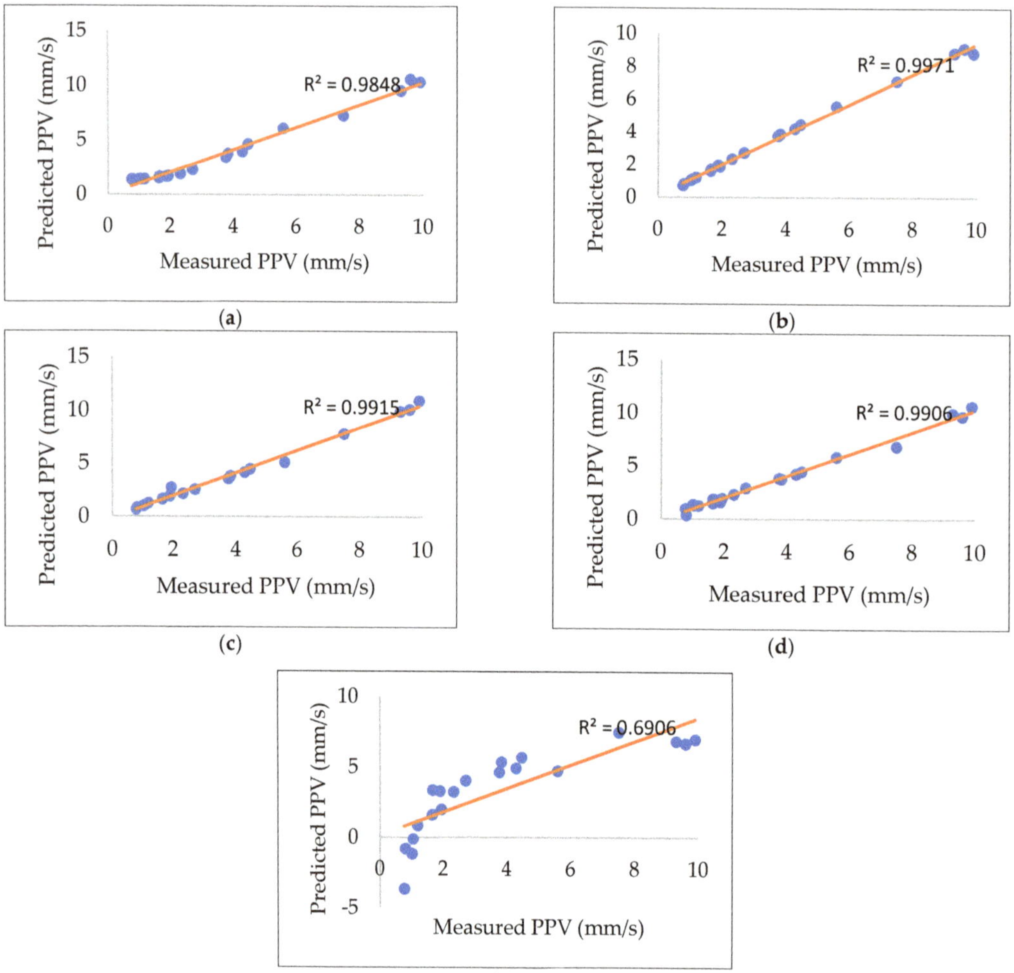

(e)

Figure 7. Comparison of Predicted and Measured PPV for: (**a**) BPNN (**b**) GPR (**c**) ELM (**d**) MARS (**e**) MVRA.

As ground vibration is one of the most unfavorable environmental effects of blasting operations which can cause serious damage to neighboring residences and structures, a precise prediction of its severity is critical to managing and lessening its incidence. The *R*, R^2 and VAF values for the GPR, MARS, BPNN, and ELM may not vary significantly, but any predictive model that delivers the most accurate prediction is of paramount relevance to the blast engineer. Hence the need to develop different models. This study found that the GPR is more accurate in forecasting ground vibration than the MARS, BPNN, ELM and MVRA and that it can be used by blast engineers to predict blast-induced ground vibration.

4.3. Sensitivity Analysis

To determine the most and least effective parameters, sensitivity analysis is performed to examine how the model responds to changes in the input variables with respect to PPV. Hence, in this study, a sensitivity analysis approach implemented in [93] was adopted. Here, while keeping the ranges of all other parameters fixed, the mean value of one of the input variables is increased (i.e., New mean = Old mean + 5% Old Mean) and subsequently the amount of changes in the predicted PPV using the GPR model is recorded. The obtained results are graphically illustrated in Figure 8.

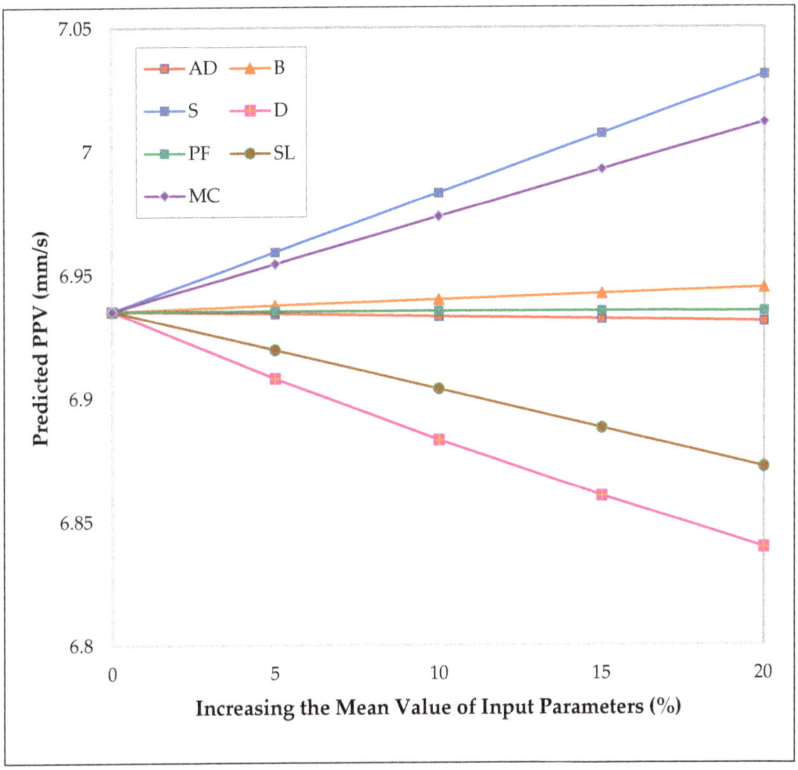

Figure 8. The Input Parameters and PPV Relationship's Strength.

As can be seen in Figure 8, increasing the mean values of spacing and maximum charge per delay, increases PPV. Furthermore, increasing the mean values of distance and stemming length decreases PPV. Increasing burden slightly increased PPV. Nevertheless, increasing values of powder factor and average hole depth did not significantly impact the values of PPV. It can thus be said that the most influential parameters that can affect PPV greatly are spacing, a maximum charge per delay, distance and stemming length.

5. Conclusions

In this paper, three AI models of GPR, ELM and BPNN were developed and applied to predict blast-induced PPV. In showing the predictive capabilities of these AI techniques, a MARS and MVRA model were also developed. In developing and evaluating these models, 101 datasets obtained from Ras Limestone Mine of Shree Cement, India were utilized. Out of the 101 datasets, 81 were utilized to create the various models, while the remaining 20 were used as test sets for the models that were developed. The input parameters in the creation of the various models were average depth (m), burden (m), spacing (m), powder factor (t/kg), the distance between the monitoring station and the blasting site (m), stemming length (m), and maximum charge per delay (kg), while the output parameter was PPV. The various developed models were then evaluated using performance metrics of R, R^2, MSE and VAF. The results obtained showed that the GPR model had the lowest MSE of 0.0903, and the highest R, R^2, and VAF values of 0.9985, 0.9971 and 99.1728% respectively, indicating that it was superior to the other models in predicting blasting-induced ground vibration. This was followed by MARS which had MSE, *R*, R^2 and VAF values of 0.1038, 0.9953, 0.9906 and 98.8692% respectively. Then ELM had an MSE of 0.1381, *R*-value of 0.9957, R^2 value of 0.9915 and VAF value of 98.5469. Then the BPNN with an MSE, *R*, R^2 and VAF of 0.1714, 0.9924, 0.9848 and 98.2273% respectively. The MVRA performed very poorly as it had, with the highest MSE of 3.2456, and lowest *R*-value of 0.8310, the R^2 value of 0.6906 and the VAF value of 66.0603%. The results obtained show that the GPR model can be utilized to forecast blast-induced ground vibration in the mining industry. The sensitivity analysis of the dataset found that spacing, a maximum charge per delay, distance and stemming length had a great influence on PPV whereas burden, powder factor and average depth had slight to no influence on PPV.

Author Contributions: Conceptualization, E.T.M., R.M.B., C.K.A., M.K.; methodology, R.M.B., C.K.A.; software, R.M.B., C.K.A.; formal analysis, R.M.B., C.K.A.; resources, E.T.M., R.M.B., C.K.A.; data curation, R.M.B. writing—original draft, M.B., E.T.M., R.M.B., C.K.A., M.K., M.M.S.S., S.K.; writing—review and editing, M.B., E.T.M., R.M.B., C.K.A., M.K., M.M.S.S., S.K.; Supervision, E.T.M., M.K., S.K.; funding acquisition, M.M.S.S. All authors have read and agreed to the published version of the manuscript.

Funding: The research is partially funded by the Ministry of Science and Higher Education of the Russian Federation under the strategic academic leadership program 'Priority 2030' (Agreement 075-15-2021-1333 dated 30 September 2021).

Institutional Review Board Statement: Not applicable.

Informed Consent Statement: Not applicable.

Data Availability Statement: The data presented in this study are available on request from the corresponding author.

Acknowledgments: Authors are thankful to Pankaj Agarwal, Assistant Vice President and Management of Shree Cement, Beawar, Rajashthan for providing data for the preparation of this paper.

Conflicts of Interest: The authors declare no conflict of interest.

Abbreviations

The following abbreviations are used in this manuscript:

Abbreviations	Explanations
ABC	Artificial bee colony
ANN	Artificial neural network
BA	Bat-inspired Algorithm
BN	Bayesian network
BBO	Biogeography-based optimization
BI	Blastability index (compressive strength/tensile strength)
BIENN	Brain-inspired emotional neural network
B	Burden
BS	Burden spacing ratio
CHAID	Chi-square automatic interaction detector
CART	Classification and regression tree
H	Distance between blasting face and monitoring point (m)
XGBoost	Extreme gradient boosting machine
ELM	Extreme learning machine
FFA	Firefly algorithm
FIS	Fuzzy inference system
FL	Fuzzy logic
GPR	Gaussian process regression
GEP	Gene expression programming
GA	Genetic algorithm
GP	Genetic programming

GOA	Grasshopper optimization algorithms
GWO	Grey wolf optimization
GMDH	Group method of data handling
HHOA	Harris hawk optimization algorithm
HD	Hole depth
ICA	Imperialistic competitive algorithm
KNN	K-nearest neighbors
LSSVM	Least square support vector machine
M5DT	M5′ decision tree
Q	Maximum charge per delay
MARS	Multivariate adaptive regression splines
ANFIS	Neuro-fuzzy inference system
NH	Number of holes
N	Number of rows
ORELM	Outlier robust ELM
PSO	Particle swarm optimization
V	Poisson's ratio
PF	Powder factor
P_v	P-wave velocity
RF	Random forest
RVR	Relevance vector regression
RSM	Response surface methodology
RD	Rock density
RQD	Rock quality designation
SaDE	Self-adaptive differential evolution
S	Spacing (m)
ST	Stemming length
SD	Subdrilling
SVM	Support vector machines
SVR	Support vector regression
VOD	Velocity of detonation
WNN	Wavelet neural network
WOA	Whale optimization algorithm
E	Young's modulus

References

1. Ak, H.; Iphar, M.; Yavuz, M.; Konuk, A. Evaluation of ground vibration effect of blasting operations in a magnesite mine. *Soil Dyn. Earthq. Eng.* **2009**, *29*, 669–676. [CrossRef]
2. Murlidhar, B.R.; Mohamad, E.T.; Armaghani, D.J. Building Information Model for Drilling and Blasting for Tropically Weathered Rock. *J. Mines Met. Fuels* **2019**, *67*, 494–500.
3. Hasanipanah, M.; Faradonbeh, R.S.; Amnieh, H.B.; Armaghani, D.J.; Monjezi, M. Forecasting blast-induced ground vibration developing a CART model. *Eng. Comput.* **2016**, *33*, 307–316. [CrossRef]
4. Yu, Z.; Shi, X.; Zhou, J.; Chen, X.; Qiu, X. Effective Assessment of Blast-Induced Ground Vibration Using an Optimized Random Forest Model Based on a Harris Hawks Optimization Algorithm. *Appl. Sci.* **2020**, *10*, 1403. [CrossRef]
5. Hasanipanah, M.; Monjezi, M.; Shahnazar, A.; Armaghani, D.J.; Farazmand, A. Feasibility of indirect determination of blast induced ground vibration based on support vector machine. *Measurement* **2015**, *75*, 289–297. [CrossRef]
6. Fouladgar, N.; Hasanipanah, M.; Amnieh, H.B. Application of cuckoo search algorithm to estimate peak particle velocity in mine blasting. *Eng. Comput.* **2016**, *33*, 181–189. [CrossRef]
7. Amiri, M.; Amnieh, H.B.; Hasanipanah, M.; Khanli, L.M. A new combination of artificial neural network and K-nearest neighbors models to predict blast-induced ground vibration and air-overpressure. *Eng. Comput.* **2016**, *32*, 631–644. [CrossRef]
8. Hasanipanah, M.; Amnieh, H.B.; Khamesi, H.; Armaghani, D.J.; Golzar, S.B.; Shahnazar, A. Prediction of an environmental issue of mine blasting: An imperialistic competitive algorithm-based fuzzy system. *Int. J. Environ. Sci. Technol.* **2017**, *15*, 551–560. [CrossRef]
9. Leskovar, K.; Težak, D.; Mesec, J.; Biondić, R. Influence of Meteorological Parameters on Explosive Charge and Stemming Length Predictions in Clay Soil during Blasting Using Artificial Neural Networks. *Appl. Sci.* **2021**, *11*, 7317. [CrossRef]
10. Lizarazo-Marriaga, J.; Vargas, C.A.; Tiria, L. A new approach to predict local site effects related to blast-induced ground vibrations. *J. Geophys. Eng.* **2018**, *15*, 1843–1850. [CrossRef]

11. Isheyskiy, V.; Marinin, M.; Dolzhikov, V. Combination of Fracturing Areas After Blasting Column Charges during Destruction of Rocks. *Int. J. Eng. Res. Technol.* **2019**, *12*, 2953–2956.
12. Duvall, W.I.; Petkof, B. *Spherical Propagation of Explosion-Generated Strain Pulses in Rock*; US Department of the Interior, Bureau of Mines: Washington, DC, USA, 1959.
13. Hidayat, R.; Cahyadi, T.A.; Winarno, E.; Saptono, S.; Koesnaryo, S. A Review of Artificial Intelligent for Prediction Ground Vibration in Blasting. In Proceedings of the 15th ReTII National Seminar, Yogyakarta, Indonesia, 27 October 2020; pp. 187–193.
14. Zadeh, L.A. Fuzzy logic, neural networks and soft computing. In *Safety Evaluation Based on Identification Approaches Related to Time-Variant and Nonlinear Structures*; Vieweg+ Teubner Verlag: Wiesbaden, Germany, 1993; pp. 320–321.
15. Fişne, A.; Kuzu, C.; Hüdaverdi, T. Prediction of environmental impacts of quarry blasting operation using fuzzy logic. *Environ. Monit. Assess.* **2011**, *174*, 461–470. [CrossRef] [PubMed]
16. Ghasemi, E.; Ataei, M.; Hashemolhosseini, H. Development of a fuzzy model for predicting ground vibration caused by rock blasting in surface mining. *J. Vib. Control* **2013**, *19*, 755–770. [CrossRef]
17. Nguyen, H. Support vector regression approach with different kernel functions for predicting blast-induced ground vibration: A case study in an open-pit coal mine of Vietnam. *SN Appl. Sci.* **2019**, *1*, 283. [CrossRef]
18. Armaghani, D.J.; Momeni, E.; Abad, S.V.A.N.K.; Khandelwal, M. Feasibility of ANFIS model for prediction of ground vibrations resulting from quarry blasting. *Environ. Earth Sci.* **2015**, *74*, 2845–2860. [CrossRef]
19. Kamali, M.; Ataei, M. Prediction of Blast Induced Ground Vibrations in Karoun III Power Plant and Dam: A Neural Network. *J. South Afr. Inst. Min. Metall.* **2010**, *110*, 481–490.
20. Mohamad, E.T.; Noorani, S.A.; Armaghani, D.J.; Saad, R. Simulation of Blasting Induced Ground Vibration by Using Artificial Neural Network. *Electron. J. Geotech. Eng.* **2012**, *17*, 2571–2584.
21. Parida, A.; Mishra, M.K. Blast Vibration Analysis by Different Predictor Approaches—A Comparison. *Procedia Earth Planet. Sci.* **2015**, *11*, 337–345. [CrossRef]
22. Arthur, C.K.; Temeng, V.A.; Ziggah, Y.Y. Novel approach to predicting blast-induced ground vibration using Gaussian process regression. *Eng. Comput.* **2020**, *36*, 29–42. [CrossRef]
23. Armaghani, D.J.; Hasanipanah, M.; Amnieh, H.B.; Mohamad, E.T. Feasibility of ICA in approximating ground vibration resulting from mine blasting. *Neural Comput. Appl.* **2016**, *29*, 457–465. [CrossRef]
24. Armaghani, D.J.; Kumar, D.; Samui, P.; Hasanipanah, M.; Roy, B. A novel approach for forecasting of ground vibrations resulting from blasting: Modified particle swarm optimization coupled extreme learning machine. *Eng. Comput.* **2020**, *37*, 3221–3235. [CrossRef]
25. Faradonbeh, R.S.; Armaghani, D.J.; Abd Majid, M.Z.; Tahir, M.M.; Murlidhar, B.R.; Monjezi, M.; Wong, H.M. Prediction of ground vibration due to quarry blasting based on gene expression programming: A new model for peak particle velocity prediction. *Int. J. Environ. Sci. Technol.* **2016**, *13*, 1453–1464. [CrossRef]
26. Hasanipanah, M.; Naderi, R.; Kashir, J.; Noorani, S.A.; Qaleh, A.Z.A. Prediction of blast-produced ground vibration using particle swarm optimization. *Eng. Comput.* **2017**, *33*, 173–179. [CrossRef]
27. Zhou, J.; Asteris, P.G.; Armaghani, D.J.; Pham, B.T. Prediction of ground vibration induced by blasting operations through the use of the Bayesian Network and random forest models. *Soil Dyn. Earthq. Eng.* **2020**, *139*, 106390. [CrossRef]
28. Choi, Y.-H.; Lee, S.S. Predictive Modelling for Blasting-Induced Vibrations from Open-Pit Excavations. *Appl. Sci.* **2021**, *11*, 7487. [CrossRef]
29. Hajihassani, M.; Armaghani, D.J.; Monjezi, M.; Mohamad, E.T.; Marto, A. Blast-induced air and ground vibration prediction: A particle swarm optimization-based artificial neural network approach. *Environ. Earth Sci.* **2015**, *74*, 2799–2817. [CrossRef]
30. Temeng, V.A.; Ziggah, Y.Y.; Arthur, C.K. A novel artificial intelligent model for predicting air overpressure using brain inspired emotional neural network. *Int. J. Min. Sci. Technol.* **2020**, *30*, 683–689. [CrossRef]
31. Murlidhar, B.R.; Armaghani, D.J.; Mohamad, E.T. Intelligence Prediction of Some Selected Environmental Issues of Blasting: A Review. *Open Constr. Build. Technol. J.* **2020**, *14*, 298–308. [CrossRef]
32. Murlidhar, B.R.; Bejarbaneh, B.Y.; Armaghani, D.J.; Mohammed, A.S.; Mohamad, E.T. Application of Tree-Based Predictive Models to Forecast Air Overpressure Induced by Mine Blasting. *Nat. Resour. Res.* **2021**, *30*, 1865–1887. [CrossRef]
33. Zhou, X.; Armaghani, D.J.; Ye, J.; Khari, M.; Motahari, M.R. Hybridization of Parametric and Non-parametric Techniques to Predict Air Over-pressure Induced by Quarry Blasting. *Nat. Resour. Res.* **2021**, *30*, 209–224. [CrossRef]
34. Zhou, J.; Koopialipoor, M.; Murlidhar, B.R.; Fatemi, S.A.; Tahir, M.M.; Armaghani, D.J.; Li, C. Use of Intelligent Methods to Design Effective Pattern Parameters of Mine Blasting to Minimize Flyrock Distance. *Nat. Resour. Res.* **2019**, *29*, 625–639. [CrossRef]
35. Han, H.; Armaghani, D.J.; Tarinejad, R.; Zhou, J.; Tahir, M.M. Random Forest and Bayesian Network Techniques for Probabilistic Prediction of Flyrock Induced by Blasting in Quarry Sites. *Nat. Resour. Res.* **2020**, *29*, 655–667. [CrossRef]
36. Murlidhar, B.R.; Kumar, D.; Armaghani, D.J.; Mohamad, E.T.; Roy, B.; Pham, B.T. A Novel Intelligent ELM-BBO Technique for Predicting Distance of Mine Blasting-Induced Flyrock. *Nat. Resour. Res.* **2020**, *29*, 4103–4120. [CrossRef]
37. Lu, X.; Hasanipanah, M.; Brindhadevi, K.; Amnieh, H.B.; Khalafi, S. ORELM: A Novel Machine Learning Approach for Prediction of Flyrock in Mine Blasting. *Nat. Resour. Res.* **2020**, *29*, 641–654. [CrossRef]
38. Nguyen, H.; Bui, X.-N.; Choi, Y.; Lee, C.W.; Armaghani, D.J. A Novel Combination of Whale Optimization Algorithm and Support Vector Machine with Different Kernel Functions for Prediction of Blasting-Induced Fly-Rock in Quarry Mines. *Nat. Resour. Res.* **2021**, *30*, 191–207. [CrossRef]

39. Ye, J.; Koopialipoor, M.; Zhou, J.; Armaghani, D.J.; He, X. A Novel Combination of Tree-Based Modeling and Monte Carlo Simulation for Assessing Risk Levels of Flyrock Induced by Mine Blasting. *Nat. Resour. Res.* **2021**, *30*, 225–243. [CrossRef]
40. Armaghani, D.J.; Hajihassani, M.; Mohamad, E.T.; Marto, A.; Noorani, S.A. Blasting-induced flyrock and ground vibration prediction through an expert artificial neural network based on particle swarm optimization. *Arab. J. Geosci.* **2014**, *7*, 5383–5396. [CrossRef]
41. Hajihassani, M.; Armaghani, D.J.; Marto, A.; Mohamad, E.T. Ground vibration prediction in quarry blasting through an artificial neural network optimized by imperialist competitive algorithm. *Bull. Eng. Geol. Environ.* **2015**, *74*, 873–886. [CrossRef]
42. Taheri, K.; Hasanipanah, M.; Golzar, S.B.; Abd Majid, M.Z. A hybrid artificial bee colony algorithm-artificial neural network for forecasting the blast-produced ground vibration. *Eng. Comput.* **2016**, *33*, 689–700. [CrossRef]
43. Shahnazar, A.; Rad, H.N.; Hasanipanah, M.; Tahir, M.M.; Armaghani, D.J.; Ghoroqi, M. A new developed approach for the prediction of ground vibration using a hybrid PSO-optimized ANFIS-based model. *Environ. Earth Sci.* **2017**, *76*, 527. [CrossRef]
44. Bayat, P.; Monjezi, M.; Rezakhah, M.; Armaghani, D.J. Artificial Neural Network and Firefly Algorithm for Estimation and Minimization of Ground Vibration Induced by Blasting in a Mine. *Nat. Resour. Res.* **2020**, *29*, 4121–4132. [CrossRef]
45. Shang, Y.; Nguyen, H.; Bui, X.-N.; Tran, Q.-H.; Moayedi, H. A Novel Artificial Intelligence Approach to Predict Blast-Induced Ground Vibration in Open-Pit Mines Based on the Firefly Algorithm and Artificial Neural Network. *Nat. Resour. Res.* **2020**, *29*, 723–737. [CrossRef]
46. Yang, H.; Hasanipanah, M.; Tahir, M.M.; Bui, D.T. Intelligent Prediction of Blasting-Induced Ground Vibration Using ANFIS Optimized by GA and PSO. *Nat. Resour. Res.* **2020**, *29*, 739–750. [CrossRef]
47. Zhang, X.; Nguyen, H.; Bui, X.N.; Tran, Q.H.; Nguyen, D.A.; Bui, D.T.; Moayedi, H. Novel Soft Computing Model for Predicting Blast-Induced Ground Vibration in Open-Pit Mines Based on Particle Swarm Optimization and XGBoost. *Nat. Resour. Res.* **2019**, *29*, 711–721. [CrossRef]
48. Chen, W.; Hasanipanah, M.; Rad, H.N.; Armaghani, D.J.; Tahir, M.M. A new design of evolutionary hybrid optimization of SVR model in predicting the blast-induced ground vibration. *Eng. Comput.* **2021**, *37*, 1455–1471. [CrossRef]
49. Arthur, C.K.; Temeng, V.A.; Ziggah, Y.Y. A Self-adaptive differential evolutionary extreme learning machine (SaDE-ELM): A novel approach to blast-induced ground vibration prediction. *SN Appl. Sci.* **2020**, *2*, 1–23. [CrossRef]
50. Fattahi, H.; Hasanipanah, M. Prediction of Blast-Induced Ground Vibration in a Mine Using Relevance Vector Regression Optimized by Metaheuristic Algorithms. *Nat. Resour. Res.* **2020**, *30*, 1849–1863. [CrossRef]
51. Yang, H.; Rad, H.N.; Hasanipanah, M.; Amnieh, H.B.; Nekouie, A. Prediction of Vibration Velocity Generated in Mine Blasting Using Support Vector Regression Improved by Optimization Algorithms. *Nat. Resour. Res.* **2020**, *29*, 807–830. [CrossRef]
52. Ding, S.; Xu, X.; Nie, R. Extreme learning machine and its applications. *Neural Comput. Appl.* **2014**, *25*, 549–556. [CrossRef]
53. Fang, Q.; Nguyen, H.; Bui, X.-N.; Nguyen-Thoi, T. Prediction of Blast-Induced Ground Vibration in Open-Pit Mines Using a New Technique Based on Imperialist Competitive Algorithm and M5Rules. *Nat. Resour. Res.* **2020**, *29*, 791–806. [CrossRef]
54. Yu, C.; Koopialipoor, M.; Murlidhar, B.R.; Mohammed, A.S.; Armaghani, D.J.; Mohamad, E.T.; Wang, Z. Optimal ELM–Harris Hawks Optimization and ELM–Grasshopper Optimization Models to Forecast Peak Particle Velocity Resulting from Mine Blasting. *Nat. Resour. Res.* **2021**, *30*, 2647–2662. [CrossRef]
55. Khandelwal, M.; Singh, T. Prediction of blast-induced ground vibration using artificial neural network. *Int. J. Rock Mech. Min. Sci.* **2009**, *46*, 1214–1222. [CrossRef]
56. Monjezi, M.; Ghafurikalajahi, M.; Bahrami, A. Prediction of blast-induced ground vibration using artificial neural networks. *Tunn. Undergr. Space Technol.* **2011**, *26*, 46–50. [CrossRef]
57. Mohammadnejad, M.; Gholami, R.; Ramezanzadeh, A.; Jalali, M.E. Prediction of blast-induced vibrations in limestone quarries using Support Vector Machine. *J. Vib. Control* **2011**, *18*, 1322–1329. [CrossRef]
58. Monjezi, M.; Mehrdanesh, A.; Malek, A.; Khandelwal, M. Evaluation of effect of blast design parameters on flyrock using artificial neural networks. *Neural Comput. Appl.* **2013**, *23*, 349–356. [CrossRef]
59. Saadat, M.; Khandelwal, M.; Monjezi, M. An ANN-based approach to predict blast-induced ground vibration of Gol-E-Gohar iron ore mine, Iran. *J. Rock Mech. Geotech. Eng.* **2014**, *6*, 67–76. [CrossRef]
60. Ghoraba, S.; Monjezi, M.; Talebi, N.; Moghadam, M.R.; Armaghani, D.J. Prediction of Ground Vibration Caused by Blasting Operations through a Neural Network Approach: A Case Study of Gol-E-Gohar Iron Mine. *Iran. J. Zhejiang Univ. Sci. A* **2015**, *10*, 1631.
61. Azimi, Y.; Khoshrou, S.H.; Osanloo, M. Prediction of blast induced ground vibration (BIGV) of quarry mining using hybrid genetic algorithm optimized artificial neural network. *Measurement* **2019**, *147*, 106874. [CrossRef]
62. Arthur, C.K.; Temeng, V.A.; Ziggah, Y.Y. Soft computing-based technique as a predictive tool to estimate blast-induced ground vibration. *J. Sustain. Min.* **2019**, *18*, 287–296. [CrossRef]
63. Hosseini, S.A.; Tavana, A.; Abdolahi, S.M.; Darvishmaslak, S. Prediction of blast-induced ground vibrations in quarry sites: A comparison of GP, RSM and MARS. *Soil Dyn. Earthq. Eng.* **2019**, *119*, 118–129. [CrossRef]
64. Jiang, W.; Arslan, C.A.; Tehrani, M.S.; Khorami, M.; Hasanipanah, M. Simulating the peak particle velocity in rock blasting projects using a neuro-fuzzy inference system. *Eng. Comput.* **2019**, *35*, 1203–1211. [CrossRef]
65. Nguyen, H.; Bui, X.-N.; Tran, Q.-H.; Le, T.-Q.; Do, N.-H.; Hoa, L.T.T. Evaluating and predicting blast-induced ground vibration in open-cast mine using ANN: A case study in Vietnam. *SN Appl. Sci.* **2019**, *1*, 125. [CrossRef]
66. Lawal, A.I.; Idris, M.A. An artificial neural network-based mathematical model for the prediction of blast-induced ground vibrations. *Int. J. Environ. Stud.* **2020**, *77*, 318–334. [CrossRef]

67. Arthur, C.K.; Temeng, V.A.; Ziggah, Y.Y. Multivariate Adaptive Regression Splines (MARS) approach to blast-induced ground vibration prediction. *Int. J. Mining Reclam. Environ.* **2020**, *34*, 198–222. [CrossRef]
68. Temeng, V.A.; Arthur, C.K.; Ziggah, Y.Y. Suitability assessment of different vector machine regression techniques for blast-induced ground vibration prediction in Ghana. *Model. Earth Syst. Environ.* **2021**, *8*, 897–909. [CrossRef]
69. Grishchenkova, E.N. Development of a Neural Network for Earth Surface Deformation Prediction. *Geotech. Geol. Eng.* **2018**, *36*, 1953–1957. [CrossRef]
70. Al-Dahidi, S.; Ayadi, O.; Adeeb, J.; Alrbai, M.; Qawasmeh, B.R. Extreme Learning Machines for Solar Photovoltaic Power Predictions. *Energies* **2018**, *11*, 2725. [CrossRef]
71. Bisoyi, S.K.; Pal, B.K. Prediction of Ground Vibration Using Various Regression Analysis. *J. Min. Sci.* **2020**, *56*, 378–387. [CrossRef]
72. Khandelwal, M.; Singh, T. Evaluation of blast-induced ground vibration predictors. *Soil Dyn. Earthq. Eng.* **2007**, *27*, 116–125. [CrossRef]
73. Ragam, P.; Nimaje, D.S. Assessment of blast-induced ground vibration using different predictor approaches-a comparison. *Chem. Eng. Trans.* **2018**, *66*, 487–492.
74. Ercins, S.; Şensöğüt, C. Performance Analysis of the Explosion Applications Realized with Electronic Ignition System at Different Times in the Same Field. *Int. J. Econ. Environ. Geol.* **2020**, *11*, 17–23.
75. Schulz, E.; Speekenbrink, M.; Krause, A. A tutorial on Gaussian process regression: Modelling, exploring, and exploiting functions. *J. Math. Psychol.* **2018**, *85*, 1–16. [CrossRef]
76. Rasmussen, C.E.; Williams, C.K.I. *Gaussian Processes for Machine Learning*; MIT Press: Cambridge, UK, 2006; pp. 105–128.
77. Rasmussen, C.E.; Nickisch, H. Gaussian Processes for Machine Learning (GPML) Toolbox. *J. Mach. Learn. Res.* **2010**, *11*, 3011–3015.
78. Mukhtar, S.M.; Daud, H.; Dass, S.C. Squared Exponential Covariance Function for Prediction of Hydrocarbon in Seabed Logging Application. In Proceedings of the AIP Conference, Depok, Indonesia, 1–2 November 2016.
79. Yang, D.; Zhang, X.; Pan, R.; Wang, Y.; Chen, Z. A novel Gaussian process regression model for state-of-health estimation of lithium-ion battery using charging curve. *J. Power Sources* **2018**, *384*, 387–395. [CrossRef]
80. Moré, J.J. The Levenberg-Marquardt Algorithm: Implementation and Theory. In *Numerical Analysis*; Watson, G.A., Ed.; Springer: Berlin/Heidelberg, Germany, 1978; pp. 105–116. [CrossRef]
81. Friedman, J.H. Multivariate Adaptive Regression Splines. *Ann. Stat.* **1991**, *19*, 1–67. [CrossRef]
82. Alexopoulos, E.C. Introduction to multivariate regression analysis. *Hippokratia* **2010**, *14*, 23–28.
83. Verma, A.K.; Sirvaiya, A. Intelligent prediction of Langmuir isotherms of Gondwana coals in India. *J. Pet. Explor. Prod. Technol.* **2016**, *6*, 135–143. [CrossRef]
84. Ding, Z.; Nguyen, H.; Bui, X.-N.; Zhou, J.; Moayedi, H. Computational Intelligence Model for Estimating Intensity of Blast-Induced Ground Vibration in a Mine Based on Imperialist Competitive and Extreme Gradient Boosting Algorithms. *Nat. Resour. Res.* **2019**, *29*, 751–769. [CrossRef]
85. Rakha, M. On the Moore–Penrose generalized inverse matrix. *Appl. Math. Comput.* **2004**, *158*, 185–200. [CrossRef]
86. Dobbin, K.K.; Simon, R.M. Optimally splitting cases for training and testing high dimensional classifiers. *BMC Med Genom.* **2011**, *4*, 31. [CrossRef]
87. Gholamy, A.; Kreinovich, V.; Kosheleva, O. *Why 70/30 or 80/20 Relation between Training and Testing Sets: A Pedagogical Explanation*; The University of Texas at El Paso: El Paso, TX, USA, 2018.
88. Codd, E.F. Further normalization of the data base relational model. *Data Base Syst.* **1972**, *6*, 33–64.
89. Ali, P.J.; Faraj, R.H.; Koya, E.; Ali, P.J.; Faraj, R.H. Data Normalization and Standardization: A Technical Report. *Mach. Learn. Tech. Rep.* **2014**, *1*, 1–6.
90. Hornik, K.; Stinchcombe, M.; White, H. Multilayer feedforward networks are universal approximators. *Neural Netw.* **1989**, *2*, 359–366. [CrossRef]
91. Sheela, K.G.; Deepa, S.N. Review on Methods to Fix Number of Hidden Neurons in Neural Networks. *Math. Probl. Eng.* **2013**, *2013*, 425740. [CrossRef]
92. Jeremiah, J.J.; Abbey, S.J.; Booth, C.A.; Kashyap, A. Results of Application of Artificial Neural Networks in Predicting Geo-Mechanical Properties of Stabilised Clays—A Review. *Geotechnics* **2021**, *1*, 147–171. [CrossRef]
93. Momeni, M.; Hadianfard, M.A.; Bedon, C.; Baghlani, A. Damage evaluation of H-section steel columns under impulsive blast loads via gene expression programming. *Eng. Struct.* **2020**, *219*, 110909. [CrossRef]

applied sciences

MDPI

Article

Application of Soft Computing Techniques for Predicting Thermal Conductivity of Rocks

Masoud Samaei [1], Timur Massalow [2], Ali Abdolhosseinzadeh [1], Saffet Yagiz [2,*] and Mohanad Muayad Sabri Sabri [3]

[1] Department of Civil Engineering, University of Tabriz, 29 Bahman Boulevard, Tabriz 5166616471, Iran
[2] Department of Mining Engineering, School of Mining and Geosciences, Nazarbayev University, Nur-Sultan 01000, Kazakhstan
[3] Peter the Great St. Petersburg Polytechnic University, 195251 Petersburg, Russia
* Correspondence: saffet.yagiz@nu.edu.kz; Tel.: +7-(7172)-705724

Abstract: Due to the different challenges in rock sampling and in measuring their thermal conductivity (TC) in the field and laboratory, the determination of the TC of rocks using non-invasive methods is in demand in engineering projects. The relationship between TC and non-destructive tests has not been well-established. An investigation of the most important variables affecting the TC values for rocks was conducted in this study. Currently, the black-boxed models for TC prediction are being replaced with artificial intelligence-based models, with mathematical equations to fill the gap caused by the lack of a tangible model for future studies and developments. In this regard, two models were developed based on which gene expression programming (GEP) algorithms and non-linear multivariable regressions (NLMR) were utilized. When comparing the performances of the proposed models to that of other previously published models, it was revealed that the GEP and NLMR models were able to produce more accurate predictions than other models were. Moreover, the high value of R-squared (equals 0.95) for the GEP model confirmed its superiority.

Keywords: thermal conductivity; geothermal systems; gene expression programming (GEP); non-linear multivariable regression (NLMR); P-wave; porosity

Citation: Samaei, M.; Massalow, T.; Abdolhosseinzadeh, A.; Yagiz, S.; Sabri, M.M.S. Application of Soft Computing Techniques for Predicting Thermal Conductivity of Rocks. *Appl. Sci.* **2022**, *12*, 9187. https://doi.org/10.3390/app12189187

Academic Editor: Ricardo Castedo

Received: 25 July 2022
Accepted: 30 August 2022
Published: 14 September 2022

1. Introduction

Due to the increase in energy prices and energy demand, energy conservation and management play a significant role in human lives and governmental policies. Heat as a form of energy is transferred more rapidly in solid mediums than in gas and liquid [1]. Therefore, knowing the ability of solid materials to transfer heat can aid in conserving energy more efficiently. Three main indices were introduced for the evaluation of the thermal behavior of solid materials [2,3] as the following:

- Thermal conductivity (TC, λ), which refers to the material's heat conduction property;
- Thermal diffusivity (TD, κ), which refers to the material's heat diffusion property;
- Specific heat capacity (C_p), which links TC and TD using density (ρ), i.e., Equation (1):

$$\kappa = \frac{\lambda}{\rho.\, C_p} \tag{1}$$

The determination of rocks' TC is of great importance in geothermal, environmental, mining, and civil engineering applications [1]. This parameter is critical for the management of geothermal reservoirs, designing power-saving walls and powerhouses, CO_2 sequestration, and underground waste disposal wells [2–6]. It can be measured using heat input and the temperature gradient of the host rock [7].

The physical and mechanical properties of rocks have a direct relation with TC as the mineral content [8–10], bulk density (ρ) [3,11], porosity ($\varnothing$) [2,12], P-wave (pressure-wave) velocity (V_p) [2,4,13], uniaxial compression strength (UCS) [2,14], saturating fluid characteristic [15], quality and geometry of the contact between the grains [16–18], discontinuities [19], temperature [20–23], and atmospheric pressure [24].

In situ measurements of TC need specialized equipment, which is very expensive and can be time-consuming [25]. However, due to the scale effect, even if the effects of the stress level, pore fluid pressure, temperature, and permeability of rocks can be considered, the laboratory values may substantially differ from those of the in situ measurements [24,26,27]. Therefore, in recent years, several researchers proposed methods to estimate the in situ TC by measuring a rock's effective properties (e.g., [2,28–31]). Measuring the uniaxial compressive strength (UCS), density, porosity, and P-wave velocity of a rock is not as difficult as measuring in situ TC. These properties can also be obtained by laboratory or indirect methods, such as using the Schmidt hammer for the determination of the UCS [24]. In this case, the following items can be found without performing TC core experiments at several depths [30]:

- Predicting and/or in-time monitoring of the heat flux;
- Extracting the temperature profile;
- Evaluating the saturation content of the formation.

Özkahraman, Selver, and Isık [2] conducted tests on rock samples that are mostly used in building constructions. They investigated the relationship between TC and rock properties, specifically for the P-wave velocity, UCS, density, and porosity. They proposed equations for the prediction of the TC regarding each of the four above-mentioned parameters, but not by simultaneously considering all parameters (multivariable equation). They found that TC has a direct relation to P-wave, UCS, and density, whereas it has an inverse relation to porosity. In another study, Yaşar, Erdoğan, and Güneyli [4] performed laboratory tests on 12 different rock samples. They found that the type of a rock's mineral composition could also affect its TC to a high degree.

In recent years, modeling using machine learning (ML) methods to better fit with actual measurements has become more popular with scientists than other modeling methods [32–37]. ML methods are applied to different issues in geotechnical engineering [38–41]. A methodology combining physical modeling and ML was proposed by Assouline et al. [42] to estimate the apparent ground thermal diffusivity at the national scale. In this methodology, a model is built with random forests (RF) using the output values from previous diffusivity estimations, as well as geological, elevation, and temperature features. The model, which exhibited an acceptable test error of 16.5%, is then used to estimate the apparent diffusivity across Switzerland. Singh, Sinha, and Singh [3] applied two ML techniques, including artificial neural networks (ANNs) and adaptive neuro-fuzzy inference (ANFIS), to predict TC by some series of datasets on rock properties (P-wave velocity, UCS, density, and porosity). The proposed models had a strong correlation coefficient between the measured and the predicted TC. Sargam et al. [43] used a multilayer perceptron (MLP) model to study concrete's thermal conductivity. A high degree of prediction accuracy was achieved by MLP. In this regard, Khandelwal [24] conducted an analysis using a feed-forward backpropagation neural network and found that the ANNs presented more accurate results than other techniques did. To predict the TC of Jalore granite, Verma et al. [44] used artificial neural networks (ANNs), linear regression, support vector regressors (SVRs), and decision tree regressors (DTRs). According to their analysis, TC was strongly correlated with density, S-wave velocity, and P-wave velocity. Moreover, it was validated by different AI tools that thermal conductivity is highly sensitive to rock's physical properties [44]. In the study done by Wang et al. [45], TC was analyzed using a convolutional neural network (CNN) and datasets of temperature fields from lattice Boltzmann method (LBM) simulations based on three-dimensional sphere-packed porous media. CNN and LBM acquired relative errors for the effective thermal conductivity of sphere-packed porous media (0.7–22.8%) and irregular porous media (3.1–16.0%). According to them, CNN is promising for the prediction of the heat transport properties of porous media with variable boundary conditions and different morphologies. For predicting geothermal gradient, thermal conductivity, the heat productivity of rocks, and the crustal/mantle heat flow, He et al. [46] used generalized linear models (GLM), deep neural networks (DNN), and gradient-boosted regression trees (GBRT). Their results showed that the DNN model, with a number of neurons multiplied by the features, performed better than the other models. The average relative prediction errors for SVM and DNN were 13.3 and 12.7%, respectively. In a hybrid SVM–DNN approach, the average relative prediction error decreased to 12.2%. Hajihassani et al. [47] developed an ANFIS and a linear multivariable regression (LMR) model using 44 datasets that were collected from the literature. Kang et al. [48] measured the thermal conductivity of various rocks in the Songliao Basin (China). The correlation between porosity, moisture content, density, P-wave velocity, and thermal conductivity were investigated. Seven prediction models were developed using extreme learning machine (ELM), support vector regression (SVR), and backpropagation neural network (BPNN) algorithms. The results demonstrated that the ELM-based model had better performance, speed, and accuracy for predicting rock thermal conductivity.

The above-developed computer-aided models (ANNs and ANFIS) are black-box methods. Some pitfalls remained in these studies [49,50]:

- The relationship among the parameters was not determined, and the most critical parameters could not be identified [51].
- These models are not usable for field uses or future studies [51].
- The possibility of over-training or over-fitting in the training stages of ANNs and ANFIS is higher than it is when using white-box methods [51,52].

In order to overcome the disadvantages discussed above and bridge the gap between ANNs and conventional experimental models, we developed mathematical equations/models using computer programs that have high confidence compared with other studies conducted in a similar field. Using multiple related inputs in this study, we developed functional relationships that can predict a specific output. Two models were proposed using a gene expression programming (GEP) algorithm and a non-linear multivariable regression (NLMR). As shown in the numerical experiments by Ferreira [53], the GEP approach can be seen as an efficient alternative to traditional machine learning approaches. The developed models were validated using statistical indices, and they were compared with the results of previously developed models.

2. Establishment of Dataset

In rock mechanics, each test induces costs, and a long period is needed for each phase of the experimental process. Therefore, it is crucial to reach the desired accuracy by testing a minimum number of samples [54]. Yamaguchi [55] sought to analyze this problem using a statistical technique called the "decision of the sample number". He tested three different kinds of igneous rocks to evaluate their compressive strength and found that a 90% confidence level could be acquired by using only ten samples. This study showed that, due to the high similarity in their physical properties, the mechanical properties of each primary type of rock (i.e., sedimentary, metamorphic, and igneous) could be similar to each other. Therefore, testing a small number of samples would be sufficient to obtain results with a high confidence level [55].

In our study, 50 datasets, including TC, UCS, density, porosity, and P-wave velocity, were taken from the literature [2–4,24]. The type of these samples was not reported, but most of them were sedimentary rocks (i.e., by referring to their P-wave velocity and porosity values). The basic descriptive statistics of the datasets are presented in Table 1.

Table 1. Descriptive statistical distribution of datasets.

Variable	Max	Min	Mean	Median	St. Deviation
TC (W/m K)	3.01	0.186	1.395	1.2628	0.762
UCS (MPa)	116.9	3.43	61.32	63.258	24.151
P-wave (m/s)	6300	1800	4486.436	4457.5105	1096.66
Density (kg/m^3)	2970	500	2508.30	2575.3305	422.851
Porosity (%)	84	0.83	5.91	2.354	12.755

2.1. Input Parameters for Models

During the preliminary steps of the analysis, using polynomial and power functions, a series of single-variable regressions were conducted to obtain more details about the relationship between TC and the independent rock properties. Table 2 summarizes the results of the single-variable regressions. As seen, the P-wave velocity and density had the highest and the lowest effects on the TC, respectively.

Table 2. Correlation coefficients of the simple regression models between the TC and the independent variables.

	P-Wave	Porosity	Density	UCS
Polynomial (1st order)	0.8249	−0.1717	0.3517	0.5463
Polynomial (2nd order)	0.8597	−0.2841	0.5058	0.5928
Power (one term)	0.8587	−0.5912	0.5113	0.5621
Power (two term)	0.8602	−0.614	0.5113	0.5942

To avoid redundancy in the future model generation, the relationships between the independent variables were investigated. As shown in Table 3, there were no redundancies between the independent parameters, and the relationships between porosity and density as well as between UCS and P-wave velocity were significant.

Table 3. Correlation coefficients of the relationships between the independent variables.

	P-Wave	Porosity	Density	UCS
P-wave	1	−0.2441	+0.3876	+0.4911
Porosity		1	−0.8094	−0.2578
Density			1	+0.4295
UCS				1

2.1.1. P-Wave Velocity

Interpreting the relationship between P-wave velocity and TC requires considering the parameters that influence P-wave velocity. Among the most-cited variables that affect P-wave velocity, mineral composition, lithology, porosity, and confining stress are cited as the main factors. Gegenhuber and Schoen [56] studied the relationship between TC and the P-wave velocity for different rock types. They found that TC and P-wave velocity had a positive and robust correlation coefficient. The mineral composition was the most effective parameter influencing their relationship. Figure 1 shows the relationship between TC and P-wave and the mineral composition of rocks. There was no information found about the mineral composition of samples in the current study. As a result, the curve fitted to TC versus P-wave velocity did not include this factor (Figure 2).The effect of lithology on seismic wave velocity in rocks was well-established by Domenico [57]. According to him, a higher seismic velocity can indicate higher quartz content, which can result in higher UCS. In addition, more significant confining stress can close microcracks in samples and improve P-wave transmission. As porosity decreases, P-wave and TC values increase. Table 3 also demonstrates this phenomenon by showing that P-wave velocity has an inverse relationship with porosity. As a result, P-wave velocity can inherently represent the most influential parameter for TC as it has a direct relationship with TC.

Furthermore, according to Freund [58], the type of pore fluid influences P-waves in porous rocks. In our study, the samples were completely dry. However, this method can be used in future studies to evaluate TC in saturated conditions.

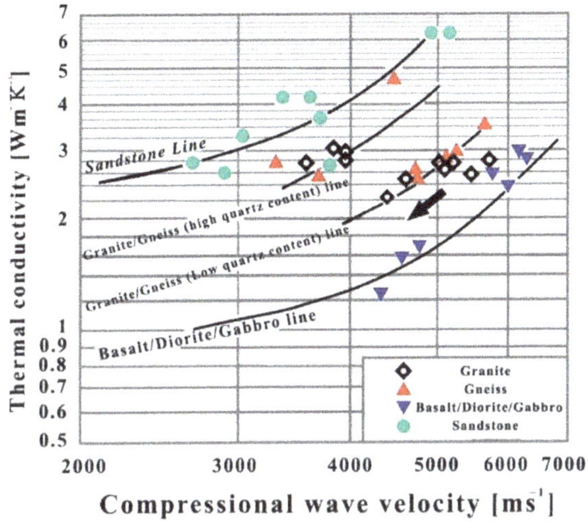

Figure 1. The points show the measured values, and the curve shows the relationship between TC and P-wave velocity. The grey arrow indicates the increase of the porosity.

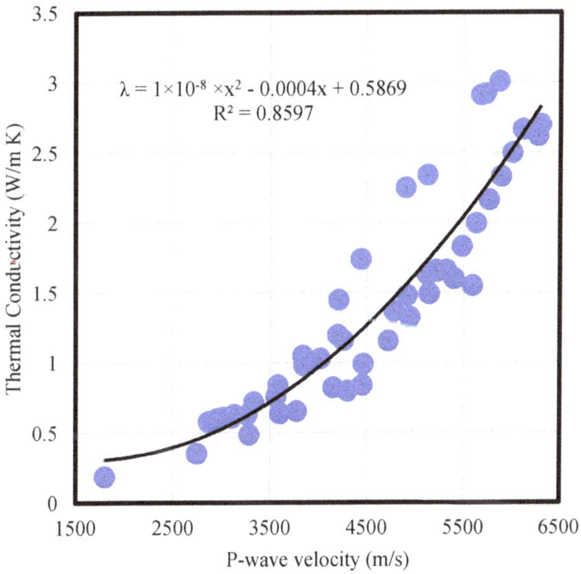

Figure 2. Measured values and relationship between TC and P-wave velocity.

2.1.2. Porosity of Rocks

As shown in Table 2, the porosity had an inverse and significant relationship with TC, which represented the more effective parameter for the TC of sedimentary rocks [59]. The void ratio of sedimentary rocks (from 1% up to 80%) could be higher than that of the volcanic and metamorphic ones (at most 1%). The wide range of substances that could fill the void spaces, such as air, water, chemical sediments, or organic matters, was the other parameter controlling TC. However, the reported datasets of the TC rocks, which were used in our study, were obtained in the laboratory, considering dry conditions and the ISRM standards.

2.1.3. Density of Rocks

Birch [60] obtained a direct relationship between P-wave and density. Horai [61] examined 166 rock-forming minerals to find an empirical equation for the TC prediction. They found an analogous relationship between increasing TC and increasing the density, as well as increasing the P-wave velocity. In our study, in addition to the results obtained by Horai (1971), another important issue was observed. Figure 3 shows the TC–density plot for silica minerals by Horai [61], whereas Figures 2 and 4 present the TC–density and TC–P-wave plots obtained in this study. The trend line of these plots reached a density of about 2500 kg/m^3, showing that there was a low gradient, while it increased for greater values of density. As observed in Figure 5, the density of rocks increased with the porosity decrease, while both of them had an extreme effect on the TC and P-wave velocity.

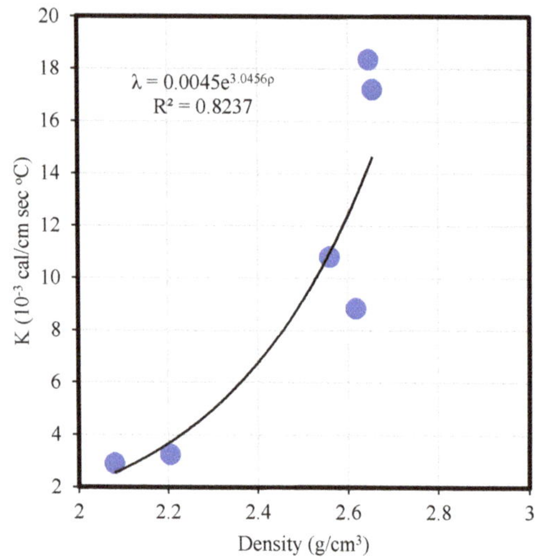

Figure 3. TC versus density for silica minerals.

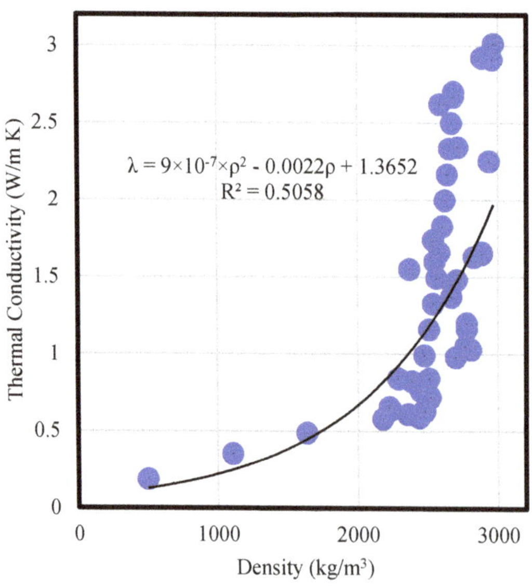

Figure 4. Measured values of and the relationship between TC and density.

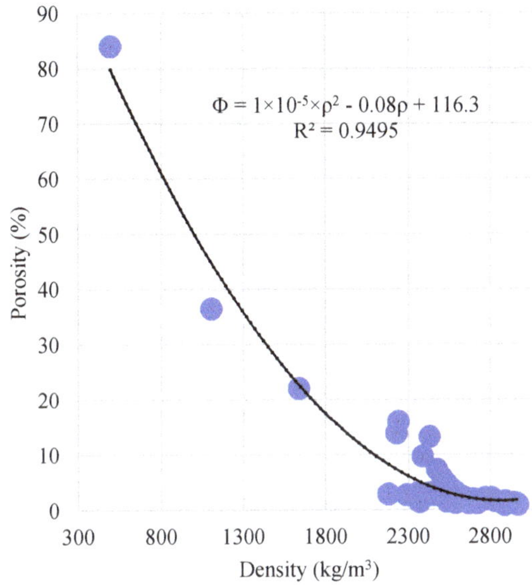

Figure 5. Measured values of and the relationship between porosity and density.

2.1.4. Uniaxial Compressive Strength

The strength of rocks is affected by their mineralogy, grain size, and porosity. Rocks with larger grain sizes and considerable porosity have lower UCS than that of other dense rocks [2]. Furthermore, an increase in quartz content will increase the strength of rocks as well as the TC (Clauser and Huenges [59] and Figure 6). Hence, the greater the UCS is, the larger the rocks' TC is (Figure 7). Sargam, Wang, and Cho [43] found that concrete mixtures with higher quartz content had higher TC and compressive strength.

Different groups of researchers (Pimienta et al. [62]; Esteban, Pimienta, Sarout, Delle, Piane, Haffen, Geraud, and Timms [30]; and El Sayed and El Sayed [12]) proposed models for TC prediction using only the P-wave velocity and porosity. Although their models had an acceptable degree of accuracy, the discussions on the role of the rock properties in TC prediction indicated that these properties were not enough. Other rock properties can also be interesting and give accurate predictions of TC (rock type, TC, UCS, density, saturating fluid characteristic, quality, and geometry of the contact between the grains).

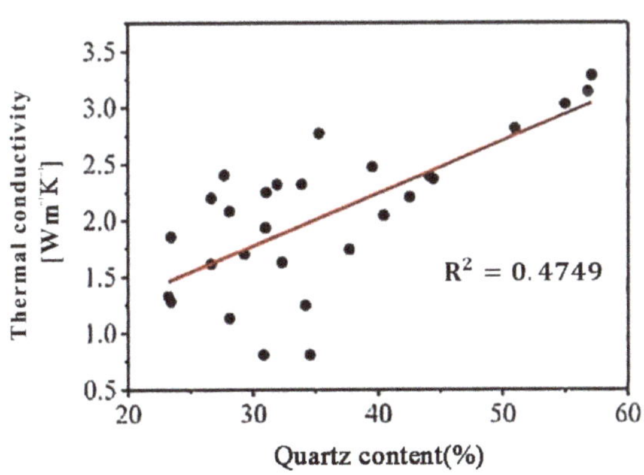

Figure 6. Thermal conductivity of basic rock-forming minerals and their compositional relationship with rocks for volcanic and sedimentary rocks.

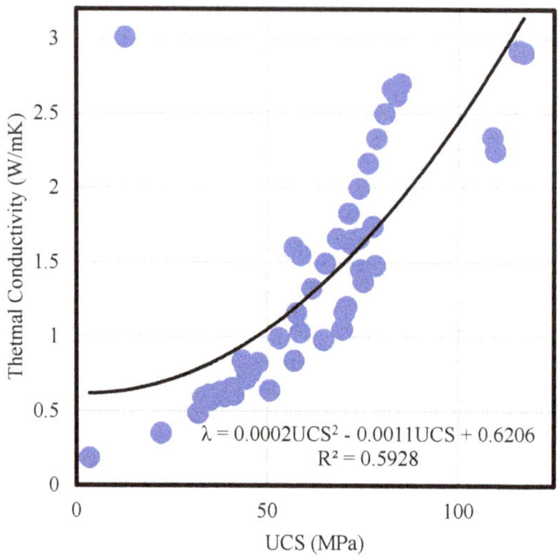

Figure 7. Measured values and relationship between TC and UCS.

3. Gene Expression Programming (GEP)

The genetic algorithm (GA) introduced by J. Holland [63] as a new stochastic optimization technique was utilized in this study, in which Darwin's theory of "survival of the fittest" was also used. This algorithm attempts to use genetic operators and a fitness function in each round of processing to optimize and classify the set of parameters that are the best solutions for the problem. Whenever the output requirements (e.g., the required proportion between measured and predicted values) are met, the algorithm is stopped [64]. A newer developed version of GA is genetic programming (GP). It was first introduced by Koza [65]. GP finds a solution for problems using variable-length sets of parameters, including mathematical functions, algebraic operators (function sets), and numbers (terminal sets). Ferreira [53] introduced gene expression programming (GEP) as a newer version of GP in which individuals are characterized as linear strings. The five central units that constitute a GEP algorithm are terminal sets, function sets, fitness functions, operators, and stop conditions [66]. Moreover, expression trees (ETs) were used to demonstrate fixed-length solutions in tree shape structures.

The fixed-length chromosome is the most obvious difference between the GEP and GP. The genomes or chromosomes are linear, symbolic strings of one or more genes with a fixed length. The genes themselves are composed of primitives (mathematical functions or variables), which are all fixed-length strings.

As shown in Figure 8, in the initial step, GEP randomly generated a series of chromosome sets that were potential solutions for the problem. In the second step, a set of chromosomes was expressed as ET. Meanwhile, ETs are interpretable as mathematical equations. Afterward, a fitness function (which is responsible for calculating the errors in predictions) evaluated the fitness of each set of parameters. If the expected results were not met, the best solutions would be selected by a selector function, and the genetic operators would combine them to generate a better set of parameters. In the following paragraphs, we will describe the most common operators.

Mutation: Since it enables immediate changes to the program output, the mutation operator is the most crucial one in the GEP algorithm. To put it another way, it changes a terminal node into a functional node and vice versa. By choosing two distinct subtrees, switching them, and then choosing another subtree, two distinct subtrees from different chromosomes are swapped. Gene inversion: Using this operator, a specific set of genes in the chromosome's head is inverted. Transposition: With the help of this operator, a chromosome's chosen portion is moved to a different location. Among the different types of transposition operators, there are the (1) insertion sequence transposition (IS), (2) root insertion sequence transposition (RIS), and (3) gene transposition.

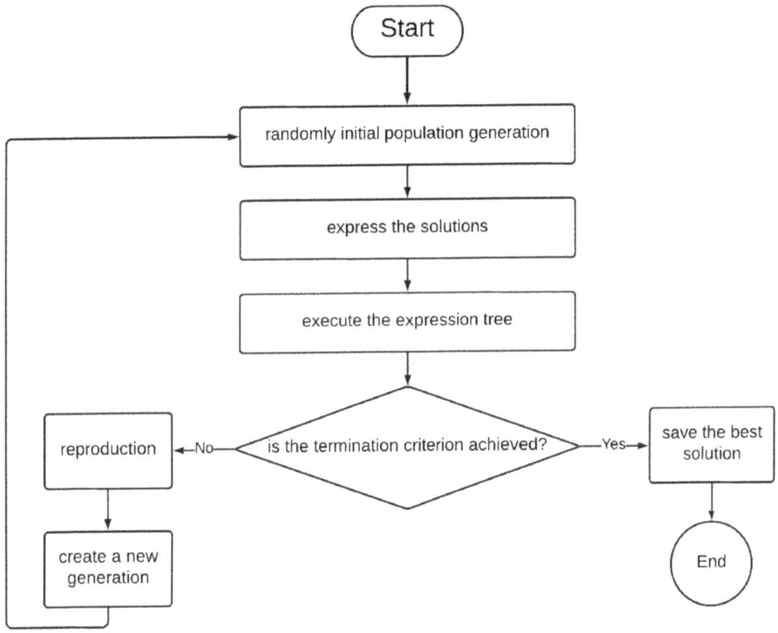

Figure 8. Flowchart of the GEP algorithm employed in our study.

Finally, if the newly created set of parameters did not have the expected fitness, the process would be repeated to reach the stopping conditions [67].

4. Results

Two empirical equations were developed in our study to predict TC using the physicomechanical properties of rocks (i.e., P-wave, porosity, density, and UCS). The first model was developed using non-linear multivariable regression (NLMR), and another model was developed using gene expression programming (GEP). To avoid overfitting during the training stages, 65% of all datasets were randomly selected for training, and the rest were used for testing the developed models. The testing results of the developed models showed how much these models could be generalized for other datasets. Furthermore, statistical indices (i.e., the coefficient of determination (R^2) and the root mean square error (RMSE)) were used to evaluate the robustness of the developed models for prediction. The R-square (Equation (2)) showed how reliable a model was for future forecasts, and the RMSE (Equation (3)) showed how much the standard deviation of the residuals was (prediction errors).

$$R^2 = 1 - \frac{\sum\left(y_{act} - y_{pre}\right)^2}{\sum\left(y_{act} - \overline{y}_{act}\right)^2} \quad (2)$$

$$RMSE = \sqrt{\frac{\sum_{i=1}^{n}\left(y_{pre} - y_{act}\right)^2}{n}} \quad (3)$$

where y_{pre} is the predicted value, y_{act} is the measured (actual) value, and $\overline{y}_{act}$ is the average of the measured values.

4.1. NLMR Model

Unlike linear regression, the NLMR model could use a wide variety of mathematical functions to find the best fitting equation between the input and output parameters [54]. However, to avoid model complexity, only polynomial and power functions were used for the model development. Based on Table 2, the NLMR functions were generated by combining non-linear single variable equations for each parameter. An optimization algorithm was used to optimize the NLMR function to achieve the best TC prediction model. The NMLR model was developed and optimized by a genetic algorithm (GA) using MATLAB software [68]. GA's general procedures are similar to GEP's, which generate random coefficients for parameters, enhance them by their operators, and try to reach a higher degree of accuracy. In order to determine the most suitable setting for the GA run, a series of trial-and-error tests were conducted. As a result, with NLMR functions, the most accurate TC prediction model was found to be Equation (4). Table 4 shows the performance indices of this model.

$$\lambda = 0.122 \times {V_p}^{-0.033} + 0.0013 \times V_p - 0.024 \times \phi^{-0.18} - 0.0177 \times \rho^{-0.0213} - 0.1171 \times UCS^{0.13} + 0.02153 \quad (4)$$

Table 4. Resulting performance indices' values of the proposed NLMR model.

Model	Data Status	R^2		RMSE	
		Training	Testing	Training	Testing
NLMR	65% of all for training 35% of all for testing	0.86	0.83	0.27	0.31

4.2. GEP Model

The GEP model was developed through the GenXPro Tools software. This software extracts the significant features of datasets, including a high number of variables, and finds relationships among them with high accuracy. Similar to the process used in Section 4.1, the same training and testing subsets were used for the GEP model's development. To develop a model for TC prediction by GEP, a simple equation in the form of $TC = f(P\text{-}wave,\ porosity,\ UCS,\ density)$ was first proposed. To acquire the best setting for the GEP model generation in the initial step, function sets and fitness functions were then chosen from the study of Zare Naghadehi et al. [69]. Afterward, several trial-and-error procedures were carried out to obtain the best settings. Having utilized these settings, the GEP-developed models obtained minimal error percentages. Table 5 presents the GEP software settings for the model generation during this study. The procedure of GEP modeling was illustrated in Section 3.

Table 5. Parameters of the GEP model.

Parameter	Value
Number of chromosomes	30
Head size	8
Number of genes	4
Linking function	Addition
Fitness function	RMSE
Mutation rate	0.044
Inversion rate	0.1
One-point recombination rate	0.3
Two-point recombination rate	0.3
Gene recombination rate	0.1
Insertion sequences transposition rate	0.1
Root insertion sequence transposition rate	0.1
Gene transposition rate	0.1
Function set	$+, -, \times, \div, \sqrt{x}, e^x$, sin, cos

The prediction of the GEP performance models was evaluated by both R-square and RMSE. Several models were developed to find a better model with the lowest RMSE and highest R-square. Table 6 gives the prediction performance of the selected GEP model.

Table 6. Resulting performance indices' values of the proposed GEP model.

Model	Data Status	R^2		RMSE	
		Training	Testing	Training	Testing
GEP	65% of all for training 35% of all for testing	0.95	0.90	0.17	0.22

The developed models are presented in terms of expression trees (ETs) or as computer codes. However, these presentations should be interpreted as a form of a mathematical equation. The extraction of the mathematical equation from ETs is an easy task. The ETs are read from left to right and bottom to the top. The ET of each gene of the GEP model is shown in Figure 9a–d, and the mathematical equation of each gene is presented as Equations (5)–(8). The genes' equations were linked, and the GEP model was generated using Equation (9).

$$\text{SubET1} = \cos\sqrt[3]{\phi \times \cos\left(\frac{\rho + \text{UCS} - V_P}{10.61 - \rho}\right)^2} \tag{5}$$

$$\text{SubET2} = \left(\cos\left(\cos\left(0.973 - \frac{V_P + \text{UCS}}{\rho - \phi}\right)\right)\right)^2 \tag{6}$$

$$\text{SubET3} = \frac{1}{\sqrt[3]{\phi^2}} + \frac{1}{\phi^3} \tag{7}$$

$$\text{SubET4} = \cos\left(\cos\left(-1 - \sqrt[3]{\tan(2\rho) - \text{UCS} + 1.71}\right)\right) \tag{8}$$

$$\lambda = \text{SubET1} + \text{SubET2} + \text{SubET3} + \text{SubET4} \tag{9}$$

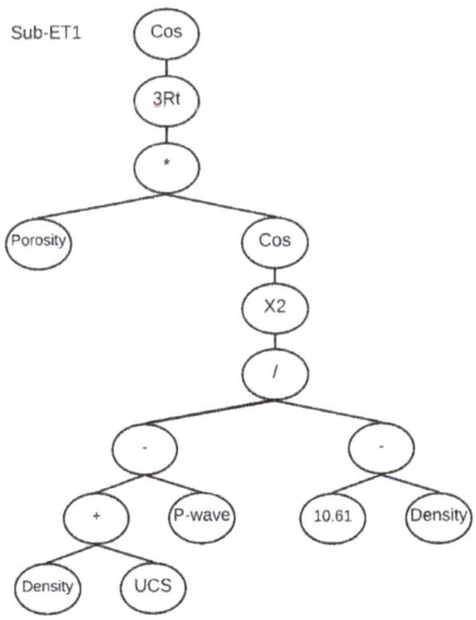

(a)

Figure 9. *Cont.*

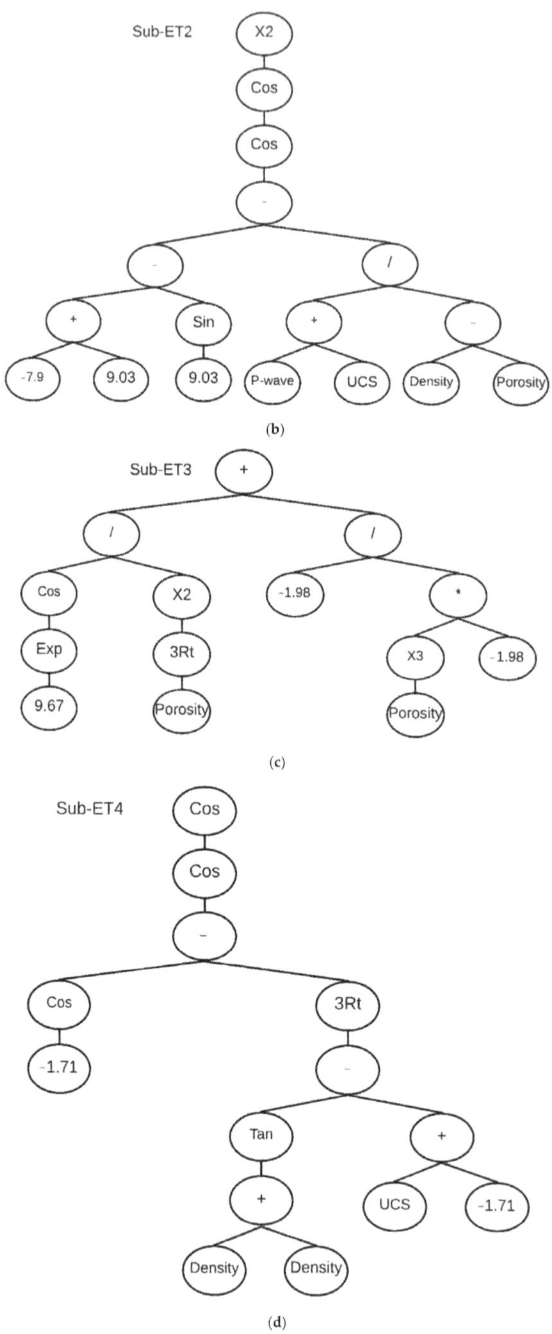

Figure 9. (**a**) Expression tree of each gene of the GEP model for TC prediction (Equation (5)). (**b**) Expression tree of each gene of the GEP model for TC prediction (Equation (6)). (**c**) Expression tree of each gene of the GEP model for TC prediction (Equation (7)). (**d**) Expression tree of each gene of the GEP model for TC prediction (Equation (8)).

A better illustration of the predicted values of TC by the GEP model against the measured TC for training and testing subsets are shown in Figures 10 and 11, respectively.

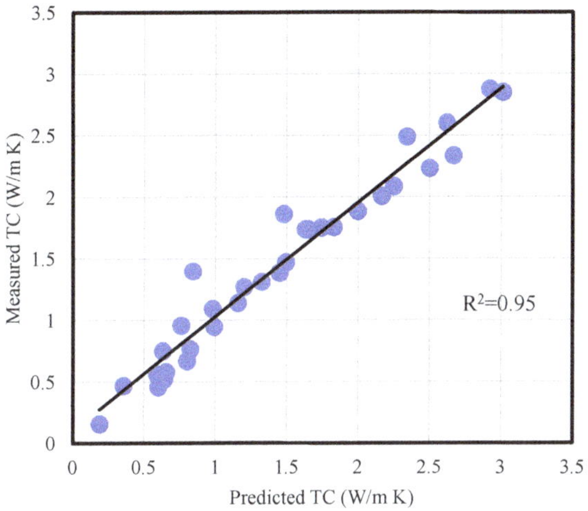

Figure 10. The correlation coefficient of the GEP model for the 65% training subsets.

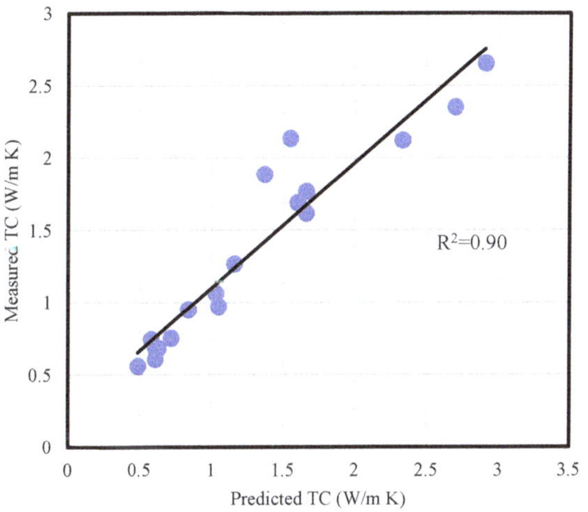

Figure 11. The correlation coefficient of the GEP model for the 35% testing subsets.

4.3. *Verification and Discussion of the Results*

To evaluate the prediction reliability of the developed models, the corresponding performance indices were compared with those of previously published studies. For the prediction of the TC using the same parameters and datasets as those used in our study, two multivariable equations were developed by Khandelwal [24] (Equation (10)) and Hajihassani, Marto, Khezri, and Kalatehjari [47] (Equation (11)); hence, these two models were chosen for the verification of the newly developed models. It is noteworthy that there are also some other efforts in the literature to predict TC by computer-aided methods using the same datasets [3,24,70,71].

$$\lambda = -1.1864 + 0.006\ \text{UCS} + 0.1493 \times \rho + 0.0134\ \phi + 0.0004\ V_p \tag{10}$$

$$\lambda = 0.00037\ V_p - 0.01653\ \phi - 0.00058\ \rho + 0.02053\ \text{UCS} - 0.06072 \tag{11}$$

These studies used some kinds of methods called black boxes, but these methods did not have the practical potential to be utilized in applications. In the meantime, as is evident from Equations (10) and (11), in the developed mathematical equations, only simple functions are used, but to represent the effect of parameters on TC, complex non-linear functions are needed. To avoid any confusion made by the selection of training and testing subsets, the performance indicators of all models were calculated again by substituting the input variables using all 50 datasets. Thus, the accuracy of the previous model could be compared exactly to the proposed one in this study. The results of the performance indicators of the proposed models for all 50 datasets are listed in Table 7.

Table 7. Performance indices and ranking of the new and the previous TC predictor models calculated for all 50 datasets.

Model	R^2	RMSE	Rank
New GEP model (Equation (9))	0.92	0.21	1
New NLMR model (Equation (4))	0.82	0.32	2
MVRA [47] (Equation (10))	0.74	0.39	3
MVRA [24,71] (Equation (11))	0.35	0.62	4

Comparing the performance indices of the proposed models and the previously published model revealed that the GEP and NLMR models produced more accurate predictions than the MVRA model did. The high value of R-square (equal to 0.95) and the low value of RMSE (equal to 0.17) confirmed the higher accuracy of the proposed GEP model. The GEP was more accurate since it used longer terms and a wider variety of mathematical functions than the NLMR one did. The existing MVRAs in the literature [24,47,71] were developed using simple non-linear variables that were regulated using commercial software. Based on the low correlation coefficients of these models, we concluded that they could not capture the complexity of the problem and the relationship between TC and the influential parameters. The NLMR model and the GEP model developed in our study were developed not only using non-linear mathematical terms to represent each parameter but also via an extensive trial and error process; thus, a soft computing approach was used to enhance the accuracy of predictions.

5. Conclusions

In our study, a literature survey was performed to the establishment of the dataset for estimating the thermal conductivity of rock via several rock properties, including the UCS, density, porosity, and P-wave velocity of rocks. Several simple variable regressions were conducted among the rocks' properties and TC. As a result of simple regressions, we found that the P-wave velocity and density had the highest and the lowest effect on the TC, respectively. Further, the relationships between porosity and density and between UCS and P-wave velocity were then considered significant and meaningful. To estimate TC via rock properties, two empirical equations were developed. First, a model was developed using NLMR, and then a second model was developed by GEP. The two statistical indices R^2 and RMSE were utilized to evaluate the robustness of the developed models in order to predict TC. While the GEP model had a higher value of the R^2 (0.95) and a lower value of RMSE (0.17), a low R^2 (0.82) and RMSE (0.32) were obtained for NLMR. A comparison of the performance indices of the proposed models and of the previously published models revealed that the GEP and NLMR models were able to produce more accurate predictions. As a result, the developed model can be used for estimating the TC of rocks, since performing TC-related tests might be time-consuming and cost-restrictive. Additionally, despite the fact that our study proposed methods and mathematical models that significantly increased prediction accuracy, there are some associated limitations. As mentioned in previous sections, a limited number of datasets were used in our study because data collection in geosciences is challenging. As a result, we recommend that future studies focus on rocks and parameter ranges that overlap only slightly with those of this study. Our study and similar studies have not quantified the texture of the rocks, which is one of the most important parameters. The data collection and analysis stages of future studies should incorporate this parameter.

Author Contributions: Conceptualization, M.S. and S.Y.; methodology, M.S. and S.Y.; software, M.S.; formal analysis, M.S., A.A. and S.Y.; resources, M.S., A.A. and S.Y.; data curation, M.S. and S.Y.; writing—original draft, M.S., T.M., A.A., S.Y. and M.M.S.S.; writing—review and editing, M.S., T.M., A.A., S.Y. and M.M.S.S.; supervision, S.Y.; funding acquisition, M.M.S.S. All authors have read and agreed to the published version of the manuscript.

Funding: The research was partially funded by the Ministry of Science and Higher Education of the Russian Federation under the strategic academic leadership program 'Priority 2030' (Agreement 075-15-2021-1333, dated 30 September 2021).

Institutional Review Board Statement: Not applicable.

Informed Consent Statement: Not applicable.

Data Availability Statement: The data presented in this study are available on request from the corresponding author.

Conflicts of Interest: The authors declare no conflict of interest.

References

1. Kardani, N.; Bardhan, A.; Samui, P.; Nazem, M.; Zhou, A.; Armaghani, D.J. A novel technique based on the improved firefly algorithm coupled with extreme learning machine (ELM-IFF) for predicting the thermal conductivity of soil. *Eng. Comput.* **2021**, *38*, 3321–3340. [CrossRef]
2. Özkahraman, H.T.; Selver, R.; Isık, E.C. Determination of the thermal conductivity of rock from P-wave velocity. *Int. J. Rock Mech. Min. Sci.* **2004**, *4*, 703–708. [CrossRef]
3. Singh, T.N.; Sinha, S.; Singh, V.K. Prediction of thermal conductivity of rock through physico-mechanical properties. *Build. Environ.* **2007**, *42*, 146–155. [CrossRef]
4. Yaşar, E.; Erdoğan, Y.; Güneyli, H. Determination of the thermal conductivity from physico-mechanical properties. *Bull. Eng. Geol. Environ.* **2008**, *67*, 219–225. [CrossRef]
5. Li, Z.-W.; Feng, X.-T.; Zhang, Y.-J.; Gong, Y.-H.; Zhu, G.-Q. Effect of mechanical damage on the thermal conductivity of granite. *Geothermics* **2020**, *53*, 1039–1051. [CrossRef]
6. Abdolhosseinzadeh, A.; Samui, P.; Samaei, M.; Garousi, A. Numerical analysis of bearing capacity of circular footing reinforced with geogrid layers. *Arab. J. Geosci.* **2022**, *15*, 750. [CrossRef]
7. *ASTM-C1045-97;* Practice for the Calculation of Thermal Transmission Properties from Steady State Heat Flux Measurements. ASTM International: West Conshohocken, PA, USA, 1990. [CrossRef]
8. Pribnow, D.; Umsonst, T. Estimation of thermal conductivity from the mineral composition: Influence of fabric and anisotropy. *Geophys. Res. Lett.* **1993**, *20*, 2199–2202. [CrossRef]
9. Hartmann, A.; Rath, V.; Clauser, C. Thermal conductivity from core and well log data. *Int. J. Rock Mech. Min. Sci.* **2005**, *42*, 1042–1055. [CrossRef]
10. Fuchs, S.; Förster, A. Rock thermal conductivity of Mesozoic geothermal aquifers in the Northeast German Basin. *Geochemistry* **2010**, *70* Suppl. 3, 13–22. [CrossRef]
11. Sundberg, J.; Back, P.-E.; Ericsson, L.O.; Wrafter, J. Estimation of thermal conductivity and its spatial variability in igneous rocks from in situ density logging. *Int. J. Rock Mech. Min. Sci.* **2009**, *46*, 1023–1028. [CrossRef]
12. El Sayed, A.M.A.; El Sayed, N.A. Thermal conductivity calculation from P-wave velocity and porosity assessment for sandstone reservoir rocks. *Geothermics* **2019**, *82*, 91–96. [CrossRef]
13. Pimienta, L.; Klitzsch, N.; Clauser, C. Comparison of thermal and elastic properties of sandstones: Experiments and theoretical insights. *Geothermics* **2018**, *76*, 60–73. [CrossRef]
14. Zhao, X.; Xu, H.; Zhao, Z.; Guo, Z.; Cai, M.; Wang, J. Thermal conductivity of thermally damaged Beishan granite under uniaxial compression. *Int. J. Rock Mech. Min. Sci.* **2019**, *115*, 121–136. [CrossRef]
15. Zimmerman, R.W. Thermal conductivity of fluid-saturated rocks. *J. Pet. Sci. Eng.* **1989**, *3*, 219–227. [CrossRef]
16. Revil, A. Thermal conductivity of unconsolidated sediments with geophysical applications. *J. Geophys. Res.* **2000**, *105*, 16749–16768. [CrossRef]
17. Jougnot, D.; Revil, A. Thermal conductivity of unsaturated clay-rocks. *Hydrol. Earth Syst. Sci.* **2010**, *14*, 91–98. [CrossRef]
18. Côté, J.; Konrad, J.-M. Assessment of structure effects on the thermal conductivity of two-phase porous geomaterials. *Int. J. Heat Mass Transf.* **2009**, *52*, 796–804. [CrossRef]
19. Li, Z.-W.; Long, M.-C.; Feng, X.-T.; Zhang, Y.-J. Thermal damage effect on the thermal conductivity inhomogeneity of granite. *Int. J. Rock Mech. Min. Sci.* **2021**, *138*, 104583. [CrossRef]

20. Vosteen, H.D.; Schellschmidt, R. Influence of temperature on thermal conductivity, thermal capacity and thermal diffusivity for different types of rock. *Phys. Chem. Earth* **2003**, *28*, 499–509. [CrossRef]
21. Urquhart, A.; Bauer, S. Experimental determination of single-crystal halite thermal conductivity, diffusivity and specific heat from −75 °C to 300 °C. *Int. J. Rock Mech. Min. Sci.* **2015**, *78*, 350–352. [CrossRef]
22. Zhao, X.; Zhao, Z.; Guo, Z.; Cai, M.; Li, X.; Li, P.; Chen, L.; Wang, J. Influence of thermal treatment on the thermal conductivity of Beishan granite. *Rock Mech. Rock Eng.* **2018**, *51*, 2055–2074. [CrossRef]
23. Chen, C.; Zhu, C.; Zhang, B.; Tang, B.; Li, K.; Li, W.; Fu, X. Effect of Temperature on the Thermal Conductivity of Rocks and Its Implication for In Situ Correction. *Geofluids* **2021**, *2021*, 6630236. [CrossRef]
24. Khandelwal, M. Prediction of thermal conductivity of rocks by soft computing. *Int. J. Earth Sci.* **2011**, *100*, 1383–1389. [CrossRef]
25. Garitte, B.; Gens, A.; Vaunat, J.; Armand, G. Thermal conductivity of argillaceous rocks: Determination methodology using in situ heating tests. *Rock Mech. Rock Eng.* **2014**, *47*, 111–129. [CrossRef]
26. Clauser, C. Thermal storage and transport properties of rocks, II: Thermal conductivity and diffusivity. In *Encyclopedia of Solid Earth Geophysics*; Springer: Dordrecht, Germany, 2011; pp. 1431–1448. [CrossRef]
27. Popov, E.; Trofimov, A.; Goncharov, A.; Abaimov, S.; Chekhonin, E.; Popov, Y.; Sevostianov, I. Technique of rock thermal conductivity evaluation on core cuttings and non-consolidated rocks. *Int. J. Rock Mech. Min. Sci.* **2018**, *108*, 15–22. [CrossRef]
28. Gu, Y.; Rühaak, W.; Bär, K.; Sass, I. Using seismic data to estimate the spatial distribution of rock thermal conductivity at reservoir scale. *Geothermics* **2017**, *66*, 61–72. [CrossRef]
29. Griffiths, C.M.; Brereton, N.R.; Beausillon, R.; Castillo, D. Thermal conductivity prediction from petrophysical data: A case study. *Geol. Soc. Lond. Spec. Publ.* **1992**, *65*, 299–315. [CrossRef]
30. Esteban, L.; Pimienta, L.; Sarout, J.; Delle Piane, C.; Haffen, S.; Geraud, Y.; Timms, N.E. Study cases of thermal conductivity prediction from P-wave velocity and porosity. *Geothermics* **2015**, *53*, 255–269. [CrossRef]
31. Boulanouar, A.; Rahmouni, A.; Boukalouch, M.; Samaouali, A.; Géraud, Y.; Harnafi, M.; Sebbani, J. Determination of thermal conductivity and porosity of building stone from ultrasonic velocity measurements. *Geomaterials* **2013**, *3*, 138–144. [CrossRef]
32. Ray, R.; Kumar, D.; Samui, P.; Roy, L.B.; Goh, A.T.C.; Zhang, W. Application of soft computing techniques for shallow foundation reliability in geotechnical engineering. *Geosci. Front.* **2021**, *12*, 375–383. [CrossRef]
33. Sarir, P.; Armaghani, D.J.; Jiang, H.; Sabri, M.M.S.; He, B.; Ulrikh, D.V. Prediction of Bearing Capacity of the Square Concrete-Filled Steel Tube Columns: An Application of Metaheuristic-Based Neural Network Models. *Materials* **2022**, *15*, 3309. [CrossRef] [PubMed]
34. Wu, D.; Wu, C. Research on the Time-Dependent Split Delivery Green Vehicle Routing Problem for Fresh Agricultural Products with Multiple Time Windows. *Agriculture* **2022**, *12*, 793. [CrossRef]
35. Chen, H.; Miao, F.; Chen, Y.; Xiong, Y.; Chen, T. A hyperspectral image classification method using multifeature vectors and optimized KELM. *IEEE J. Sel. Top. Appl. Earth Obs. Remote Sens.* **2021**, *14*, 2781–2795. [CrossRef]
36. Zhou, X.; Ma, H.; Gu, J.; Chen, H.; Deng, W. Parameter adaptation-based ant colony optimization with dynamic hybrid mechanism. *Eng. Appl. Artif. Intell.* **2022**, *114*, 105139. [CrossRef]
37. An, Z.; Wang, X.; Li, B.; Xiang, Z.; Zhang, B. Robust visual tracking for UAVs with dynamic feature weight selection. *Appl. Intell.* **2022**, 1–14. [CrossRef]
38. Dai, L.; Wu, X.; Zhou, M.; Ahmad, W.; Ali, M.; Sabri, M.M.S.; Salmi, A.; Ewais, D.Y.Z. Using Machine Learning Algorithms to Estimate the Compressive Property of High Strength Fiber Reinforced Concrete. *Materials* **2022**, *15*, 4450. [CrossRef]
39. Hajihassani, M.; Jahed Armaghani, D.; Kalatehjari, R. Applications of particle swarm optimization in geotechnical engineering: A comprehensive review. *Geotech. Geol. Eng.* **2018**, *36*, 705–722. [CrossRef]
40. Zangoei, A.; Monjezi, M.; Armaghani, D.J.; Mehrdanesh, A.; Ahmadian, S. Prediction and optimization of flyrock and oversize boulder induced by mine blasting using artificial intelligence techniques. *Environ. Earth Sci.* **2022**, *81*, 359. [CrossRef]
41. Huat, C.Y.; Moosavi, S.M.H.; Mohammed, A.S.; Armaghani, D.J.; Ulrikh, D.V.; Monjezi, M.; Hin Lai, S. Factors influencing pile friction bearing capacity: Proposing a novel procedure based on gradient boosted tree technique. *Sustainability* **2021**, *13*, 11862. [CrossRef]
42. Assouline, D.; Mohajeri, N.; Gudmundsson, A.; Scartetzzini, J.-L. Combining fourier analysis and machine learning to estimate the shallow-ground thermal diffusivity in Switzerland. In Proceedings of the IGARSS 2018—2018 IEEE International Geoscience and Remote Sensing Symposium, Valencia, Spain, 22–27 July 2018; pp. 1144–1147.
43. Sargam, Y.; Wang, K.; Cho, I.H. Machine learning based prediction model for thermal conductivity of concrete. *J. Build. Eng.* **2021**, *34*, 101956. [CrossRef]
44. Verma, A.; Jha, M.K.; Gautam, P.; Mishra, A.K.; Vardhan, H.; Singh, S.K. Prediction of thermal conductivity and damage in Indian Jalore granite for design of underground research laboratory. *Neural Comput. Appl.* **2021**, *33*, 13183–13192. [CrossRef]
45. Wang, M.; Wang, H.; Yin, Y.; Rahardja, S.; Qu, Z. Temperature field prediction for various porous media considering variable boundary conditions using deep learning method. *Int. Commun. Heat Mass Transf.* **2022**, *132*, 105916. [CrossRef]
46. He, J.; Li, K.; Wang, X.; Gao, N.; Mao, X.; Jia, L. A Machine Learning Methodology for Predicting Geothermal Heat Flow in the Bohai Bay Basin, China. *Nat. Resour. Res.* **2022**, *31*, 237–260. [CrossRef]
47. Hajihassani, M.; Marto, A.; Khezri, N.; Kalatehjari, R. Indirect measure of thermal conductivity of rocks through adaptive neuro-fuzzy inference system and multivariate regression analysis. *Meas. J. Int. Meas. Confed.* **2015**, *67*, 71–77. [CrossRef]

48. Kang, J.; Yu, Z.; Wu, S.; Zhang, Y.; Gao, P. Feasibility analysis of extreme learning machine for predicting thermal conductivity of rocks. *Environ. Earth Sci.* **2021**, *80*, 455. [CrossRef]
49. Barkhordari, M.S.; Armaghani, D.J.; Sabri, M.M.S.; Ulrikh, D.V.; Ahmad, M. The Efficiency of Hybrid Intelligent Models in Predicting Fiber-Reinforced Polymer Concrete Interfacial-Bond Strength. *Materials* **2022**, *15*, 3019. [CrossRef]
50. Huang, J.; Zhou, M.; Sabri, M.M.S.; Yuan, H. A novel neural computing model applied to estimate the dynamic modulus (dm) of asphalt mixtures by the improved beetle antennae search. *Sustainability* **2022**, *14*, 5938. [CrossRef]
51. Tu, J.V. Advantages and disadvantages of using artificial neural networks versus logistic regression for predicting medical outcomes. *J. Clin. Epidemiol.* **1996**, *49*, 1225–1231. [CrossRef]
52. Huang, J.; Zhou, M.; Yuan, H.; Sabri, M.M.S.; Li, X. Prediction of the compressive strength for cement-based materials with metakaolin based on the hybrid machine learning method. *Materials* **2022**, *15*, 3500. [CrossRef]
53. Ferreira, C. Gene Expression Programming: A New Adaptive Algorithm for Solving Problems. *arXiv Preprint* **2001**, arXiv:cs/0102027.
54. Samaie, M.; Ranjbarnia, M.; Zare Naghadehi, M. Prediction of the Rock Brittleness Index Using Nonlinear Multivariable Regression and the CART Regression Tree. *J. Civ. Environ. Eng.* **2018**, *48*, 33–40.
55. Yamaguchi, U. The number of test-pieces required to determine the strength of rock. *Int. J. Rock Mech. Min. Sci. Geomech. Abstr.* **1970**, *7*, 209–227. [CrossRef]
56. Gegenhuber, N.; Schoen, J. New approaches for the relationship between compressional wave velocity and thermal conductivity. *J. Appl. Geophys.* **2012**, *76*, 50–55. [CrossRef]
57. Domenico, S.N. Rock lithology and porosity determination from shear and compressional wave velocity. *Geophysics* **1984**, *49*, 1188–1195. [CrossRef]
58. Freund, D. Ultrasonic compressional and shear velocities in dry clastic rocks as a function of porosity, clay content, and confining pressure. *Geophys. J. Int.* **1992**, *108*, 125–135. [CrossRef]
59. Clauser, C.; Huenges, E. Thermal conductivity of rocks and minerals. In *Rock Physics & Phase Relations: A Handbook of Physical Constants*; John Wiley & Sons, Inc.: New York, , NY, USA, 1995; Volume 3, pp. 105–126. [CrossRef]
60. Birch, F. The velocity of compressional waves in rocks to 10 kilobars: 2. *J. Geophys. Res.* **1961**, *66*, 2199–2224. [CrossRef]
61. Horai, K.i. Thermal conductivity of rock-forming minerals. *J. Geophys. Res.* **1971**, *76*, 1278–1308. [CrossRef]
62. Pimienta, L.; Sarout, J.; Esteban, L.; Piane, C.D. Prediction of rocks thermal conductivity from elastic wave velocities, mineralogy and microstructure. *Geophys. J. Int.* **2014**, *197*, 860–874. [CrossRef]
63. Holland, J.H. *Adaptation in Natural and Artificial Systems: An Introductory Analysis with Applications to Biology, Control, and Artificial Intelligence*; MIT Press: Cambridge, MA, USA, 1992. Available online: https://books.google.com/books?hl=en&lr=&id=5EgGaBkwvWcC&oi=fnd&pg=PR7&dq=63.+Holland,+J.+Adaptation+in+naturna+and+artificial+systems:+An+introductory+analysis+with+application+to+biology.+Control+Artif.+Intell.+1975&ots=mJlk05IInw&sig=oarnmkRQ8ErijpQz8 (accessed on 29 August 2022).
64. Chambers, L.D. *The Practical Handbook of Genetic Algorithms: Applications*; Chapman and Hall/CRC: Boca Raton, FL, USA, 2000.
65. Koza, J.R.; Koza, J.R. *Genetic Programming: On the Programming of Computers by Means of Natural Selection*; MIT Press: Cambridge, MA, USA, 1992; Volume 1.
66. Bayat, P.; Monjezi, M.; Mehrdanesh, A.; Khandelwal, M. Blasting pattern optimization using gene expression programming and grasshopper optimization algorithm to minimise blast-induced ground vibrations. *Eng. Comput.* **2021**, *38*, 3341–3350. [CrossRef]
67. Samaei, M.; Ranjbarnia, M.; Nourani, V.; Naghadehi, M.Z. Performance prediction of tunnel boring machine through developing high accuracy equations: A case study in adverse geological condition. *Measurement* **2019**, *152*, 107244. [CrossRef]
68. MathWorks, I. *MATLAB and Statistics Toolbox Release 2012b*; The MathWorks, Inc.: Natick, MA, USA, 2012.
69. Zare Naghadehi, M.; Samaei, M.; Ranjbarnia, M.; Nourani, V. State-of-the-art predictive modeling of TBM performance in changing geological conditions through gene expression programming. *Measurement* **2018**, *126*, 46–57. [CrossRef]
70. Rostami, A.; Masoudi, M.; Ghaderi-Ardakani, A.; Arabloo, M.; Amani, M. Effective thermal conductivity modeling of sandstones: SVM framework analysis. *Int. J. Thermophys.* **2016**, *37*, 59. [CrossRef]
71. Khandelwal, M. Application of an expert system to predict thermal conductivity of rocks. *Neural Comput. Appl.* **2012**, *21*, 1341–1347. [CrossRef]

applied sciences

MDPI

Article

Inverse Modeling of Seepage Parameters Based on an Improved Gray Wolf Optimizer

Yongkang Shu [1], Zhenzhong Shen [1,2], Liqun Xu [1,*], Junrong Duan [1], Luyi Ju [1] and Qi Liu [3]

1 College of Water Conservancy and Hydropower Engineering, Hohai University, Nanjing 210098, China
2 State Key Laboratory of Hydrology-Water Resources and Hydraulic Engineering, Hohai University, Nanjing 210098, China
3 Datang Hydropower Science & Technology Research Institute Co., Ltd., Nanning 530007, China
* Correspondence: xlq@hhu.edu.cn

Abstract: The seepage parameters of the dam body and dam foundation are difficult to determine accurately and quickly. Based on the inverse analysis, a Gray Wolf Optimizer (GWO) was introduced into this study to search the target hydraulic conductivity. A novel approach for initialization, a polynomial-based nonlinear convergence factor, and weighting factors based on Euclidean norms and hierarchy were applied to improve GWO. The practicability and effectiveness of Improved Gray Wolf Optimizer (IGWO) were evaluated by numerical experiments. Taking Kakiwa dam located on the Muli River of China as a case, an inversion analysis for seepage parameters was accomplished by adopting the proposed optimization algorithm. The simulated hydraulic heads and seepage volume agree with measurements obtained from piezometers and measuring weir. The steady seepage field of the dam was analyzed. The results indicate the feasibility of IGWO in determining the seepage parameters of Kakiwa dam.

Keywords: inverse analysis; hydraulic conductivities; Gray Wolf Optimizer

Citation: Shu, Y.; Shen, Z.; Xu, L.; Duan, J.; Ju, L.; Liu, Q. Inverse Modeling of Seepage Parameters Based on an Improved Gray Wolf Optimizer. *Appl. Sci.* **2022**, *12*, 8519. https://doi.org/10.3390/app12178519

Academic Editors: Danial Jahed Armaghani, Yixia Zhang, Pijush Samui, Ahmed Hussein Kamel Ahmed Elshafie and Aydin Azizi

Received: 14 July 2022
Accepted: 24 August 2022
Published: 25 August 2022

1. Introduction

In hydraulic engineering, seepage parameters of dams and dam foundations change with operating time and loading conditions. The changes in seepage parameters weaken the strength of the structure and lead to failure. Seepage analysis is commonly used to monitor the working conditions of dams and dam foundations for the safety of hydraulic projects [1–5]. The hydraulic conductivity, a key parameter in seepage analysis, is closely related to the accuracy of the analysis results. Minimized error between the simulated and actual values of hydraulic conductivity could improve the reliability of the analysis. In-situ tests have been proven to be helpful in determining the hydraulic conductivity of dams and dam foundations. The hydraulic conductivity determined by in-situ testing agrees with the actual value when the test samples are small. However, this method is time-consuming and costly when there are large quantities of models. Another method to solve this problem is inverse analysis. The inverse research based on monitoring data and numerical simulation results demonstrates economy and efficiency. The essence of the inverse analysis is to determine the hydraulic conductivity by measurements and simulated results. In the inversion analysis, optimization algorithms are widely applied for iterative search over a range of hydraulic conductivity values. The optimal hydraulic conductivity is determined by iteration while minimizing the objective function.

Considering the repetition of the iterative process, optimization algorithms are widely used to improve efficiency and accuracy in the searching process of the inverse problem. For example, on the basis of the Radial Basis Function (RBF) neural network optimized by Particle Swarm Optimization (PSO), Chi et al. [6] constructed an inverse model for the permeability coefficient of a high core rockfill dam; Combining error Back-Propagation Neural Network (BPNN) and Genetic Algorithm (GA), Deng and Lee [7] proposed an inverse analysis method for determining the displacements. This method was successfully applied in the displacement identification of the lock profile of the Three Gorges Project, which led to reasonable results. Zhao et al. [8] developed the differential evolution (DE) algorithm to determine soil parameters in the field of deep excavation, which improved the stability of the backtracking parameters. Simulated annealing [9–12] and ant colony optimization [13–17] have also been extensively used in the inverse problem. Significantly, much progress has been made in the research field of seepage because of optimization algorithms [18–25]. Tan et al. [26] proposed a biological immune mechanism-based quantum particle swarm optimization (IQPSO) algorithm to solve the inversion problem of seepage parameters. Based on back propagation neural network (BPNN) and genetic algorithm (GA), Zhou et al. [27] developed a new approach for inverse modeling of the transient groundwater flow in dam foundations, which improved the uniqueness and reliability of the inversed results and made tractable the large-scale inverse problems in engineering practices. Zhang et al. [28] proposed an inverse analysis model by using the genetic algorithm (GA) and finite element analysis technology, to solve the calcium leaching problems.

Although optimization algorithms are frequently employed for inverse problems, they suffer from low accuracy, slow convergence, and poor robustness. The Gray Wolf Optimizer (GWO) proposed by Mirjalili [29] has been shown to be efficient and intelligent in engineering optimization. Mirjalili [29] compared the performance of GWO with Particle Swarm Optimization (PSO), Gravity Search Algorithm (GSA), Differential Evolution (DE), Evolutionary Programming (EP), and Evolutionary strategy (ES). The results demonstrate that the GWO can provide very competitive results compared to these well-known metaheuristics. It has been extensively adopted in various fields due to its simple structure, fewer parameters, and easy coding implementation. However, it tends to converge to locally optimal solutions. In addition, suboptimal values could result from completely randomized initial populations. Therefore, strategies of improvement are proposed as necessary.

Generally, there are three main strategies to improve the GWO, including adjustments of initial populations, convergence factor, and formula of a location update [30]. Pradhan et al. [31] combined the concept of opposition with GWO, initially providing a uniform population for the algorithm. Long et al. [32] introduced the theory of good point set to population initialization, which improved the homogeneity of the population. Based on this, Long et al. [33] considered the dynamics of the iterative process and proposed an equation of the convergence factor based on the number of iterations. This exponential function simulates the iterative process and balances the local and global search to a certain extent. Mittal et al. [34] described the decay process of the parameter by an exponential function, which leads to improved accuracy for GWO. Salgotra et al. [35] applied the spiral property from the whale optimization algorithm (WOA) to the GWO, which solves the premature convergence in the evolutionary algorithm. Mostafa et al. [36] introduced variational operators to update the location of the individual in GWO and improved the algorithm's performance. Gupta and Deep [37] adopted a random wandering strategy to enhance the accuracy of the algorithm.

A great deal of research has been conducted on the improvement of GWO. However, local and global search, along with homogenization and randomization, cannot be relatively balanced by these improvements. GWO based on these strategies is still limited in terms of efficiency and accuracy. Therefore, there remains potential for improvement across these dimensions.

Based on the evolution of the GWO, three strategies are proposed to ensure the accuracy and efficiency of the algorithm. Initial populations of semi-uniform and semi-random were proposed for rational initialization. A polynomial-based nonlinear convergence factor was applied to maintain a balance between global and local search. Weighting factors based on Euclidean norms and hierarchy were given to dynamically update the wolves' positions for jumping out of the local optimum at the late stage of the iterative search [38,39]. The Improved Gray Wolf Optimizer (IGWO) has been proven to be effective through numerical experiments. This algorithm was applied for the inversion model of the Kakiwa Dam located in Sichuan Province, China. The fitness function is constructed by measurements of piezometers and measuring weir. The finite element method was used to simulate the seepage process under the assumption of steady flow. The free tetrahedral grid is used to construct the two-dimensional mesh of the dam. The finite element calculation is carried out in COMSOL Multiphysics, while the iterative control and data extraction are implemented in MATLAB. The objective hydraulic conductivity and the corresponding fitness value were determined when the maximum number of iterations was achieved. The seepage field at the dam site was also presented.

2. Improved Gray Wolf Optimizer

2.1. Overview of Gray Wolf Optimizer

The Gray Wolf Optimizer (GWO) is a new group intelligence algorithm considering gray wolves' hierarchy and group hunting. There is a strict hierarchy in the gray wolf population. The population is classified into four levels of status in accordance with the fitness values of individuals. Wolves in the first level of the population are responsible for making decisions and leading the group in hunting. Wolves in the second level take responsibility for helping to manage the group. Wolves in the third level of the group obey the orders of the first two levels of gray wolves. All the remaining populations are set at the fourth level.

The hunting process tends to be taken as group action in gray wolf populations, which could be summarized by tracking, encircling, and attacking. The rank in the wolf pack changes dynamically with the individual fitness value in the hunting process. The fitness value can be considered the distance between a wolf and its prey. This means that the closer the distance to the target, the higher the level of the wolf.

Let α, β, and γ represent the three dominant wolves in rank order. The mathematical model of gray wolf hunting is established as described in Equation (1).

$$\begin{cases} d = |C \cdot X_p(t) - X(t)| \\ X(t+1) = X(t) - A \cdot d \end{cases} \tag{1}$$

where d is the perceived distance between the gray wolf and the prey; t represents the number of iterations. C denotes the coefficient vector; $X_p(t)$ and $X(t)$ are the positions of the wolf and the prey, respectively; $X(t+1)$ means the position of the wolf after the iteration; A stands for the coefficient vector.

The expressions for A and C are shown in Equation (2).

$$\begin{cases} A = a \cdot (2m_1 - 1) \\ C = 2m_2 \end{cases} \tag{2}$$

where a is the convergence factor; m_1 and m_2 are both random numbers between 0 and 1.

It is assumed that the first three levels of gray wolves have a better perception of the location of the prey. The populations in the fourth layer decide the direction and distance of the next movement according to the positions of the three dominant wolves. The wolf's position in the fourth rank is updated according to Equation (3).

$$X(t+1) = \frac{1}{3}\sum_{i=1}^{3} X_i \tag{3}$$

where $X(t+1)$ denotes the position of the wolf φ after update; $X_i(i = 1,2,3)$ represent the position vectors of the wolf α, β, and γ, respectively.

The position vectors of α, β, and γ are expressed by Equation (4).

$$\begin{cases} X_1 = |X_\alpha - A_1 \cdot d_1| \\ X_2 = |X_\beta - A_2 \cdot d_2| \\ X_3 = |X_\gamma - A_3 \cdot d_3| \end{cases} \quad (4)$$

where $X_i (i = 1, 2, 3)$ stand for the position vectors of the wolf α, β, and γ, respectively; $X_j (j = \alpha, \beta, \gamma)$ represent the prey position perceived by α, β, and γ, respectively. $A_i (i = 1, 2, 3)$ are the coefficient vectors; $d_i (i = 1, 2, 3)$ mean the distance between the three dominant wolves and the prey.

$d_i (i = 1, 2, 3)$ could be expressed as Equation (5) follows.

$$\begin{cases} d_1 = |C_1 \cdot X_\alpha - X(t)| \\ d_2 = |C_2 \cdot X_\beta - X(t)| \\ d_3 = |C_3 \cdot X_\gamma - X(t)| \end{cases} \quad (5)$$

where $C_i (i = 1, 2, 3)$ are the coefficient vectors; $X(t)$ is the wolf's position in the t-th iteration.

2.2. *Strategies of Improvement*

2.2.1. Initial Populations of Semi-Uniform and Semi-Random

The instability of the solution could be increased by completely randomized initial populations, which leads to unstable results. A novel approach for initializing populations was presented to balance uniformity and randomness. The solution range is divided into intervals equidistantly according to the population size. The j-th interval can be expressed by Equation (6).

$$\Delta_j = \left[lb + \frac{ub - lb}{S}(j-1), lb + \frac{ub - lb}{S} j \right] (j = 1, 2 \ldots, S) \quad (6)$$

where Δ_j is the j-th interval; lb and ub are the upper and lower bounds of the solution set, respectively; S is the population size.

Generate a random initial solution in each interval to ensure randomness. The initial populations are uniformly distributed in the solution space without losing randomness. The expression of the initial solution is given in Equation (7).

$$R_j = lb + \frac{ub - lb}{S} j + rand() \cdot \frac{ub - lb}{S} (j = 1, 2 \ldots, S) \quad (7)$$

where R_j is the initial solution of the j-th interval; lb and ub are the upper and lower bounds of the solution set, respectively; S is the population size; $rand()$ is a random real number between 0 and 1.

2.2.2. Polynomial-Based Nonlinear Convergence Factor

The way to search for prey is determined by the coefficient vector A. The gray wolf can be in any position between the current individual and the prey at the next moment when $|A| < 1$, indicating that the next position of the wolf will be closer to the location of the prey. This is considered a local search. When $|A| > 1$, the next location of the wolf will be further away from the prey than the current location. Gray wolves tend to search over a wider area, which is considered a global search. The positions of gray wolves change with a rapid convergence speed. In this case, the search step size of the gray wolf becomes smaller, thus achieving a refined search. A varies dynamically with convergence factor a. Considering good symmetry and smoothness, a function based on a third-degree polynomial was used to fit the convergence factor a. The expression for a is shown in Equation (8).

$$a(t) = r_1 \left(\frac{t}{T}\right)^3 + r_2 \left(\frac{t}{T}\right)^2 + r_3 \frac{t}{T} + r_4 \quad (8)$$

where a is the convergence factor; $r_i(i = 1, 2, 3, 4)$ denote the real-valued parameters. t represents the number of iterations. T is the maximum number of iterations.

The constraints are given here, as presented in Equation (9).

$$\begin{cases} a(0) = 2, a(T) = 0, a\left(\frac{T}{2}\right) = 1 \\ a'(0) < 0, a'(T) < 0 \end{cases} \tag{9}$$

Thus, a could be indicated by Equation (10).

$$\begin{array}{c} a(t) = \frac{4+2r_3}{T^3}t^3 - \frac{6+3r_3}{T^2}t^2 + r_3\frac{t}{T} + 2 \\ -3.0 \times 10^{-3} < r_3 < 0 \end{array} \tag{10}$$

Figure 1 shows the nonlinear convergence factor evolution at different values of r_3. The maximum number of iterations is set to 500. The three values of r_3 are -2×10^{-3}, -1×10^{-3}, and -2×10^{-4}. The convergence factor values for the three curves decrease with the number of iterations. With the increase of iterations, the cut-off point between global and local search is reached when $T_0 = 250$. Global and local searches could be equally divided and effectively balanced in this condition. In addition, the curve corresponding to $r_3 = -2 \times 10^{-3}$ is lower than the other two curves at the early search stage and higher at the late search, indicating its focus on local search and adequate step size. Considering the drawback of converging to the local optimum in GWO, the curve corresponding to $r_3 = -2 \times 10^{-3}$ was chosen to ensure the property of jumping out of the local optimum in the study.

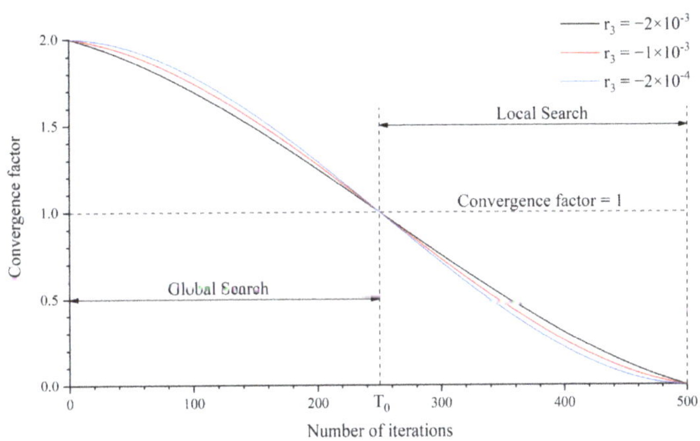

Figure 1. Nonlinear convergence factor based on third-degree polynomials.

2.2.3. Weighting Factors Based on Euclidean Norm and Hierarchy

In the GWO strategy, the position of the gray wolf is updated by the average formula. One limitation of this strategy is that the leadership of the wolf located in the first rank is not considered. Another weakness is that the weighting factors are kept constant as the iteration proceeds. Weighting factors based on Euclidean norms and hierarchy are proposed to overcome this problem, as shown in Equation (11).

$$\begin{cases} \rho_1 = \frac{\|X_1\|}{\|X_1\| + \|X_2\| + \|X_3\|} \\ \rho_2 = \frac{\|X_2\|}{\|X_1\| + \|X_2\| + \|X_3\|} \\ \rho_3 = \frac{\|X_3\|}{\|X_1\| + \|X_2\| + \|X_3\|} \end{cases} \tag{11}$$

Here ρ_i $(i = 1, 2, 3)$ are the weighting factors of the first-three level wolves, respectively; X_i $(i = 1, 2, 3)$ are the Euclidean norms of the first three levels, respectively.

The weights of the wolf α, β, and γ are multiplied by 0.6, 0.3, and 0.1 to reinforce the wolf pack hierarchy. The formula for updating the location of the gray wolf can be improved, as Equation (12) states.

$$X(t+1) = \frac{6\rho_1 \cdot X_1 + 3\rho_2 \cdot X_2 + \rho_3 \cdot X_3}{10} \tag{12}$$

where ρ_i $(i = 1, 2, 3)$ are the weighting factors of the first-three level wolves, respectively; X_i $(i = 1, 2, 3)$ represent the position vectors of the wolf α, β, and γ respectively.

2.3. Numerical Experiment of Algorithm Performance

A numerical experiment was performed to verify the effectiveness of IGWO. Six typical functions were selected for simulation in the experiments, including Sphere, Rosenbrock, Quartic, Rastrigin, Ackley, and Griewank. Table 1 shows the mathematical expressions, dimensions, and search ranges of these typical functions. Sphere, Rosenbrock, and Quartic are single-peak functions. Especially Quartic is a multidimensional flat bottom function with random disturbances. The single-peak functions are mainly applied to determine the accuracy of IGWO. Rastrigin, Ackley, and Griewank are multi-peaked functions that tend to cause the algorithm to converge to a locally optimal solution. The performance to jump out of the local optimum could be tested reasonably for IGWO. In addition, the experimental results of IGWO, SGWO [33], and GWO are compared in the simulation.

Table 1. Test functions in the numerical experiment.

Test Function	Mathematical Expression	Dimension	Search Range
Sphere	$\sum_{i=1}^{D} x_i^2$	30	[−100, 100]
Rosenbrock	$\sum_{i=1}^{D-1} \left[100\left(x_{i+1} - x_i^2\right)^2 + (x_i - 1)^2\right]$	30	[−30, 30]
Quartic	$\sum_{i=1}^{D} ix_i^4 + random[0, 1]$	30	[−1.28, 1.28]
Rastrigin	$\sum_{i=1}^{D-1} \left[x_i^2 - 10cos(2\pi x_i) + 10\right]$	30	[−5.12, 5.12]
Ackley	$-20exp\left(-0.2\sqrt{\frac{1}{30}\sum_{i=1}^{D} x_i^2}\right) - exp\left(\frac{1}{30}\sum_{i=1}^{D} cos2\pi x_i\right) + 20 + e$	30	[−32, 32]
Griewank	$\frac{1}{4000}\sum_{i=1}^{D} x_i^2 - \prod_{i=1}^{D} cos\left(\frac{x_i}{\sqrt{i}}\right) + 1$	30	[−600, 600]

To ensure a fairness, the population size is 30, and the maximum number of iterations is 500 for IGWO, SGWO, and GWO. The three algorithms were performed 30 times independently for each function, and the average values were taken as the simulation results. The results of the numerical experiments are given in Table 2. The optimal values simulated by IGWO are closer to the theoretical optimal solution than the results of the other two algorithms. Among them, the simulation results of Sphere, Rastrigin, and Griewank are equal to the theoretical values, revealing the high accuracy of IGWO.

Figure 2 presents the convergence curves of the six test functions with IGWO, SGWO, and GWO. The convergence curves of all three algorithms continue to decrease as the number of iterations increases. In particular, the convergence curve of IGWO decreases significantly faster than the corresponding curves of the other two algorithms, indicating the progress of IGWO in terms of running time. Furthermore, for the multi-peaked test function, the convergence curve of IGWO continues to decrease while the search of the other two algorithms converges to a local optimum solution. This result demonstrates the ability to jump out of the local search and converge to the global optimum value in IGWO.

Table 2. Results of the numerical experiments.

Test Function	Optimization Algorithm	Simulated Optimum Value	Theoretical Optimum Value
	IGWO	0	
Sphere	SGWO	6.45×10^{-33}	0
	GWO	1.34×10^{-26}	
	IGWO	2.89×10	
Rosenbrock	SGWO	2.70×10	0
	GWO	2.72×10	
	IGWO	2.04×10^{-4}	
Quartic	SGWO	2.83×10^{-4}	0
	GWO	1.40×10^{-3}	
	IGWO	0	
Rastrigin	SGWO	5.68×10^{-14}	0
	GWO	1.71×10^{-12}	
	IGWO	4.44×10^{-15}	
Ackley	SGWO	1.51×10^{-14}	0
	GWO	1.11×10^{-13}	
	IGWO	0	
Griewank	SGWO	1.16×10^{-2}	0
	GWO	2.84×10^{-2}	

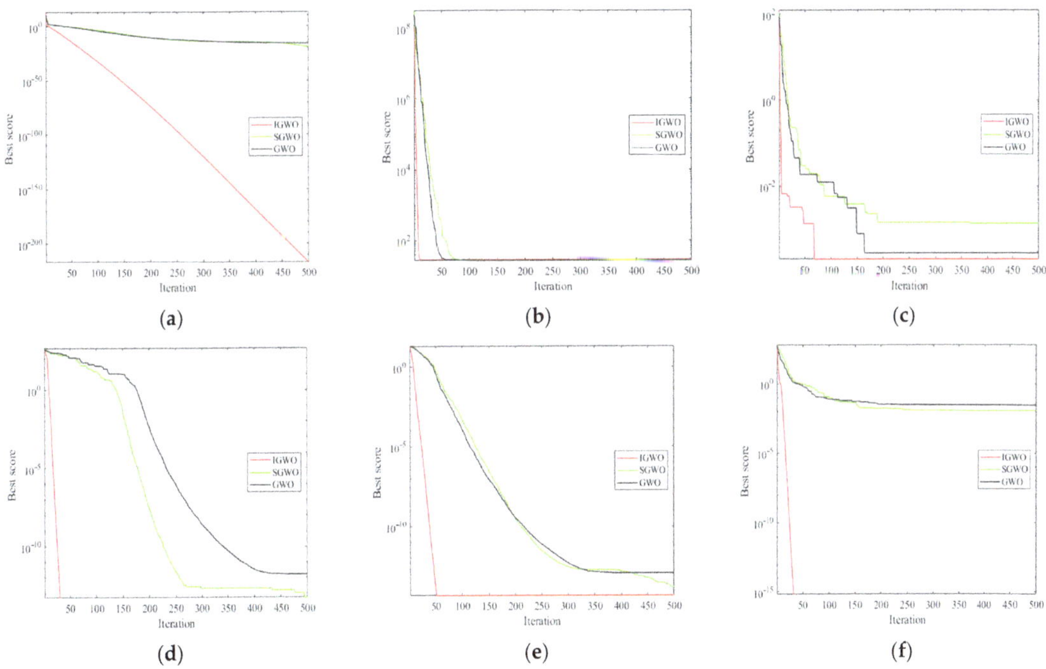

Figure 2. Convergence curves of the test function with the three optimization algorithms: (**a**) Sphere; (**b**) Rosenbrock; (**c**) Quartic; (**d**) Rastrigin; (**e**) Ackley; (**f**) Griewank.

3. Inverse Model of Seepage Parameters

3.1. The Objective Function

The aim of the inverse model is to determine the hydraulic conductivity for each partition while minimizing the value of the objective function. The objective function was constructed by hydraulic head and leakage in this paper, improving the reliability of the simulation results. Suppose that the hydraulic conductivity of each medium is isotropic. Denote by $\boldsymbol{K} = [k_1, k_2, \ldots, k_n]$ the combination of hydraulic conductivity, in which k_i represents the hydraulic conductivity of the ith media. The number of piezometers and measuring weirs are indicated by m and n, respectively. $\boldsymbol{H} = [H_1, H_2, \ldots, H_M]$ is expressed as a sequence of hydraulic head measurements. Similarly, $\boldsymbol{Q} = [Q_1, Q_2, \ldots, Q_N]$ is a series of leakage volume measurements. $H_i(\boldsymbol{K})$ and $Q_j(\boldsymbol{K})$ are the simulated hydraulic head and leakage volume by finite element method. The mathematical model for the inverse problem is established, as shown in Equation (13).

$$\begin{gathered} minf = \left(\sum_{i=1}^{M} \frac{\|H_i(\boldsymbol{K}) - H_i\|_2^2}{\|H_i\|_2^2} \right)^{\frac{1}{2}} + w \left(\sum_{i=1}^{N} \frac{\|Q_j(\boldsymbol{K}) - Q_j\|_2^2}{\|Q_j\|_2^2} \right)^{\frac{1}{2}} \\ \text{s.t. } \boldsymbol{K}_{min} \le \boldsymbol{K} \le \boldsymbol{K}_{max} \end{gathered} \tag{13}$$

Here $\boldsymbol{K}_{min}$ and $\boldsymbol{K}_{max}$ are the lower and upper bounds of hydraulic conductivity values, respectively. The range of hydraulic conductivity can be roughly determined by geological data and engineering experience. w is a weight factor for balancing the hydraulic head and the leakage volume. In this paper, the simulated leakage value is estimated by the flow rate and area of the overwater cross-section. It is suggested that the value of the weighting factor is set small considering an error between the simulated value and the measurement at the shoulder part of the dam. Zhou [23] compared the relative errors of hydraulic head and leakage volume at different weights. The results show that the simulated values are in good agreement with the measurements, and the minimum value of relative error is reached at the condition of $w = 0.02$. The finding was applied in this paper.

The objective combination of hydraulic conductivity was obtained by the searching process. The search process was accelerated by IGWO. The objective combination of hydraulic conductivity was applied to simulate the seepage field of the dam during operation.

3.2. Procedure of the Inversion Model

The procedure of the inversion model could be summarized by specific steps. The steps are as follows.

Step 1: Set initial parameters of IGWO. The number of search agents S, the maximum number of iterations T, and the bounds of hydraulic conductivity values, $\boldsymbol{K}_{min}$ and $\boldsymbol{K}_{max}$, are determined initially.

Step 2: Initialize the population. Equation (1) is applied for initialization, ensuring that the initial populations are uniformly distributed in the solution space.

Step 3: Calculate the fitness of individual gray wolves. The three gray wolves with the top fitness values are selected as α, β, and γ.

Step 4: Update the position. Determine the distance of the gray wolf from the three dominant wolves, respectively. Orient the location and calculate the weighting factors of the three wolves. The position of the gray wolf at the fourth level is updated according to Equation (12).

Step 5: Iterative Judgment. Determines whether the maximum number of iterations has been achieved. If not, skip to the third step. Otherwise, end the iterative procedure. Output the objective hydraulic conductivity and the corresponding fitness value.

Step 6: Positive verification. The combination of the target hydraulic conductivity is substituted into the finite element model for positive verification. Compare the calculated and monitored values of hydraulic head and leakage volume and evaluate the reasonableness of the simulation.

Figure 3 presents the flow chart of the model.

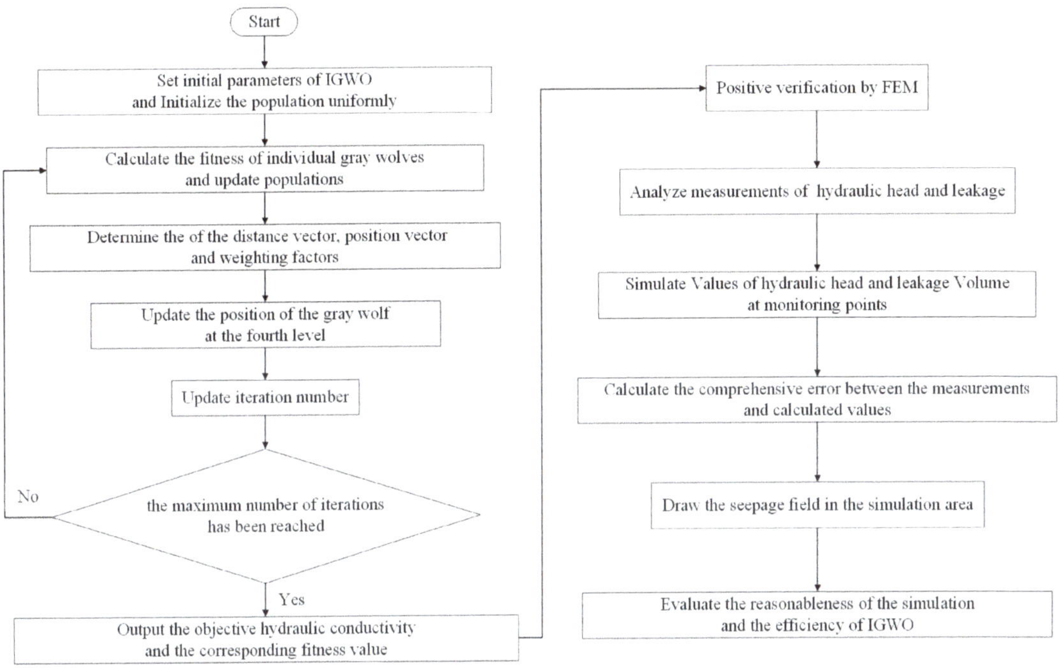

Figure 3. Flow chart of the inverse model based on IGWO.

4. A Casebook Study

4.1. Project Overview

Located on the Muli River in the Sichuan Province of China, the Kakiwa Hydropower Station is a project focused on power generation and ecological preservation. The location of the Kakiwa Dam is indicated in Figure 4.

Figure 4. The location of the Kakiwa Dam.

A concrete panel rockfill dam is selected as the barrage in the pivot project, with a maximum height of 171 m. The crest width of the dam is 11 m. The dam mainly consists of a concrete face slab, blanket area, cushion area, transition area, rockfill area, drainage area, ballast area, and grout curtain. The normal storage level is 2850.00 m, the calibration flood level is 2852.20 m, and the dead water level is 2800.00 m. Figure 5 shows the maximum cross-section of the dam body.

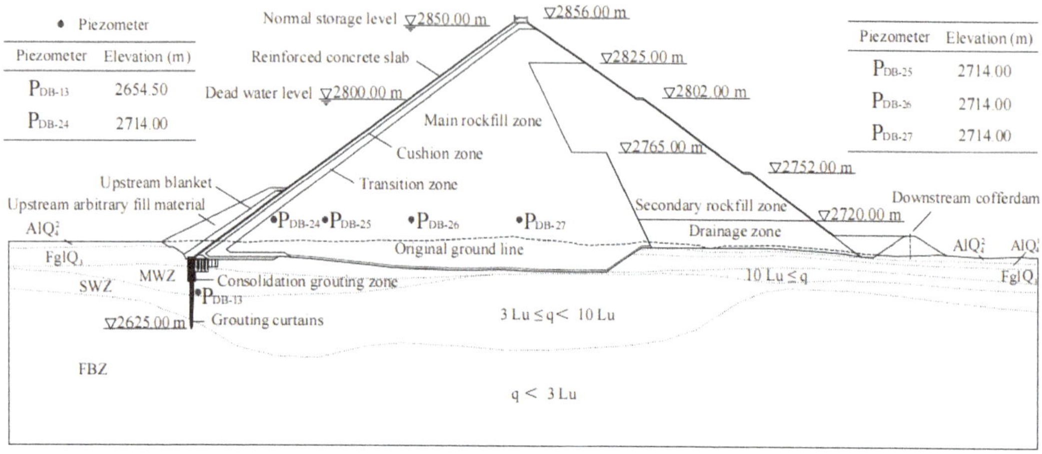

Figure 5. The max cross-section of the dam body.

A total of 37 piezometers were installed for seepage monitoring of the dam body and dam foundation. Among them, the piezometer P_{DB-13} is located downstream of the curtain. $P_{DB-24} - P_{DB-27}$ are installed near the original ground line. Figure 5 presents the locations and elevations of these piezometers. The water measuring weir is installed at the downstream cofferdam axis of the dam.

The hydraulic conductivities of the five media, including the grout curtain, the top cover layer, the second cover layer, the moderately weathered zone, the slightly weathered zone, and the fresh bedrock zone, are limited to reasonable ranges and needed to be optimized. The parameter ranges are given in the results of the simulation. In addition, the hydraulic conductivities of other media are indicated in Table 3.

Table 3. Hydraulic conductivity of stationary medium.

Material	Hydraulic Conductivity (m/s)
Upstream face slab	1.00×10^{-6}
Bedding material	9.90×10^{-4}
Transition material	7.50×10^{-2}
Main rock-fill zone	8.70×10^{-1}
Secondary rock-fill zone	9.80×10^{-1}
Upstream blanket	1.00×10^{-5}

4.2. Analysis of Monitoring Data

Figure 6 shows the monitoring data of the piezometers around the grout curtain and the water measuring weirs during the operation period. The hydraulic head on the downstream side of the curtain is relatively consistent with the upstream water level. The hydraulic head measured by the piezometer lags behind the upstream head, which is called the hysteresis effect. The measured hydraulic head rises less than the upstream water level. In addition, the value of the piezometer P_{DB-24} is approximately 2776 m, and the difference among the values of piezometers P_{DB-24} to P_{DB-27} is not significant, indicating the efficiency of the impermeable curtain.

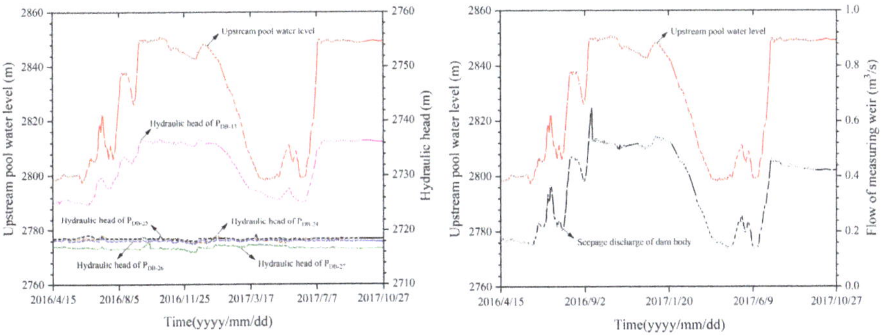

(**a**) (**b**)

Figure 6. Measurements of hydraulic head and seepage discharge during storage period: (**a**) hydraulic head; (**b**) seepage volume.

Similarly, the seepage volume is consistent with the upstream water level. The value of seepage volume at the dam body is relatively small, with a stable value of 0.41 m^3/s in the operation period.

It is assumed that the seepage field is stable for simplicity. A period with a slight variation of upstream and downstream reservoir levels and long duration was selected for the inverse model, which could minimize the seepage lag effect to a certain extent. As seen in Figure 6, the upstream pool level is relatively stable from 10 July 2017, to 27 October 2017, with values between 2849.55 m and 2850.06 m. Therefore, this period was chosen for the inversion model.

4.3. Computation Model

IGWO was used for iterative search. The maximum number of iterations is 200, and the initial population size is 30. The simulation is performed 20 times independently by IGWO, and the average values of simulated hydraulic conductivity are taken as the target results.

The multi-physics field simulation software COMSOL Multiphysics was applied in this simulation. A two-dimensional finite element mesh was created for inverse modeling each medium's hydraulic conductivity. The seepage properties of the dam body and foundation were analyzed. The free tetrahedral grid is applied to build the mesh, composed of 11,635 domain elements and 14,400 vertices. The maximum and minimum widths of the cells are 20 m and 5 m, respectively. A right-hand cartesian coordinate system is constructed with the x-axis pointing to the downstream reservoir and the z-axis pointing to the sky vertically. The upstream and downstream are both extended by 2.5 times the dam's height on the x-axis. The depth of the foundation is taken as 400 m. In addition, the upstream and downstream water levels are, respectively, set at 2850 m and 2702 m. The two-dimensional finite element mesh is presented in Figure 7.

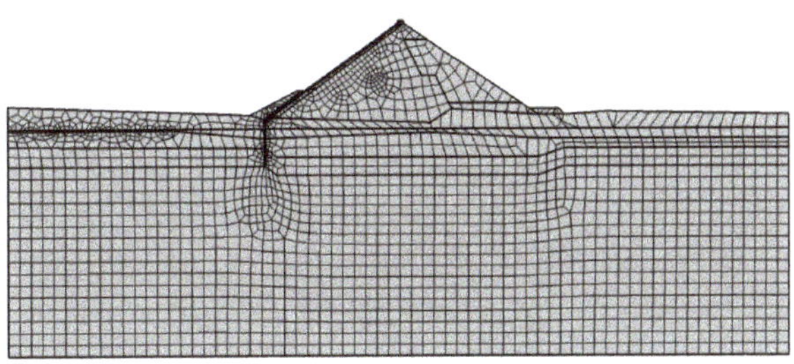

Figure 7. 2D FE mesh of Kakiwa dam.

4.4. Results of the Simulation

4.4.1. Hydraulic Conductivity

The hydraulic conductivity of each medium was determined by IGWO. The hydraulic conductivity of each medium at the dam site is given in Table 4. All the results are within the corresponding search range.

Table 4. Hydraulic conductivity of each medium determined by IGWO.

Material	Hydraulic Conductivity (m/s)	Search Range (m/s)
Grout curtain	3.00×10^{-5}	1.00×10^{-6}–1.00×10^{-4}
The top cover layer	5.33×10^{-2}	1.00×10^{-3}–1.00×10^{-1}
The second cover layer	2.67×10^{-3}	1.00×10^{-4}–1.00×10^{-2}
Moderately-weathered zone	5.50×10^{-4}	1.00×10^{-5}–1.00×10^{-3}
Slightly weathered zone	1.10×10^{-4}	1.00×10^{-5}–1.00×10^{-3}
Fresh bedrock zone	3.10×10^{-5}	1.00×10^{-6}–1.00×10^{-4}

4.4.2. Hydraulic Head

The results of hydraulic conductivity were substituted into the finite element model for positive analysis to verify the reasonableness of this simulation. The simulated values of the hydraulic head and leakage volume at the monitoring points were obtained and compared with the corresponding measurements. Absolute and relative errors of hydraulic head and leakage were calculated. A contour of the hydraulic head in the dam site area was also predicted.

The mathematical expression for the relative error of the hydraulic head is given in Equation (14):

$$\delta_H = \frac{|H_i - H|}{\Delta H} \times 100\% \quad (14)$$

where δ_H is the relative error of the hydraulic head; H_i and H represent the simulated and measured values of the hydraulic head, respectively. ΔH denotes the difference in water level between the upstream and downstream sides, taken as 148 m.

Table 5 compares the calculated and measured hydraulic head values at the monitoring points. The calculated values at the monitoring points are relatively close to the measurements. Among them, the simulated values at the piezometers $P_{DB-24} - P_{DB-27}$ show a very high consistency with the corresponding measurements. The maximum value of the absolute error −1.41 m, and the maximum value of the relative error is 0.95%. Meanwhile, the value of the hydraulic head decreases with the increase of seepage distance.

Table 5. Comparison between the calculated and measured hydraulic head values.

Monitoring Points	Measured Hydraulic Head (m)	Simulated Hydraulic Head (m)	Absolute Error (m)	Relative Error (%)
P_{DB-13}	2736.17	2735.74	−0.43	0.29
P_{DB-24}	2718.67	2717.51	−1.16	0.78
P_{DB-25}	2718.28	2717.16	−1.12	0.76
P_{DB-26}	2718.21	2716.80	−1.41	0.95
P_{DB-27}	2716.61	2716.24	−0.37	0.25

4.4.3. Leakage Volume of the Dam Foundation

The seepage volume is estimated by the flow velocity and the overflow surface. The formula for calculating leakage volume is given in Equation (15).

$$Q_i = \iint vBdxdz \quad (15)$$

where Q_i is the simulated seepage volume; v means the flow rate at the vertical spillway surface; B represents the length of the dam taken as 355 m.

The relative error for the leakage volume is determined by Equation (16).

$$\delta_Q = \frac{|Q_i - Q|}{Q} \times 100\% \tag{16}$$

where δ_Q is the relative error of the dam body leakage; Q_i and Q are the simulated and measured values of the dam body leakage, respectively.

Table 6 shows the comparison of the calculated and measured values of the dam body leakage during the stable upstream water level. The simulated leakage values are in good agreement with the actual measurements, showing the accuracy of the IGWO strategy and the reliability of the simulation.

Table 6. Comparison between the simulated and measured values of the dam body leakage.

Leakage	Measured Values (m^3/s)	Simulated Values (m^3/s)	Absolute Error (m^3/s)	Relative Error (%)
Dam body	0.40	0.38	−0.02	5.00

The leakage measurements are averaged over the simulation period. The relative error of the calculated seepage volume is 5.00%, demonstrating the positive performance of IGWO in the simulation.

Figure 8 presents the contour map of the hydraulic head at the dam site. The distribution of contours is in accordance with seepage characteristics. The results show the accuracy of the simulation and the reasonableness of the calculated hydraulic conductivity.

Figure 8. Distribution of seepage contours at the maximum section of the dam.

5. Discussion

The hydraulic conductivities of the dam and dam foundation change with operation and loading, affecting the effectiveness and safety of the hydraulic project. An Improved Gray Wolf Optimizer for solving the problem that hydraulic conductivity is not easily determined, is introduced in this paper.

The objective hydraulic conductivity was determined based on IGWO. The obtained hydraulic conductivities of the dam and dam foundation were applied in the finite element positive analysis. The results show the effectiveness of IGWO in determining hydraulic conductivity. It is suggested that IGWO could be used to obtain reasonable simulation results in similar inverse problems.

The hydraulic conductivity of each medium in the dam body and dam foundation simulated in this paper was obtained with a stable upstream water level. In fact, the upstream water level of a reservoir varies continuously with the operating conditions and purpose. The hydraulic conductivity will also change with the monitored values. It is worth noting that IGWO is still applicable under this condition.

6. Conclusions

Aiming at balancing the local and global search of GWO, along with uniformity and stochasticity, IGWO was proposed in this paper. The improvement of IGWO in accuracy and running time was indicated in numerical experiments. IGWO was used in the inverse modeling of the hydraulic conductivity of the Kakiwa dam. The hydraulic head and leakage were used to set up the objective function. Errors in the hydraulic head and leakage were calculated. The steady seepage field was analyzed in the application case. The main conclusions of this paper are given as follows.

(1) The performance of IGWO is improved due to the three strategies. A novel approach for initialization contributes to populations of semi-uniform and semi-random. The polynomial-based nonlinear convergence factor is selected to keep the equilibrium of the local and global search. Weighting factors based on Euclidean norms and hierarchy helps to update the position of the wolf dynamically.

(2) A numerical experiment was conducted to demonstrate the performance of IGWO. The optimal values obtained by IGWO are closer to the theoretical solution than the results of the other two algorithms. The results of the experiment demonstrate the feasibility and efficiency of IGWO.

(3) The target combination of hydraulic conductivity was obtained by IGWO. The values of hydraulic conductivities were substituted into the finite element model. The values of the hydraulic head and leakage quantity at the corresponding measurement points were obtained. The maximum values of absolute and relative errors of the hydraulic head were—1.41 m and 0.95%, respectively. The absolute and relative errors of the seepage volume were—0.02 m^3/s and 5.00%, respectively. The results of the application case show that the inversion model and the algorithm are reliable and efficient.

Author Contributions: Conceptualization, Y.S. and L.X.; methodology, Z.S.; software, Y.S.; validation, J.D. and L.J.; formal analysis, L.X.; investigation, J.D.; resources, Z.S.; data curation, L.J. and Q.L.; writing—original draft preparation, Y.S.; writing—review and editing, Y.S.; visualization, J.D. and Q.L.; supervision, Z.S.; project administration, Z.S.; funding acquisition, L.X. All authors have read and agreed to the published version of the manuscript.

Funding: This research was funded by the "National Key R&D Program of China, grant number 2019YFC1510802".

Institutional Review Board Statement: Not applicable.

Informed Consent Statement: Not applicable.

Data Availability Statement: Not applicable.

Conflicts of Interest: The authors declare no conflict of interest.

References

1. Liu, M.; Zhong, J.Q.; Rao, X.B.; Liu, J.; Wang, W.J. Seepage monitoring analysis and safety assessment for Yatang dam. *J. Yangtze River Sci. Res. Inst.* **2007**, *24*, 30–33. (In Chinese)
2. Chu, X.W.; Xu, M.; Wang, Z.M. Stability analysis of seepage on the accumulation dam of a phosphogypsum tailings. *J. Eng. Geol.* **2016**, *24*, 661–667. (In Chinese)
3. Xu, J.B.; Wei, W.; Bao, H.; Zhang, K.; Lan, H.; Yan, C.; Sun, W. Failure models of a loess stacked dam: A case study in the ansai area (China). *Bull. Eng. Geol. Environ.* **2020**, *79*, 1009–1021. [CrossRef]
4. Jiang, Z.X.; He, J.P. Detection model for seepage behavior of earth dams based on data mining. *Math. Probl. Eng.* **2018**, *2018*, 8191802. [CrossRef]
5. Fan, X.F.; Wu, Z.Y.; Liu, L.; Wen, Y.; Yu, S.; Zepeng, Z.; Li, Z. Analysis of sluice foundation seepage using monitoring data and numerical simulation. *Adv. Civ. Eng.* **2019**, *2019*, 2850916.

6. Chi, S.; Ni, S.; Liu, Z. Back analysis of the permeability coefficient of a high core rockfill dam based on a RBF neural network optimized using the PSO algorithm. *Math. Probl. Eng.* **2015**, *2015*, 1–15. [CrossRef]
7. Deng, J.H.; Lee, C.G. Displacement back analysis for a steep slope at the Three Gorges Project site. *Int. J. Rock Mech. Min.* **2001**, *38*, 259–268. [CrossRef]
8. Zhao, B.D.; Zhang, L.L.; Jeng, D.S.; Wang, J.H.; Chen, J.J. Inverse analysis of deep excavation using differential evolution algorithm. *Int. J. Numer. Anal. Met.* **2015**, *39*, 115–134. [CrossRef]
9. Ortiz-Aleman, C.; Martin, R. Inversion of electrical capacitance tomography data by simulated annealing: Application to real two-phase gas-oil flow imaging. *Flow Meas. Instrum.* **2005**, *16*, 157–162. [CrossRef]
10. Tabatabaeenejad, A.; Moghaddam, M. Inversion of subsurface properties of layered dielectric structures with random slightly rough interfaces using the method of simulated annealing. *IEEE Trans. Geosci. Remote Sens.* **2009**, *47*, 2035–2046. [CrossRef]
11. Song, G.R.; Hong, G.F.; Lu, Y.; Xu, Y.; Qin, D.; Wu, B.; He, C. Inversion of elastic constants of anisotropic (100) silicon based on surface wave velocity by acoustic microscopy using Particle Swarm-Based-Simulated annealing optimization. *J. Nondestruct. Eval.* **2015**, *34*, 43. [CrossRef]
12. Netto, A.; Dunbar, J. 3-D constrained inversion of gravimetric data to map the tectonic terranes of southeastern Laurentia using simulated annealing. *Earth Planet. Sc. Lett.* **2019**, *513*, 12–19. [CrossRef]
13. Shi-chang, L.; Qing-sheng, Z.; Zhe, Y.; Hao-lan, Y. Ant colony optimization for nonlinear AVO inversion of network traffic allocation optimization. *Expert Syst. Appl.* **2010**, *37*, 8343–8347. [CrossRef]
14. Lin, P.; Liu, X.L.; Chen, H.X.; Kim, J. Ant colony optimization analysis on overall stability of high arch dam basis of field monitoring. *Sci. World J.* **2014**, *2014*, 483243. [CrossRef] [PubMed]
15. Gao, W. Displacement back analysis for underground engineering based on immunized continuous ant colony optimization. *Eng. Optimiz.* **2016**, *48*, 868–882. [CrossRef]
16. Ghorbani, E.; Moosavi, M.; Hossaini, M.F.; Assary, M.; Golabchi, Y. Determination of initial stress state and rock mass deformation modulus at lavarak HEPP by back analysis using ant colony optimization and multivariable regression analysis. *Bull. Eng. Geol. Environ.* **2021**, *80*, 429–442. [CrossRef]
17. Li, Z.X.; Rao, S.W. Frequency domain soil parameters inversion of horizontally multilayered earth model with considering high-frequency field. *IET Gener. Transm. Dis.* **2018**, *12*, 5690–5699. [CrossRef]
18. He, J.; Chen, S.H.; Shahrour, I. A revised solution of equivalent permeability tensor for discontinuous fractures. *J. Hydrodyn.* **2012**, *24*, 711–717. [CrossRef]
19. Virbulis, J.; Bethers, U.; Saks, T.; Sennikovs, J.; Timuhins, A. Hydrogeological model of the baltic artesian basin. *Hydrogeol. J.* **2013**, *21*, 845–862. [CrossRef]
20. Yu, H.; Wang, X.; Ren, B.; Zeng, T.; Lv, M.; Wang, C. An efficient bayesian inversion method for seepage parameters using a data-driven error model and an ensemble of surrogates considering the interactions between prediction performance indicators. *J. Hydrol.* **2022**, *604*, 127235. [CrossRef]
21. Xu, L.Q.; Li, L. Inversion Analysis of seepage parameters based on improved shuffled frog leaping algorithm. *Math. Probl. Eng.* **2021**, *2021*, 6536294. [CrossRef]
22. Shi, Z.; Wu, Z.; Gu, C.; Chen, B.; Zhang, H.; Yin, W.; Wu, B. Calculation methods for the permeability coefficient of concrete face rockfill dam with cracks. *Adv. Civ. Eng.* **2019**, *2019*, 6571092. [CrossRef]
23. Zhang, H.Y.; Song, Z.Y.; Peng, P.; Sun, Y.; Ding, Z.; Zhang, X. Research on seepage field of concrete dam foundation based on artificial neural network. *Alex. Eng. J.* **2021**, *60*, 1–14. [CrossRef]
24. Kumar, R.; Pal, S.K.; Gupta, P.K. Water seepage mapping in an underground coal-mine barrier using self-potential and electrical resistivity tomography. *Mine Water Environ.* **2021**, *40*, 622–638. [CrossRef]
25. Shu, Y.; Shen, Z.; Xu, L.; Zhang, K.; Yang, C. Inversion analysis of impervious curtain permeability coefficient using calcium leaching model, extreme learning machine, and optimization algorithms. *Appl. Sci.* **2022**, *12*, 3272. [CrossRef]
26. Tan, J.; Xu, L.; Zhang, K.; Yang, C. A biological immune mechanism-based quantum PSO algorithm and its application in back analysis for seepage parameters. *Math. Probl. Eng.* **2020**, *2020*, 2191079. [CrossRef]
27. Zhou, C.B.; Liu, W.; Chen, Y.F.; Hu, R.; Wei, K. Inverse modeling of leakage through a rockfill dam foundation during its construction stage using transient flow model, neural network and genetic algorithm. *Eng. Geol.* **2015**, *187*, 183–195. [CrossRef]
28. Zhang, K.; Shen, Z.; Xu, L.; Shu, Y.; Yang, C. Inverse modeling of grout curtain hydraulic conductivity evolution considering the calcium leaching effect. *Mathematics* **2022**, *10*, 381. [CrossRef]
29. Mirjalili, S.; Mirjalili, S.M.; Lewis, A. Grey wolf optimizer. *Adv. Eng. Softw.* **2014**, *69*, 46–61. [CrossRef]
30. Zhang, X.F.; Wang, X.Y. Comprehensive review of grey wolf optimization algorithm. *Comput. Sci.* **2019**, *46*, 30–38. (In Chinese)
31. Pradhan, M.; Roy, P.K.; Pal, T. Oppositional based grey wolf optimization algorithm for economic dispatch problem of power system. *Ain Shams Eng. J.* **2018**, *9*, 2015–2025. [CrossRef]
32. Long, W.; Zhao, D.Q.; Xu, S.J. Improved grey wolf optimization algorithm for constrained optimization problem. *J. Comput. Appl.* **2015**, *35*, 2590–2595. (In Chinese)
33. Long, W.; Wu, T.B. Improved grey wolf optimization algorithm coordinating the ability of exploration and exploitation. *Control. Decis.* **2017**, *32*, 1749–1757. (In Chinese)
34. Yin, L.; Sun, Z. Distributed multi-objective grey wolf optimizer for distributed multi-objective economic dispatch of multi-area interconnected power systems. *Appl. Soft Comput. J.* **2022**, *117*, 108345. [CrossRef]

35. Salgotra, R.; Singh, U.; Singh, S.; Mittal, N. A hybridized multi-algorithm strategy for engineering optimization problems. *Knowl.-Based Syst.* **2021**, *217*, 106790. [CrossRef]
36. Mostafa, A.; Fouad, A.; Houseni, M.; Allam, N.; Hassanien, A.E.; Hefny, H.; Aslanishvili, I. A hybrid grey wolf based segmentation with statistical image for CT liver images. In *International Conference on Advanced Intelligent Systems and Informatics*; Springer: Cham, Switzerland, 2016; pp. 846–855.
37. Gupta, S.; Deep, K. A novel random walk Grey Wolf Optimizer. *Swarm Evol. Comput.* **2019**, *44*, 101–112. [CrossRef]
38. Malik, M.; Mohideen, E.R.; Ali, L. Weighted distance Grey wolf optimizer for global optimization problems. In Proceedings of the 2015 IEEE International Conference on Computational Intelligence and Computing Research, Tamilnadu, India, 10–12 December 2015.
39. Panda, M.; Das, B. Grey Wolf Optimizer and Its Applications: A Survey. In Proceedings of the 3rd International Conference on Microelectronics, Computing and Communication Systems, Ranchi, India, 12–13 May 2018; Nath, V., Mandal, J.K., Eds.; Springer: Singapore, 2019; pp. 179–194.

applied sciences

MDPI

Article

Analysis of Malicious Node Identification Algorithm of Internet of Vehicles under Blockchain Technology: A Case Study of Intelligent Technology in Automotive Engineering

Jing Chen [1,*], Tong Li [2] and Rui Zhu [3]

1 School of Information, Yunnan University, Kunming 650500, China
2 School of Big Data, Yunnan Agricultural University, Kunming 650201, China
3 School of Software, Yunnan University, Kunming 650091, China
* Correspondence: cjing@mail.ynu.edu.cn

Abstract: False messages sent by malicious or selfish vehicle nodes will reduce the operation efficiency of the Internet of Vehicles, and can even endanger drivers in serious cases. Therefore, it is very important to detect malicious vehicle nodes in the network in a timely manner. At present, the existing research on detecting malicious vehicle nodes in the Internet of Vehicles has some problems, such as difficulties with identification and a low detection efficiency. Blockchain technology cannot be tampered with or deleted and has open and transparent characteristics. Therefore, as a shared distributed ledger in decentralized networking, blockchain can promote collaboration between transactions, processing and interaction equipment, and help to establish a scalable, universal, private, secure and reliable car networking system. This paper puts forward a block-network-based malicious node detection mechanism. Using blockchain technology in a car network for malicious node identification algorithm could create a security scheme that can ensure smooth communication between network vehicles. A consensus on legal vehicle identification, message integrity verification, false message identification and malicious vehicle node identification form the four parts of the security scheme. Based on the public–private key mechanism and RSA encryption algorithm, combined with the malicious node identification algorithm in the Internet of Vehicles, the authenticity of the vehicle's identity and message is determined to protect the vehicle's security and privacy. First, a blockchain-based, malicious node detection architecture is constructed for the Internet of vehicles. We propose a malicious node identification algorithm based on the blockchain consensus mechanism. Combined the above detection architecture with the consensus mechanism, a comprehensive and accurate verification of vehicle identity and message authenticity is ensured, looking at the four aspects of vehicle identification, accounting node selection, verification of transmission message integrity and identification of the authenticity of transmission messages. Subsequently, the verification results will be globally broadcast in the Internet of Vehicles to suppress malicious behavior, further ensure that reliable event messages are provided for the driver, improve the VANET operation environment, and improve the operation efficiency of the Internet of Vehicles. Comparing the proposed detection mechanism using simulation software, the simulation results show that the proposed blockchain-based trust detection mechanism can effectively improve the accuracy of vehicle node authentication and identification of false messages, and improve network transmission performance in the Internet of Vehicles environment.

Keywords: blockchain technology; intelligent technology; internet of vehicles; malicious nodes; identification algorithm

Citation: Chen, J.; Li, T.; Zhu, R. Analysis of Malicious Node Identification Algorithm of Internet of Vehicles under Blockchain Technology: A Case Study of Intelligent Technology in Automotive Engineering. *Appl. Sci.* **2022**, *12*, 8362. https://doi.org/10.3390/app12168362

Academic Editors: Pijush Samui, Aydin Azizi, Ahmed Hussein Kamel Ahmed Elshafie, Yixia Zhang and Danial Jahed Armaghani

Received: 22 May 2022
Accepted: 8 August 2022
Published: 21 August 2022

1. Introduction

Establishing an intelligent transportation system using the Internet is a wise choice. This kind of intelligent system is used in the field of transportation: using the real-time interactions between cars, trains and boat vehicles to coordinate data on their trajectory and running state can to help alleviate the frequent road safety accidents, improve the robustness of network security, etc. As a more advanced system in the field of transportation, the intelligent transportation system has good prospects for the future development of intelligent transportation. The intelligent transportation system could combine computer science, sensors, the Internet of Things and AI technology and be applied to the delivery of goods, service management and vehicle production. This would further strengthen the relationship between vehicles, roads and users, accelerate improvements in driving safety, reduce road jams and reduce the energy losses caused by the construction of a new transportation management system. The connection of the vehicle, the driver and the road is the core role of the intelligent transportation system. At the same time, the intelligent transportation system can also provide the corresponding service management agency. According to the analysis of relevant research conclusions, the use of intelligent transportation system can provide more intelligent road information for locomotives, and can also promote the vehicles can easily obtain the state information in front of any road, and make more effective use of road infrastructure and other resources. A large number of scholars in the field have shown that, with the use of intelligent transportation system technology, traffic jams in the next 20 years will be reduced to 40%, the existing road blockage problems will be effectively solved, the traffic accident rate could be reduced by 8%, and the resulting accident deaths would be reduced by 30~70%, which is of great significance for the healthy growth in domestic transportation and stable economic development. In recent years, the safety problems with Internet of Vehicles systems have gradually become a research hotspot. The Internet of Vehicles has the characteristics of complex dynamic topology changes, rapid vehicle movement, and unreliable transmission, which means that both the internal and external network are faced with security threats, such as malicious attacks. This means that solving the network security problems has become a great challenge. In recent years, VANET has been proposed as the basis of ITS to improve traffic efficiency and ensure the safety of vehicles and drivers. As VANET is characterized by its dynamic topology, high mobility, and variability, it is vulnerable to various attacks originating from malicious nodes. Malicious nodes in the Internet of Vehicles will broadcast and forward false traffic warning messages for selfish purposes, which will lead to traffic congestion, threaten people's lives, damage the entire VANET network function and affect its performance. They also discard received messages or refuse to help other vehicle nodes to forward messages. To enable the vehicle to operate normally and communicate on the road, it is necessary to detect false information in a timely fashion. Therefore, ensuring VANET security has become a pervasive area of research, and issues related to identifying malicious nodes and creating messages remain the focus of VANET security research.

2. Application of Regional Chain Technology in the Internet of Vehicles

The deep integration of the Internet of Things, computers, mobile communication and other technologies with intelligent transportation promotes the wide application of vehicular communication and computing equipment in vehicles. This also transforms vehicles from traditional vehicles to mobile devices with computing communication abilities. This makes communication between vehicles possible, leading to the birth and rapid rise of Internet of Vehicles technology. The Internet of Vehicles is essentially a dynamic communication system for communication between vehicles and mobile public networks. The Internet obtains the data that are shared with the vehicle, and analyzes data on the vehicle and road, vehicle and the driver, the driver and the driver, and the relationship between the driver and the third-party service providers. According to the analysis, the existence of the above relationship can be used to solve the urban traffic problems, and thus help to manage the urban intelligent transportation. In recent years, the safety problem of Internet of Vehicles systems has gradually become a research hotspot. The Internet of Vehicles has the characteristics of complex dynamic topology changes, fast vehicle movement, and unreliable transmission; therefore, both the internal and external network are faced with security threats such as malicious attacks, which mean that solving the network security problems has become a great challenge.

The network environment of the Internet of Vehicles is very complex and changeable. The authentication and trust detection methods of on-board and network (VANET) technology are not applicable to the Internet of Vehicles network environment, because this can only handle a small number of simple event requirements, and cannot effectively store, query and trace big data on vehicles. With the development of blockchain technology as a new technology solution, the problems regarding storing and querying big data on the Internet of Vehicles can be solved. The blockchain core technology consensus mechanism can ensure the security and reliability of data transmission in the network. In actual intelligent transportation applications, recording transactions' traceability through the blockchain can effectively and strongly ensure behavior traceability and responsibility backtracking, achieve fairness and justice, and promote the efficient operation of various affairs. In practice, the security risks in the Internet of Vehicles are still very large, so we need to pay closer attention to the harm caused by malicious node attacks. The rational use of blockchain technology is of great significance to the process of detecting malicious nodes in the Internet of Vehicles. The vehicle status can be globally broadcast to fundamentally suppress malicious behavior, thus ensuring that the driver is provided with reliable event messages.

Therefore, to reduce the security threat problem and network security problems, and protect the security and privacy of vehicles, it is necessary to improve the malicious node identification of the Internet of Vehicles, as well as to reasonably determine vehicle identity and the authenticity of the information. Based on this, this research chose to use a malicious node identification algorithm, focusing on the implementation of malicious node identification algorithms in the Internet of Vehicles under regional chain technology, and judge the main value of malicious node identification algorithms in the Internet of Vehicles.

3. Research Status Quo

Many domestic and foreign researchers have put forward corresponding detection and identification methods to resolve network security problems in the Internet of Vehicles.

This paper applies blockchain technology to the Internet of Vehicles to detect and identify malicious nodes, and solve the problems in the traditional Internet of Vehicles trust model. Abboud, K., Omar, H.A., et al. [1] proposed a method using RFID technology, which is implemented by verifying the vehicle's interactions with the cloud while driving on the road. According to the statistics, the vehicle will use electronic tags to send the data requiring authentication to the cloud storage, and the cloud storage will authenticate the vehicle data after receiving the information, and broadcast the information in the cloud network. Lu Zhongmei, Chen Wei, et al. [2] proposed a vehicle privacy protection method that combines the certificates issued to a vehicle by a trusted agency and pseudonyms authorized by a trusted agency to build an intelligent privacy authentication system to jointly authenticate a vehicle's identity. However, due to the frequent communication problems in the Internet of vehicles, the certification requirements cannot be met efficiently and in a timely manner. These schemes enable the vehicle to communicate on the road according to the pseudonym produced by the authority, and authenticate the identity of the vehicle through interaction with the RSU. However, these schemes are relatively dependent on the RSU, which leads to the problem of excessive RSU workload [3–5].

The application of blockchain in the Internet of Vehicles derives from the following research conclusions. VANET traditional certification and trust detection methods have many problems, which make it difficult to adapt them to the complex networking network environment. Traditional security mechanisms can only deal with a few simple events, cannot store vehicle big data, and struggle with problems regarding queries and traceability. The rise of blockchain technology means that it could be a way to solve the problems with new technology for networking security [6]. Arushi A, Kumar YS, et al. [7] propose a certified and secure data transmission algorithm to ensure that real information is communicated between nodes. The authors also introduced blockchain to some of the connected car services to improve efficiency. However, the vehicle needs to be registered at a centralized authority, and the system still has a single-point-of-failure problem. Wagner M, Mcmillin B, et al. [8] proposed a new blockchain architecture with local, physically verified transactions. With this new architecture, they proposed a protocol that protects vehicle self-organized networks (VANET) that do not regularly communicate with the RSU or other infrastructure components. In addition, they proposed a way to overcome the real-time challenges of applying Bitcoin's blockchain to disconnection by changing the transaction validation mechanism and blockchain management process, and adding a trusted CA.

In sum, the security risks in the Internet of Vehicles are great, and the harm caused by malicious node attacks cannot be ignored. Existing methods to reduces these risks in the Internet of Vehicles have been effective in their target applications, but they also face several technical challenges, including the need to improve detection accuracy and enhance privacy protection. Some scholars have applied blockchain technology to the Internet of Vehicles to solve these security and trust problems. Blockchain technology leads to decentralization and means that information cannot be tampered with, and the existing blockchain technology is used to detect malicious nodes. However, the accuracy of security messages and the malicious node detection rate need to be further improved. Therefore, the reasonable use of blockchain technology to detect malicious nodes in networks requires further research and exploration [8].

4. Application of Blockchain Technology in the Internet of Vehicles

4.1. Blockchain Technology

Blockchain technology is essentially a distributed database that stores a large amount of data, with specific transactions on each block, and notes the time when messages occur. They are connected in chronological order to form a chain structure [9]. A block in a blockchain is simply called a list of records, while a blockchain can be regarded as a record chain composed of many blocks, forming a public ledger that records a lot of encrypted data. The ledger is publicly shared among individual users. Each transaction is recorded into a new block, in which the recorded data are unchangeable and time-linked [10]. Finally, together, these new blocks form a complete blockchain in chronological order. The blockchain provides secure shared databases, ledgers, and transaction logs without being managed by central trusts [11–13]. The consistency and synchronization of the data in a blockchain are achieved through a consensus mechanism, in which a group of participants in a distrust peer network collaborate in a fully transparent way and only accept valid transactions [14]. However, Bitcoin was originally designed without taking privacy into consideration. By viewing the ledger, any public key can be used to trace back to one's real identity [15].

4.2. Structure of the Malicious Node Identification Algorithm

The ways in which vehicles communicate with each other using VANET can be divided into the following three main categories of message transmission: beacon messages, early warning messages, and entertainment messages [16–19]. Warning messages have the highest priority level among the three messages, mainly because they may threaten the performance of the Internet of Vehicles, and even affect the personal safety of drivers and passengers. In essence, the main function of early warning messages is to send road safety warning messages to the vehicle as soon as possible when an emergency occurs, making them is very important for safe driving [20]. This study mainly focuses on the authenticity of traffic early warning messages in VANET. The warning message report table is shown in Table 1.

Table 1. Warning message report form.

Event ID	Event Type	TTL_{Event} (Lifecycle of the Warning Message)	Ran_{msg} (Transmission Range of the Alert Message)
I	Traffic congestion	$Th_{t\text{-}I}$	$Th_{d\text{-}I}$
II	Traffic accident	$Th_{t\text{-}II}$	$Th_{d\text{-}II}$
III	Road construction	$Th_{t\text{-}III}$	$Th_{d\text{-}III}$
IV	Road icing	$Th_{t\text{-}IV}$	$Thd_{\text{-}IV}$

Emergencies are divided into the following four categories: traffic jams, traffic accidents, road construction, road icing [21]. The lifetime and transmission range of the alert message for each emergency are recorded. After the analysis, different types of early warning messages were found to have different life cycles and transmission ranges for the four categories of traffic jams, traffic accidents, road construction and road icing. The scheme structure diagram expression is shown in Figure 1.

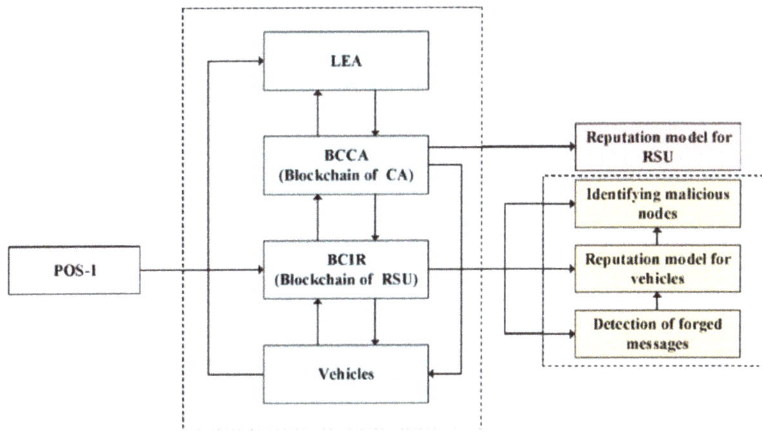

Figure 1. Schematic structure diagram.

4.3. Authentication Algorithm

There are several authentication algorithms: authentication based on shared key, authentication based on biological features, and public key encryption algorithm. Different authentication methods also have different security levels. This paper adopted the authentication method based on the shared key, which means that the server side and the user share one or one set of passwords. When the user requires authentication, the user submits a password that is jointly owned by the user and the server, either by entering the password or through a device that holds the password.

A set of vehicles in the Internet of Vehicles are set to V, obtaining:

$$V = \{V_1, V_2, V_3, \ldots V_i, V_j \ldots V_n\}$$

In the above formula, any vehicle source sending a message node is represented by V_j (source sending message node), and V_i (target receiving message node) represents the target receiving the message node. After V_j sends a message to V_i, V_i carefully verifies the integrity of the message, and then produces a series of interactive behaviors with other entities in the network [22,23]. If the alert message issued by V_j is acquired by V_i, then V sends out the collected messages. The information is sent to the nearest roadside unit (RSU) in the communication range, and the RSU must identify the information and determine whether the information is legal. After the data package sent by V_i is received by RSU, the first step in judging the message's credibility is make an accurate judgment of whether

V_i has a legal identity [24–29]. When combined with CA (authentication center in the network), the detailed authentication verification process is as follows:

Step 1. For the RSU within the vehicle communication range, vehicle V_i transmits the request, based on which the message received from vehicle V_j is recognized [30].

Step 2. After the RSU receives the request, a pseudonym containing vehicle V_i is sent with a PID_v instead. A random number (L) is reported to the nearest surrounding CA for authentication. L is randomly produced by a linear congruence generator (LCG). The LCG calculation is described as follows [30]:

$$\begin{cases} L_0 = \mathrm{d} \\ L_{\mathrm{r}+1} = (A \times L_r + Z)\mathrm{mod}(M) \end{cases}$$

In this formula, d represents the seed value, the current system time is its set initial value, the increment value is Z, multiplier is A, modulus size is M, and Z and M are prime. Generally, M will take the power square of 2.

Step 3. If the CA receives information from the RSU, the RSU will conduct a traceability search of the real situation of vehicle V_i using vehicle mapping, according to the pseudonym of vehicle V_i in the first step, and obtain the real vehicle identity information accordingly. An audit result (report) is produced after the search. In the audit results, if the false vehicle information is backed up in the results, then vehicle V_i is legal; if there is no record, then vehicle V_i is an illegal vehicle. Then, a random key, L, is encrypted using a private key (PR_{CA}) to generate a session key K_n. After the encryption of a random key L, a private key (PR_{CA}) is used to generate a session key, K_n. After using K_n for report encryption, $E_{K_n}[report]$ is obtained. After completing the above work, the RSU receives an encrypted message. By taking the result of the public key (PK_j) encryption under the RSU, the corresponding generating encryption function is as follows (E represents the encryption function, and the encrypted ciphertext is C):

$$E : C = E_{PB_{RSU}}[E_{PB_{RSU}}[K_n] || E_{k_n}[report]]$$

Step 4. When the inspection results after CA encryption are transmitted to the RSU, the RSU uses the private key (PR_{RSU}) to decrypt the generated ciphertext C, corresponding to $E_{PB_{Rsu}}[K_n]$ and $E_{K_n}[report]$. Based on the above, the RSU key is used to decrypt $E_{PB_{Rsu}}[K_n]$, and the obtained session key is K_n. After the same operation, the report views the audit results of CA, to make the final judgment on whether vehicle V_i is legal. The detailed decryption function is as follows (D represents the decryption function):

$$D : [E_{PB_{RSU}}[K_n] || E_{k_n}[report]] = D_{PB_{RSU}}[C]$$

After completing the above four steps, the RSU can make an accurate judgment regarding the legitimacy of vehicle V_i.

4.4. POS Consensus Algorithm Improvement

Compared with other traditional consensus algorithms, the main purpose of the nodes in BCIR is to fake the message verification using computer technology. Therefore, this study set up a new consensus mechanism POS1 algorithm for VANET: (1) If RSU wants to participate in the selection of bookkeeping node, it will register with CA the first time the selection of a new bookkeeping node is initiated, and will submit a part of the deposit according to the standard. This can become a candidate and enter the next step of bookkeeping selection. (2) All RSU applications are accepted by the CA, and the full energy value of the RSU will be updated. The algorithm updates are as follows:

Part 1. No matter which RSU has its own initial energy value in the blockchain network, in VANET, the RUS performs the very diverse work. For example, it actively participates in campaign billing nodes; assists CA in verifying vehicle identity information; forwards packets and assists in verifying vehicle messages; broadcasts messages; etc. Due to the differences the behavior of the participating network, the energy value of the RSU will also change. For example, in VNAET, when RSU broadcasts information, the energy will undergo a changing consumption trend. To give RSU enough encouragement, RSU needs to receive enough energy return, and the reward should be greater more than the amount of energy consumed to continue smoothly.

Part 2. The size of the RSU registration number needed to participate in the accounting node election is the variable, where any RSU corresponds to ITS own k value (energy value). If the number is 0, then $k = 0$.

Part 3. In the new round with several bookkeeping choices, CA records any registered RSU as R_n; the corresponding k is added 1 on the original basis to obtain the energy value of $(k + 1)$.

Part 4. If the energy value before R_n registration is set to Egy_{0-l}, then the deposit corresponding to R_n registration is ΔEgy_{0-l}, corresponding to the energy value that is to be reduced. The calculation is as follows:

$$\Delta Egy_{0-l} = \frac{k}{2(k+1)} Egy_{0-l}$$

Part 5. The latest RSU energy value Egy_{n-1} obtained after successful registration and deposit is as follows:

$$Egy_{\mathrm{n}-l} = Egy_{0-l} - \Delta Egy_{0-l}$$

Part 6. As long as the RSU is registered, the above steps should be followed before updating the latest energy value.

(3) For the latest energy values of all current RSU candidates, CA needs to create energy value statistics. According to the set threshold Th_{Egy}, if the latest energy value Egy_{n-1} is lower than Th_{Egy}, then CA will eliminate the RSU that should have been Egy_{n-1} the first time, and the RSU will withdraw from the election. If the latest energy value Egy_{n-1} is greater than Th_{Egy}, then the RSU continues and accepts the next election.

(4) The equity value corresponding to the RSU that is already owned in the election lineup should be calculated at this stage. The equity essentially refers to the assets or energy owned by the node. In other words, if the RSU behavior is very active, more interest will be obtained. If the behavior of the RSU is very negative, fewer benefits will arise.

Part 1. Based on the blockchain conditions, we set the exit time of the equity proof mechanism (POS) as ΔT_{POS}. At every period of exit time, the accounting function will be stimulated accordingly. If the RSU needs to select the accounting node but the interval ΔT_{POS} time has not met the standard, then, during this process, the RSU will submit the application to the CA the first time, and the application content will immediately trigger a bookkeeping election immediately. At this time, the CA in the network will be the first to elect a new accounting node.

Part 2. As there are obvious differences between the network behaviors in which each RSU participates, the corresponding energy values also show a constantly changing trend. CA needs to record each RSU, that is, to record the energy value corresponding to the RSU in each stage from the first round to the last round. The energy value corresponding to the RSU of each round is set to Egy_{x}, and the energy value of the RSU differs at different moments. Therefore, the energy value of RSU cannot be uniformly estimated for a certain round, that is, the energy value does not have a certain value. In the current study, the energy value of the RSU in each round was considered as the energy value of the last moment corresponding to the RSU in this round.

Part 3. While participating in the election, each RSU is counted by CA from the first registration election. RSU participates in and initiates many elections in the latest election, and the number of elections is recorded as J. Following the above steps, CA will calculate the energy value corresponding to the RSU in round J. In addition, the corresponding equity value in each RSU election is recorded as Stake_{R-1}, and Stake_{R-1}'s calculation formula is as follows (in the formula, a represents the return growth rate in the POS algorithm mechanism, and the constant a generally takes a value of 0.05; Egy_{x} means the corresponding energy value for each round of RSU, which is set to Egy_{x}):

$$\text{Stake}_{R-1} = \sum_{x=1}^{J} Egy_x \times (1 + a\%)^J$$

Part 4. If round F is set as a particular number, five rounds are generally selected during the calculation process, and the energy value of the corresponding RSU is in a stable state. Then, in the process of RSU equity value calculation, we the J value corresponding to RSU is chosen as zero. The corresponding J value should not be considered until the corresponding energy value of RSU changes. This is because, if the RSU does not participate in the network transaction for a very long time, the corresponding energy value will be in a stable and constant state; in other words, the energy value does not change. The F round is set as a specific round, and the equity value and J value are considered, aiming to stimulate RSU activity, and stimulate RSU to more actively participate in the election and network behavior.

Part 5. Following the above steps, CA calculates the equity value of all candidate RSU candidates and determines that the accounting node of this round is the maximum equity value corresponding to RSU. After determining the new accounting node, CA will immediately update the energy value corresponding to the RSU. By setting the energy value of the RSU after the J election as Egy_{J}, the energy value of the RSU after the (J − 1) election is recorded as $Egy_{\mathrm{J}-1}$; the refund proportion of the deposit is expressed by P_{r}; the reward issued by the successful RSU accounting node is recorded as $\Delta Egy_{\mathrm{reword}}$; the attenuation coefficient is $\mathrm{e}^{\frac{1}{\Delta T_{J-1-J}}}$. The correct update process is as follows:

$$Egy_{\mathrm{J}} = Egy_{\mathrm{J}-1} \cdot \mathrm{e}^{\frac{1}{\Delta T_{J-1-J}}} + \Delta Egy_{\mathrm{consume}}.P_r + \Delta Egy_{\mathrm{reword}}$$

To express the RSU revenue in the form of an increased energy value, the actual reward is calculated as follows:

$$\Delta Egy_{\mathrm{reword}} = \frac{1 - \Delta Egy_{J-1}}{2}$$

CA will also reset the J value of the accounting function RSU. After this round, the J value of the RSU is counted again. When the RSU obtains the bookkeeping function, CA will also reset ITS J value, and the J value of the RSU will start counting again after this round. The following algorithm details the incentive consensus mechanism (see Algorithm 1).

Algorithm 1 Consensus mechanism POS-I

Input: Egy_0, k, $\mathrm{Th}_{\mathrm{Egy}}$, $\Delta\mathrm{T}_{\mathrm{POS}}$, a, J, Pr;
Output: committer peer;
1: RSU sends a request to CA:
2: BCCA initializes an election for selecting committer peer;
3: **for** all RSU participating in the election **do**
4: R_{l} submITS $\Delta Egy_{\mathrm{consume}-1} = \frac{k}{2(k+1)} Egy_{o-l}$ as deposit;
5: Calculate $Egy_{\mathrm{e}-1} = Egy_{o-l} - \Delta Egy_{\mathrm{consume}-1}$;
6: **if** $Egy_{\mathrm{e}-1} < \mathrm{Th}_{Egy}$ Then;
7: R_{l} cannot participate in the election;
8: **else**
9: R_{l} is regarded as a candidate;
10: **end if**
11: Calculate $\mathrm{Stake}_{\mathrm{R}-1} = \sum_{x=1}^{J} Egy_x \times (1 + \mathrm{a}\%)^J$;
12: **end for**
13: Selecting the node whose has $\mathrm{Max}_{\mathrm{stake}}$ as the committer peer;
14: Calculate $\Delta Egy_{\mathrm{reword}} = \frac{1-Egy_{J-l}}{2}$
15: Calculate $Egy_J = Egy_{J-1} \times \mathrm{e}^{\frac{1}{\Delta T_{J-1-J}}} + \Delta Egy_{consume} \times \mathrm{Pr} + \Delta Egy_{reword}$,
16: Output committer peer.

5. Results Analysis of the Malicious Node Identification Algorithm of Blockchain Technology

5.1. Simulation Software Setting

The experimental simulation verifies the performance of the false message detection mechanism on the ONE simulation platform. ONE is a discrete-time engine open-source simulation platform written in the Java language. It is a simulation software that is mainly used to design and evaluate the routing mechanisms of data forwarding and message communication in the Internet of Vehicles. The ONE simulation platform contains a variety of simulation modules, which will be updated during each simulation process to realize the whole simulation function. The main functions of the ONE simulation platform include: mobile modeling of nodes, sending, receiving and processing routing messages, node communication, visual representation of results, etc. This platform has various different communication protocols to simulate the process of message transmission and generate the trajectory information during the transmission process. All ONE's components are independent of each other. ONE simulates the network simulation environment to test and optimize the security scheme performance. In the simulation experiments, a motion model based on the shortest path map was used in the ONE to simulate vehicle behavior on the road. The model initially places nodes in random locations, but selects a specific destination for all nodes in the map. Vehicle nodes are not arbitrary and include various types: emergency vehicles (police cars and ambulances), vehicles with fixed lines (buses and trams), and randomly distributed vehicles (private cars and taxis). There are two scenarios in the simulation experiments: low flow density and high flow density. Each simulation was run 20 times on average. In the experiment, the number of malicious vehicles that sent forged or forged messages in VANET ranged from 10% to 45%. When malicious vehicles exceed 45%, VANET's efficiency will dramatically decline, and it will be unable to provide any reliable services.

- Suppose an attacker cannot attack more than half the vehicles on the network.
- Authorities and the RSU are equipped with customized hardware with high computing abilities.
- Certification bodies and RSU are equipped with custom hardware, with a much higher computing power than general computers.
- As long as the public or private keys are not stolen, encryption technology can be used to provide secure communication channels between entities.

5.2. Evaluating Indicator

According to the requirements of the Internet of Vehicle malicious node identification algorithm, we selected the following indicators to verify the performance and results of the algorithm.

The first indicator is the false alarm rate (*FAR*), detecting the number of messages identified as false using the N_i representative, and detecting the number of false messages identified as true information using the N_j representative, from which *FAR* can be obtained:

$$FAR = \frac{N_i}{N_j}$$

The second indicator is the missed detection rate (*MDR*), where the number of detected true messages identified as false messages is replaced by N_{miss}, while the number of detected messages identified as true messages is N. From this, the *MDR* is obtained:

$$MDR = \frac{N_{miss}}{N}$$

Simulation experiments were carried out to verify the results of the malicious node identification algorithm of the Internet of Vehicles, and *FAR* and *MDR* were used to evaluate the results of the Internet of Vehicles' trust evaluation algorithm. The independent variable taken by the experiment is the proportion of malicious vehicle nodes in all vehicle networks, and the results of the malicious vehicle node identification algorithm are evaluated for low-traffic-density and high-traffic-density conditions.

Whether the traffic density is low or high, the malicious node identification algorithm is less than 20%, and the *FAR* decreases as the percentage increases. At the same time, a significant trend towards decreased *MDR* also occurs with increasing percentage. Details are shown below. In the presence of a false information attack and black hole attack, the anti-attack performance of the malicious message detection algorithm is analyzed using several details: delivery rate (*Dr*), average end-to-end delay (*Ad*), and network overhead (*Or*).

First, *Dr* is calculated. The actual meaning of *Dr* is the ratio of the number of messages that are successfully sent to the target node or the designated location to the total number of messages that are generated and sent by the source vehicle nodes in the network. The calculation formula is as follows:

$$Dr = \frac{N_a}{N_t}$$

In the formula, the total number of messages generated and sent by the source vehicle node is N_t, the number of messages successfully sent to the target vehicle or designated location is N_a, and there is a proportional relationship between the communication quality between the nodes and *Dr*. In other words, the higher the *Dr*, the higher the communication quality, meaning that the overall network performance is very good. If the *Dr* is lower, then the communication quality is lower, meaning that the overall network performance is very poor.

Second, *Ad* is calculated. *Ad* is a very important indicator when evaluating the anti-attack performance of malicious message detection algorithms; by using the Ad evaluation method, how the additional overhead of security measures increases latency in the routing process can be described. The actual meaning of *Ad* is the average time taken to pass a message between two nodes

$$Ad = \frac{\sum_{n-1}^{N} T_n}{N}$$

In this formula, the total number of sent messages is recorded as N; the total time taken to send each message during N messages is $\sum_{n-1}^{N} T_n$. After analysis, an inverse relationship was found between *Ad* and the overall simulation performance; that is, if *Ad* is shorter, the corresponding performance is better, and if *Ad* is longer, the corresponding performance is worse.

Finally, *Or* is calculated. *Or* is another very important indicator when evaluating the anti-attack performance of the malicious message detection algorithm. This refers to the ratio of the total number of nodes involved showing the target node the total number of nodes involved in forwarding the message to the nodes required for transmission. The calculation formula is as follows:

$$Or = \frac{Num_{relay}}{Num_{total}}$$

In the formula, the number of nodes involved in forwarding the packet is recorded as Num_{relay}; the number of all nodes required during the transmission is recorded as Num_{total}. After analysis, an inverse relationship was found between *Or* and performance. That is, the smaller *Or*, the better the network performance of the corresponding algorithm, the fewer network resources that are occupied and the lower the number of additional nodes that are occupied. However, the larger the *Or*, the worse the corresponding algorithm network performance, indicating that more network resources are occupied and more additional nodes are occupied.

5.3. Simulation Software and Simulation Environment

5.3.1. Simulation Software

Simulation software: the ONE

The ONE is a discrete-time engine open-source simulation platform written in the Java language. This is a simulation software, which is mainly used to design and evaluate the routing mechanism of data forwarding and message communication in the Internet of Vehicles. The real-time interactive graphical interface of the ONE simulation is shown in Figure 2.

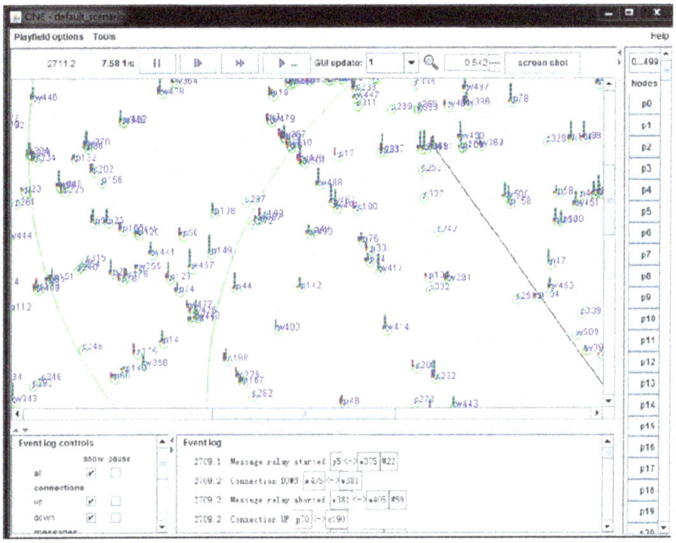

Figure 2. Interactive graphical interface of the emulator.

The emulator involves the classification and number of node types in the network, the path model of node movement, and the number of messages generated by the interaction, which are all displayed in real time in the interface. The ONE simulation platform contains a variety of simulation modules, which will be updated in each simulation process to realize the whole simulation function. The main functions of the ONE simulation platform include: mobile modeling of nodes, sending, receiving and processing routing messages, node communication, visual representation of results, etc. In this simulation platform, various different communication protocols simulate message transmission, and generate the trajectory information during the transmission process. All of ONE's components are independent of each other. The main structure and modules of the ONE simulation platform are shown in Figure 3.

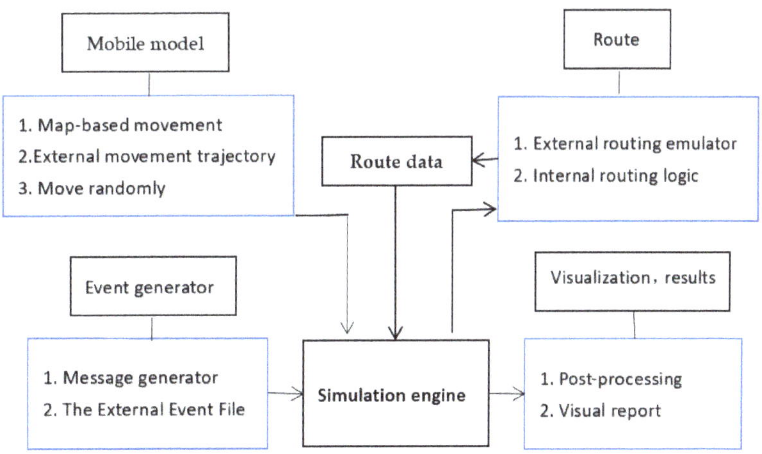

Figure 3. Simulation platform architecture.

5.3.2. Simulation Environment

ONE is used to simulate the network simulation environment to test and optimize the performance of the security scheme. In simulation experiments, a motion model based on the shortest map path was used in ONE to simulate vehicle behavior on the road. The model initially places nodes at random locations, but selects a specific destination for all nodes in the map and generates the shortest path from the start to end point using the Dijkstra algorithm. This experiment was implemented on the Helsinki city map, and a scene size of 4500 m × 3400 m was selected for the simulation experiments, as shown in Figure 4.

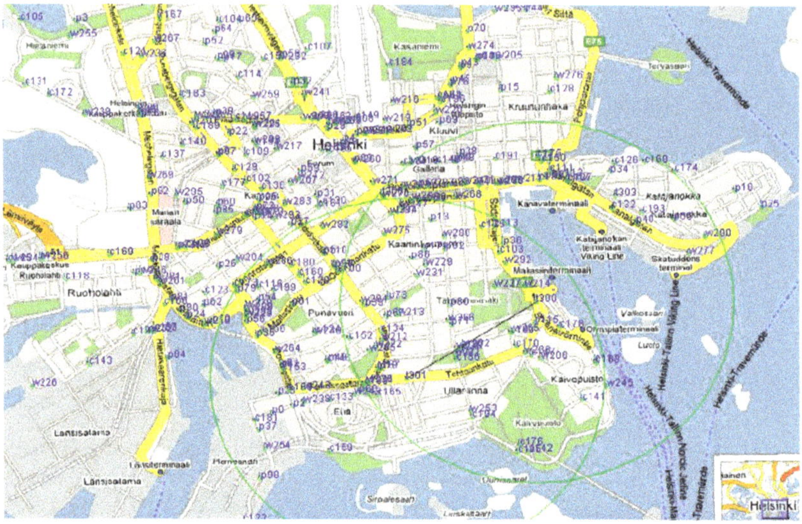

Figure 4. City map of Helsinki, Finland.

In the simulation experiment, the vehicle nodes communicate messages through the IEEE 802.11p communication protocol. The specific experimental simulation parameters are shown in Table 2.

Table 2. Experimental simulation parameters are set.

The Parameter Name	Parameter Values
Simulation Scene Range (m^2)	4500 × 3400
Simulation Time (s)	0–43,200
Vehicle mobility model	Shortest path movement model
Vehicle node type grouping (group)	11
Total number of vehicle nodes (individual)	200–600
Vehicle communication range (m)	10
Node speed (m/s)	2
RSU quantity (s)	10
Vehicle Cache Size (M)	40
Packet lifecycle (min)	15

There are two scenarios in the simulation experiments: low flow density and high flow density. Each simulation was run 20 times on average. In the experiment, the number of the malicious vehicles that sent forged messages or forged messages in VANET ranged from 10% to 45%. When malicious vehicles exceed 45%, VANET's efficiency will dramatically decline, and it will be unable to provide any reliable services. Therefore, this extreme case is not considered in this paper.

The locations of all nodes are shown in Figure 5.

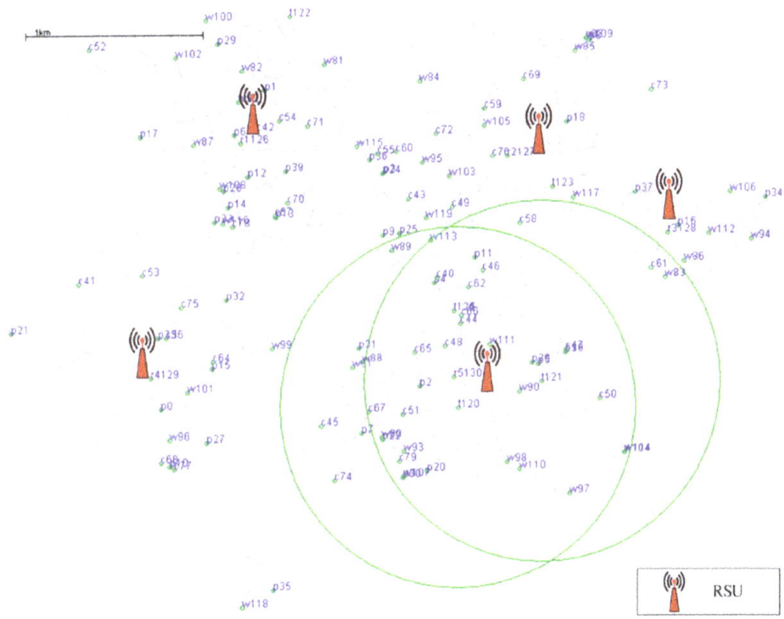

Figure 5. The location of all nodes.

To smooth the simulation experiment, some necessary assumptions in the theoretical algorithm and simulation experiments are presented as the basis of the proposed scheme.

- Suppose that an attacker cannot attack more than half of the vehicles in the network.
- Authorities and the RSU are equipped with customized hardware with high computing abilities.
- Certification bodies and RSU are equipped with custom hardware with a much higher computing power than general computers.
- As long as the public or private keys are not stolen, encryption technology can be used to provide secure communication channels between entities.

5.4. Simulation Results

To objectively and effectively analyze the malicious node detection mechanism algorithm based on blockchain, the simulation experiment is divided into two groups, looking at the number of vehicles for low-flow-density and high-flow-density scenarios. The results of each experiment are the average of the data obtained from 20 simulations. The simulation experiment takes the proportion of malicious vehicle nodes in the network as the independent variable and analyzes the algorithm's performance under low and high traffic flow densities. The experimental simulation results are as follows (See Table 3).

Overall, high-density scenarios outperform low-density scenarios using this algorithm. The proposed algorithm has a low *FAR* at both high and low densities. This is because this algorithm uses the blockchain to jointly verify the legitimacy of the sending nodes, determine the integrity and reliability of the VANET transmission messages, and greatly improve the identification rate of malicious messages.

Dr, *Ad* and *Or* were used to evaluate and analyze the performance, and each experiment was the average of the data obtained from 20 simulations. The simulation experiment also takes the proportion of all vehicles of malicious vehicle nodes in the network as the independent variable and analyzes the algorithm's performance in the case of low traffic density and high flow density, respectively. The experimental simulation results are as follows (See Table 4).

Table 3. The *FAR* versus *MDR* comparison results.

Number of Malicious Vehicle Nodes in the Network	*FAR*		*MDR*	
	Low Flow Density	High Flow Density	Low Flow Density	High Flow Density
15%	0.156	0.112	0.100	0.105
25%	0.168	0.125	0.125	0.119
35%	0.171	0.135	0.129	0.120
45%	0.182	0.152	0.130	0.125

Table 4. Comparative results of the D, A, and O.

Number of Malicious Vehicle Nodes in the Network	*Dr* (%)		*Ad* (s)		*Or* (KB/S)	
	Low Flow Density	High Flow Density	Low Flow Density	High Flow Density	Low Flow Density	High Flow Density
10%	70.02	94.25	0.102	0.118	29.52	129.20
15%	69.52	90.21	0.161	0.156	40.20	134.56
20%	68.65	88.68	0.175	0.167	43.25	142.30
25%	67.25	86.98	0.201	0.210	50.20	149.52
30%	66.82	86.70	0.214	0.234	53.29	150.29
35%	66.30	84.02	0.226	0.301	58.63	167.98
40%	65.48	80.20	0.238	0.365	61.30	165.32
45%	65.00	78.68	0.262	0.402	67.05	172.02

In general, comprehensive experimental performance analysis can conclude with an increase in experimental simulation time and the gradual increase of malicious nodes in the network. If faced with the experimental simulation time and the high density of malicious vehicle nodes, the above algorithm can still play a significant role, and it has been verified by many scholars.

6. Conclusions

After a series of performance tests and verification of the results of the malicious node identification algorithm based on blockchain technology, the algorithm proposed in this paper can identify vehicle information in either low- or high-density vehicle scenarios. Based on the judgment of *FAR*, *MDR* and other indicators, the Internet of Vehicles' malicious node identification algorithm based on blockchain technology has a very significant false alarm rate and missing detection rate when identifying vehicle information. In this paper, *Dr*, *Ad*, *Or* and other indicators were selected to reveal malicious node identification algorithm's performance in the Internet of Vehicles. In general, the computing system based on blockchain technology is feasible, and ITS network performance is superior to other algorithms.

In the face of the increasingly prominent security needs and the complex diversity of malicious node attacks in the Internet of Vehicles, to further improve the detection of malicious nodes, the following perspectives could be the focus of future research work. First, the malicious node recognition algorithm model could be optimized to improve the system availability. The algorithm presented here is a result of the limited experimental conditions. A transportation service platform should be considered to evaluate the performance of this plan in real-world transportation networks. Second, the evaluation indicators should be increased to improve system comprehensiveness. The application of blockchain technology in car networking applications is relatively low at present. Although the application of blockchain technology in the field of car networking has shown a good effect in this paper, the article used in the network evaluation index is not comprehensive. In future, other security metrics will need to be performance evaluated to integrate blockchain technology more comprehensively with automotive networks. Third, traffic scenarios should be enriched to improve the robustness of the system.

This blockchain, car networking, malicious-node-detection mechanism used a ONE simulation platform simulation experiment, looking at high-density and low-density scenario car scenarios to analyze the networking and malicious node identification algorithm performance. The malicious node detection performance of *FAR* and *MDR*, for algorithm models *Dr*, *Ad*, *Or* were also looked at, to conduct a comparative analysis of the algorithm performance. The final experimental results show that the proposed trust detection mechanism is feasible, and the ITS network performance is superior to that of other algorithms.

The blockchain-based malicious node detection mechanism proposed in this paper has a very significant application effect. The aim was build a model in more diversified traffic scenarios to test the security and trust mechanism, which can further improve the detection rate of malicious nodes and optimize the algorithm performance. In the future, this mechanism can be applied to richer traffic scenarios to improve its expansion and robustness.

Author Contributions: Conceptualization, J.C.; Project administration, T.L.; Resources, T.L.; Writing—original draft, J.C.; Writing—review & editing, R.Z. All authors have read and agreed to the published version of the manuscript.

Funding: Authors received funding from Yunnan Province's major science and technology special plan project "Research and application demonstration of key technologies of blockchain serving key industries": 202002AD080002.

Institutional Review Board Statement: Not applicable.

Informed Consent Statement: Not applicable.

Data Availability Statement: Not applicable.

Conflicts of Interest: The authors declare no conflict of interest.

References

1. Abboud, K.; Omar, H.A.; Zhuang, W. Inter working of DSRC and Cellular Network Technologies for V2X Communications: A Survey. *IEEE Trans. Veh. Technol.* **2016**, *65*, 9457–9470. [CrossRef]
2. Zhongmei, L.; Wei, C.; Jie, W.; Haitao, Y. Very low latency and high reliability communication of Internet of Vehicles: Status and Outlook. *Signal Process.* **2019**, *35*, 1773–1783.
3. Nasrollahi, M.; Fathi, M.R. Modeling Big Data Enablers for Service Operations Management. In *Big Data and Blockchain for Service Operations Management*; Springer: Cham, Switzerland, 2022; pp. 49–94.
4. Ravari, S.S.M.; Fathi, M.R.; Mohammadi, M.; Bandarian, R. Investigating the concept of effectiveness in technology development projects in a research and technology organizations; evaluating eight technology development projects in the Research Institute of Petroleum Industry (RIPI). *Pet. Bus. Rev.* **2020**, *4*, 21–41. [CrossRef]
5. Yu, H.; Zhao, C.; Li, S.; Wang, Z.; Zhang, Y. Pre-Work for the Birth of Driver-Less Scraper (LHD) in the Underground Mine: The Path Tracking Control Based on an LQR Controller and Algorithms Comparison. *Sensors* **2021**, *21*, 7839. [CrossRef]
6. Parsajoo, M.; Armaghani, D.J.; Mohammed, A.S.; Khari, M.; Jahandari, S. Tensile strength prediction of rock material using non-destructive tests: A comparative intelligent study. *Transp. Geotech.* **2021**, *31*, 100652. [CrossRef]
7. Arushi, A.; Kumar, Y.S. Block Chain Based Security Mechanism for Internet of Vehicles (IoV). In Proceedings of the 3rd International Conference on Internet of Things and Connected Technologies (ICIoTCT), Jaipur, India, 26–27 March 2018; pp. 267–272.
8. Wagner, M.; Mcmillin, B. Cyber-Physical Transactions: A Method for Securing VANETs with Blockchains. In Proceedings of the IEEE Pacific Rim International, Symposium on Dependable Computing, Taipei, Taiwan, 4–7 December 2018; pp. 64–73.
9. Yu, H.; Li, S. The Function Design for the Communication-Based Train Control (CBTC) System: How to Solve the Problems in the Underground Mine Rail Transportation? *Appl. Syst. Innov.* **2021**, *4*, 31. [CrossRef]
10. Campanile, L.; Iacono, M.; Marulli, F.; Mastroianni, M. Designing a GDPR compliant blockchain-based IoV distributed information tracking system. *Inf. Process. Manag.* **2021**, *58*, 102511. [CrossRef]
11. Chen, W.; Yuan, L.; Wei, L. Progress of Internet of Vehicles Industry and Key Technology Analysis. *Zte Technol.* **2020**, *26*, 5–11.

12. Lin, L.; Lu, L.; Yuming, G. Analysis of Internet of Vehicles Communication Standardization and Industry Development. *Telecommun. Sci.* **2020**, *36*, 15–26.
13. Rongyue, Y.; Pengzhou, Z.; Qing, S. Research and Outlook of Intelligent Internet of Vehicles based on 5G Technology. *Telecommun. Sci.* **2020**, *36*, 106–114.
14. Jiahui, Q.; Zhichao, Z.; Xiaobo, L.; Yu, X.; Chao, C.; Liu, L. Research and Application of Internet of Vehicles Technology based on MEC. *Telecommun. Sci.* **2020**, *36*, 45–55.
15. Junhao, Y.; Zongpu, J.; Dongying, L. Intelligent Mine Internet of Vehicles System Architecture and Key Technologies. *Coal Sci. Technol.* **2020**, *48*, 249–254.
16. Rundong, W.; Wanwei, L.; Xiuliang, M.; Wenjun, Y. Summary of Mutual trust Certification and safe Communication of Internet of Vehicles. *Comput. Sci.* **2020**, *47*, 1–9.
17. Xiaolong, X.; Zijie, F.; Lianyong, Q.; Wanchun, D.; Qiang, H.; Yucong, D. Distributed service uninstallation method based on deep reinforcement learning in the edge computing environment of Internet of Vehicles. *J. Comput. Sci.* **2021**, *44*, 2382–2405.
18. Li, S.; Wang, G.; Yu, H.; Wang, X. Engineering Project: The Method to Solve Practical Problems for the Monitoring and Control of Driver-Less Electric Transport Vehicles in the Underground Mines. *World Electr. Veh. J.* **2021**, *12*, 64. [CrossRef]
19. Momeni, E.; Nazir, R.; Armaghani, D.J.; Maizir, H. Prediction of pile bearing capacity using a hybrid genetic algorithm-based ANN. *Measurement* **2014**, *57*, 122–131. [CrossRef]
20. Tao, S.; Xiuhua, L.; Hui, L.; Junhao, W.; Qingyu, X.; Jie, C. Summary of research on the car network security encryption authentication technology in the era of big data. *Comput. Sci.* **2022**, *49*, 340–353.
21. Xiao, W.; Tingting, T.; Shuangshuang, H.; Dongpu, C.; Feiyue, W. Parallel vehicle networking: Intelligent vehicle network management and control based on ACP. *J. Autom.* **2018**, *44*, 1391–1404.
22. Xinghua, L.; Cheng, Z.; Ying, C.; Huilin, Z.; Jian, W. A Review of Internet of Vehicles Security. *J. Inf. Secur.* **2019**, *4*, 17–33.
23. Wang, R.; Deng, X.; Xu, Z.; Zhao, X. Review of Simulation Test and Evaluation Technology of Internet of Vehicles. *Comput. Appl. Res.* **2019**, *36*, 1921–1926, 1939.
24. Paschek, D.; Mocan, A.; Draghici, A. Industry 5.0–The Expected Impact of Next Industrial Revolution. In *Thriving on Future Education, Industry, Business and Society, Proceedings of the MakeLearn and TIIM International Conference, Piran, Slovenia, 15–17 May 2019*; ToKnowPress: Bangkok, Thailand, 2019.
25. Hasanipanah, M.; Monjezi, M.; Shahnazar, A.; Armaghani, D.J.; Farazmand, A. Feasibility of indirect determination of blast induced ground vibration based on support vector machine. *Measurement* **2015**, *75*, 289–297. [CrossRef]
26. Garg, S.; Singh, A.; Aujla, G.S.; Kaur, S.; Batra, S.; Kumar, N. A Probabilistic Data Structures-Based Anomaly Detection Scheme for Software-Defined Internet of Vehicles. *IEEE Trans. Intell. Transp. Syst.* **2020**, *22*, 3557–3566. [CrossRef]
27. Junejo, M.H.; Ab Rahman AA, H.; Shaikh, R.A.; Mohamad Yusof, K.; Memon, I.; Fazal, H.; Kumar, D. A Privacy-Preserving Attack-Resistant Trust Model for Internet of Vehicles Ad Hoc Networks. *Sci. Program.* **2020**, *2020*, 1–21. [CrossRef]
28. Lamba, K.; Singh, S.P. Modeling Big Data Enablers for Operations and Supply Chain Management. *Int. J. Logist. Manag.* **2018**, *29*, 629–658. [CrossRef]
29. Chen, J.M.; Li, T.T.; Panneerselvam, J. TMEC: A Trust Management Based on Evidence Combination on Attack-Resistant and Collaborative Internet of Vehicles. *IEEE Access* **2018**, *7*, 148913–148922. [CrossRef]
30. Farooq, S.M.; Hussain, S.M.S.; Ustun, T.S. A Survey of Authentication Techniques in Vehicular Ad-Hoc Networks. *IEEE Intell. Transp. Syst. Mag.* **2021**, *13*, 39–52. [CrossRef]

applied sciences

MDPI

Article

Multifunctional Models, Including an Artificial Neural Network, to Predict the Compressive Strength of Self-Compacting Concrete

Kawan Ghafor

Civil Engineering Department, College of Engineering, University of Sulaimani, Sulaymaniyah 46001, Iraq; kawan.ghafor@univsul.edu.iq; Tel.: +96-4770-1562-324

Abstract: In this study, three different models were developed to predict the compressive strength of SCC, including the nonlinear relationship (NLR) model, multiregression model (MLR), and artificial neural network. Thus, a set of 400 data were collected and analyzed to evaluate the effect of seven variables that have a direct impact on the CS, such as water to cement ratio (w/c), cement content (C, kg/m^3), gravel content (G, kg/m^3), sand content (S, kg/m^3), fly ash content, (FA, kg/m^3), superplasticizer content (SP, kg/m^3), and curing time (t, days) up to 365 days. Several statistical assessment parameters, such as the coefficient of determination (R^2), root mean squared error (RMSE), mean absolute error (MAE), and scatter index (SI), were used to assess the performance of the predicted models. Depending on the statistical analysis, the median percentage of superplasticizers for the production of SCC was 1.33%. Furthermore, the percentage of fly ash inside all mixes ranged from 0 to 100%, with 1 to 365 days of curing and sand content ranging from 845 to 1066 kg/m^3. The results indicated that ANN performed better than other models with the lowest SI values. Curing time has the most impact on forecasts for the CS of SCC modified with FA.

Keywords: SCC; compressive strength; fly ash; statistical analysis; modeling

Citation: Ghafor, K. Multifunctional Models, Including an Artificial Neural Network, to Predict the Compressive Strength of Self-Compacting Concrete. *Appl. Sci.* **2022**, *12*, 8161. https://doi.org/10.3390/app12168161

Academic Editors: Pijush Samui, Aydin Azizi, Ahmed Hussein Kamel Ahmed Elshafie, Yixia Zhang and Danial Jahed Armaghani

Received: 5 July 2022
Accepted: 29 July 2022
Published: 15 August 2022

1. Introduction

Self-consolidating concrete or self-compacting concrete (SCC) is a concrete mix that has low yield stress, high deformability, good segregation resistance (prevents separation of particles in the mix), and moderate viscosity (necessary to ensure uniform suspension of solid particles during transportation, placement (without external compaction), and until the concrete sets). In everyday terms, when poured, SCC is an extremely fluid mix with the following distinctive practical features: it flows very easily within and around the formwork, can flow through obstructions and around corners ("passing ability"), is close to self-leveling (although not self-leveling), does not require vibration or tamping after pouring, and follows the shape and surface texture of a mold (or form) very closely once set. As a result, pouring SCC is much less labor-intensive than standard concrete mixes. Once poured, SCC is usually similar to standard concrete in terms of its setting and curing time (gaining strength) and strength. SCC does not require a high proportion of water to become fluid and SCC may contain less water than standard concretes. Instead, SCC gains its fluid properties from an unusually high proportion of fine aggregate, such as sand (typically 50%), combined with superplasticizers (additives that ensure particles disperse and do not settle in the fluid mix) and viscosity-enhancing admixtures (VEAs) [1,2]. The introduction of SCC represents a major technological advance, leading to a better quality of concrete produced and a faster and more economical concrete construction process [3]. When there is a lot of reinforcement in the structure members, using an SCC is the perfect choice for structural components. A superplasticizer and other mineral admixtures can increase the SCC's high flowability and stability [4].

Fly ash (FA) is a siliceous or aluminosilicate particle that has been finely divided. When moisture is present, it chemically combines with the calcium hydroxide generated when Portland cement hydrates to create a compound with cementing properties [5]. As a byproduct of burning pulverized coal, FA is a finely split residue produced in coal-fired power plants. The coal is crushed and blown into the burning chamber, where it burns right away to heat the boiler tubes. Regarding the influence of fly ash on self-compacting concrete (SCC) properties, a study indicates that a high volume of fly ash can be used in SCC to produce high-strength and low-shrinkage concrete. Subhan Ahmad et al. compared hardened properties of normal concrete (NC) and self-compacting concrete (SCC) [6]. While most of the impurities deposited in coal during its formation, such as clay, shale, quartz, and feldspar, typically fuse and stay suspended in the flue gas, the volatile matter and carbon are burned off. The importance of a low water–cement ratio for enhancing the durability of concrete has long been accepted. Low water content leads to low workability of the fresh concrete; if this concrete is not properly compacted, the durability of the structures will be impaired. In the mid-1970s, superplasticizers were introduced to the market to produce rhinoplastic concretes [7,8]. The flaming material is swiftly transferred to cooler regions, where it is formed into sphere-shaped particles. Lighter ash particles (FA) stay suspended in exhaust gases while heavier ash particles fall to the bottom of the burning chamber. Mechanical separators, electrostatic precipitators, or bag filters are used to remove these FA particles [9]. Previous research has shown that using mineral admixtures such as fly ash in concrete mixes can improve the slump without increasing expense while lowering the amount of superplasticizer needed to achieve equal slump flow when compared to concrete prepared only with Portland cement, and the presence of the fly ash lessens the heat of hydration and increases the durability [10]. Due to its finer and more rounded particles than cement, fly ash can be used to replace cement in the creation of self-compacting concrete. Fly ash could operate as a filler to cover the inter-particle spaces that cement is unable to occupy by making use of these physical features [11].

SCC consists of similar components to conventionally vibrated concrete: cement, aggregates, water, mineral, and chemical admixtures. The passing ability and segregation resistance of SCC are achieved by the decrease in coarse aggregate content and increases in the powder quantities. The superplasticizers (high-range water reducers) are responsible for the high fluidity of the concrete mix, while the powder and viscosity-modifying agents lead to better stability and cohesion, reducing bleeding and segregation of the mix.

Mineral admixtures are used as extra-fine material; the most used is the limestone filler which represents a chemically inert by-product of limestone crushers. The addition of limestone powder improves the particle packing by filling the small pores between cement grains and augmenting the water retention of fresh mixes. The compressive strength is calculated by the failure load divided by the cross-sectional area resisting the load and reported megapascals (MPa). Concrete hardens and acquires strength as it hydrates. This process occurs rapidly at first and slows down as time goes by; it continues over a long period.. This study investigated the compressive strength at 28 days, an age considered for design purposes when a substantial percentage of hydration has taken place. The cube and cylinder strength of SCC and conventional vibrated concrete was determined at the same w/c ratio and cement content [12,13]. To ascertain the compressive strength of SCC modified with FA, several cylindrical or cubical specimens are currently produced and tested over various curing times in the practice field. Generally speaking, work on the construction site should not continue until the results of the compressive strength test at a given age, particularly 28 days, are available. Due to this, testing takes longer than expected and is more expensive [14]. Construction sites are also delayed as a result. Fly ash can replace 40% of the cement in self-compacting concrete to increase its strength to 65 MPa [10]. The term "binder content" refers to the cement and any additional pozzolanic elements that may be included in the mixture to enhance specific SCC characteristics. FA is additionally one of the most reactive pozzolanic materials. With an increase in cement or binder content, the compressive strength of SCC was typically noticeably enhanced [4,15].

When the amount of cement was increased from 360 to 450 kg/m^3 at the same water content, the compressive strength of SCC increased from 48 and 52 MPa to roughly 52 and 60 MPa at 28 and 90 days, respectively [16–18].

The w/c or w/b ratio, similar to normal concrete, is one of the key factors affecting the CS of SCC at different curing ages. Compared to regular concrete, SCC frequently has a lower w/b ratio. When the w/b ratio is lower, a higher percentage of cement and binder ingredients produces a higher CS and a more homogeneous matrix [4]. The w/c or w/b ratio, similar to normal concrete, is one of the key factors affecting the CS of SCC at different curing ages. Compared to normal concrete, SCC frequently has a lower w/b ratio. When the w/b ratio is lower, a higher proportion of cement and binder ingredients produces a higher CS and a more homogeneous matrix. When the w/b ratio was decreased from 0.45 to 0.35 up to 56 days of curing, the CS of SCC was significantly improved at all curing ages [19]. The w/b ratio affects the CS of regular concrete more so than it does the CS of SCC [20].

One of the main advantages SCC has over other special concrete varieties is that it does not need any special curing procedures. Similar to regular concrete, SCC is cured using the same procedure. According to earlier studies [4,21], the most crucial factor affecting the CS of SCC is the age of the concrete samples after the examination. Longer curing times produced better compressive strength for the same mix proportions because of the faster rate of hydration. The CS of SCC was increased from 13 MPa to 54 MPa by extending the curing time from one day to 28 days, as demonstrated by Corinaldesi and Moriconi [22]. According to Sahmaran et al. [23], the 7-day CS of SCC was 55.9 MPa. However, when the curing age of SCC was raised to 180 days, this value increased to 71 MPa. SP's high cement and binder contents make it impossible to make SCC without it. The presence of SP improves workability. Although SP is added to SCC to enhance its rheology, different dosages can also affect compressive strength. The CS of SCC increased by 30% after 28 days of curing as the SP dosage increased from 5.5 kg/m^3 to 8.25 kg/m^3 [24]. Aggregates comprise 60 and 70 percent of SCC's volume, significantly impacting the material's rheological and hardened performance. Compared to conventional concrete, SCC typically contains less coarse material [4]. According to earlier studies, the CS of SCC was less affected by the fine aggregate content than the coarse aggregate content.

Additionally, compared to coarse aggregates with a maximum size of 10 mm, those with a maximum size of 20 mm had a higher CS of SCC [25]. Artificial neural networks have been widely used in many studies, such as predicting the piezoelectric effect of the plate of engineering structures in vibration and noise reduction. This study employed an artificial neural network (ANN) model to explore the piezoelectric patch size and thickness's effect on a plate's first-order natural frequency and displacement amplitude. With the finite element method (FEM), a rectangular plate actuated by a piezoelectric patch was analyzed with various patch sizes. The FEM data were later used to build an ANN model. The dynamic response of the plate was predicted by the ANN model and validated with FEM in terms of first-order natural frequency and displacement amplitude. Case studies showed that the ANN model could accurately predict both natural frequency and displacement amplitude with patch length, width, and thickness input. When the information of the ANN model was simplified to patch size and thickness or the patch's volume, the accuracy worsened. The patch size and thickness's influences on the first-order natural frequency were coupled, and the maximal and minimal values were predicted based on the ANN model. The purpose of this research is to develop and apply the artificial neural network (ANN) with an extreme learning machine (ELM) to forecast the gross domestic product (GDP) growth rate. The economic growth forecasting was analyzed based on the value of agriculture, manufacturing, industry, and services added to GDP. The results were compared with the ANN with the backpropagation (BP) learning approach since BP could be considered a conventional learning methodology. The reliability of the computational models was assessed based on simulation results and several statistical indicators. The results showed that the ANN with ELM learning methodology could be applied effectively in applications of GDP forecasting [23–27].

Since compressive strength depends on various characteristics and mixture proportions, more sophisticated methodologies should be used to lessen the need for laboratory experiments and provide engineers with more convenient tools and numerical equations for predicting experimental results. An excellent strategy is to use soft computing techniques. The main benefit of using these techniques is the development of workarounds and solutions for linear or nonlinear problems where mathematical models cannot readily show the relationship between the pertinent factors in the problem [27–30]. A methodical multiscale model was developed to calculate the CS of concrete made with significant amounts of fly ash [28]. Four hundred fifty different experimental datasets were used to create a model. M5P trees, linear, nonlinear, and multilogistic regression were developed as five distinct qualification modeling methods. It was concluded that the ANN, M5P tree, and multilogistic regression models could predict the CS of large-volume fly ash concrete wells based on higher R^2 and lower RMSE and MAE values [29].

Despite the wide use of FA in SCC mixes in the earlier studies, an accurate model could not be identified for effective use by the construction industry. As a result, it was attempted in this study to analyze and quantify the effects of a wide range of mixture proportions, including FA content, cement content, w/b ratio, SP content, fine aggregate content, and coarse aggregate content, on the CS of SCC up to 365 days of curing. Different model approaches, such as nonlinear regression, multiregression, and ANN models, were used to predict the CS of SCC modified by FA using 400 data samples from the literature. The main goals are to (i) conduct statistical analysis and assess how mixture compositions such as FA content, C content, G content, w/c ratio, SP content, and curing time affect the CS of SCC; (ii) examine and design a reliable model for estimating the CS among all models (NLR, MLR, and ANN models) using appropriate statistical measures; and (iii) reduce the overall cost by designing mixes with any necessary specifications without the need for trail mixes.

2. Materials and Methods

The collected data from various literature for this paper were 400; inserting all data into an Excel sheet, they were sorted into two groups. The largest group comprised two-thirds of the collected data and was named as the training dataset and used to develop the models. The other group, one-third of the dataset, was used to test the proposed models. As mentioned earlier, the main objective of this study is to design models to predict the compressive strength of SCC. Table 1 includes the detailed ranges and information about each variable studied in the paper, which was water to cement ratio (w/c), cement content (C, kg/m^3), gravel content (G, kg/m^3), sand content (S, kg/m^3), fly ash content, (FA, kg/m^3), superplasticizer content (SP, kg/m^3), and curing time (t, days). All mentioned independent parameters were used to evaluate and estimate the compressive strength and compare the results with the measured strength. The procedure of the current study is simply designed in Figure 1.

Table 1. Datasets used in the modeling process.

Ref.	Fly Ash, FA (%)	Cement (kg/m^3)	Water/Cement Ratio (%)	Curing Time (Days)	Superplasticizer Content (%)	Fine Aggregate (kg/m^3)	Coarse Aggregate (kg/m^3)	Slump Flow Diameter (mm)	Compressive Strength (MPa)
[1]	0–40	365–468	0.4–0.6	14, 28, 56	2.2	803–918	778–879	652–772	35.7–69
[3]	35–90	183–317	0.38–0.72	7, 28, 90	0.2–1	476–1066	837	555–790	6.2–74.2
[5]	15–35	355–465	0.41–0.44	7, 28, 356	1.8–2	910	590	603–673	22.7–61.2
[28]	0–60	161–326	0.35–0.5	1, 7, 28	0.5–1.9	650–866	155–251	570–650	6.1–48.3
[31]	0–30	420–600	0.33–0.46	7, 28, 90	2	900	750	650–720	65.6–85.3
[32]	0–30	315–480	0.4, 0.45	7, 28	1.5–2.8	890	810	650–695	39–52

Table 1. *Cont.*

Ref.	Fly Ash, FA (%)	Cement (kg/m³)	Water/Cement Ratio (%)	Curing Time (Days)	Superplasticizer Content (%)	Fine Aggregate (kg/m³)	Coarse Aggregate (kg/m³)	Slump Flow Diameter (mm)	Compressive Strength (MPa)
[33]	0–60	180–550	0.33, 0.44	7, 28, 91	1	640–890	780–924	600–806	9.6–77.5
[34]	0–55	225–500	0.35	3, 7, 28	1.5,1.6	908–967	652–694	630–700	40–78
[35]	50–70	221–369	0.29	7, 28, 90	2–3.33	579	703	650–780	10.4–27
[36]	-	160–280	0.33–0.42	1, 7, 28	0.1–0.6	808–1066	900	570–770	5–52
[37]	0–70	150–500	0.28–0.35	7–365	1.33	902–967	597–639	665–775	14.9–75.6
[38]	0–60	180–550	0.32, 0.44	28, 90	0.78–3	686–826	829–935	670–730	30–91.1
[39]	15–75	355–465	0.41–0.62	28–365	1.2–2	640–910	590	590–690	18–59.4
[40]	-	375	0.47	28	1	779–825	700–746	-	22.8–31.46
[41]	0–60	161–336	0.35–0.50	1, 7, 28	0.4–3.8	739–866	843–1118	570–650	4.9–48.3
[42]	60	180,450	0.37	1, 7, 28, 90	0.1–0.48	827–850	827–850	561	2.67–70.8
[43]	0–75	142–570	0.33	28	-	765–835	765–835	-	30–66
[44]	0–80	100–500	0.36	1, 7, 28, 56	0.7	751–874	876	657–712	4.55–84
[45]	-	310–622	0.38	28	0.56–1	780	720	-	38–59
[46]	0–100	115–539	0.33–0.72	7, 28, 90, 120	0.65–1.1	743	924	700–730	16–84.1
Remarks	Ranged from 0–100%	Ranged from 100–620 kg/m³	Varied from 0.28–0.72	Varied from 1–365 days	Ranged from 0.1–3.33%	Varied from 476–1024 kg/m³	Ranged from 590–1118 kg/m³	Varied from 561–806 mm	Ranged from 2.67–91.1 MPa

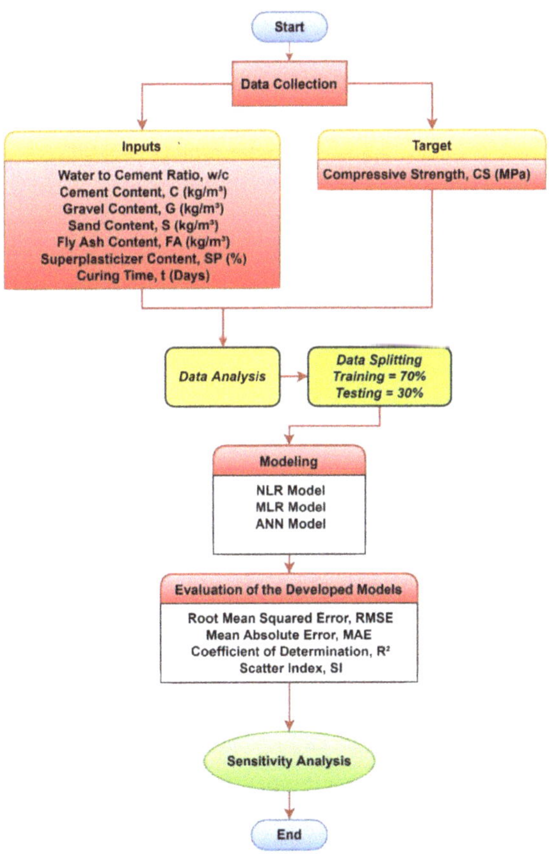

Figure 1. Current study methodology flowchart diagram.

3. Statistical Evaluation

This part of the paper is about the statistical analysis of the dependent and independent variables separately to determine whether a strong relationship exists between each variable and the compressive strength of SCC modified with FA or not (Figure 2). Thus, the plot of all considered parameters, including water to cementitious material ratio, cement content, gravel content, sand content, fly ash content, superplasticizer content, and curing time with compressive strength, was prepared and analyzed (Figures 2–4). The detail of the analysis results for each variable is explained as follows.

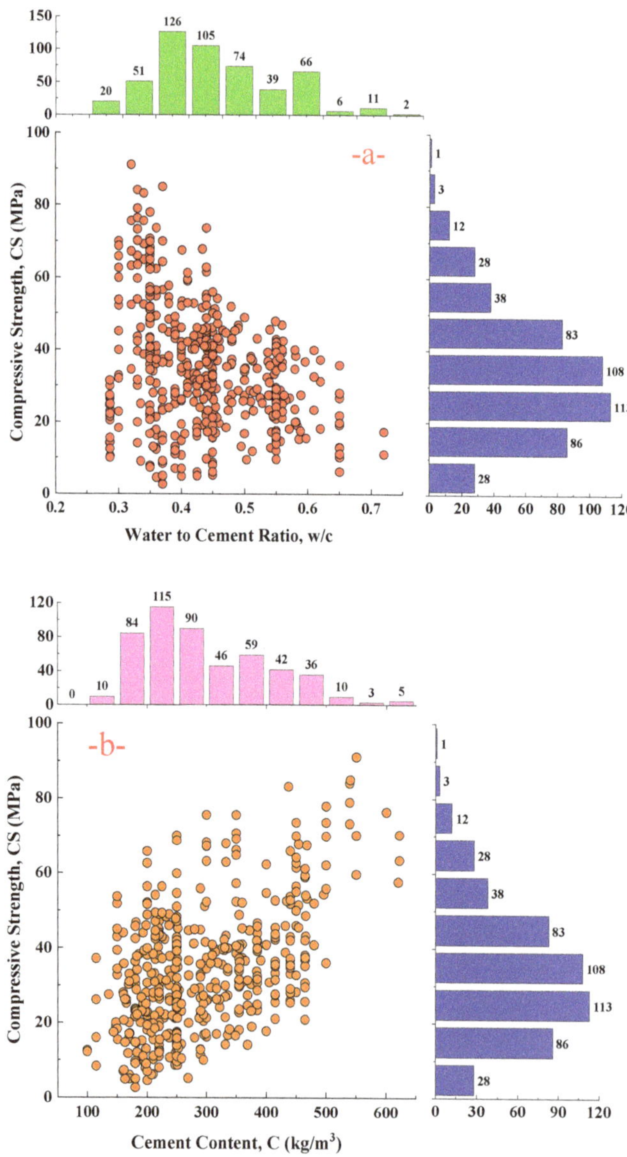

Figure 2. *Cont.*

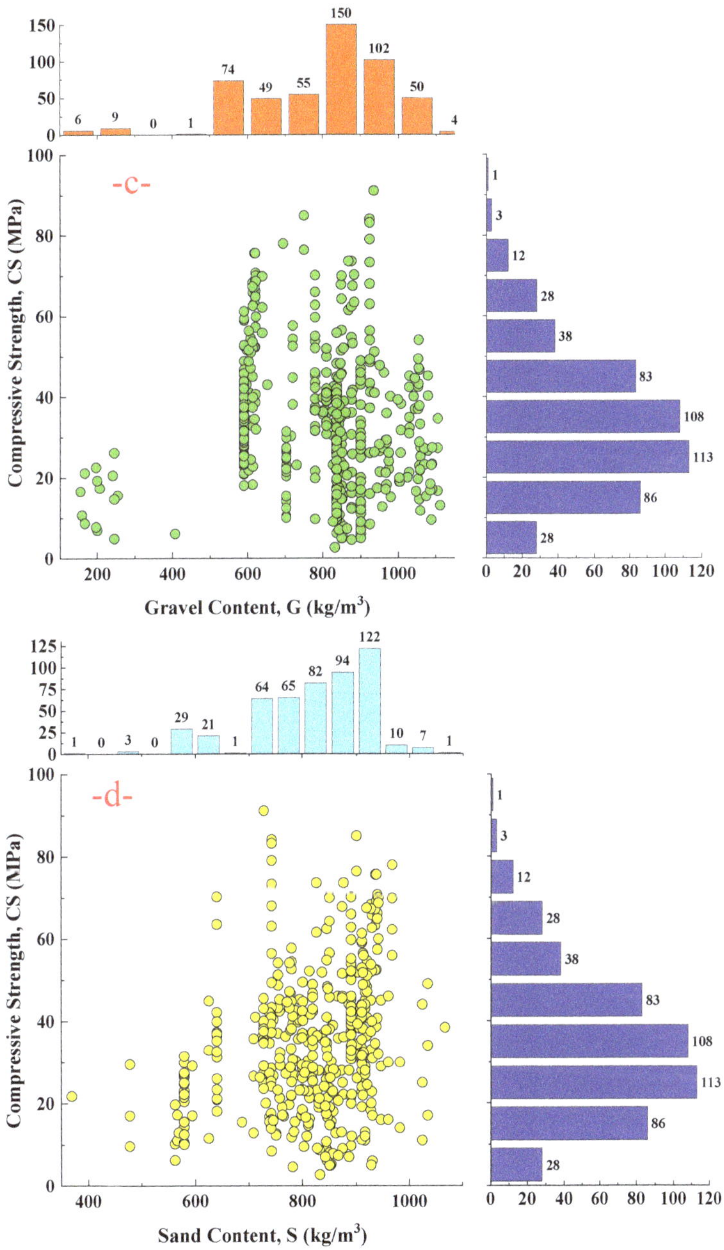

Figure 2. *Cont.*

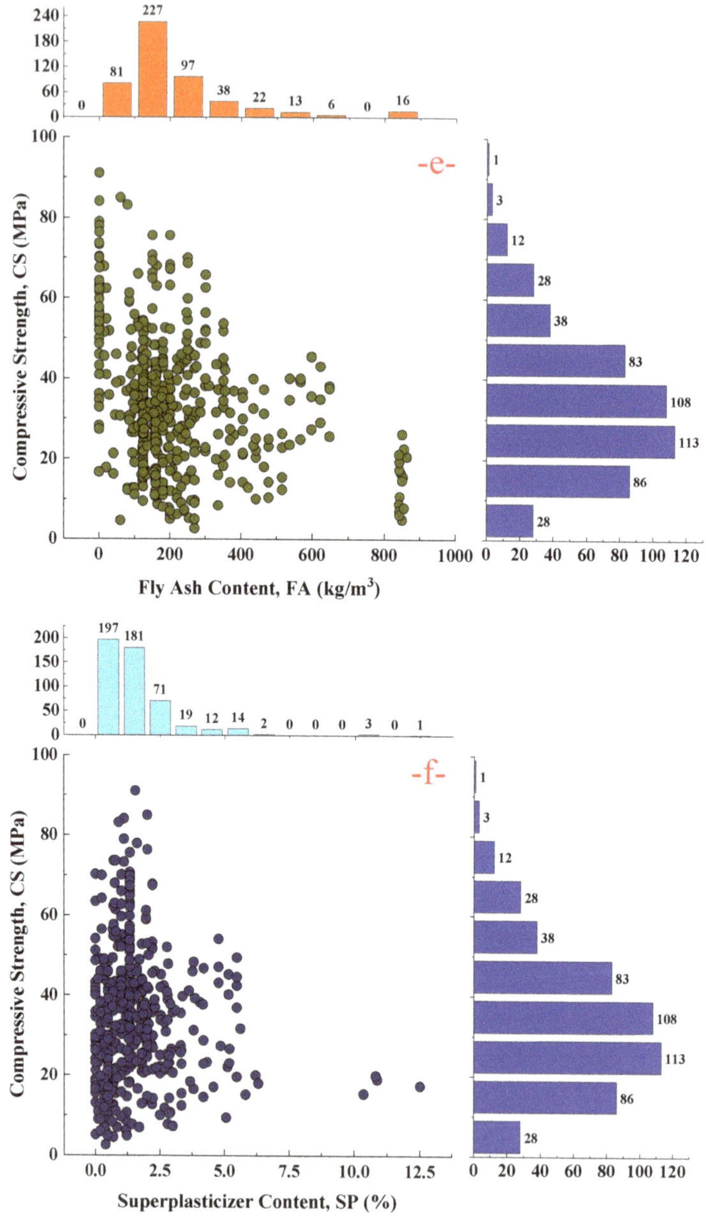

Figure 2. *Cont.*

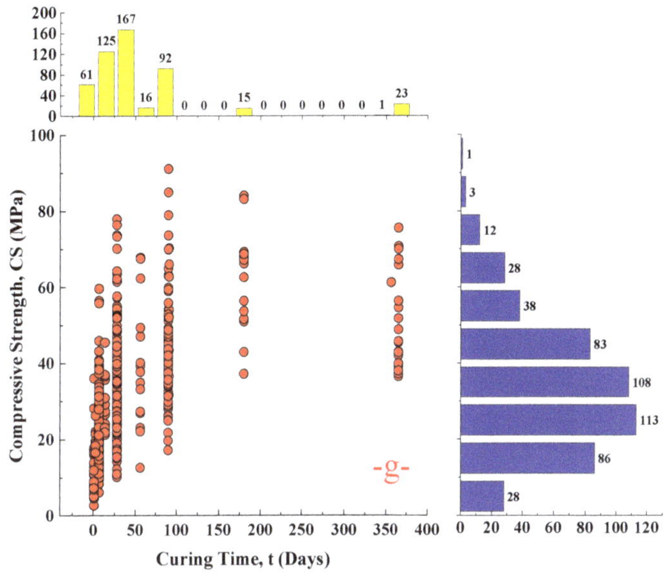

Figure 2. Marginal plot for compression strength with (**a**) w/c, (**b**) cement content, (**c**) gravel content, (**d**) sand content, (**e**) fly ash content, (**f**) superplasticizer content, and (**g**) curing time.

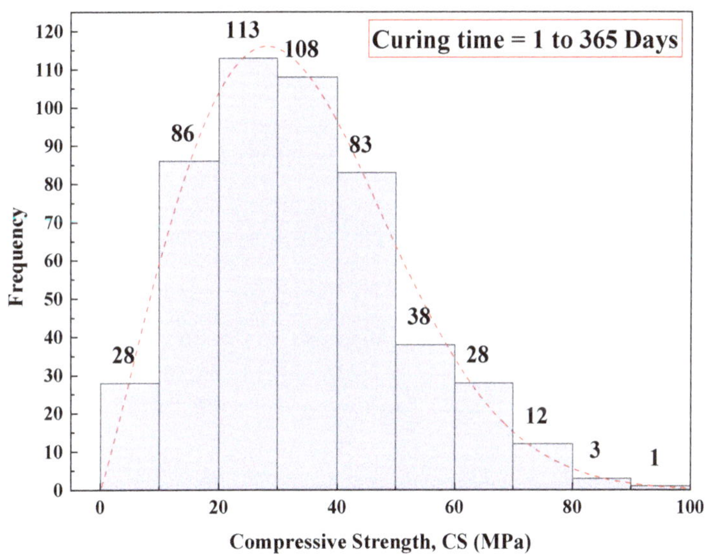

Figure 3. The self-compacting concrete's compressive strength histogram after being modified with fly ash.

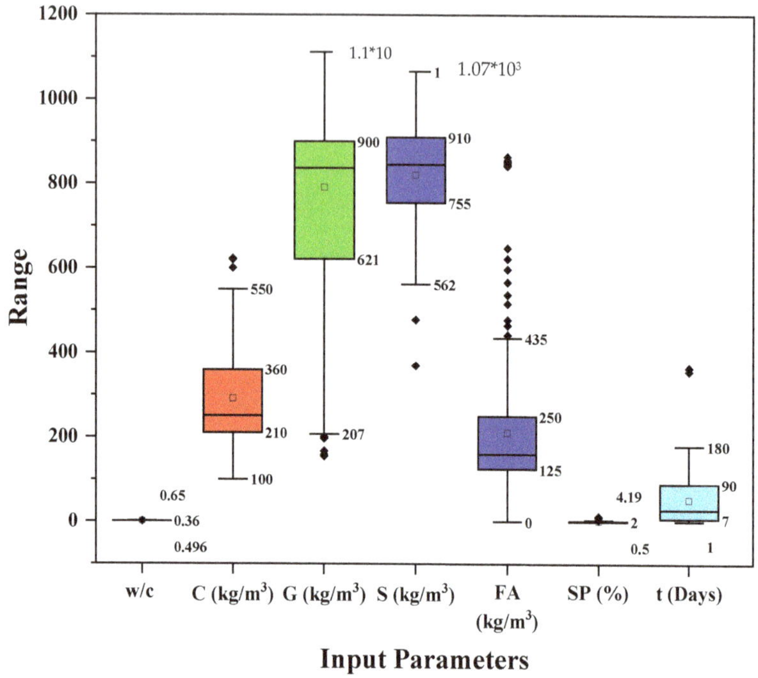

Figure 4. Boxplot for model input parameters.

3.1. Water/Cement Ratio

The w/c ratio of SCC mixtures ranged from 0.28 to 0.72, with a median of 0.42, according to the total amount of data gathered. According to the statistical analysis, the other variables had the following values: 0.0075 for variance, 0.088 for standard deviation, 0.59 for skewness, and 0.24 for kurtosis. Figure 2 illustrates the connection between compressive strength and w/c ratio using the w/c ratio histogram (a) (Table 2).

Table 2. The statistical analysis of the collected datasets.

Variable	St. Dev	Variance	Minimum	Median	Maximum	Skewness	Kurtosis
w/c	0.088	0.00775	0.28	0.42	0.72	0.59	−0.24
C, (kg/m^3)	108.76	11,828.96	100	258	622	0.66	−0.35
G, (kg/m^3)	196.06	38,441.25	590	837	1118	−1.20	2.21
S, (kg/m^3)	106.3	11,298.6	476	845	1066	−0.93	1.18
FA, (kg/m^3)	189.98	36,092	0	160	905	2.00	4.16
SP, (kg/m^3)	1.461	2.1346	0	1.33	12.5	2.76	13.07
t, (days)	86.74	7524.2	1	28	365	2.69	6.73
CS, (MPa)	18.61	346.6	2.67	34.9	91.1	0.48	−0.38

3.2. Cement Content

According to the data gathered, the cement content of SCC mixtures ranged from 100 to 622 kg/m^3, with a median value of 258 kg/m^3. Based on the statistical analysis, the other variables had the following values: variance of 11,828.96, the standard deviation of 108.76, skewness of 0.66, and kurtosis of −0.35 (Figure 2b) (Table 2).

3.3. Gravel Content

Based on the total collected data, the SCC mixtures' gravel content ranged from 590 to 1118, with an average of 837. The other variables' values were as follows: variance of 38,441 standard deviation of 196.06.34, and skewness and kurtosis of −1.20 and −2.21, respectively. The relationship between compressive strength and gravel content with the histogram of gravel content is reported in Figure 2c (Table 2).

3.4. Sand Content

Sand content varied from 476 to 1066 kg/m^3, with an average value of 845 kg/m^3. According to the statistical analysis, the other variables had the following values: a variance of 11,298 a standard deviation of 106.3, and skewness and kurtosis values of −0.93 and 1.18, respectively (d) (Table 2).

3.5. FA Content

The FA content of SCC mixtures ranged from 0 to 905 kg/m^3, with a median of 160 kg/m^3, according to the complete dataset. According to the statistical analysis, the other variables had the following values: variance of 36,092 standard deviation of 189.98, skewness of 2, and kurtosis of 4.16. Figure 2 reports the relationship between compressive strength and FA content (e) (Table 2).

3.6. SP Content

The SP contents ranged from 0 to 12.5 kg/m^3, with a median value of 1.33 kg/m^3. According to the statistical analysis, the other variables had the following values: variance of 2.134, standard deviation of 1.46, skewness of 2.76, and kurtosis of 13.07 (Figure 2f) (Table 2).

3.7. Curing Time

The median curing time in the data was 28 days, and the curing time ranged from 1 to 365 days. The statistical analysis revealed the following values for the other variables: variance of 7524 the standard deviation of 86.74, skewness, and kurtosis values of 2.69 and 6.73, respectively (Figure 2g) (Table 2).

3.8. Compressive Strength

The compressive strength of SCC mix modified with FA ranged from 2.67 to 91.1 MPa, with a median of 34.9 MPa, according to the total amount of data gathered. According to the statistical analysis, the other variables had the following values: variance of 346.64, the standard deviation of 18.61, skewness of 0.48, and kurtosis of −0.38. Figure 3 displays the SCC compressive strength histogram.

4. Modeling

From the results of the analyzed data in Section 3, no direct relationship was obtained according to the correlation matrix (Figure 5). Therefore, three models were predicted to evaluate each parameter's influence on the strength of self-compacting concrete modified with FA. All models were assessed using several standard evaluation criteria to compare all models and select the most accurate ones, such as highest R2, lower MAE, RMSI, and SI values. The models proposed in this study are used to predict the compressive strength of FA-self-compacted concrete and select the best model, which gives a better estimation of compressive strength than the measured compressive strength from the experimental data. The collected datasets were randomly split into training, testing, and validating [34]. The training dataset was used to train the LR, NLR, and MLR models and obtain the optimal weights and biases, while the testing dataset was used to confirm the fulfillment of the proposed models. The comparison among model predictions was made based on the following assessment criteria: the model should be scientifically valid, and it should give less error between the measured and predicted data, a lower RMSE, OBJ, and SI, and a higher R^2 value.

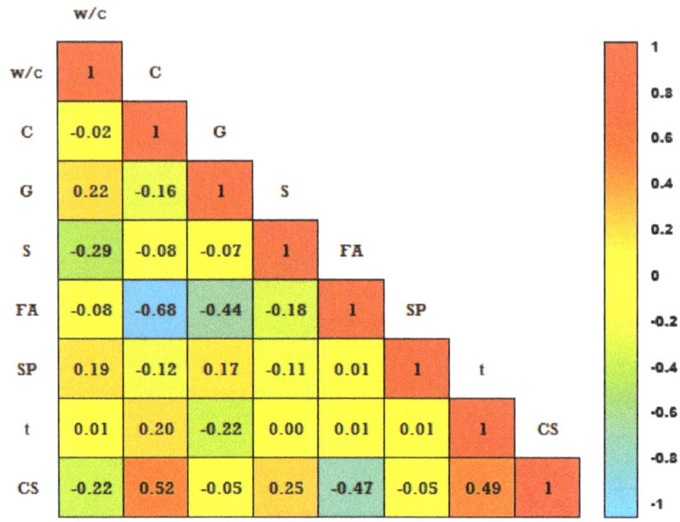

Figure 5. Correlation matrix for the model input parameters and target.

4.1. Nonlinear Model (LR)

This study is focused on developing a model that assesses the impact of the greatest number of parameters on the compressive strength of SCC modified with FA, as was previously mentioned. This study's general technique for assessing compressive strength [47–49] is nonlinear regression.

$$CS = a * \left(\frac{w}{c}\right)^{b} * (C)^{c} * (G)^{d} * (S)^{e} * (FA)^{f} * (SP)^{g} * (t)^{h} + i * \left(\frac{w}{c}\right)^{j} * (C)^{k} * (G)^{l} * (S)^{m} * (SP)^{n} * (t)^{o} + p \quad (1)$$

where w/c denotes the water to cement ratio (percent), C is the cement content (kg/m^3), G represents the gravel content (kg/m^3), S is the sand content (kg/m^3), SP is the superplasticizer content (kg/m^3), FA is the fly ash content (kg/m^3), t is the curing time (days), and a to p are model inputs.

4.2. Multimodel (MLR)

MLR is used when the criterion variable has more than two phases. Since it may reveal the relationship between a nominal dependent variable and two or more independent variables (Equation (2)), MLR is comparable to multiple linear regression [50].

$$CS = a * \left(\frac{w}{c}\right)^{b} (C)^{c} G^{d} S^{e} FA^{f} SP^{g} t^{h} \quad (2)$$

A limitation of Equation (2) is that it cannot be used to predict the compressive strength of SCC in the absence of FA. In this model, the FA content ought to be greater than 0 (the constraint of Equation (2) is FA content larger than 0 percent). The (a, b, c, d, e, f, g, and h) are model parameters.

4.3. Artificial Neural Network (ANN)

A feed-forward neural network (FNN) is the opposite of an ANN [51–53]. Its input, output, and hidden layers are displayed in Figure 6. The input layer receives the signal to be evaluated. The input layer predicts and sort's output. Data travel from input to output, much like an ANN's feed-forward network. Trial and error iterations assisted in determining the model's optimal number of hidden layers to lower error and boost R^2. In order to obtain the lowest RMSE, MAE, and R^2 values, a single hidden layer with six neural networks was selected through trial and error due to the complexity of the equation for multiple hidden layers.

$$\beta_n = a_n(w/c) + b_n(C) + c_n(G) + d_n(S) + e_n(FA) + f_n(SP) + g_n(t) + i_n \quad (3)$$

$$CS = \frac{Node_1}{1+e^{-\beta 1}} + \frac{Node_2}{1+e^{-\beta 2}} + \cdots + \frac{Node_n}{1+e^{-\beta n}} + Threshold \quad (4)$$

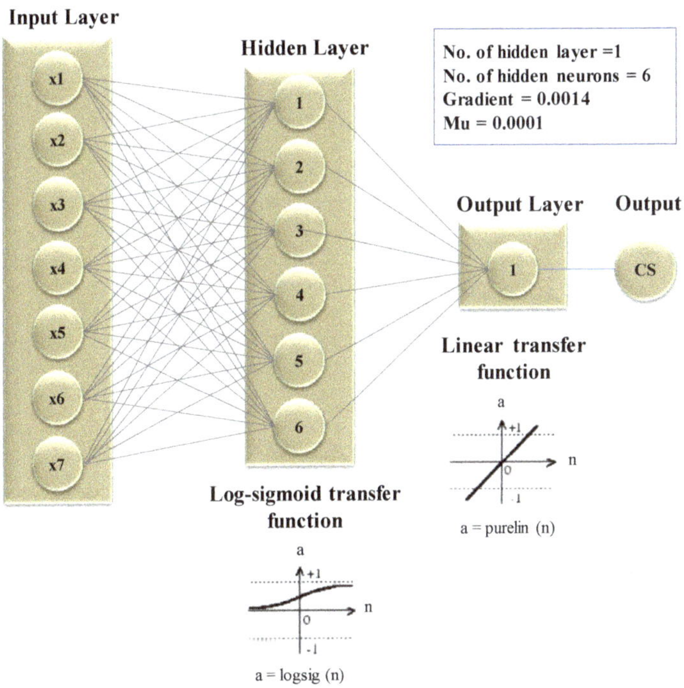

Figure 6. The ideal ANN structure has six hidden neurons and one hidden layer.

5. Assessment Criteria for the Developed Models

The coefficient of determination (R^2), root mean squared error (RMSE), mean absolute error (MAE), and scatter index (SI) was used to assess the accuracy and efficacy of the proposed models. Their equations are:

$$R^2 = \left[\frac{\sum_{p=1}^{p}(t_p - t')(y_p - y')}{\sqrt{\left[\sum_{p=1}^{p}(t_p - t')^2\right]\left[\sum_{p=1}^{p}\left(y_p - y'\right)^2\right]}} \right]^2, \quad (5)$$

$$RMSE = \sqrt{\frac{\sum_{p=1}^{p}\left(y_p - t_p\right)^2}{p}}, \quad (6)$$

$$MAE = \frac{\sum_{p=1}^{p}\left|(y_p - t_p)\right|}{p}, \quad (7)$$

$$\text{SI} = \frac{\text{RMSE}}{t'}. \quad (8)$$

The predicted and measured path pattern values are denoted by y_p and t_p, respectively, and the average value of the measured and predicted values is denoted by y' and t'. The terms tr and tst, which stand for training and testing datasets, denote the number of patterns (collected data) in each set. When evaluating any equation or model, one is the ideal value for R^2, but the lowest value is preferable for other parameters (MAE and RMSE), and zero is the ideal SI value. However, the SI parameter has several limitations, and when it is greater than 0.3, it indicates that the models are performing poorly.

6. Analysis and Output

Slump Flow Diameter

The test that is most frequently used to gauge SCC flow performance is the slump flow test. During this test, the slump flow diameter of the SCC mix is measured. When using the EFNARC SCC criterion, the slump flow diameter should meet one of the classifications listed in the standard. A slump flow diameter of 550 to 650 mm is considered in slump flow class 1, 650 to 750 mm is considered in slump flow class 2, and 750 to 850 mm is considered to be in slump flow class 3 [54]. Any slump flow diameter that is both less than 550 mm and more than 850 mm cannot be classified as SCC, according to EFNARC guidelines. With a median slump flow diameter of 672 mm and minimum and maximum slump flow diameters of 550 mm and 850 mm, respectively, all of the mixes can be regarded as SCC, per 401 data (Table 1) from prior studies. Figure 7 depicts the relationship between the slump flow diameter and the CS of self-compacted concrete [4].

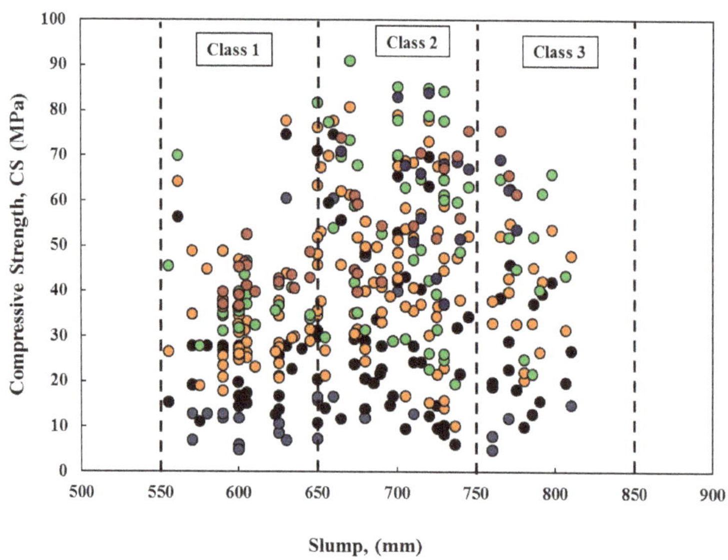

Figure 7. The relationship between the SCC mixture's compressive strength and slump flow diameter.

7. Nonlinear Regression Model

Figure 8 depicts the relationship between SCC's measured and anticipated compressive strength modified with FA for training and the testing dataset. The sum of error squares and the least square method was optimized to determine the value of each parameter in the existing model. The outcome of every experiment is represented by Equation (9).

$$CS = 0.11\left(\frac{w}{b}\right)^{-1.196}(C)^{0.109}(G)^{0.628}(S)^{0.068}(FA)^{-0.007}(SP)^{0.006}(t)^{0.104} - 41.608\left(\frac{w}{b}\right)^{-1.682}(C)^{-0.239}(G)^{0.79}(S)^{-0.812}(t)^{-0.015} + 0.0001. \quad (9)$$

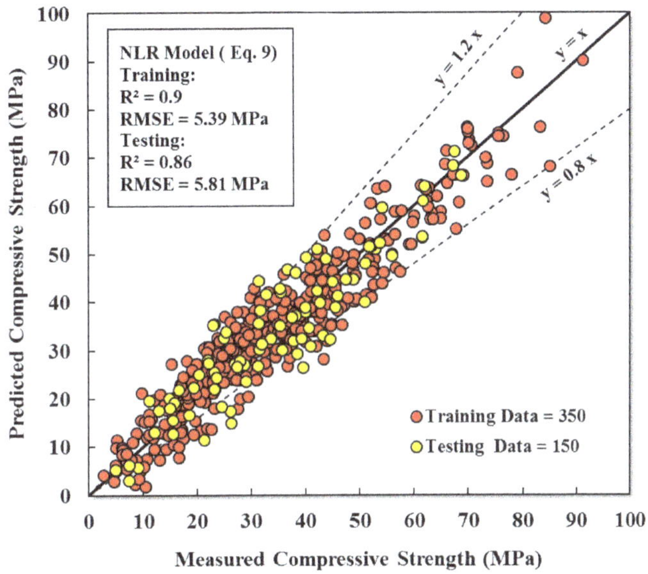

Figure 8. Comparison between measured and predicted compressive strength for NLR model.

According to the equation above, S has the least impact on the compressive strength of SCC modified with FA, while W has the greatest. As shown in Figure 9, the SI values for the developed model are 0.158 and 0.175 for the training and testing datasets, respectively. The values for assessment parameters such as R^2, RMSE, and MAE are 0.90, 5.39 MPa, and 4.35 MPa for the testing dataset.

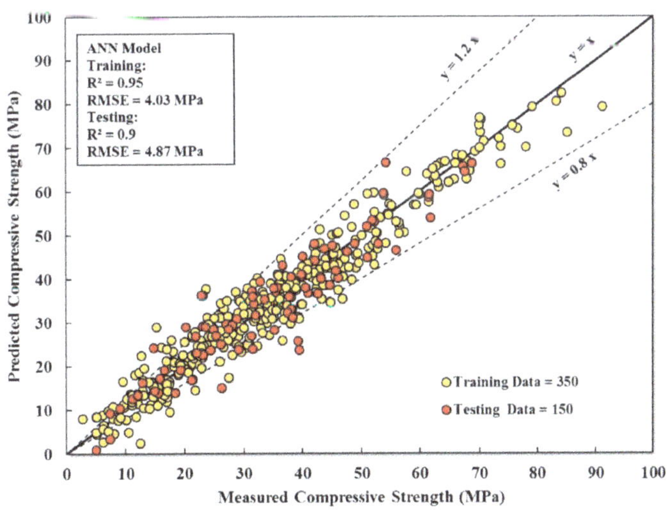

Figure 9. Comparison between measured and predicted compressive strength for the ANN model.

8. Multiregression Model (MLR)

Figure 8 uses training and testing datasets for SCC modified with FA to compare the predicted compressive strength to the actual compressive strength. Following this model, the w/c ratio and sand content concentration are the most important factors affecting the compressive strength of SCC. Equation (10) can be used to represent the multi regression model with various variable parameters (Figure 10):

$$CS = 4.56 * 10^{-6}\left(\frac{w}{b}\right)^{-0.642}(C)^{-0.487}(G)^{0.624}(S)^{1.146}(FA + 0.0001)^{-0.011}(SP)^{0.008}(t)^{0.224}. \quad (10)$$

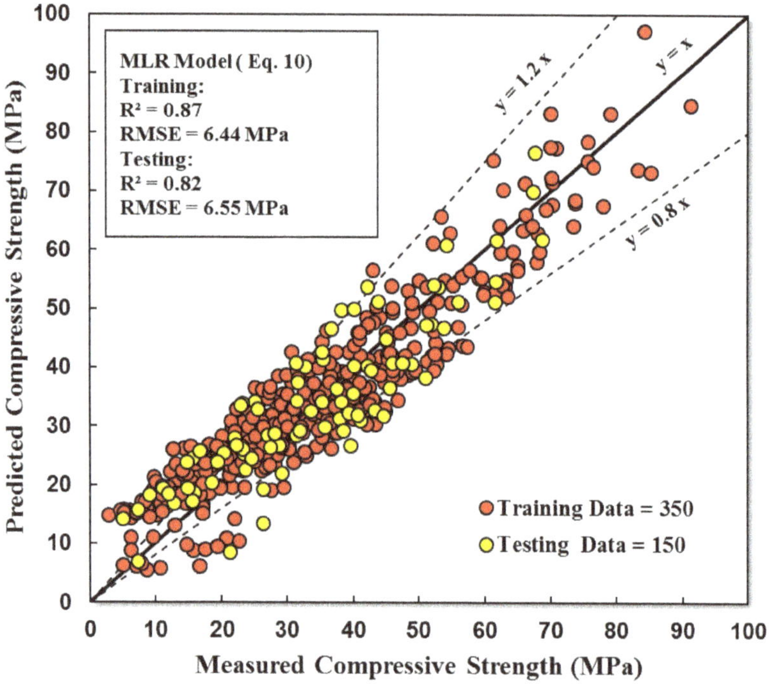

Figure 10. Comparison between measured and predicted compressive strength for MLR model.

9. Artificial Neural Network (ANN)

The network was fed both training and test data to predict the compressive strength values for the appropriate input parameters (Figure 10). An ANN model was created iteratively (such as the number of hidden layer neurons, learning rate, momentum, and iteration). Nineteen neural nets were used in this study's hidden layer system. The training period is 50,000 s, the learning rate is 0.1, and the momentum is 0.1. The quantity of epochs is a hyperparameter that controls how frequently the learning algorithm may process the training dataset. The higher the R^2, the lower the RMSE and, the lower the MAE is as the error is reduced, the more epochs there are. Figure 10 illustrates the basic idea of producing data using an ANN model by plotting the projected compressive strength vs. the actual value.

									w/b		
−0.516	−2.554	−5.391	−4.137	−5.155	0.837	−3.108	−3.38946		C		β_1
−2.950	−2.648	−2.776	−3.033	−9.037	−1.516	−0.625	−5.86885		G		β_2
30.737	−11.951	−10.236	−19.384	−13.954	32.126	2.856	47.38619	×	S	=	β_3
−0.507	0.047	−7.706	−5.574	−5.971	5.024	2.979	8.12332		FA		β_4
61.384	−198.964	183.340	104.406	804.041	67.087	−278.128	570.3129		SP		β_5
0.730	−0.299	−1.200	−0.649	−0.729	0.452	32.931	34.86496		t		β_6
									1		

$$CS = \frac{-1.112}{1+e^{-\beta_1}} + \frac{0.601}{1+e^{-\beta_2}} - \frac{0.222}{1+e^{-\beta_3}} + \frac{0.529}{1+e^{-\beta_4}} - \frac{0.179}{1+e^{-\beta_5}}\frac{1.801}{1+e^{-\beta_6}} - 2.071 \quad (11)$$

The R^2, RMSE, and MAE values for this model are, respectively, 0.95, 4.03 MPa, and 3.04 MPa.

10. Comparison between Developed Models

As mentioned in earlier sections, the effectiveness of the suggested new models was evaluated using four statistical parameters (R^2, MAE, RMSE, and SI). Compared to the NLR and MLR models, the ANN model had the lowest RMSE and MAE values and the highest R^2 values. The coefficient of determination for all models using training and test sets is shown in Figure 9. Figure 9 demonstrates that the ANN model's predicted and measured compressive strength values are much more comparable, demonstrating the ANN model's superior performance to other models.

The SI assessment parameter values for the proposed models during the training and testing phases are shown in Figure 9. The SI values for all models and phases (training and testing), as displayed in Figure 9, were less than 0.2, indicating excellent performance for all models. Additionally, the SI values for the ANN model and the other performance parameters are lower compared to other models. The ANN model has smaller SI value values across all phases compared to the NLR model, for instance, 208 percent lower in training and 68 percent lower in testing datasets.

11. Sensitivity Investigation

The most important factor affecting the CS of SCC combinations was identified and evaluated using a model sensitivity test. As a quick method, the ANN model was used to examine sensitivity. Sensitivity analysis was performed, with each training dataset having only one input variable extracted. Each training dataset's evaluation criteria, such as R2, RMSE, and MAE, were developed separately. According to the data, the curing time, cement content, and the proportion of water to cement are the factors most strongly influencing SCC compressive strength. The CS was thus greatly improved by lengthening the curing time and cement content. Figure 11 depicts the effect of mixed proportions on the CS of SCC. The capacity of models to predict the compression strength of SCC with FA for the training and test dataset was thoroughly evaluated by the following metrics: coefficient of determination (R^2), root mean square error (RMSE), scatter index (SI), and mean absolute error (MAE). However, the R^2 value, also known as the coefficient of determination, has been observed to be the best for model assessment [4]. R^2 values from 0.70 to 0.80 indicate satisfactory results, values from 0.80 to 0.95 indicate good model prediction, and values above 0.95 indicate excellent prediction, while values below 0.60 indicate unsatisfactory results [4,13]. In addition, sensitivity analysis was also carried out. The results of these analyses are presented in detail in the text (Figure 12).

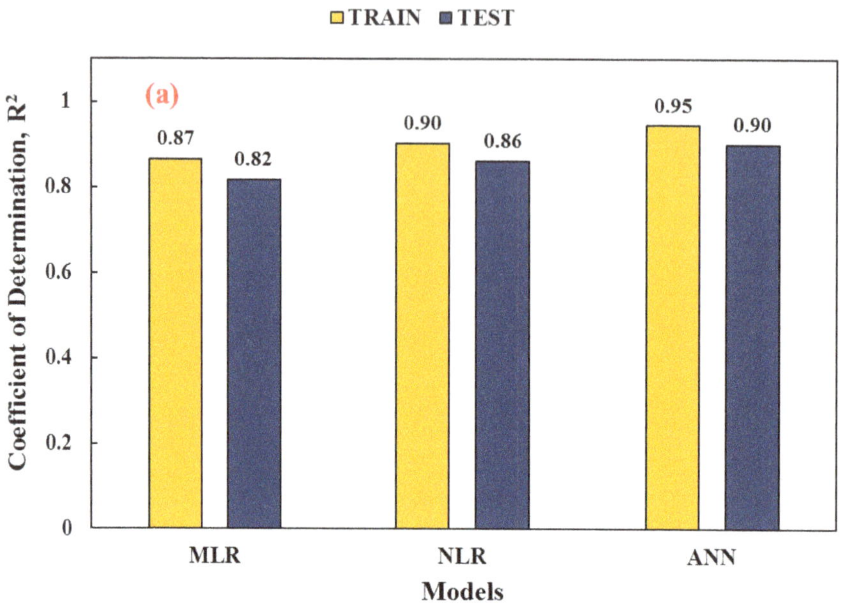

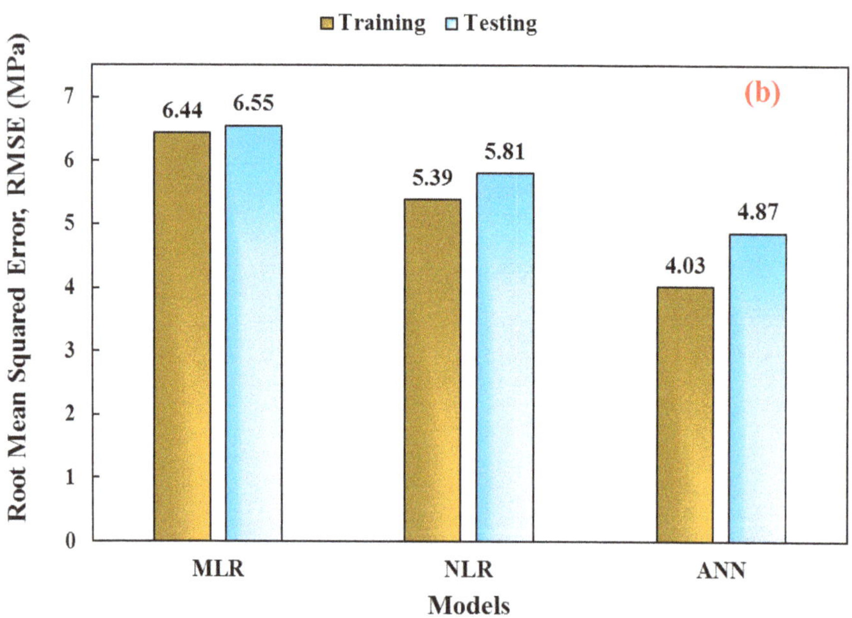

Figure 11. *Cont.*

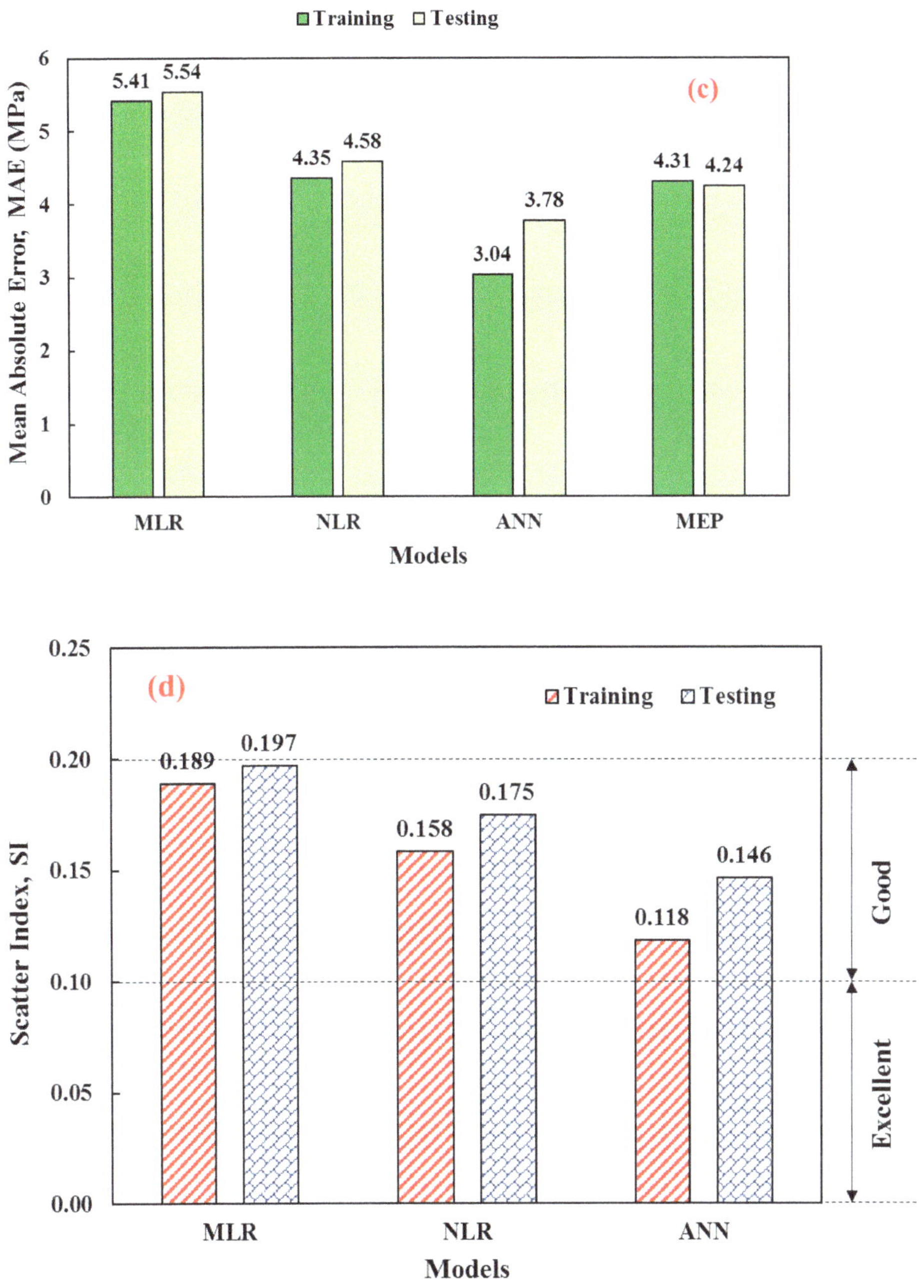

Figure 11. The developed models based on (**a**) coefficient of determination, (**b**) root mean squared error, (**c**) mean absolute error, (**d**) scatter index.

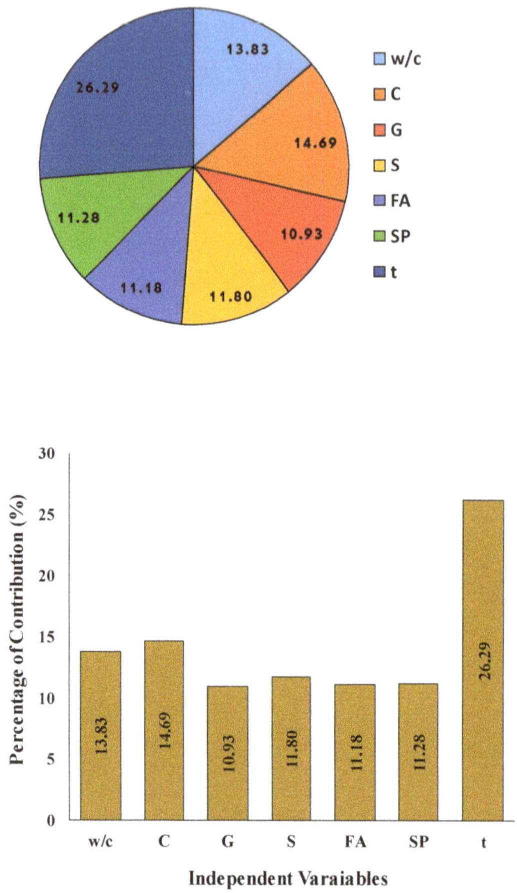

Figure 12. Sensitivity analysis to investigate the effect of input variables on the CS of CSS using the ANN model.

12. Conclusions

Based on the analysis and simulation of data from earlier studies to forecast the compressive strength of SCC for 400 different mix proportions, the following conclusions were made:

1. Based on data collected from the literature, the fly ash enhanced the compressive strength of normal concrete as a partial replacement for cement. Depending on the statistical analysis, the median percentage of superplasticizers for the production of SCC was 1.33%. Furthermore, the percentage of fly ash inside all mixes ranged from 0 to 100%, with 1 to 365 days of curing and sand content ranging from 845 to 1066 kg/m^3.
2. This study developed the NLR, MLR, and ANN models to predict the compressive strength of SCC mixes. The ANN model outperformed other models in the training and testing datasets, with higher R^2 values, and lower RMSE, MAE, and SI values.
3. The SI values for all models and phases (training and testing) were less than 0.2, indicating good performance. Furthermore, compared to the NLR model, the ANN model has smaller SI values in all phases, for example, 60% lower in training and 35% lower in the testing dataset.

4. Using several evaluation criteria, including the root mean square error (RMSE), the coefficient of determination (R^2), the SI, and the mean absolute error (MAE). The sequence of ANN models was the best model provided in this research based on data acquired from the literature and produced a higher R^2 and lower MAE and RMSE.
5. The sensitivity analysis test was performed in order to check the most effective dependent variables on independent variables' output performance. The results indicated that the most effective parameters causing the output result were curing time and fly ash content.
6. The box plot for the proposed models indicated that the ANN model had better centered mean square error and standard deviation performance.
7. The NLR model as a reliable mathematical model can predict the compressive strength of self-compacted concrete with a high coefficient of determination value.
8. The overall findings and analysis showed that it was possible to effectively modify SCC's strength and other properties by producing replacement SCC with cement containing 35% fly ash. However, percentages more than 35% might be detrimental to SCC's performance.
9. The sensitivity analysis showed that the curing time is the key input parameter that influences the improvement rate and forecasts the compressive strength of SCC.

Funding: This research received no external funding.

Institutional Review Board Statement: Not applicable.

Informed Consent Statement: Not applicable.

Data Availability Statement: The data are available upon request.

Acknowledgments: College of Engineering, the University of Sulaimani supported this study.

Conflicts of Interest: The authors declare no conflict of interest.

References

1. Satish, K.; Kumar, S.; Rai, B. Self-compacting concrete using fly ash and silica fumes as pozzolanic material. *J. Eng. Technol.* **2017**, *6*, 394–407.
2. Okamura, H. Self-compacting high-performance concrete. *Concr. Int.* **1997**, *19*, 50–54.
3. Sonebi, M. Medium strength self-compacting concrete containing fly ash: Modelling using factorial experimental plans. *Cem. Concr. Res.* **2004**, *34*, 1199–1208. [CrossRef]
4. Faraj, R.H.; Mohammed, A.A.; Mohammed, A.; Omer, K.M.; Ahmed, H.U. Systematic multiscale models to predict the compressive strength of self-compacting concretes modified with nanosilica at different curing ages. *Eng. Comput.* **2021**, 1–24. [CrossRef]
5. Siddique, R. Properties of self-compacting concrete containing class F fly ash. *Mater. Des.* **2011**, *32*, 1501–1507. [CrossRef]
6. Toporov, D. *Combustion of Pulverised Coal in a Mixture of Oxygen and Recycled Flue Gas*; Elsevier: Amsterdam, The Netherlands, 2014.
7. Khalil, N.; Hassan, E.; Shakdofa, M.; Farahat, M. Beneficiation of the huge waste quantities of barley and rice husks as well as coal fly ashes as additives for Portland cement. *J. Ind. Eng. Chem.* **2014**, *20*, 2998–3008. [CrossRef]
8. Meshram, P.; Purohit, B.K.; Sinha, M.K.; Sahu, S.K.; Pandey, B.D. Demineralization of low grade coal—A review. *Renew. Sustain. Energy Rev.* **2015**, *41*, 745–761. [CrossRef]
9. Domone, P.; Hsi-Wen, C. Testing of binders for high performance concrete. *Cem. Concr. Res.* **1997**, *27*, 1141–1147. [CrossRef]
10. Sunarmasto; Kristiawan, S.A. Effect of Fly Ash on Compressive Strength and Porosity of Self-Compacting Concrete. *Appl. Mech. Mater.* **2015**, *754–755*, 447–451. [CrossRef]
11. Faraj, R.H.; Sherwani, A.F.H.; Jafer, L.H.; Ibrahim, D.F. Rheological behavior and fresh properties of self-compacting high strength concrete containing recycled PP particles with fly ash and silica fume blended. *J. Build. Eng.* **2021**, *34*, 101667.
12. Neville, A.M.; Brooks, J.J. *Concrete Technology*; Longman Scientific & Technical: London, UK, 1987; pp. 242–246.
13. Neville, A.M. *Properties of Concrete*; Longman: London, UK, 1995; Volume 4.
14. Shariati, M.; Mafipour, M.S.; Ghahremani, B.; Azarhomayun, F.; Ahmadi, M.; Trung, N.T.; Shariati, A. A novel hybrid extreme learning machine–grey wolf optimizer (ELM-GWO) model to predict compressive strength of concrete with partial replacements for cement. *Eng. Comput.* **2020**, *38*, 757–779. [CrossRef]
15. Silvestre, J.; de Brito, J. Review on concrete nanotechnology. *Eur. J. Environ. Civ. Eng.* **2016**, *20*, 455–485. [CrossRef]
16. Karamoozian, A.; Karamoozian, M.; Ashrafi, H. Effect of Nano Particles on Self Compacting Concrete: An Experimental Study. *Life Sci. J.* **2013**, *10*, 95–101.

17. Larsen, O.; Naruts, V.; Aleksandrova, O. Self-compacting concrete with recycled aggregates. *Mater. Today Proc.* **2019**, *19*, 2023–2026. [CrossRef]
18. Ghasemi, M.; Mousavi, S.R. Studying the fracture parameters and size effect of steel fiber-reinforced self-compacting concrete. *Constr. Build. Mater.* **2019**, *201*, 447–460. [CrossRef]
19. Dinakar, P.; Manu, S. Concrete mix design for high strength self-compacting concrete using metakaolin. *Mater. Des.* **2014**, *60*, 661–668. [CrossRef]
20. Ahmadi, M.A.; Alidoust, O.; Sadrinejad, I.; Nayeri, M. Development of mechanical properties of self compacting concrete contain rice husk ash. *Int. J. Comput. Inf. Syst. Sci. Eng.* **2007**, *1*, 259–262.
21. Quercia, G.; Spiesz, P.; Hüsken, G.; Brouwers, H. SCC modification by use of amorphous nano-silica. *Cem. Concr. Compos.* **2014**, *45*, 69–81. [CrossRef]
22. Corinaldesi, V.; Moriconi, G. Characterization of self-compacting concretes prepared with different fibers and mineral additions. *Cem. Concr. Compos.* **2011**, *33*, 596–601. [CrossRef]
23. Yu, P.; Fu, X.; Fan, M. An Artificial Neural Network Model for Flexoelectric Actuation and Control of Beams. In Proceedings of the ASME 2021 International Mechanical Engineering Congress and Exposition, Virtual Online, 1–5 November 2021; Volume 7A: Dynamics, Vibration, and Control. p. V07AT07A049.
24. Min, H.; Zhang, J.; Fan, M. Size Effect of a Piezoelectric Patch on a Rectangular Plate with the Neural Network Model. *Materials* **2021**, *14*, 3240. [CrossRef]
25. Jin, L.; Li, S.; Yu, J.; He, J. Robot manipulator control using neural networks: A survey. *Neurocomputing* **2018**, *285*, 23–34. [CrossRef]
26. Mair, C.; Kadoda, G.; Lefley, M.; Phalp, K.; Schofield, C.; Shepperd, M.; Webster, S. An investigation of machine learning based prediction systems. *J. Syst. Softw.* **2000**, *53*, 23–29. [CrossRef]
27. Milačić, L.; Jović, S.; Vujović, T.; Miljković, J. Application of artificial neural network with extreme learning machine for economic growth estimation. *Phys. A Stat. Mech. Its Appl.* **2017**, *465*, 285–288. [CrossRef]
28. Jamei, M.; Mohammed, A.S.; Ahmadianfar, I.; Sabri, M.M.S.; Karbasi, M.; Hasanipanah, M. Predicting Rock Brittleness Using a Robust Evolutionary Programming Paradigm and Regression-Based Feature Selection Model. *Appl. Sci.* **2022**, *12*, 7101. [CrossRef]
29. Sor, N.A.H. The effect of superplasticizer dosage on fresh properties of self-compacting lightweight concrete produced with coarse pumice aggregate. *J. Garmian Univ.* **2018**, *5*, 190–209.
30. Madandoust, R.; Ranjbar, M.M.; Ghavidel, R.; Shahabi, S.F. Assessment of factors influencing mechanical properties of steel fiber reinforced self-compacting concrete. *Mater. Des.* **2015**, *83*, 284–294. [CrossRef]
31. Ghafor, K.; Mahmood, W.; Qadir, W.; Mohammed, A. Effect of Particle Size Distribution of Sand on Mechanical Properties of Cement Mortar Modified with Microsilica. *ACI Mater. J.* **2020**, *117*, 47–60. [CrossRef]
32. Salih, A.; Rafiq, S.; Mahmood, W.; Al-Darkazali, H.; Noaman, R.; Ghafor, K.; Qadir, W. Systemic multi-scale approaches to predict the flowability at various temperature and mechanical properties of cement paste modified with nano-calcium carbonate. *Constr. Build. Mater.* **2020**, *262*, 120777. [CrossRef]
33. Salih, A.; Rafiq, S.; Mahmood, W.; Ghafor, K.; Sarwar, W. Various simulation techniques to predict the compressive strength of cement-based mortar modified with micro-sand at different water-to-cement ratios and curing ages. *Arab. J. Geosci.* **2021**, *14*, 411. [CrossRef]
34. EFNARC, F. *Specification and Guidelines for Self-Compacting Concrete*; European Federation of Specialist Construction Chemicals and Concrete System: Farnham, UK, 2002.
35. Rafiq, S.K. Modeling and statistical assessments to evaluate the effects of fly ash and silica fume on the mechanical properties of concrete at different strength ranges. *J. Build. Pathol. Rehabil.* **2020**, *5*, 441. [CrossRef]
36. Arivalagan, S. Experimental Analysis of Self Compacting Concrete Incorporating Different Range of High-Volumes of Class F Fly Ash. *Sch. J. Eng. Technol.* **2013**, *1*, 104–111.
37. Krishnapal, P.; Yadav, R.K.; Rajeev, C. Strength characteristics of self compacting concrete containing fly ash. *Res. J. Eng. Sci. ISSN* **2013**, *2278*, 9472.
38. Abdalhmid, J.M.; Ashour, A.F.; Sheehan, T. Long-term drying shrinkage of self-compacting concrete: Experimental and analytical investigations. *Constr. Build. Mater.* **2019**, *202*, 825–837. [CrossRef]
39. Bingöl, A.F.; Tohumcu, İ. Effects of different curing regimes on the compressive strength properties of self compacting concrete incorporating fly ash and silica fume. *Mater. Des.* **2013**, *51*, 12–18. [CrossRef]
40. Ahmed, H.U.; Mahmood, L.J.; Muhammad, M.A.; Faraj, R.H.; Qaidi, S.M.; Sor, N.H.; Mohammed, A.S.; Mohammed, A.A. Geopolymer concrete as a cleaner construction material: An overview on materials and structural performances. *Clean. Mater.* **2022**, *5*, 100111. [CrossRef]
41. Patel, R. Development of Statistical Models to Simulate and Optimize Self-Consolidating Concrete Mixes Incorporating High Volumes of Fly Ash. Master's Thesis, Ryerson University, Toronto, ON, Canada, 2004; p. 1802.
42. Şahmaran, M.; Yaman, İ.Ö.; Tokyay, M. Transport and mechanical properties of self consolidating concrete with high volume fly ash. *Cem. Concr. Compos.* **2009**, *31*, 99–106. [CrossRef]
43. Güneyisi, E.; Gesoğlu, M.; Özbay, E. Strength and drying shrinkage properties of self-compacting concretes incor-porating multi-system blended mineral admixtures. *Constr. Build. Mater.* **2010**, *24*, 1878–1887. [CrossRef]
44. Siddique, R.; Aggarwal, P.; Aggarwal, Y. Influence of water/powder ratio on strength properties of self-compacting concrete containing coal fly ash and bottom ash. *Constr. Build. Mater.* **2012**, *29*, 73–81. [CrossRef]

45. Dhiyaneshwaran, S.; Ramanathan, P.; Baskar, I.; Venkatasubramani, R. Study on durability characteristics of self-compacting concrete with fly ash. *Jordan J. Civ. Eng.* **2013**, *7*, 342–353.
46. Bouzoubaâ, N.; Lachemi, M. Self-compacting concrete incorporating high volumes of class F fly ash: Preliminary results. *Cem. Concr. Res.* **2001**, *31*, 413–420. [CrossRef]
47. Nikbin, I.; Beygi, M.; Kazemi, M.; Amiri, J.V.; Rahmani, E.; Rabbanifar, S.; Eslami, M. A comprehensive investigation into the effect of aging and coarse aggregate size and volume on mechanical properties of self-compacting concrete. *Mater. Des.* **2014**, *59*, 199–210. [CrossRef]
48. Wang, J.; Mohammed, A.S.; Macioszek, E.; Ali, M.; Ulrikh, D.V.; Fang, Q. A Novel Combination of PCA and Machine Learning Techniques to Select the Most Important Factors for Predicting Tunnel Construction Performance. *Buildings* **2022**, *12*, 919. [CrossRef]
49. Mohammed, A.; Rafiq, S.; Sihag, P.; Kurda, R.; Mahmood, W. Soft computing techniques: Systematic multiscale models to predict the compressive strength of HVFA concrete based on mix proportions and curing times. *J. Build. Eng.* **2020**, *33*, 101851. [CrossRef]
50. Salih, A.; Rafiq, S.; Sihag, P.; Ghafor, K.; Mahmood, W.; Sarwar, W. Systematic multiscale models to predict the effect of high-volume fly ash on the maximum compression stress of cement-based mortar at various water/cement ratios and curing times. *Measurement* **2020**, *171*, 108819. [CrossRef]
51. Mohammed, A.; Rafiq, S.; Mahmood, W.; Al-Darkazalir, H.; Noaman, R.; Qadir, W.; Ghafor, K. Artificial Neural Network and NLR techniques to predict the rheological properties and compression strength of cement past modified with nanoclay. *Ain Shams Eng. J.* **2020**, *12*, 1313–1328. [CrossRef]
52. Mohammed, A.; Mahmood, W. Statistical Variations and New Correlation Models to Predict the Mechanical Behavior and Ultimate Shear Strength of Gypsum Rock. *Open Eng.* **2018**, *8*, 213–226. [CrossRef]
53. Mahmood, W.; Mohammed, A. Hydraulic Conductivity, Grain Size Distribution (GSD) and Cement Injectability Limits Predicted of Sandy Soils Using Vipulanandan Models. *Geotech. Geol. Eng.* **2020**, *38*, 2139–2158. [CrossRef]
54. Akeed, M.H.; Qaidi, S.; Faraj, R.H.; Mohammed, A.S.; Emad, W.; Tayeh, B.A.; Azevedo, A.R. Ultra-high-performance fiber-reinforced concrete. Part II: Hydration and microstructure. *Case Stud. Constr. Mater.* **2022**, *17*, e01289. [CrossRef]

applied sciences

MDPI

Article

Predicting Rock Brittleness Using a Robust Evolutionary Programming Paradigm and Regression-Based Feature Selection Model

Mehdi Jamei [1], Ahmed Salih Mohammed [2], Iman Ahmadianfar [3], Mohanad Muayad Sabri Sabri [4], Masoud Karbasi [5] and Mahdi Hasanipanah [6,*]

1 Faculty of Engineering, Shohadaye Hoveizeh Campus of Technology, Shahid Chamran University, Dasht-e Azadegan, Susangerd 6155634899, Iran; m.jamei@shhut.ac.ir
2 Civil Engineering Department, College of Engineering, University of Sulaimani, Sulaymaniyah 46001, Iraq; ahmed.mohammed@univsul.edu.iq
3 Department of Civil Engineering, Behbahan Khatam Alanbia University of Technology, Behbahan 6361663973, Iran; im.ahmadian@gmail.com
4 Peter the Great St. Petersburg Polytechnic University, St. Petersburg 195251, Russia; mohanad.m.sabri@gmail.com
5 Water Engineering Department, Faculty of Agriculture, University of Zanjan, Zanjan 4537138791, Iran; m.karbasi@znu.ac.ir
6 Institute of Research and Development, Duy Tan University, Da Nang 550000, Vietnam
* Correspondence: hasanipanahmahdi@duytan.edu.vn

Abstract: Brittleness plays an important role in assessing the stability of the surrounding rock mass in deep underground projects. To this end, the present study deals with developing a robust evolutionary programming paradigm known as linear genetic programming (LGP) for estimating the brittleness index (BI). In addition, the bootstrap aggregate (Bagged) regression tree (BRT) and two efficient lazy machine learning approaches, namely local weighted linear regression (LWLR) and KStar approach, were examined to validate the LGP model. To the best of our knowledge, this is the first attempt to estimate the BI through the LGP model. A tunneling project in Pahang state, Malaysia, was investigated, and the requirement datasets were measured to construct the proposed models. According to the results from the testing phase, the LGP model yielded the best statistical indicators (R = 0.9529, RMSE = 0.4838, and I_A = 0.9744) for modeling BI, followed by LWLR (R = 0.9490, RMSE = 0.6607, and I_A = 0.9400), BRT (R = 0.9433, RMSE = 0.6875, and I_A = 0.9324), and KStar (R = 0.9310, RMSE = 0.7933, and I_A = 0.9095), respectively. In addition, the sensitivity analysis demonstrated that the dry density factor demonstrated the most effective prediction of BI.

Keywords: rock brittleness; linear genetic programming; bagged regression tree; lazy machine learning method

Citation: Jamei, M.; Mohammed, A.S.; Ahmadianfar, I.; Sabri, M.M.S.; Karbasi, M.; Hasanipanah, M. Predicting Rock Brittleness Using a Robust Evolutionary Programming Paradigm and Regression-Based Feature Selection Model. *Appl. Sci.* **2022**, *12*, 7101. https://doi.org/10.3390/app12147101

Academic Editor: Arcady Dyskin

Received: 19 April 2022
Accepted: 19 June 2022
Published: 14 July 2022

1. Introduction

The brittleness of rock should be measured as the main property of rock mass in any ground excavation project. It is important to properly consider the brittleness of the rock to design structures of geotechnical engineering, particularly structures constructed on the rock mass. For example, engineers can use the information on rock brittleness to assess the wellbore performance quality and stability of a hydraulic fracturing job [1–3]. Furthermore, such information can be used to regulate the mechanical properties of shale rocks well. Meanwhile, Young's modulus and strength of these properties can be defined using certain parameters such as the volumetric fraction of strong minerals [4–6].

One of the reasons for different disasters due to rock mechanics, such as rock bursts, is brittleness [7–9]. The literature shows that brittleness can be an effective and significant factor that can predict tunnel boring machines (TBMs) and road header performance [10,11].

Moreover, this property can effectively define the excavation effectiveness of drilling as a parameter highly affecting coal mining processes [3,12]. Therefore, measuring rock brittleness is necessary for any ground excavation project [7]. Although all of the above facts had been explained, Altindag [13] argued that there was no consensus on measurement and definition of standards for this brittleness. On the other hand, Yagiz [12] argues that rock brittleness is affected by different properties of the rock. Some researchers have described the relationship between brittleness and ductility inversion or the lack of ductility [14]. Ramsey [15] defined brittleness as the lack of cohesion in rock particles. Brittleness was defined by Obert and Duvall [16] as the inclination of a material, such as rock or cast iron, to split. There are normally six characteristics of highly brittle rock: a large compressive-to-tensile strength ratio, a large interior friction angle, the production of small particles, failure under an insignificant force, high firmness, and producing completely developed characteristics after hardness lab experiments [16].

The relationship between the rock's uniaxial compressive and tensile strengths is a significant subject in rock brittleness index (BI) studies [17–19]. Nevertheless, the relationship between BI and other rock properties such as Poisson's ratio, internal friction angle, hardness, elasticity modulus, etc. is limited in the literature [20,21]. There has not enough capability to estimate BI in these models due to them using one or two dependent parameters [12,22].

Rock brittleness can be approximated using empirical formulas proposed by several studies [20,23,24]. Alternatively, multi-input and single-input predictive methods such as multiple and simple linear regression can be used to predict the BI value of rock [22,24]. However, despite a higher accuracy than the existing simple regression [17,25], they sometimes cannot accurately describe complex systems' behavior since they are not always robust enough [26]. Furthermore, rock BI cannot be predicted due to the insufficient accuracy level of these models [22]. Recently, many researchers have applied machine learning (ML) methods and metaheuristic algorithms to solve engineering and science problems [27–33].

Despite some researchers confirming that ML techniques could be used to solve problems in engineering fields, studies with a focus on the prediction of rock BI have not used different ML techniques yet. Kaunda and Asbury [34] used Poisson's ratio, velocity, and elastic modulus to apply a neural network (NN) method. Yagiz and Gokceoglu [17] formed a fuzzy system and conducted a multiple regression method to estimate rock BI by using different input parameters such as Brazilian tensile strength (BTS). Their findings demonstrated the effective application of the fuzzy system to estimate BI. Koopialipoor et al. [25] suggested some models that predict rock BI value. The proposed equations were developed by hybridizing the firefly and ANN algorithm into a single model. Another study by Khandelwal et al. [22] tested the feasibility of a genetic programming model to predict the brittleness level of intact rocks. Multiple input variables such as unit weight, BTS, and UCS were employed to estimate the rock mass BI. Jahed Armaghani et al. [3] offered different support vector machine methods for BI prediction. In their study, different kernels were used to implement SVM methods. They indicated the effectiveness of proposed SVM methods in the BI prediction field. In another study, Yagiz et al. [28] predicted BI values through a differential evolution (DE) algorithm using 48 datasets. With this aim, they employed DE to develop linear and nonlinear models. They demonstrated an acceptable application of the DE algorithm in predicting BI. Recently, comprehensive study was conducted by Sun et al. [8] to predict BI using several efficient machine learning methods such as SVM and Chi-square automatic interaction detector methods. According to their results, the proposed models could predict BI with good performance.

This study aims to assess the applicability of a novel evolutionary programming paradigm (LGP) for estimating BI to enhance the accuracy of BI simulation compared to the previous study [3]. Three advanced machine learning methods (bootstrap aggregate (Bagged) regression tree (BRT), local weighted linear regression (LWLR), and KStar models) were implemented for evaluation of the predictive performance of the LGP approach.

To the best of our knowledge, all implemented models have not yet been used in rock mechanics-based soft computing research. Here, as the novelty, best subset analysis was employed to identify the best input combination, and the results of obtained models were validated using several metrics, a graphical tool, and error analysis. In addition, an efficient sensitivity analysis was conducted to determine the most influential features in BI modeling.

2. Materials and Methods

2.1. Materials

2.1.1. Field Investigation

A tunneling project in Pahang state, Malaysia, was used to extract the data used in this study. Additional information regarding the field study can be found in Jahed Aramaghani et al. [3]. Three tunnel boring machines (TBMs) were used to excavate 35 km of the tunnel, and drilling and blasting techniques were used to excavate the rest of the tunnel [3]. Although most of the excavated rocks consisted of granite (based on the mentioned techniques), there were metamorphic and some sedimentary rocks in the geological units. The research team collected a total of 120 granite block samples from the tunnel face at different tunnel distances and several locations, and the tests were performed by transferring these block samples to the rock mechanics laboratory. Then, the procedure suggested by the ISRM [35] was applied to prepare the block rock samples for each planned test. Laboratory tests—including UCS, point load, density, the Schmidt hammer, BTS, and p-wave—were planned and conducted on the samples in the experimental program. Then, to model this study, the obtained results were considered. As mainly suggested by the literature, the BI values were calculated as BI = UCS/BTS, and then set as the output. The related inputs of the model included the p-wave velocity (V_p), point load strength index (Is_{50}), dry density (D), and Schmidt hammer rebound number (R_n). In Figures 1 and 2, BTS and UCS tests were conducted on the samples and their failures, respectively.

Figure 1. Failure of a sample under a BTS test [3].

In this study, 85 data points were collected to model the BI; 75% (64 data points) of the data was allocated for the training dataset, with the rest for the testing dataset. The descriptive statistics of all features and target variables are tabulated in Table 1. The skewness ([0.116, 0.7339]) and kurtosis ([−0.76, 0.3369]) range of variables demonstrate that both criteria fall in an acceptable range ([−2, 2]) [36,37]. Thus, it can be inferred that all datasets have a fairly normal distribution, which is a good indication for modeling rock brittleness with data-driven methods.

Figure 2. UCS test (**a**) before and (**b**) after failure [3].

Table 1. Descriptive statistics of all variables used in the modeling.

Parameters	R_n	V_p (m/s)	Dry Density (g/cm^3)	Is_{50} (MPa)	BI
Minimum	20	2910	2.38	0.8722	10.12
Maximum	59	7943	2.75	6.59	16.75
Mean	37.16	4975	2.536	3.441	12.61
Std. Deviation	10.12	1199	0.079	1.118	1.554
Range	39	5033	0.37	5.718	6.626
Skewness	0.3951	0.2449	0.1161	0.1294	0.7339
Kurtosis	−0.76	−0.605	−0.3473	0.3369	0.2216

The predictive and target parameters for decreasing the computational cost and complexity of prediction, normalized in range of [0, 1], are expressed in the following formula:

$$x_{nor} = \frac{x - x_{min}}{x_{max} - x_{min}} \quad (1)$$

where the x_{nor} is the normalized value and x_{max}, x_{min}, and x are the maximum, minimum, and original values of the modeling dataset, respectively.

2.1.2. Feature Selection Process

Best feature selection is one of the most crucial stages for building a predictive model based on a data-driven model; it has a key role in the accuracy and reliability of developed models. Best subsect regression analysis [38] is one of the most popular schemes for identifying the best input features based on linear regression modeling. In this approach, six metrics (mean square error (MSE), correlation coefficient (R), adjusted R^2, Mallows coefficient (Cp) [39], Akaike (AIC) [40], and Amemiya (PC) [41]) have been computed for choosing the best input combination [38] (see Table 2). The possible tree combination demonstrates that the last case includes all input parameters and has the highest R^2 (0.817) and lowest Mallows, Akaike, and Amemiya (MSE = 0.463, Cp = 5 AIC = −60.552, and PC = 0.21); as such, this case can be identified as the best combination for modeling BI. Thus, the functional relationship between the chosen features and target can be expressed as follows:

$$\mathrm{BI} = \mathrm{f}(\mathrm{R_n}, \mathrm{V_p}, \mathrm{D}, \mathrm{I_{S50}}) \quad (2)$$

Table 2. Best subset analysis for selecting the optimum input combination.

Number of Variables	Variables	MSE	R^2	Adjusted R^2	Mallows' Cp	Akaike's AIC	Amemiya's PC
2	V_p/D	0.652	0.736	0.730	36.387	−33.419	0.276
3	V_p, D, I_{S50}	0.530	0.788	0.781	15.611	−50.109	0.227
4	R_n, V_p, D, I_{S50}	0.463	0.817	0.808	5.000	−60.552	0.201

2.2. Methods

2.2.1. Linear Genetic Programming (LGP)

The LGP is a novel variant of the GP model proposed by Koza [42]. The LGP model is a version of the tree-based GP model with linear instruction. A comparison between the structure of the LGP and GP models is displayed in Figure 3. In the LGP, each program is described by using a parameter-length sequence of C language instructions. The instructions of LGP model include arithmetic operations (+, −, ÷, ×), conditional branches (if $x[i] \leq y[l]$), and function calls (exp(x), x, sin, cos, tan) [43]. Each function consists of an assignment to a parameter $x[i]$, which simplifies the utilization of multiple outputs in the LGP model. Table 3 reports the functional set and operation parameters employed in the GP. The main steps of the LGP can be described as follows:

A. **Initialization**: Creating the initial population randomly (programs), and then calculating the fitness function of each program.

B. **Main operators:**

   (1) **Tournament selection**: This operator randomly selects several individuals from the population. Two individuals with the best fitness functions are chosen from these individuals, and two others as the worst solutions [43].

   (2) **Crossover operator**: This operator is applied to combine some elements of the best solutions with each other to create two new solutions (individuals).

   (3) **Mutation operator**: Mutation is used to create two new individuals by transforming each of the best solutions.

C. **Elitist mechanism**: The worst solutions are replaced with transformed solutions based on this mechanism.

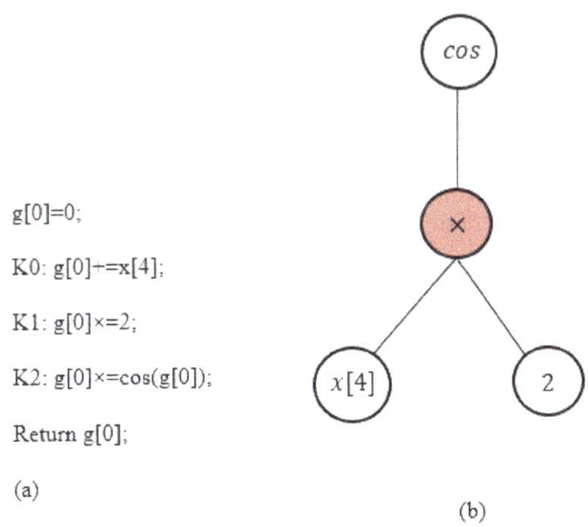

$$y = g[0] = cos(3 \times x[4])$$

Figure 3. Comparison between (**a**) GP and (**b**) LGP structure.

Table 3. The characteristics and setting parameters of proposed AI-based approaches.

Models	Setting of Parameter	
LGP	Function set	+, −, ×, ÷, √, power, sin, cos
	Population size	300
	Mutation frequency %	85
	Crossover frequency %	50
	Number of replication	10
	Block mutation rate %	20
	Instruction mutation rate %	20
	Instruction data mutation rate %	60
	Homologous crossover %	90
	Program size	64–256
LWLR	• μ = 4	
KStar	• Global blend = 30	
BRT	• Function: "Bag", Learning cycles = 50, MinLeafSize = 1	

2.2.2. Local Weighted Linear Regression (LWLR)

The LWLR method is an advanced version of the multiple linear regression (MLR) model developed by Atkeson et al. [44]. LWLR is able to improve MLR performance significantly. To illustrate the LWLR model, consider the following model:

$$z_{mk} = \alpha_{ko} + \sum_{m=1}^{M} \alpha_{km} x_{km} + \varepsilon_k \tag{3}$$

In the above model, z_{mk} is a dependent variable that can be calculated based on at least two independent variables (x_k). α is the regression coefficient calculated by the least-squares (LS) method, M is the number of data, and ε is the random error.

In the LWLR method, a weight function describes the relationship between input and output data. The fitness function of the LWLR model can be expressed by the following equation [44–46]:

$$F = \frac{1}{2M} \sum_{m=1}^{M} w_m (z_{om} - z_m)^2 \tag{4}$$

where w is the regression weight, z_o is the observed data, and z is the data obtained from the model. The above equation can also be expressed in the form of the following matrix:

$$F = (X\alpha - Z)^T W (X\alpha - Z) \tag{5}$$

By solving the above equation for α, we obtain

$$\alpha = \left(X^T W X\right)^{-1} X^T W Z \tag{6}$$

where X is the matrix of input training dataset, W denotes the weight matrix, and Z is the vector of data obtained from the model. A kernel function can be used instead of the weight matrix in the LWLR model [47,48]. In the present study, the RBF function was used as the kernel in the LWLR model. The RBF kernel equation is defined as follows:

$$w_{ik} = \exp\left(-\mu (x_i - x_k)^2\right) a \tag{7}$$

where μ is a positive number as a kernel variable and $(x_i - x_k)$ is the difference between point i and k [49]. It should be mentioned that the main setting parameter of LWLR model can be optimized by a trial and error procedure.

2.2.3. KStar Model

The KStar algorithm is a lazy learner method introduced by Machine [50]. This method is an instance-based (IB) algorithm with a fast learning capability. Generally, the IB requires only one instance for each group to create successful estimations. In this method, the distance between various instances is considered by the complexity of transforming an instance into another [51]. The KS employs an entropy-based distance function for the regression.

Considering a transformation and instance as V and I, respectively, the instance maps to other instances utilizing $i : I \to I$ which belong to V ($i \in V$). For mapping instances to themselves, a parameter called the distinct member (μ) is used, where $\mu(\alpha) = \alpha$. This parameter is used to determine all prefix codes from V^*. V^* comprises members which describe a one-to-one transformation to V. Provided that the Pf is a probability function on V^*, the probability of all paths from n to m is defined as

$$P^*\left(\frac{m}{n}\right) = \sum P(v) \tag{8}$$

where v is the value of set V. Then, the K^* function can be expressed as

$$K^*\left(\frac{m}{n}\right) = -\log_2 P^*\left(\frac{m}{n}\right) \tag{9}$$

If the examples are real numbers, then it is possible to demonstrate that $P^*\left(\frac{m}{n}\right)$ is dependent solely on the absolute difference between m and n. Therefore, it can be defined as

$$K^*\left(\frac{m}{n}\right) = K^*(i) = \frac{1}{2}\log_2\left(2e - e^2\right) - \log_2(e) + i\left[\log_2(i - e) - \log_2\left(1 - \sqrt{2e - e^2}\right)\right] \tag{10}$$

where $i = |m - n|$ and e denotes the model parameter, whose possible values range from 0 to 1. As a result, the distance between two points is equivalent to their absolute difference. Furthermore, for real numbers, the assumption is that the real space is underlain by a discrete space with extremely short distances between discrete instances. The first thing that has to be done is to evaluate those expressions in their limit as the variable e becomes closer and closer to 0. Thus, we obtain

$$P^*(i) = \sqrt{e/2} \cdot e^{-i\sqrt{2e}} \tag{11}$$

The likelihood of generating an integer with a value between i and $I + i$ can be expressed as a probability density function (PDF) as follows:

$$P^*(i) = \sqrt{e/2} \cdot e^{-i\sqrt{2e}} \cdot \Delta i \tag{12}$$

To obtain the PDF over the real numbers, $x/x_0 = i\sqrt{2e}$ can be adjusted in aspects of a real value x.

$$P^*(x) = \frac{1}{2x_0} e^{\frac{-x}{x_0}} dx \tag{13}$$

where x_0, the mean predicted value for x across the distribution P, must be suitable for practical purposes. A number between n_o and N is picked in the KStar method, which selects x_0 as the training instance with the lowest distance from m. It should be noted that the KS model was developed in this study by utilizing open-source WEKA software. The main parameter of the KS model is the global blend (GB: n), which is determined by using the trial-and-error method.

2.2.4. Bootstrap Aggregate (Bagged) Regression Tree (BRT)

Bagging (bootstrap aggregating) is one of the learning methods of the ensemble learning model [52]. In this method, the training data series is divided into *N* new training data series by the bootstrap sampling method, and a weak learner is used to train *N* datasets. In the bootstrap sampling method, random sampling is performed by replacement, which means that some of the training series data may be repeated, and some may be omitted. In the bagging regression tree (BRT) method, each of the *N* training data series is learned by a tree regression model. The final result is obtained by averaging the output of the *N* tree models (Figure 4). In the tree regression method, the results of each tree individually have high variance and low bias. Averaging the results of *N* trees reduces the variance of the model, increases the accuracy, and prevents overfitting of the model. The performance of the BRT method depends on the correct choice of the number of trees (*N*). To determine the optimal value of *N*, out-of-bag (OOB) error estimation curves can be used. Usually, two-thirds of the data series are used in model training by bootstrapping. One-third of the remaining data that does not enter the training phase in each tree is called out-of-bag (OOB) observations. OOB observations are used to estimate the prediction error. The error value of the obtained OOB observations is a good criterion for model error validation. In the present study, the fit ensemble function in MATLAB software was used to build a bagged regression tree model.

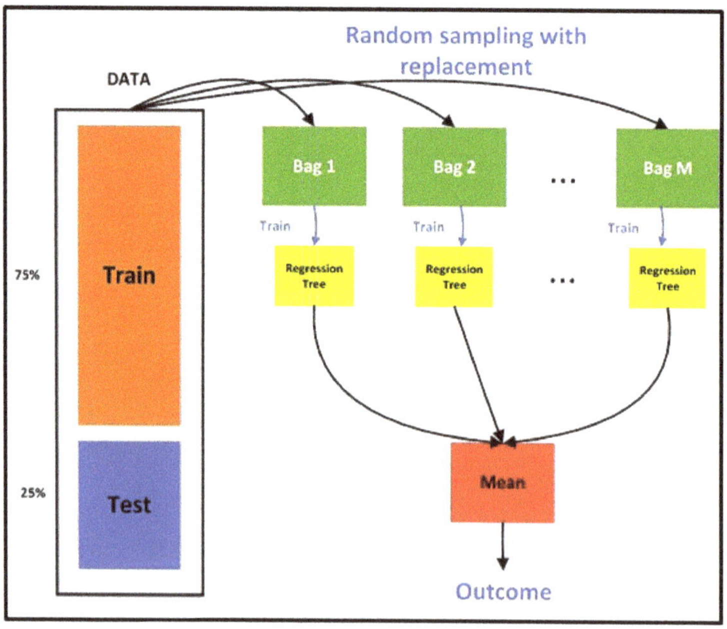

Figure 4. Training procedure of bagged regression tree (BRT).

3. Statistical Criteria for Evaluation of Models

To check the precision of the proposed models, different statistical criteria including *R*, root mean square error (RMSE), mean absolute percentage error (MAPE), Scatter Index (SI), and Willmott's agreement Index (IA) were employed [53–66].

1. Correlation coefficient (R) can be expressed as

$$R = \frac{\sum_{i=1}^{N}\left(BI_{p,i} - \overline{BI}_{p}\right).\left(BI_{o,i} - \overline{BI}_{o}\right)}{\sqrt{\sum_{i=1}^{N}\left(BI_{p,i} - \overline{BI}_{p}\right)^{2}\sum_{i=1}^{N}\left(BI_{o,i} - \overline{BI}_{o}\right)^{2}}},\ 0 < R < 1 \tag{14}$$

2. Root mean square error (RMSE) can be expressed as

$$\text{RMSE} = \left(\frac{1}{N}\sum_{i=1}^{N}\left(BI_{o,i} - BI_{p,i}\right)^2\right)^{0.5} \tag{15}$$

3. Mean absolute percentage error is defined as

$$\text{MAPE} = \frac{100}{N}\sum_{i=1}^{N}\frac{\left|BI_{o,i} - BI_{p,i}\right|}{BI_{o,i}} \tag{16}$$

4. Scatter Index can be expressed as

$$\text{SI} = \text{RMSE}/\overline{BI}_o \tag{17}$$

5. Willmott's agreement Index [49] can be expressed as

$$I_A = \frac{\sum_{i=1}^{N}\left(BI_{o,i} - BI_{p,i}\right)^2}{\sum_{i=1}^{N}\left(\left|BI_{o,i} - \overline{BI}_o\right| + \left|BI_{o,i} - \overline{BI}_o\right|\right)^2},\ 0 < I_A < 1 \tag{18}$$

where BI_o is observed value; BI_p is predicted value; $\overline{BI}_o$ and $\overline{BI}_p$ are average values of observed and predicted data, respectively; and N is the number of data.

4. Results and Discussion

The LGP model is provided based on free software called "Discipulus"; its setting parameters are listed in Table 3. In addition, to provide the BRT model, the "bag" method of the "fitresemble" function of the Machine Learning Toolbox of MATLAB 2019 was implemented. The setting parameters for the BRT model are tabulated in Table 3, which were optimized to avoid overfitting by using a trial-and-error procedure [67,68]. The kernel variable in the LWLR model was adopted through a trial-and-error process, leading to a value of 0.4. To provide the KStar model, the global blend—as a crucial parameter of the model—was optimized using a grid search scheme, leading to a value of 30. Figure 5 demonstrates the road map of predicting the procedure of BI parameters using provided AI models.

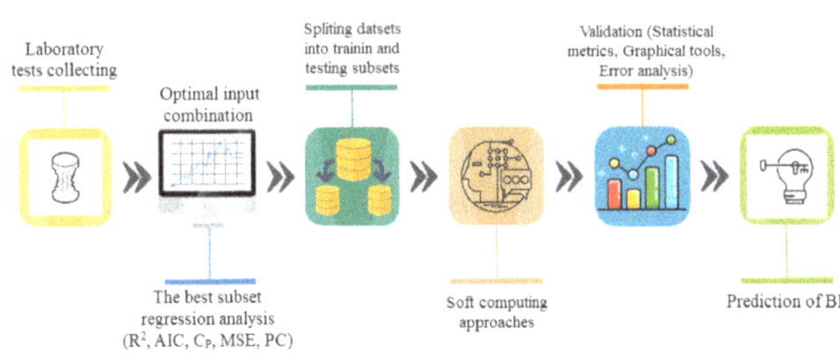

Figure 5. The road map of predicting BI using machine learning approaches.

This paper examines the LGP approach to predict the brittleness index (BI) based on four input variables: R_n, V_p, D, and Is_{50}. Also, two lazy machine learning models (namely LWLR and KStar) and a tree decision-based model (BRT) were measured to evaluate the outcome of the LGP approach. Figure 6 depicts the regression tree constructed from the BRT model, in which the terminal nodes or leafs identify the response of prediction. Table 4 presents the modeling results obtained by all models in the training and testing phases. The quantitative results in the training phase indicate that the KStar model (R = 0.9984, RMSE = 0.0865, MAPE = 0.2564, and I_A = 0.9992) is superior to the BRT (R = 0.9459, RMSE = 0.5297, and MAPE = 3.1569), LWLR (R = 0.9252, RMSE = 0.5960, and MAPE = 3.4088), and LGP (R = 0.9248, RMSE = 0.5867, and MAPE = 3.6279) models. Testing results show that the LGP approach exhibits the best efficiency for BI prediction by having the highest correlation coefficient (R = 0.9529) and lowest metrics error (RMSE = 0.4838 and MAPE = 3.2155), followed by LWLR (R = 0.9490, RMSE = 0.6607, and MAPE = 4.1549), BRT (R = 0.9433, RMSE = 0.6875, and MAPE = 4.3884), and KStar (R = 0.9310, RMSE = 0.9733, and MAPE = 5.0573), respectively. A scatter plot of each model, as a powerful graphical tool, is depicted in Figure 7 for comparison between predicted and observed values of BI. Careful examination of the scatters indicates that the LGP approach—due to the closest distribution of predicted points to the 1:1 line—demonstrates better performance than the other AI methods for whole data. The LWLR and BRT models, with acceptable accuracy and similar predictive performance, are ranked as the second and third best models, respectively. KStar, despite the remarkable performance in the training phase (R = 0.9984), is identified as the weakest method due to the highest dispersion of testing predicted points.

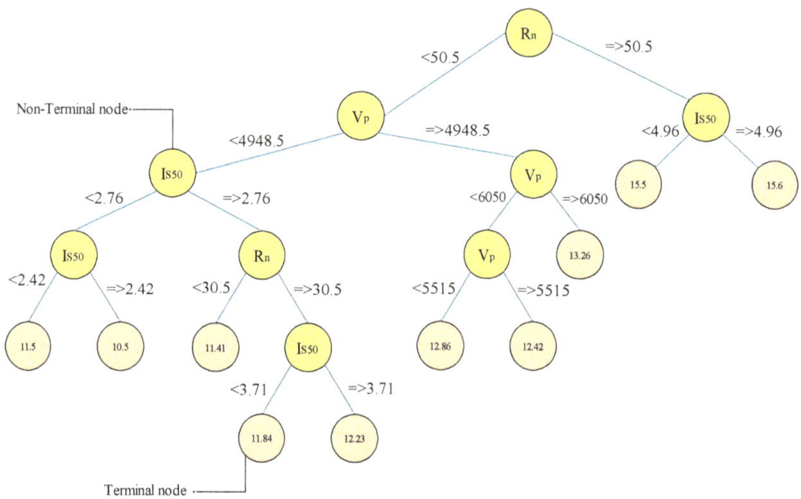

Figure 6. Decision trees of BRT model for prediction of BI.

Table 4. Quantitative evaluation of AI base approaches for predicting BI.

	Metrics	LGP	K-Star	BRT	LWLR
Training	R	0.9248	0.9984	0.9459	0.9252
	RMSE	0.5867	0.0865	0.5297	0.5960
	MAPE%	3.6279	0.2564	3.1569	3.4088
	SI	0.0463	0.0068	0.0418	0.0470
	I_A	0.9560	0.9992	0.9628	0.9531
	St.D	1.3339	1.5195	1.2640	1.2828
Testing	R	0.9529	0.9310	0.9433	0.9490
	RMSE	0.4838	0.7933	0.6875	0.6607
	MAPE%	3.2155	5.0573	4.3884	4.1549
	SI	0.0389	0.0638	0.0553	0.0532
	I_A	0.9744	0.9095	0.9324	0.9400
	St.D	1.5059	1.0861	1.1116	1.1686

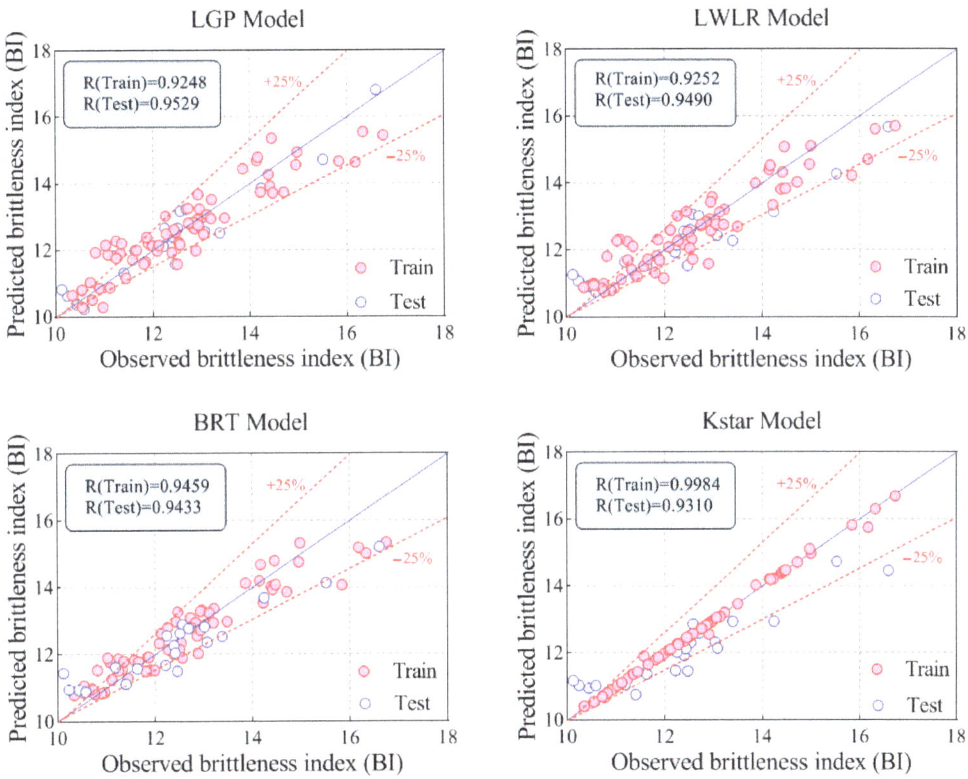

Figure 7. Comparison of four soft computing approaches and observed BI using scatter plot.

In the next graphical validation stage, half violin plots for all datasets are featured to show the distribution of quantitative data across several predicted values levels compared to observed ones. The underlying distribution of the models has been estimated using a smooth kernel density function by showing attractive benchmark points, namely the median and quartiles depicted in Figure 8. It is abundantly clear that KStar and BRT have closer Q25% values (11.366 and 11.562, respectively) to the observed values (11.395) compared with the LWLR and LGP approaches, whereas the LGP and LWLR Q75% values (13.251 and 13.146, respectively) exhibit better agreement with the observed values (13.21). Given the arrangement of the datasets, it is evident that the first Q25% is filled into the training data. Regarding KStar, the remarkable performance in training and disappointment in the testing phase implies that overfitting occurred in this paradigm.

The trend variation of BI plots in both training and testing modes is shown in Figure 9. The results indicate that the LGP model can properly capture the nonlinear behavior of BI in both triaging and testing datasets, and is capable of demonstrating promising predictive performance compared to other models. Complete error analysis was performed to evaluate the performance of the proposed predictive methods in BI estimation. According to Figure 10, the KStar (RDB = 5.52%) and LWLR (RDB = 21.51%) models are identified as having the best and worst predictive performance, as indicated from the lowest and highest relative deviation bands in the training stage, respectively. Furthermore, LGP with the lowest RDB (14.40%) and KStar with the highest RDB (23.41%) have yielded the most promising and weakest results in forecasting BI in testing mode, respectively.

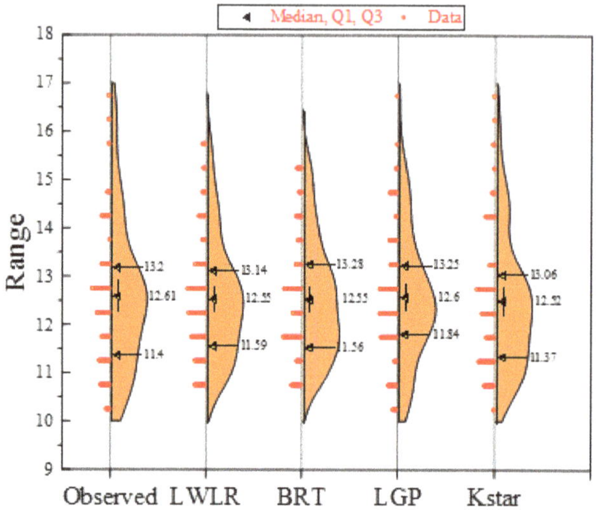

Variable	LWLR	BRT	Kstar	LGP	Observed
$Q_{0.25}$	11.594	11.562	11.366	11.843	11.395
Median	12.552	12.553	12.521	12.607	12.610
$Q_{0.75}$	13.146	13.287	13.062	13.252	13.21
IQR	1.572	1.725	1.696	1.409	1.815

Figure 8. Performance assessment of predicted and observed BI values using half violin plots.

As a final error assessment, the cumulative distribution function (CDF) of the absolute percentage of relative error (APRE) for the testing dataset was considered. Figure 11 indicates that for more than 80% of testing datasets in predicting BI values, the APRE values of LGP, BRT, LWLR, and KStar are less than 5%, 7.01%, 7.65%, and 7.80%, respectively. Eventually, it can be inferred that the LGP model, as the main novelty of this research, is superior to all proposed AI models for accurately predicting BI. The KStar approach, despite its amazing performance in the training phase, yielded the weakest results in the test phase among all models, which means that this method may not work properly for unseen data. The KStar model cannot be identified as an efficient predictive method for BI prediction due to overfitting. Thus, LGP and LWLR were identified as the best and second-best predictive models. The BRT model—ranking third, with predictive performance close to LWLR—yielded the admitted results for the prediction of BI values. It is worth noting that although KStar in this study showed unfavorable performance in testing mode, the accuracy of its results is far better than the results of previous research [3]. In the literature, some studies have predicted BI by using different machine learning methods. Yagiz et al. [69] used the genetic algorithm (GA) and particle swarm optimization (PSO) to predict BI. According to their results, the values of R^2 ranged between 0.851 and 0.932. In another study, Koopialipoor et al. [25] predicted BI through a combination of ANN and firefly algorithm, yielding prediction results with an R^2 of 0.896. In the present study, BI has been predicted with better performance (R^2 of 0.953) from the LGP model. This indicates the effectiveness of the model proposed in this study compared to aforementioned models used in the literature. According to the objectives of this study, the uncertainty of the data has not been investigated. Given great importance, uncertainty of data and results of machine learning-based methods could be considered as the subject of future research. Also, the models presented in the current study generally suffer from a lack of laboratory data. Therefore, in the future, it is necessary to examine the accuracy of presented methods with a greater number of datasets.

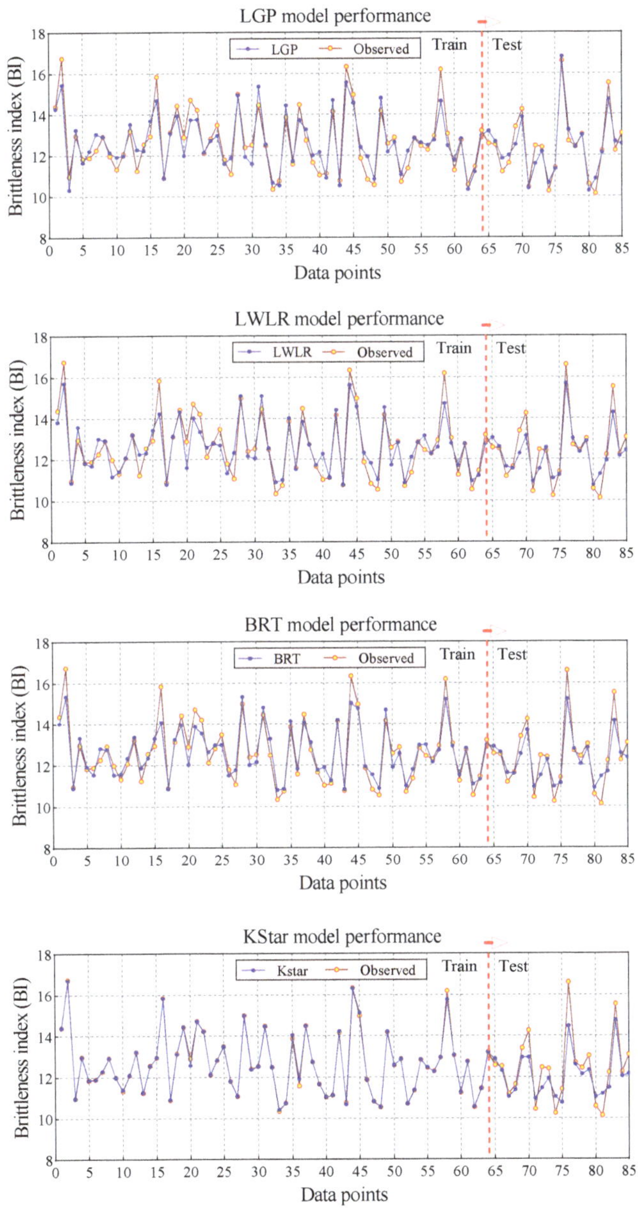

Figure 9. The trend physical plot for comparison between the observed and predicted BI values.

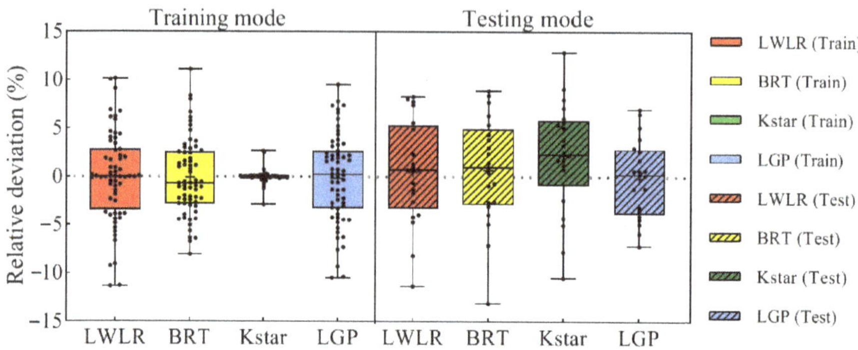

Figure 10. Box plots for the relative deviation (%) distribution of all predictive models in testing and training.

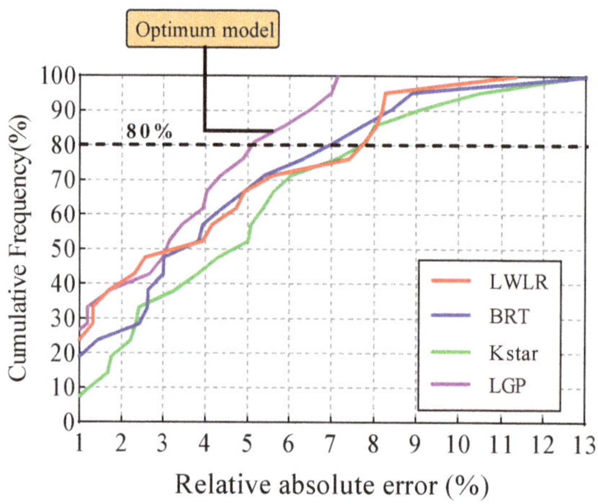

Figure 11. The cumulative frequency percentage versus the relative absolute error (%) for LWLR, BRT, KStar, and LGP models for the testing dataset.

5. Sensitivity Analysis

For more effective use of the AI methods, recognizing the effective parameters is an essential issue. One of the most widely used techniques for sensitivity analysis (SA) is consecutive elimination of the input variables and executing the AI model for all created situations. This research used the LGP model as the best model to implement the SA. Table 5 lists the SA results for five modes of combining inputs. The results demonstrate that Dry Density, with the lowest R (0.9081) and highest RMSE (0.8027) and MAPE (5.4642), is the most efficient input variable to estimate the brittleness index (BI). In addition, the V_p (R = 0.9163 and RMSE = 0.7944) ranks second, followed by Is_{50} (R = 0.9169 and RMSE = 0.7861) and R_n (R = 0.9273 and RMSE = 0.6959). A spider plot based on the six statistical criteria for all combining inputs is displayed in Figure 12. According to this figure, the combination with eliminating the dry density variable (i.e., all-dry density), showing the lowest R and I_A and highest RMSE and MAPE, has the greatest impact on the accuracy of predicting BI. It should be mentioned that some feature selection methods such as Boruta-random forest can be utilized to specify the influential parameters, which has great ability to capture the non-linear interaction between the predictors and target. This aim can be considered as an alternative of classical sensitivity analysis.

Table 5. The sensitivity analysis results for all possible situations.

Metric	All-R_n	All-V_p	All-Dry Density	All-Is_{50}	All
R	0.9273	0.9163	0.9081	0.9169	0.9433
RMSE	0.6959	0.7944	0.8027	0.7861	0.6875
MAPE	4.4592	5.1695	5.4642	5.0433	4.3884
SI	0.0560	0.0639	0.0646	0.0633	0.0553
I_A	0.9318	0.9018	0.9004	0.9049	0.9324
St.D	1.6277	1.6277	1.6277	1.6277	1.6277
Rank	4.0000	3.0000	1.0000	2.0000	-

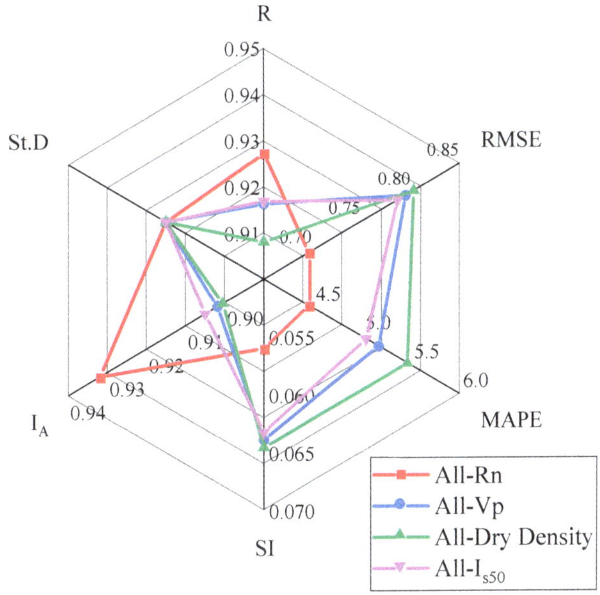

Figure 12. The influence input variables ranking for estimation of BI value.

6. Conclusions

Precise estimation of BI is necessary for any ground excavation project, and this issue requires the application of appropriate prediction models. With this in view, several advanced ML methods, including LGP, BRT, LWLR, and KStar models, were proposed to estimate BI. In this regard, a database collected from a tunneling project in Pahang state, Malaysia, was used, using four input parameters (V_p, Is_{50}, D, and R_n) and BI as the output parameter. In the modeling processes, 64 and 21 datasets, respectively, were used for training and testing phases. Finally, the models' accuracy was compared using several statistical criteria such as R and RSME. The findings of this study can be summarized as follows:

1. Based on the results, all developed models' performance capacity was suitable and acceptable. Accordingly, all proposed models can be used with confidence for future research on predictions of other issues in the field of rock mechanics.
2. Among the proposed models, the KStar (R = 0.9984 and RMSE = 0.0865) model predicted BI with the best performance in the training phase, while the best performance for the testing phase was achieved by the LGP (R = 0.9529 and RMSE = 0.4838) model. In addition, both LWLR (R = 0.9490 and RMSE = 0.6607) and BRT (R = 0.9433 and RMSE = 0.6875), ranking second and third, respectively, lead to desired results for modeling BI values.

3. The authors recommend increasing the accuracy of BI modeling as a possible future study, examining the ensemble of stacked models to integrate the advantages of standalone data-driven models.
4. Sensitivity analysis demonstrated that dry density (*D*) was the most influential parameter with respect to BI.

Author Contributions: Conceptualization, M.H. and A.S.M.; methodology, M.J., I.A., M.M.S.S. and M.K.; validation, M.J. and I.A.; investigation, A.S.M. and M.K.; writing—original draft, M.H., A.S.M., M.J., I.A., M.M.S.S. and M.K., writing—review and editing, M.H., A.S.M., M.J., I.A., M.M.S.S. and M.K.; supervision, M.H.; funding acquisition, M.M.S.S. All authors have read and agreed to the published version of the manuscript.

Funding: This research was partially funded by the Ministry of Science and Higher Education of the Russian Federation, under the strategic academic leadership program 'Priority 2030' (Agreement 075-15-2021-1333, dated 30 September 2021).

Institutional Review Board Statement: Not applicable.

Informed Consent Statement: Not applicable.

Data Availability Statement: Not applicable.

Conflicts of Interest: The authors declare no conflict of interest.

References

1. Rickman, R.; Mullen, M.J.; Petre, J.E.; Grieser, W.V.; Kundert, D. A practical use of shale petrophysics for stimulation design optimization: All shale plays are not clones of the Barnett Shale. In Proceedings of the SPE Annual Technical Conference and Exhibition, Denver, CO, USA, 21–24 September 2008.
2. Miskimins, J.L. The impact of mechanical stratigraphy on hydraulic fracture growth and design considerations for horizontal wells. *Bulletin* **2012**, *91*, 475–499.
3. Jahed Armaghani, D.; Asteris, P.G.; Askarian, B.; Hasanipanah, M.; Tarinejad, R.; Huynh, V.V. Examining hybrid and single SVM models with different kernels to predict rock brittleness. *Sustainability* **2020**, *12*, 2229. [CrossRef]
4. Hajiabdolmajid, V.; Kaiser, P. Brittleness of rock and stability assessment in hard rock tunnelling. *Tunn. Undergr. Space Technol.* **2003**, *18*, 35–48. [CrossRef]
5. Rybacki, E.; Reinicke, A.; Meier, T.; Makasi, M.; Dresen, G. What controls the mechanical properties of shale rocks?–Part I: Strength and Young's modulus. *J. Pet. Sci. Eng.* **2015**, *135*, 702–722. [CrossRef]
6. Rybacki, E.; Meier, T.; Dresen, G. What controls the mechanical properties of shale rocks?—Part II: Brittleness. *J. Pet. Sci. Eng.* **2016**, *144*, 39–58. [CrossRef]
7. Singh, S.P. Brittleness and the mechanical winning of coal. *Min. Sci. Technol.* **1986**, *3*, 173–180. [CrossRef]
8. Sun, D.; Lonbani, M.; Askarian, B.; Jahed Armaghani, D.; Tarinejad, R.; Pham, B.T.; Huynh, V.V. Investigating the Applications of Machine Learning Techniques to Predict the Rock Brittleness Index. *Appl. Sci.* **2020**, *10*, 1691. [CrossRef]
9. Zhou, J.; Guo, H.; Koopialipoor, M.; Armaghani, D.J.; Tahir, M.M. Investigating the effective parameters on the risk levels of rockburst phenomena by developing a hybrid heuristic algorithm. *Eng. Comput.* **2020**, *37*, 1679–1694. [CrossRef]
10. Yagiz, S. Utilizing rock mass properties for predicting TBM performance in hard rock condition. *Tunn. Undergr. Space Technol.* **2008**, *23*, 326–339. [CrossRef]
11. Ebrahimabadi, A.; Goshtasbi, K.; Shahriar, K.; Cheraghi Seifabad, M. A model to predict the performance of roadheaders based on the Rock Mass Brittleness Index. *J. S. Afr. Inst. Min. Metall.* **2011**, *111*, 355–364.
12. Yagiz, S. Assessment of brittleness using rock strength and density with punch penetration test. *Tunn. Undergr. Space Technol.* **2009**, *24*, 66–74. [CrossRef]
13. Altindag, R. Assessment of some brittleness indexes in rock-drilling efficiency. *Rock Mech. Rock Eng.* **2010**, *43*, 361–370. [CrossRef]
14. Morley, A. *Strength of Material, Longmans*, 11th ed.; Green: London, UK, 1954.
15. Ramsay, J.G. *Folding and Fracturing of Rocks*; Mc Graw Hill B. Co.: New York, NY, USA, 1967; Volume 568.
16. Obert, L.; Duvall, W.I. *Rock Mechanics and the Design of Structures in Rock*; Wiley: Hoboken, NJ, USA, 1967.
17. Yagiz, S.; Gokceoglu, C. Application of fuzzy inference system and nonlinear regression models for predicting rock brittleness. *Expert Syst. Appl.* **2010**, *37*, 2265–2272. [CrossRef]
18. Wang, Y.; Watson, R.; Rostami, J.; Wang, J.Y.; Limbruner, M.; He, Z. Study of borehole stability of Marcellus shale wells in longwall mining areas. *J. Pet. Explor. Prod. Technol.* **2014**, *4*, 59–71. [CrossRef]
19. Meng, F.; Zhou, H.; Zhang, C.; Xu, R.; Lu, J. Evaluation methodology of brittleness of rock based on post-peak stress–strain curves. *Rock Mech. Rock Eng.* **2015**, *48*, 1787–1805. [CrossRef]
20. Hucka, V.; Das, B. Brittleness determination of rocks by different methods. *Int. J. Rock Mech. Min. Sci. Geomech. Abstr.* **1974**, *11*, 389–392. [CrossRef]

21. Lawn, B.R.; Marshall, D.B. Hardness, toughness, and brittleness: An indentation analysis. *J. Am. Ceram. Soc.* **1979**, *62*, 347–350. [CrossRef]
22. Khandelwal, M.; Faradonbeh, R.S.; Monjezi, M.; Armaghani, D.J.; Bin Abd Majid, M.Z.; Yagiz, S. Function development for appraising brittleness of intact rocks using genetic programming and non-linear multiple regression models. *Eng. Comput.* **2017**, *33*, 13–21. [CrossRef]
23. Altindag, R. The role of rock brittleness on analysis of percussive drilling performance. In Proceedings of the 5th National Rock Mechanics, Isparta, Turkey, 30–31 October 2000; pp. 105–112.
24. Nejati, H.R.; Moosavi, S.A. A new brittleness index for estimation of rock fracture toughness. *J. Min. Environ.* **2017**, *8*, 83–91.
25. Koopialipoor, M.; Noorbakhsh, A.; Noroozi Ghaleini, E.; Jahed Armaghani, D.; Yagiz, S. A new approach for estimation of rock brittleness based on non-destructive tests. *Nondestruct. Test. Eval.* **2019**, *34*, 354–375. [CrossRef]
26. Armaghani, D.J.; Koopialipoor, M.; Marto, A.; Yagiz, S. Application of several optimization techniques for estimating TBM advance rate in granitic rocks. *J. Rock Mech. Geotech. Eng.* **2019**, *11*, 779–789. [CrossRef]
27. Dehghan, S.; Sattari, G.H.; Chelgani, S.C.; Aliabadi, M.A. Prediction of uniaxial compressive strength and modulus of elasticity for Travertine samples using regression and artificial neural networks. *Min. Sci. Technol.* **2010**, *20*, 41–46. [CrossRef]
28. Yagiz, S.; Yazitova, A.; Karahan, H. Application of differential evolution algorithm and comparing its performance with literature to predict rock brittleness for excavatability. *Int. J. Min. Reclam. Environ.* **2020**, *34*, 672–685. [CrossRef]
29. Ahmadianfar, I.; Jamei, M.; Chu, X. A novel hybrid wavelet-locally weighted linear regression (W-LWLR) model for electrical conductivity (EC) prediction in water surface. *J. Contam. Hydrol.* **2020**, *232*, 103641. [CrossRef]
30. Jiang, H.; Mohammed, A.S.; Kazeroon, R.A.; Sarir, P. Use of the Gene-Expression Programming Equation and FEM for the High-Strength CFST Columns. *Appl. Sci.* **2021**, *11*, 10468. [CrossRef]
31. Asteris, P.G.; Rizal, F.I.M.; Koopialipoor, M.; Roussis, P.C.; Ferentinou, M.; Armaghani, D.J.; Gordan, B. Slope Stability Classification under Seismic Conditions Using Several Tree-Based Intelligent Techniques. *Appl. Sci.* **2022**, *12*, 1753. [CrossRef]
32. Massalov, T.; Yagiz, S.; Adoko, A.C. Application of Soft Computing Techniques to Estimate Cutter Life Index Using Mechanical Properties of Rocks. *Appl. Sci.* **2022**, *12*, 1446. [CrossRef]
33. Qian, Y.; Aghaabbasi, M.; Ali, M.; Alqurashi, M.; Salah, B.; Zainol, R.; Moeinaddini, M.; Hussein, E.E. Classification of Imbalanced Travel Mode Choice to Work Data Using Adjustable SVM Model. *Appl. Sci.* **2021**, *11*, 11916. [CrossRef]
34. Kaunda, R.B.; Asbury, B. Prediction of rock brittleness using nondestructive methods for hard rock tunnelling. *J. Rock Mech. Geotech. Eng.* **2016**, *8*, 533–540. [CrossRef]
35. Hatheway, A.W. The complete ISRM suggested methods for rock characterization, testing and monitoring; 1974–2006. *Environ. Eng. Geosci.* **2009**, *15*, 47–48. [CrossRef]
36. Geroge, D.; Mallery, P. *SPSS for Windows Step by Step: A Simple Guide and Reference*; Allyn and Bacon: Boston, MA, USA, 2003.
37. Nie, N.H.; Bent, D.H.; Hull, C.H. *SPSS: Statistical Package for the Social Sciences*; McGraw-Hill: New York, NY, USA, 1975.
38. Kobayashi, M.; Sakata, S. Mallows' Cp criterion and unbiasedness of model selection. *J. Econom.* **1990**, *45*, 385–395. [CrossRef]
39. Gilmour, S.G. The interpretation of Mallows's Cp-statistic. *J. R. Stat. Soc. Ser. D Stat.* **1996**, *45*, 49–56.
40. Akaike, H. A new look at the statistical model identification. *IEEE Trans. Automat. Contr.* **1974**, *19*, 716–723. [CrossRef]
41. Claeskens, G.; Hjort, N.L. *Model Selection and Model Averaging*; Cambridge Books: Cambridge, UK, 2008.
42. Koza, J.R. *Genetic Programming: On the Programming of Computers by Means of Natural Selection*; MIT Press: Cambridge, MA, USA, 1992.
43. Gandomi, A.H.; Mohammadzadeh, D.; Pérez-Ordóñez, J.L.; Alavi, A.H. Linear genetic programming for shear strength prediction of reinforced concrete beams without stirrups. *Appl. Soft Comput.* **2014**, *19*, 112–120. [CrossRef]
44. Atkeson, C.G.; Moore, A.W.; Schaal, S. Locally Weighted Learning for Control. *Artif. Intell. Rev.* **1997**, *11*, 75–113. [CrossRef]
45. Jamei, M.; Ahmadianfar, I. Prediction of scour depth at piers with debris accumulation effects using linear genetic programming. *Mar. Georesources Geotechnol.* **2020**, *38*, 468–479. [CrossRef]
46. Ahmadianfar, I.; Jamei, M.; Chu, X. Prediction of local scour around circular piles under waves using a novel artificial intelligence approach. *Mar. Georesources Geotechnol.* **2019**, *39*, 44–55. [CrossRef]
47. Wang, J.; Yu, L.C.; Lai, K.R.; Zhang, X. Locally weighted linear regression for cross-lingual valence-arousal prediction of affective words. *Neurocomputing* **2016**, *194*, 271–278. [CrossRef]
48. Pourrajab, R.; Ahmadianfar, I.; Jamei, M.; Behbahani, M. A meticulous intelligent approach to predict thermal conductivity ratio of hybrid nanofluids for heat transfer applications. *J. Therm. Anal. Calorim.* **2020**, *146*, 611–628. [CrossRef]
49. Ahmadianfar, I.; Bozorg-Haddad, O.; Chu, X. Gradient-based optimizer: A new metaheuristic optimization algorithm. *Inf. Sci. (NY)* **2020**, *540*, 131–159. [CrossRef]
50. Machine, P. Practical Machine Learning Tools and Techniques. In *Data Mining*; Elsevier: Amsterdam, The Netherlands, 2011.
51. Williams, T.P.; Gong, J. Predicting construction cost overruns using text mining, numerical data and ensemble classifiers. *Autom. Constr.* **2014**, *43*, 23–29. [CrossRef]
52. Breiman, L. Bagging predictors. *Mach. Learn.* **1996**, *24*, 123–140. [CrossRef]
53. Hasanipanah, M.; Monjezi, M.; Shahnazar, A.; Jahed Armaghanid, D.; Farazmand, A. Feasibility of indirect determination of blast induced ground vibration based on support vector machine. *Measurement* **2015**, *75*, 289–297. [CrossRef]
54. Parsajoo, M.; Armaghani, D.J.; Mohammed, A.S.; Khari, M.; Jahandari, S. Tensile strength prediction of rock material using. *Transp. Geotech.* **2021**, *31*, 100652. [CrossRef]

55. non-destructive tests: A comparative intelligent study. *Transp. Geotech.* **2021**, *31*, 100652. [CrossRef]
56. Asteris, P.G.; Mamou, A.; Hajihassani, M.; Hasanipanah, M.; Koopialipoor, M.; Le, T.T.; Kardani, N.; Armaghani, D.J. Soft computing based closed form equations correlating L and N-type Schmidt hammer rebound numbers of rocks. *Transp. Geotech.* **2021**, *29*, 100588. [CrossRef]
57. Pham, B.T.; Nguyen, M.D.; Nguyen-Thoi, T.; Ho, L.S.; Koopialipoor, M.; Quoc, N.K.; Armaghani, D.J.; Van Le, H. A novel approach for classification of soils based on laboratory tests using Adaboost, Tree and ANN modeling. *Transp. Geotech.* **2020**, *27*, 100508. [CrossRef]
58. Zhou, J.; Qiu, Y.; Zhu, S.; Jahed Armaghani, D.; Khandelwal, M.; Mohamad, E.T. Estimating TBM advance rate in hard rock condition using XGBoost and Bayesian optimization. *Undergr. Space* **2020**, *6*, 506–515. [CrossRef]
59. Harandizadeh, H.; Jahed Armaghani, D.; Hasanipanah, M.; Jahandari, S. A novel TS Fuzzy-GMDH model optimized by PSO to determine the deformation values of rock material. *Neural Comput. Appl.* **2022**, *in press*. [CrossRef]
60. Hasanipanah, M.; Amnieh, H.B. A fuzzy rule-based approach to address uncertainty in risk assessment and prediction of blast-induced flyrock in a quarry. *Nat. Resour. Res.* **2020**, *29*, 669–689. [CrossRef]
61. Hasanipanah, M.; Amnieh, H.B. Developing a new uncertain rule-based fuzzy approach for evaluating the blast-induced backbreak. *Eng. Comput.* **2020**, *37*, 1879–1893. [CrossRef]
62. Hasanipanah, M.; Keshtegar, B.; Thai, D.K.; Troung, N.-T. An ANN adaptive dynamical harmony search algorithm to approximate the flyrock resulting from blasting. *Eng. Comput.* **2020**, *38*, 1257–1269. [CrossRef]
63. Hasanipanah, M.; Meng, D.; Keshtegar, B.; Trung, N.T.; Thai, D.K. Nonlinear models based on enhanced Kriging interpolation for prediction of rock joint shear strength. *Neural Comput. Appl.* **2020**, *33*, 4205–4215. [CrossRef]
64. Hasanipanah, M.; Zhang, W.; Jahed Armaghani, D.; Rad, H.N. The potential application of a new intelligent based approach in predicting the tensile strength of rock. *IEEE Access* **2020**, *8*, 57148–57157. [CrossRef]
65. Zhu, W.; Nikafshan Rad, H.; Hasanipanah, M. A chaos recurrent ANFIS optimized by PSO to predict ground vibration generated in rock blasting. *Appl. Soft. Comput.* **2021**, *108*, 107434. [CrossRef]
66. Hasanipanah, M.; Jamei, M.; Mohammed, A.S.; Nait Amar, M.; Hocine, O.; Khedher, K.M. Intelligent prediction of rock mass deformation modulus through three optimized cascaded forward neural network models. *Earth Sci. Inform.* **2022**, *in press*. [CrossRef]
67. Babyak, M.A. What you see may not be what you get: A brief, nontechnical introduction to overfitting in regression-type models. *Psychosom. Med.* **2004**, *66*, 411–421.
68. Hill, T.; Lewicki, P. *Statistics: Methods and Applications: A Comprehensive Reference for Science, Industry, and Data Mining*; StatSoft, Inc.: Tulsa, OK, USA, 2006.
69. Yagiz, S.; Ghasemi, E.; Adoko, A.C. Prediction of Rock Brittleness Using Genetic Algorithm and Particle Swarm Optimization Techniques. *Geotech. Geol. Eng.* **2018**, *36*, 3767–3777. [CrossRef]

applied sciences

MDPI

Article

Adaptive Salp Swarm Algorithm for Optimization of Geotechnical Structures

Mohammad Khajehzadeh [1], Amin Iraji [2,*], Ali Majdi [3], Suraparb Keawsawasvong [4] and Moncef L. Nehdi [5,*]

1 Department of Civil Engineering, Anar Branch, Islamic Azad University, Anar 77419-43615, Iran; mohammad.khajehzadeh@gmail.com
2 Engineering Faculty of Khoy, Urmia University of Technology, Urmia 57166-93188, Iran
3 Department of Building and Construction Techniques, Al-Mustaqbal University College, Hillah 51001, Iraq; alimajdi@mustaqbal-college.edu.iq
4 Department of Civil Engineering, Thammasat School of Engineering, Thammasat University, Bangkok 10200, Thailand; ksurapar@engr.tu.ac.th
5 Department of Civil Engineering, McMaster University, Hamilton, ON L8S 4M6, Canada
* Correspondence: a.iraji@uut.ac.ir (A.I.); nehdim@mcmaster.ca (M.L.N.); Tel.: +1-905-525-9140 (ext. 23824) (M.L.N.)

Abstract: Based on the salp swarm algorithm (SSA), this paper proposes an efficient metaheuristic algorithm for solving global optimization problems and optimizing two commonly encountered geotechnical engineering structures: reinforced concrete cantilever retaining walls and shallow spread foundations. Two new equations for the leader- and followers-position-updating procedures were introduced in the proposed adaptive salp swarm optimization (ASSA). This change improved the algorithm's exploration capabilities while preventing it from converging prematurely. Benchmark test functions were used to confirm the proposed algorithm's performance, and the results were compared to the SSA and other effective optimization algorithms. A Wilcoxon's rank sum test was performed to evaluate the pairwise statistical performances of the algorithms, and it indicated the significant superiority of the ASSA. The new algorithm can also be used to optimize low-cost retaining walls and foundations. In the analysis and design procedures, both geotechnical and structural limit states were used. Two case studies of retaining walls and spread foundations were solved using the proposed methodology. According to the simulation results, ASSA outperforms alternative models and demonstrates the ability to produce better optimal solutions.

Keywords: salp swarm optimizer; spread foundation; retaining structures; economic design

Citation: Khajehzadeh, M.; Iraji, A.; Majdi, A.; Keawsawasvong, S.; Nehdi, M.L. Adaptive Salp Swarm Algorithm for Optimization of Geotechnical Structures. *Appl. Sci.* **2022**, *12*, 6749. https://doi.org/10.3390/app12136749

Academic Editors: Pijush Samui, Aydin Azizi, Arcady Dyskin, Ahmed Hussein Kamel Ahmed Elshafie, Yixia Zhang and Danial Jahed Armaghani

Received: 23 April 2022
Accepted: 28 June 2022
Published: 3 July 2022

1. Introduction

The objective function in most engineering problems is non-convex and discontinuous, with a large number of design variables. As a result, traditional deterministic optimization techniques based on mathematical principles may struggle to find a global optimum solution due to local optima trapping. The use of powerful metaheuristic optimization algorithms for obtaining a global optimum to overcome this limitation is of interest, and metaheuristic algorithms have proven to be an excellent alternative for solving complex problems in recent decades [1–6].

The most common geo-structures in practical application are reinforced concrete retaining walls and spread footings, which have received considerable attention in recent studies [7,8]. These structures are commonly used and typically involve a large volume of material. In the past, the initial anticipated dimensions of retaining structures were tested for stability and other building code requirements. If the dimensions were insufficient to meet the constraints, they were adjusted until all the requirements were met. During this time-consuming, iterative process, the cost of construction was not taken into account. In the optimum design of these structures, the dimensions that provide the lowest cost and weight of construction while meeting all the design requirements are automatically determined.

Several metaheuristic algorithms for geotechnical engineering problems have recently been developed and are widely used. Despite the fact that metaheuristic methods can produce acceptable results, no algorithm outperforms another in solving all the optimization problems. Furthermore, the objective function in most geotechnical engineering optimization problems, such as shallow foundations, retaining structures, and pile optimization, is discontinuous and has a large number of design variables. As a result, several research projects have been launched in order to improve the performance and efficiency of the existing metaheuristics. Some of these are modified particle swarm optimizations [9,10], modified harmony search algorithms [11], modified gravitational search algorithms [12], modified sine cosine algorithms [13], improved salp swarm algorithms [14], modified ant colony optimizations [15], modified teaching–learning-based optimizations [16], improved tunicate swarm algorithms [17], and modified wild horse optimizations [18]. According to the effectiveness of the metaheuristics and their modified versions, these methods have been widely used to solve several geotechnical engineering problems, as presented in Table 1.

Table 1. Application of metaheuristic algorithms for geotechnical engineering problems.

Author, Year	Reference	Optimization Method	Application
Goh, 2000	[19]	Genetic algorithm	Locate the critical circular slip surface in slope stability analysis
Zolfaghari, Heath, and McCombie, 2005	[20]	Genetic algorithm	Search for critical noncircular failure surface in slope stability analysis
Cheng et al., 2007	[1]	Particle swarm optimization	Analyze two-dimensional slope stability
Cheng et al., 2008	[11]	Improved harmony search algorithm	Analyze slope stability
Chan, Zhang, and Ng, 2009	[21]	Hybrid genetic algorithms	Optimize pile groups
Kahatadeniya, Nanakorn, and Neaupane, 2009	[22]	Ant colony optimization	Determine the critical failure surface of earth slope
Khajehzadeh et al., 2011	[23]	Modified particle swarm optimization	Optimize design of spread footing and retaining wall
Camp and Akin, 2012	[24]	Big bang–big crunch optimization	Optimize design of retaining wall
Camp and Assadollahi, 2013	[25]	Hybrid big bang–big crunch algorithm	Optimize CO_2 and cost of reinforced concrete footings
Khajehzadeh et al., 2013	[26]	Hybrid firefly algorithm	Multi-objective optimization of foundations
Kang, Li, and Ma, 2013	[27]	Artificial bee colony algorithm	Locate the critical slip surface in slope stability analysis
Khajehzadeh, Taha, and Eslami, 2014	[12]	Hybrid adaptive gravitational search algorithm	Multi-objective optimization of retaining walls
Kashani, Gandomi, and Mousavi, 2016	[28]	Imperialistic competitive algorithm	Locate the critical slip surface of earth slope
Gordan et al., 2016	[29]	Particle swarm optimization and neural network	Predict seismic slope stability
Gandomi and Kashani, 2017	[7]	Accelerated particle swarm optimization, firefly algorithm, Levy-flight krill herd, whale optimization algorithm, ant lion optimizer, grey wolf optimizer, moth–flame optimization algorithm, and teaching–learning-based optimization algorithm	Minimize construction cost of shallow foundation
Aydogdu, 2017	[30]	Biogeography-based optimization algorithm	Optimize cost of retaining wall

Table 1. *Cont.*

Author, Year	Reference	Optimization Method	Application
Gandomi et al., 2017	[31]	Genetic algorithm, differential evolution, evolutionary strategy, and biogeography-based optimization	Analyze slope stability
Mahdiyar et al., 2017	[32]	Monte Carlo simulation technique	Assess safety of slope
Gandomi, Kashani, and Zeighami, 2017	[2]	Interior search algorithm	Optimize retaining wall
Chen et al., 2019	[33]	Hybrid imperialist competitive algorithm and artificial neural network	Predict safety factor values of retaining walls
Koopialipoor et al., 2019	[34]	Imperialist competitive algorithm, genetic algorithm, particle swarm optimization, and artificial bee colony combined with artificial neural network	Predict slope stability under static and dynamic conditions
Yang et al., 2019	[35]	Neural network system	Design retaining wall structures based on smart and optimal systems
Xu et al., 2019	[36]	Hybrid artificial neural network and ant colony optimization	Assess dynamic conditions of retaining wall structures
Himanshu and Burman, 2019	[37]	Particle swarm optimization	Determine critical failure surface considering seepage and seismic loading
Kalemci et al., 2020	[38]	Grey wolf optimization algorithm	Optimize retaining walls
Kaveh, Hamedani, and Bakhshpoori, 2020	[39]	Eleven metaheuristic algorithms	Optimize design of cantilever retaining walls
Kashani et al., 2020	[4]	Differential algorithm, evolution strategy, and biogeography-based optimization algorithm	Optimize design of shallow foundation
Sharma, Saha, and Lohar, 2021	[40]	Hybrid butterfly and symbiosis organism search algorithm	Optimize retaining wall
Kaveh and Seddighian, 2021	[41]	Black hole mechanics optimization, firefly algorithm, evolution strategy, and sine cosine algorithm	Optimize slope critical surfaces considering seepage and seismic effects
Temur, 2021	[42]	Teaching–learning-based optimization	Optimize retaining wall
Li and Wu, 2021	[43]	Improved salp swarm algorithm	Locate critical slip surface of slopes
Khajehzadeh, Keawsawasvong, et al., 2022	[44]	Hybrid tunicate swarm algorithm and pattern search	Seismic analysis of earth slope
Arabali et al., 2022	[45]	Adaptive tunicate swarm algorithm	Optimize construction cost and CO_2 emissions of shallow foundation
Khajehzadeh, Keawsawasvong, and Nehdi, 2022	[46]	Artificial neural network combined with rat swarm optimization	Predict the ultimate bearing capacity of shallow foundations and their optimum design
Khajehzadeh, Kalhor, et al., 2022	[47]	Adaptive sperm swarm optimization	Optimize design of retaining structures under seismic load
Kashani et al., 2022	[48]	Multi-objective particle swarm optimization, multi-objective multiverse optimization and Pareto envelope-based selection algorithm	Multi-objective optimization of mechanically stabilized earth retaining wall

A new meta-heuristic algorithm called the salp swarm algorithm (SSA) simulates salp fish swarming in deep waters [49]. Section 2 contains more information on the SSA's motivation and mathematical modelling. The SSA in its basic model can be extended or hybridized with another algorithm to produce better answers for future problems, similar to other metaheuristic approaches [14,43,50].

This paper presents an adaptive salp swarm algorithm (ASSA) for optimization by introducing new position-updating equations for leader and follower salps. This change significantly improves the algorithm's performance and convergence speed. A set of well-known standard benchmark functions from the literature is used to validate the effectiveness of the proposed approach. Furthermore, numerical geotechnical structure optimization tests are used to investigate the proposed method's performance and efficiency.

2. Salp Swarm Algorithm

A salp is a type of marine animal in the Salpidae family. It has a cylindrical structure with apertures at the ends similar to those of a jellyfish, which move and eat by pumping water through internal feeding filters in their gelatinous bodies. The salp swarm algorithm (SSA), a population-based optimization technique, was developed by Mirjalili et al. [49]. The salp chain can be used to calculate the SSA's behavior while hunting for optimal feeding sources (i.e., the target of this swarm is a food position in the search space called FP). To mathematically model salp chains, they are sampled into two groups: followers and leaders. The salp at the head of the chain is known as the leader, while the others are known as followers. The swarm is led by the leader of these salps, and the followers follow in his footsteps. The chain begins with a leader, who is followed by the followers to guide their movements.

Similar to other swarm-based algorithms, the salp location is specified in a n-dimensional search space, where n is the number of variables in a given problem. As a result, the positions of all the salps are recorded in a two-dimensional matrix known as X, as shown in Equation (1):

$$X_i = \begin{bmatrix} x_1^1 & x_2^1 & \dots & x_d^1 \\ x_1^2 & x_2^2 & \dots & x_d^2 \\ \vdots & \vdots & \dots & \vdots \\ x_1^n & x_2^n & \dots & x_d^n \end{bmatrix} \tag{1}$$

The fitness of each salp is then determined in order to define which salp has the best fitness. It is also supposed that the swarm's goal is a food position called FP in the search area.

The following equation can be used by the leader salp to change positions:

$$x_i^1 = \begin{cases} FP_i + r_1((ub_i - lb_i)r_2 + lb_i) & r_3 \geq 0 \\ FP_i - r_1((ub_i - lb_i)r_2 + lb_i) & r_3 < 0 \end{cases} \tag{2}$$

where x_i^1 denotes the first salp's position in the ith dimension, and FP_i denotes the food position in the ith dimension. The lower and upper bounds of the ith dimension are represented by lb_i and ub_i, respectively, and the coefficient r_1 is calculated with Equation (3):

$$r_1 = 2e^{-\left(\frac{4t}{t_{max}}\right)^2} \tag{3}$$

In addition, the random numbers r_2 and r_3 are between 0 and 1. The maximum number of iterations is t_{max}, and the current iteration is t. It is worth noting that the r_1 coefficient is critical in a SSA because it balances exploration and exploitation throughout the search. The following equations are used to change the positions of the followers:

$$x_i^j = \frac{1}{2}\left(x_i^j + x_i^{j-1}\right) \tag{4}$$

where $j \geq 2$. In case some agents transfer outside of the search area, Equation (6) shows how to move salps back into the search area if they leave it:

$$x_i^j = \begin{cases} lb_i & if\ x_i^j \leq lb_i \\ ub_i & if\ x_i^j \geq ub_i \\ x_i^j & otherwise \end{cases} \tag{5}$$

The pseudocode of the SSA is shown in Algorithm 1.

Algorithm 1. Salp swarm algorithm

```
Initialize the salp population x_i (i = 1, 2, ... , n) considering lb_i and ub_i
while t ≤ t_max
      Calculate the fitness of each search agent (salp)
      Put the best search agent as FP (Food position)
      Update r_1 by Equation (3)
           for each salp (x_i)
                if i = 1
                     Update the position of the leading salp by Equation (2)
                else
                     Update the position of the follower salp by Equation (4)
                end
           end
      Amend the salps based on the upper and lower bounds of variables
   Calculate the fitness of each search agent FP
Update the food position
t = t + 1
end
return the food position FP and its best fitness
```

3. Adaptive Salp Swarm Algorithm

Even though the SSA has the capability to generate acceptable results in comparison to other well-known techniques [49], the obtained results of the SSA are prone to becoming stuck in a local optimum, making it unsuitable for very complex problems with multiple local optima [43].

The leading salp adjusts its location in the SSA in response to the food situation (i.e., the position of the best salp in the whole population), as observed in Equation (2). The SSA algorithm updates the location of the leader salp around a single point at each incarnation pass, and other salps (followers) follow the leader. If the algorithm fails to recover because it lacks knowledge of the food position (FP), the algorithm fails. In other words, once an algorithm converges, it loses its ability to explore and then becomes inactive. As a result of this mechanism, the SSA algorithm becomes locked at local minimum points. In light of these circumstances, an adaptive version of the SSA (ASSA) is proposed to address the aforementioned flaw, while also increasing the algorithm's search capability and flexibility.

In the proposed ASSA, half of the population is considered as leaders, and the remaining salps are followers, which improves the algorithm's performance and exploring capabilities. The following equation is then used to update the position of the leader salps:

$$x_i^j = \begin{cases} x_i^j + r_1\left(FP_i - x_i^j\right) & r_3 \geq 0.5 \\ x_i^j - r_1\left(FP_i - x_i^j\right) & r_3 < 0.5 \end{cases} \tag{6}$$

The leaders adjust their positions in response to the state of the food source, as well as their previous position, as shown in Equation (6).

This process increases exploration while also allowing the SSA to conduct a more powerful global search across the entire search space. To improve the proposed ASSA's search efficiency, the followers update their positions according to the following equation:

$$x_i^j = rand^2\left(x_i^j + x_i^{j-1}\right) \tag{7}$$

In addition, in the suggested ASSA, at each iterative process, the worst salp with the highest objective function value is replaced with a completely random salp. The flowchart of the proposed ASSA is shown in Figure 1.

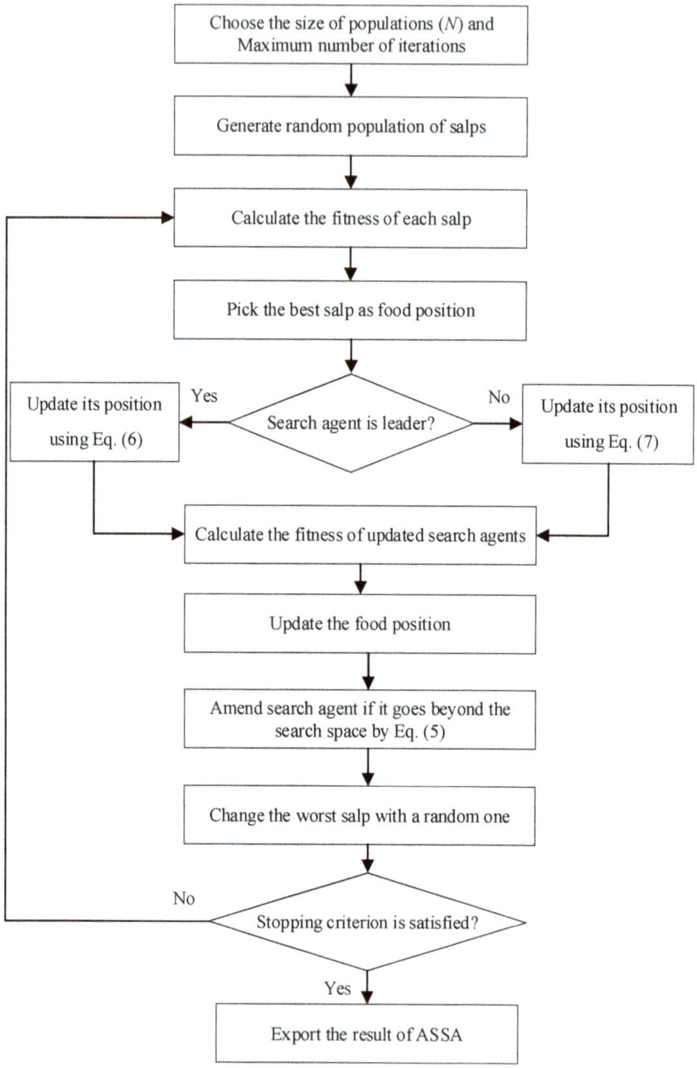

Figure 1. Flowchart of ASSA.

4. Model Verification

A set of numerical reference test functions was used in this section to compare and confirm the achievement and effectiveness of the proposed adaptive salp swarm algorithm (ASSA). In the empirical evidence literature, these functions have commonly been used to determine the performance of optimizers [51,52].

The mathematical models and characteristics of these test functions are shown in Tables 2 and 3. This standard set was divided into two categories: (1) unimodal functions with a single global best for testing the algorithm convergence pace and enslavement ability and (2) multimodal functions with multiple local minimums and a global ideal for testing an algorithm's local optima avoidance and exploratory capacity. MATLAB R2020b was used to create the suggested algorithms. All these functions should be minimized. Furthermore, all the functions had dimensions of 30. Three-dimensional drawings of these benchmark functions are illustrated in Figures 2 and 3.

The proposed ASSA was compared to the original SSA, as well as to some well-known optimization methods, such as the genetic algorithm developed by [53], the particle swarm optimization (PSO) proposed by [54], the firefly algorithm (FA) introduced by [55], the multiverse optimizer (MVO) developed by [56], and the tunicate swarm algorithm (TSA) introduced by [52]. For all methodologies, the sizes of the solutions (N) and the maximum number of iterations (t_{max}) were set to 30 and 1000, respectively, in order to make fair comparisons between them.

Because the results of a single run of a metaheuristic method are stochastic, they may be incorrect. As a result, statistical analysis should be performed in order to provide a fair comparison and evaluate an algorithm's efficacy. To address this issue, 30 runs for the mentioned methods were performed, with the results presented in Tables 4 and 5.

Table 2. Description of unimodal benchmark functions.

Function	Range	f_{min}	n (Dim)
$F_1(X) = \sum_{i=1}^{n} x_i^2$	$[-100,\ 100]^n$	0	30
$F_2(X) = \sum_{i=1}^{n} \lvert x_i \rvert + \prod_{i=1}^{n} \lvert x_i \rvert$	$[-10,\ 10]^n$	0	30
$F_3(X) = \sum_{i=1}^{n} \left(\sum_{j=1}^{i} x_j \right)^2$	$[-100,\ 100]^n$	0	30
$F_4(X) = \max_i \{ \lvert x_i \rvert,\ 1 \le i \le n \}$	$[-100,\ 100]^n$	0	30
$F_5(X) = \sum_{i=1}^{n-1} \left[100\left(x_{i+1} - x_i^2\right)^2 + (x_i - 1)^2 \right]$	$[-30,\ 30]^n$	0	30
$F_6(X) = \sum_{i=1}^{n} ([x_i + 0.5])^2$	$[-100,\ 100]^n$	0	30
$F_7(X) = \sum_{i=1}^{n} i x_i^4 + random[0,1)$	$[-1.28,\ 1.28]^n$	0	30

Table 3. Description of multimodal benchmark functions.

Function	Range	f_{min}	n (Dim)
$F_8(X) = \sum_{i=1}^{n} -x_i \sin\left(\sqrt{\lvert x_i \rvert}\right)$	$[-500,\ 500]^n$	$428.9829 \times n$	30
$F_9(X) = \sum_{i=1}^{n} \left[x_i^2 - 10\cos(2\pi x_i) + 10 \right]$	$[-5.12,\ 5.12]^n$	0	30
$F_{10}(X) = -20\exp\left(-0.2\sqrt{\frac{1}{n}\sum_{i=1}^{n} x_i^2}\right) - \exp\left(\frac{1}{n}\sum_{i=1}^{n}\cos(2\pi x_i)\right) + 20 + e$	$[-32,\ 32]^n$	0	30
$F_{11}(X) = \frac{1}{4000}\sum_{i=1}^{n} x_i^2 - \prod_{i=1}^{n}\cos\left(\frac{x_i}{\sqrt{i}}\right) + 1$	$[-600,\ 600]^n$	0	30
$F_{12}(X) = \frac{\pi}{n}\left\{10\sin(\pi y_1) + \sum_{i=1}^{n-1}(y_i - 1)^2\left[1 + 10\sin^2(\pi y_{i+1})\right] + (y_n - 1)^2\right\} + \sum_{i=1}^{n} u(x_i, 10,\ 100,\ 4)$; $y_i = 1 + \frac{x_{i+4}}{4}$ $u(x_i, a,\ k,\ m) = \begin{cases} k(x_i - a)^m & x_i > a \\ 0 & a < x_i < a \\ k(-x_i - a)^m & x_i < -a \end{cases}$	$[-50,\ 50]^n$	0	30
$F_{13}(X) = 0.1\{\sin^2(3\pi x_1) + \sum_{i=1}^{n}(x_i - 1)^2\left[1 + \sin^2(3\pi x_i + 1)\right] + (x_n - 1)^2\left[1 + \sin^2(2\pi x_n)\right]\} + \sum_{i=1}^{n} u(x_i, 5,\ 100,\ 4)$	$[-50,\ 50]^n$	0	30

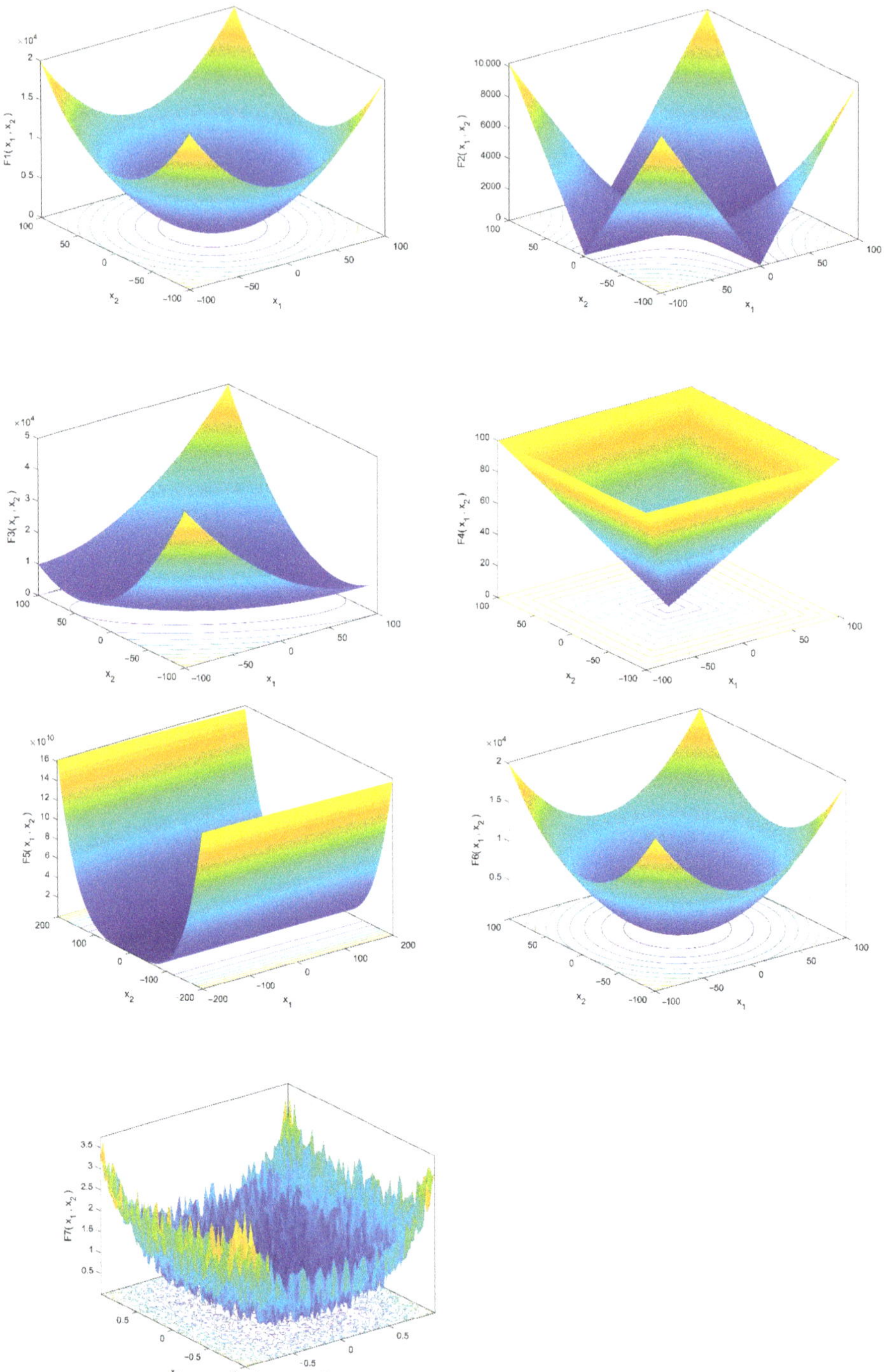

Figure 2. 3-D versions of unimodal benchmark functions.

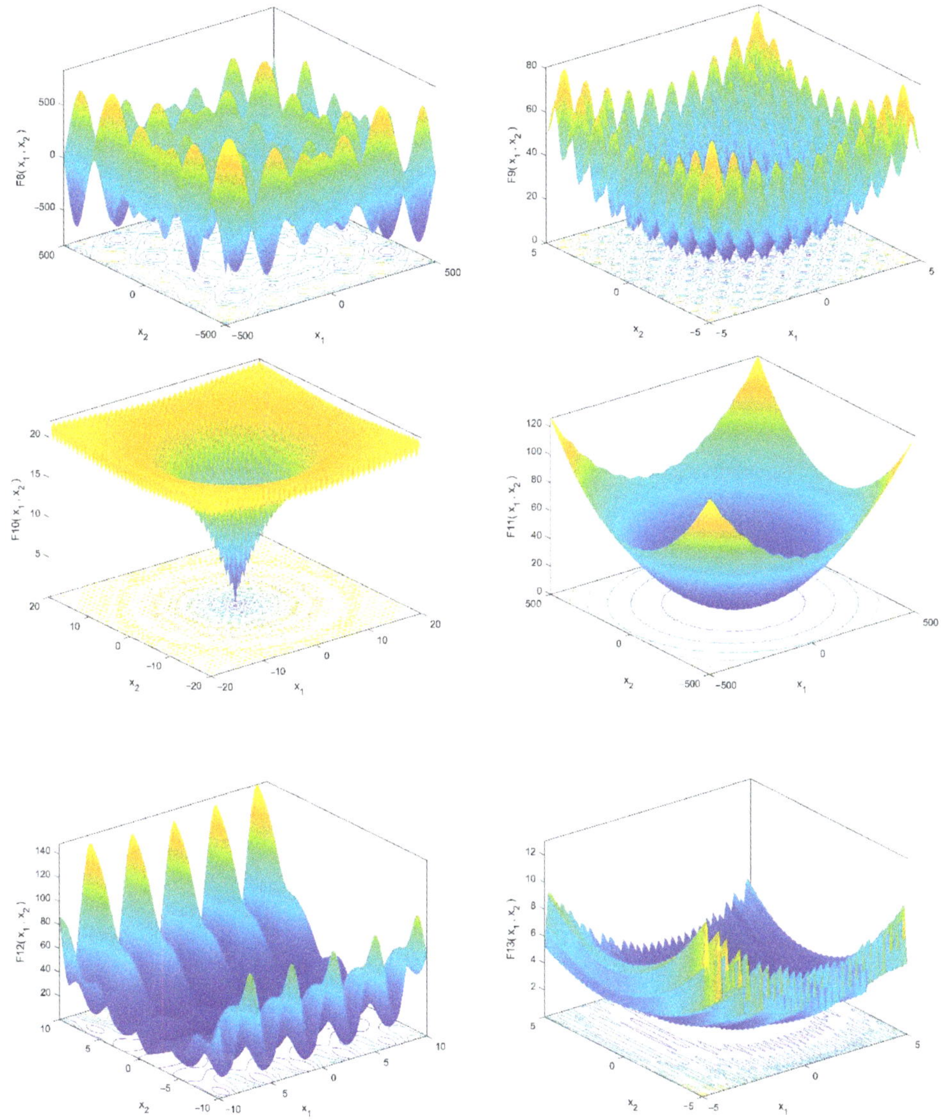

Figure 3. 3-D versions of multimodal benchmark functions.

Table 4. Comparison of different methods in solving unimodal test functions.

F	Index	ASSA	SSA	GA	PSO	FA	MVO	TSA
F_1	Mean	$\mathbf{2.23 \times 10^{-227}}$	3.29×10^{-7}	1.95×10^{-12}	4.98×10^{-9}	7.11×10^{-3}	2.81×10^{-1}	8.31×10^{-56}
	Std.	**0.00**	5.92×10^{-7}	2.01×10^{-11}	1.40×10^{-8}	3.21×10^{-3}	1.11×10^{-1}	1.02×10^{-58}
F_2	Mean	$\mathbf{5.96 \times 10^{-105}}$	1.911	6.53×10^{-18}	7.29×10^{-4}	4.34×10^{-1}	3.96×10^{-1}	8.36×10^{-35}
	Std.	$\mathbf{1.91 \times 10^{-104}}$	1.614	5.10×10^{-17}	1.84×10^{-3}	1.84×10^{-1}	1.41×10^{-1}	9.86×10^{-35}
F_3	Mean	$\mathbf{3.27 \times 10^{-180}}$	1.50×10^{3}	7.70×10^{-10}	14.0	1.66×10^{3}	43.1	1.51×10^{-14}
	Std.	**0.00**	7.07×10^{2}	7.36×10^{-9}	7.13	6.72×10^{2}	8.97	6.55×10^{-14}
F_4	Mean	$\mathbf{1.56 \times 10^{-104}}$	2.44×10^{-5}	91.7	6.00×10^{-1}	1.11×10^{-1}	8.80×10^{-1}	1.95×10^{-5}
	Std.	$\mathbf{3.47 \times 10^{-105}}$	1.89×10^{-5}	56.7	1.72×10^{-1}	4.75×10^{-2}	2.50×10^{-1}	4.49×10^{-4}
F_5	Mean	$\mathbf{2.56 \times 10^{-1}}$	1.36×10^{2}	5.57×10^{2}	49.3	79.7	1.18×10^{2}	28.4
	Std.	$\mathbf{4.78 \times 10^{-1}}$	1.54×10^{2}	41.6	38.9	73.9	1.43×10^{2}	8.40×10^{-1}
F_6	Mean	$\mathbf{3.76 \times 10^{-7}}$	5.72×10^{-7}	3.15×10^{-1}	6.92×10^{-2}	6.94×10^{-3}	2.02×10^{-2}	3.67
	Std.	$\mathbf{1.23 \times 10^{-7}}$	2.44×10^{-7}	9.98×10^{-2}	2.87×10^{-2}	3.61×10^{-3}	7.43×10^{-3}	3.35×10^{-1}
F_7	Mean	$\mathbf{2.71 \times 10^{-6}}$	8.82×10^{-5}	6.79×10^{-4}	8.94×10^{-2}	6.62×10^{-2}	5.24×10^{-2}	1.80×10^{-3}
	Std.	$\mathbf{2.33 \times 10^{-6}}$	6.94×10^{-5}	3.29×10^{-3}	2.06×10^{-2}	4.23×10^{-2}	1.37×10^{-2}	4.62×10^{-4}

Table 5. Comparison of different methods in solving multimodal test functions.

F	Index	ASSA	SSA	GA	PSO	FA	MVO	TSA
F_8	Mean	$\mathbf{-1.21 \times 10^{4}}$	-7.46×10^{3}	-5.11×10^{3}	-6.01×10^{3}	-5.85×10^{3}	-6.92×10^{3}	-7.89×10^{3}
	Std.	$\mathbf{4.89 \times 10^{2}}$	6.34×10^{2}	4.37×10^{2}	1.30×10^{3}	1.61×10^{3}	9.19×10^{2}	599.2
F_9	Mean	**0.00**	55.45	1.23×10^{-1}	47.2	15.1	1.01×10^{2}	151.4
	Std.	**0.00**	18.27	41.1	10.3	12.5	18.9	35.87
F_{10}	Mean	$\mathbf{8.88 \times 10^{-16}}$	2.84	5.31×10^{-11}	3.86×10^{-2}	4.58×10^{-2}	1.15	2.409
	Std.	**0.00**	6.58×10^{-1}	1.11×10^{-10}	2.11×10^{-1}	1.20×10^{-2}	7.87×10^{-1}	1.392
F_{11}	Mean	**0.00**	2.29×10^{-1}	3.31×10^{-6}	5.50×10^{-3}	4.23×10^{-3}	5.74×10^{-1}	7.7×10^{-3}
	Std.	**0.00**	1.29×10^{-1}	4.23×10^{-5}	7.39×10^{-3}	1.29×10^{-3}	1.12×10^{-1}	5.7×10^{-3}
F_{12}	Mean	$\mathbf{2.31 \times 10^{-5}}$	6.82	9.16×10^{-8}	1.05×10^{-2}	3.13×10^{-4}	1.27	6.373
	Std.	$\mathbf{2.46 \times 10^{-5}}$	2.72	4.88×10^{-7}	2.06×10^{-2}	1.76×10^{-4}	1.02	3.458
F_{13}	Mean	$\mathbf{1.44 \times 10^{-4}}$	21.31	6.39×10^{-2}	4.03×10^{-1}	2.08×10^{-3}	6.60×10^{-2}	2.897
	Std.	$\mathbf{1.95 \times 10^{-4}}$	16.99	4.49×10^{-2}	5.39×10^{-1}	9.62×10^{-4}	4.33×10^{-2}	6.43×10^{-1}

Tables 4 and 5 show that, for all the functions, the ASSA could provide better solutions in terms of mean value of the objective functions than the conventional SSA, as well as the other optimization techniques. The results also showed that the mean and standard deviation of the ASSA were significantly lower than those of the other strategies, indicating that the algorithm was stable. The ASSA outperformed both the standard method and alternative optimization approaches, according to the findings.

In order to obtain significant effectiveness between two or more algorithms, a nonparametric Wilcoxon's rank sum test is often used [57]. In this study, a pairwise comparison was performed using the best results from 30 runs of each algorithm. The Wilcoxon's rank sum test returned the p-value, the sum of the positive ranks (R+), and the sum of the negative ranks (R−). Table 6 shows the results of the Wilcoxon's rank sum test for all the benchmark functions. The p-value is the smallest level of significance for detecting differences. In this study, the level of significance was set at 0.05 ($\alpha = 0.05$). If the p-value was smaller than 0.05, it meant that the better result achieved by the best method in each pairwise comparison was statistically significant and was not obtained by chance. However, there was no significant difference between the two examined methods if the p-value was greater than 0.05. Such a result is indicated with "NA" in the "win" rows of Table 6. In addition, if the R+ was greater than the R−, the ASSA had a better performance than the alternative technique. Otherwise, the ASSA had a poor performance, and the other approach had a better performance [58].

Table 6. Results of Wilcoxon's rank sum test for benchmark functions.

Fun.	Index	ASSA vs. SSA	ASSA vs. GA	ASSA vs. PSO	ASSA vs. FA	ASSA vs. MVO	ASSA vs. TSA
F_1	p-val.	2.0×10^{-6}	2.0×10^{-6}	2.0×10^{-6}	2.0×10^{-6}	2.0×10^{-6}	2.0×10^{-6}
	R+	465	465	465	465	465	465
	R−	0.0	0.0	0.0	0.0	0.0	0.0
	Win	ASSA	ASSA	ASSA	ASSA	ASSA	ASSA
F_2	p-val.	2.0×10^{-6}	2.0×10^{-6}	2.0×10^{-6}	2.0×10^{-6}	2.0×10^{-6}	2.0×10^{-6}
	R+	465	465	465	465	465	465
	R−	0.0	0.0	0.0	0.0	0.0	0.0
	Win	ASSA	ASSA	ASSA	ASSA	ASSA	ASSA
F_3	p-val.	2.0×10^{-6}	2.0×10^{-6}	2.0×10^{-6}	2.0×10^{-6}	2.0×10^{-6}	2.0×10^{-6}
	R+	465	465	465	465	465	465
	R−	0.0	0.0	0.0	0.0	0.0	0.0
	Win	ASSA	ASSA	ASSA	ASSA	ASSA	ASSA
F_4	p-val.	2.0×10^{-6}	2.0×10^{-6}	2.0×10^{-6}	2.0×10^{-6}	2.0×10^{-6}	2.0×10^{-6}
	R+	465	465	465	465	465	465
	R−	0.0	0.0	0.0	0.0	0.0	0.0
	Win	ASSA	ASSA	ASSA	ASSA	ASSA	ASSA
F_5	p-val.	2.0×10^{-6}	2.0×10^{-6}	2.0×10^{-6}	2.0×10^{-6}	2.0×10^{-6}	2.0×10^{-6}
	R+	465	465	465	465	465	465
	R−	0.0	0.0	0.0	0.0	0.0	0.0
	Win	ASSA	ASSA	ASSA	ASSA	ASSA	ASSA
F_6	p-val.	6.0×10^{-6}	2.0×10^{-6}	2.0×10^{-6}	2.0×10^{-6}	2.0×10^{-6}	2.0×10^{-6}
	R+	453	465	465	465	465	465
	R−	12	0.0	0.0	0.0	0.0	0.0
	Win	ASSA	ASSA	ASSA	ASSA	ASSA	ASSA
F_7	p-val.	6.0×10^{-6}	2.0×10^{-6}	2.0×10^{-6}	2.0×10^{-6}	2.0×10^{-6}	2.0×10^{-6}
	R+	453	465	465	465	465	465
	R−	12	0.0	0.0	0.0	0.0	0.0
	Win	ASSA	ASSA	ASSA	ASSA	ASSA	ASSA
F_8	p-val.	2.0×10^{-6}	2.0×10^{-6}	2.0×10^{-6}	2.0×10^{-6}	2.0×10^{-6}	2.0×10^{-6}
	R+	465	465	465	465	465	465
	R−	0.0	0.0	0.0	0.0	0.0	0.0
	Win	ASSA	ASSA	ASSA	ASSA	ASSA	ASSA
F_9	p-val.	2.0×10^{-6}	2.0×10^{-6}	2.0×10^{-6}	2.0×10^{-6}	2.0×10^{-6}	2.0×10^{-6}
	R+	465	465	465	465	465	465
	R−	0.0	0.0	0.0	0.0	0.0	0.0
	Win	ASSA	ASSA	ASSA	ASSA	ASSA	ASSA
F_{10}	p-val.	2.0×10^{-6}	2.0×10^{-6}	2.0×10^{-6}	2.0×10^{-6}	2.0×10^{-6}	2.0×10^{-6}
	R+	465	465	465	465	465	465
	R−	0.0	0.0	0.0	0.0	0.0	0.0
	Win	ASSA	ASSA	ASSA	ASSA	ASSA	ASSA
F_{11}	p-val.	2.0×10^{-6}	2.0×10^{-6}	2.0×10^{-6}	2.0×10^{-6}	2.0×10^{-6}	2.0×10^{-6}
	R+	465	465	465	465	465	465
	R−	0.0	0.0	0.0	0.0	0.0	0.0
	Win	ASSA	ASSA	ASSA	ASSA	ASSA	ASSA
F_{12}	p-val.	2.0×10^{-6}	2.0×10^{-6}	2.0×10^{-6}	2.0×10^{-6}	2.0×10^{-6}	2.0×10^{-6}
	R+	465	0.0	465	465	465	465
	R−	0.0	465	0.0	0.0	0.0	0.0
	Win	ASSA	GA	ASSA	ASSA	ASSA	ASSA
F_{13}	p-val.	2.0×10^{-6}	2.0×10^{-6}	2.0×10^{-6}	2.0×10^{-6}	2.0×10^{-6}	2.0×10^{-6}
	R+	465	465	465	465	465	465
	R−	0.0	0.0	0.0	0.0	0.0	0.0
	Win	ASSA	ASSA	ASSA	ASSA	ASSA	ASSA
	Superior /Inferior/NA	13/0/0	12/1/0	13/0/0	13/0/0	13/0/0	13/0/0

According to the findings of the Wilcoxon's rank sum test in Table 6, the pairwise comparison of the ASSA and the SSA in the optimization of thirteen test functions demonstrated that the new approach outperformed the original method in all thirteen cases. Similarly, in the other pairwise comparisons, the ASSA provided better results for the majority of the test suite. As a result of the nonparametric statistical analysis, the ASSA created much better answers and performed significantly better than the other techniques.

5. Foundation Optimization

A shallow spread foundation, as an essential geotechnical structure, must safely and reliably support the superstructure, guarantee stability against soil-bearing capacity failings and excessive settlement, and reduce concrete stresses. Aside from these design criteria, spread footings must meet a number of other criteria: they must have enough shear and moment capacities in both the long and short dimensions; the load-carrying capacity of the foundation must not be surpassed; and the reinforcing steel configuration must meet all building code criteria [59]. The foundation optimization problem requires determining the objective function, layout constraint, and design variables, which are discussed in the following subsections.

5.1. Objective Function

The total cost of the spread footing was the study's objective function, which can be expressed mathematically as follows:

$$f(X) = C_c V_c + C_e V_e + C_b V_b + C_f A_f + C_s W_s \quad (8)$$

In Equation (8), C_c, C_e, C_b, C_f, and C_s are the unit costs of concrete, excavation, backfill, formwork, and reinforcement, respectively. The unit costs considered here are listed in Table 7 [60].

Table 7. Spread footing assembly unit cost [60].

Item	Symbol	Unit	Unit Cost (USD)
Excavation	C_e	m^3	25.16
Formwork	C_f	m^2	51.97
Reinforcement	C_s	kg	2.16
Concrete	C_c	m^3	173.96
Backfill	C_b	m^3	3.97

5.2. Design Variables

Figure 4 depicts the design features for the given model. The design variables were divided into two categories: those that described geometric dimensions and those that described steel reinforcement. As shown in Figure 4, there were four spatial design variables that reflected the foundation dimensions: the foundation's length (Y_1), the width (Y_2), the thickness (Y_3), and the embedment's depth (Y_4). The steel reinforcement also had two design variables: the longitudinal reinforcement (Y_5) and the transverse reinforcement (Y_6).

5.3. Design Constraints

While optimizing and designing a reinforced concrete footing, both structural and geotechnical limit states should be considered. Two different geotechnical limit states are the bearing capacity of the surrounding geo-material and the permitted settlement of the footing. The shear capacity of the footing (one- and two-way shear), flexural capacity, and reinforcement limitation are all structural limit states. The structural limit states are investigated using ACI 318-11 specifications [59]. Service loads are commonly used to satisfy geotechnical limit states. Even so, factored loads can be used for structural limit states. Table 8 provides a list of both structural and geotechnical limit states.

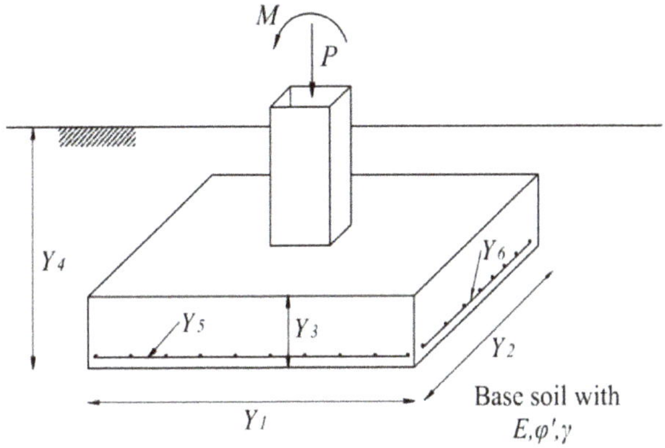

Figure 4. Design variables of the footing.

Table 8. Design constraints of spread footing.

Failure Mode	Constraint
Bearing capacity	$q_{max} \leq \frac{q_{all}}{FS}$
Settlement of foundation	$\delta \leq \delta_{all}$ $\delta = \frac{P(1-\mu^2)}{k_z E\sqrt{Y_1 Y_2}}$ $k_z = -0.0017\left(\frac{Y_2}{Y_1}\right)^2 + 0.0597\left(\frac{Y_2}{Y_1}\right) + 0.9843$
Eccentricity	$e \leq Y_1/6$
One-way (wide beam) shear	$V_u \leq \frac{1}{6}\varnothing_v\sqrt{f\prime_c}bd$
Two-way (punching) shear	$V_u \leq$ $min\left\{\left(1+\frac{2}{\beta_c}\right)/6, \left(\frac{\alpha_s d}{b_0}+2\right)/12, \frac{1}{3}\right\}\varnothing_v\sqrt{f\prime_c}bd$
Bending moment	$M_u \leq \varnothing_M A_s f_y\left(d-\frac{a}{2}\right)$
Minimum and maximum reinforcements	$\rho_{min}bd \leq A_s \leq \rho_{max}bd$
Limitation of depth of embedment	$0.5 \leq Y_4 \leq 2$

All the parameters presented in Table 8 are defined in Table 9.

Table 9. Definition of parameters of Table 7.

Parameter	Definition
q_{ult}	ultimate bearing capacity of the foundation soil
$q_{\max}$	maximum contact pressure at the interface between the bottom of a foundation and the underlying soil
δ_{all}	allowable settlement of foundation
δ	immediate settlement of foundation
ϕ_V	shear strength reduction factor equal to 0.75
f'_c	compression strength of concrete
b_0	perimeter of critical section taken at $d/2$ from face of column
b	width of the section
β_c	ratio of long side to short side of column section
α_s	is equal to 40 for interior columns
M_u	bending moment
ϕ_M	flexure strength reduction factor equal to 0.9
A_s	cross-sectional area of steel reinforcement
f_y	yield strength of steel
ρ_{min}	minimum reinforcement ratio
ρ_{max}	maximum reinforcement ratio

6. Retaining Structure Optimization

Reinforced concrete retaining walls are structures that are built to withstand lateral soil pressure as the land elevation changes. The retaining structure design process necessitates several considerations, such as structural dimensions, material characteristics, and needed reinforcement. Generally, the designer's experience plays a critical role in the cost-effective and safe design of these structures. However, the optimum design of retaining walls is independent of user experience, and the results satisfy both safety and economy.

6.1. Objective Functions

In the case of retaining structure optimization, the total construction cost of the retaining wall was considered as an objective function that incorporated the cost of materials, as well as labor and installation costs, that could be represented as follows:

$$f(X) = C_c V_c + C_e V_e + C_b V_b + C_f A_f + C_s W_s \quad (9)$$

In Equation (9), C_c, C_e, C_b, C_f, and C_s are the unit costs of concrete, excavation, backfill, formwork, and reinforcement, respectively. Table 10 presents the unit construction of a retaining structure [61].

Table 10. Basic prices considered in the analysis.

Item	Unit	Unit Cost (USD/m)
Excavation	m^3	11.41
Foundation formwork	m^2	36.82
Stem formwork	m^2	37.08
Reinforcement	kg	1.51
Concrete in foundations	m^3	104.51
Concrete in stem	m^3	118.05
Backfill	m^3	38.10

6.2. Design Variables

Figure 5 depicts the retaining wall model's cross-section, design variables, and external load. As shown in this diagram, the dimensions of the retaining wall are represented by five geometric design variables: the heel width, represented by X_1; the top stem thickness, represented by X_2; the bottom stem thickness, represented by X_3; the toe width, represented by X_4; and the base slab thickness, represented by X_5. Three additional design features are included in the steel reinforcement of the various sections of the retaining wall. The vertical steel reinforcement in the stem is designated as X_6, the horizontal steel reinforcement in the toe is designated as X_7, and the horizontal steel reinforcement in the heel is designated as X_8. B is the foundation's base width, H is the wall's total height, and H' is the stem's height.

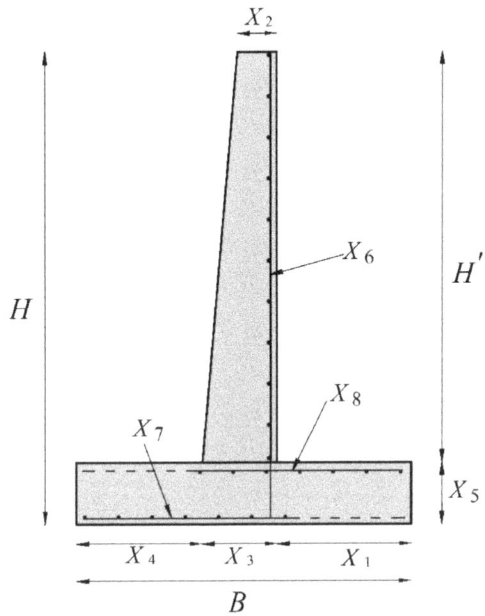

Figure 5. Design variables of the retaining structure.

6.3. Design Constraints

Figure 6 depicts the general forces acting on the retaining wall. Table 11 summarizes and presents the various design constraints that were taken into account when optimizing the concrete retaining wall.

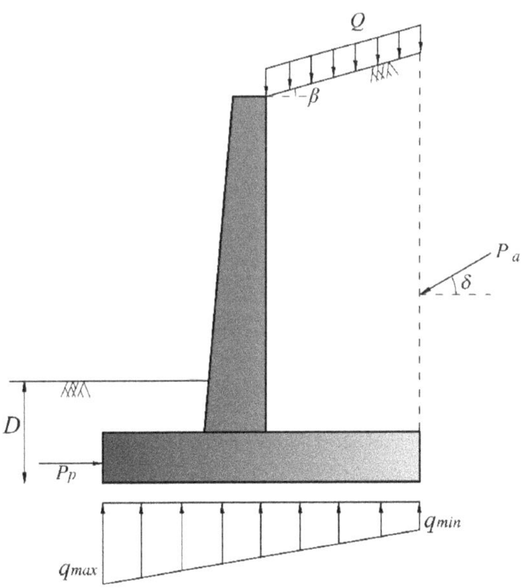

Figure 6. Forces acting on the retaining wall.

Table 11. Failure modes of retaining wall.

Failure Mode	Constraints
Sliding stability	$FS_S \leq (\Sigma F_R / \Sigma F_d)$
Overturning stability	$FS_O \leq (\Sigma M_R / \Sigma M_O)$
Bearing capacity	$FS_b \leq (q_{ult} / q_{max})$
Eccentricity failure	$e \leq (B/6)$ $e = \frac{b}{2} - \frac{\sum M_R - \sum M_O}{\sum V}$
Toe shear	$V_{ut} \leq V_{nt}$
Toe moment	$M_{ut} \leq M_{nt}$
Heel shear	$V_{uh} \leq V_{nh}$
Heel moment	$M_{uh} \leq M_{nh}$
Shear at bottom of stem	$V_{us} \leq V_{ns}$
Moment at bottom of stem	$M_{us} \leq M_{ns}$
Deflection at top of stem	$(1/150) \times H' \leq \delta_{max}$

All the parameters presented in Figure 6 and Table 11 are defined in Table 12.

Table 12. Definition of parameters of Figure 6 and Table 10.

Parameter	Definition
β	backfill slop angle
D	depth of soil in front of the wall
Q	distributed surcharge load
P_a	active earth pressure
P_p	passive earth pressure
FS_S	required factor of safety against sliding
FS_O	required factor of safety against overturning
FS_b	required factor of safety against bearing capacity
$\sum F_R$	sum of the horizontal resisting forces
$\sum F_d$	sum of the horizontal driving forces
$\sum M_R$	sum of the moments of forces that tends to resist overturning about the toe
$\sum M_O$	sum of the moments of forces that tends to overturn the structure about the toe
$\sum V$	sum of the vertical forces due to the weight of wall
V_{ut}	ultimate shearing force of toe
V_{uh}	ultimate shearing force of heel
V_{us}	ultimate shearing force of stem
V_n	nominal shear strength of concrete
M_{ut}	ultimate bending moment of toe
M_{uh}	ultimate bending moment of heel
M_{us}	ultimate bending moment of toe stem
M_n	nominal flexural strength of concrete
δ_{max}	maximum deflection at the top of the stem

In addition to the constraints mentioned above, the design variables had practical lower and upper values [62]. Table 13 summarizes the lower and upper boundaries of the design variables.

Table 13. Upper bound and lower bound for design variables of retaining wall.

Description	Lower Bound	Upper Bound
Width of footing	$B_{\min} = 0.4\,H$	$B_{\max} = 0.7\,H$
Thickness of base slab	$X_{5\min} = H/12$	$X_{5\max} = H/10$
Width of toe	$X_{4\min} = 0.4\,H/3$	$X_{4\max} = 0.7\,H/3$
Stem thickness at top	$X_{2\min} = 20$ cm	-
Steel reinforcement ratio	$\rho_{min} = max\left\{\frac{1.4}{f_y}, 0.25\frac{\sqrt{f'_c}}{f_y}\right\}$	$\rho_{max} = 0.85\beta_1\frac{f'_c}{f_y}\left(\frac{600}{600+f_y}\right)$

7. Design Examples

This section investigates numerical problems of geotechnical structures in order to evaluate the ASSA performance. To address the current inquiry, a MATLAB code was developed to computerize the design approach based on the ACI 318-11 specifications, as stated earlier [59].

In order to consider the constraints and transform a constrained optimization to an unconstrained problem, a penalty function method was used in this paper:

$$F(X) = f(X) + r\sum_{i=1}^{p} max\{0, g_i(X)\}^l \tag{10}$$

where $F(X)$ is the penalized objective function, $f(X)$ is the problem's original objective function presented in (8) and (9), and $g(X)$ is the problem's constraints presented in mboxcreftabref:applsci-1718195-t007,tabref:applsci-1718195-t010 for the spread footing and retaining wall, respectively. r is a penalty factor considered equal to 1000, l is the power of the penalty function considered equal to 2, and p is the total number of constraints.

To demonstrate the efficacy of the proposed technique, the findings were compared to state-of-the-art algorithms such as particle swarm optimization (PSO) and the firefly algorithm (FA) in the following cases. The maximum number of iterations in any algorithm was assumed to be 1000. Because of the stochastic behavior of the metaheuristics in the following experiments, all the algorithms were run 30 times, and the best results of the analyses for the minimum cost obtained by each method are reported.

7.1. Spread Footing Optimization

The first two design examples were concerned with the best design for a dry sand inner surface spread footing. Table 14 lists the other input parameters for such case studies.

Table 14. Input parameters for design examples 1 and 2.

Parameter	Unit	Value for Example 1	Value for Example 2
Effective friction angle of base soil	degree	35	30
Unit weight of base soil	kN/m^3	18.5	18.5
Young's modulus	MPa	50	35
Poisson's ratio	–	0.3	0.3
Vertical dead load (D)	kN	2000	4200
Vertical live load (L)	kN	1000	2100
Moment (M)	kN-m	0.0	850
Concrete cover	cm	7.0	7.0
Yield strength of reinforcing steel	MPa	400	400
Compressive strength of concrete	MPa	30	28
Factor of safety for bearing capacity	–	3.0	3.0
Allowable settlement of footing	m	0.04	0.04

The presented procedure solved the problem by combining all the previously mentioned algorithms. Tables 15 and 16 show the best results of the analyses for the lowest cost.

Table 15. Optimization results for design example 1.

Design Variable	Unit	Optimum Values ASSA	Optimum Values SSA	Optimum Values FA	Optimum Values PSO
(Y_1)	cm	169.5	158.3	155.3	169.4
(Y_2)	cm	218.8	248.5	253.1	219.2
(Y_3)	cm	57.5	58.1	58.2	60
(Y_4)	cm	200	158.2	200	200
(Y_5)	cm^2	39.58	48.2	49.65	37.75
(Y_6)	cm^2	25.13	21.74	20.94	23.91
Objective function	USD	1091	1098	1162	1108

Table 16. Optimization result for design example 2.

Design Variable	Unit	Optimum Values ASSA	Optimum Values SSA	Optimum Values FA	Optimum Values PSO
(Y_1)	cm	153	153.1	159.3	153.2
(Y_2)	cm	833.4	833.2	819.1	837.6
(Y_3)	cm	80.6	80.6	82.4	80.8
(Y_4)	cm	200	200	200	200
(Y_5)	cm^2	277.1	277.2	256.8	278.1
(Y_6)	cm^2	20.54	21.1	24.7	20.6
Objective function	USD	4512	4520	4650	4544

Tables 15 and 16 show that the optimization findings computed by the proposed ASSA were lower than those calculated by the conventional SSA and other approaches, indicating that the new method was effective. Table 15 shows that, contrary to popular belief that the best shape for a footing under vertical load is square, a rectangular footing provided a more cost-effective design.

7.2. Retaining Structure Optimization

The optimal design of two retaining walls with heights of 4 and 6 m was the subject of this section. Table 17 lists the other parameters that were required for this example.

Table 17. Input parameters for design examples 3 and 4.

Parameter	Unit	Value for Example 3	Value for Example 4
Height of stem	m	4.0	6
Internal friction angle of retained soil	degree	36	36
Internal friction angle of base soil	degree	0.0	34
Unit weight of retained soil	kN/m^3	17.5	17.5
Unit weight of base soil	kN/m^3	18.5	18.5
Unit weight of concrete	kN/m^3	23.5	24
Cohesion of base soil	kPa	125	100
Depth of soil in front of wall	m	0.5	0.75
Surcharge load	kPa	20	30
Backfill slop	degree	10	15

Table 17. *Cont.*

Parameter	Unit	Value for Example 3	Value for Example 4
Concrete cover	cm	7.0	7.0
Yield strength of reinforcing steel	MPa	400	400
Compressive strength of concrete	MPa	21	28
Shrinkage and temporary reinforcement percent	-	0.002	0.002
Factor of safety for overturning stability	-	1.5	1.5
Factor of safety against sliding	-	1.5	1.5
Factor of safety for bearing capacity	-	3.0	3.0

Tables 18 and 19 show the results of the assessments for the examples with the lowest cost.

Table 18. Optimization results for design example 3.

Design Variable	Unit	Optimum Values ASSA	Optimum Values SSA	Optimum Values FA	Optimum Values PSO
(X_1)	m	0.7233	0.6947	0.6948	0.6436
(X_2)	m	0.2	0.2	0.2	0.25
(X_3)	m	0.4674	0.5	0.5	0.55
(X_4)	m	0.7778	0.7778	0.7778	0.7778
(X_5)	m	0.2727	0.2727	0.2723	0.2727
(X_6)	cm^2/m	6.67	6.66	6.66	6.66
(X_7)	cm^2/m	6.75	6.75	6.75	6.75
(X_8)	cm^2/m	6.75	6.75	6.75	6.75
Objective function	USD/m	822.73	827.02	860.42	848.17

Table 19. Optimization results for design example 4.

Design Variable	Unit	Optimum Values ASSA	Optimum Values SSA	Optimum Values FA	Optimum Values PSO
(X_1)	m	1.423	1.391	1.459	1.444
(X_2)	m	0.25	0.25	0.246	0.249
(X_3)	m	0.531	0.532	0.466	0.517
(X_4)	m	0.755	0.772	0.773	0.774
(X_5)	m	0.331	0.374	0.352	0.339
(X_6)	cm^2/m	25.38	25.64	32.21	27.52
(X_7)	cm^2/m	6.78	6.75	6.75	7.02
(X_8)	cm^2/m	7.94	7.47	7.57	8.39
Objective function	USD/m	1631.2	1643.1	1668.4	1653.9

Tables 18 and 19 show that, when compared to the traditional SSA and other methods, the ASSA may be able to provide a better solution by calculating lower values of the objective functions. It can be observed that the ASSA's best price was relatively lower than that of the SSA and significantly lower than that of the PSO and FA. However, additional experiments revealed that increasing the maximum number of iterations reduced the distinctions between the algorithm results. The fact that the best solution was found in the first iteration was due to the effective modification of the algorithm proposed in this study. The modified algorithm demonstrated a much-enhanced efficacy.

In order to determine the statistical significance of the comparative results between the considered algorithms in all the design examples, a nonparametric Wilcoxon's rank sum test was performed between the results. In this regard, utilizing the results obtained from 30 runs of each method, a pairwise comparison was conducted. According to the results of the Wilcoxon's rank sum test in Table 20, the pairwise comparison between the ASSA and the SSA revealed that, in the optimization of four design examples, the new method had superior performances in three cases. In addition, for design example 3, both methods were statistically equivalent. Similarly, in the other pairwise comparisons, the ASSA provided better results. Therefore, the nonparametric statistical analysis proved that the ASSA generated significantly better solutions and, comparatively, had a superior performance over the other algorithms.

Table 20. Results of Wilcoxon's rank sum test for design examples.

Example No.	Index	ASSA vs. SSA	ASSA vs. FA	ASSA vs. PSO
Ex. 1	*p*-val.	6.0×10^{-6}	1.73×10^{-6}	1.73×10^{-6}
	R+	453	465	465
	R−	12	0.0	0.0
	Win	ASSA	ASSA	ASSA
Ex. 2	*p*-val.	0.012	1.73×10^{-6}	1.73×10^{-6}
	R+	354	465	465
	R−	111	0.0	0.0
	Win	ASSA	ASSA	ASSA
Ex. 3	*p*-val.	0.106	1.73×10^{-6}	1.73×10^{-6}
	R+	311	465	465
	R−	154	0.0	0.0
	Win	NA	ASSA	ASSA
Ex. 4	*p*-val.	1.73×10^{-6}	1.73×10^{-6}	1.73×10^{-6}
	R+	465	465	465
	R−	0.0	0.0	0.0
	Win	ASSA	ASSA	ASSA
	Superior /Inferior/NA	3/0/1	4/0/0	4/0/0

8. Conclusions

The primary objective of this study was to introduce an adaptive version of the salp swarm algorithm (ASSA). Two new equations for the leader- and follower-updating positions were introduced to improve the proposed ASSA's search and discovery abilities. In the standard SSA, the leading salp modifies its position based on a single point, which is the food location. However, due to a lack of knowledge about the real position of the food location, the algorithm may be locked at the local optimum. To overcome this weakness and to improve the exploration ability of the algorithm, in the proposed ASSA, half of the population was considered as leaders, which adjusted their positions not only based on the food location but also based on their previous positions. In addition, instead of the constant value considered in an SSA for follower-position-updating, in the ASSA, a random value was proposed. In addition, at each iteration of the optimization process, the ASSA replaced the worst salp, yielding the highest fitness value with a randomly generated salp. A statistical analysis was carried out in order to make an accurate assessment of the new algorithm's performance. The proposed method was shown to perform admirably in terms of accuracy, stability, and robustness when tested on some well-known unimodal and multimodal test functions. The paper's second goal was to automate a cost-effective design process for spread foundations and retaining walls. A computer program in Matlab was developed to reduce the cost of retaining structures and spread footings. On four case studies of these structures, the proposed method was compared to a classical SSA and some state-of-the-art metaheuristic algorithms. Given the final results, it was demonstrated that the ASSA outperformed the other techniques and should be able to provide better optimal results. The new method concurrently satisfied geotechnical and structural limit states while simultaneously providing a cost-effective design.

Author Contributions: M.K.: methodology, software, and data curation. A.I.: investigation and writing—original draft preparation. A.M.: conceptualization and methodology. S.K.: resources and writing—original draft preparation. M.L.N.: supervision, project administration, validation, funding acquisition, and final draft preparation. All authors have read and agreed to the published version of the manuscript.

Funding: This research received no external funding.

Institutional Review Board Statement: Not applicable.

Informed Consent Statement: Not applicable.

Data Availability Statement: The data that support the findings of this study are available from the corresponding author upon request.

Conflicts of Interest: This research work abides by the highest standards of ethics, professionalism, and collegiality. All authors have no explicit or implicit conflict of interest of any kind related to this manuscript.

References

1. Cheng, Y.M.; Li, L.; Chi, S.-C.; Wei, W. Particle swarm optimization algorithm for the location of the critical non-circular failure surface in two-dimensional slope stability analysis. *Comput. Geotech.* **2007**, *34*, 92–103. [CrossRef]
2. Gandomi, A.; Kashani, A.; Zeighami, F. Retaining wall optimization using interior search algorithm with different bound constraint handling. *Int. J. Numer. Anal. Methods Geomech.* **2017**, *41*, 1304–1331. [CrossRef]
3. Khajehzadeh, M.; Taha, M.R.; El-Shafie, A.; Eslami, M. Search for critical failure surface in slope stability analysis by gravitational search algorithm. *Int. J. Phys. Sci.* **2011**, *6*, 5012–5021.
4. Kashani, A.R.; Gandomi, M.; Camp, C.V.; Gandomi, A.H. Optimum design of shallow foundation using evolutionary algorithms. *Soft Comput.* **2020**, *24*, 6809–6833. [CrossRef]
5. Almazán-Covarrubias, J.H.; Peraza-Vázquez, H.; Peña-Delgado, A.F.; García-Vite, P.M. An Improved Dingo Optimization Algorithm Applied to SHE-PWM Modulation Strategy. *Appl. Sci.* **2022**, *12*, 992. [CrossRef]
6. Agresta, A.; Baioletti, M.; Biscarini, C.; Caraffini, F.; Milani, A.; Santucci, V. Using Optimisation Meta-Heuristics for the Roughness Estimation Problem in River Flow Analysis. *Appl. Sci.* **2021**, *11*, 10575. [CrossRef]
7. Gandomi, A.H.; Kashani, A.R. Construction cost minimization of shallow foundation using recent swarm intelligence techniques. *IEEE Trans. Ind. Inform.* **2017**, *14*, 1099–1106. [CrossRef]
8. Nigdeli, S.M.; Bekdaş, G.; Yang, X.-S. Metaheuristic optimization of reinforced concrete footings. *KSCE J. Civ. Eng.* **2018**, *22*, 4555–4563. [CrossRef]
9. Eslami, M.; Shareef, H.; Mohamed, A.; Khajehzadeh, M. Optimal location of PSS using improved PSO with chaotic sequence. In Proceedings of the International Conference on Electrical, Control and Computer Engineering 2011 (InECCE), Kuantan, Malaysia, 21–22 June 2011; pp. 253–258.
10. Delice, Y.; Aydoğan, E.K.; Özcan, U.; İlkay, M.S. A modified particle swarm optimization algorithm to mixed-model two-sided assembly line balancing. *J. Intell. Manuf.* **2017**, *28*, 23–36. [CrossRef]
11. Cheng, Y.; Li, L.; Lansivaara, T.; Chi, S.; Sun, Y. An improved harmony search minimization algorithm using different slip surface generation methods for slope stability analysis. *Eng. Optim.* **2008**, *40*, 95–115. [CrossRef]
12. Khajehzadeh, M.; Taha, M.R.; Eslami, M. Multi-objective optimisation of retaining walls using hybrid adaptive gravitational search algorithm. *Civ. Eng. Environ. Syst.* **2014**, *31*, 229–242. [CrossRef]
13. Ji, Y.; Tu, J.; Zhou, H.; Gui, W.; Liang, G.; Chen, H.; Wang, M. An adaptive chaotic sine cosine algorithm for constrained and unconstrained optimization. *Complexity* **2020**, *2020*, 6084917. [CrossRef]
14. Hegazy, A.E.; Makhlouf, M.; El-Tawel, G.S. Improved salp swarm algorithm for feature selection. *J. King Saud Univ. Comput. Inf. Sci.* **2020**, *32*, 335–344. [CrossRef]
15. Gao, W. Modified ant colony optimization with improved tour construction and pheromone updating strategies for traveling salesman problem. *Soft Comput.* **2021**, *25*, 3263–3289. [CrossRef]
16. Duan, D.; Poursoleiman, R. Modified teaching-learning-based optimization by orthogonal learning for optimal design of an electric vehicle charging station. *Util. Policy* **2021**, *72*, 101253. [CrossRef]
17. Li, L.-L.; Liu, Z.-F.; Tseng, M.-L.; Zheng, S.-J.; Lim, M.K. Improved tunicate swarm algorithm: Solving the dynamic economic emission dispatch problems. *Appl. Soft Comput.* **2021**, *108*, 107504. [CrossRef]
18. Ali, M.H.; Kamel, S.; Hassan, M.H.; Tostado-Véliz, M.; Zawbaa, H.M. An improved wild horse optimization algorithm for reliability based optimal DG planning of radial distribution networks. *Energy Rep.* **2022**, *8*, 582–604. [CrossRef]
19. Goh, A. Search for critical slip circle using genetic algorithms. *Civ. Eng. Environ. Syst.* **2000**, *17*, 181–211. [CrossRef]

20. Zolfaghari, A.R.; Heath, A.C.; McCombie, P.F. Simple genetic algorithm search for critical non-circular failure surface in slope stability analysis. *Comput. Geotech.* **2005**, *32*, 139–152. [CrossRef]
21. Chan, C.M.; Zhang, L.; Ng, J.T. Optimization of pile groups using hybrid genetic algorithms. *J. Geotech. Geoenviron. Eng.* **2009**, *135*, 497–505. [CrossRef]
22. Kahatadeniya, K.S.; Nanakorn, P.; Neaupane, K.M. Determination of the critical failure surface for slope stability analysis using ant colony optimization. *Eng. Geol.* **2009**, *108*, 133–141. [CrossRef]
23. Khajehzadeh, M.; Taha, M.R.; El-Shafie, A.; Eslami, M. Modified particle swarm optimization for optimum design of spread footing and retaining wall. *J. Zhejiang Univ. Sci. A* **2011**, *12*, 415–427. [CrossRef]
24. Camp, C.V.; Akin, A. Design of retaining walls using big bang–big crunch optimization. *J. Struct. Eng.* **2012**, *138*, 438–448. [CrossRef]
25. Camp, C.V.; Assadollahi, A. CO_2 and cost optimization of reinforced concrete footings using a hybrid big bang-big crunch algorithm. *Struct. Multidiscip. Optim.* **2013**, *48*, 411–426. [CrossRef]
26. Khajehzadeh, M.; Taha, M.R.; Eslami, M. A new hybrid firefly algorithm for foundation optimization. *Natl. Acad. Sci. Lett.* **2013**, *36*, 279–288. [CrossRef]
27. Kang, F.; Li, J.; Ma, Z. An artificial bee colony algorithm for locating the critical slip surface in slope stability analysis. *Eng. Optim.* **2013**, *45*, 207–223. [CrossRef]
28. Kashani, A.R.; Gandomi, A.H.; Mousavi, M. Imperialistic competitive algorithm: A metaheuristic algorithm for locating the critical slip surface in 2-dimensional soil slopes. *Geosci. Front.* **2016**, *7*, 83–89. [CrossRef]
29. Gordan, B.; Jahed Armaghani, D.; Hajihassani, M.; Monjezi, M. Prediction of seismic slope stability through combination of particle swarm optimization and neural network. *Eng. Comput.* **2016**, *32*, 85–97. [CrossRef]
30. Aydogdu, I. Cost optimization of reinforced concrete cantilever retaining walls under seismic loading using a biogeography-based optimization algorithm with Levy flights. *Eng. Optim.* **2017**, *49*, 381–400. [CrossRef]
31. Gandomi, A.; Kashani, A.; Mousavi, M.; Jalalvandi, M. Slope stability analysis using evolutionary optimization techniques. *Int. J. Numer. Anal. Methods Geomech.* **2017**, *41*, 251–264. [CrossRef]
32. Mahdiyar, A.; Hasanipanah, M.; Armaghani, D.J.; Gordan, B.; Abdullah, A.; Arab, H.; Majid, M.Z.A. A Monte Carlo technique in safety assessment of slope under seismic condition. *Eng. Comput.* **2017**, *33*, 807–817. [CrossRef]
33. Chen, H.; Asteris, P.G.; Jahed Armaghani, D.; Gordan, B.; Pham, B.T. Assessing dynamic conditions of the retaining wall: Developing two hybrid intelligent models. *Appl. Sci.* **2019**, *9*, 1042. [CrossRef]
34. Koopialipoor, M.; Jahed Armaghani, D.; Hedayat, A.; Marto, A.; Gordan, B. Applying various hybrid intelligent systems to evaluate and predict slope stability under static and dynamic conditions. *Soft Comput.* **2019**, *23*, 5913–5929. [CrossRef]
35. Yang, H.; Koopialipoor, M.; Armaghani, D.J.; Gordan, B.; Khorami, M.; Tahir, M. Intelligent design of retaining wall structures under dynamic conditions. *Steel Compos. Struct. Int. J.* **2019**, *31*, 629–640.
36. Xu, C.; Gordan, B.; Koopialipoor, M.; Armaghani, D.J.; Tahir, M.; Zhang, X. Improving performance of retaining walls under dynamic conditions developing an optimized ANN based on ant colony optimization technique. *IEEE Access* **2019**, *7*, 94692–94700. [CrossRef]
37. Himanshu, N.; Burman, A. Determination of critical failure surface of slopes using particle swarm optimization technique considering seepage and seismic loading. *Geotech. Geol. Eng.* **2019**, *37*, 1261–1281. [CrossRef]
38. Kalemci, E.N.; İkizler, S.B.; Dede, T.; Angın, Z. Design of reinforced concrete cantilever retaining wall using Grey wolf optimization algorithm. *Structures* **2020**, *23*, 245–253. [CrossRef]
39. Kaveh, A.; Hamedani, K.B.; Bakhshpoori, T. Optimal design of reinforced concrete cantilever retaining walls utilizing eleven meta-heuristic algorithms: A comparative study. *Period. Polytech. Civ. Eng.* **2020**, *64*, 156–168. [CrossRef]
40. Sharma, S.; Saha, A.K.; Lohar, G. Optimization of weight and cost of cantilever retaining wall by a hybrid metaheuristic algorithm. *Eng. Comput.* **2021**, 1–27. [CrossRef]
41. Kaveh, A.; Seddighian, M.R. Optimization of Slope Critical Surfaces Considering Seepage and Seismic Effects Using Finite Element Method and Five Meta-Heuristic Algorithms. *Period. Polytech. Civ. Eng.* **2021**, *65*, 425–436. [CrossRef]
42. Temur, R. Optimum design of cantilever retaining walls under seismic loads using a hybrid TLBO algorithm. *Geomech. Eng.* **2021**, *24*, 237–251.
43. Li, S.; Wu, L. An improved salp swarm algorithm for locating critical slip surface of slopes. *Arab. J. Geosci.* **2021**, *14*, 359. [CrossRef]
44. Khajehzadeh, M.; Keawsawasvong, S.; Sarir, P.; Khailany, D.K. Seismic Analysis of Earth Slope Using a Novel Sequential Hybrid Optimization Algorithm. *Period. Polytech. Civ. Eng.* **2022**, *66*, 355–366. [CrossRef]
45. Arabali, A.; Khajehzadeh, M.; Keawsawasvong, S.; Mohammed, A.H.; Khan, B. An Adaptive Tunicate Swarm Algorithm for Optimization of Shallow Foundation. *IEEE Access* **2022**, *10*, 39204–39219. [CrossRef]
46. Khajehzadeh, M.; Keawsawasvong, S.; Nehdi, M.L. Effective hybrid soft computing approach for optimum design of shallow foundations. *Sustainability* **2022**, *14*, 1847. [CrossRef]
47. Khajehzadeh, M.; Kalhor, A.; Tehrani, M.S.; Jebeli, M. Optimum design of retaining structures under seismic loading using adaptive sperm swarm optimization. *Struct. Eng. Mech.* **2022**, *81*, 93–102.
48. Kashani, A.R.; Camp, C.V.; Azizi, K.; Rostamian, M. Multi-objective optimization of mechanically stabilized earth retaining wall using evolutionary algorithms. *Int. J. Numer. Anal. Methods Geomech.* **2022**, *46*, 1433–1465. [CrossRef]

49. Mirjalili, S.; Gandomi, A.H.; Mirjalili, S.Z.; Saremi, S.; Faris, H.; Mirjalili, S.M. Salp Swarm Algorithm: A bio-inspired optimizer for engineering design problems. *Adv. Eng. Softw.* **2017**, *114*, 163–191. [CrossRef]
50. Zhao, X.; Yang, F.; Han, Y.; Cui, Y. An opposition-based chaotic salp swarm algorithm for global optimization. *IEEE Access* **2020**, *8*, 36485–36501. [CrossRef]
51. Zervoudakis, K.; Tsafarakis, S. A mayfly optimization algorithm. *Comput. Ind. Eng.* **2020**, *145*, 106559. [CrossRef]
52. Kaur, S.; Awasthi, L.K.; Sangal, A.; Dhiman, G. Tunicate swarm algorithm: A new bio-inspired based metaheuristic paradigm for global optimization. *Eng. Appl. Artif. Intell.* **2020**, *90*, 103541. [CrossRef]
53. Holland, J.H. Genetic algorithms. *Sci. Am.* **1992**, *267*, 66–73. [CrossRef]
54. Kennedy, J.; Eberhart, R. Particle swarm optimization. In Proceedings of the ICNN'95-International Conference on Neural Networks, Perth, Australia, 27 November–1 December 1995; pp. 1942–1948.
55. Yang, X.S. Firefly algorithms for multimodal optimization. *Lect. Notes Comput. Sci.* **2009**, *5792*, 169–178.
56. Mirjalili, S.; Mirjalili, S.M.; Hatamlou, A. Multi-verse optimizer: A nature-inspired algorithm for global optimization. *Neural Comput. Appl.* **2016**, *27*, 495–513. [CrossRef]
57. Derrac, J.; García, S.; Molina, D.; Herrera, F. A practical tutorial on the use of nonparametric statistical tests as a methodology for comparing evolutionary and swarm intelligence algorithms. *Swarm Evol. Comput.* **2011**, *1*, 3–18. [CrossRef]
58. Toz, M. Chaos-based Vortex Search algorithm for solving inverse kinematics problem of serial robot manipulators with offset wrist. *Appl. Soft Comput.* **2020**, *89*, 106074. [CrossRef]
59. ACI Committee 318. *ACI 318-11 Building Code Requirements for Structural Concrete*; American Concrete Institute: Farmington Hills, MI, USA, 2011.
60. Wang, Y.; Kulhawy, F.H. Economic design optimization of foundations. *J. Geotech. Geoenviron. Eng.* **2008**, *134*, 1097–1105. [CrossRef]
61. Yepes, V.; Gonzalez-Vidosa, F.; Alcala, J.; Villalba, P. CO_2-optimization design of reinforced concrete retaining walls based on a VNS-threshold acceptance strategy. *J. Comput. Civ. Eng.* **2012**, *26*, 378–386. [CrossRef]
62. Bowles, J. *Foundation Analysis and Design*; McGraw-Hill: New York, NY, USA, 1982.

applied sciences

MDPI

Article

Estimation of Blast-Induced Peak Particle Velocity through the Improved Weighted Random Forest Technique

Biao He [1], Sai Hin Lai [1,*], Ahmed Salih Mohammed [2], Mohanad Muayad Sabri Sabri [3,*] and Dmitrii Vladimirovich Ulrikh [4]

[1] Department of Civil Engineering, Faculty of Engineering, Universiti Malaya, Kuala Lumpur 50603, Malaysia; s2005282@siswa.um.edu.my
[2] Civil Engineering Department, College of Engineering, University of Sulaimani, Sulaymaniyah 46001, Iraq; ahmed.mohammed@univsul.edu.iq
[3] Peter the Great St. Petersburg Polytechnic University, St. Petersburg 195251, Russia
[4] Department of Urban Planning, Engineering Networks and Systems, Institute of Architecture and Construction, South Ural State University, 76, Lenin Prospect, Chelyabinsk 454080, Russia; ulrikhdv@susu.ru
* Correspondence: laish@um.edu.my (S.H.L.); mohanad.m.sabri@gmail.com (M.M.S.S.)

Abstract: Blasting is one of the primary aspects of the mining operations, and its environmental effects interfere with the safety of lives and property. Therefore, it is essential to accurately estimate the environmental impact of blasting, i.e., peak particle velocity (PPV). In this study, a regular random forest (RF) model was developed using 102 blasting samples that were collected from an open granite mine. The model inputs included six parameters, while the output is PPV. Then, to improve the performance of the regular RF model, five techniques, i.e., refined weights based on the accuracy of decision trees and the optimization of three metaheuristic algorithms, were proposed to enhance the predictive capability of the regular RF model. The results showed that all refined weighted RF models have better performance than the regular RF model. In particular, the refined weighted RF model using the whale optimization algorithm (WOA) showed the best performance. Moreover, the sensitivity analysis results revealed that the powder factor (PF) has the most significant impact on the prediction of the PPV in this project case, which means that the magnitude of the PPV can be managed by controlling the size of the PF.

Keywords: blasting; ground vibration; PPV prediction; random forest; whale optimization algorithm

Citation: He, B.; Lai, S.H.; Mohammed, A.S.; Sabri, M.M.S.; Ulrikh, D.V. Estimation of Blast-Induced Peak Particle Velocity through the Improved Weighted Random Forest Technique. *Appl. Sci.* **2022**, *12*, 5019. https://doi.org/10.3390/app12105019

Academic Editor: Ricardo Castedo

Received: 12 April 2022
Accepted: 11 May 2022
Published: 16 May 2022

1. Introduction

Blasting is an economical method of rock excavation in mining and civil engineering, and produces a series of adverse environmental effects such as blasting vibration [1–3], flying rocks [4–6], and back beak [7–9]. Among these adverse effects, the harm caused by blasting vibration is quite serious. For example, the surrounding structures can be damaged or fail because of excessive structural vibration produced by ground vibration during blasting [10,11]. Therefore, it is indispensable to predict the magnitude of blast vibration accurately.

The standard base parameter for assessing the magnitude of the blast-induced ground vibration is the peak particle velocity (PPV) [12]. Many studies on PPV prediction have been implemented, such as the proposed empirical formulas, multiple linear and nonlinear regression methods, and machine learning (ML) methods. Among these methods, the empirical formulas for the PPV prediction are easy to construct, but these formulas do not have enough accuracy since few factors are considered to implement the prediction task [13,14]. Moreover, some empirical formulas are usually designed for a given location with distinct geological parameters and field morphology, indicating that these formulas are limited and unable to predict the PPV at other blasting locations [15] accurately. Concerning the multiple linear and nonlinear regression methods, several scholars have demonstrated that these methods are capable of handling high-dimensional problems, which means that the effects of multiple factors can be considered simultaneously when predicting PPV [15,16]. However, many studies prove that the accuracy of statistical models is inferior compared to the ML methods [17–20].

In recent years, a large number of ML techniques have been applied in different areas of geotechnics, such as tunnel construction and risk assessment [21–28], soil classification [29], pile technology [30,31], materials properties [32–35], slope stability [36,37], blasting environmental issues [38,39], pillar stability prediction [40], and rock material properties [41–44], which reveal the favorable application prospects of the ML techniques. Similarly, as for the research on PPV prediction, to compensate for the shortcomings of empirical formulas and statistical models, there has been a strong inclination toward using the ML techniques in predicting the PPV because of their feasible ability to handle multidimensional nonlinear problems. For instance, Zhang et al. [1] used five machine learning techniques (i.e., classification and regression trees (CART), chi-squared automatic interaction detection (CHAID), random forest (RF), artificial neural network (ANN), and support vector machine (SVM) to predict the PPV caused by mine blasting. Their research utilized five parameters, including maximum charge per delay (MC), stemming (ST), distance from the measuring station to the blast face (DI), powder factor (PF), and hole depth (HD), to develop the ML models. The results showed that the RF had a superior capability in predicting PPV compared with the other four techniques. Lawal [45] developed an ANN-based formula to predict the PPV using two inputs, i.e., the distance from the monitoring point to the blast face and the explosive charge per delay. This study attempts to construct an interpretable formula through the ANN to identify the effect of inputs on the PPV. Rana et al. [46] compared the performance of two AI techniques (e.g., ANN and decision tree (DT)) for forecasting the blast-induced PPV. In the study, eight input parameters, i.e., total charge, number of holes, hole diameter, distance from blasting face, hole depth, tunnel cross section, the maximum charge per delay, and charge per hole, were used to design the ML models. The results indicated that the precision of the DT model performed better than the ANN model. Some relevant studies for predicting the PPV are shown in Table 1.

Table 1. Research on blast-induced PPV prediction.

Techniques	Input Variables	Number of Samples	Studies
RF, CART, CHAID, SVM, ANN	MC, HD, ST, PF, DI	102	Zhang et al. [1]
ANN	DI, MC	100	Lawal, A.I. [45]
GEP, ANFIS, SCA-ANN	DI, MC, ρ, SRH	100	Lawal, A.I. et al. [47]
MPSO-ANN	DI, MC	137	BUI Xuan-Nam et al. [48]
ANN, DT	TC, A, MC, NH, H, DI, HD, CPH	137	Rana et al. [46]
ANFIS	DI, MC	44	Iphar et al. [49]
ANN	HD, ST, DI, MC	182	Monjezi et al. [50]
FIS	B, S, ST, N, MC, DI	120	Ghasemi et al. [51]
ICA-ANN	BS, ST, MC, DI, V_p, E	95	Hajmassani et al. [52]
ICA	MC, DI	73	Jahed Armaghani et al. [53]
ANN, FIS	MC, DI	162	Mohamed [54]

Note: MC—maximum charge per delay; HD—hole depth; ST—stemming; PF—powder factor; DI—distance from the measuring station to the blast face; ρ—rock density; SRH—Schmidt rebound hardness; TC—total charge; A—tunnel cross section area; NH—number of holes; H—hole diameter; CPH—charge per hole; B—burden; S—spacing; N—number of raw; BS—burden to spacing; E—Young modulus; V_p—p-wave velocity; CART—classification and regression trees; CHAID—chi-squared automatic interaction detection; RF—random forest; ANN—artificial neural network; SVM—support vector machine; MPSO—modified particle swarm optimization; GEP—gene expression programming; ANFIS—adaptive neuro-fuzzy inference system; SCA—sine cosine algorithm; DT—decision tree; ICA—imperialist competitive algorithm; FIS—fuzzy logic.

In light of the above, the good performance and adaptability of the ML models in predicting the PPV have been proven gradually, and hence a large number of studies of ML models for predicting the PPV emerged. Furthermore, the ML models with better adaptability and higher prediction accuracy used for the prediction of blast-induced PPV need to be further developed and improved.

In the present study, we utilized a classic ML model termed the RF to conduct the PPV prediction task. After designing the regular RF model using the datasets collected from a quarry mine, we proposed five categories for modifying the regular RF model's weights and, as a result, making better predictions. The designed weighting frameworks include the improved weighted RF model based on the prediction accuracy of decision trees, as well as the optimized weights of decision trees obtained by three metaheuristic algorithms, i.e., whale optimization algorithm (WOA), gray wolf optimization (GWO), and tunicate swarm algorithm (TSA). Subsequently, four evaluation metrics were used to validate the performance of the developed models. Finally, a sensitivity analysis was conducted to identify the predominant factors for PPV prediction in this engineering case.

2. Project Description and Data Collection

In the present study, a granite mine in Penang state, Malaysia, was selected as the subject to research PPV prediction, and thus its blasting operations were investigated. Granite is the most common rock type found in the study area. The top layer is generally less than three feet thick and consists mostly of sandy clay with humus and tree roots. Explosive operations are rather common in this mine, and these operations are repeated at various intervals. Blasting at this location aims to create aggregates for various building projects with annual capacity ranging from 500–700 thousand tons per year, and large quantities of explosives are used in explosive operations. For example, in holes with sizes ranging from 76 mm to 89 mm, explosives weighing 856 kg to 9420 kg are commonly utilized. Concerning the present study, 102 blasting operations were recorded, including the design details of the blast parameters and the PPV values obtained by the Vibra ZEB seismographer. The measured and recorded blast parameters involve the number of holes, hole diameter, hole depth, burden, spacing, stemming length, subdrilling, total charge, powder factor, the maximum charge per delay, and distance from the blast-face to the measuring points. Among them, six parameters were selected as the inputs that will be used to implement the ML modeling, which is in accord with previously published works [1,48,49]. The simple statistics of the six parameters are shown in Table 2, including their max/min values, mean values, and standard deviations. It can be seen from Table 2 that the distance from the seismograph to the explosion site was about 285 m–531 m, and the range of the PPV is between 0.13 and 11.05 mm/s.

Table 2. Statistic of the data collected from the study area.

Parameters	Symbol	Unit	Type	Max	Min	Mean	Std. Dev.
Burden to spacing	BS	-	Input	0.92	0.7	0.819	0.004
Hole depth	HD	m	Input	23.17	5.23	14.115	15.973
Stemming	ST	m	Input	3.6	1.9	2.630	0.157
Powder factor	PF	kg/m^3	Input	0.94	0.23	0.654	0.034
Max charge per delay	MC	kg	Input	305.6	45.8	179.623	4246.587
Distance	DI	m	Input	531	285	379.520	5100.269
Peak particle velocity	PPV	mm/s	Output	11.05	0.13	5.337	9.267

3. Methods

3.1. RF

The RF approach is a machine learning model proposed by Leo Breiman in 2001 [55]. Because of its special algorithmic mechanism and efficient performance, the RF model has attracted much attention from researchers in various research fields. The RF model is an approach that integrates many base learners (also called decision trees) through integrated learning. It uses bagging and bootstrap techniques to train the decision trees, which solves the problem of insufficient performance of individual decision trees when dealing with complex data. At the same time, the RF is also a nonparametric classification or regression method, so it does not require prior knowledge when processing data.

The schematic diagram of the design of the RF and improved RF models is depicted in Figure 1. In the process of building decision trees, the RF model generates more randomness as the number of decision trees in the forest increases. Instead of selecting the optimal value step by step like a decision tree, the RF uses its ability of random selection and the voting mechanism of the decision tree to find the optimal value quickly. This property allows the RF model to have better classification or regression performance and a strong generalization ability of the learning system. For the regression task, when predicting the unknown output value, every single decision tree yields a predicted value, and the final value is the average of all decision trees, as shown in Equation (1). In this process, each decision tree occupies the same weight:

$$y_{pre} = \sum_{i=1}^{t} weight_i \times Tree_i(x) \quad (1)$$

where x is the input variable, y_{pre} denotes the predicted value corresponding to the input x, t is the amount of the constructed decision trees, $\{Tree_1, Tree_2, \ldots, Tree_t\}$ represent the set of decision trees in a forest, and $weight_i$ is the weight of each decision tree. In the current case (i.e., for the regular RF model), the value of weight is obtained as follows:

$$weight_i = \frac{1}{t} \quad (2)$$

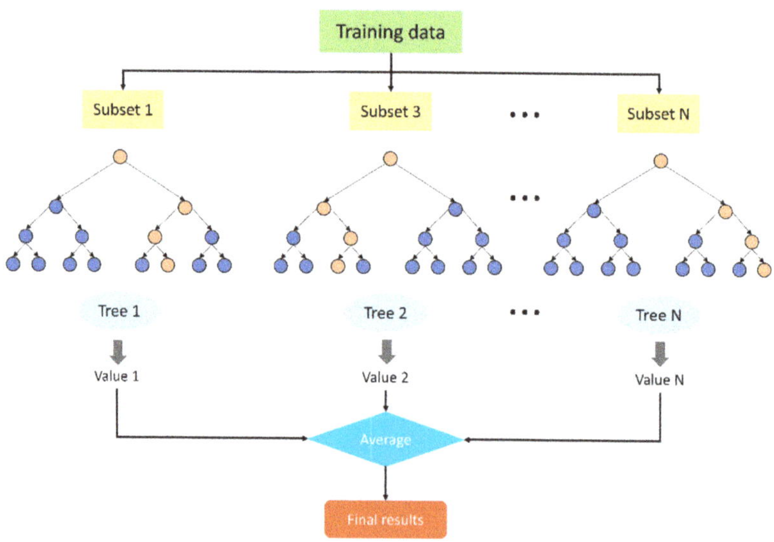

Figure 1. Schematic diagram of the RF model.

In light of the above, it can be inferred that the number of decision trees is a crucial hyperparameter that governs the predictive capability of the RF model. Therefore, a key task in this paper is to determine the optimal number of decision trees. Moreover, to avoid possible overfitting of the RF model, another hyperparameter termed the maximum depth of decision trees must also be tuned. These two critical steps will be discussed in more detail in a later section. In addition, the datasets used for constructing the RF model are randomly split into two parts; that is, 80% of datasets are used for training the RF model, and 20% of datasets are used for validating the performance of the built models. Simultaneously, a fourfold cross-validation procedure is applied in this work when implementing model training procedures. Note that fourfold cross-validation means that the datasets are divided into four equal parts, one of which is used to test the model, and the rest are used to train the model in each modeling session. A total of four modeling sessions are performed, and the average of the four modeling sessions is used as the final result.

3.2. GWO

The gray wolf optimization (GWO) algorithm is a new swarm intelligence optimization algorithm proposed by Mirialili et al. in 2014, which is based on the simulation of the hierarchical mechanism and predatory behavior of the gray wolf population in nature. Moreover, the optimization of the GWO algorithm is achieved through the process of wolf stalking, encircling, chasing, and attacking prey [56]. Gray wolves are top carnivores, and their lifestyles are mostly gregarious, thus constituting a hierarchical pyramid in the gray wolf population, with a strict hierarchical management system, as shown in Figure 2.

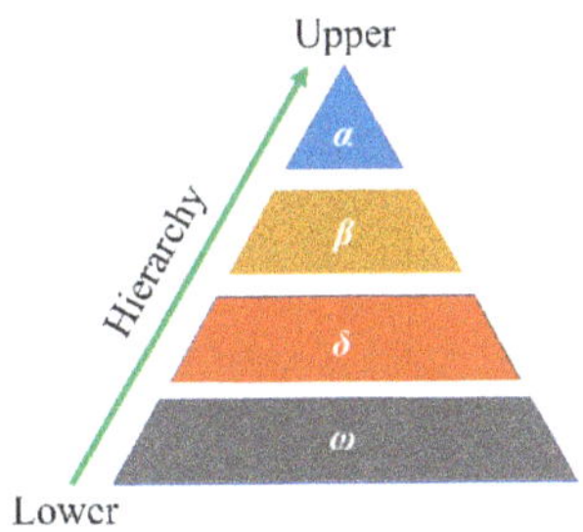

Figure 2. Hierarchy of wolf in GWO algorithm.

The first level of the pyramid is the head of the population, called α, which is mainly responsible for all the decision-making matters of the population. The second pyramid level is called β, which assists α in making management decisions. The third level of the pyramid is δ, which is mainly responsible for scouting, sentry, hunting, and guarding. Furthermore, the bottom level of the pyramid is called ω, which is mainly responsible for coordinating the relationship within the population. The hierarchy of gray wolves plays a crucial role in achieving prey hunting. The predation process is led by α. At first, the wolves search, track, and approach the prey in a team pattern; then, the wolves encircle the prey from all directions, and when the encirclement is small enough and perfect, the wolves will attack from β and δ, which are closest to the prey, under the command of α. The mathematical models of this process are explained below.

First, the encirclement of the prey by the gray wolves during predation can be characterized by the following equation:

$$D = \left|C \times X_p(t) - X(t)\right| \tag{3}$$

where $X_p(t)$ and $X(t)$ denote the position of the prey and the position of the wolves during the t-th iteration, respectively, and C is the coefficient, which is computed by the following equation:

$$C = 2r_1 \tag{4}$$

Here r_1 is the random value in the interval [0, 1].

Then, the equation for updating the position of the gray wolf in search space is as follows:

$$X(t+1) = X_p(t) - A \times D \tag{5}$$

where A is the convergence factor. A can be computed by the following equation:

$$A = 2a \times r_2 - a \tag{6}$$

$$a = \left(2 - 2 \times \left(\frac{t}{t_{max}}\right)\right) \tag{7}$$

where r_2 is the random value in the interval [0, 1], and t_{max} denotes the maximum iterations.

After that, when the gray wolf determines the position of the prey, the wolf α will lead β and δ to initiate the pursuit behavior. In the wolf pack, α, β, and δ are the closest to the prey. The positions of these three wolves can be used to determine the location of the prey. The mathematical description is as follows:

$$D_\alpha = |C_1 \times X_\alpha(t) - X(t)| \tag{8}$$

$$D_\beta = |C_2 \times X_\beta(t) - X(t)| \tag{9}$$

$$D_\delta = |C_3 \times X_\delta(t) - X(t)| \tag{10}$$

$$X_1 = X_\alpha - A_1 \times D_\alpha \tag{11}$$

$$X_2 = X_\beta - A_2 \times D_\beta \tag{12}$$

$$X_3 = X_\delta - A_3 \times D_\delta \tag{13}$$

$$X_p(t+1) = \frac{X_1 + X_2 + X_3}{3} \tag{14}$$

Using Equations (8)–(13) can obtain the distance between the prey and the α, β, and δ, and then using Equation (14) identifies the direction of movement of individual gray wolves toward the prey.

3.3. WOA

The whale optimization algorithm (WOA) is a novel swarm intelligence optimization algorithm proposed by Mirjalili and Lewis in 2016 [57]. The WOA is based on the simulation of the hunting behavior of humpback whales in nature, and it optimizes the searching process by mimicking the behavior of searching, encircling, pursuing, and attacking the prey by the humpback whales. Whales are considered the largest mammals globally, with adults growing up to 30 m long and weighing 180 tons. Whales have a unique feeding behavior, which is the bubble-net method, as shown in Figure 3. The method is divided into two stages: upward spiral and double circulation. Based on this special foraging behavior, the WOA algorithm is obtained.

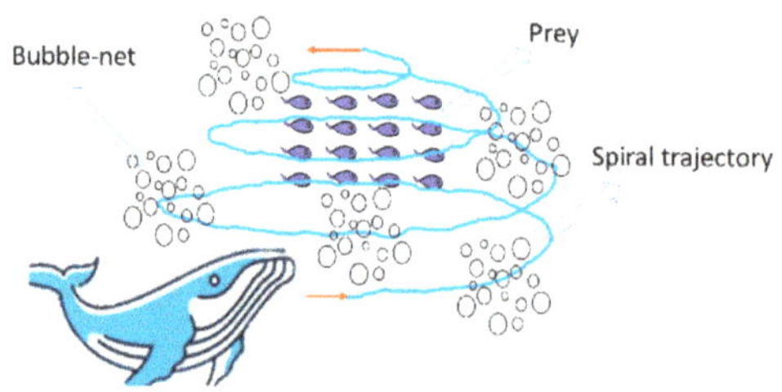

Figure 3. Bubble-net attacking method of whales.

In the WOA, it is assumed that the number of populations is N and the dimension of the search space is d, and the position of the i-th whale in the dimensional space can be expressed as $X_i = \left(X_1^d, X_2^d, X_3^d, \ldots, X_N^d\right)$. The position of the prey in the search space corresponds to the optimal global solution. Whales can identify the position of the prey and then encircle it since there is no a priori knowledge of the position of the optimal global solution in the search space before solving the optimization problem. Therefore, in the WOA, assuming that the whale at the optimal position in the current population is the prey and all other whale individuals in the population encircle the optimal whale, the mathematical model of the process is as follows:

$$X(t+1) = X_p(t) - A \times \left| C \times X_p(t) - X(t) \right| \tag{15}$$

where t is the current iteration, $X(t)$ denotes the position of whales, $X_p(t)$ denotes the position of prey, and A and C are the coefficients that can be defined as follows:

$$A = 2a \times rand_1 - a \tag{16}$$

$$C = 2 \times rand_2 \tag{17}$$

Here $rand_1$ and $rand_2$ represent the random values in the interval [0, 1], and a is the convergence factor that decreases linearly from 2 to 0 as the number of iterations increases.

To describe the bubble-net attacking behavior of whales with a mathematical model, two different methods were designed in the WOA, namely, the shrinking encircling mechanism and the spiral updating position. Among them, the shrinking encircling mechanism is implemented by the linear reduction in the convergence factor a in Equation (16). In the method of spiral updating position, the spiral motion of the whale is simulated to capture the prey, and its mathematical model is shown as follows:

$$X(t+1) = D' \times e^{bl} \times \cos(2\pi l) + X_p(t) \tag{18}$$

where $D' = \left| X_p(t) - X(t) \right|$ denotes the distance between the prey and the i-th whale, b signifies a constant for controlling the shape of the logarithmic spiral, and l is a random value in the interval $[-1, 1]$. Noted that, the whales move within a constricted circle and simultaneously follow a spiral path towards their prey.

In addition to the bubble-net attacking method, whales will also randomly search for prey. Individual whales search randomly according to each other's position, and the mathematical model can be expressed as follows:

$$X(t+1) = X_{rand}(t) - A \times |C \times X_{rand}(t) - X(t)| \tag{19}$$

where X_{rand} represents a position vector of an individual search agent randomly chosen from the current population.

3.4. TSA

The tunicates are small, net-like organisms that are found throughout the sea. They live solitary or parasitic lives and can locate food in the ocean. Based on the biological inspiration of the predatory behavior of tunicates, Kaur et al. proposed a swarm intelligence optimization algorithm termed tunicate swarm algorithm (TSA), which is inspired by the jet-propelled migration mechanism and the intelligent foraging behavior of tunicates in the ocean [58]. The mathematical models of this algorithm are explained below.

(1) Initialization

Similar to most metaheuristic algorithms, the TSA starts to execute the optimization process by initializing the tunicate population, i.e., by initializing the positions of every tunicate ($\vec{A_0}$) in the search space, as shown in the following equation:

$$\vec{A_0} = \vec{A_{min}} + rand\left(\vec{A_{max}} - \vec{A_{min}} \right) \tag{20}$$

where $\vec{A_{min}}$ and $\vec{A_{max}}$ denote the upper and lower limits of the search space, respectively, and *rand*() signifies the random value in the interval [0, 1].

(2) Avoid conflicts between search agents

To avoid conflicts between the individuals when implementing the searching task, the TSA utilizes vector $\vec{A}$ to calculate the position of new search agents, which can be illustrated by the following equations:

$$\vec{A} = \frac{\vec{G}}{\vec{M}} \tag{21}$$

$$\vec{G} = c_1 + c_2 - \vec{F} \tag{22}$$

$$\vec{M} = \lfloor P_{min} + c_3 \times P_{max} - P_{min} \rfloor \tag{23}$$

where $\vec{A}$ denotes the new positions of the search agents, $\vec{G}$ denotes the gravity, $\vec{M}$ denotes the interaction forces between the tunicates, $\vec{F}$ signifies the current advection in the deep sea, and $\vec{F} = 2c_3$, c_1, c_2, and c_3 are the random value in the interval [0, 1].

(3) Move to the best neighbor

After avoiding conflicts between the individuals, the search agents will move towards the position of the best neighbor, which the following equation can interpret:

$$\vec{P_d} = \left| \vec{F_s} - r \times \vec{P_p}(t) \right| \tag{24}$$

where $\vec{P_d}$ denotes the distance between the food and the search agents, $\vec{F_s}$ represents the position of the food, $\vec{P_p}(t)$ signifies the positions of the search agents during the t-th iteration, and r is the random value in the interval [0, 1].

(4) Move towards the best individual

The mathematical description of the movement of the tunicate population towards the position of the optimal search agent is as follows:

$$\vec{P_p}(t^*) = \begin{cases} \vec{F_s} + \vec{A} \times \vec{P_d}, & r \geq 0.5 \\ \vec{F_s} - \vec{A} \times \vec{P_d}, & r < 0.5 \end{cases} \tag{25}$$

where $\vec{P_p}(t^*)$ is the position of the search agent closest to the target food.

(5) Swarm behavior

TSA is implemented to mimic the tunicates' swarm behavior by saving the first two optimal solutions and updating the other tunicates' positions according to the first two optimal solutions. The swarm behavior of the tunicates is mathematically described as follows:

$$\vec{P_p}(t+1) = \frac{\vec{P_p}(t) + \vec{P_p}(t+1)}{2 + c_1} \tag{26}$$

where $\vec{P_p}(t+1)$ denotes the positions of the search agent when implementing the $(t + 1)$-th iteration.

3.5. Improved RF Models

Although the RF model has good performance and has been applied in many areas, there is still some room for improvement. For the random forest approach, in addition to sampling with the replacement for constructing training datasets, a random number of features are picked each time to lower the degree of correlation between individual produced decision trees. The end outcome is a simple average computed by each created decision tree of the forest [55]. Although the RF shows remarkable performance in some regression or classification tasks, it seems that some improvements for the combined way of the base learners (i.e., decision trees) can be made to achieve a better prediction of the RF. Decision trees exhibit different predictive capabilities because of the bootstrap replicates and random feature picking. However, in the regular RF, each decision tree is assigned with the same weights, which seems unreasonable and can be further improved. Therefore, some techniques such as weighting the decision trees based on their predictive capabilities are proposed to optimize the regular RF. The accuracy of decision trees is considered the indicator that can amend the weights of decision trees instead of using the same weight for each tree [59]. In this paper, two indicators, i.e., the coefficient of determination (R^2) and the root mean squared error (RMSE) of decision trees, are used as benchmarks for improving the weights of decision trees. For example, the first category is that decision trees with higher R^2 will be assigned relatively higher weights, and vice versa. Another case is that decision trees with lower RMSE will be assigned relatively higher weights, and vice versa. Furthermore, we also put forward using three metaheuristic algorithms (i.e., GWO, WOA, and TSA) to search for the best weights of decision trees and then compare the effect of their improvements on the regular RF model.

To sum up, there are a total of five methods that are leveraged to refine the weights of decision trees in a forest. Before applying these methods, the first task that needs to be carried out is the parameter tuning of the RF model, which is to establish the optimal RF model based on the training dataset. Subsequently, the task is to use the five categories mentioned above to amend the weights of decision trees. This part will be elaborated hereinafter. Moreover, the schematic diagram of the design of the RF and improved RF models is depicted in Figure 4.

3.5.1. Improved Weights Based on the Accuracy

First, let $\{Tree_1(x),\ Tress_2(x),\ Tree_3(x),\ \cdots,\ Tree_t(x)\}$ represent the set of decision trees. For the regular RF model, the final predicted values can be obtained through Equation (27), which indicates that the weight of every single decision tree is equal to $1/t$.

$$y_{pre} = \sum_{i=1}^{t} \frac{1}{t} \times Tree_i(x) \tag{27}$$

where t is the number of decision trees in a forest.

For the improved weights based on R^2, the weight of every single decision is obtained through Equation (28), and the final predicted values can be obtained using Equation (29). The key step of this approach is to compute the weights of decision trees. After constructing the optimal RF model based on the training sets, the accuracy (i.e., R^2) of each decision tree can be obtained. Then, according to Equation (28), all decision trees will be assigned a normalized weight, and the larger the term R^2, the larger the weight assigned, which means that these decision trees with large weights will have a greater role in the final prediction. Equations (28) and (29) are:

$$\text{weight\_R}^2{}_i = \frac{R_i^2}{\sum_{i=1}^{t} R_i^2} \tag{28}$$

$$y_{pre} = \sum_{i=1}^{t} \text{weight\_R}^2{}_i \times Tree_i(x) \tag{29}$$

Similarly, for the improved weights based on the RMSE, the weight of every single decision is obtained through Equation (30), and the final predicted values can be obtained using Equation (31). The weights of the decision trees are obtained by first taking the inverse of the RMSE and then performing a normalized calculation, which means that the larger the RMSE of a decision tree, the smaller the weight it is assigned to, and vice versa. Equations (30) and (31) are:

$$\text{weight\_RMSE}_i = \frac{\frac{1}{\text{RMSE}_i}}{\sum_{i=1}^{t}\left(\frac{1}{\text{RMSE}_i}\right)} \tag{30}$$

$$y_{pre} = \sum_{i=1}^{t} \text{weight\_RMSE}_i \times Tree_i(x) \tag{31}$$

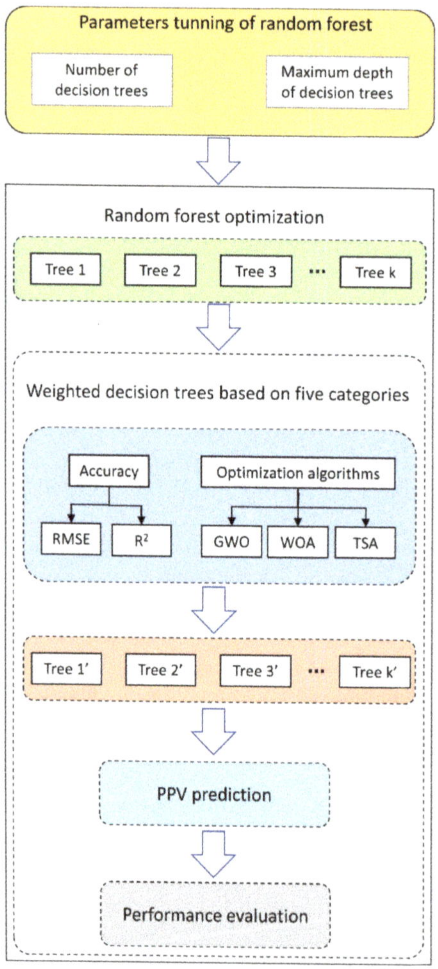

Figure 4. Flowchart of the design of the RF and improved RF models.

3.5.2. Improved Weights Based on Three Metaheuristic Algorithms

In this section, we consider the weights of decision trees in a forest as an unknown space of high dimensionality to be solved. After determining the number of trees in a forest, the dimension of the space to be solved is equal to the number of trees. Then, the GWO, WOA, and TSA algorithms are used to capture the optimal weights in the unknown space. To achieve this goal, it is assumed that the independent variables to be solved are set to $\{x_1, x_2, x_3, \cdots, x_t\}$, where t is the number of trees and the range of x is in the interval [0, 1], and then the standardized weights are computed as follows:

$$\text{weight\_optmization}_i = \frac{x_i}{\sum_{i=1}^{t} x_i} \tag{32}$$

where $\text{weight\_optmization}_i$ represents the weights of decision trees optimized by the metaheuristic algorithms, and it is clear that $\sum_{i=1}^{t} \text{weight\_optimization}_i = 1$.

After that, the RMSE is considered the criteria used to evaluate each result's performance. In other words, the fitness function that needs to be solved by the metaheuristic algorithms is the RMSE function. Moreover, the L2 regularization (also named penalty terms) is also incorporated in the RMSE function, which avoids overfitting. Therefore, the fitness function consists of two parts: the RMSE function and the L2 regularization, which is presented by Equation (34):

$$y_{pre} = \sum_{i=1}^{t} \text{weight\_optmization}_i \times Tree_i(x) \tag{33}$$

$$\text{Fitness function} = \sqrt{\frac{1}{m}\sum_{i=1}^{m}\left(y_i - y_{pre}\right)^2} + \gamma \times \sqrt{\sum_{i=1}^{t} \text{weight\_optmization}_i{}^2} \tag{34}$$

where y_{pre} denotes the predicted PPV computed by the improved RF model based on optimization algorithms, m is the number of training samples, and γ is the coefficient of the L2 regularization, whose value is set to 0.08 in this paper according to the trial-and-error method.

For the GWO, WOA, and TSA, the parameters that need to be set are the swarm sizes and the number of iterations. The appropriate selection of these parameters can effectively and quickly lead to optimal results. Therefore, after constructing the model several times, the swarm sizes set to each optimization algorithm are 50, 100, 150, and 200, respectively, and the number of iterations is set at 1000.

The GWO, WOA, and TSA optimization techniques can be used to improve the RF model in the following way:

(1) Data preparation: randomly divide the raw data into a training set (80% of raw data) and a testing set (20% of raw data);
(2) Initialization: Initialize the swarm size, iterations, as well as some necessary parameters of the three optimization algorithms;
(3) Fitness evaluation: Calculate the fitness value of the population, evaluate its fitness, and then save the best fitness value before starting the next iteration;
(4) Update parameters: Update the fitness value based on the outcome of each iteration, which aims to capture the ideal solutions;
(5) Suspension conditions check: When the optimal fitness value no longer changes, or the maximum number of iterations is reached, the optimal solutions of the weights of decision trees are obtained.

3.6. Criteria for Evaluation

An important task to be conducted after the modeling is to evaluate each model's accuracy and generalization ability. As mentioned previously, in this work, 80% of the data samples were assigned as training that was used to train the RF and improve the RF models, while the rest 20% of the data samples were assigned as testing data to verify the performance of the RF and improved RF models. To evaluate the performance of the built models, four metrics were used in this work, i.e., the coefficient of determination (R^2), the root mean squared error (RMSE), the mean absolute error (MAE), and the variance account for (VAF). These evaluation indicators are defined below.

The square of the correlation between the predicted and measured values is represented by R^2. The RMSE characterizes the standard deviation of the fitting error between the predicted values and the measured values. The MAE indicates the mean absolute error between the predicted values and measured values. The VAF describes the prediction performance by comparing the standard deviation of the fitting error with the standard deviation of the actual value. The following equations were utilized to calculate the R^2, RMSE, MAE, and VAF values:

$$R^2 = 1 - \frac{\sum_{i=1}^{n}(y_i - \hat{y}_i)^2}{\sum_{i=1}^{n}(y_i - \overline{y})^2} \tag{35}$$

$$\text{RMSE} = \sqrt{\frac{1}{n}\sum_{i=1}^{n}(y_i - \hat{y}_i)^2} \tag{36}$$

$$\text{MAE} = \frac{1}{n}\sum_{i=1}^{n}|y_i - \hat{y}_i| \tag{37}$$

$$\text{VAF} = \left(1 - \frac{var(y_i - \hat{y}_i)}{var(y_i)}\right) \times 100 \tag{38}$$

where y_i, $\hat{y}_i$, and $\overline{y}$ denote the measured, predicted, and mean values of the PPV, respectively. When the predicted and measured PPV values are precisely the same, R^2 is 1, the RMSE is 0, the MAE is 0, and the VAF is 100 (%).

4. Results and Discussion

4.1. Parameter Tuning of the RF Model

The main purpose of this section is to determine the optimal hyperparameters of the RF model, including the number of decision trees and the maximum depth of decision trees. In addition, before conducting the model training, the original dataset was standardized to eliminate the negative effect of magnitude between the data and speed up the model's training. The formula for standardization of data is given as follows:

$$X_i^* = \frac{X_i - \mu_i}{\sigma_i},\ i = 1, 2, \ldots, 6 \tag{39}$$

where X_i denotes the data samples belonging to feature i, μ_i and σ_i denote the mean and standard deviation of X_i, respectively, and X_i^* signifies the standardized datasets that are prepared for constructing the RF model.

The parameters tuning of RF consist of two steps. First, the scale of the RF is optimized to determine the number of decision trees in a forest. In this regard, the number of trees increases from 1 to 200 with 1 increment each time. Meanwhile, other parameters are set to default values. Figure 5 shows the performance of the RF model with respect to the number of trees. Intuitively, it can be seen that the mean squared error decreases when the number of trees increases from 1 to 50, and then with the increase in the number of trees, the mean squared error only shows small fluctuations. The results show that the minimum mean squared error is reached when the number of trees is 51. Consequently, the scale of the RF model is set to 51 decision trees. Furthermore, using the testing set to validate the performance of the RF model when the number of trees is 51, the results show that the R^2, RMSE, MAE, and VAF values on the testing set are 0.915, 0.275, 0.224, and 93.504, respectively.

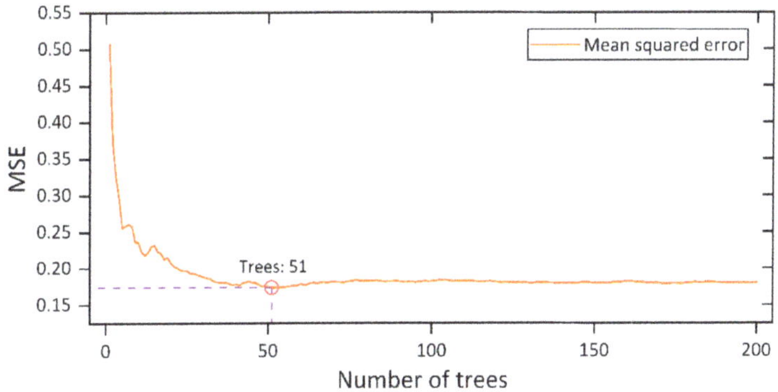

Figure 5. MSE with respect to the number of trees in a forest.

After that, the optimal maximum depth of decision trees was determined. The number of the maximum depth increases from 1 to 10 with 1 increment each time, while the number of decision trees is 51 currently. Figure 6 depicts the performance of the RF model concerning the maximum depth of decision trees. It can be found that the minimum mean squared error is reached when the maximum depth is set to 7. When the maximum depth values are larger than 7, there is no apparent decrease in the mean squared error. At the same time, the performance evaluation results of the current RF model on the testing set are R^2 of 0.923, the RMSE of 0.262, the MAE of 0.209, and the VAF of 93.994, respectively. Thus, according to the aforementioned results, the maximum depth is set to 7. Now, the RF model has been successfully constructed, and the later work uses five techniques to optimize the weights of decision trees, aiming to improve the performance of the current RF model.

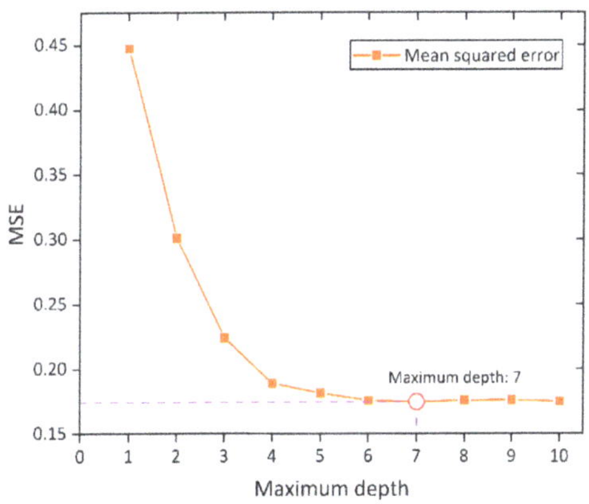

Figure 6. MSE with respect to the maximum depth of trees in a forest.

4.2. Improved RF-Based Models

In this section, five methods were used to optimize the regular RF model, namely the refined weights of decision trees based on their predictive capabilities that are characterized by R^2 and the RMSE, and the refined weights of decision trees based on the optimization solution of three metaheuristic algorithms (i.e., GWO, WOA, and TSA). The main goal of the five techniques is to obtain the optimal weight set of decision trees that can minimize the generalization error of the improved RF models. Moreover, to validate the models' performance, each model's accuracy and error on the training set and testing set are both utilized. In this way, the optimal improved RF model can be effectively obtained.

Figure 7 presents the calculation results of the three metaheuristic algorithms on the training set. It can be seen that there are some differences in the calculation results due to the swarm sizes. For example, for the RF-WOA model, the final fitness value is 200 swarms sizes < 150 swarm sizes < 100 swarm sizes < 50 swarm sizes; for the RF-GWO model, the final fitness value is 100 swarm sizes < 200 swarm sizes < 150 of swarm sizes < 50 swarm sizes; and for the RF-TSA model, the final fitness value is 150 of swarm sizes < 200 swarm sizes < 50 of swarm sizes < 100 swarm sizes. It should be noted that the above results are obtained based on the training set. For a clearer identification of the performance of each model, it should be combined with its performance on the testing set. For this purpose, a scoring evaluation method was employed to select the best model [60]. The principle of this method is that the higher the score, the better the performance, and vice versa. Consequently, the final scoring of the RF-WOA, RF-GWO, and RF-TSA models with different swarm sizes on both training and test sets is obtained in Tables 3–5. According to the results in Tables 3–5, it can be concluded that the best RF-WOA model with a total score of 26 is obtained when the swarm sizes are set to 100, the best RF-GWO model with a total score of 28 is obtained when the swarm sizes are set to 200, and the best RF-TSA model with a total score of 27 is obtained when the swarm sizes are set to 50.

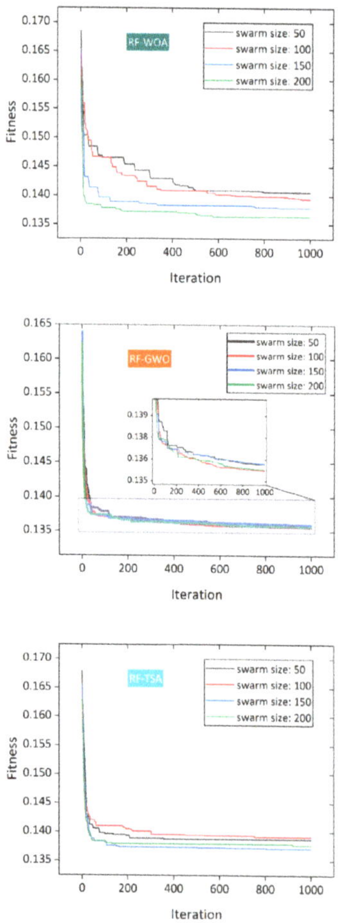

Figure 7. Various RF-based metaheuristic models based on different swarm sizes.

Table 3. Performance of different swarm sizes of the RF-WOA models on predicting the PPV.

Models		Training								Testing								Score Summation
	Swarm Size	R^2	Score	RMSE	Score	MAE	Score	VAF	Score	R^2	Score	RMSE	Score	MAE	Score	VAF	Score	
RF-WOA	50	0.986	4	0.119	1	0.094	2	98.584	1	0.932	4	0.246	4	0.187	4	94.723	3	23
	100	0.986	4	0.118	2	0.093	3	98.603	2	0.932	4	0.246	4	0.188	3	95.032	4	26
	150	0.986	4	0.117	3	0.092	4	98.612	3	0.927	2	0.254	2	0.191	2	94.430	2	22
	200	0.986	4	0.116	4	0.092	4	98.647	4	0.928	3	0.253	3	0.196	1	94.394	1	24

Table 4. Performance of different swarm sizes of the RF-GWO models on predicting the PPV.

Models		Training								Testing								Score Summation
	Swarm Size	R^2	Score	RMSE	Score	MAE	Score	VAF	Score	R^2	Score	RMSE	Score	MAE	Score	VAF	Score	
RF-GWO	50	0.986	4	0.116	4	0.092	4	98.632	1	0.930	4	0.250	4	0.192	3	94.516	3	27
	100	0.986	4	0.116	4	0.092	4	98.633	2	0.929	3	0.251	3	0.192	3	94.475	1	24
	150	0.986	4	0.116	4	0.092	4	98.637	4	0.929	3	0.252	2	0.193	2	94.488	2	25
	200	0.986	4	0.116	4	0.093	3	98.635	3	0.929	3	0.251	3	0.191	4	94.528	4	28

Table 5. Performance of different swarm sizes of the RF-TSA models on predicting the PPV.

Models		Training								Testing								Score Summation
	Swarm Size	R^2	Score	RMSE	Score	MAE	Score	VAF	Score	R^2	Score	RMSE	Score	MAE	Score	VAF	Score	
RF-TSA	50	0.986	4	0.118	2	0.093	3	98.598	2	0.931	4	0.248	4	0.191	4	94.743	4	27
	100	0.986	4	0.118	2	0.094	2	98.582	1	0.924	1	0.260	1	0.201	2	93.823	1	14
	150	0.986	4	0.116	4	0.093	3	98.635	4	0.927	2	0.255	2	0.191	4	94.483	2	25
	200	0.986	4	0.117	3	0.092	4	98.619	3	0.929	3	0.251	3	0.195	3	94.569	3	26

Table 6. Performance of the RF and improved RF models on predicting the PPV.

Models	Training								Testing								Score Summation
	R^2	Score	RMSE	Score	MAE	Score	VAF	Score	R^2	Score	RMSE	Score	MAE	Score	VAF	Score	
RF-WOA (100)	0.986	6	0.118	5	0.093	6	98.603	5	0.932	6	0.246	6	0.188	6	95.032	6	46
RF-GWO (200)	0.986	6	0.116	6	0.093	6	98.635	6	0.929	4	0.251	4	0.191	5	94.528	4	41
RF-TSA (50)	0.986	6	0.118	5	0.093	6	98.598	4	0.931	5	0.248	5	0.191	5	94.743	5	41
RF	0.973	3	0.164	2	0.123	3	97.298	1	0.923	1	0.262	1	0.209	2	93.994	1	14
RF-R^2	0.974	4	0.161	3	0.121	4	97.380	2	0.924	2	0.261	2	0.207	3	94.078	2	22
RF-RMSE	0.975	5	0.156	4	0.119	5	97.530	3	0.925	3	0.258	3	0.203	4	94.219	3	30

After obtaining the best model for individual metaheuristic algorithms in their respective model comparisons, the next step is to perform further comparative analysis of them and the two previously mentioned improved weighting methods based on the accuracy (i.e., R^2 and the RMSE) to identify the optimal improved RF model. Concerning the improved weights based on the R^2 and RMSE of every single decision tree, the weights of the decision trees are computed through Equations (28) and (30), respectively. For the metaheuristic algorithms, the RF-WOA with 100 swarm sizes, the RF-GWO with 200 swarm sizes, and the RF-TSA with 50 swarm sizes are chosen because of their best performance. At the same time, the regular RF model is also used for comparative analysis in this section. Then, the results of the performance of these six models are tabulated in Table 6. As can be seen from Table 6, the model with the highest score is the RF-WOA, whose scoring is 46, followed by the RF-GWO and RF-TSA, whose scorings are both 41. As for the RF-RMSE and RF-R^2, their scores were 30 and 22, respectively, which indicates the performance of the RF-RMSE is slightly better than that of the RF-R^2. In general, these five models all outperform the regular RF model with a scoring of 14. Moreover, the ranking of these five models is as follows: RF-WOA > RF-GWO = RF-TSA > RF-RMSE > RF-R^2 > RF.

To intuitively observe the differences among these models, we divide the four evaluation metrics into two groups, one for the VAF and $R^2 \times 100$, where larger values indicate better model performance, and the other one for the RMSE and MAE, where smaller values indicate better models. As depicted in Figure 8, for the group of the VAF and $R^2 \times 100$ (the left one in Figure 8), the accuracies of the RF-GWO, RF-WOA, and RF-TSA are quite close and are all better than the other three models (i.e., RF-RMSE, RF-R^2, and RF), whereas, for the group of the RMSE and MAE (the right one in Figure 8), the error of the RF-GWO is slightly lower than those of the RF-WOA and RF-TSA. Likewise, the errors of the three metaheuristic-based RFs are all lower than those of the RF-RMSE, RF-R^2, and RF. Moreover, it can be concluded that all of the refined weights RF models have a good performance on the training set compared with the regular RF model. As for the testing set, the results are shown in Figure 9. For the group of the VAF and $R^2 \times 100$ (the left one in Figure 9), the ranking of the accuracy of these six models is shown as RF-WOA > RF-TSA > RF-GWO > RF-RMSE > RF-R^2 > RF. The same situation occurs in the group of the RMSE and MAE (the right one in Figure 9), which indicates the ranking of error of these six models is RF-WOA < RF-TSA < RF-GWO < RF-RMSE < RF-R^2 < RF. Therefore, it can be inferred that the RF-WOA has a better generalization ability than other models, which can be verified as it has a better performance on the testing set. Additionally, we can also conclude that all of the refined weights RF models have a good performance on the testing set compared with the regular RF model, which means the refined techniques proposed in this paper can effectively enhance the performance of the regular RF model.

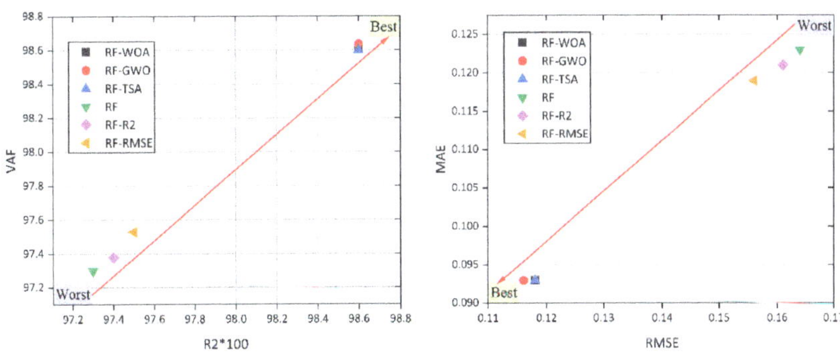

Figure 8. Results of the evaluation metrics of the developed models on the training set.

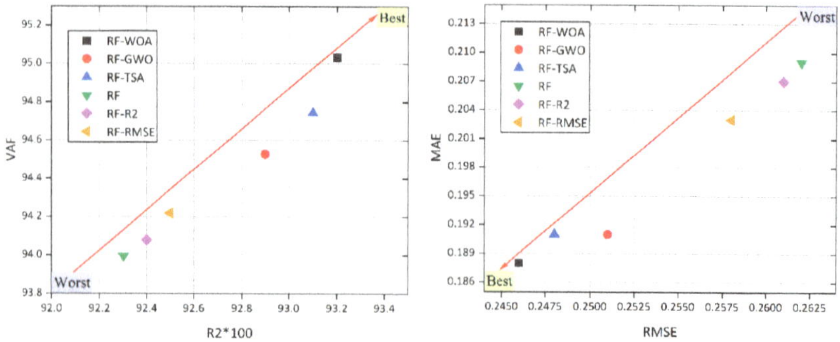

Figure 9. Results of the evaluation metrics of the developed models on the testing set.

In a different way, the Taylor diagram is presented in this work, which may be used to graphically summarize how closely a set of patterns fits observations [61]. The similarities between the patterns and observations are quantified by utilizing their correlation coefficients, centered root-mean-square errors, and standard deviations, as shown in Equation (40) [62]:

$$E'^2 = \sigma_p^2 + \sigma_a^2 - 2 \times \sigma_p \times \sigma_a \times R \tag{40}$$

where E' is the centered root-mean-square error between the predicted and measured values, σ_p^2 and σ_a^2 are the variances of the predicted and measured values, respectively, and R is the correlation coefficient between the predicted and measured values.

Figures 10 and 11 depict how closely the six developed models match the training and testing sets, respectively, as references. In the Taylor diagram, the standard deviation is shown by the distance between the circles representing the models and the origin point on the x-axis, and the ticks on the clockwise arc represent the correlation coefficient. The actual PPV values are shown by the star-shaped point 'REF' on the x-axis, and the distance between the other circles and the point 'REF' reflects the centered RMSE (i.e., the grey arc).

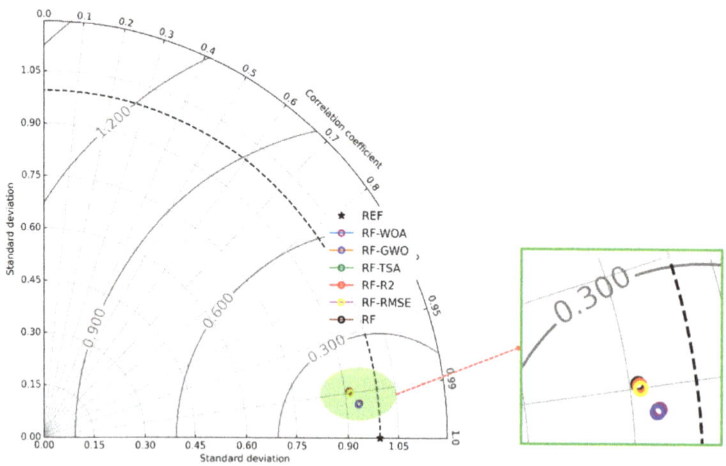

Figure 10. Taylor diagram of the training results for the six developed models.

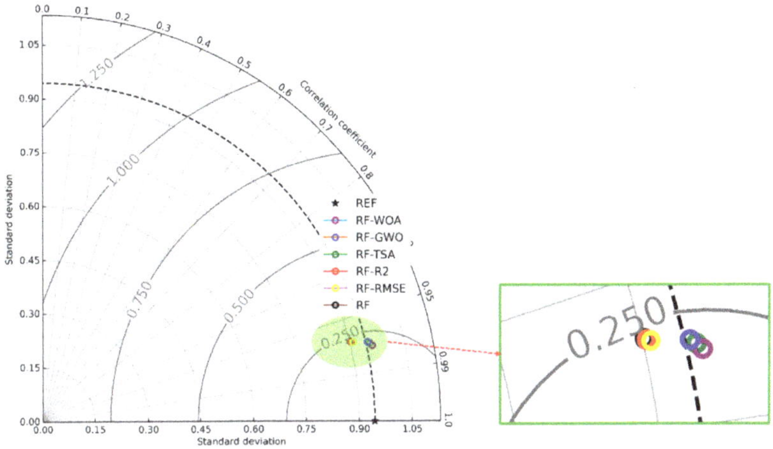

Figure 11. Taylor diagram of the testing results for the six developed models.

In Figure 10, it can be found that the standard deviations of all six models are less than the standard deviations of the measured values, and the standard deviations of the RF-WOA model, RF-GWO model, and RF-TSA model are closer to the standard deviation of the actual values compared with the models of the RF, RF-R^2, and RF-RMSE. Moreover, the same applies to the correlation coefficient, i.e., the correlation coefficients of the RF-WOA model, RF-GWO model, and RF-TSA model are closer to that of the actual values, but they differ little from each other. For the centered RMSE, the results indicate that the RF-WOA model, RF-GWO model, and RF-TSA model have smaller errors than those of the RF, RF-$R^{2,}$ and RF-RMSE models. From a holistic perspective, the calculation results of the three metaheuristic algorithms on the training set are closer to the actual PPV values, while the RF-R^2 and RF-RMSE models perform slightly better on the training set than the RF model.

In Figure 11, it can be inferred that the standard deviations of the three metaheuristic algorithms on the testing set are close to the standard deviation of the actual PPV and slightly greater. With regard to the centered RMSE and correlation coefficient, the calculation results of the RF-WOA models are better than that of the RF-TSA and RF-GWO models. Overall, based on the distance between these circles and the point 'REF', it can be determined that the ranking of the model's superiority on the testing set is as RF-WOA > RF-TSA > RF-GWO > RF-RMSE > RF-R^2 > RF. Compared with the results on the training set, the results on the testing set can more clearly reflect the performance differences among these models, especially for the results obtained by the metaheuristic algorithms.

To sum up, the RF-WOA is the optimal model that is recommended for refining the weights of decision trees of the RF model in this study. Compared with the regular RF model, for the training set, the RF-WOA model increases the R^2 value of the RF from 0.973 to 0.986 and the VAF of the RF from 97.298 to 98.603 and simultaneously reduces the RMSE of the RF model from 0.164 to 0.118 and the MAE of the RF from 0.123 to 0.093. For the testing set, the RF-WOA model increases the R^2 value of the RF from 0.923 to 0.932 and the VAF value of the RF from 93.994 to 95.032 and simultaneously reduces the RMSE of the RF model from 0.262 to 0.246 and the MAE of the RF from 0.209 to 0.188. Finally, the measured and predicted PPV by the RF-WOA model on both training and test sets are depicted in Figures 12 and 13, respectively.

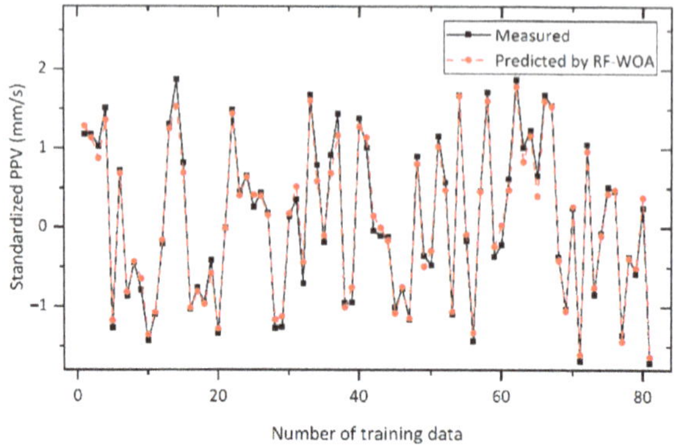

Figure 12. Measured and predicted PPV of the RF-WOA model on the training set.

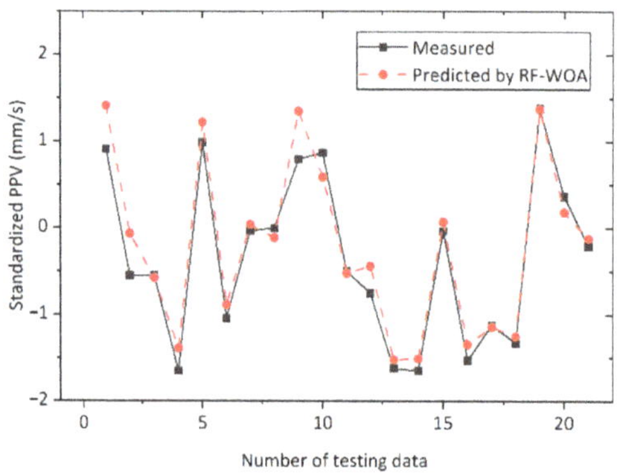

Figure 13. Measured and predicted PPV of the RF-WOA model on the testing set.

4.3. Sensitivity Analysis of Predictor Variables

Owing to the possible hazards of the PPV that are caused by blasting operations, it is indispensable to clarify the major factor affecting the PPV. Therefore, in this section, further analysis is conducted to identify the significant correlation of predictor variables that are used to predict the PPV. As previously stated, six variables, i.e., BS, HD, ST, PF, MC, and DI, were used to develop the RF models in this study. Based on the attribute of the RF model, a significant correlation of each input variable can be obtained. The (normalized) total decrease in the criterion brought by a feature is used to calculate the input variable's relevance with the target variable, which is also known as the Gini importance [56,57]; the higher it is, the more significant the input variable. In this way, the significant correlation of the input variables with PPV can be determined, as shown in Figure 14. It can be revealed that the most significant predictor variable is the PF, followed by the BS, DI, and HD. The least significant parameters include ST and MC. Accordingly, we will focus more on the relationship between the PF and PPV.

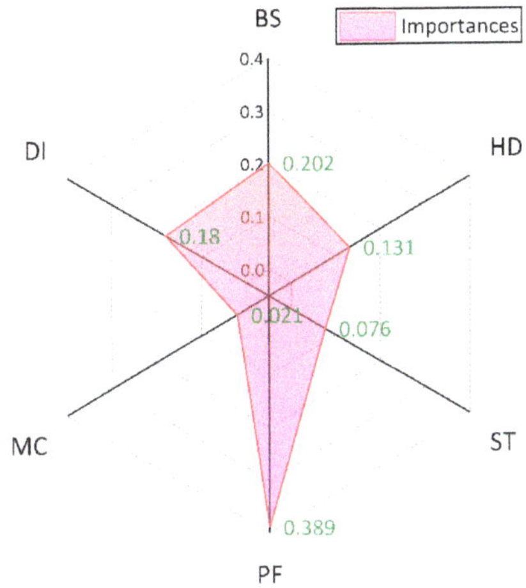

Figure 14. Significant correlation of the input variables with PPV.

To obtain a clearer picture of how the PF affects the PPV, partial dependency plots (PDPs) are utilized to achieve this goal. The PDPs illustrate the intuitive relationship between the target response and an input feature of interest and simultaneously marginalize the values of other input features. The PDPs visualize the average influence of the input feature on the target response, which can reveal a homogeneous connection between the feature and the target response [63]. Figure 15 shows the partial dependence between the PPV and PF. From an overall perspective, as the PF gradually increased from 0.23 to 0.94, the PPV gradually increased from 3.13 to 6.94. The process of the PF causing the PPV changes can be divided into three stages. The first stage is that when PF increases from 0.23 to 0.46, there is no significant change in the PPV, and the PPV reaches a minimum value when the PF equals 0.46. The second stage is that when the PF increases from 0.46 to 0.63, the PPV shows a dramatic increase from 3.01 to 6.27 in an exponential-like trend. The third stage is when the PF increases from 0.63 to 0.94, and the PPV increases from 6.27 to 6.94, but the amplification is significantly smaller compared with that in the second stage. From this, managing the PPV through curbing the size of the PF is a potentially effective means.

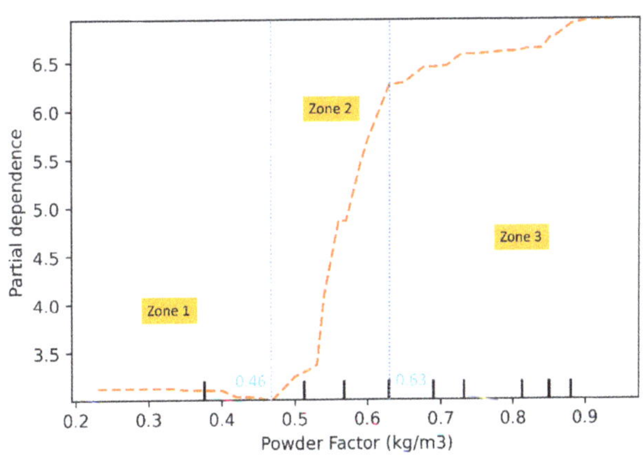

Figure 15. The partial dependency between the PPV and PF.

4.4. Comparision with the Published Works

As for the datasets used in the present paper, it has also been used in several published articles, e.g., references [1,64,65]. This section aims to compare the superiority of the model proposed in this paper with the existing models in the published papers.

Reference [64] proposed two models used for the PPV prediction, i.e., the nonlinear multiple regression (NLMR) and the gene expression programming (GEP). Likewise, 102 datasets of blasting operations were randomly divided into 80% for the training set and 20% for the testing set. The input parameters, including BS, HD, ST, PF, MC, and DI, are the same as the input parameters used in the present study. The results of the GEP and NLMP models are presented in Table 7. Reference [1] also used the same datasets to research the PPV prediction. In the research, a special case is that a feature selection technique was used to filter out unimportant input parameters before conducting the modeling. In this way, five variables except for BS, including HD, ST, PF, MC, and DI, were chosen as the input parameters, which is the main difference from the current paper. In their study, five ML models, i.e., CART, CHAID, RF, ANN, and SVM, were developed using the new datasets after feature selection. Furthermore, the datasets used for developing the models were randomly split into two parts, i.e., 70% of datasets for training the models and 30% of datasets for validating the performance of the models. The performance results of these five models are presented in Table 7. Reference [65] also utilized the same datasets to develop two ML models, i.e., RF and Bayesian Network (BN). Similar to Reference [1], feature selection was also used to filter features of low importance before building the ML model. After that, the input parameters that have been identified consist of DI, PF, HD, ST, and MC, in addition to BS.

Moreover, the datasets were randomly split into two parts, i.e., 70% of datasets for training the models and 30% of datasets for validating the performance of the models. The performance results of these two models are presented in Table 7. Overall, according to the results in Table 7, the improved RF model (i.e., the RF-WOA) proposed in this paper can significantly outperform the models proposed in the published papers, which proves the improved RF model is good enough and robust to predict the PPV caused by mining blasting. Likewise, this shows that it is reasonable and feasible to amend the weights of the decision trees of the RF model as proposed in this paper.

Table 7. Results of the evaluation metrics of the models that used the same datasets.

Reference	Models	Training				Testing			
		R^2	RMSE	MAE	VAF	R^2	RMSE	MAE	VAF
Present paper	RF-WOA	0.986	0.118	0.093	98.603	0.932	0.246	0.188	95.032
[64]	GEP	0.914	0.920	0.755	91.304	0.874	0.963	0.851	87.107
	NLMR	0.829	1.365	1.125	80.878	0.790	1.498	1.221	69.261
[1]	RF	0.940	0.770	0.620	92.970	0.830	1.460	1.190	82.170
	CART	0.670	1.670	1.320	67.030	0.560	2.390	1.840	54.600
	CHAID	0.910	0.860	0.540	91.300	0.680	1.900	1.470	67.790
	ANN	0.890	0.960	0.750	89.140	0.840	1.410	1.130	83.710
	SVM	0.880	1.020	0.770	88.480	0.850	1.500	1.170	84.540
[65]	RF	0.930	-	-	-	0.903	-	-	-
	BN	0.930	-	-	-	0.871	-	-	-

5. Conclusions

This paper utilized the RF and improved weighted RF models to predict the blast-induced PPV. Thus, a dataset of a total of 102 samples collected from an open granite mine was used to develop the target regular RF model. The input parameters used for modeling included BS, HD, ST, PF, MC, and DI, while the output was the PPV. Then, five techniques, i.e., refined weights based on the accuracy (R^2 and the RMSE) of decisions trees as well as the optimization results of three metaheuristics algorithms (WOA, GWO, and TSA), were employed to enhance the performance of the regular RF model by reassigning the new weights of decision trees. The results concluded that the optimal hyperparameters of the regular RF model are 51 decision trees and 7 of maximum depth. Subsequently, the performance evaluation results of the five weighted RF models showed that all improved RF models outperformed the regular RF model. Moreover, the RF-WOA model shows the best performance among these five models, as evidenced by the fact that the RF-WOA model has the R^2 values of 0.986 and 0.932, the RMSE values of 0.118 and 0.246, the MAE values of 0.093 and 0.188, and the VAF values of 98.603 and 95.032 for the training and testing sets, respectively. Additionally, compared with the models developed in published articles that also used the same dataset, the RF-WOA model still shows the best accuracy, proving that the RF-WOA model developed in this study has better performance, adaptability, and robustness. Furthermore, the sensitivity analysis revealed that the PF model shows a great significant correlation with the PPV prediction, and the results of the PDPs indicated that when the PF is less than 0.46, the PPV maintains a small fluctuation. At the same time, when the PF exceeds 0.46, there is a significant increasing trend of PPV, and when the PF exceeds 0.63, the increasing trend of the PPV gradually shrinks, which suggests that governing the magnitude of the PPV by managing the PF is an effective and practical measure for the case of this project.

Author Contributions: Conceptualization, B.H. and A.S.M.; methodology, B.H., S.H.L.; software, B.H. and M.M.S.S.; validation, B.H., A.S.M.; formal analysis, B.H. and S.H.L.; investigation, B.H.; writing—original draft preparation, B.H., M.M.S.S., S.H.L., A.S.M. and D.V.U.; writing—review and editing, B.H., M.M.S.S., S.H.L., A.S.M. and D.V.U.; supervision, S.H.L., A.S.M. and M.M.S.S.; funding acquisition, M.M.S.S. All authors have read and agreed to the published version of the manuscript.

Funding: The research is partially funded by the Ministry of Science and Higher Education of the Russian Federation under the strategic academic leadership program 'Priority 2030' (Agreement 075-15-2021-1333 dated 30 September 2021).

Institutional Review Board Statement: Not applicable.

Informed Consent Statement: Not applicable.

Data Availability Statement: The data are available upon request.

Acknowledgments: Authors of this study wish to express their appreciation to the Universiti Malaya for supporting this study and making it possible.

Conflicts of Interest: The authors declare no conflict of interest.

References

1. Zhang, H.; Zhou, J.; Armaghani, D.J.; Tahir, M.M.; Pham, B.T.; Huynh, V. Van A Combination of Feature Selection and Random Forest Techniques to Solve a Problem Related to Blast-Induced Ground Vibration. *Appl. Sci.* **2020**, *10*, 869. [CrossRef]
2. Hasanipanah, M.; Monjezi, M.; Shahnazar, A.; Armaghani, D.J.; Farazmand, A. Feasibility of indirect determination of blast induced ground vibration based on support vector machine. *Measurement* **2015**, *75*, 289–297. [CrossRef]
3. Yu, Z.; Shi, X.; Zhou, J.; Gou, Y.; Huo, X.; Zhang, J.; Armaghani, D.J. A new multikernel relevance vector machine based on the HPSOGWO algorithm for predicting and controlling blast-induced ground vibration. *Eng. Comput.* **2020**, *38*, 905–1920. [CrossRef]
4. Guo, H.; Zhou, J.; Koopialipoor, M.; Armaghani, D.J.; Tahir, M. Deep neural network and whale optimization algorithm to assess flyrock induced by blasting. *Eng. Comput.* **2021**, *37*, 173–186. [CrossRef]
5. Zhou, J.; Koopialipoor, M.; Murlidhar, B.R.; Fatemi, S.A.; Tahir, M.M.; Armaghani, D.J.; Li, C. Use of intelligent methods to design effective pattern parameters of mine blasting to minimize flyrock distance. *Nat. Resour. Res.* **2020**, *29*, 625–639. [CrossRef]
6. Guo, H.; Nguyen, H.; Bui, X.-N.; Armaghani, D.J. A new technique to predict fly-rock in bench blasting based on an ensemble of support vector regression and GLMNET. *Eng. Comput.* **2019**, *37*, 421–435. [CrossRef]
7. Yu, Q.; Monjezi, M.; Mohammed, A.S.; Dehghani, H.; Armaghani, D.J.; Ulrikh, D.V. Optimized support vector machines combined with evolutionary random forest for prediction of back-break caused by blasting operation. *Sustainability* **2021**, *13*, 12797. [CrossRef]
8. Sayadi, A.; Monjezi, M.; Talebi, N.; Khandelwal, M. A comparative study on the application of various artificial neural networks to simultaneous prediction of rock fragmentation and backbreak. *J. Rock Mech. Geotech. Eng.* **2013**, *5*, 318–324. [CrossRef]

9. Faramarzi, F.; Farsangi, M.A.E.; Mansouri, H. An RES-based model for risk assessment and prediction of backbreak in bench blasting. *Rock Mech. Rock Eng.* **2013**, *46*, 877–887. [CrossRef]
10. Afeni, T.B.; Osasan, S.K. Assessment of noise and ground vibration induced during blasting operations in an open pit mine—A case study on Ewekoro limestone quarry, Nigeria. *Min. Sci. Technol.* **2009**, *19*, 420–424. [CrossRef]
11. He, B.; Armaghani, D.J.; Lai, S.H. A Short Overview of Soft Computing Techniques in Tunnel Construction. *Open Constr. Build. Technol. J.* **2022**, *16*, 1–6. [CrossRef]
12. Lawal, A.I.; Kwon, S.; Kim, G.Y. Prediction of the blast-induced ground vibration in tunnel blasting using ANN, moth-flame optimized ANN, and gene expression programming. *Acta Geophys.* **2021**, *69*, 161–174. [CrossRef]
13. Rai, R.; Shrivastva, B.K.; Singh, T.N. Prediction of maximum safe charge per delay in surface mining. *Trans. Inst. Min. Metall. Sect. A Min. Technol.* **2005**, *114*, 227–232. [CrossRef]
14. Hasanipanah, M.; Bakhshandeh Amnieh, H.; Khamesi, H.; Jahed Armaghani, D.; Bagheri Golzar, S.; Shahnazar, A. Prediction of an environmental issue of mine blasting: An imperialistic competitive algorithm-based fuzzy system. *Int. J. Environ. Sci. Technol.* **2018**, *15*, 1–10. [CrossRef]
15. Hasanipanah, M.; Faradonbeh, R.S.; Amnieh, H.B.; Armaghani, D.J.; Monjezi, M. Forecasting blast-induced ground vibration developing a CART model. *Eng. Comput.* **2017**, *33*, 307–316. [CrossRef]
16. Ram Chandar, K.; Sastry, V.R.; Hegde, C.; Shreedharan, S. Prediction of peak particle velocity using multi regression analysis: Case studies. *Geomech. Geoengin.* **2017**, *12*, 207–214. [CrossRef]
17. Khandelwal, M.; Singh, T.N. Prediction of blast induced ground vibrations and frequency in opencast mine: A neural network approach. *J. Sound Vib.* **2006**, *289*, 711–725. [CrossRef]
18. Lawal, A.I.; Idris, M.A. An artificial neural network-based mathematical model for the prediction of blast-induced ground vibrations. *Int. J. Environ. Stud.* **2020**, *77*, 318–334. [CrossRef]
19. Parida, A.; Mishra, M.K. Blast Vibration Analysis by Different Predictor Approaches—A Comparison. *Procedia Earth Planet. Sci.* **2015**, *11*, 337–345. [CrossRef]
20. Xue, X.; Yang, X. Predicting blast-induced ground vibration using general regression neural network. *JVC/J. Vib. Control* **2014**, *20*, 1512–1519. [CrossRef]
21. Yang, H.; Wang, Z.; Song, K. A new hybrid grey wolf optimizer-feature weighted-multiple kernel-support vector regression technique to predict TBM performance. *Eng. Comput.* **2020**. [CrossRef]
22. Yang, H.; Song, K.; Zhou, J. Automated Recognition Model of Geomechanical Information Based on Operational Data of Tunneling Boring Machines. *Rock Mech. Rock Eng.* **2022**, *55*, 1499–1516. [CrossRef]
23. Armaghani, D.J.; Mohamad, E.T.; Narayanasamy, M.S.; Narita, N.; Yagiz, S. Development of hybrid intelligent models for predicting TBM penetration rate in hard rock condition. *Tunn. Undergr. Sp. Technol.* **2017**, *63*, 29–43. [CrossRef]
24. Armaghani, D.J.; Koopialipoor, M.; Marto, A.; Yagiz, S. Application of several optimization techniques for estimating TBM advance rate in granitic rocks. *J. Rock Mech. Geotech. Eng.* **2019**, *11*, 779–789. [CrossRef]
25. Zhou, J.; Yazdani Bejarbaneh, B.; Jahed Armaghani, D.; Tahir, M.M. Forecasting of TBM advance rate in hard rock condition based on artificial neural network and genetic programming techniques. *Bull. Eng. Geol. Environ.* **2020**, *79*, 2069–2084. [CrossRef]
26. Li, Z.; Yazdani Bejarbaneh, B.; Asteris, P.G.; Koopialipoor, M.; Armaghani, D.J.; Tahir, M.M. A hybrid GEP and WOA approach to estimate the optimal penetration rate of TBM in granitic rock mass. *Soft Comput.* **2021**, *25*, 11877–11895. [CrossRef]
27. Zhou, J.; Chen, C.; Wang, M.; Khandelwal, M. Proposing a novel comprehensive evaluation model for the coal burst liability in underground coal mines considering uncertainty factors. *Int. J. Min. Sci. Technol.* **2021**, *31*, 14. [CrossRef]
28. Zhou, J.; Li, X.; Mitri, H.S. Classification of rockburst in underground projects: Comparison of ten supervised learning methods. *J. Comput. Civ. Eng.* **2016**, *30*, 4016003. [CrossRef]
29. Pham, B.T.; Nguyen, M.D.; Nguyen-Thoi, T.; Ho, L.S.; Koopialipoor, M.; Kim Quoc, N.; Armaghani, D.J.; Le, H. Van A novel approach for classification of soils based on laboratory tests using Adaboost, Tree and ANN modeling. *Transp. Geotech.* **2021**, *27*, 100508. [CrossRef]
30. Armaghani, D.J.; Harandizadeh, H.; Momeni, E.; Maizir, H.; Zhou, J. *An Optimized System of GMDH-ANFIS Predictive Model by ICA for Estimating Pile Bearing Capacity*; Springer: Cham, The Netherlands, 2022; Volume 55, ISBN 0123456789.
31. Huat, C.Y.; Moosavi, S.M.H.; Mohammed, A.S.; Armaghani, D.J.; Ulrikh, D.V.; Monjezi, M.; Hin Lai, S. Factors Influencing Pile Friction Bearing Capacity: Proposing a Novel Procedure Based on Gradient Boosted Tree Technique. *Sustainability* **2021**, *13*, 11862. [CrossRef]
32. Huang, J.; Sun, Y.; Zhang, J. Reduction of computational error by optimizing SVR kernel coefficients to simulate concrete compressive strength through the use of a human learning optimization algorithm. *Eng. Comput.* **2021**. [CrossRef]
33. Huang, J.; Kumar, G.S.; Ren, J.; Zhang, J.; Sun, Y. Accurately predicting dynamic modulus of asphalt mixtures in low-temperature regions using hybrid artificial intelligence model. *Constr. Build. Mater.* **2021**, *297*, 123655. [CrossRef]
34. Asteris, P.G.; Lourenço, P.B.; Roussis, P.C.; Adami, C.E.; Armaghani, D.J.; Cavaleri, L.; Chalioris, C.E.; Hajihassani, M.; Lemonis, M.E.; Mohammed, A.S. Revealing the nature of metakaolin-based concrete materials using artificial intelligence techniques. *Constr. Build. Mater.* **2022**, *322*, 126500. [CrossRef]
35. Mahmood, W.; Mohammed, A.S.; Asteris, P.G.; Kurda, R.; Armaghani, D.J. Modeling Flexural and Compressive Strengths Behaviour of Cement-Grouted Sands Modified with Water Reducer Polymer. *Appl. Sci.* **2022**, *12*, 1016. [CrossRef]

36. Asteris, P.G.; Rizal, F.I.M.; Koopialipoor, M.; Roussis, P.C.; Ferentinou, M.; Armaghani, D.J.; Gordan, B. Slope Stability Classification under Seismic Conditions Using Several Tree-Based Intelligent Techniques. *Appl. Sci.* **2022**, *12*, 1753. [CrossRef]
37. Cai, M.; Koopialipoor, M.; Armaghani, D.J.; Thai Pham, B. Evaluating Slope Deformation of Earth Dams due to Earthquake Shaking using MARS and GMDH Techniques. *Appl. Sci.* **2020**, *10*, 1486. [CrossRef]
38. Zhou, J.; Dai, Y.; Khandelwal, M.; Monjezi, M.; Yu, Z.; Qiu, Y. Performance of Hybrid SCA-RF and HHO-RF Models for Predicting Backbreak in Open-Pit Mine Blasting Operations. *Nat. Resour. Res.* **2021**, *30*, 4753–4771. [CrossRef]
39. Zhou, J.; Qiu, Y.; Khandelwal, M.; Zhu, S.; Zhang, X. Developing a hybrid model of Jaya algorithm-based extreme gradient boosting machine to estimate blast-induced ground vibrations. *Int. J. Rock Mech. Min. Sci.* **2021**, *145*, 104856. [CrossRef]
40. Li, C.; Zhou, J.; Armaghani, D.J.; Li, X. Stability analysis of underground mine hard rock pillars via combination of finite difference methods, neural networks, and Monte Carlo simulation techniques. *Undergr. Sp.* **2021**, *6*, 379–395. [CrossRef]
41. Parsajoo, M.; Armaghani, D.J.; Mohammed, A.S.; Khari, M.; Jahandari, S. Tensile strength prediction of rock material using non-destructive tests: A comparative intelligent study. *Transp. Geotech.* **2021**, *31*, 100652. [CrossRef]
42. Asteris, P.G.; Mamou, A.; Hajihassani, M.; Hasanipanah, M.; Koopialipoor, M.; Le, T.-T.; Kardani, N.; Armaghani, D.J. Soft computing based closed form equations correlating L and N-type Schmidt hammer rebound numbers of rocks. *Transp. Geotech.* **2021**, *29*, 100588. [CrossRef]
43. Momeni, E.; Armaghani, D.J.; Hajihassani, M.; Amin, M.F.M. Prediction of uniaxial compressive strength of rock samples using hybrid particle swarm optimization-based artificial neural networks. *Measurement* **2015**, *60*, 50–63. [CrossRef]
44. Huang, J.; Zhang, J.; Gao, Y. Intelligently predict the rock joint shear strength using the support vector regression and Firefly Algorithm. *Lithosphere* **2021**, *2021*, 2467126. [CrossRef]
45. Lawal, A.I. An artificial neural network-based mathematical model for the prediction of blast-induced ground vibration in granite quarries in Ibadan, Oyo State, Nigeria. *Sci. African* **2020**, *8*, e00413. [CrossRef]
46. Rana, A.; Bhagat, N.K.; Jadaun, G.P.; Rukhaiyar, S.; Pain, A.; Singh, P.K. Predicting Blast-Induced Ground Vibrations in Some Indian Tunnels: A Comparison of Decision Tree, Artificial Neural Network and Multivariate Regression Methods. *Min. Metall. Explor.* **2020**, *37*, 1039–1053. [CrossRef]
47. Lawal, A.I.; Kwon, S.; Hammed, O.S.; Idris, M.A. Blast-induced ground vibration prediction in granite quarries: An application of gene expression programming, ANFIS, and sine cosine algorithm optimized ANN. *Int. J. Min. Sci. Technol.* **2021**, *31*, 265–277. [CrossRef]
48. Bui, X.N.; Nguyen, H.; Nguyen, T.A. Artificial Neural Network Optimized by Modified Particle Swarm Optimization for Predicting Peak Particle Velocity Induced by Blasting Operations in Open Pit Mines. *Inz. Miner.* **2021**, *1*, 79–90. [CrossRef]
49. Iphar, M.; Yavuz, M.; Ak, H. Prediction of ground vibrations resulting from the blasting operations in an open-pit mine by adaptive neuro-fuzzy inference system. *Environ. Geol.* **2008**, *56*, 97–107. [CrossRef]
50. Monjezi, M.; Ghafurikalajahi, M.; Bahrami, A. Prediction of blast-induced ground vibration using artificial neural networks. *Tunn. Undergr. Sp. Technol.* **2011**, *26*, 46–50. [CrossRef]
51. Ghasemi, E.; Ataei, M.; Hashemolhosseini, H. Development of a fuzzy model for predicting ground vibration caused by rock blasting in surface mining. *J. Vib. Control* **2013**, *19*, 755–770. [CrossRef]
52. Hajihassani, M.; Jahed Armaghani, D.; Marto, A.; Tonnizam Mohamad, E. Ground vibration prediction in quarry blasting through an artificial neural network optimized by imperialist competitive algorithm. *Bull. Eng. Geol. Environ.* **2014**, *74*, 873–886. [CrossRef]
53. Armaghani, D.J.; Hasanipanah, M.; Amnieh, H.B.; Mohamad, E.T. Feasibility of ICA in approximating ground vibration resulting from mine blasting. *Neural Comput. Appl.* **2018**, *29*, 457–465. [CrossRef]
54. Mohamed, M.T. Performance of fuzzy logic and artificial neural network in prediction of ground and air vibrations. *Int. J. Rock Mech. Min. Sci.* **2011**, *39*, 425–440.
55. Breiman, L. Random Forests. *Mach. Learn.* **2001**, *45*, 5–32. [CrossRef]
56. Mirjalili, S.; Mirjalili, S.M.; Lewis, A. Grey wolf optimizer. *Adv. Eng. Softw.* **2014**, *69*, 46–61. [CrossRef]
57. Mirjalili, S.; Lewis, A. The whale optimization algorithm. *Adv. Eng. Softw.* **2016**, *95*, 51–67. [CrossRef]
58. Kaur, S.; Awasthi, L.K.; Sangal, A.L.; Dhiman, G. Tunicate Swarm Algorithm: A new bio-inspired based metaheuristic paradigm for global optimization. *Eng. Appl. Artif. Intell.* **2020**, *90*, 103541. [CrossRef]
59. Li, H.B.; Wang, W.; Ding, H.W.; Dong, J. Trees Weighting Random Forest method for classifying high-dimensional noisy data. In Proceedings of the 2010 IEEE 7th International Conference on E-Business Engineering, Shanghai, China, 10–12 November 2010; pp. 160–163.
60. Zorlu, K.; Gokceoglu, C.; Ocakoglu, F.; Nefeslioglu, H.A.; Acikalin, S. Prediction of uniaxial compressive strength of sandstones using petrography-based models. *Eng. Geol.* **2008**, *96*, 141–158. [CrossRef]
61. Taylor, K.E. Summarizing multiple aspects of model performance in a single diagram. *J. Geophys. Res. Atmos.* **2001**, *106*, 7183–7192. [CrossRef]
62. Taylor, K.E. Taylor Diagram Primer—Working Paper. 2005. Available online: http://www.atmos.albany.edu/daes/atmclasses/atm401/spring_2016/ppts_pdfs/Taylor_diagram_primer.pdf (accessed on 10 May 2022).
63. Friedman, J.H. Greedy function approximation: A gradient boosting machine. *Ann. Stat.* **2001**, *29*, 1189–1232. [CrossRef]
64. Shirani Faradonbeh, R.; Jahed Armaghani, D.; Abd Majid, M.Z.; MD Tahir, M.; Ramesh Murlidhar, B.; Monjezi, M.; Wong, H.M. Prediction of ground vibration due to quarry blasting based on gene expression programming: A new model for peak particle velocity prediction. *Int. J. Environ. Sci. Technol.* **2016**, *13*, 1453–1464. [CrossRef]

65. Zhou, J.; Asteris, P.G.; Armaghani, D.J.; Pham, B.T. Prediction of ground vibration induced by blasting operations through the use of the Bayesian Network and random forest models. *Soil Dyn. Earthq. Eng.* **2020**, *139*, 106390. [CrossRef]

applied sciences

MDPI

Article

Ultra-Scratch-Resistant, Hydrophobic and Transparent Organosilicon-Epoxy-Resin Coating with a Double Cross-Link Structure

Zeyu Qiu [1], Haofeng Lin [1], Longlong Zeng [1], Yunfeng Liang [1], Chunhong Zeng [2] and Ruijiang Hong [1,*]

[1] Guangdong Provincial Key Laboratory of Photovoltaic Technology, Institute for Solar Energy Systems, School of Physics, Sun Yat-sen University, Guangzhou 510006, China; chouzy@mail2.sysu.edu.cn (Z.Q.); linhf27@mail2.sysu.edu.cn (H.L.); zengllong@mail2.sysu.edu.cn (L.Z.); liangyf36@mail2.sysu.edu.cn (Y.L.)
[2] School of Materials, Sun Yat-sen University, Guangzhou 510275, China; zengchh7@mail.sysu.edu.cn
* Correspondence: hongruij@mail.sysu.edu.cn

Abstract: In this paper, an ultra-scratch-resistant, hydrophobic and transparent coating was fabricated by the sol–gel method using (3-Glycidyloxypropyl) triethoxysilane (GPTES) and curing agents. When the silanol was condensated, the ring-opening reaction of the epoxy groups also took place, which formed a double-cross-linked network (Si–O–Si and R_3N). This network structure restricted the molecule chains from being twisted or dislocated, resulting in a great improvement of the abrasion resistance of the coating. A pencil hardness grade up to 8H was obtained. The coating also showed excellent stability after being soaked in pH = 2 and pH = 12 solutions, seawater and acetone, respectively. In addition, a water contact angle of 121° was obtained by post-treatment with hexamethyldisilazane (HMDS). The average transmittance of the coating reached to 90% in the wavelength range of 400~800 nm, nearly identical to the glass substrate. With multiple desirable properties and a simple fabrication process, this low-cost coating shows great potential in many practical applications.

Keywords: scratch-resistant; hydrophobic; GPTES; transparent; sol–gel

Citation: Qiu, Z.; Lin, H.; Zeng, L.; Liang, Y.; Zeng, C.; Hong, R. Ultra-Scratch-Resistant, Hydrophobic and Transparent Organosilicon-Epoxy-Resin Coating with a Double Cross-Link Structure. *Appl. Sci.* **2022**, *12*, 4854. https://doi.org/10.3390/app12104854

Academic Editors: Pijush Samui, Aydin Azizi, Ahmed Hussein Kamel Ahmed Elshafie, Yixia Zhang and Danial Jahed Armaghani

Received: 21 April 2022
Accepted: 10 May 2022
Published: 11 May 2022

1. Introduction

Multifunctional transparent coatings are closely related to our lives and are categorized as one of the hot topics of many researchers [1,2]. While maintaining transparency, the coating is also endowed with multiple properties, such as self-cleaning [3,4], anti-fogging [5,6], oil–water separation [7], anti-corrosion [8–10], anti-reflection [11,12] and self-healing [13–15]. There are usually two ways to achieve self-cleaning, the super-hydrophilic surface and the super-hydrophobic surface. When water droplets come into contact with the coating, they will immediately be spread over the entire surface to form a continuous water film, thereby taking away the dirt on the surface [16]. The smooth surface that reduces the scattering of light makes the coating transparent and anti-fogging [5]. On the other hand, when the water droplets come into contact with the coating with low surface tension, the water droplets will keep their spherical shape and roll off quickly from the surface [17]. The preparation of superhydrophobic surfaces often relies on two factors, micro or nano-scaled hierarchical structures [18] and low surface energy materials [19]. In addition, a transparent coating is often used as a protective coating to protect metallic substrates from corrosion, because the dense and inert coating separates the metal from the environment [20]. Furthermore, anti-reflective coatings are often used in solar cells [12]. The refractive index of the antireflective coatings is lower than that of the substrate. Through the interference of light, it effectively reduces the reflection loss of incident light and improves the efficiency of solar cells [21]. What is more, the damaged structure can be recovered spontaneously by self-healing materials with external stimuli [15]. Self-healing coatings repair damage in two ways, inherent reversible noncovalent interactions [22–27] and dynamic covalent bonds [28–32].

However, these coatings are often soft and susceptible to scratches and abrasion. Scratches may make the surface rough, which will greatly cause the scattering of light, thereby reducing the transparency of the coating [33]. In addition, self-cleaning, anti-corrosion and other functions may also be affected, which severely restricts its application in daily life. Although the self-healing coating can relieve the effects of scratches to a certain extent, it cannot totally eliminate the damage of scratches [15]. Therefore, scratch resistance is also an important point for transparent coatings. Zhang et al. prepared a bilayer antireflective coating with the top layer of ultra-low refractive index from fully dispersing nano-silica particles by mixing HMDS and achieved an average transmittance of 99.90% in the visible region [34]. However, the coating with an ultra-low refractive index was usually rather soft, easily scratched. Mousavi et al. fabricated a transparent scratch-resistant coating through the direct oxidation of Al-coated glass [35]. After annealing at 600 °C, the pencil hardness of the coating increased to 9H due to the hard Al_2O_3 particles. However, the transmittance of the coating declined from 90% to 75%. Hua Zhou et al. prepared durable and superhydrophobic fabric coatings through simple mixtures of fluorinated silica nanoparticles and polydimethylsiloxane (PDMS) and showed that the water contact angle only decreased from 170° to 150° after 28,000 cycles of abrasion under 12 kPa [36]. However, the transparency of the coating was not mentioned.

In general, the epoxy resin needs to be solidified to increase its hardness [37–39] through a ring-opening reaction to form an organic network. (3-Glycidyloxypropyl) Trimethoxysilane (GLYMO) and (3-Glycidoxypropyl) Triethoxysilane (GPTES) contain epoxy groups, as silane coupling agents, often used as surface hardening agents [37]. Zhi et al. provided a method to fabricate a durable superhydrophobic antireflection coating via introducing an organic network from KH560 and octadecylamine (ODA) [40]. In detail, the coating resisted scratches of a 4H pencil and the transmittance was 93%, which represented a 3% improvement of the uncoated substrate. Omer Kesmez et al. reported a hybrid organic–inorganic photocatalytic nanocomposite film, composed of Ce-doped TiO_2 nanoparticles and TEOS, GPTES, 1H, 1H, 2H, 2H-perfluorooctyl triethoxysilane [41]. This coating exhibited good transparency and the pencil hardness was >9H. Therefore, silane coupling agents containing epoxy groups can enhance mechanical damage resistance.

In this work, we prepare an ultra-scratch-resistant and hydrophobic polymer coating, based on a double-cross-link structure from GPTES and a curing agent, diethylenetriamine (DETA) or m-Xylylenediamine (MXDA). It protected the substrate from the scratches of an 8H pencil without deteriorating its transparency. The fabrication process of the coating, sol–gel method, is a simple and cost-effective thin film preparation method. It can be found that the coating was relatively durable after being soaked in different corrosive liquids. From the results of thermogravimetric analysis, it is also demonstrated that this polymer coating with a wide working temperature and hydrophobicity provided the possibility for practical applications on metal and/or wood surface.

2. Experimental Procedure

2.1. Materials

(3-Glycidyloxypropyl) Triethoxysilane (GPTES), hexamethyldisilazane (HMDS), tetraethyl orthosilicate (TEOS), methyltriethoxysilane (MTES), diethylenetriamine (DETA), m-xylylenediamine (MXDA) and ammonia (25~28%) were purchased from Shanghai Aladdin Biochemical Technology Co., Ltd., Shanghai, China. Anhydrous ethanol (EtOH), acetone, sodium hydroxide (NaOH) and sodium chloride (NaCl) were purchased from Tianjin Zhiyuan Chemical Reagent Co., Ltd., Tianjin, China. All the reagents used in this work were not purified further. High purity water was prepared by a Purescience water purification system.

2.2. Preparation of Coatings

2.2.1. Preparation of the DETA-Organosilicon-Epoxy-Resin (DETA-OSER) Coating and MXDA-Organosilicon-Epoxy-Resin (MXDA-OSER) Coating

GPTES (5.5 mL) was mixed with the EtOH (43.0 mL) with stirring for 30 min. Then, high purity water (1.0 mL) and DETA (0.8 mL) or MXDA (1.3 mL) were added to the solution and the mixture was stirred for 4 h. After stirring, the solution was transferred to a cool place to age for 7 days. After 7 days, the solution formed a sol. Then, EtOH (50 mL) was added to dilute the sol for later dip-coating.

Glass and silicon wafer were used as substrates to prepared samples for transmittance and refractive index measurement directly. The glass substrates were cleaned in an ultrasonic bath with high purity water, ethanol and acetone, respectively. After that, they were dried in the baker at 60 °C prior to dip-coating. The DETA-OSER coatings were dip-coated on glass substrates with a withdrawal rate of 1.5 mm/s. Then, the samples were immersed in HMDS for 3 days to obtain hydrophobic surface. Finally, the *DETA-OSER*-coated glasses were annealed in a muffle furnace at 160 °C for 1.5 h, and the MXDA-*OSER*-coated glasses were annealed in a muffle furnace at 155 °C for 1.5 h.

2.2.2. Preparation of the NH_3-Organosilicon-Epoxy-Resin (NH_3-OSER) Coating

GPTES (5.5 mL) was mixed with the EtOH (43.0 mL) with stirring for 30 min. Then, high purity water (1.0 mL) and ammonia (0.9 mL) were added to the solution and the mixture was stirred for 4 h. After stirring, the solution was transferred to a cool place to age for 7 days. After 7 days, the solution formed a sol. Then, EtOH (50 mL) was added to dilute the sol for later dip-coating.

The NH_3-OSER coatings were dip-coated on the cleaned glass substrates at the withdrawal rate 1.5 mm/s. Then, the samples were immersed in HMDS for 3 days to obtain a hydrophobic surface. Finally, the coated glasses were annealed in a muffle furnace at 160 °C for 1.5 h.

2.2.3. Preparation of the TEOS/MTES (TM) Coatings

TEOS (tetraethyl orthosilicate) (2.5 mL) and MTES (methyltriethoxysilane) (7.3 mL) were mixed with the EtOH (36.0 mL) with stirring for 30 min. Then, high purity water (1.3 mL) and ammonia (0.9 mL) were added to the solution and the mixture was stirred for 1 h. After stirring, the solution was transferred to a cool place to age for 7 days. After 7 days, the solution formed a sol. Then, EtOH (50 mL) was added to dilute the sol for later dip-coating.

The TM coatings were dip-coated on the cleaned glass substrates at the withdrawal rate 1.5 mm/s. Then, the samples were immersed in HMDS for 3 days to obtain a hydrophobic surface. Finally, the coated glasses were annealed in a muffle furnace at 160 °C for 1.5 h.

2.3. Characterization

The optical transmittance spectra of the coated glasses were measured by using a UV–VIS-NIR spectrophotometer (Hitachi U-4100, Tokyo, Japan) at room temperature. The refraction index and film thickness were measured by a spectroscopic ellipsometry (SENTECH SE800PV, Berlin, Germany). The surface morphologies and Young's modulus of the coatings were determined by atomic force microscope (Dimension Fastscan, Bruker, Billerica, MA, USA). The scratch resistance was evaluated by the pencil hardness test. The pencil hardness test was operated according to ASTM D 3363-2005. The optical microscope images of the scratches were taken by metallurgical microscope (LEICA DM2500M). FTIR spectra were recorded on the infrared spectrometer (Vertex70 Hyperion3000, with a diamond crystal plane (single reflection) Attenuated Total Reflection (ATR) attachment), with a resolution of 4 cm^{-1} and range of 4000~400 cm^{-1}, to measure the possible groups on the coatings. Thermogravimetric analysis was performed by using thermogravimetry (TG209F1 libra) from 30 °C to 710 °C at a rate of 10 °C/min in the air to measure the decomposition temperature of the coating. The water contact angles were measured at room temperature by an optical contact angle system (OCA 20, Dataphysics) with a droplet volume of 5 μL.

3. Results and Discussion

3.1. Formation Mechanism of the Double-Cross-Link Structure

The simple fabrication process of the coatings is schematically illustrated in Figure 1. For convenience, GPTES, DETA and MXDA are replaced by simple graphics in Figure 1a. In general, Si–O–C_2H_5 can be catalyzed by acids and alkalis to hydrolyze and produce silanol. At the same time, silanol can also be catalyzed to condense and produce Si–O–Si [4]. Aliphatic polyamine can also catalyze the hydrolysis and condensation of Si–O–C_2H_5, as discussed in the next section. Low-temperature epoxy groups ring-opening reaction and organic network formation can be achieved by the use of amine curing agents. In principle, each active hydrogen in an amine is capable of opening and linking to one epoxy groups [42]. That is, 1 mol of DETA react with 5 mol of GPTES and 1 mol of MXDA react with 4 mol of GPTES. Therefore, as shown in Figure 1b, DETA and MXDA not only catalyze the hydrolysis of Si–O–C_2H_5 and the condensation of silanol, but react with GPTES to form the double-cross-link network. As the reaction proceeds, the molecular chains gradually expand to achieve the much larger networks in Figure 1c,d (both amorphous from XRD, not present).

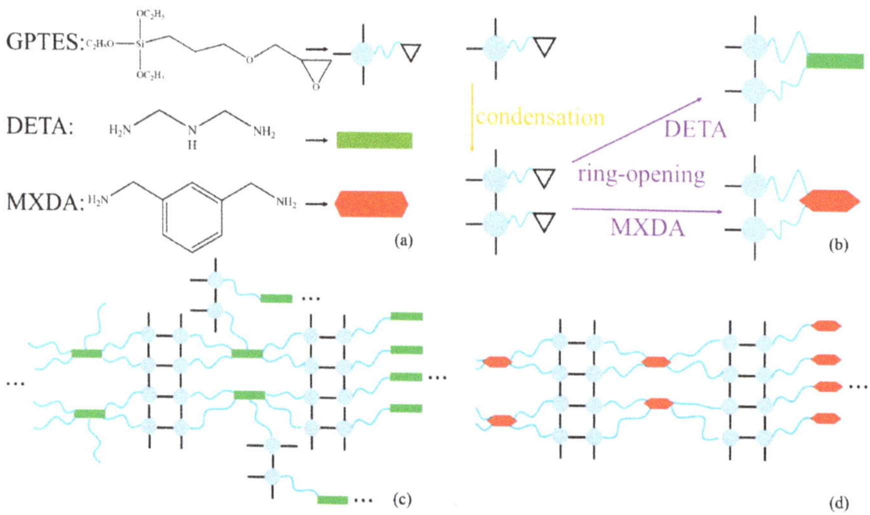

Figure 1. Schematic illustration of the OSER coating. (**a**) Simplification of reactant. (**b**) Reaction process. (**c**) DETA-OSER network. (**d**) MXDA-OSER network.

For the sake of verifying whether the cross-linked network was achieved, the FTIR spectrum was applied to infer the possible groups in coatings. Figure 2 shows FTIR spectrums of NH_3-OSER, MXDA-OSER, DETA-OSER and GPTES. Absorption peaks at ~2920, ~2853 and ~1457 cm^{-1} were observed corresponding to the C–H asymmetric, symmetrical stretching vibration and the in-plane deformation vibration, respectively [43]. This indicated that all the samples contained methylene. Moreover, the characteristic peak shown at ~3292 cm^{-1} could be attributed to the stretching vibration of hydroxyl [44], including hydrogen bonds, which indicated that the Si–O–C_2H_5 in three coatings have hydrolyzed, whereas GPTES have not. Since the characteristic peak of O–H is very close to N–H, this might also suggest the existence of –NH_2. The characteristic peak shown at ~910 cm^{-1} could be attributed to the vibration of epoxy. This indicated incompletely reacted epoxy groups. In addition, in terms of the characteristic peak shown at ~1094 cm^{-1}, which represented the Si–O stretching vibration [45], it could be confirmed that GPTES contained unhydrolyzed Si–O–C_2H_5 and this peak corresponds to the cross-link networks of Si–O–Si in the other three samples. Furthermore, the presence of amino was observed at ~1649 and ~1025 cm^{-1}, and was assigned to the symmetric N–H bending modes of –NH_2 groups and C–N stretching modes of R_3N, respectively [45]. In other words, the open-ring reaction occurred between the GPTES and DETA or MXDA. On the contrary, GPTES was only catalyzed by ammonia to hydrolyze, but not ring-open. In summary, GPTES achieved the single Si–O–Si cross-link with the help of ammonia, but DETA and MXDA both catalyzed hydrolysis and condensation, and reacted with epoxy groups to produce the double cross-links Si–O–Si and R_3N.

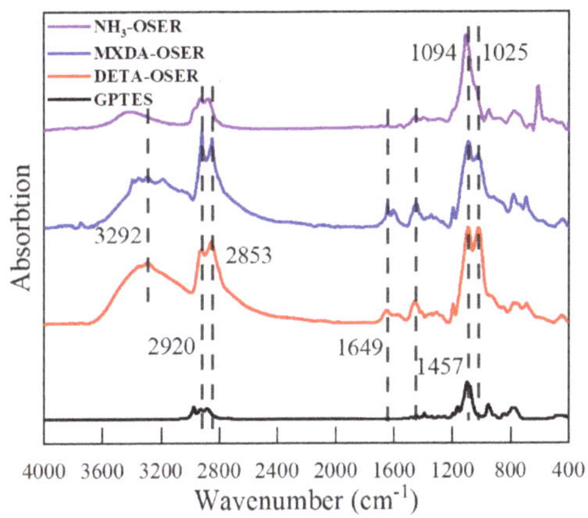

Figure 2. FTIR spectra of NH_3-OSER, MXDA-OSER, DETA-OSER and GPTES.

3.2. Mechanical Property of Coatings

Scratch resistance is an important property of coatings, especially the ones for optical applications. The hardness of the coatings was assessed by a pencil hardness test on the basis of the ASTM D3363 standard [46], using pencils ranging from 6B (the softest) to 9H (the hardest). As shown in Figure 3f, with MTES as the silane coupling agent, the TM coating had the softest pencil hardness < 6B, due to its low refractive index and high porosity. In fact, the hydrogen bonds among the methyl-embedded particles are weakened and the extent of cross-link is greatly reduced, leading to a high porosity [33]. The NH_3-OSER coating, using GPTES as the silane coupling agent, was also soft, which showed a pencil hardness < 3B, as Figure 3e. That means the single Si–O–Si cross-link is not strong enough to achieve ultra-scratch resistance. However, as shown in Figure 3a, there were only minor scratches on DETA-OSER, caused by an 8H pencil, yet it suffered evident scratch damage by 9H in Figure 3b. What is more, MXDA-OSER was also ultra-scratch resistant, absolutely none scratches on its surface as Figure 3c showed. In summary, owing to the double cross-links, the hardness of coatings is greatly enhanced to 8H. GPTES achieved the cross-link structure of Si–O–Si through hydrolysis and condensation because of ammonia, but this single cross-link was not very strong. The extent of the cross-link between the molecular chains is relatively weak, and there are still the possibility of slippage and dislocation under external force tearing. When introducing the curing agent, molecular chains are double cross-linked to each other, which strengthens the stability and robustness of the network and increases the relative molecular mass. Macroscopically, these make the polymer coatings rather hard, up to 8H. In fact, the extension of the double cross-link decides the hardness of the coating. As mentioned, each mole of active hydrogen in the amines react with one mole of epoxy group (H:epoxy = 1:1), theoretically. However, there is always a dynamic equilibrium in organic reaction, that is, not every epoxy group goes through a ring-opening reaction. In order to improve the conversion of epoxy groups, excessive curing agent is supposed to be added to the solution. According to Table 1, when the proportion of hydrogen increased, the pencil hardness of both coatings also increased, proving that the excessive curing agent made the ring-opening reaction more thorough, and then the double cross-link network was strengthened. In detail, when the curing agent was less ($\leq$1:1), the curing efficiency of DETA was higher, and the hardness reached to 5H. In other words, most GPTES was cured by DETA with a low concentration. As for MXDA, its small amount had a small increase in hardness. When H:epoxy = 2:1, the hardness of two coatings increased to 8H. However, the sol soon becomes a gel because of excessive curing in about 8 days. This is because the size of the cross-linked networks continues to expand as the aging time increases. Macroscopically, the fluidity of the sol is continuously weakened, and finally becomes a gel, making it impossible to go through the dip-coating process. So, the sol (epoxy:H = 1:2) needs to be diluted to slow down the growth of the cross-linked networks to prolong its life. According to experiments, when the concentration of sol was diluted to half, the sol was kept in a fluidized condition after 180 days.

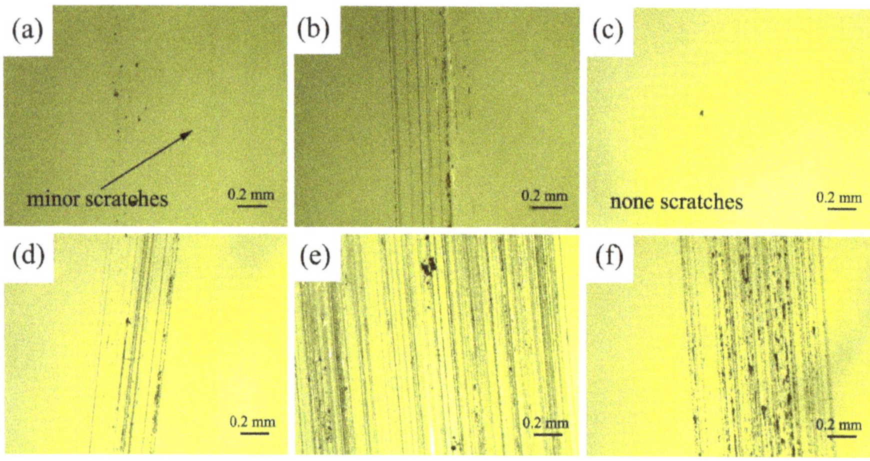

Figure 3. Optical microscope images of the scratches on various coatings. (**a**) DETA-OSER, scratched with an 8H pencil. (**b**) DETA-OSER, scratched with a 9H pencil. (**c**) MXDA-OSER, scratched with an 8H pencil. (**d**) MXDA-OSER, scratched with a 9H pencil. (**e**) NH_3-OSER, scratched with a 3B pencil. (**f**) TM, scratched with a 6B pencil.

Table 1. Pencil hardness of DETA-OSER and MXDA-OSER in different molar ratios: H:epoxy.

	0.5	**1.0**	**1.5**	**2.0**
DETA-OSER	2H	5H	7H	8H
MXDA-OSER	3B	H	6H	8H

3.3. Morphology and Optical Property of Coatings

In order to confirm the hardness quantitatively and figure out the surface morphologies of the coatings, the coatings' Young's modulus and surface roughness were determined by atomic force microscope. Young's modulus describes the ability of a solid material to resist deformation. That is, the higher Young's modulus of the coating, the stronger its ability to resist bending and the greater its hardness. The black lines marked in Figure 4 are the center line average, which represents the average of Young's modulus of the coatings. The results shown in Figure 4 revealed that MXDA-OSER obtained the highest value of ~10.8 GPa, while TM obtained a minimum value of ~1.5 GPa. This result was approximately consistent with the scratch-resistance observation shown in Figure 3. The stability of the double cross-links structure was verified. Figure 5 showed the surface morphologies of the coatings. The corresponding root-mean-square deviation roughness (Rq) is given in

Figure 5. Spherical clusters on the surface could be observed as shown in Figure 5c,d. The roughness of NH_3-OSER and TM (Rq = 2.024, 7.651 nm) was much higher than that of the DETA-OSER and MXDA-OSER (Rq = 0.316 nm, 0.274 nm). The result reflects that the DETA-OSER and MXDA-OSER had extremely smooth surfaces, which was attributed to their tightly linked molecular chains with double-cross-link structures. The smooth surface also greatly reduced the light scattering, resulting in a high transmittance of visible light. It is noticed that the scratch resistance of TM and NH_3-OSER was much weaker. Therefore, the further investigation on these two coatings will not be carried out.

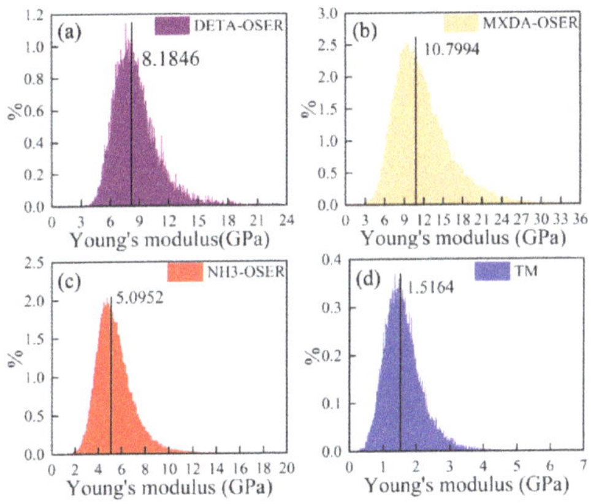

Figure 4. Histograms of the Young's modulus distribution of the coatings. (**a**) DETA-OSER. (**b**) MXDA-OSER. (**c**) NH_3-OSER. (**d**) TM.

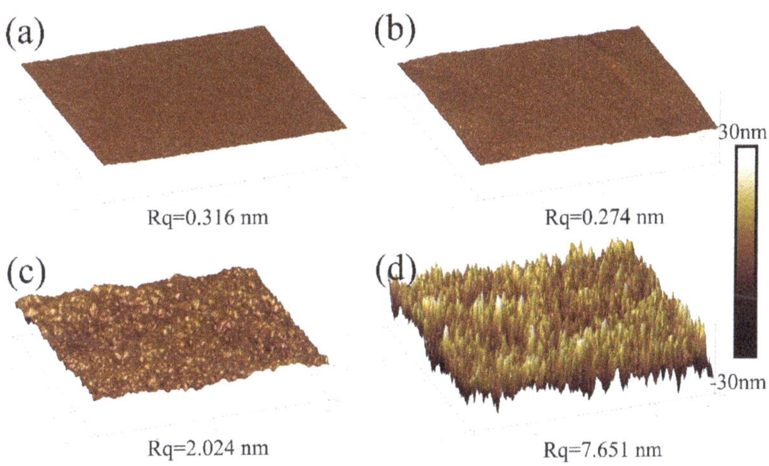

Figure 5. The 3D surface morphology of the coatings. (**a**) DETA-OSER. (**b**) MXDA-OSER. (**c**) NH_3-OSER. (**d**) TM.

According to Figure 6, the transmittance of the DETA-OSER was a little higher than that of bare glass. This slightly difference is caused by the refractive index. The refractive index and thickness of the coatings and glass, measured by spectroscopic ellipsometry, are given in Table 2. The result indicates that DETA-OSER can also be used as an antireflective coating. However, when using MXDA as a curing agent, phenyl was introduced into the molecular chains, which increased the density of particles in the molecular chains to a certain extent. Nevertheless, the large phenyl enhances the rigidity of the molecular chains. Consequently, chains are more difficult to deform, curl and shift, which makes the coating extremely hard and ultra-scratch-resistant. Back to Figure 3c, there were no scratches, scratched with an 8H pencil, whereas in DETA-OSER, without phenyl, there were minor scratches on it. In a word, MXDA-OSER was more scratch resistant than DETA-OSER, but at the cost of a slight decrease in transmittance.

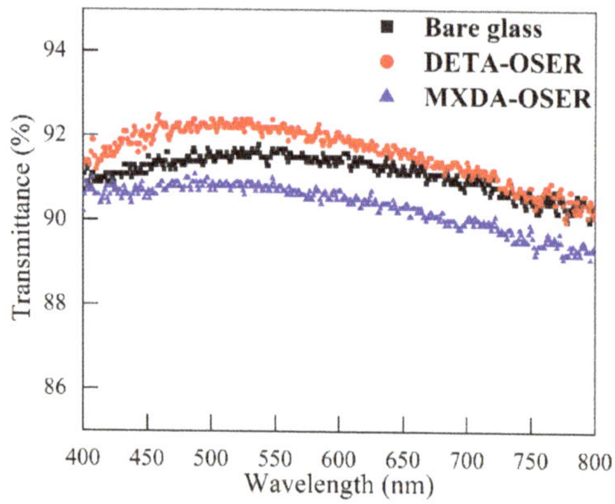

Figure 6. Transmittance spectra of the DETA-OSER, MXDA-OSER and bare glass.

Table 2. Refractive index and thickness of the DETA-OSER, MXDA-OSER and bare glass.

	Refractive Index (±0.003)	Thickness
DETA-OSER	1.493	103 nm
Bare glass	1.512	700 μm
MXDA-OSER	1.525	114 nm

3.4. Durability in Different Environments of Coatings

The transmittance of coatings often decreases because of dust in practical applications. The coating with self-cleaning ability can effectively reduce the influence of dust on transmittance. Owing to being soaked in HMDS, the coating got the self-cleaning ability. Si–CH_3 in HMDS was transferred to the surface of the coating, making the surface hydrophobic. When the water droplet was dropped on the surface, the contact angle was up to 121°, which is shown in Figure 7b. However, in Figure 7a, the contact angle of the unsoaked DETA-OSER was just 69°, which was lower than the soaked one. It indicated that the hydrophobic groups were successfully grafted onto the surface. Additionally, this happened to MXDA-OSER. Generally, being soaked in the HMDS reduces the surface energy and increases the contact angle of water droplets, which endows the coating with self-cleaning ability.

The durability of the coating is a key technical concern. In order to study its chemical stability in different environments, DETA-OSER and MXDA-OSER were soaked in pH = 2 and pH = 12 solutions, seawater (3.5% NaCl aqueous solution) and organic solvent (anhydrous acetone), respectively. According to Tables 3 and 4, DETA-OSER and MXDA-OSER showed to be insoluble in acetone because their polarity was weak owing to the large molecular chains. Besides, DETA-OSER and MXDA-OSER, to a certain extent, could resist the erosion of the acid and seawater, but the hardness of DETA-OSER decreased slightly. The possible reason is that the molecular chains with phenyl are more inert due to the steric hindrance of phenyl. In addition, both DETA-OSER and MXDA-OSER could be soaked in an aqueous alkali only for a short time. The reason for hardness decline is that the Si–O–Si framework reacts with NaOH to produce the soluble Na_2SiO_3. Gradually, the corrosion of strong alkali destroyed the cross-link structure.

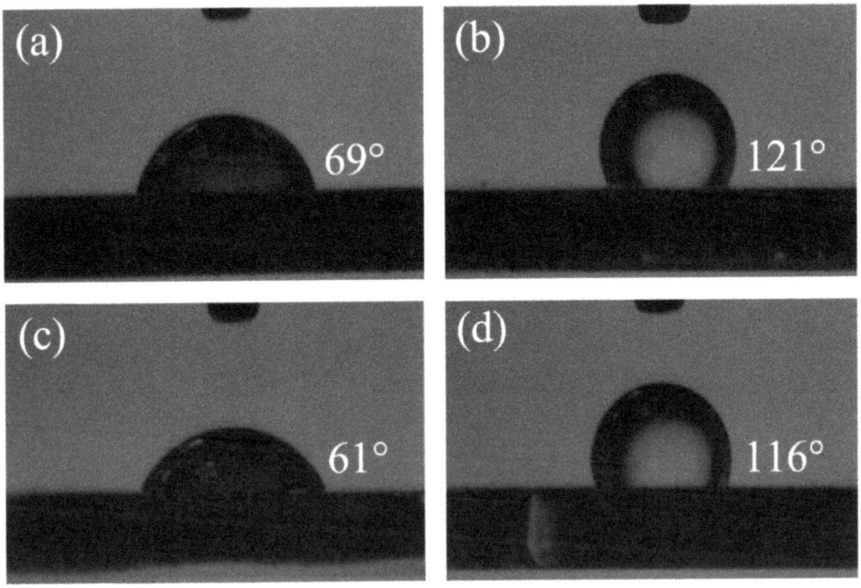

Figure 7. Contact angle of a water droplet on the surface of (**a**) DETA-OSER, before being soaked in HMDS. (**b**) DETA-OSER, after being soaked in HMDS. (**c**) MXDA-OSER, before being soaked in HMDS. (**d**) MXDA-OSER, after being soaked in HMDS.

Table 3. The pencil hardness grade of DETA-OSER being soaked in different solutions.

	1 Day	3 Days	5 Days	7 Days
pH = 2	8H	8H	8H	7H
pH = 12	8H	<6B	<6B	<6B
Seawater	8H	8H	8H	6H
Acetone	8H	8H	8H	8H

Table 4. The pencil hardness grade of MXDA-OSER being soaked in different solutions.

	1 Day	3 Days	5 Days	7 Days
pH = 2	8H	8H	8H	8H
pH = 12	8H	3B	<6B	<6B
Seawater	8H	8H	8H	8H
Acetone	8H	8H	8H	8H

To study the thermal behavior of materials, the thermal-oxidative decomposition processes of samples were investigated. According to the DTG curves shown in Figure 8, a maximum value with 160 °C and 13.4 min was observed. When the temperature was lower than 160 °C, the water adsorbed on the surface and started to evaporate, and the free Si–OH and the unreacted C_2H_5–O–Si also began to condense to produce a Si–O–Si cross-link structure, which reduced the mass during this time. In other words, annealing at 160 °C made the cross-link more thorough and then increased the hardness of the coating. When the temperature was higher than 160 °C, the rate of mass decrease began to increase, which meant that the polymer coating began to decompose intensely. Therefore, the DETA-OSER coating has the highest working temperature of 160 °C. Similarly, a maximum value of 155 °C was obtained at 12.7 min for MXDA-OSER coating. Meanwhile, it is observed that the DTG curves fluctuated greatly in the high-temperature region (>160 °C). This was due to the fact that the coatings begun to thermal decompose, producing gases such as CO_x, NO_x, NH_3 and alkanes with a different decomposition temperature and time.

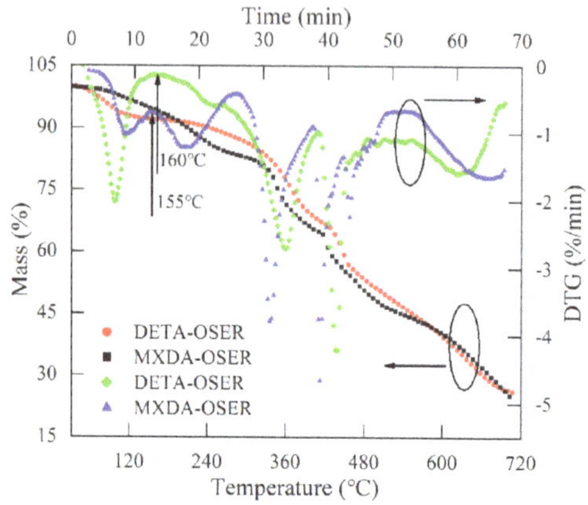

Figure 8. TGA and DTG curves of the samples of DETA-OSER and MXDA-OSER in air. The samples were made of the aged sol after drying the solvent at 60 °C, and they were not cured at a high temperature.

3.5. *The Coating on Different Substrates*

OSER can be applied on various substrates to protect the surface. The influence of the substrate on the scratch resistance was investigated. As shown in Figure 9, there were evident scratches on the coatings with a 9H pencil, but no scratches with an 8H, which meant that both DETA-OSER and MXDA-OSER were still ultra-scratch-resistant, even on iron substrates with a rough surface. Besides, owing to their inert and hydrophobic properties, OSER could be used as a protective coating to prevent the substrate from being scratched or becoming wet. Additionally, due to its high transparency, it had almost no effect on the pattern of the substrate.

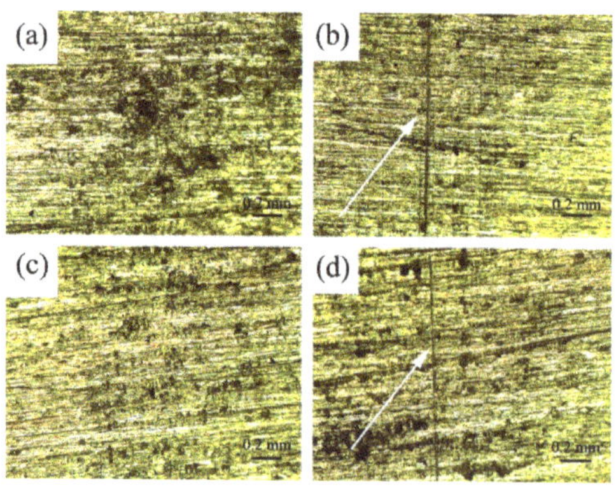

Figure 9. Surface morphologies of the OSERs on iron substrates after pencil hardness test. (**a**) DETA-OSER, scratched with an 8H pencil. (**b**) DETA-OSER, scratched with a 9H pencil. (**c**) MXDA-OSER, scratched with an 8H pencil. (**d**) MXDA-OSER, scratched with a 9H pencil.

4. Conclusions

In summary, we demonstrated a robust, hydrophobic and transparent coating based on organosilicon-epoxy resin (OSER). Aliphatic polyamines, as catalysts and reactants at the same time, make GPTES hydrolyze and condense as well as react with epoxy groups, thereby forming a double-cross-link structure. The double cross-links allow the coatings to resist scratching, macroscopically. The coating shows excellent scratch resistance and good transparency. Besides, its durable and hydrophobic properties prevent the substrate from becoming wet by many solutions. The cost-effective coating exhibits great potential value in commercial applications.

Author Contributions: Conceptualization, Z.Q. and H.L.; methodology, Z.Q. and H.L.; formal analysis, Z.Q., L.Z. and Y.L.; investigation, Z.Q.; data curation, Z.Q.; writing—original draft preparation, Z.Q.; writing—review and editing, C.Z. and R.H.; visualization, Z.Q.; supervision, R.H.; project administration, R.H.; funding acquisition, R.H. All authors have read and agreed to the published version of the manuscript.

Funding: This research supported by Guangzhou Municipal Science and Technology Bureau (grant no. 201804020031) and Guangdong Basic and Applied Basic Research Foundation (grant nos. 2020A1515010057 and 2021A1515010626).

Acknowledgments: The authors gratefully acknowledge financial support from Guangzhou Municipal Science and Technology Bureau (grant no. 201804020031) and Guangdong Basic and Applied Basic Research Foundation (grant nos. 2020A1515010057 and 2021A1515010626).

Conflicts of Interest: The authors declare no conflict of interest.

References

1. Deng, X.; Mammen, L.; Butt, H.-J.; Vollmer, D. Candle Soot as a Template for a Transparent Robust Superamphiphobic Coating. *Science* **2012**, *335*, 67–70. [CrossRef] [PubMed]
2. Deng, X.; Mammen, L.; Zhao, Y.; Lellig, P.; Müllen, K.; Li, C.; Butt, H.-J.; Vollmer, D. Transparent, Thermally Stable and Mechanically Robust Superhydrophobic Surfaces Made from Porous Silica Capsules. *Adv. Mater.* **2011**, *23*, 2962–2965. [CrossRef] [PubMed]
3. Zhu, T.; Cheng, Y.; Huang, J.; Xiong, J.; Ge, M.; Mao, J.; Liu, Z.; Dong, X.; Chen, Z.; Lai, Y. A transparent superhydrophobic coating with mechanochemical robustness for anti-icing, photocatalysis and self-cleaning. *Chem. Eng. J.* **2020**, *399*, 125746. [CrossRef]
4. Cai, S.; Zhang, Y.L.; Zhang, H.L.; Yan, H.W.; Lv, H.B.; Jiang, B. Sol–Gel Preparation of Hydrophobic Silica Antireflective Coatings with Low Refractive Index by Base/Acid Two-Step Catalysis. *ACS Appl. Mater. Interfaces* **2014**, *6*, 11470–11475. [CrossRef]
5. Chevallier, P.; Turgeon, S.; Sarra-Bournet, C.; Turcotte, R.; Laroche, G. Characterization of Multilayer Anti-Fog Coatings. *ACS Appl. Mater. Interfaces* **2011**, *3*, 750–758. [CrossRef]
6. Brown, P.S.; Atkinson, O.D.L.A.; Badyal, J.P.S. Ultrafast Oleophobic-Hydrophilic Switching Surfaces for Antifogging, Self-Cleaning, and Oil-Water Separation. *ACS Appl. Mater. Interfaces* **2014**, *6*, 7504–7511. [CrossRef]
7. Chen, C.; Weng, D.; Mahmood, A.; Chen, S.; Wang, J. Separation Mechanism and Construction of Surfaces with Special Wettability for Oil/Water Separation. *ACS Appl. Mater. Interfaces* **2019**, *11*, 11006–11027. [CrossRef]
8. Mohammadkhani, R.; Ramezanzadeh, M.; Saadatmandi, S.; Ramezanzadeh, B. Designing a dual-functional epoxy composite system with self-healing/barrier anti-corrosion performance using graphene oxide nano-scale platforms decorated with zinc doped-conductive polypyrrole nanoparticles with great environmental stability and non-toxicity. *Chem. Eng. J.* **2019**, *382*, 122819.
9. Cho, E.C.; Chang-Jian, C.W.; Chen, H.C.; Chuang, K.S.; Zheng, J.H.; Hsiao, Y.S.; Huang, J.H. Robust Multifunctional Superhydrophobic Coatings with Enhanced Water/Oil Separation, Self-Cleaning, Anti-Corrosion, and Anti-biological adhesion. *Chem. Eng. J.* **2017**, *314*, 347–357. [CrossRef]
10. Ye, Y.; Zhang, D.; Liu, T.; Liu, Z.; Pu, J.; Liu, W.; Wang, L. Superior corrosion resistance and self-healable epoxy coating pigmented with silanzied trianiline-intercalated graphene. *Carbon* **2018**, *142*, 164–176. [CrossRef]
11. Siddique, R.H.; Gomard, G.; Hoelscher, H. The role of random nanostructures for the omnidirectional anti-reflection properties of the glasswing butterfly. *Nat. Commun.* **2015**, *6*, 6909. [CrossRef]
12. Zhang, J.; Ai, L.; Lin, S.; Lan, P.; Lu, Y.; Dai, N.; Tan, R.; Fan, B.; Song, W. Preparation of humidity, abrasion, and dust resistant antireflection coatings for photovoltaic modules via dual precursor modification and hybridization of hollow silica nanospheres. *Sol. Energy Mater. Sol. Cells* **2019**, *192*, 188–196. [CrossRef]
13. Zhou, H.; Wang, H.; Niu, H.; Gestos, A.; Lin, T. Robust, Self-Healing Superamphiphobic Fabrics Prepared by Two-Step Coating of Fluoro-Containing Polymer, Fluoroalkyl Silane, and Modified Silica Nanoparticles. *Adv. Funct. Mater.* **2013**, *23*, 1664–1670. [CrossRef]
14. Dong, X.; Gao, S.; Huang, J.; Li, S.; Zhu, T.; Cheng, Y.; Zhao, Y.; Chen, Z.; Lai, Y. A self-roughened and biodegradable superhydrophobic coating with UV shielding, solar-induced self-healing and versatile oil-water separation ability. *J. Mater. Chem. A* **2019**, *7*, 2122–2128. [CrossRef]
15. Liang, B.; Zhong, Z.; Jia, E.; Zhang, G.; Su, Z. Transparent and Scratch-Resistant Antifogging Coatings with Rapid Self-Healing Capability. *ACS Appl. Mater. Interfaces* **2019**, *11*, 30300–30307. [CrossRef]
16. Banerjee, S.; Dionysiou, D.D.; Pillai, S.C. Self-cleaning applications of TiO_2 by photo-induced hydrophilicity and photocatalysis. *Appl. Catal. B* **2015**, *176*, 396–428. [CrossRef]

17. Latthe, S.S.; Sutar, R.S.; Kodag, V.S.; Bhosale, A.K.; Kumar, A.M.; Sadasivuni, K.K.; Xing, R.; Liu, S. Self-cleaning superhydrophobic coatings: Potential industrial applications. *Prog. Org. Coat.* **2019**, *128*, 52–58. [CrossRef]
18. Chattopadhyay, S.; Huang, Y.; Jen, Y.; Ganguly, A.; Chen, K.-H.; Chen, L. Anti-reflecting and photonic nanostructures. *Mater. Sci. Eng. R Rep.* **2010**, *69*, 1–35. [CrossRef]
19. Zhang, X.-X.; Xia, B.-B.; Ye, H.-P.; Zhang, Y.-L.; Xiao, B.; Yan, L.-H.; Lv, H.-B.; Jiang, B. One-step sol–gel preparation of PDMS–silica ORMOSILs as environment-resistant and crack-free thick antireflective coatings. *J. Mater. Chem.* **2012**, *22*, 13132–13140. [CrossRef]
20. Figueira, R.B.; Fontinha, I.R.; Silva, C.J.R.; Pereira, E.V. Hybrid Sol-Gel Coatings: Smart and Green Materials for Corrosion Mitigation. *Coatings* **2016**, *6*, 12. [CrossRef]
21. Chi, F.; Liu, D.; Wu, H.; Lei, J. Mechanically robust and self-cleaning antireflection coatings from nanoscale binding of hydrophobic silica nanoparticles. *Sol. Energy Mater. Sol. Cells* **2019**, *200*, 109939. [CrossRef]
22. Wang, Z.; van Andel, E.; Pujari, S.P.; Feng, H.; Dijksman, J.A.; Smulders, M.M.J.; Zuilhof, H. Water-repairable zwitterionic polymer coatings for anti-biofouling surfaces. *J. Mater. Chem. B* **2017**, *5*, 6728–6733. [CrossRef]
23. Wang, C.; Wu, H.; Chen, Z.; McDowell, M.T.; Cui, Y.; Bao, Z. Self-healing chemistry enables the stable operation of silicon microparticle anodes for high-energy lithium-ion batteries. *Nat. Chem.* **2013**, *5*, 1042–1048. [CrossRef]
24. Burnworth, M.; Tang, L.; Kumpfer, J.R.; Duncan, A.J.; Beyer, F.L.; Fiore, G.L.; Rowan, S.J.; Weder, C. Optically healable supramolecular polymers. *Nature* **2011**, *472*, 334–337. [CrossRef]
25. Nakahata, M.; Takashima, Y.; Yamaguchi, H.; Harada, A. Redox-responsive self-healing materials formed from host–guest polymers. *Nat. Commun.* **2011**, *2*, 511. [CrossRef]
26. Armaghani, D.J.; Mohamad, E.T.; Narayanasamy, M.S.; Narita, N.; Yagiz, S. Development of hybrid intelligent models for predicting TBM penetration rate in hard rock condition. *Tunn. Undergr. Space Technol.* **2017**, *63*, 29–43. [CrossRef]
27. Momeni, E.; Armaghani, D.J.; Hajihassani, M.; Amin, M.F.M. Prediction of uniaxial compressive strength of rock samples using hybrid particle swarm optimization-based artificial neural networks. *Measurement* **2015**, *60*, 50–63. [CrossRef]
28. Lu, Y.-X.; Guan, Z. Olefin Metathesis for Effective Polymer Healing via Dynamic Exchange of Strong Carbon–Carbon Double Bonds. *J. Am. Chem. Soc.* **2012**, *134*, 14226–14231. [CrossRef]
29. Amamoto, Y.; Otsuka, H.; Takahara, A.; Matyjaszewski, K. Self-Healing of Covalently Cross-Linked Polymers by Reshuffling Thiuram Disulfide Moieties in Air under Visible Light. *Adv. Mater.* **2012**, *24*, 3975–3980. [CrossRef]
30. Ji, S.; Cao, W.; Yu, Y.; Xu, H. Visible-Light-Induced Self-Healing Diselenide-Containing Polyurethane Elastomer. *Adv. Mater.* **2015**, *27*, 7740–7745. [CrossRef]
31. Cash, J.J.; Kubo, T.; Bapat, A.P.; Sumerlin, B.S. Room-Temperature Self-Healing Polymers Based on Dynamic-Covalent Boronic Esters. *Macromolecules* **2015**, *48*, 2098–2106. [CrossRef]
32. Hasanipanah, M.; Monjezi, M.; Shahnazar, A.; Jahed Armaghani, D.; Farazmand, A. Feasibility of indirect determination of blast induced ground vibration based on support vector machine. *Measurement* **2015**, *75*, 289–297. [CrossRef]
33. Zhang, Y.; Zhao, C.; Wang, P.; Ye, L.; Luo, J.; Jiang, B. A convenient sol–gel approach to the preparation of nano-porous silica coatings with very low refractive indices. *Chem. Commun.* **2014**, *50*, 13813–13816. [CrossRef] [PubMed]
34. Zhang, S.; Xiao, P.; Wang, P.; Luo, J.; Jiang, B. Spherical-chain silica with super-hydrophobic surface and ultra-low refractive index for multi-functional broadband antireflective coatings. *Sol. Energy* **2020**, *207*, 1222–1230. [CrossRef]
35. Mousavi, H.; Jilavi, M.H.; Koch, M.; Arzt, E.; De Oliveira, P.W. Development of a Transparent Scratch Resistant Coating through Direct Oxidation of Al-Coated Glass. *Adv. Eng. Mater.* **2017**, *19*, 1600617. [CrossRef]
36. Zhou, H.; Wang, H.; Niu, H.; Gestos, A.; Wang, X.; Lin, T. Fluoroalkyl Silane Modified Silicone Rubber/Nanoparticle Composite: A Super Durable, Robust Superhydrophobic Fabric Coating. *Adv. Mater.* **2012**, *24*, 2409–2412. [CrossRef] [PubMed]
37. Jin, F.-L.; Li, X.; Park, S.-J. Synthesis and application of epoxy resins: A review. *J. Ind. Eng. Chem.* **2015**, *29*, 1–11. [CrossRef]
38. Memon, H.; Liu, H.; Rashid, M.A.; Chen, L.; Jiang, Q.; Zhang, L.; Wei, Y.; Liu, W.; Qiu, Y. Vanillin-Based Epoxy Vitrimer with High Performance and Closed-Loop Recyclability. *Macromolecules* **2020**, *53*, 621–630. [CrossRef]
39. Tian, Y.; Wang, Q.; Shen, L.; Cui, Z.; Kou, L.; Cheng, J.; Zhang, J. A renewable resveratrol-based epoxy resin with high Tg, excellent mechanical properties and low flammability. *Chem. Eng. J.* **2020**, *383*, 123124. [CrossRef]
40. Zhi, J.; Zhang, L.-Z. Durable superhydrophobic surface with highly antireflective and self-cleaning properties for the glass covers of solar cells. *Appl. Surf. Sci.* **2018**, *454*, 239–248. [CrossRef]
41. Kesmez, Ö. Preparation of hybrid nanocomposite coatings via sol-gel method for hydrophobic and self-cleaning properties. *J. Mol. Struct.* **2020**, *1205*, 127572. [CrossRef]
42. Davis, S.R.; Brough, A.R.; Atkinson, A. Formation of silica/epoxy hybrid network polymers. *J. Non-Cryst. Solids* **2003**, *315*, 197–205. [CrossRef]
43. Grancaric, A.M.; Colleoni, C.; Guido, E.; Botteri, L.; Rosace, G. Thermal behaviour and flame retardancy of monoethanolamine-doped sol-gel coatings of cotton fabric. *Prog. Org. Coat.* **2017**, *103*, 174–181. [CrossRef]
44. Nazir, T.; Afzal, A.; Siddiqi, H.M.; Saeed, S.; Dumon, M. The influence of temperature and interface strength on the microstructure and performance of sol–gel silica–epoxy nanocomposites. *Polym. Bull.* **2011**, *67*, 1539–1551. [CrossRef]
45. Zhang, Y.; Zhang, X.; Ye, H.; Xiao, B.; Yan, L.; Jiang, B. A simple route to prepare crack-free thick antireflective silica coatings with improved antireflective stability. *Mater. Lett.* **2012**, *69*, 86–88. [CrossRef]
46. *ASTM D3363*; Standard Test Method for Film Hardness by Pencil Test. ASTM International: West Conshohocken, PA, USA, 2020. Available online: https://www.astm.org/Standards/D3363.htm (accessed on 15 September 2021).

applied sciences

MDPI

Article

Soil Classification from Piezocone Penetration Test Using Fuzzy Clustering and Neuro-Fuzzy Theory

Joon-Shik Moon [1], Chan-Hong Kim [2] and Young-Sang Kim [3,*]

[1] Department of Civil Engineering, Kyungpook National University, Daegu 41566, Korea; orangedreamer@gmail.com
[2] Korea Mine Rehabilitation and Mineral Resources Corporation, Wonju 26464, Korea; chahkim@komir.or.kr
[3] Department of Civil Engineering, Chonnam National University, Gwangju 61186, Korea
* Correspondence: geoyskim@jnu.ac.kr; Tel.: +82-62-530-1654

Abstract: The advantage of the piezocone penetration test is a guarantee of continuous data, which are a source of reliable interpretation of the target soil layer. Much research has been carried out for several decades, and several classification charts have been developed to classify in situ soil from the cone penetration test result. Even though most present classification charts or methods were developed on the basis of data which were compiled over many countries, they should be verified to be feasible for local country. However, unfortunately, revision of those charts is quite difficult or almost impossible even though a chart provides misclassified soil class. In this research, a new method for developing soil classification model is proposed by using soft computing theory—fuzzy C-mean clustering and neuro-fuzzy theory—as a function of 5173 piezocone penetration test (PCPT) results and soil boring logs compiled from 17 local sites around Korea. Feasibility of the proposed soil classification model was verified from the viewpoint of accuracy of the classification result by comparing the classification results not only for data which were used for developing the model but also new data, which were not included in developing the model with real boring logs, other fuzzy computing classification models, and Robertson's charts. The biggest advantage of the proposed method is that it is easy to make the piezocone soil classification system more accurate by updating new data.

Keywords: piezocone; soil classification; fuzzy C-means clustering; neuro-fuzzy

Citation: Moon, J.-S.; Kim, C.-H.; Kim, Y.-S. Soil Classification from Piezocone Penetration Test Using Fuzzy Clustering and Neuro-Fuzzy Theory. *Appl. Sci.* **2022**, *12*, 4023. https://doi.org/10.3390/app12084023

Academic Editors: Danial Jahed Armaghani, Yixia Zhang, Pijush Samui, Ahmed Hussein Kamel Ahmed Elshafie and Aydin Azizi

Received: 21 March 2022
Accepted: 12 April 2022
Published: 15 April 2022

1. Introduction

Underground information is a main factor to be considered during the construction and design phases. In particular, stratigraphy is essential for economical design of a foundation because most construction projects are carried out at the deposit layer on bedrock. Boring logs from subsurface exploration at a constant interval along the project area are the only source of data. They are drawn from some resources such as the penetration rate, soil color, and driller's experience, which is dependent upon their career and, thus, may could not always reflect the nature of the ground. Therefore, penetration tests, such as the cone penetration test (CPT), piezocone penetration test (PCPT), and standard penetration test (SPT), have been used together. PCPT has an advantage in the view of continuity and standardization, even when evaluating interbedded thin layers from thick deposit layers. Research on soil classification from CPT results was commenced by Begemann [1]. Furthermore, Douglas and Olsen [2] developed a new soil classification chart using electric cone penetration test results. After introducing the piezocone which can measure pore pressure readings, many researchers including Robertson et al. [3], Robertson [4], and Jefferies and Davis [5] developed various types of soil classification charts and/or techniques. However, most classification charts provide only soil behavior type, while local engineers who are familiar with Unified Soil Classification System (USCS) have trouble with understanding relevant results. In addition, the adopted charts and methods sometimes give different soil classification results for the same input parameters, and even two charts developed by one researcher may lead to different soil types. To complement the weakness of the chart type soil classification method, and considering the fuzziness of the ground, there has been progress in studies on soft computing. Pradhan [6] developed fuzzy membership functions on the basis of Robertson et al.'s chart [3], and Zhang and Tumay [7] suggested a fuzzy soil classification method according to Douglas and Olsen's chart [2]. On the other hand, Hegazy and Mayne [8] introduced a clustering method as a function of normalized cone resistance, Q_t, and pore pressure ratio, B_q. Clustering methods can give soil classifications between upper and lower soil data but not the soil type of each soil datum. On the other hand, soil classification by fuzzy theory can provide soil type to each soil datum and has the advantage of simply being updated for newly acquired soil data. However, it also has a problem that the classification result is highly dependent on the fuzzy membership function. As described before, Pradhan [6] and Zhang and Tumay [7] developed fuzzy membership functions on the basis of charts such as those proposed by Robertson et al., and Douglas and Olson, respectively. Therefore, these fuzzy classifications seemingly remain unable to reflect local soil type. Recently, machine learning has been used to classify soils from CPT data [9–13] and to successfully estimate soil and design parameters [9,13]. Rauter and Tschuchnigg [14] suggested a machine learning classifier based on a support vector machine, artificial neural network, and random forest to predict soil classes according to Oberhollenzer et al. [15] and soil behavior types according to Robertson [16–18]. They showed that machine learning algorithms can classify soils on the basis of grain size distribution and the updated soil behavior classification from Robertson (i.e., SBT, SBTn, ModSBTn). However, since they used cone tip resistance q_c, sleeve friction f_s, total vertical stress σ_v, and static pore pressure u_o as input variables, their model can still be improved by adopting pore pressure parameters such as B_q.

In this study, a new soil classification method was developed using the neuro-fuzzy technique, in which the membership function was developed by a neural network and not adjusted by the trial-and-error method to present classification charts. Moreover, input variables and relevant soil types were determined on the basis of proximity between compiled soil data using the fuzzy C-mean clustering (FCM) method and not by the developer's experience. To show the feasibility of the proposed model, new PCPT results which were not included in the soil database were classified using the proposed neuro-fuzzy model and compared with Robertson et al.'s chart classification, Pradhan's fuzzy classification, Zhang and Tumay's fuzzy classification, and the Unified Soil Classification System (USCS).

2. Soil Classification Method for CPT and PCPT

2.1. Soil Classification Charts

Figure 1 shows Robertson et al. [3]'s classification charts as a function of q_t, R_f, and B_q. Their definitions are as follows:

$$q_t = q_c + (1-a)u_{bt}, \tag{1}$$

$$R_f = \frac{f_s}{q_c} \times 100(\%), \tag{2}$$

$$B_q = \frac{(u_{bt} - u_o)}{(q_t - \sigma_{vo})}, \tag{3}$$

where q_t is the corrected cone tip resistance, q_c is the measured cone tip resistance, u_{bt} is the penetration-induced pore pressure measured behind the cone tip, a is the unequal area ratio, R_f is the friction ratio, f_s is the sleeve friction, B_q is the pore pressure ratio, u_o is the static pore pressure before cone penetration, and σ_{vo} is the total stress. Robertson's charts have been widely used, and their feasibility was verified by several researchers.

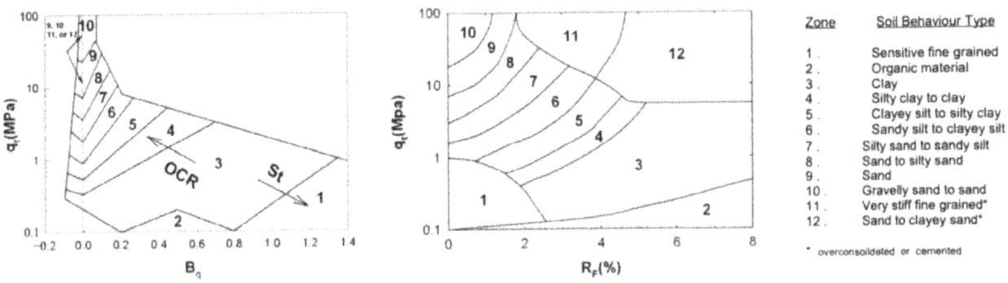

Figure 1. Soil classification charts by Robertson et al. [3].

2.2. Soil Classification Using Fuzzy Theory

Natural phenomena are known not to be decided in absolute terms such as 0 or 1. Zadeh [19] introduced "soft computing", the concept of fuzzy theory to describe nature's ambiguousness. This theory can present intermediate values using the fuzzy membership function. Various soft computing methods have been suggested after Zadeh [19], and studies on soil classification from CPT and PCPT using soft computing are summarized below.

2.2.1. Pradhan's Study

Pradhan [6] suggested a soil classification method using fuzzy theory. He developed fuzzy membership functions for input variables q_t, $F_r(= f_s/q_t)$, and B_q according to Robertson et al. [2]. However, soil types were only classified into "clay", "silt", and "sand". The maximum grade for membership functions was limited to 0.8 when considering uncertainty in soil classification. Figure 2 shows the membership functions of three soil sets in terms of q_t, B_q, and F_r respectively. The reader is referred to Pradhan [6] for the detailed expression of fuzzy membership functions.

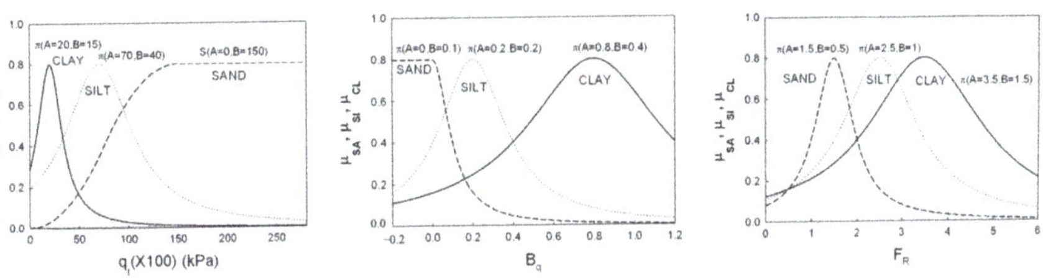

Figure 2. Fuzzy membership functions of soil sets for three parameters, q_t, F_r, and B_q.

Three soil fuzzy sets for clay, silt, and sand were named CL, SI, and SA, as defined in Equation (4). μ_{CL}, μ_{SI}, and μ_{SA} represent the summation of membership function values from each chart for clay, silt, and sand, respectively.

$$\begin{aligned} \text{Clayey soil}: \ \mathrm{CL} &= \sum \mu_{CL}(a_i)/a_i (i = 1,2,3), \\ \text{Silty soil}: \ \mathrm{SI} &= \sum \mu_{SI}(a_i)/a_i (i = 1,2,3), \\ \text{Sandy soil}: \ \mathrm{SA} &= \sum \mu_{SA}(a_i)/a_i (i = 1,2,3), \end{aligned} \tag{4}$$

where $a_1 = q_t$, $a_2 = F_r$, and $a_3 = B_q$.

2.2.2. Zhang and Tumay's Study

Zhang and Tumay [7] grouped soil into three types, i.e., HPC (highly probable clay), HPM (highly probable mixed soil), and HPS (highly probable sand) on the basis of the Unified Soil Classification System (USCS) and used q_c and R_f as input variables. They suggested new fuzzy membership functions $\mu_C(U)$, $\mu_m(U)$, and $\mu_s(U)$ as shown in Figure 3, with an intermediate soil classification index (U) empirically based on Douglas and Olsen [2]'s chart.

$$U = \frac{(a_1X - a_2Y + b_1)(c_1X - c_2Y + d_1)}{(c_1X - c_2Y + d_1)^2 + (c_2X + c_1Y + d_2)^2} - \frac{(a_2X + a_1Y + b_2)(c_2X + c_1Y + d_2)}{(c_1X - c_2Y + d_1)^2 + (c_2X + c_1Y + d_2)^2}, \quad (5)$$

where $X = 0.1539R_f + 0.8870 \log q_c - 3.35$, and $Y = -0.2957R_f + 0.4617 \log q_c - 0.37$.

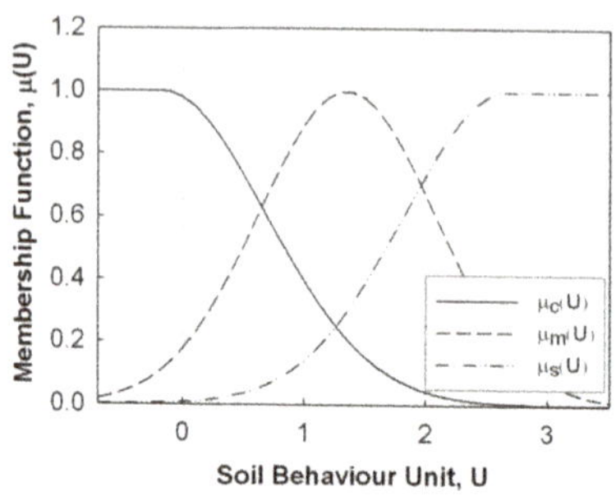

Figure 3. Fuzzy membership functions developed by Zhang and Tumay [7].

Since Pradhan [6] and Zhang and Tumay [7] developed fuzzy membership functions using a trial-and-error method according to the charts of Robertson et al. [3] and Douglas and Olsen [2], respectively, their method may result in very similar results to the soil classification result presented in the original chart. To overcome this, in this study, fuzzy memberships were determined using a neural network process.

3. Fuzzy Clustering and Neuro-Fuzzy Modeling

3.1. Database

The database for this study was built from 17 local sites of South Korea, as shown in Figure 4, along the coastal line, and six sites were used for verification of the model. The closed circle indicates the location where training data for training the fuzzy membership function were extracted, and the open square indicates the location where the verification data for verifying the completed fuzzy soil classification system were obtained. Table 1 summarizes the site location, number of PCPTs, and soil type classified by USCS. The measured values from piezocone penetration tests—q_c, f_s, u_{bt}—were averaged within the interval of 5 to 10 cm and picked up at the same depth where SPT and undisturbed samples were taken. The database contained 5173 data points in total. Table 2 shows the classification results of the database into six categories following USCS. A huge number of clayey soil (CH, CL) samples were, included while the number of silty soil and sandy soil samples was relatively small due to the difficulty in soil sampling with a thin wall tube sampler.

Figure 4. Seventeen local sites for this study.

Table 1. Included number of PCPTs and USCS classes for each site.

Sites		Nos.	Soil Type by USCS
Gyeonggi	Pyeongtaek	2	CL, SP, SW
	Siheung	3	CL, SM, SP
	Ilsan	1	SM, SP
	Incheon	3	CL, ML
Chungnam	Seocheon	4	CL, SM, SP
	Asan	1	CL, SM
Jeonnam	Yeongam	1	CH, SP, SW
	Gwangyang	1	CL, CH, ML, SP
Jeonbuk	Kunjang	9	CH, CL, ML
Gyeongnam	Yangsan	4	CL, SM
	Tongyeong	2	CH
	Hadong	2	CL, CH, MH, SP
	Ulsan	2	SM, SP-SC
	Yongwon	4	CH
	Cheonseong	4	CH
	Gaduk	4	CH
Kangwon	Naegok	2	CL, MH, SP, SW

Table 2. Number of soil data points and corresponding PCPT data range.

USCS	Nos.	q_t (MPa)	f_s (MPa)	u_{bt} (MPa)
CH	2746	0.108 to 1.250	0.0001 to 0.030	0.051 to 0.683
CL	1861	0.028 to 6.520	0.0001 to 0.052	0.002 to 1.088
MH	36	0.608 to 1.640	0.003 to 0.028	0.018 to 0.255
ML	284	0.436 to 6.818	0.004 to 0.096	−0.092 to 0.562
SM, SP-SC	148	0.217 to 160.328	0.006 to 5.740	−0.960 to 3.256
SP, SW	98	1.232 to 36.263	0.005 to 0.857	−0.168 to 0.528

3.2. FCM (Fuzzy C-Means) Clustering Algorithm

Before developing the neuro-fuzzy model for PCPT-based soil classification, a grouping procedure was carried out to establish the unique structure between soil behavior type and PCPT input parameters, as well as to determine the appropriate number of soil types in the database and the input variables. Generally, some techniques exist to find structures in the database and divide them into small groups. An unsupervised learning strategy, clustering, can be used for that purpose, with the FCM (fuzzy C-means) algorithm being most widely used. This algorithm searches for fuzzy divisions $\overline{F} = \{\overline{F_1}, \overline{F_2}, \cdots, \overline{F_c}\}$ to minimize the function, as expressed in Equation (6), when a dataset composed of n items is divided into c clusters.

$$J_m(U, V : X) = \sum_{i=1}^{c} \sum_{k=1}^{n} (\mu_{ik})^m \|X_k - V_i\|^2, \tag{6}$$

where $V = \{V_1, V_2, \cdots, V_c\}$ is the set of c central vectors, and $\|X_k - V_i\|$ is the geometric distance between data X_k and the center of the ith cluster. In addition, μ_{ik} is the grade of cluster $\overline{F}_i$ including data X_k and satisfies Equation (7) in the element of the fuzzy partition matrix $U = [\mu_{ij}]$ with the size of $(c \times n)$.

$$\mu_{ik} \in [0,1], \sum_{i=1}^{c} \mu_{ik} = 1 \tag{7}$$

The procedure of the FCM clustering algorithm is summarized below.

① Assume partition number c $(2 \le c \le n)$ and fuzziness of partition m.
② Select initial values of fuzzy partition matrix, $U(t)$. Random values are assumed for satisfying Equation (6).
③ Calculate the center of cluster V using Equation (8).

$$V_i^{(t+1)} = \frac{\sum_{i=1}^{c} \left(\mu_{ik}{}^{(t)}\right)^m X_k}{\sum_{i=1}^{c} \mu_{ik}{}^{(t)}}, m > 1, i = 1, \cdots, c \tag{8}$$

④ Recompose fuzzy partition matrix using Equation (9).

$$\mu_{ik} = \frac{1}{\sum_{j=1}^{c} \left(\frac{|X_k - V_i|^2}{|X_k - V_j|^2} \right)^{1/(m-1)}}, i = 1, \cdots, c, k = 1, \cdots, n \tag{9}$$

⑤ Complete the procedure if $|U(t+1) - U(t)| < \delta$. Otherwise, repeat phase ④. Here, δ is assumed to be 10^{-3}.
⑥ Repeat phases①to⑤and decide optimized partition number, as well as c and m values.

To determine the substructure of the compiled database and optimized input parameters, the success rate of any clusters and input parameters was evaluated after combining PCPT parameters, hydrostatic pressure, and total vertical stress, as shown in Table 3. First, the database was divided into 3–6 clusters to determine the optimized clusters of soil type according to combined parameters. Outputs were clay (CH, CL), silt (MH, ML), and sand (SM, SP, SP-SC, SW) for three clusters, clay (CH, CL), silt (MH, ML), sand with fine grained soil (SM, SP-SC), and relatively coarse sand (SP, SW) for four clusters, clay (CH, CL), silt with high liquid limit (MH), silt with low liquid limit (ML), sand with fine grained soil (SM, SP-SC), and coarse sand (SP, SW) for five clusters, and CH, CL, MH, ML, sand with fine grained soil (SM, SP-SC), and coarse sand (SP, SW) for six clusters. A total of 5 (input parameters) × 4 (clusters) were considered. The success rates of FCM clustering were evaluated as presented in Table 3. When all data points were concentrated in a specific cluster and appropriate clustering was not possible, is the success rate was regarded as "bad". According to the results, the maximum success rate was 74% when q_t, R_f, and B_q were used as input parameters and three output clusters were selected. Furthermore, the m value representing the fuzziness of the partition was optimized, as given in Figure 5. Success rates were increased to m = 4 and seemed to converge after m = 4. Thus, m = 4 was adopted for this study.

Table 3. Success rate of FCM clustering for selected input parameters and specified clusters.

Input Parameters	Success Rate (%)			
	Three Clusters	Four Clusters	Five Clusters	Six Clusters
q_t, f_s, u_{bt}	71	61	46	Bad
q_t, R_f, B_q	74	60	48	42
q_t, f_s, Δu *	70	60	69	42
q_t, f_s, u_{bt}, σ_{vo}	71	53	Bad	52
q_t, f_s, u_{bt}, R_f, B_q, σ_{vo}	70	58	48	Bad

* $\Delta u = u_{bt} - u_o$.

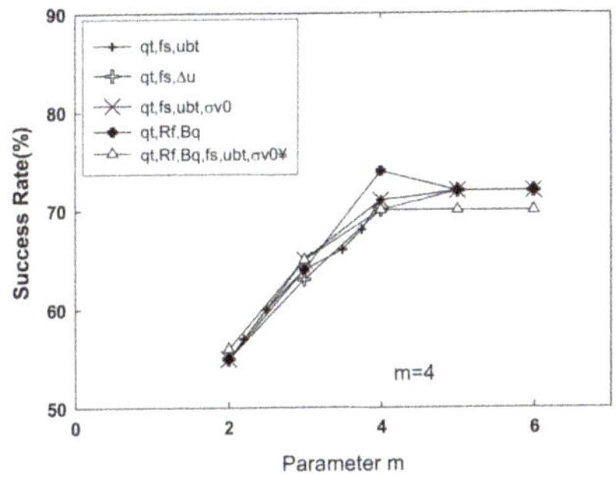

Figure 5. Change in success rate with respect to m value.

3.3. *Neuro-Fuzzy Algorithm*

After the FCM clustering described in Section 3.2, a neuro-fuzzy model was developed on the basis of the optimum number of clusters (three, i.e., clay, silt, and sand) and input variables (q_t, R_f, B_q). The fuzzy technique has the advantage of presenting data or events which cannot be numerically expressed, whereas the selection of the fuzzy membership function for inference is not always objective and precise. The decisions of fuzzy membership functions by Pradhan [6] and Zhang and Tumay [7] were dependent upon subjective trial-and-error methods or experimental methods. Thus, any revision or supplement considering local characteristics may not be easily considered even though it is necessary. However, the neuro-fuzzy method combining a neutral network and the fuzzy method has the merits of both techniques and is expected to overcome the previously mentioned defects. The neural network has the advantage of facing variations of data, but the input data are in numeric form. On the other hand, the fuzzy technique allows presenting numeric data using a membership function. Thus, the techniques are complementary. Moreover, the neuro-fuzzy technique can objectively decide membership functions and can be easily updated if needed through determining the optimized membership functions using the neural network algorithm. Figure 6 shows a schematic diagram of the ANFIS neuro-fuzzy model with two input parameters.

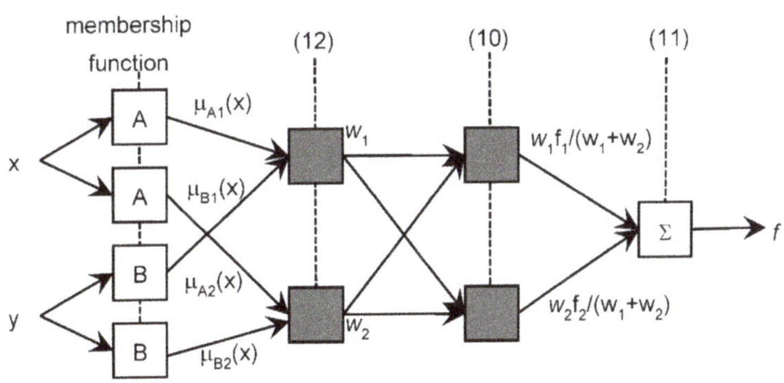

Figure 6. Procedure of ANFIS neuro-fuzzy model.

The general neuro-fuzzy model uses membership functions in the input and output phases. In this case, more time is required to complete calculation and convergence. Thus, the ANFIS (adaptive network-based fuzzy inference system) adopts a first-order function in the output instead of a membership function. The relevant procedure of the ANFIS model is summarized below.

① The neuro-fuzzy output shown in Figure 6 is defined as a first-order function as shown in Equation (10) if two input parameters are assumed. Here, p, q, and r are constants to be decided after neural network training, while x and y are input parameters, which are PCPT indices.

$$\begin{aligned} f_1 &= p_1x + q_1y + r_1, \\ f_2 &= p_2x + q_2y + r_2, \\ f &= f_1 + f_2 \end{aligned} \tag{10}$$

② Total outputs in the system considering weighting factors w_1 and w_2 are given by Equation (11). Weighting factors are calculated using Equation (12) after evaluating each fuzzy membership function for given input parameters. Here, μ_{A1} corresponds to the finalized fuzzy membership function of x, while μ_{B1} corresponds to the finalized fuzzy membership function y.

$$f = \frac{w_1f_1 + w_2f_2}{w_1 + w_2}. \tag{11}$$

$$w_1 = \mu_{A1}\mu_{B1}, w_2 = \mu_{A2}\mu_{B2}. \tag{12}$$

③ The training procedure is completed after optimizing the parameters (p, q, and r) of the first-order function and of membership functions to minimize output error, e, defined by output f_k and estimation T_k.

$$e = \frac{1}{2}\sum (f_k - T_k)^2. \tag{13}$$

Commercial soft computing package, Matlab was used to complete the training procedure when the least square error in Equation (13) was within the target error value, $\varepsilon = 0.01$ or when the maximum training loop attained 200. If any of the predetermined conditions were not satisfied, is the outcome was regarded as "bad". Figure 7 shows the commonly used candidate membership functions for the input process—i.e., triangular, Gaussian, bell-shapes, and sigmoidal (S-shaped) membership functions.

Membership function	Shape	Membership function	Shape
Triangular	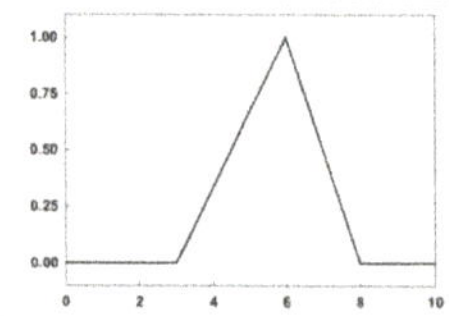	Gauss	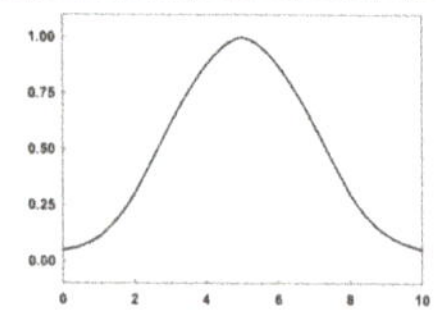
Bell	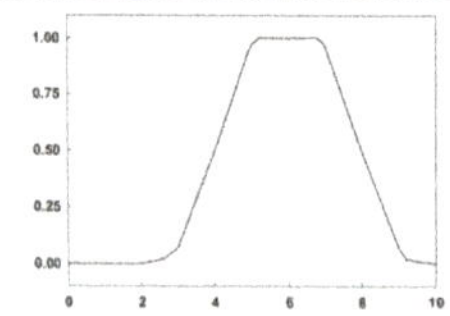	Sigmoidal	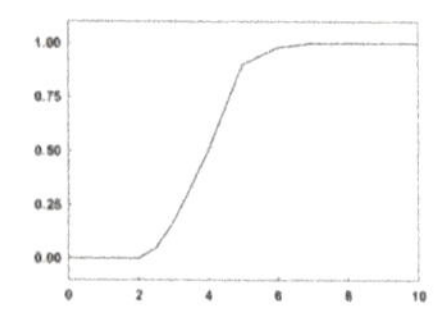

Figure 7. Shapes of fuzzy membership functions for input parameters.

Tables 4–7 show the various neuro-fuzzy analysis results to find the best combination of fuzzy membership functions for input variables. The success rate for each class, which is defined by the match with the soil type in the database, and the averaged success rates per each combination are presented. As shown in the tables, the success rate generally ranged from 70% to 79%. Among the results, the maximum success rate was 79.09% when the triangular membership function, Gaussian membership function, and sigmoidal membership function were selected as the membership functions for q_t, R_f, and B_q respectively. The optimized shapes of each membership function after training are shown in Figure 8.

Table 4. Success rate when using a triangular membership function for q_t.

Selected Fuzzy Membership Functions			Success Rate (%)			
q_t (MPa)	R_f	B_q	Clay	Silt	Sand	Average
Triangular	Triangular	Triangular	99.88	60.31	75.61	78.60
	Triangular	Gaussian	99.43	57.19	72.76	76.46
	Triangular	Bell	99.58	55.31	76.83	77.24
	Triangular	Sigmoidal	99.63	57.81	75.61	77.68
	Gaussian	Triangular	99.85	55.06	80.49	78.80
	Gaussian	Gaussian	99.43	57.19	73.58	76.73
	Gaussian	Bell	99.63	54.38	77.24	77.08
	Gaussian	Sigmoidal	99.58	57.19	80.49	79.09
	Bell	Triangular	99.38	58.44	73.98	77.27
	Bell	Gaussian	99.48	55.94	75.61	77.01
	Bell	Bell	99.63	54.06	78.46	77.38
	Bell	Sigmoidal	99.60	55.94	80.89	78.81
	Sigmoidal	Triangular	99.43	57.5	73.98	76.97
	Sigmoidal	Gaussian	99.48	56.25	75.61	77.11
	Sigmoidal	Bell	99.58	53.13	77.24	76.65
	Sigmoidal	Sigmoidal	99.58	55.31	80.89	78.59

Table 5. Success rate when using a Gaussian membership function for q_t.

Selected Fuzzy Membership Functions			Success Rate (%)			
q_t (MPa)	R_f	B_q	Clay	Silt	Sand	Average
Gauss	Triangular	Triangular	99.60	57.19	73.98	76.92
	Triangular	Gaussian	99.65	56.25	73.98	76.63
	Triangular	Bell	99.43	51.25	72.36	74.35
	Triangular	Sigmoidal	99.48	57.19	71.95	76.21
	Gaussian	Triangular	Bad	Bad	Bad	Bad
	Gaussian	Gaussian	99.65	57.50	76.42	77.86
	Gaussian	Bell	99.43	52.50	76.02	75.98
	Gaussian	Sigmoidal	99.48	55.63	78.05	77.72
	Bell	Triangular	99.13	54.69	72.36	75.39
	Bell	Gaussian	99.65	57.50	76.83	77.99
	Bell	Bell	99.43	52.50	74.80	75.58
	Bell	Sigmoidal	99.50	55.63	78.05	77.73
	Sigmoidal	Triangular	99.13	55.63	67.07	73.94
	Sigmoidal	Gaussian	99.58	55.63	76.02	77.08
	Sigmoidal	Bell	99.48	51.88	72.76	74.71
	Sigmoidal	Sigmoidal	99.50	55.31	76.42	77.08

Table 6. Success rate when using a bell-shaped membership function for q_t.

Selected Fuzzy Membership Functions			Success Rate (%)			
q_t (MPa)	R_f	B_q	Clay	Silt	Sand	Average
Bell	Triangular	Triangular	98.93	53.13	63.41	71.82
	Triangular	Gaussian	99.60	55.00	72.76	75.79
	Triangular	Bell	99.50	53.13	73.58	75.40
	Triangular	Sigmoidal	99.43	57.19	69.51	75.38
	Gaussian	Triangular	Bad	Bad	Bad	Bad
	Gaussian	Gaussian	99.73	57.50	77.64	78.29
	Gaussian	Bell	99.58	55.31	78.46	77.78
	Gaussian	Sigmoidal	99.45	55.31	77.24	77.33
	Bell	Triangular	99.80	54.06	80.89	78.25
	Bell	Gaussian	99.60	55.94	75.61	77.05
	Bell	Bell	99.50	54.69	75.61	76.60
	Bell	Sigmoidal	99.48	54.38	76.42	76.76
	Sigmoidal	Triangular	98.98	55.63	65.04	73.22
	Sigmoidal	Gaussian	99.58	54.06	76.83	76.82
	Sigmoidal	Bell	99.45	50.31	71.14	73.63
	Sigmoidal	Sigmoidal	99.48	54.69	72.36	75.51

Table 7. Success rate when using a sigmoidal membership function for q_t.

Selected Fuzzy Membership Functions			Success Rate (%)			
q_t (MPa)	F_R	B_q	Clay	Silt	Sand	Average
Sigmoidal	Triangular	Triangular	99.50	59.38	74.39	77.76
	Triangular	Gaussian	99.63	52.81	73.17	75.20
	Triangular	Bell	99.58	47.81	75.20	74.20
	Triangular	Sigmoidal	99.53	53.75	75.20	76.16
	Gaussian	Triangular	99.50	60.31	76.42	78.74
	Gaussian	Gaussian	99.60	53.44	74.80	75.95
	Gaussian	Bell	99.58	51.88	76.42	75.96
	Gaussian	Sigmoidal	99.55	52.81	76.83	76.40
	Bell	Triangular	99.53	59.06	78.46	79.02
	Bell	Gaussian	99.55	53.44	74.39	75.79
	Bell	Bell	99.58	52.50	74.80	75.63
	Bell	Sigmoidal	99.55	52.81	75.61	75.99
	Sigmoidal	Triangular	99.55	59.06	78.05	78.89
	Sigmoidal	Gaussian	99.60	52.50	77.24	76.45
	Sigmoidal	Bell	99.55	53.75	74.80	76.03
	Sigmoidal	Sigmoidal	99.55	52.81	73.98	75.45

Input Variable	q_t (MPa)	R_f	B_q
Optimized shape of membership	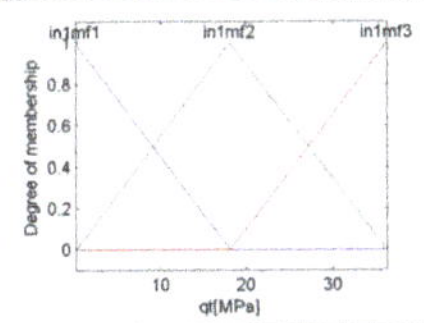	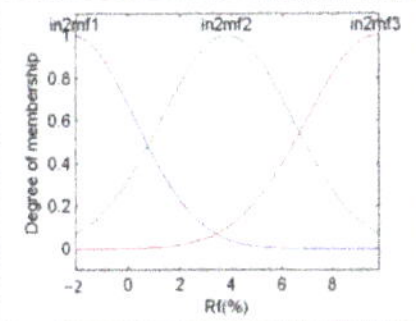	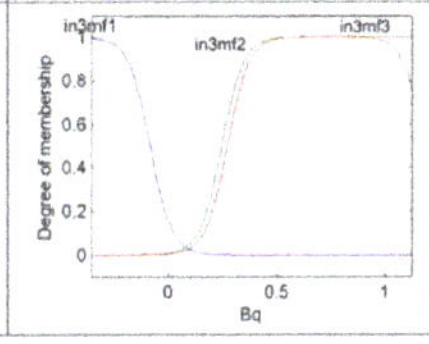

Figure 8. Optimized Shapes of fuzzy membership functions after training.

4. Verification of Suggested Neuro-Fuzzy Model

Verifications were performed with additional PCPT results which were not included in the training data, and the prediction results were compared to relevant boring logs. Piezocone tests for verification were carried out at Busan, Gyeongnam, and Jeonnam along the southern coast of Korea, Ulsan on the eastern coast of Korea, and Incheon on the western coast of Korea, as shown in Figure 4. Representative soil layers were three sites for clay, one for silt, and one for sand. For comparison, predictions from Pradhan's [6] and Zhang and Tumay's [7] methods using fuzzy theory and Robertson et al.'s chart [3] are also presented with the results of the newly suggested neuro-fuzzy model in this study. Soil classification results by Robertson et al. [3] were mainly related to soil behavior type, and their zones on the two charts were revised for the simplicity, i.e., clay for zone 3, silt for zones 4 and 5, sand for zones 8 and 9, and silt or sand for zones 6 and 7. Other zones not mentioned here are seldom found in South Korea (Kim et al., [20]). Indices for the results were as follows: 1 for clay, 2 for silt, and 3 for sand. In addition, zones 6 and 7 on Robertson's charts were marked as 2.5 and 0 for unclassified types.

4.1. Busan New Port Site

This site is located on the sea, and two PCPT penetrations were carried out. Soil layers were clay and a mixture of clay, sand, and gravel from the top of seabed. Thin silt lenses were found at the upper part of the deposit due to variations in seawater level. Laboratory test results from the undisturbed sample revealed the clay layer as highly compressible "CH". Water contents ranged from 51.5% to 75.3%, liquid limits ranged from 67.8% to 101.9%, and plastic limits ranged from 27.4% to 34.8%. Piezocone test results, boring logs, passing #200 sieve, water contents, and Atterberg limits are shown in Figures 9 and 10. From the results, Pradhan's method misclassified upper clay to some depth as silt, while Zhang and Tumay's method gave a better prediction of the narrow silt layer between clay at GL-16–17 m in Figure 9, but failed to predict the lower clay layer in Figure 10 as silt. Robertson's $q_t - B_q$ chart also provided satisfactory prediction when compared to boring logs, but had unclassified zones at the upper clay in both cases. On the other hand, the proposed neuro-fuzzy model from this study successfully classified the interbedded silt layer in Figure 9 and the clay layer in Figure 10. From the results, it was found that the proposed model provided more consistent classification with boring logs than others.

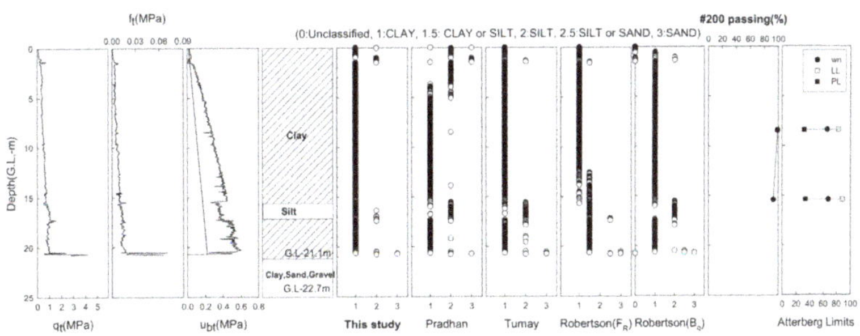

Figure 9. Verification results at Busan new port site-1.

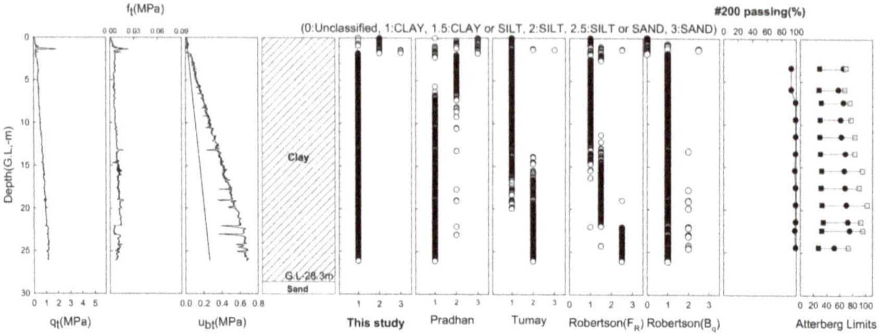

Figure 10. Verification results at Busan new port site-2.

4.2. Yangsan Site

According to the boring log, silty sand was distributed from surface to G.L-2.4 m and was layered by clay to G.L-11.7 m. A silt layer was interbedded in the clay layer at GL-5 m. A piezocone test was performed up to G.L-10.0 m. USCS results from undisturbed samples showed upper silty sand "CH" or "CL". Water contents ranged from 38.6% to 72.7%, liquid limits ranged from 36.0% to 68.6%, and plastic limits ranged from 17.0% to 26.9%. The classification results from every method are shown in Figure 11. Zhang and Tumay's method misclassified the silt layer and interbedded silt layer as clay due to the negative pore pressure measured at these layers. Kim et al. [20] reported this phenomenon whereby the precision of Zhang and Tumay's method is relatively low when negative pore pressure is measured because it does not incorporate the pore pressure index. Pradhan's method predicted the upper silty sand and mid-silty layer well, but failed to classify some upper clay right below the silty sand (circled zone) into silt. Robertson et al.'s estimation showed an unexpected result that it gave mainly unclassified points in the B_q chart. However, the proposed method succeeded in classifying the upper silty sand and mid-silty layer, as well as the clay layer.

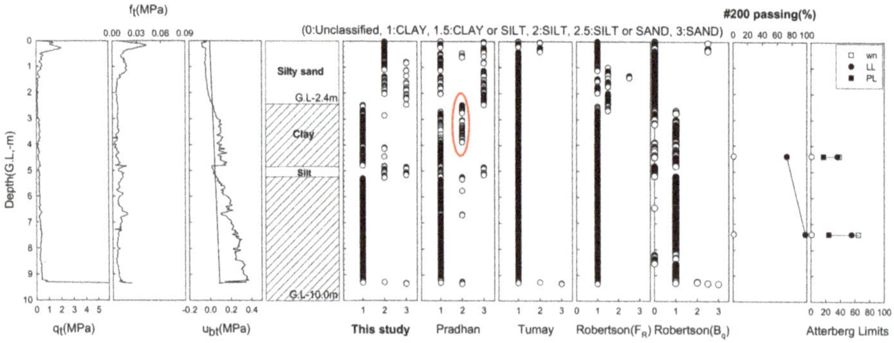

Figure 11. Verification results at Yangsan site.

4.3. Busan New Port Support Area Site

An 8 m thick intentional sand embankment above the clay layer was performed as a preloading for the purpose of accelerating consolidation as shown in Figure 12. Piezocone tests were performed to identify the bottom of the sand layer during the consolidation. Embanked sand was classified as "SM" from USCS. According to the results, Zhang and Tumay's method and the proposed neuro-fuzzy model yielded good agreement with the boring log. However, Robertson et al.'s classification from the B_q chart gave unclassified points from 3 m to 7 m, as seen for the Yangsan site. Pradhan's method also revealed low applicability when the pore pressure by PCPT was similar to hydrostatic pressure because its membership functions were derived from Robertson et al.'s charts.

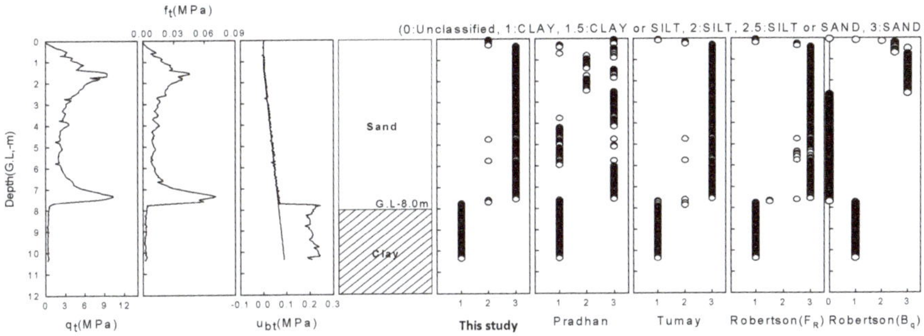

Figure 12. Verification results at Pusan new port support area site.

4.4. Jeonnam Dojang Port Site

This site is located on the southwestern coast of Korea on the sea. The top of the seabed was covered by silty sand up to G.L-1.8 m and layered by clay up to the depth of G.L-9.8 m as shown in Figure 13. A mixture of gravel and sand was distributed below clay. A casing tube was initially driven to G.L-0.5 m, and a piezocone test was carried out up to G.L-10 m. Soil classification results according to USCS from undisturbed samples in the clay layer were mainly "CH" or "CL". Water contents ranged from 29.59% to 65.27%, liquid limits ranged from 36.1% to 76.2%, and plastic limits ranged from 20.8% to 30.8%. Zhang and Tumay's method misclassified top silty sand mainly as clay because it did not consider negative pore pressure, as explained previously. Furthermore, Pradhan's method misclassified the upper part of "CH" or "CL" by USCS as silt. However, the neuro-fuzzy model classified clay and silty sand except at about 1 m thickness, but this may have occurred during the ground investigation considering the 1 m interval of SPT. The two charts proposed by Robertson et al. [3] failed to correctly detect silty sand. Moreover, the R_f chart and B_q chart gave different soil types in silty sand, which could confuse engineers.

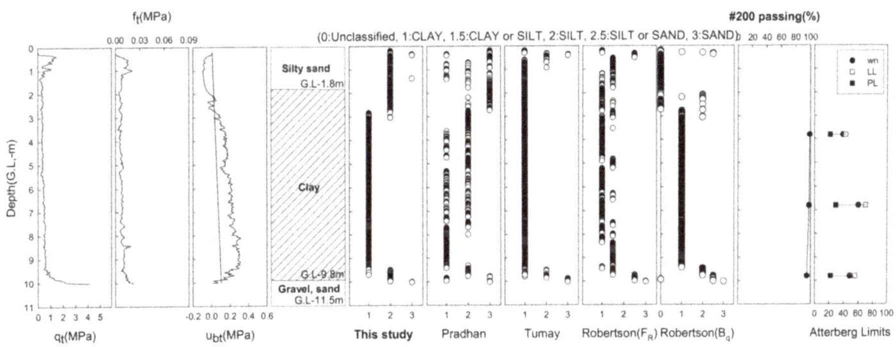

Figure 13. Verification results at Jeonnam Dojang new port site.

4.5. Incheon Trade Center Site

Incheon is located on the northwestern coast of Korea, where strong tidal action exists. The maximum tide difference is almost 9 m, and fine-grained soil is almost "ML". According to the boring log, fill reclaimed up to G.L-8.5 m and was layered by clayey silt up to G.L-19 m. Sandy silt and silty sand existed below the clayey silt. A piezocone test was performed to a depth of G.L-18.6 m through a fill into a casing tube. Soil classification results according to USCS from undisturbed samples in the clayey silt layer were mainly "ML" due to tidal action, except for "CL" at around G.L-17 m to 18 m. Water contents ranged from 28.2% to 32.8%, liquid limits ranged from 34.2% to 38.9%, and plastic limits ranged from 21.9% to 29.5%. According to the results shown in Figure 14, Zhang and Tumay's method uniformly estimated all layers as silt and sand. The B_q chart proposed by Robertson et al. [3] also classified clayey silt into silt or sand. Some misclassification was observed when considering the classification of USCS as "ML" or partially "CL". Pradhan's method estimated the mixture of clay, silt, and sand with depth and did not show any difference in detecting the thin clay layer at GL-17 m. The suggested neuro-fuzzy model also classified the clayey silt layer into mainly silt and succeeded in detecting the thin clay layer at around 17 m with "CL".

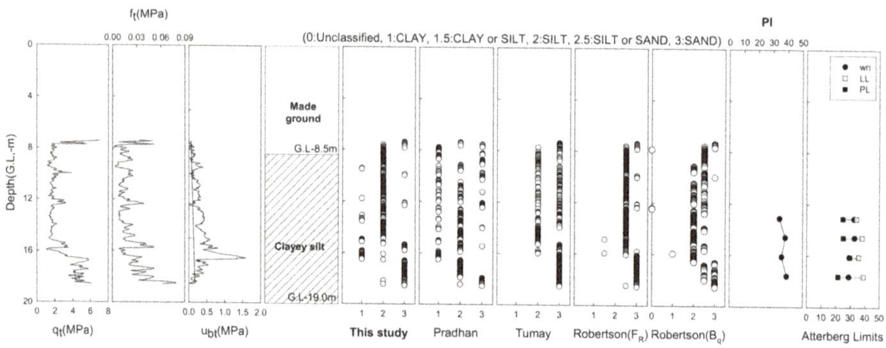

Figure 14. Verification results at Incheon trade center site.

4.6. Ulsan Southern Breakwater Site

This site is located on the southeastern coast of Korea, where the seawater depth is approximately 30 m. Clay was distributed to G.L-17.0 m and layered by sand or gravel. A casing tube was initially driven to G.L-3.7 m, and a piezocone test was carried out to G.L-16.7 m. The soil classification results of clay layer according to USCS from undisturbed samples were mainly "CH" through all depths, except for "CL" at G.L-16.0 to 16.8 m. Water contents of "CH" ranged from 70.1% to 90.8%, liquid limits ranged from 76.7% to 96.4%, and plastic limits ranged from 31.9% to 36.4%, whereas the water content of "CL" was about 38.1%, with a liquid limit of 41.2% and plastic limit of 21.9%. According to the classification results shown in Figure 15, all methods correctly classified the clay layer. However, Zhang and Tumay's method seemed to misclassify sand/gravel into silt since it did not appropriately reflect the variation of pore pressure, as mentioned above.

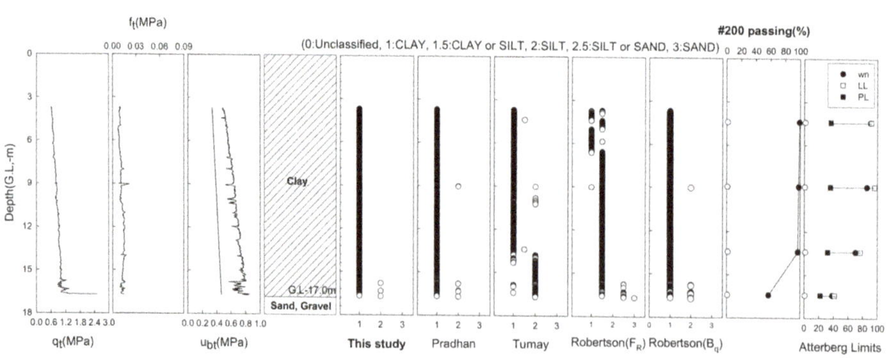

Figure 15. Verification results at Ulsan southern breakwater site.

5. Conclusions

A new soil classification system using FCM clustering and the neuro-fuzzy technique with piezocone test results was developed, and the main findings are summarized as follows:

(1) FCM clustering of the local database suggested that three input parameters of q_t, R_f, and B_q combined with three soil groups, i.e., clay, silt, and sand presented the highest success rate of 74.0%, and the m value representing the partition was optimized as 4.
(2) The neuro-fuzzy model was developed on the basis of the FCM clustering results with three input parameters q_t, R_f, and B_q and three output classes, i.e., clay, silt, and sand. The training procedure was performed with a total of 5173 data points using various combinations of fuzzy membership functions. As a result, a maximum success rate of 79.09% was shown when the triangular membership function for q_t, Gaussian membership function for R_f, and sigmoidal membership function for B_q were applied.
(3) Zhang and Tumay's method revealed low applicability when the penetration pore pressure by piezocone was negative or the same as the hydrostatic pressure since this method does not consider the pore pressure as an input parameter. The two charts presented by Robertson et al. sometimes failed to classify the upper reclaimed sand layer and interbedded sand or silt layer. Since Pradhan's method was adjusted to best match Robertson's diagram, both methods tended to yield essentially similar soil classification results. However, since Pradhan's method expressed a single soil classification using the overlapping fuzzy membership in Robertson's two diagrams and limited the maximum value of the fuzzy membership to 0.8, the soil classification result from Pradhan did not always match that of Robertson et al.
(4) The suggested neuro-fuzzy model matched well with boring logs and provided a better agreement with the classification in Korea. In addition, it has strong advantages in terms of revising or updating the model when the database is supplemented with new data.

Author Contributions: Conceptualization, Y.-S.K. and C.-H.K.; methodology, Y.-S.K. and C.-H.K.; validation, Y.-S.K. and J.-S.M.; investigation, C.-H.K.; writing—original draft preparation, Y.-S.K. and C.-H.K.; writing—review and editing, Y.-S.K. and J.-S.M.; supervision, Y.-S.K.; project administration, Y.-S.K.; funding acquisition, Y.-S.K. All authors have read and agreed to the published version of the manuscript.

Funding: This research was funded by the Ministry of Oceans and Fisheries, Korea, grant number 20180323.

Institutional Review Board Statement: Not applicable.

Informed Consent Statement: Not applicable.

Data Availability Statement: Not applicable.

Conflicts of Interest: The authors declare no conflict of interest.

References

1. Begemann, H.K. The friction jacket cone as an aid in determining the soil profile. In Proceedings of the 6th International Conference on Soil Mechanics and Foundation Engineering, Montreal, QC, Canada, 8–15 September 1965; Volume 1, pp. 17–20.
2. Douglas, B.J.; Olsen, R.S. Soil classification using electric cone penetrometer. Cone Penetration testing and Experience. In Proceedings of the ASCE National Convention, St. Louis, MO, USA, 26–30 October 1981; pp. 209–217.
3. Robertson, P.K.; Campanella, R.G.; Gillespie, D.; Greig, J. Use of Piezometer Cone Data. In Proceedings of the ASCE Specialty Conference In Situ '86: Use of In Situ Test in Geotechnical Engineering, Blacksburg, VA, USA, 23–25 June 1986; pp. 1263–1280.
4. Robertson, P.K. Soil classification using the cone penetration test. *Can. Geotech. J.* **1990**, *27*, 151–158. [CrossRef]
5. Jefferies, M.G.; Davies, M.P. Soil classification by the cone penetration test: Discussion. *Can. Geotech. J.* **1991**, *28*, 173–176. [CrossRef]
6. Pradhan, T.B.S. Soil identification using Piezocone Data by Fuzzy Method. *Soils Found.* **1998**, *38*, 255–262. [CrossRef]
7. Zhang, Z.; Tumay, M.T. Statistical to Fuzzy Approach toward CPT Soil Classification. *J. Geotech. Geoenvironmental Eng.* **1999**, *125*, 179–186. [CrossRef]

8. Hegazy, Y.A.; Mayne, P.W. Objective Site Characterization using Clustering of Piezocone Data. *J. Geotech. Geoenvirnomental Eng.* **2002**, *128*, 986–996. [CrossRef]
9. Tsiaousi, D.; Travasarou, T.; Drosos, V.; Ugalde, J.; Chacko, J. Machine Learning Applications for Site Characterization Based on CPT Data. In *Geotechnical Earthquake Engineering and Soil Dynamics V: Slope Stability and Landslides*; Laboratory Testing and In Situ Testing; American Society of Civil Engineers: Reston, VA, USA, 2018; pp. 461–472.
10. Reale, C.; Gavin, K.; Librić, L.; Jurić-Kaćunić, D. Automatic classification of fine-grained soils using CPT measurements and Artificial Neural Networks. *Adv. Eng. Inform.* **2018**, *36*, 207–215. [CrossRef]
11. Wang, H.; Wang, X.; Wellmann, J.F.; Liang, R.Y. A Bayesian unsupervised learning approach for identifying soil stratification using cone penetration data. *Can. Geotech. J.* **2019**, *56*, 1184–1205. [CrossRef]
12. Kurup, P.U.; Griffin, E. Prediction of Soil Composition from CPT Data Using General Regression Neural Network. *J. Comput. Civ. Eng.* **2006**, *20*, 281–289. [CrossRef]
13. Zhang, W.; Wu, C.; Zhong, H.; Li, Y.; Wang, L. Prediction of undrained shear strength using extreme gradient boosting and random forest based on Bayesian optimization. *Geosci. Front.* **2021**, *12*, 469–477. [CrossRef]
14. Rauter, S.; Tschuchnigg, F. CPT Data Interpretation Employing Different Machine Learning Techniques. *Geosciences* **2021**, *11*, 265. [CrossRef]
15. Oberhollenzer, S.; Premstaller, M.; Marte, R.; Tschuchnigg, F.; Erharter, G.H.; Marcher, T. Cone penetration test dataset Premstaller Geotechnik. *Data Brief* **2021**, *34*, 106618. [CrossRef] [PubMed]
16. Robertson, P.K. Interpretation of cone penetration tests—A unified approach. *Can. Geotech. J.* **2009**, *46*, 1337–1355. [CrossRef]
17. Robertson, P.K. Soil Behaviour Type from the CPT: An Update. In Proceedings of the 2nd International Symposium on Cone Penetration Testing, Huntington Beach, CA, USA, 9–11 May 2010.
18. Robertson, P. Cone penetration test (CPT)-based soil behaviour type (SBT) classification system—An update. *Can. Geotech. J.* **2016**, *53*, 1910–1927. [CrossRef]
19. Zadeh, L.A. Fuzzy sets. *Inf. Control* **1965**, *8*, 338–353. [CrossRef]
20. Kim, C.H.; Im, J.C.; Kim, Y.S. Study on Applicability of CPT Based Soil Classification Chart. *J. KSCE* **2008**, *28*, 293–301. (In Korean)

applied sciences

MDPI

Article

A Design of CGK-Based Granular Model Using Hierarchical Structure

Chan-Uk Yeom [1] and Keun-Chang Kwak [2,*]

[1] Center for IT-Bio Convergence System Agriculture, Chonnam University, Gwangju 61186, Korea; walt18@naver.com
[2] Department of Electronics Engineering IT-Bio Convergence System, Chosun University, Gwangju 61452, Korea
* Correspondence: kwak@chosun.ac.kr; Tel.: +82-062-230-6086

Abstract: In this paper, we propose context-based GK clustering and design a CGK-based granular model and a hierarchical CGK-based granular model. Existing fuzzy clustering generates clusters using Euclidean distances. However, there is a problem in that performance decreases when a cluster is created from data with strong nonlinearity. To improve this problem, GK clustering is used. GK clustering creates clusters using Mahalanobis distance. In this paper, we propose context-based GK (CGK) clustering, which adds a method that considers the output space in the existing GK clustering, to create a cluster that considers not only the input space but also the output space. there is. Based on the proposed CGK clustering, a CGK-based granular model and a hierarchical CGK-based granular model were designed. Since the output of the CGK-based granular model is in the form of a context, it has the advantage of verbally expressing the prediction result, and the CGK-based granular model with a hierarchical structure can generate high-dimensional information granules, so meaningful information with high abstraction value granules can be created. In order to verify the validity of the method proposed in this paper, as a result of conducting an experiment using the concrete compressive strength database, it was confirmed that the proposed methods showed superior performance than the existing granular models.

Keywords: granular model; incremental granular model; interval-based fuzzy c-means clustering; coverage; specificity; performance index

Citation: Yeom, C.-U.; Kwak, K.-C. A Design of CGK-Based Granular Model Using Hierarchical Structure. *Appl. Sci.* **2022**, *12*, 3154. https://doi.org/10.3390/app12063154

Academic Editor: Vincent A. Cicirello

Received: 17 February 2022
Accepted: 18 March 2022
Published: 19 March 2022

1. Introduction

In In the field of artificial intelligence, an inference engine is a system component that applies logical rules to a knowledge base to infer new information, where the first inference engines are expert systems. Conventional expert systems comprise knowledge bases and inference engines. A knowledge base stores information about the actual world, and an inference engine applies logical rules to the knowledge base and new inferred knowledge. In this process, each piece of new information in the knowledge base can generate additional rules from the inference engine. These expert systems include fuzzy inference systems. Fuzzy inference systems are the core units of fuzzy logic system, which perform decision making as a basic task and employ logical gates such as "OR", "AND", and "IF-THEN" rules to generate the required decision rules.

Fuzzy inference systems are broadly divided into Mamdani and Sugeno types. Mamdani-type inference systems are created by combining a series of language control rules obtained from experts, and the output of each rule has a fuzzy set form. Because they have an intuitive and easily understood rule base, they are suitable in fields that employ expert systems that are created from the expert knowledge of humans, such as medical diagnoses. Sugeno-type inference systems are also called Takagi-Sugeno-Kang inference systems, and they use single output membership functions, which are a form of linear function, of a constant or an input value. Sugeno-type inference systems include a defuzzification process, and rather than calculating the center of a 2D region, they adopt a weighted average or weighted sum of several data points; hence, they have the advantage of exhibiting a higher computational efficiency than Mamdani-type inference systems. These fuzzy inference systems are used in various forecasting fields and are actively being studied [1–9]. A previous study [10] proposed a Fuzzy convolutional neural network (F-CNN) that combines fuzzy inference with a CNN to predict traffic flow, which is a core part of predicting traffic volume. Yeom [11] proposed adaptive neuro-fuzzy inference system (ANFIS), which has an incremental structure and adopts context-based fuzzy clustering. Parsapoor [12] proposed brain emotional learning-based fuzzy inference system (BELFIS) to predict solar activity. Kannadasan [13] proposed an intelligent prediction model for predicting performance indices such as surface roughness and geometric tolerance in computer numerical control (CNC) operations, which plays an important role in machine product manufacturing. Guo [14] proposed a model called backpropagation-based (BP) kernel function Granger causality, which adopts symmetry geometry to embed dimensions and fuzzy inference systems for time-series predictions; in addition, this model was utilized to examine the causal relationships between brain regions. Hwang [15] proposed a motion cue-based fuzzy inference system to predict the normal walking speeds of sudden pedestrian movements at the initial walking stage when the heel is lifted.

Neural network expert systems are expert systems that mimic human intelligence by combining artificial neural networks (ANNs) and expert systems. In conventional expert systems, human inference methods are designed using decision trees and logical inferences, while ANNs focus on the structure and learning capacity of the human brain and reflect this in their knowledge expression. If these two systems are combined, the process of deriving results can be confirmed by the expert system, while learning can be performed by the ANN without user intervention. Accordingly, it is possible to create a system that is capable of more effective inferences than existing individual systems. The following studies on such neural network expert systems have been conducted [16–20]. Liu [21] proposed recurrent self-evolving fuzzy neural network (RSEFNN), which adopts online gradient descent learning rules to solve brainwave regression problems in brain dynamics, to predict driving fatigue. Dumas [22] proposed prediction neural network (PNN), which is based on fully connected neural networks and CNNs, and is used for internal image prediction. Lin [23] proposed an embedded backpropagation neural network comprising two hidden layers for earthquake magnitude prediction.

The aforementioned fuzzy inference systems and ANNs have different processes and solve various prediction problems. In addition, studies are being conducted on solving problems by combining two or more different methods, rather than using one method. Inference systems that combine different methods are called hybrid systems, and the granular computing (GrC) [24,25] method is adopted as a method for constructing hybrid systems. GrC is a computing theory related to the processing of information objects, called "information granules" (IG), that occur during the process of extracting knowledge from data and information, as well as abstractifying the data.

In the computing performed in general-used fuzzy inference systems, ANNs, and deep learning methods, the model output appears in a crisp form or as numbers. If the model output is in a crisp form or a number with a clear value, the numerical error relative to the actual output value can be calculated; however, difficulties occur when the difference between the model and actual output is expressed linguistically. However, in GrC, the model output is expressed in a soft form or as a fuzzy set; hence, GrC is effective at handling and processing data and information that are uncertain, incomplete, or with vague boundaries. In the actual world, people mainly use linguistic expressions rather than numerical expressions, and the brain, which makes inferences in uncertain and incomplete environments, utilizes linguistic values instead of numerical values to perform inferences and make decisions. Accordingly, GrC can represent the process by which humans think and make decisions. The following studies on GrC have been conducted [26–29]. Zhu [30] proposed a novel approach that develops and analyzes a granular input space and designed a granular model accordingly. Truong [31] proposed fuzzy possibilistic C-means (FPCM) clustering and a GrC method to solve anomalous value-detection problems. Zuo [32] proposed three types of granular fuzzy regression-domain adaptative methods, to apply GrC to transfer learning. Hu [33] proposed a method that adopts GrC to granularize fuzzy rule-based models and assess the proposed models. Zhao [34] made long-term predictions about energy systems in the steel industry by designing a granular model based on IGs created via fuzzy clustering. By analyzing the aforementioned research, it has become possible to create IGs that are generated via GrC, and to use these to design a granular model (GM), as well as calculate soft form output and express it linguistically. In addition, performance evaluation methods are proposed to evaluate the prediction performance of soft form output. However, studies are required to improve the prediction performance of granular models by creating optimal IGs, including methods for generating IGs and setting their form and size.

Conventional fuzzy clustering creates circle-shaped clusters starting at the cluster's center in the input space. However, when the input space's data exhibits geometric features, a problem emerges in which the clustering is not properly performed. To address this problem, Gustafuson-Kessel (GK) clustering is employed, as it can generate clusters while considering the geometric features of the data. This study proposes context-based GK (CGK) clustering, which considers both the input space and also the output space during existing GK clustering to generate geometrically-shaped clusters. This study also designed a CGK-based granular model that utilizes the proposed context-based GK clustering to generate context-shaped IGs in the output space and geometrically-shaped IGs in the input space. In addition, to resolve the problem of geometric increases in the numbers of rules when large amounts of data are adopted, this study proposes a CGK-based granular model with a hierarchical structure that combines the CKG-based granular model and the normal prediction model into an aggregate structure, such that meaningful rules can be generated. The remainder of this paper is organized as follows. Section 2 describes fuzzy clustering and GK clustering, while Section 3 describes IGs, existing granular models, the proposed context-based GK clustering, and the CGK-based granular model. Section 4 describes the hierarchical CGK-based granular model that is combined into an aggregate structure, and Section 5 verifies the validity of the proposed method by analyzing its performance using prediction-related benchmarking data. Finally, Section 6 presents this paper's conclusions and future research plans.

2. Data Clustering

Clustering is the task of placing data sets into clusters, such that the data in the same cluster are more mutually similar than data in other clusters. It is mainly used in data search and analysis, and as a data analysis method, it is adopted in various fields such as image analysis, bioinformatics, pattern recognition, and machine learning. Because the concept of clustering cannot be precisely defined, various clustering algorithms exist. These include connectivity based clustering, centroid based clustering, distribution based clustering, density based clustering, and grid based clustering, while a typical clustering method is fuzzy clustering.

2.1. Fuzzy Clustering

Fuzzy clustering is a method that was developed by Dunn and improved by Bezdek [35], which exhibits the feature of allowing the given data to belong to two or more clusters. In non-fuzzy clustering, the given data can only belong to exactly one cluster; hence, it is divided into separate clusters. In fuzzy clustering, data can belong to two or more clusters according to the membership values. For example, a banana can be yellow or green (non-fuzzy clustering criteria, or it can be yellow and green (fuzzy clustering criteria). Here, certain parts of the entire banana can be yellow, and they can be green. The banana can belong to green (green = 1), and it can belong to yellow (yellow = 0.5) and green (green = 0.5), which is not yellow (yellow = 0). The membership values can be between zero and one, while the sum of the membership values is 1. Membership values are assigned to the given data. These membership values numerically indicate the extent to which the data belongs to each cluster. If the data has a low membership value, it can be known that it is on the edge of the cluster; conversely, if it has a high membership value, it can be deduced that it is in the center part of the cluster.

Fuzzy clustering can be generalized by the following formulas.

$$J_m = \sum_{i=1}^{N} \sum_{k=1}^{c} u_{ki}^m d^2(x_i, v_k) \tag{1}$$

$$\sum_{k=1}^{c} u_{ki} = 1, \forall i \in \{1, 2, \ldots, N\} \tag{2}$$

where $X = \{x_1, x_2, \ldots, x_N\} \in R^{N \times D}$ and $x_i \in R^{1 \times D}$ represents the data and data items, respectively. N denotes the number of data items, and c is the number of clusters, which is $2 \leq c \leq N$. $u_{ki} \in R$ represents the membership value of the kth x_i, and $m \in Z^+$ is the fuzzification coefficient for the fuzzy membership values.

The cluster center and fuzzy membership function are obtained via an iterative process by minimizing Equation (1) according to the constraint conditions defined in Equation (2). Therefore, the objective function is modified using Lagrange multipliers and expressed as:

$$J_m = \sum_{i=1}^{N} \left(\sum_{k=1}^{c} u_{ki}{}^m d^2(x_i, v_k) + \lambda_i \left(1 - \sum_{k=1}^{c} u_{ki}\right)\right) \tag{3}$$

where λ_i denotes the Lagrange multiplier. Therefore, the clustering problem involves identifying the cluster center set $v^* = \{v_k{}^*, \forall k \in \{1, 2, \ldots, c\}\}$ and the fuzzy membership function set $U^* = \{u_{ki}{}^*, \forall k \in \{1, 2, \ldots, c\}, \forall i \in \{1, 2, \ldots, N\}\}$ by minimizing Equation (3). The minimization of cluster centers $v_k{}^*$ can be obtained via Equation (4), and the minimization of the fuzzy membership functions can be obtained using Equation (5), which are expressed as:

$$v_k{}^* = \frac{\sum_{i=1}^{N} u_{ki}{}^m x_i}{\sum_{i=1}^{N} u_{ki}{}^m} \tag{4}$$

$$u_{ki}{}^* = \frac{1}{\sum_{j=1}^{c} \left(\frac{d^2(x_i, v_k)}{d^2(x_i, v_j)}\right)^{\frac{1}{(m-1)}}} \tag{5}$$

Equations (4) and (5) are computed repeatedly to obtain the final cluster centers and fuzzy membership functions.

2.2. Fuzzy Clustering That Considers the Output Space

The aforementioned fuzzy clustering is a clustering that considers the features of the data in the input space. A fuzzy clustering that considers the output space generates clusters by considering both the features of the data in the input space and also the similarity and features of the data in the output space. This clustering type includes context-based fuzzy C-means (CFCM) clustering and interval-based fuzzy C-means (IFCM) clustering [36], which differ according to how the output space is divided. Figure 1. illustrates the fuzzy clustering that considers the output space. In Figure 1a. triangle-shaped contexts (fuzzy sets), which are IGs, are created in the output space, while clusters that correspond to each context are created in the input space. In Figure 1b interval-shaped IGs are created in the output space, and clusters that correspond to each interval are created in the input space.

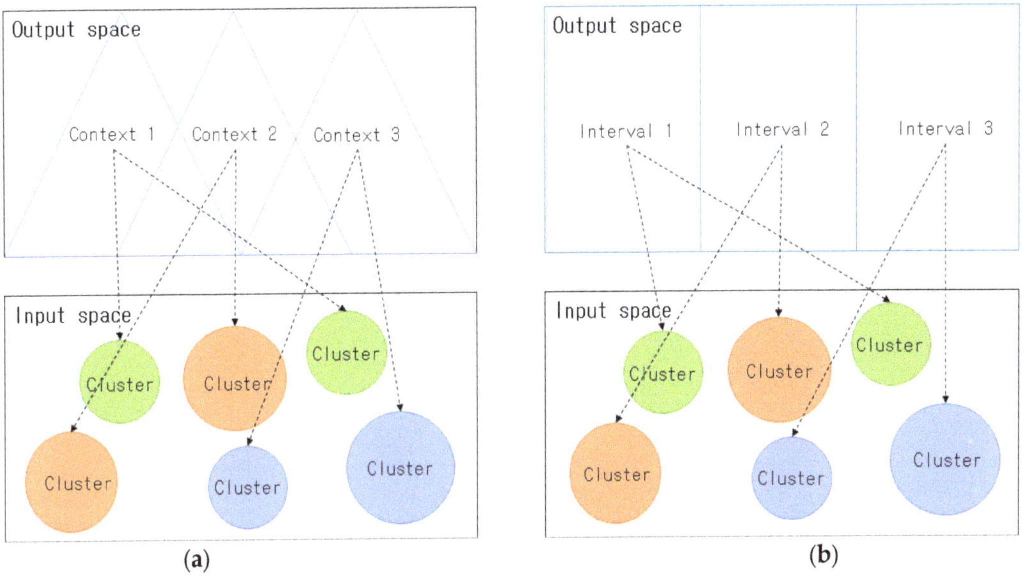

Figure 1. Context-based fuzzy clustering and interval-based fuzzy clustering concept: (**a**) context-based fuzzy clustering; (**b**) interval-based fuzzy clustering.

In normal fuzzy clustering, clusters are created using only the Euclidean distance between the cluster centers and the data in the input space, without considering the features of the data in the output space. However, in context-based fuzzy clustering, triangle-shaped contexts (fuzzy sets) are created in the output space using the method proposed by Pedrycz [37,38], while clusters are created via fuzzy clustering in each context; hence, clusters can be created in a more sophisticated manner than in conventional fuzzy clustering. Figure 2a presents normal fuzzy clustering, and Figure 2b shows clusters that were created in context-based fuzzy clustering by considering the features of the output space.

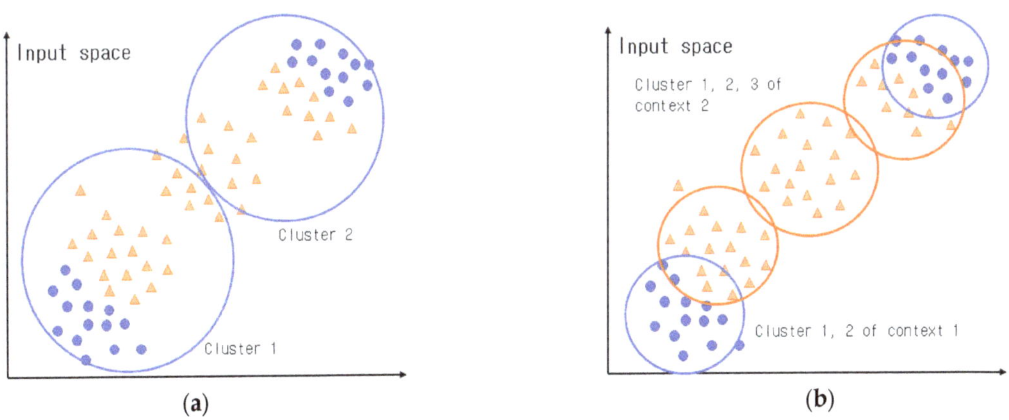

Figure 2. Comparison of clusters created in fuzzy clustering and context-based fuzzy clustering: (**a**) presents normal fuzzy clustering; (**b**) context-based fuzzy clustering.

As illustrated in Figure 2, fuzzy clustering creates clusters using the distance between the cluster centers and the data in the input space without considering the properties of the data in the output space. In contrast, context-based fuzzy clustering creates clusters by considering the properties of the data in the output space; hence, it can create clusters more efficiently than conventional fuzzy clustering.

The context of the data in the output space can be expressed as $D : T \rightarrow [0,\ 1]$. D represents all of the data in the output space. Here, it is assumed that the context for the given data can be adopted. $f_k = T(d_k)$ represents the extent to which the kth data belongs in the context created in the output space. f_k can be a value between zero and one, and the requirements for the membership matrix are as expressed in Equation (6) owing to the aforementioned properties.

$$U(f) = \left\{ eqalignu_{ik} \in [0,1] \mid \sum_{i=1}^{c} u_{ik} = f_k \,\forall\, k \text{ and } 0 < \sum_{k=1}^{N} u_{ik} < N \right\} \tag{6}$$

$$u_{ik} = \frac{f_k}{\sum_{j=1}^{c} \left(\frac{\|x_k - c_i\|}{\|x_k - c_j\|} \right)^{\frac{2}{m-1}}} \tag{7}$$

The membership matrix U updated by Equation (6) can be expressed as Equation (7). Here, m is the fuzzification coefficient, and $m = 2$ is generally used. For the contexts, the output space is uniformly divided into fuzzy set forms, while the degree of membership f_k is obtained. Usually, the output space is divided uniformly; however, it can be divided flexibly according to a Gaussian probability distribution according to the features of the data. The sequence in which the context-based fuzzy clustering is performed is presented below.

[Step 1] Select the number of contexts that can be expressed linguistically and the number of clusters that can be created in each context, and then initialize the membership matrix U with arbitrary values between zero and one. The numbers of the contexts and clusters can be set as the same number, or different values can be set by the user.

[Step 2] Divide the output space uniformly into fuzzy set forms and create fixed-sized contexts that can be expressed linguistically. In addition, a Gaussian probability distribution can be used to flexibly divide the output space and create contexts of different sizes.

[Step 3] Use Equation (8) to calculate the centers of the clusters in each context.

$$c_i = \frac{\sum_{k=1}^{N} {u_{ik}}^m x_k}{\sum_{k=1}^{N} {u_{ik}}^m} \tag{8}$$

[Step 4] Use Equations (9) and (10) to calculate the objective function. Here, the calculated value is compared to the previous objective function value, and the above process is repeated, provided it is greater than the threshold value that was set, or the process ends if it is less than the threshold value.

$$J = \sum_{i=1}^{c} \sum_{k=1}^{N} {u_{ik}}^m {d_{ik}}^2 \tag{9}$$

$$\left| J^h - J^{h-1} \right| \leq \varepsilon \tag{10}$$

where d_{ik} denotes the Euclidean distance between the kth data and ith cluster center, and h repress nets the number of iterations.

[Step 5] Equation (7) is adopted to update the membership function U, and Step 3 is performed.

2.3. GK Clustering

Regardless of the data in the input space belonging to a cluster, the cluster is normally determined by the distance between the data and the center of each cluster. As described in Section 1, fuzzy clustering adopts Euclidean distance to create clusters. Euclidean distance is primarily used when circle-shaped clusters are created, and it has the problem of being unable to create clusters that are not circle-shaped. To resolve this problem, GK clustering was proposed [39–41], as it can create geometrically-shaped clusters. GK clustering employs

Mahalanobis distance, rather than Euclidean distance, to calculate the distance between cluster centers and data. Figure 3 illustrates clusters that were created in fuzzy and GK clustering, and Equation (11) presents the Mahalanobis distance.

$$d_{GK}{}^2(x_k, v_i) = \| x_k - v_i \|_{A_i}{}^2 = (x_k - v_i)^T A_i (x_k - v_i) \tag{11}$$

where $d_{GK}{}^2$ denotes the square of the distance between the ith cluster's center v_i and the kth data x_k, while A_i is the variance matrix of the ith cluster. In GK clustering, Equation (12) is used to calculate the variance matrix A_i in Equation (11).

$$A_i = \frac{\sum_{k=1}^{N} u_{ik}{}^m (x_k - v_i)(x_k - v_i)^T}{\sum_{k=1}^{N} u_{ik}{}^m} \tag{12}$$

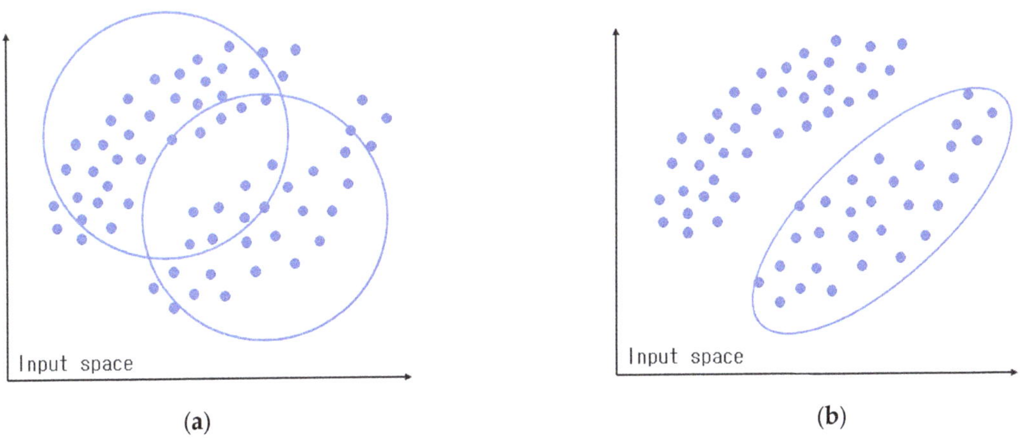

Figure 3. Comparison of clusters created with fuzzy and GK clustering: (**a**) fuzzy clustering; (**b**) GK clustering.

The variance matrix that is calculated using Equation (12) is adopted when calculating the distance between the cluster center and the data in Equation (13):

$$D_{GK}{}^2 = (x_k - v_i)^T [\rho_i \, det(A_i)^{\frac{1}{N}} A_i{}^{-1}](x_k - v_i) \tag{13}$$

where ρ_i denotes the volume of each cluster. When A_i is calculated in Equation (13), the matrix may become zero if the number of data is insufficient; hence, the minimum value is limited using Equation (14).

$$(1 - \gamma)A_i + \gamma \, det(A_i)^{\frac{1}{N}} I \rightarrow A_i \tag{14}$$

where A_i is the variance matrix that is calculated using all data, while I and γ denote the unit matrix and weight value constant, respectively. The eigen value and eigen vector can be calculated from the variance matrix. The calculated maximum eigen value is used to limit the minimum eigen value, such that the shape of the cluster can be maintained geometrically.

3. IG Creation and Granular Model Design

3.1. Creating Rational IG

Computing and inferences in GrC are centered on IGs, which are considered fundamental concepts and algorithms, rather than being centered on numbers. IGs are a core element in GrC because they play an important role in knowledge representation and processing [42–44]. Although IGs created using various types of clustering are relatively limited, they can reflect the general structure of some original data. Original data comprising numbers cannot depict the features and connections in the data, but IGs make this possible. Rational IG creation is focused on using the original data to create meaningful IGs. To create rational IGs, two requirements must be satisfied: coverage and specificity. Figure 4. Presents the coverage and specificity in IGs.

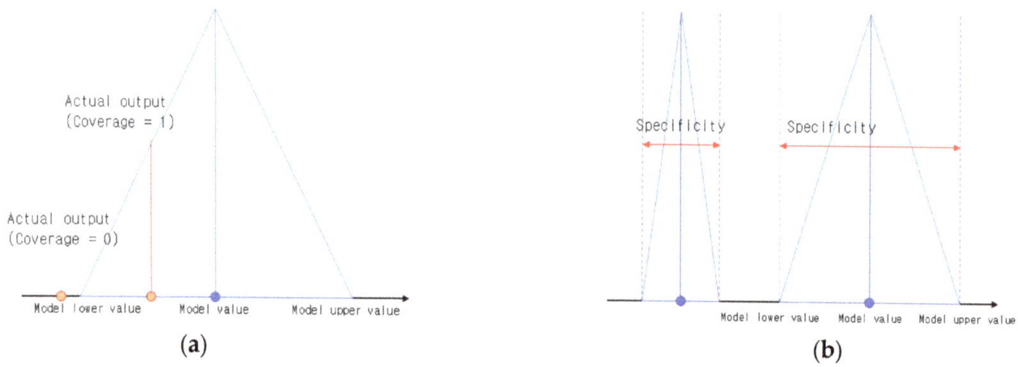

Figure 4. Concepts of coverage and specificity for creating rational IGs: (**a**) concept of coverage; (**b**) concept of specificity.

Coverage refers to whether the target data is included in the formed IG. In other words, it shows how much of the overall target data has accumulated within the IG's range, including the extent of the accumulation. The more data that accumulates in the IG, the higher the coverage value. This can verify the validity of the IG, and the model may be better in terms of modeling functions. incl, which is the degree of inclusion and is specified according to the form in which the IG Y_k is created. When Y_k is in context form, incl has a value close to one when y_k is included in $Y_k = [y_k^-, y_k^+]$, and it has a value close to zero when it is not included. In other words, coverage can be adopted to count the number that includes the data y_k in the granularized output of the granular model, while an average value can be calculated for all data. Ideally, the coverage has a value that is close to one, and all data is included in the granular model's output.

$$Coverage = \frac{1}{N}\sum_{k=1}^{N} incl\ (y_k, Y_k) \tag{15}$$

Specificity represents how specifically and semantically the IG Y_k can be described. In general, the specificity of a given IG Y_k must satisfy Equation (16). In other words, the IG must be created with as much detail as possible, and each IG must have a meaning that can be described. When an IG is in context form, the specificity becomes higher as the interval, i.e., the distance between the upper and lower bounds, becomes narrower. If the IG Y_k is reduced to point form, the specificity arrives at a value close to one.

$$\text{if } Y_k \subset Y_k' \text{ then } specificity(Y_k) \succ specificity(Y_k'), \text{ and } specificity(\{y\}) = 1 \tag{16}$$

$$Specificity = \frac{1}{N}\sum_{k=1}^{N} exp\left(-\left|y_k^+ - y_k^-\right|\right) \tag{17}$$

A continuous decreasing function of the interval length can be considered instead of the exponential function used in Equation (17). Coverage and specificity can be adopted to evaluate the IG's validity and the granular model's prediction performance. In other words, the granular model can be evaluated by considering the coverage and specificity of the IG, and a method that can simultaneously maximize coverage and specificity should be determined. These two properties have a trade-off relationship. This implies that the higher the coverage value, the lower the specificity value. Rational IGs can be represented by Equation (18), and this is called the PI.

The PI plays an important role in evaluating the model's accuracy and clarity, and various methods for evaluating model performance have been developed. General performance evaluation methods include root-mean-square error (RMSE) and mean absolute percentage error (MAPE). RMSE evaluates performance by subtracting the model's predicted values from the actual predicted values, calculating the mean of the squares, and squaring the obtained value. MAPE evaluates performance by subtracting the model's predicted value from the actual output value and dividing by the model's predicted value. These performance evaluation methods are mainly used when the model's output value is a numerical value. However, in the case of granular models comprise IGs, the model output is not a numerical value but an IG; hence, it is difficult to evaluate the model using general performance evaluation methods. To address this issue, studies are actively being conducted on adopting coverage and specificity as performance evaluation methods for granular models [45–49]. The higher the PI, the more meaningful the IG, and granular models with excellent performance can be designed.

$$Performance\ index = coverage(\varepsilon) \cdot specificity(\varepsilon) \tag{18}$$

The PI value obtained from a granular model can be adopted to represent the relationship between coverage and specificity as coordinates, and the changes in model performance, which are related to changes in the PI value, can be observed. Figure 5 illustrates the trade-off relationship between coverage and specificity. If coverage approaches zero, specificity approaches one, and the shape of the IG approaches a point. It can be observed that as the coverage increases, the size of the IG increases, but the specificity decreases.

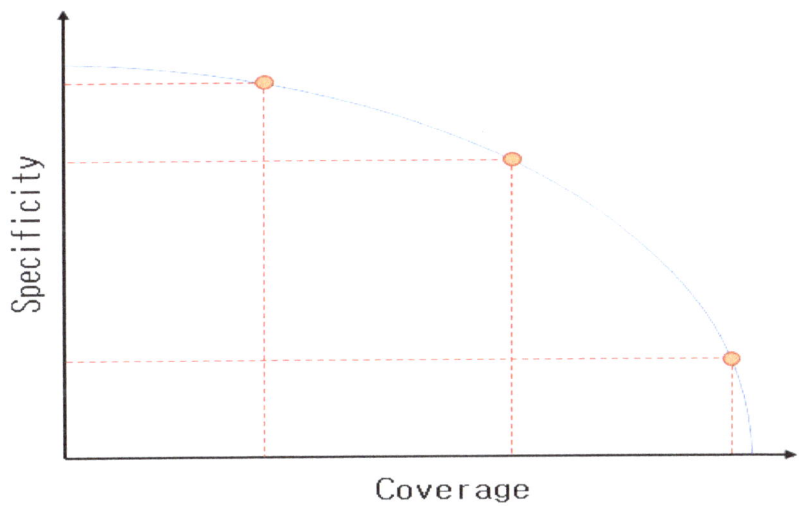

Figure 5. Trade-off relationship between IG coverage and specificity.

3.2. Fuzzy-Based Granular Model

Because the inference values of fuzzy rule-based inference systems used in various real-world fields of application are numeric values, there are limitations to describing these results linguistically. Fuzzy granular models, which are designed based on IGs that are created using fuzzy clustering, can express and process knowledge because their output values are IGs. Fuzzy granular models are created by granularizing a predetermined level of information in the data included in A. Owing to the granular properties of the data, granularized output is created from the variables of an existing fuzzy model with numerical input and output. This is based on the rational IG creation method described in Section 3.2. The IGs used in the fuzzy granular model exhibit the shapes of the fuzzy sets. The IG's level of granularization is assumed to be $\varepsilon(\varepsilon \in [0, 1])$. The granularization level creates the IG a_{i0} with a fuzzy set shape by allowing IGs of the given level $\varepsilon(\varepsilon \in [0, 1])$, which can be described as shown below, due to $a_{i0}, a_{i1}, a_{i2}, \ldots, a_{iN}$, which represent the data in each rule's output space.

$$G(a_{i0}) = [min(a_{i0}(1 - e), a_{i0}(1 + e)), max(a_{i0}(1 - e), a_{i0}(1 + e))] = [a_{i0}{}^{+}, a_{i0}{}^{-}] = A_{i0} \quad (19)$$

Using the same method, the IGs $A_{i1}, A_{i2}, \ldots, A_{iN}$ are created by granularizing $a_{i1}, a_{i2}, \ldots, a_{iN}$, which represent the data in the output space. A general fuzzy granular model divides the output space uniformly to create triangle-shaped contexts and clusters in each context. The fuzzy granular model's output value Y is expressed in context form, and each fuzzy rule regarding the input x_k creates the following IG output:

$$\text{if } x_k \in \Omega_i, \text{ then } Y_{ik} = f_i(x_k, A_i) = A_{i0} \oplus A_{i1} \otimes x_{k1} \oplus A_{i2} \otimes x_{k2} \oplus \ldots \oplus A_{iN} \otimes x_{kN} \quad (20)$$

The following method is used to calculate Y_k, which is the IG output in context form that was created based on all fuzzy rules.

$$Y_k = \sum_{i=1}^{c} \Omega_i(x_k) \oplus Y_k \quad (21)$$

where $\oplus$, $\otimes$ represent the completed addition and multiplication operations for each IG, respectively. Figure 6. Presents the structure of the fuzzy granular model.

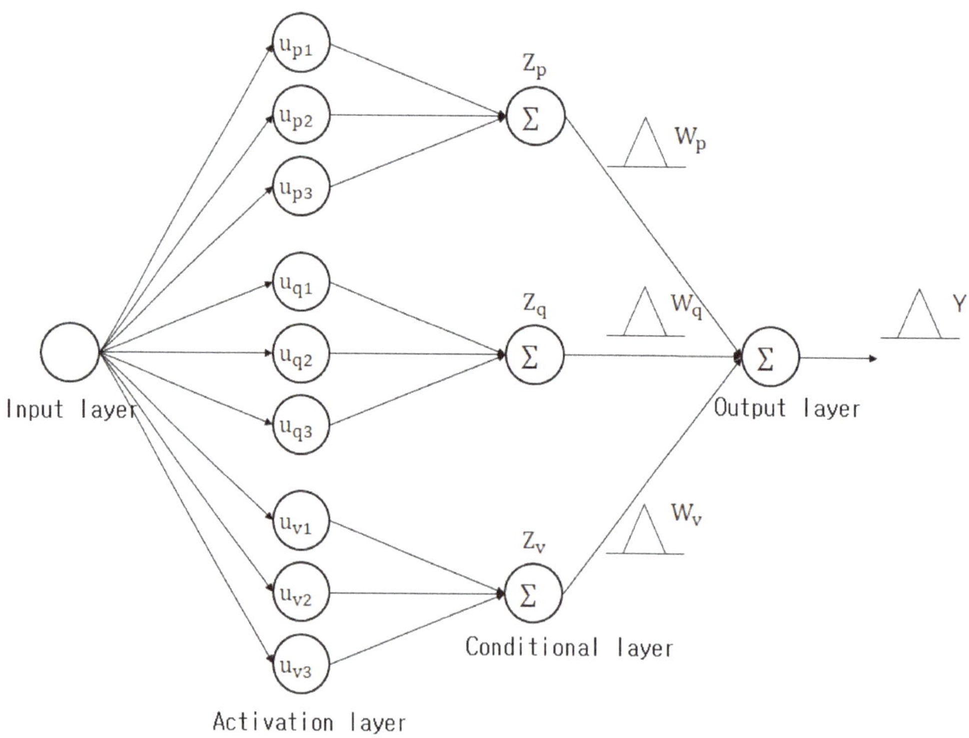

Figure 6. Structure of a fuzzy-based granular model.

3.3. CGK Clustering

CGK clustering is a clustering method that considers the output space. It creates clusters based on the correlations between the data in the input and output spaces by considering the output space in conventional GK clustering. It is assumed that there are data with two features. The data above can be depicted in red and blue according to the dependent variable. Figure 7. Presents the data with two features.

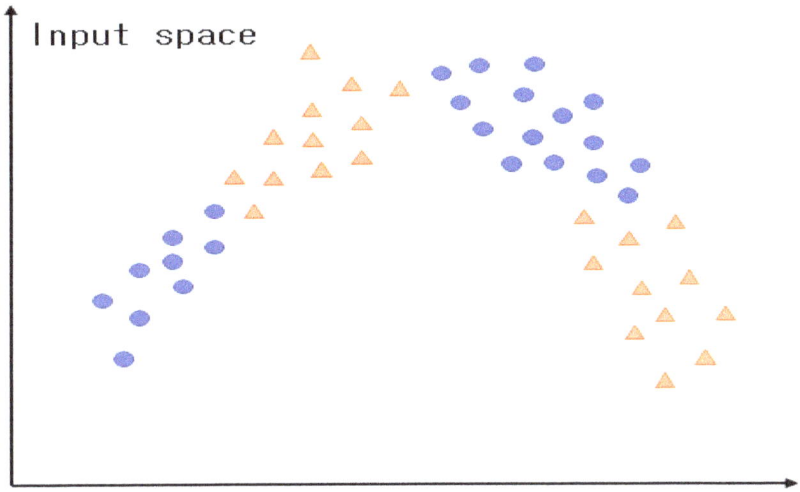

Figure 7. Data set with two features in the output space.

Figure 8a presents clusters created via normal GK clustering. In Figure 8a, it can be observed that the features of the data in the input space were considered when creating the clusters; however, the features of the output space were not considered. Figure 8b presents clusters created via CGK clustering that consider the output space. As illustrated in this figure, clusters are created by considering both the input and output spaces; hence, the features of the data in the output space can be preserved, and more efficient clusters can be created than in normal GK clustering. Figure 9. Illustrates the concept of CGK that considers the output space.

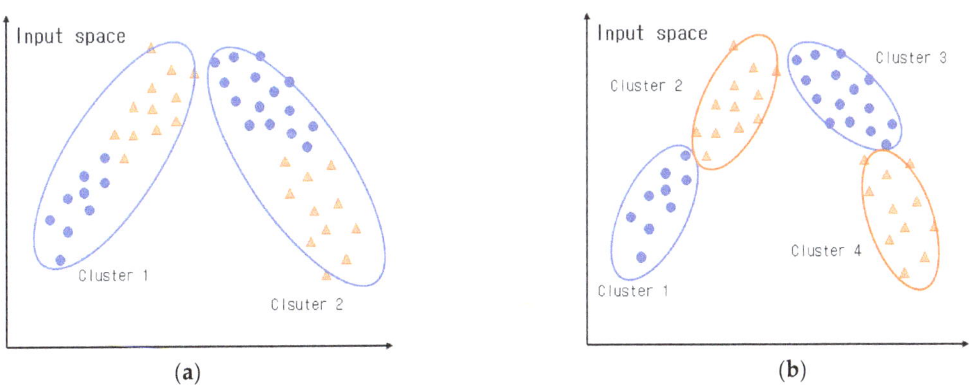

Figure 8. Comparison of clusters created in GK and CGK clustering: (**a**) general GK clustering; (**b**) context-based GK clustering.

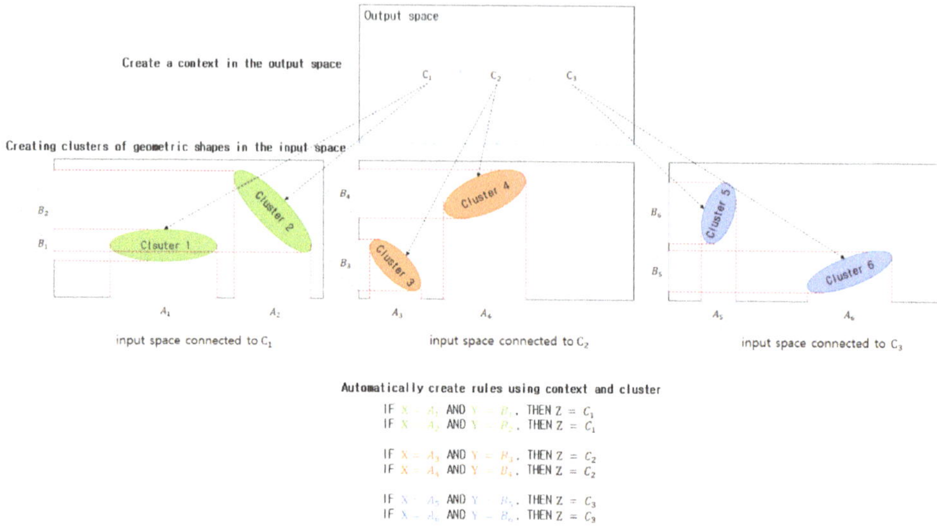

Figure 9. CGK clustering concept.

The context regarding the data in the output space can be expressed as expressed in Equation (22). Here, D denotes the data in the output space. If it is assumed that a context-shaped IG is adopted for the given data in the output space, $f_k = T(d_k)$ represents the degree to which the context created in the output space belongs to the kth data.

$$D : T \rightarrow [0,\ 1] \tag{22}$$

Fuzzy clustering adopts Euclidean distance to create clusters, while GK clustering improves upon this by creating clusters with Mahalanobis distance using Equation (11).

Where A_i is a matrix with $det(A_i) = \rho_i$, which is a fixed constant for each i. Because fuzzy clustering uses Euclidean distance, it exhibits excellent performance for only problems that create circle-shaped clusters. To circumvent this disadvantage, GK clustering adopts $d_{GK}{}^2(x_k,\ v_i)$ to extend the Euclidean distance of fuzzy clustering, such that clusters with various geometric shapes can be created, and it allows the distance standard to adapt to local areas. The objective functions are expressed in Equations (23)–(25).

$$J_m{}^{GK}(\mu,\ v) = \sum_{k=1}^{n} \sum_{i=1}^{c} \mu_{ik}{}^m d_{GK}{}^2(x_k,\ v_i) \tag{23}$$

$$v_i = \frac{\sum_{k=1}^{n} \mu_{ik}{}^m x_k}{\sum_{k=1}^{n} \mu_{ik}{}^m} \tag{24}$$

$$\mu_{ik} = \frac{\| x_k - v_i \|_{A_i}{}^{\frac{-2}{(m-1)}}}{\sum_{j=1}^{c} \| x_k - v_j \|_{A_j}{}^{\frac{-2}{(m-1)}}} \tag{25}$$

Equations (23)–(25) are repeated in each context generated in the output space to create geometrically-shaped clusters. Below is the sequence in which context-based GK clustering is performed.

[Step 1] The number of contexts that can be expressed linguistically and the number of clusters that are created in each context are selected, as well as ε. Here, ε sets the degree of the geometric shape, and a value greater than zero must be selected. The membership function U is initialized with values between zero and one. The numbers of contexts and clusters can be set to be the same, or they can be set differently.

[Step 2] Context-shaped IGs with fixed sizes can be created by uniformly dividing the output space, while context-shaped IGs with different sizes can be created by via a Gaussian probability distribution.

[Step 3] Equation (24) is adopted to calculate the centers of the clusters in the contexts in the output space and a membership matrix.

[Step 4] Equations (23) and (26) are adopted to calculate an objective function, and the aforementioned process is repeated if the calculated value is greater than the previous objective-function value. Conversely, if the calculated value is less than the previous objective-function value, the above process ends.

$$\| \mu^t - \mu^{t-1} \| \le \varepsilon \tag{26}$$

3.4. CGK-Based Granular Model Design

GK granular models are designed to adopt CGK clustering that considers the output space, to create context-shaped IGs in the output space and create geometrically-shaped clusters in each context. Figure 10. presents the structure of a GK granular model in which three contexts are created in the output space and three clusters are created in each context. As illustrated in the figure, there are conditional and conclusion variables. The conclusion variables represent the context-shaped IGs that are created in the output space, while the conditional variables represent the centers of the clusters that are created in each context, i.e., IGs that are created in the input space. As mentioned above, a uniform creation method and a flexible creation method can be adopted to create the contexts in the output space. The GK granular model's final output value Y is calculated using Equation (27).

$$Y = \sum\nolimits_{\oplus} W_t \otimes z_t \tag{27}$$

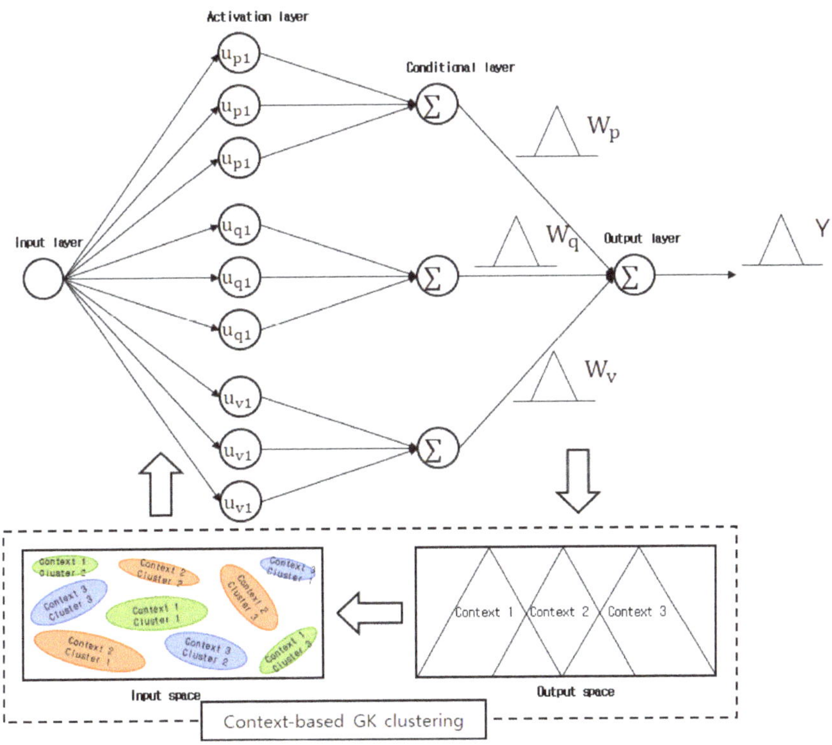

Figure 10. CGK-based granular model structure.

Here, the addition and multiplication symbols $\oplus$, $\otimes$ represent the completed addition and multiplication operations for the IGs, respectively. Fuzzy sets are created during the process of handling the GK granular model conditions. At this point, the clusters created via CGK clustering can be represented by the GK granular model's hidden layer. The area between the hidden and output layers is expressed as a context that can be described linguistically. The sum, which is the GK granular model's final output, can be expressed using all contexts as expressed in Equation (28):

$$Y = (z_{11} \otimes A_1 \oplus z_{12} \otimes A_1 \oplus \ldots \oplus z_{1n1} \otimes A_1) \oplus (z_{21} \otimes A_2 \oplus z_{22} \otimes A_2 \oplus \ldots \oplus z_{2n2} \otimes A_2) \oplus \ldots (z_{c1} \otimes A_c \oplus z_{c2} \otimes A_c \oplus \ldots \oplus z_{cnc} \otimes Ac) \quad (28)$$

The GK granular model's final output is expressed as a triangle-shaped context, and it can be expressed as a fuzzy set:

$$Y_i = \left(y_i{}^{-},\, y_i,\, y_i{}^{+}\right) \quad (29)$$

where $y_i{}^{-}$, y_i, and $y_i{}^{+}$ denote the GK granular model's lower bound, model, and upper bound values, respectively, and they refer to each of the triangle-shaped context's points. The lower bound, model, and upper bound values can be expressed by Equations (30)–(32):

$$y_i{}^{-} = \left(z_{11}a_1 + z_{12}a_1{}^{-} + \ldots + z_{1n1}a_1{}^{-}\right) + \ldots \left(z_{c1}a_c{}^{-} + z_{c2}a_c{}^{-} + \ldots + z_{cnc}a_c{}^{-}\right) \quad (30)$$

$$y_i = \left(z_{11}a_1 + z_{12}a_1 + \ldots + z_{1n1}a_1\right) + \ldots \left(z_{c1}a_c + z_{c2}a_c + \ldots + z_{cnc}a_c\right) \quad (31)$$

$$y_i{}^{+} = \left(z_{11}a_1 + z_{12}a_1{}^{+} + \ldots + z_{1n1}a_1{}^{+}\right) + \ldots \left(z_{c1}a_c{}^{+} + z_{c2}a_c{}^{+} + \ldots + z_{cnc}a_c{}^{+}\right) \quad (32)$$

When CGK clustering is performed, the membership matrix U can be expressed as values between zero and 1, while the membership matrix's requirements can be expressed as:

$$U(f) = \left\{ eqalign u_{ik} \in [0,\, 1] \mid \sum_{i=1}^{c} u_{ik} = f_k \; \forall\, k \text{ and } 0 < \sum_{k=1}^{N} u_{ik} < N \right\} \quad (33)$$

Here, the contexts are created by uniformly or flexibly dividing the output space into fuzzy set shapes. The GK granular model's structure is as follows. In the input layer, data is received and enters the GK granular model. The activation layer is the cluster activation step in which clusters that correspond to the contexts that were created in the output space are created in the input space. The conditional layer performs conditional clustering in each context. The activation and conditional layers are connected, and the data information is adopted in GK clustering when a context is provided. The GK granular model is focused on the activation and conditional layers. The contexts are connected to the GK clustering in the conditional layer, and fuzzy sets are created by considering the features of the data in the input space. A specified number of clusters is created in each context, and the total number of nodes in the output layer is the same as the number of contexts. The final output values that are added up in the output layer are represented as a triangle-shaped context.

4. Granular Model Design with a Hierarchical Structure

4.1. CGK-Based Granular Model Design with a Hierarchical Structure

As the number of input variables for a fuzzy system and granular model increase, the number of rules increase geometrically. Large rule bases have the problem of reducing the computation efficiency of fuzzy systems and granular models. In addition, they make it difficult to understand the action of granular models, and complicate the adjustment of rules and membership functions. The possibility of generalizing fuzzy systems and granular models with large rule bases is minimized because various prediction-related fields of application provide limited amounts of data. To resolve these problems, rather than using a single fuzzy system and a single granular model, it is possible to design a granular model with a hierarchical structure in which these are mutually connected. Because the granular model is arranged in a hierarchical tree structure, the tree is called the hierarchical structure. The output of the low-level granular models in the hierarchical structure is adopted as the input for the high-level granular models. Granular models with hierarchical structures are computationally more efficient than single granular models with the same number of inputs, and are also designed with a simple structure [50–52].

Hierarchical structures that can be used in various prediction-related fields of application include incremental, aggregated, and cascaded structures. Figure 11. Presents each type of hierarchical structure. In incremental structures, input variables are combined in several stages, while output values are calculated at several levels. As illustrated in Figure 11, the granular model $GM_i{}^n$ is built with a 3-stage structure. Here, i is the nth level's granular model. In an incremental granular model, when i is one, it means that there is one fuzzy inference system on each level. The nth level's ith granular model's jth input is called $x_{ij}{}^n$, while the nth level's ith granular model's kth output is called $y_{ik}{}^n$. When the input variables on each of the levels of an incremental granular model are selected, their ranks are determined according to their degrees of contribution to the final output value. The input variable with the highest degree of contribution is usually used on the lowest level; conversely, the input variable with the lowest degree of contribution is adopted on the highest level. In other words, low-rank input values depend on high-rank input values.

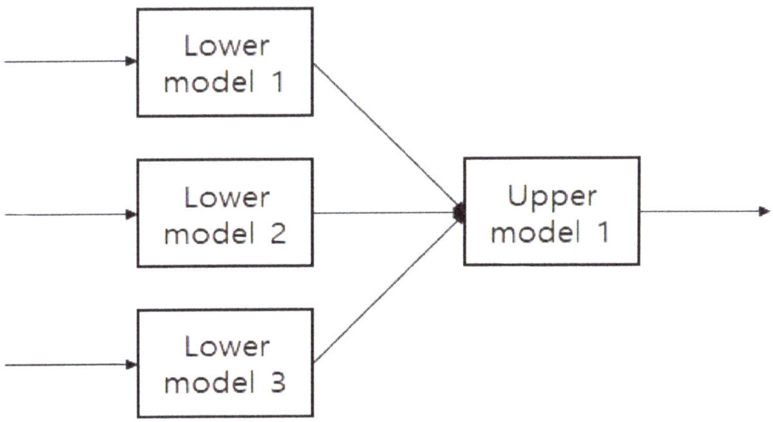

Figure 11. Granular model with a hierarchical structure.

In an aggregate structure, the original data's input variables are used on the lowest level, and the output of the low-level granular model that receives each input variable as the input is inputted in the high-level granular models, and the obtained results are combined. For example, the granular model $GM_{i_n}{}^n$ is built with two stages as, illustrated in Figure 11. i_n is the index of the granular model on the nth level. The input variables in aggregated granular models are grouped for performing specific decision making. For example, an autonomous robot's search task combines two tasks: searching while avoiding collisions with obstacles, and arriving at the goal. To perform the search task, the granular model adopts input variables related to obstacles. To perform the task of arriving at the goal, input variables related to the robot's current position and movement direction are employed. Aggregated granular models can be modified to design parallel aggregated granular models that directly add up the outputs of low-level granular models to calculate their final output.

A cascaded structure is a structure that combines the aforementioned incremental structure with the hierarchical structure, and it is suitable for systems that include both correlated input variables and non-correlated input variables. It has a form in which the correlated input variables are grouped into an aggregated structure and the non-correlated input variables are added as an incremental structure.

4.2. CGK-Based Granular Model Design with an Aggregated Structure

This paper presents a design for a granular model that adopts an aggregated structure. When an aggregated structure is built, rather than using low-level and high-level granular models, the low level comprises LR (LR) models, neural network models, and radial basis function networks, and each prediction model's output is adopts as the input for the high-level fuzzy granular model to calculate the final output.

An LR model [53] models the linear correlations between input and output variables. Figure 12 shows the concept of linear regression. Simple linear regression models are based on explanatory variables, while multiple linear regression models are based on two or more explanatory variables. Linear regression models estimate unknown parameters in the data. A linear regression model can be expressed as:

$$y_i = \beta_1 x_{i1} + \ldots + \beta_p x_{ip} + e_i = {x_i}^T \beta + e_i, \quad i = 1, 2, \ldots, n \tag{34}$$

where β_i and p denote each independent variable's coefficient and the number of parameters estimated by the linear regression model, respectively. T indicates transposition, while ${x_i}^T\beta$ represents the inner product of x_i and β. Furthermore, e_i is the error term, and it represents the error variables. This refers to the error between the dependent and independent variables.

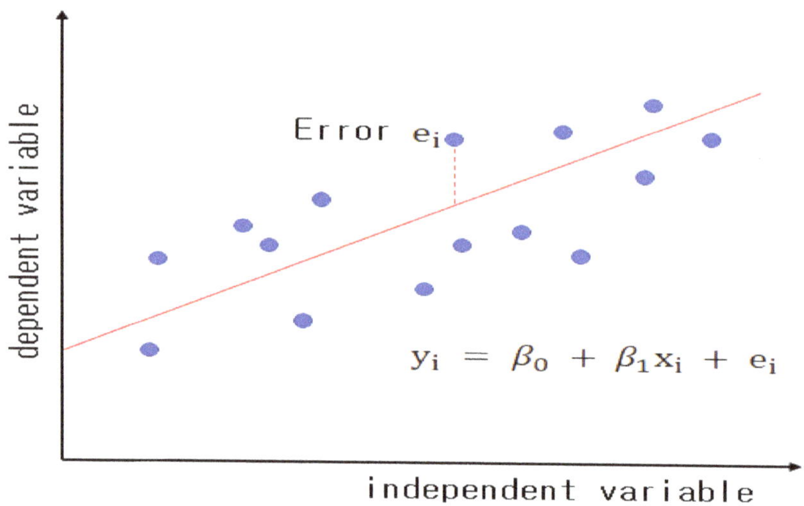

Figure 12. Linear regression model concept.

Neural networks [54] are algorithms that are created with inspiration from biological neural networks in cognitive science and machine learning. These models can solve problems by altering the strength of the connections between synapses via the learning performed by the nodes that constitute the neural network by combining synapses. Figure 13. Presents the structure of a simple neural network. A simple neural network consists of an input layer, hidden layer, and output layer. The input layer inputs the data's input variables into the neural network, and the number of input variables must equal the number of input layer nodes. Usually, no calculation is performed in the input layer, and the layer simply performs the role of passing the values on. The hidden layer is between the input and output layers. If there are two or more hidden layers, it is called a multi-layer neural network. The output layer calculates the neural network's output. To achieve this, it adopts an activation function that is suitable for the problem to be solved.

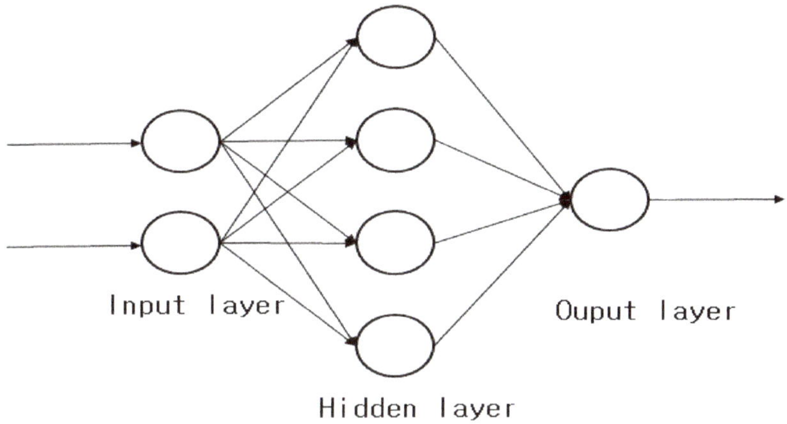

Figure 13. Neural network structure.

A radial basis function network [55,56] is a neural network that adopts a radial basis function, instead of a sigmoid function, as the activation function in a conventional neural network structure. Figure 14. Presents the structure of a radial basis function network. A radial basis function network has a simple structure because there is only one hidden layer and the form of the output is linear; therefore, weight value calculations can be performed efficiently.

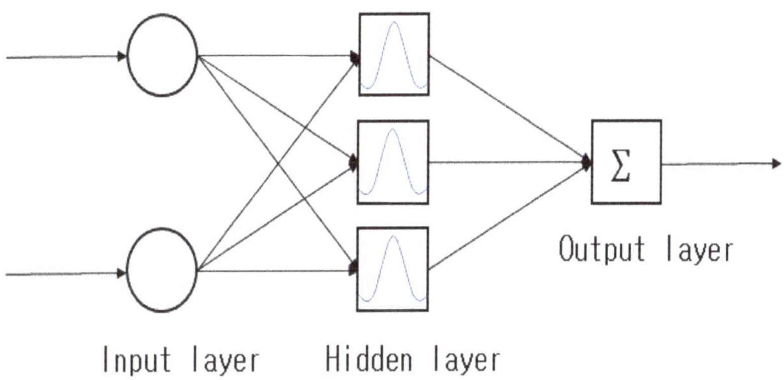

Figure 14. Radial basis function network structure.

The outputs from a linear regression model, neural network, and radial basis function network are combined and adopted as the input of a high-level fuzzy granular model. The high-level fuzzy granular model determines the number of contexts created in the output space and the number of clusters created in the input spaces, and then creates the IGs. Accordingly, the final output of the hierarchical structure is calculated. Existing fuzzy granular models are limited in creating meaningful IGs in the input and output spaces when processing large-scale data, and they have the problem of long computation times. In contrast, the fuzzy granular model with a hierarchical structure proposed in this study has the advantages of being able to create meaningful IGs from large-scale data and reduce processing times by taking the data created by combining the output from the low-level linear regression model, neural network, and radial basis function network, as well as adopting it as the input of the high-level fuzzy granular model. Here, if the clustering used by the granular model is context-based fuzzy clustering, the model is a fuzzy granular model with a hierarchical structure, and if the clustering is context-based GK clustering, the model is a GK granular model with a hierarchical structure. Figure 15. Presents the structure of a granular model with a hierarchical structure.

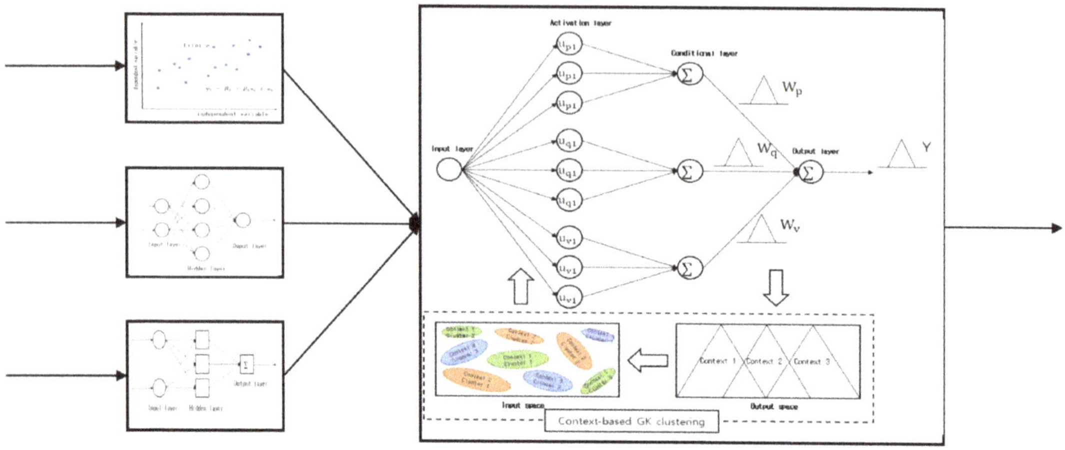

Figure 15. Structure of a CGK granular model with an aggregated structure.

5. Experiment and Results Analysis

To examine the validity of the CGK-based Granular Model (CGK-GM), which is the method proposed in this study, as well as the CGK granular model with an aggregated structure (AGM), the experiments were performed through a concrete compressive strength database [57] which are benchmarking databases used in the field of forecasting. For convenience in the experiments and the results analysis, the two proposed granular models are labeled CGK-GM, AGM. The databases used in the experiments are presented as follows. The concrete compressive strength database was collected by Tsinghua University in Taiwan, and it comprises 1030 instances and 9 variables. The input variables include the cement, fly ash, blast furnace slag, water, superplasticizer, coarse aggregate, fine aggregate, and time. The output variable is the concrete's compressive strength.

Experimental Method and Results

In this study, the prediction performance of the granular model was evaluated by the Performance Index (PI) method, which is a performance evaluation method that is suitable for IGs and granular models, rather than the generally used evaluation methods. As expressed in Equation (18). The experiment method is presented as follows. Each database was divided into 50% learning data and 50% validation data, normalized to values between zero and one and then used in the experiments. The numbers of contexts (P) and clusters (C) in conventional GM, the proposed CGK-GM, and AGM were increased from 2 to 6 in increments of 1 during the experiments, while the fuzzification coefficient was fixed at 2. In addition, the experiments were performed using the uniform and flexible method of creating contexts.

The following shows the results of the concrete compressive strength prediction experiment. Table 1 shows the prediction performance of the existing GM that created the context uniformly, and Table 2 shows the prediction performance of the GM that created the context flexibly. Figure 16 shows the output value and actual output value of the existing GM, and Figure 17 shows the performance index value of the existing GM for the verification data. In Figure 16, the x-axis represents the number of verification data for the concrete compressive strength, and the y-axis represents the concrete compressive strength value. The black solid line is the actual concrete compressive strength value, and the red dotted line shows the output value of the existing GM. As shown in the figure, it can be confirmed that the GM output value predicts a large change in the actual output value, but does not predict a small change. In Figure 17, the *x*-axis represents the number of clusters created in the input space, and the *y*-axis represents the number of contexts created in the output space. The *z*-axis shows the performance index values for the verification data. As shown in the figure, it can be seen that when the number of contexts is 6 and the number of clusters is 6 when the contexts are created equally, the performance index value is 0.4276, which is the best.

Table 1. Performance index of GM that created context evenly.

P \ C	2	3	4	5	6
2	0	0	0	0	0
3	0	0	0	0	0
4	0.3165	0.3175	0.3191	0.3184	0.3191
5	0.4039	0.4184	0.4223	0.4175	0.4204
6	0.4031	0.4136	0.4183	0.4159	0.4276

Table 2. Performance index of GM that created context flexibly.

P \ C	2	3	4	5	6
2	0	0	0	0	0
3	0.0097	0.0093	0.0097	0.0093	0.0090
4	0.2931	0.2947	0.2953	0.2945	0.2948
5	0.3897	0.3962	0.3962	0.3946	0.3960
6	0.4022	0.4120	0.4183	0.4212	0.4230

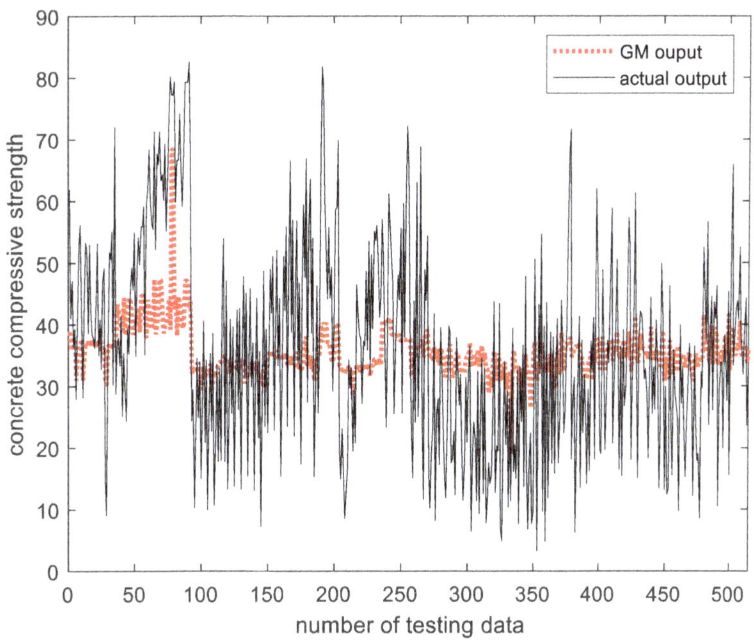

Figure 16. Comparison of the output value of the existing GM with the actual output value (context is created equally, the number of contexts = 6, the number of clusters = 6).

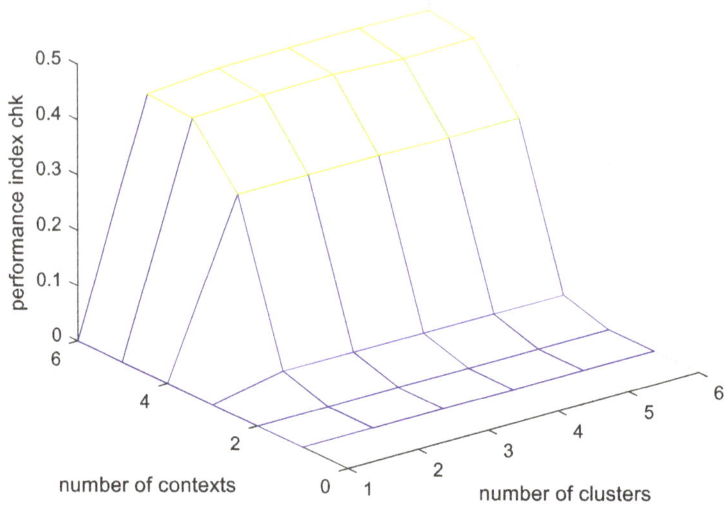

Figure 17. GM's performance index for validation data (context equally generated, number of contexts = 6, number of clusters = 6).

Table 3 lists the prediction performance of the CGK-GM in which the contexts are uniformly generated, and Table 4 lists the prediction performance of the CGK-GM in which the contexts are flexibly generated. Figure 18 shows the output and actual output values of CGK-GM, and Figure 19 shows the predictive performance of CGK-GM using performance indicators. As shown in Figure 18, it can be seen that the CGK-GM output value predicts only a large change in the actual output value, but predicts the actual output value more similarly than the conventional GM output value. Figure 19. Shows the performance index values of CGK-GM for the verification data. As shown in the figure, when the number of contexts is 6 and the number of clusters is 4 when the contexts are equally created, the performance index value is 0.4700, which is the best.

Table 3. Performance index of CGK-GM that created context evenly.

P \ C	2	3	4	5	6
2	0	0	0	0	0
3	0.0204	0.0204	0.0204	0.0205	0.0206
4	0.3328	0.3315	0.3308	0.3295	0.3315
5	0.4409	0.4379	0.4350	0.4350	0.4300
6	0.4618	0.4629	0.4700	0.4618	0.4606

Table 4. Performance index of CGK-GM that created context flexibly.

P \ C	2	3	4	5	6
2	0	0	0	0	0
3	0.0295	0.0241	0.0282	0.0259	0.0233
4	0.3060	0.3070	0.3084	0.3070	0.3052
5	0.4182	0.4254	0.4183	0.4178	0.4162
6	0.4662	0.4640	0.4569	0.4466	0.4449

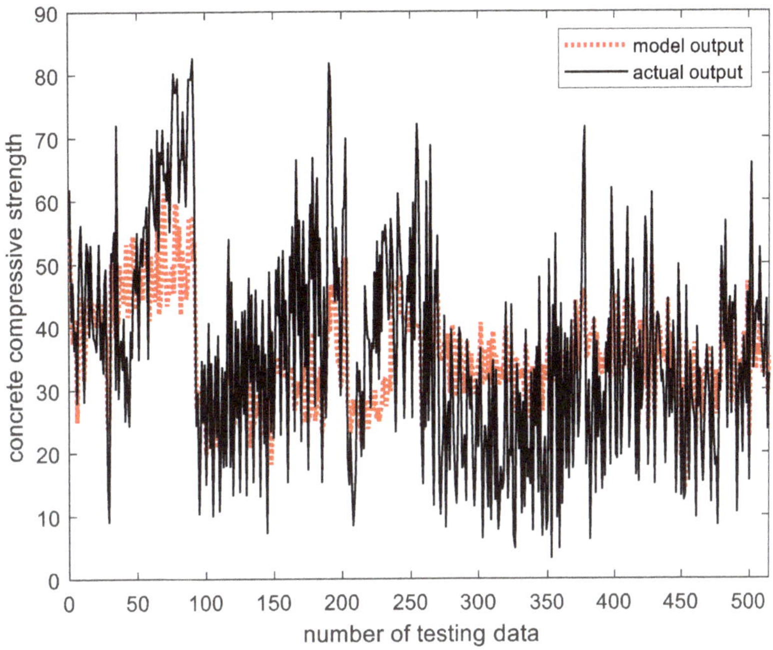

Figure 18. Comparison of the output value of the existing CGK-GM with the actual output value (context is created equally, the number of contexts = 6, the number of clusters = 4).

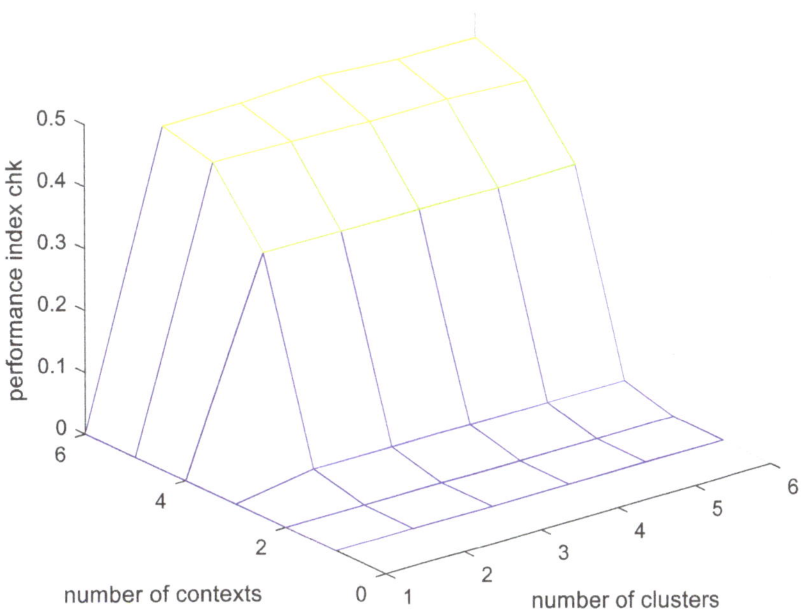

Figure 19. CGK-GM's performance index for validation data (context equally generated, number of contexts = 6, number of clusters = 4).

Table 5 lists the prediction performance of the AGM in which the context is equally generated, and Table 6 lists the prediction performance of the AGM in which the context is flexibly generated. Figure 20 shows the output value and actual output value of AGM, and Figure 21 shows the predictive performance of AGM using performance indicators. As shown in Figure 20 it can be confirmed that the output value of AGM similarly predicts the actual output value with strong nonlinear characteristics. Figure 21 shows the performance index values of CGK-GM for the verification data. As shown in the figure, when the number of contexts is 6 and the number of clusters is 4 when the contexts are created equally, the performance index value is 0.5208, which is the best.

Table 5. Performance index of AGM that created context evenly.

P \ C	2	3	4	5	6
2	0	0	0	0	0
3	0.0201	0.0204	0.0203	0.0202	0.0203
4	0.3194	0.3221	0.3302	0.3315	0.3295
5	0.4548	0.4587	0.4548	0.4488	0.4637
6	0.5125	0.5149	0.5208	0.5184	0.5196

Table 6. Performance index of AGM that created context flexibly.

P \ C	2	3	4	5	6
2	0	0	0	0	0
3	0.0481	0.0437	0.0446	0.0391	0.0451
4	0.3071	0.3023	0.3136	0.3187	0.3129
5	0.4344	0.4348	0.4380	0.4471	0.4418
6	0.4884	0.4981	0.4921	0.5074	0.5028

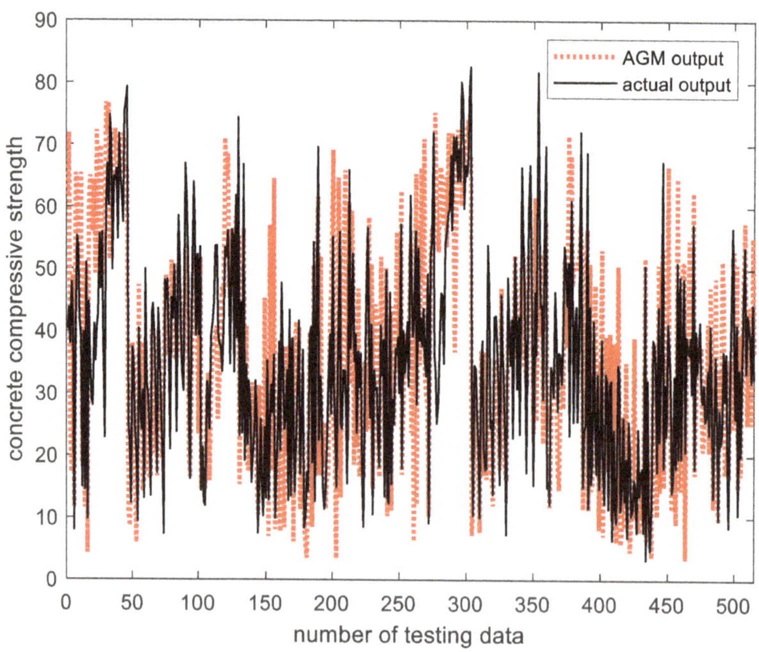

Figure 20. Comparison of the output value of the existing AGM with the actual output value (context is created equally, the number of contexts = 6, the number of clusters = 6).

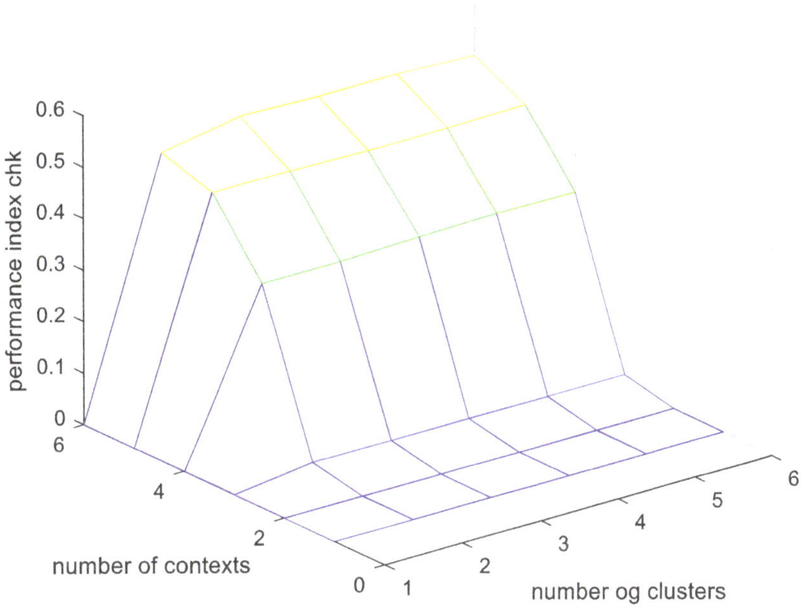

Figure 21. AGM's performance index for validation data (context equally generated, number of contexts = 6, number of clusters = 6).

Table 7 shows the experimental results of concrete compressive strength prediction. As shown in the table, it can be seen that the existing GM creates the contexts equally and the performance index value for the verification data is 0.4276 when the context is 6 and the cluster is 6. In this paper, it was confirmed that the proposed methods, CGK-GM and AGM, show better prediction performance than the conventional GM when the contexts are created equally. As an additional experiment, a house price prediction experiment using the Boston house price database [58] was performed. Table 8 shows the experimental results of Boston house price prediction. As a result of the experiment, the existing GM created the context equally in the output space, and when there were 6 contexts and 4 clusters, a value of 0.5431 was obtained. It was confirmed that the proposed method, CGK-GM, generates contexts evenly and shows better performance at 0.5502 when there are 6 contexts and 5 clusters. It can be seen that AGM flexibly creates a context in the output space and shows better prediction performance than the previous model at 0.5870 when there are 3 contexts and 6 clusters. As a result of conducting an experiment using two databases, it was confirmed that the shape of the context created in the output space affects the performance according to the characteristics of the data.

Table 7. Experimental Results of Predicting Concrete Compressive Strength.

Model Types	IG Type	Num. of Contexts	Num. of Clusters	Training PI	Testing PI
GM	Uniform	6	6	0.4311	0.4276
CGK-GM	Uniform	6	4	0.4743	0.4700
AGM	Uniform	6	4	0.5287	0.5208

Table 8. Experimental Results of Predicting Boston Home Price.

Model Types	IG Type	Num. of Contexts	Num. of Clusters	Training PI	Testing PI
GM	Uniform	6	4	0.5410	0.5431
CGK-GM	Uniform	6	5	0.5476	0.5502
AGM	Flexible	6	3	0.5842	0.5870

6. Conclusions

In this paper, we proposed a CGK-based particle model using context-based GK clustering and a CGK-based particle model with a hierarchical structure. Conventional fuzzy clustering generates clusters by calculating the distance between the center of the cluster and each data using the Euclidean distance. However, there is a problem in that the performance decreases when the data has geometric characteristics. To improve this problem, GK clustering is used. GK clustering uses Mahalanobis distance to calculate the distance between the center of the cluster and each data to generate a geometrical cluster. This paper proposes context-based GK clustering that considers the output space in the existing GK clustering and creates a cluster that considers not only the input space but also the output space. Using the proposed CGK clustering, we designed a CGK-based particle model (CGK-GM) and a CGK-based particle model with aggregated structure (AGM). The advantages of the proposed CGK-based particle model can be summarized as follows.

First, unlike the existing new network, it is possible to automatically generate an explanatory meaningful fuzzy IF-THEN rule that can be expressed verbally by generating information particles in the input space and output space from numerical input and output data. Second, unlike the existing fuzzy clustering, it is effective to process numerical input/output databases with geometric features because it is possible to create a geometrical cluster. Third, meaningful information particles with high abstraction values can be generated by combining the general prediction models, such as linear regression model, neural network, and radiative basis function neural network, with the CGK-based particle model proposed in this paper.

To verify the feasibility of the proposed method, an experiment was conducted using the concrete compressive strength data, a benchmarking database. To evaluate the performance of each particle model, we used a performance index using the scalability and specificity that we consider when generating rational information particles. As a result of the experiment, it was confirmed that the proposed methods were superior to the existing particle models.

In the future, based on the rational information particle generation principle, we plan to conduct research on generating various types of information particles and optimally allocating information particles created in the input space and output space.

Author Contributions: Conceptualization, C.-U.Y. and K.-C.K.; Methodology, C.-U.Y. and K.-C.K.; Software, C.-U.Y. and K.-C.K.; Validation, K.-C.K.; Formal Analysis, C.-U.Y. and K.-C.K.; Investigation, C.-U.Y. and K.-C.K.; Resources, K.-C.K.; Data Curation, C.-U.Y.; Writing-Original Draft Preparation, C.-U.Y.; Writing-Review and Editing, C.-U.Y. and K.-C.K.; Visualization, C.-U.Y.; Supervision, K.-C.K.; Project Administration, K.-C.K.; Funding Acquisition, K.-C.K. All authors have read and agreed to the published version of the manuscript.

Funding: This research was supported by the Basic Science Research Program through the National Research Foundation of Korea (NRF) funded by the Ministry of Education (No.2018R1D1A1B07044907) (No. 2017R1A6A1A03015496).

Conflicts of Interest: The authors declare no conflict of interest.

References

1. Cazarez, R.L.U.; Diaz, N.G.; Equigua, L.S. Multi-layer adaptive fuzzy inference system for predicting student performance in online higher education. *IEEE Lat. Am. Trans.* **2021**, *19*, 98–106. [CrossRef]

2. Yeom, C.U.; Kwak, K.C. Performance comparison of ANFIS models by input space partitioning methods. *Symmetry* **2018**, *10*, 700. [CrossRef]
3. Ishak, K.E.H.K.; Ayoub, M.A. Predicting the efficiency of the oil removal from surfactant and polymer produced water by using liquid-liquid hydrocyclone comparison of prediction abilities between response surface methodology. *IEEE Access* **2019**, *7*, 179605–179619. [CrossRef]
4. Elbaz, K.; Shen, S.L.; Sun, W.J.; Yin, Z.Y.; Zhou, A. Prediction model of shield performance during tunneling via incorporating improved particle swarm optimization into ANFIS. *IEEE Access* **2020**, *8*, 39659–39671. [CrossRef]
5. Saucedo, J.A.M.; Hemanth, J.D.; Kose, U. Prediction of electroencephalogram time series with electro-search optimization algorithm trained adaptive neuro-fuzzy inference system. *IEEE Access* **2019**, *7*, 15832–15844. [CrossRef]
6. Alreshoodi, M.; Danish, E.; Woods, J.; Fernando, A.; Alwis, C.D. Prediction of perceptual quality for mobile video using fuzzy inference systems. *IEEE Trans. Consum. Electron.* **2015**, *61*, 546–554. [CrossRef]
7. Sahoo, S.; Panda, S.K.; Pratihar, D.K.; Mukhopadhyay, S. Prediction of step length using neuro-fuzzy approach suitable for prosthesis control. *IEEE Trans. Instrum. Meas.* **2020**, *69*, 5658–5665. [CrossRef]
8. Loung, T.H.L.; Tran, M.T.; Tran, T.N.; Le, H.S.; Nguyen, L.G.; Vo, T.N.N.; Pham, V.H. A new complex fuzzy inference system with fuzzy knowledge graph and extensions in decision making. *IEEE Access.* **2020**, *8*, 164899–164921.
9. Ai, C.; Jia, L.; Hong, M.; Zhang, C. Short-term road speed forecasting based on hybrid RBF neural network with the aid of fuzzy system-based techniques in urban traffic flow. *IEEE Access* **2020**, *8*, 69461–69470. [CrossRef]
10. An, J.; Fu, L.; Chen, W.; Zhen, J. A novel fuzzy-based convolutional neural network method to traffic flow prediction with uncertain traffic accident information. *IEEE Access* **2019**, *7*, 20708–20722. [CrossRef]
11. Yeom, C.U.; Kwak, K.C. Adaptive neuro-fuzzy inference system predictor with an incremental tree structure based on a context-based fuzzy clustering approach. *Appl. Sci.* **2020**, *10*, 8495. [CrossRef]
12. Parsapoor, M.; Bilstrupo, U.; Svensson, B. Forecasting solar activity with computational intelligence models. *IEEE Access* **2018**, *6*, 70902–70909. [CrossRef]
13. Kannadasan, K.; Edla, D.R.; Yadav, M.H.; Bablani, A. Intelligent-ANFIS model for predicting measurement of surface roughness and geometric tolerances in three-axis CNC milling. *IEEE Trans. Instrum. Meas.* **2020**, *69*, 7683–7694. [CrossRef]
14. Guo, H.; Zeng, W.; Shi, Y.; Deng, J.; Zhao, L. Kernel granger causality based on back propagation neural network fuzzy inference system on fMRI data. *IEEE Trans. Neural Syst. Rehabil. Eng.* **2020**, *28*, 1049–1058. [CrossRef]
15. Hwang, Y.; Kang, B.; Kim, W. Motion cue-based sudden pedestrian behavior prediction using fuzzy inference. *IEEE Access* **2021**, *9*, 135245–135255. [CrossRef]
16. Ding, Y.; Yang, S.; Chen, Y.; Long, Q.; Wei, J. Explaining and predicting mobile government microblogging services participation behaviors a SEM-neural network method. *IEEE Access* **2019**, *7*, 39600–39611. [CrossRef]
17. Ibragimov, B.; Toesca, D.A.S.; Yuan, Y.; Koong, A.C.; Chang, D.T.; Xing, L. Neural networks for deep radiotherapy dose analysis and prediction of liver SBRT outcomes. *IEEE J. Biomed. Health Inform.* **2019**, *23*, 1821–1833. [CrossRef]
18. Jia, T.; Yan, P. Predicting citywide road traffic flow using deep spatiotemporal neural networks. *IEEE Trans. Intell. Transp. Syst.* **2020**, *22*, 3101–3111. [CrossRef]
19. Wu, Q.; Jiang, Z.; Hong, K.; Liu, H.; Yang, L.T.; Ding, J. Tensor-based recurrent neural network and multi-modal prediction with Its applications in traffic network management. *IEEE Trans. Netw. Serv. Manag.* **2021**, *18*, 780–792. [CrossRef]
20. Bouda, M.; Krishna, G.; Krystofik, J.; Oda, S.; Palacharla, P. Towards vendor-agnostic real-time optical network design with extended Kalman state estimation and recurrent neural network machine learning Invited. *J. Opt. Commun. Netw.* **2021**, *13*, B21–B34. [CrossRef]
21. Liu, Y.T.; Lin, Y.Y.; Wu, S.L.; Chuang, C.H.; Lin, T.C. Brain dynamics in predicting driving fatigue using a recurrent self-evolving fuzzy neural network. *IEEE Trans. Neural Netw. Learn. Syst.* **2015**, *27*, 347–360. [CrossRef] [PubMed]
22. Dumas, T.; Roumy, A.; Guillemot, C. Context-adaptive neural network-based prediction for image compression. *IEEE Trans. Image Processing* **2019**, *29*, 679–693. [CrossRef] [PubMed]
23. Lin, J.W.; Chao, C.T.; Chiou, J.S. Determining neuronal number in each hidden layer using earthquake catalogues as training data in training an embedded back propagation neural network for predicting earthquake magnitude. *IEEE Access* **2018**, *6*, 52582–52597. [CrossRef]
24. Pedrycz, W.; Vasilakos, A.V. Linguistic models and linguistic modeling. *IEEE Trans. Syst. Man Cybern. Part B* **1999**, *29*, 745–757. [CrossRef] [PubMed]
25. Zhou, P.; Xu, Z.; Song, C.; Shao, Z. Long-term hybrid prediction method based on multiscale decomposition and granular computing for oxygen supply network. *Comput. Chem. Eng.* **2021**, *153*, 107442. [CrossRef]
26. Hmouz, R.A.; Pedrycz, W.; Balamash, A.S.; Morfeq, A. Hierarchical system modeling. *IEEE Trans. Fuzzy Syst.* **2017**, *26*, 258–269. [CrossRef]
27. Pedrycz, W.; Homenda, W.; Jastrzebska, A.; Yu, F. Information granules and granular models selected design investigations. In Proceedings of the 2020 IEEE International Conference on Fuzzy Systems (FUZZ-IEEE), Glasgow, UK, 19–24 July 2020.
28. Liu, Y.; Zhao, J.; Wang, W.; Pedrycz, W. Prediction intervals for granular data streams based on evolving type-2 fuzzy granular neural network dynamic ensemble. *IEEE Trans. Fuzzy Syst.* **2020**, *29*, 874–888. [CrossRef]
29. Wang, S.; Pedrycz, W. Robust granular optimization a structured approach for optimization under integrated uncertainty. *IEEE Trans. Fuzzy Syst.* **2014**, *23*, 1372–1386. [CrossRef]

30. Zhu, X.; Pedrycz, W.; Li, Z. A development of granular input space in system modeling. *IEEE Trans. Cybern.* **2019**, *51*, 1639–1650. [CrossRef]
31. Truong, H.Q.; Ngo, L.T.; Pedrycz, W. Advanced fuzzy possibilistic C-means clustering based on granular computing. In Proceedings of the 2016 IEEE International Conference on Systems, Man, and Cyberneticx, Budapest, Hungary, 9–12 October 2016.
32. Zuo, H.; Zhang, G.; Pedrycz, W.; Behbood, V.; Lu, J. Granular fuzzy regression domain adaptation in Takagi-Sugeno fuzzy models. *IEEE Trans. Fuzzy Syst.* **2017**, *26*, 847–858. [CrossRef]
33. Hu, X.; Pedrycz, W.; Wang, X. Granular fuzzy rule-Bbased models a study in a comprehensive evaluation and construction of fuzzy models. *IEEE Trans. Fuzzy Syst.* **2016**, *25*, 1342–1355. [CrossRef]
34. Zhao, J.; Han, Z.; Pedrycz, W.; Wang, W. Granular model of long-term prediction for energy system in steel industry. *IEEE Trans. Cybern.* **2015**, *46*, 388–400. [CrossRef]
35. Bezdek, J.C. *Pattern Recognition with Fuzzy Objective Function Algorithms*; Springer: Berlin/Heidelberg, Germany, 1981; ISBN 0-306-40671-31.
36. Silva, L.; Moura, R.; Canuto, A.M.P.; Santiago, R.H.N.; Bedregal, B. An interval-based framework for fuzzy clustering applications. *IEEE Trans. Fuzzy Syst.* **2015**, *23*, 2174–2187. [CrossRef]
37. Pedrycz, W. Conditinal fuzzy C-means. *Pattern Recognit. Lett.* **1996**, *17*, 625–632. [CrossRef]
38. Pedrycz, W. Conditional fuzzy clustering in the design of radial basis function neural networks. *IEEE Trans. Neural Netw.* **1998**, *9*, 601–612. [CrossRef] [PubMed]
39. Gustafuson, D.E.; Kessel, W.C. Fuzzy clustering with a fuzzy covariance matrix. In Proceedings of the 1978 IEEE Conference on Decision and Control inclusing the 17th Symposium on Adaptive Precesses, San Diego, CA, USA, 10–12 January 1979.
40. Babuska, R. *Fuzzy Modeling for Control*; Kluwer Academic Publishers: Dordrecht, The Netherlands, 1998.
41. Babuska, R.; van der Veen, P.J.; Kaymak, U. Improved covariance estimation for Gustafuson-Kessel clustering. In Proceedings of the 2002 IEEE International Conference on Fuzzy Systems, Honolulu, HI, USA, 12–17 May 2002; pp. 1081–1085.
42. Nguyen, T.T.; Pham, X.C.; Liew, A.W.C.; Pedrycz, W. Aggregation of classifiers a justifiable information granularity approach. *IEEE Trans. Cybern.* **2018**, *49*, 2168–2177. [CrossRef] [PubMed]
43. Zhang, B.; Pedrycz, W.; Wang, X.; Gacek, A. Design of interval type-2 information granules based on the principle of justifiable granularity. *IEEE Trans. Fuzzy Syst.* **2020**, *29*, 3456–3469. [CrossRef]
44. Wang, D.; Pedrycz, W.; Li, Z. Granular data aggregation an adaptive principle of the justifiable granularity approach. *IEEE Trans. Cybern.* **2018**, *49*, 417–426. [CrossRef]
45. Wang, D.; Pedrycz, W.; Li, Z. A two-phase development of fuzzy rule-based model and their analysis. *IEEE Access* **2019**, *7*, 80328–80341. [CrossRef]
46. Pedrycz, W.; Wang, X. Designing fuzzy sets with the use of the parametric principle of justifiable granularity. *IEEE Trans. Fuzzy Syst.* **2015**, *24*, 489–496. [CrossRef]
47. Zhang, B.; Pedrycz, W.; Fayek, A.R.; Gacek, A.; Dong, Y. Granular aggregation of fuzzy rule-based models in distributed data environment. *IEEE Trans. Fuzzy Syst.* **2020**, *29*, 1297–1310. [CrossRef]
48. Lu, W.; Wang, D.; Pedrycz, W.; Zhang, L.; Yang, J.; Liu, X. Granular fuzzy modeling for multidimensional numeric data A layered approach based on hyperbox. *IEEE Trans. Fuzzy Syst.* **2018**, *27*, 775–789. [CrossRef]
49. Zhang, B.; Dong, Y.; Feng, X.; Pedrycz, W. Maximum fuzzy consensus feedback mechanism with minimum cost and private interest in group decision-making. *IEEE Trans. Fuzzy Syst.* **2020**, *29*, 2689–2700. [CrossRef]
50. Wang, L.; Han, Z.; Pedrycz, W.; Zhao, J. A granular computing-based hybrid hierarchical method for construction of long-term prediction intervals for gaseous system of steel industry. *IEEE Access* **2020**, *8*, 63538–63550. [CrossRef]
51. Pedrycz, W.; Hmouz, R.A.; Balamash, A.S.; Morfeq, A. Hierarchical granular clustering an emergence of information granules of higher type and higher order. *IEEE Trans. Fuzzy Syst.* **2015**, *23*, 2270–2283. [CrossRef]
52. Siddique, N.; Adeli, H. *Computational Intelligence: Synergies of Fuzzy Logic, Neural Networks and Evolutionary Computing*; John Wiley & Sons Ltd.: Oxford, UK, 2013.
53. Freedman, D.A. *Statistical Models: Theory and Practice*; Cambrige University Press: Cambrige, UK, 2009.
54. Siegelmann, H.T.; Sontag, E.D. Turing computability with neural nets. *Appl. Math. Lett.* **1991**, *4*, 77–80. [CrossRef]
55. Broomhead, D.S.; Lowe, D. Multivariable functional interpolation and adaptive networks. *Complex Syst.* **1988**, *2*, 321–355.
56. Broomhead, D.S.; Lowe, D. *Radial Basis Functions, Multi-Variable Functional Interpolation and Adaptive Networks*; Technical Report; RSRE (Royal Signals & Radar Establishment): Malvern, UK, 1988; Volume 4148.
57. UCI Machine Learning Repository, Concrete Compressive Strength Data Set. Available online: https://archive.ics.uci.edu/ml/datasets/Concrete+Compressive+Strength (accessed on 16 February 2022).
58. Available online: https://www.cs.toronto.edu/~{}delve/data/boston/bostonDetail.html (accessed on 16 February 2022).

applied sciences

MDPI

Article

Development of Open-Assistant Environment for Integrated Operation of 3D Bridge Model and Engineering Document Information

Sang I. Park [1,2], Bong-Geun Kim [3], Wonhui Goh [4] and Goangseup Zi [4,*]

[1] Department of Civil, Environmental and Architectural Engineering, University of Colorado, Boulder, CO 80309, USA; sang.i.park@colorado.edu
[2] Research Institute for Safety Performance, Korea Authority of Land & Infrastructure Safety, Jinju 52856, Korea
[3] Taesung SNI Singapore Branch, Singapore 208652, Singapore; bgkim@tssni.com
[4] School of Civil, Environmental and Architectural Engineering, Korea University, Seoul 02841, Korea; wonhee0689@korea.ac.kr
* Correspondence: g-zi@korea.ac.kr; Tel.: +82-2-3290-3324

Abstract: This study proposes a method for assistant environments to integrate 3D bridge model information and engineering document fragments. The engineering document content varies depending on the process. Therefore, we accept a loose coupling concept to support the independence of each information set instead of using a specific data model for effective integration. The engineering document is translated into an Extensible Markup Language (XML)-based structured format based on the explicit and apparent semantic structure of the document. An extended industry foundation classes (IFC) schema is proposed to manage the bridge information model, as well as document fragments. An information document (iMapDoc) is proposed to manage interim data to connect a 3D digital model, an IFC model, and engineering document fragments. Document fragments on a specific component in the 3D bridge model are retrieved to validate the developed integrated assistant module.

Keywords: 3D bridge model; IFC-based bridge model; engineering document; document fragment; integrated operation

Citation: Park, S.I.; Kim, B.-G.; Goh, W.; Zi, G. Development of Open-Assistant Environment for Integrated Operation of 3D Bridge Model and Engineering Document Information. *Appl. Sci.* **2022**, *12*, 2510. https://doi.org/10.3390/app12052510

Academic Editors: Danial Jahed Armaghani, Yixia Zhang, Pijush Samui, Ahmed Hussein Kamel Ahmed Elshafie and Aydin Azizi

Received: 21 December 2021
Accepted: 23 February 2022
Published: 28 February 2022

1. Introduction

A certain level of knowledge is required to understand and use construction information presented in the form of engineering documents, such as structural calculation documents. Therefore, an environment that can provide information regarding various types, such as documents or 3D digital models, depending on the intended purpose must be established; this is because completing a construction project is achieved through the collaboration of participants with various knowledge backgrounds. Tatum [1] emphasized the role of a three-dimensional (3D) model-based environment in promoting efficiency in the communication of all architect, engineering, and construction (AEC) project participants, and we agree with his opinion. Building information modeling (BIM) is a specialized application field of 3D model-based integration from an information point of view; furthermore, it is expected to be one of the most important applications for the integrated operation of BIM and document-type information in the AEC field. Data in BIM and documents must be managed, as well as linked or aligned, appropriately to achieve an effective integration of BIM and engineering documents in a system or environment.

The information of model objects is managed internally by the tool itself in the case of closed BIM, which generates a model in a proprietary format using BIM authoring tools. Access to model data is possible within the software or within an application programming interface (API) scope. End-users in the practical field often base their preference on closed BIM from the perspective of information management because model visualization and object generation/editing are performed within one platform [2]. However, this method may cause reliability issues, owing to restriction to full data access, platform dependency, and non-guaranteed transparency of internal processes [3]. In open BIM represented by industry foundation classes (IFC), access to all model object information and modification is enabled [4] according to the standard data access interface (SDAI) standardized by the International Organization for Standardization (ISO) [5]. This method, however, requires advanced programming knowledge to control the IFC physical file (IPF) directly for modifying geometries, and there are instability issues in the process of importing the IPF from the BIM tools.

No standardized method is available for data access, unlike in BIM, because the content included in the construction engineering document is in a denormalized form; hence, various relevant studies are being conducted. Marking-up on contents can be interpreted by a machine, although semi-automatic or manual preprocessing is required for generating mark-up [6]. BIM-based automated compliance checking (ACC), the core task of which is to identify the semantic meaning of codes or regulations [3], proceeds in a similar manner to that of BIM construction document connection. To identify the semantic meaning from documents, specific rules were assigned to the plain documents to extract the content [7,8], or an ontology model was used [9–11]. Recently, research that extracts meaning from regulations through natural language processing or machine learning has been actively conducted [12–14]. These studies focus on the "latent semantic structure" categorized by Wang et al. [15]. The connection between BIM and regulation focuses on mapping the attributes extracted from each model (BIM and document), rather than document fragments or elements from the ACC perspective [16–18]. This results in relatively high costs for extracting the required information, as well as inefficiency if the documents are used as non-geometric reference information instead of a regulatory review. If engineering documents are used as references, then the approach is effective when the "explicit or apparent semantic structure" is used to reconstruct construction documents [19–21].

Choi et al. [22] has performed BIM-document integration by adding the necessary rules to the IFC entities or properties. Opitz et al. [23] linked document contents stored in a repository with an IFC model via separate link elements. As summarized in Table 1, however, previous research mainly focuses on the perspective of document information extraction, and so do not consider the information connection considering the IFC schema or do not pay much attention to the relationship with the 3D object.

Herein, we studied methods that can link engineering documents as reference information with the bridge information model, and subsequently developed an integrated open-assistant environment. Furthermore, we developed a structuralizing method using the explicit and apparent semantic structure of unstructured plain document contents by improving the work of Kim et al. [20]. We adopted the extended IFC schema proposed by Park et al. [24] for bridge information modeling and described how to manage document information based on IFC. An Autodesk Revit-based add-in module was developed to generate and manage a bridge information model using the adopted IFC schema. An interim mapping document (iMapDoc) generated during the process can serve as an excellent bridge for a seamless interface between the closed and open BIM, as well as providing relevant document fragments for each model component, even when the 3D model objects are modified.

Table 1. Overview of related research.

Previous Study	Extraction of 3D Model Info.	Extraction of Doc. Info.	Integration, Mapping Method	Limitation
Hjelseth and Nisbet [6]	X	RASE [1] methodology	Not proposed	Integration method between 3D model and doc. Info. was not proposed.
Zhong et al. [9]	X	CQIEOntology [2] (Manual)	Not proposed	Integration method between 3D model and doc. Info. was not proposed.
Choi et al. [22]	IFC	X	IFC user-defined property sets (PSETs)	Regulation codes must be mapped into IFC PSETs manually.
Opitz et al. [23]	IFC schema	X	SQL and BIMfit Model Query	The document content should already be stored in DB in a fragile state.
Beach et al. [7]	IFC	Extended RASE (RASE + XML tag)	Experts performed the mapping between the code fragments and IFC entities.	Mapping was performed manually.
Zhou and El-Gohary [10]	IFC	Rule-based OBIE [3] algorithm	IFC–SIE [4]–logic facts transformation	OBIE algorithm highly depends on specific knowledge domain (building energy conservation codes).
Sydora and Stroulia [8]	IFC	Rule Language (manual)	IFC-based automatic mapping	The information should be organized to interpret into rule language manually.

[1] RASE: requirement, applicability, selection, and exceptions. [2] CQIEOntology: quality inspection and evaluation ontology. [3] OBIE: ontology-based information extraction. [4] SIE: semantic information element.

2. Document Analysis and Translation to Structured Format

2.1. Translation Process from Unstructured Engineering Document to XML Document

An engineering document saved as a plain text file is used as an input file to eliminate errors that may occur when recognizing letters from visualized documents. Figure 1 illustrates the entire process of generating a structured Extensible Markup Language (XML) document from a text document referred to by Kim et al. [20]. As depicted in Figure 1, the input engineering document is translated into an XML document through three main steps, according to the subtitle structure. The first step in performing translation to the structured XML document is to store each sentence of the contents sequentially by classifying strings for the types of heading symbols, headings, subtitles, content, and references into a temporary table based on the engineering document model. The types of heading symbols are used to identify the hierarchy of the contents. The temporary table is rearranged after identifying the sentence structure using the existing data stored in the temporary table. The hierarchical information of the contents is identified using the rearranged data connected to the types of heading symbols and the tree structure of the document. Finally, the XML file is generated using the information saved in the temporary table and hierarchical information.

2.2. Content Analysis of Engineering Documents

We defined several notations to describe the content of the engineering documents efficiently, as follows:

1. The string $S = s_1, s_2, \cdots, s_n$ represents a set of characters with a finite sequence; here, $s_k \in \Psi$, $1 \leq k \leq n$, and Ψ is the set of all the characters, including space, in a given document;
2. In $A ::= B$, the symbol '::=' implies that A can be expressed as B;
3. The symbol '|' represents 'or'.

The text information comprising an engineering document can be separated using finite string sets with sequences; the string set S^i of the ith row can be expressed as Equation (1), referring to Kim et al. [20].

$$S^i ::= H^i|C^i|H^iC^i|H^iR^i|C^iR^i|H^iC^iR^i \tag{1}$$

where H^i is a string set for the subtitle, $H^i = s_1, s_2, \cdots, s_l$; C^i is a string set for the document content, $C^i = s_{l+1}, s_{l+2}, \cdots, s_m$, $R_i (= s_{m+1}, s_{m+2}, \cdots, s_n)$ represents the string set for reference, and $0 \leq l \leq m \leq n$.

Figure 2 depicts the content analysis algorithm for extracting the document structure based on Equation (1). More definitions and processes are available in the previous research by Kim et al. [20].

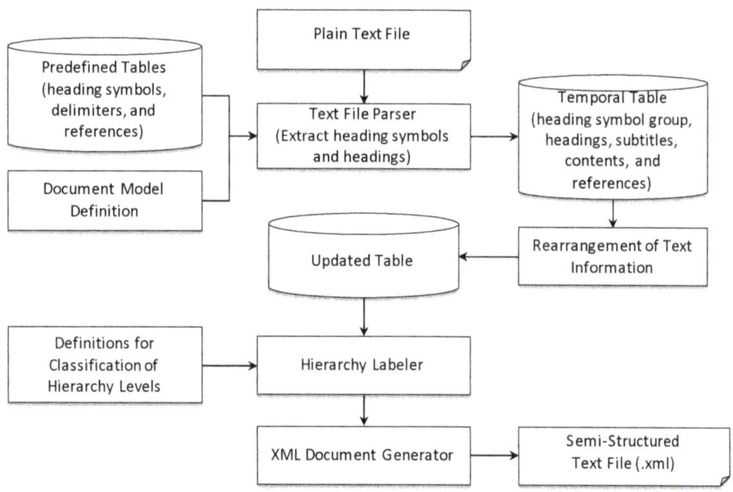

Figure 1. Document translation process for text information.

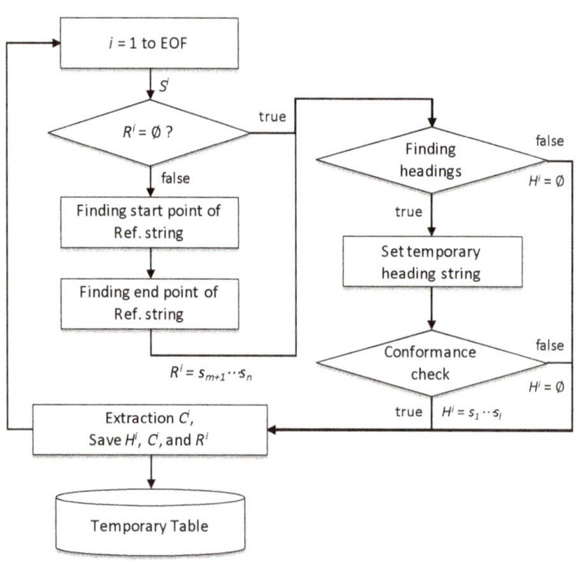

Figure 2. Algorithm for content analysis to extract document structure.

2.3. Identification of Bullet-Form Text Strings

To identify the syntax's meaning, which comprises the text string of the engineering documents, the temporary table constructed through the component extraction algorithm of the document is used, as described in Section 2.2. TS^i, the string set of a random ith row, can be expressed as shown in Equation (2).

$$TS^i = \left\{HS^i_{ID}, hs^i, hc^i, C^i, R^i\right\} \tag{2}$$

where HS^i_{ID} is the unique group number including the heading hs^i; hc^i is the string for the title except the heading symbol; and C^i and R^i are text strings for the content and reference, respectively, as defined in Equation (1). We regarded the ith row to $i + n$th row of the temporary table contents, when they satisfy all conditional expressions shown in Equation (3a), as the bullet-form text strings. Here, $n = 1, 2, 3, \cdots$.

$$HS^i_{ID} = HS^{i+n}_{ID} \tag{3a}$$

$$hs^{i+\alpha} \in BS^d \quad (\alpha = 0, 1, 2, \cdots \leq n) \tag{3b}$$

$$C^i = \varnothing \wedge C^{i+n} = \varnothing \tag{3c}$$

In Equation (3b), BS^d is the set of heading symbol groups with the depth d, where d is a natural number exceeding 1, as defined by users. In Equation (3c), $\varnothing$ means the set with no elements and $\wedge$ logical conjunction or meeting in a lattice. The identified bullet-form text strings according to Equation (3a) are assigned to the $i - 1$th document content (C) using Equation (4); subsequently, the contents in the temporary table are rearranged.

$$C^{i-1}_{new} = C^{i-1}_{old} + \sum_{j=1}^{n}\left(nl + hs^j + hc^j\right) \tag{4}$$

where C^{i-1}_{new} is the text string for the newly updated $i - 1$th document content, C^{i-1}_{old} is the string text for the $i - 1$th document content before rearranging the content, and nl is the character that represents a new line.

2.4. Identification of Hierarchical Structure of Subtitles

The problem of translating unstructured document contents into a tree-shaped hierarchical structure can be described by estimating the depth of the headings of the corresponding contents. The hierarchical information of the headings is identified by comparing the number of heading symbol groups. We used the results from a previous study that provided a generalized solution to this problem [20]. Using the algorithm described above, the unstructured document content was regenerated into a structured document format (Figure 3) using the developed translator, as shown in Figure 4. The tags of the XML element indicate the content items describing the function of the sentence, and the heading symbols identified for hierarchical classification are expressed as the *header* attributes of the element. If the element should refer to other documents or codes, the references appear in the *reference* attribute of the element. The text data are expressed as parsed character data (PCDATA) of XML. The core functions of the translator shown in Figure 3 are combined into the integrated module to manage the bridge and document model, which will be explained later.

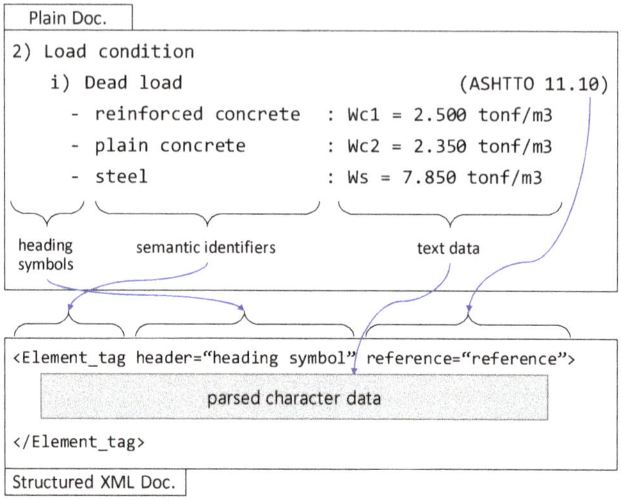

Figure 3. Mapping relationship between a plain and XML document.

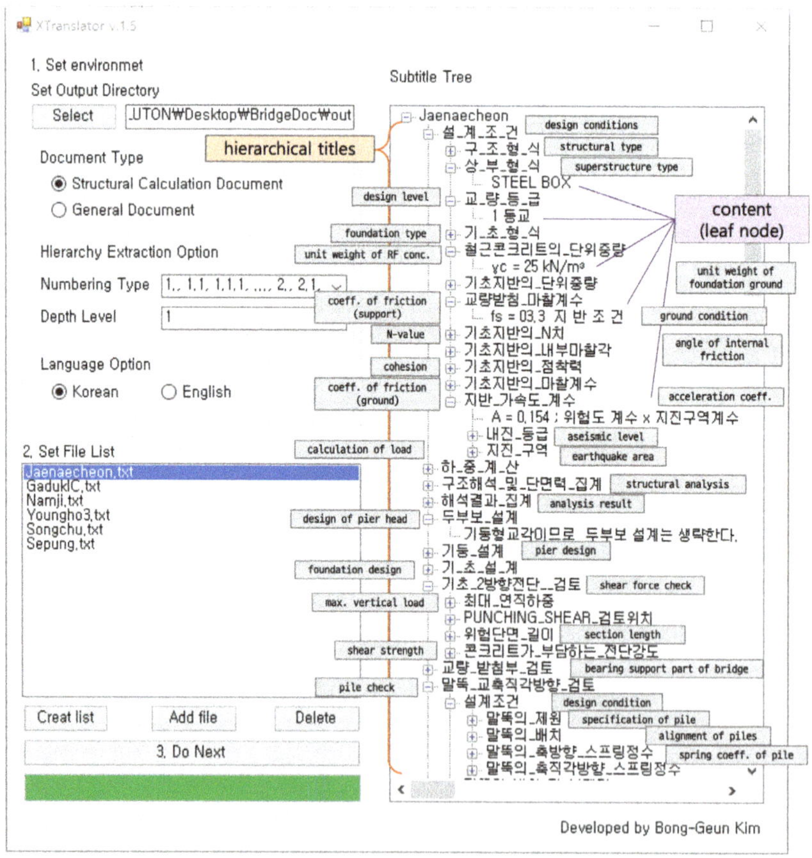

Figure 4. Module developed to translate plain document to XML document.

2.5. Performance Evaluation of Document Translation Module

We used precision and recall, which are widely used in the information retrieval field for a performance evaluation of the component extraction algorithm of a document. Precision and recall use the number of true positive (*TP*), true negative (*TN*), false positive (*FP*), and false negative (*FN*) values; *TP* means that the extracted title sentence is correct, *TN* means that the extracted content sentence is correctly recognized, and *FP* means that the algorithm misrecognizes a sentence as a title. *FN* means that the title sentence is not recognized as the title sentence. The equations used to measure the precision and recall are as follows.

$$Precision = \frac{TP}{TP + FP} \tag{5}$$

$$Recall = \frac{TP}{TP + FN} \tag{6}$$

The hierarchies recognized by the proposed algorithm in this study are relative classification among heading symbols; the recognized results of the hierarchical classification of the preceding items affect the following items. Therefore, the module performance of this part was checked following these equations:

$$GAH = \frac{C^G}{TP} \tag{7}$$

$$PAH = \frac{C^P}{TP} \tag{8}$$

where GAH means 'generalized accuracy for hierarchy labeler' and PAH means 'precise accuracy for hierarchy labeler'. C^G is the number of results from which relative hierarchical classification was performed correctly among TPs, and C^P is the number of results from which absolute hierarchical classification was performed correctly among TPs. Therefore, GAH and PAH represent ratio values corresponding to C^G and C^P, respectively.

The proposed algorithm performance, including the content extracting and hierarchical classification processes, was evaluated as the following equations.

$$GAA = \frac{C^G}{TP + FN} \tag{9}$$

$$PAA = \frac{C^P}{TP + FN} \tag{10}$$

where GAA means 'generalized accuracy for application module' and PAA means 'precise accuracy for application module'.

Figure 5 shows the results of applying Equations (5)–(10) to 20 bridge engineering documents with an average number of sentences of 4814 and the number of title sentences to be extracted as 433. The lowest and mean values of PAA are 97.36% and 99.47%, respectively, so it can be judged that the performance as a content item extraction algorithm for BIM-based integration is sufficient.

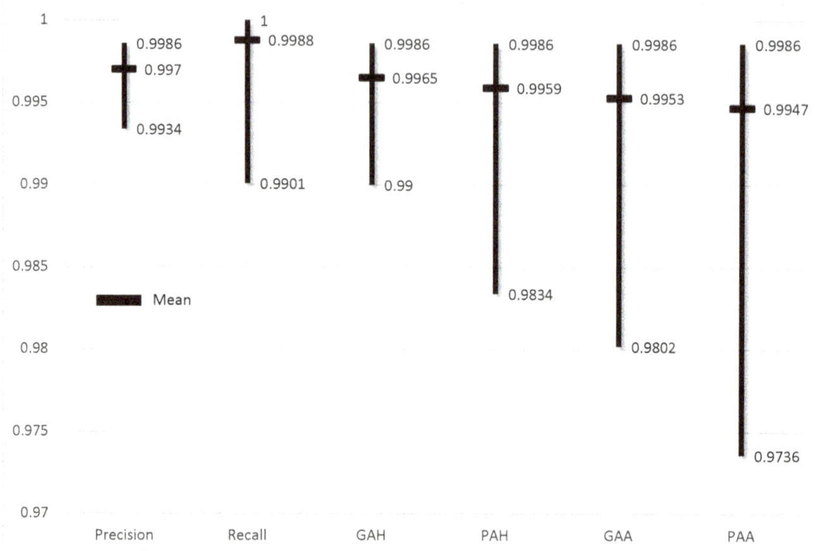

Figure 5. Performance evaluations based on precision, recall, GAH, PAH, GAA, and PAA.

3. IFC-Based Bridge Information Modeling with Document Metadata

3.1. IFC Schema for Bridge Model and Document Information

The most recent official IFC released by buildingSMART International (bSI) is version IFC4.0.2.1 [25]. IFC4.0.2.1 has opened up the possibility of future extensions for civil infrastructures (*IfcCivilElement*), although they will be removed in a future version. The unofficial version currently being developed (IFC4.3RC4) includes alignments (*IfcAlignment*), roads (*IfcRoad*), railways (*IfcRailway*), ports (*IfcMarineFacility*), and bridges (*IfcBridge*). Regarding the bridge structure, IFC4.3RC3 includes the *IfcBridge* entity in order to represent the spatial information of the bridge as a subtype of *IfcFacility*. *IfcBridge* has an enumeration type to embody a spatial function for the subspace of the bridge [26]. More specific spatial components constituting the bridge structure can be managed using the *IfcBridgePartTypeEnum* type of *IfcFacilityPart*. The representation of the physical element of the bridge can be used in the subtypes of *IfcBuiltElement*.

We used the schema developed by Park et al. [24] instead of IFC4.0.2.1 or IFC4.3RC4 for the following reasons:

- IFC4.3RC4 is a schema under development that has not yet been officially released. Therefore, current BIM authoring tools, such as Autodesk Revit, cannot handle the information generated by IFC4.3RC4;
- Information on the bridge structure components covered by IFC4.3RC4 is limited; it does not define bridge-specific and bridge-related attributes that should be treated as entity and attribute level.

IfcBridgeAddMeshfree, proposed by Park et al. [24], extends additional entities focusing on the bridge structure and bridge components (see Figure 6). The detailed elements of bridges regarded as enumeration types in IFC4.3RC4 are also defined as entities.

In the *IfcBridgeAddMeshfree* schema, *IfcBridge* is used to manage the spatial information of the bridge structure itself, *IfcBridgeSpan* is used to segregate the spatial information based on the bridge's along the road, and *IfcBridgeSpacePart* is used for the transverse direction of the bridge. In order to represent the physical object of the bridge, it was categorized into girder part, slab part, abutment and pier part, and detailed member component part. Each of these contain enumeration types for detailed types or functions.

External document-related information can be managed through the *IfcDocumentInformation* entity in the IFC schema. The *IfcDocumentInformation* can capture metadata, such as the document name, document purpose, and/or revision information of an external document, but not the document content. We linked each content fragment of a document corresponding to a model component rather than linking an entire document for an entire 3D model because *IfcDocumentInformation* can be associated with all IFC objects through the *IfcRelAssociatesDocument* entity. The implementation of *IfcDocumentInformation*-related information will be described later.

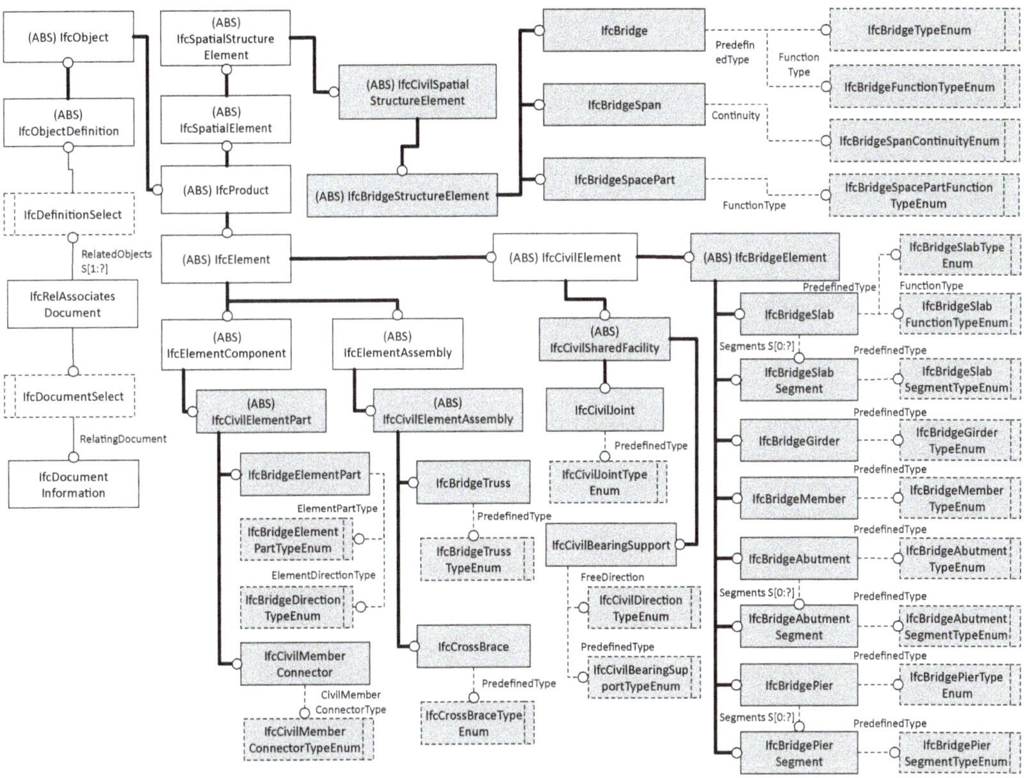

Figure 6. EXPRESS-G diagram to manage bridge model and document meta data; the shaded boxes denote extended entities for the bridge structure.

3.2. Assistant Module for Integrated Management of Bridge Model and Document Information

We developed a module to manage 3D model information, reflect the extended IFC-based bridge, and enable information retrieval for engineering documents related to bridge segments. The module was based on the Revit API provided by Autodesk [27], a representative BIM authoring tool. This enabled the integration of the extended IFC and engineering document retrieval in the Revit environment. We used the original functions provided by Revit to generate a 3D geometric model (Figure 7a). We modeled the components using general or building elements corresponding to bridge components referred by the concept of Park et al.'s [2] previous research for bridge components, such as bridge piers, that Revit cannot provide. As a note, to connect with the extended IFC entities for the bridge structure, each component should be created as one object. As a preliminary procedure, the user should directly generate spatial objects suitable for bridge and bridge components to apply IFC entities (see Figure 7b). This module suggests an appropriate IFC entity by parsing the EXPRESS files of the extended IFC schema for the spatial object, as depicted in Figure 7c. All 3D physical objects of the bridge collected automatically through the Revit API can be mapped with an appropriate IFC entity using the interface shown in Figure 7d. The appropriate IFC entity can be suggested through the basic information included in the object, such as the object's family name and description. If the object information has a specific code or words included in the product breakdown structure (PBS) document developed in this study, then the physical object can be mapped precisely with the corresponding IFC entity using this PBS document. The PBS document comprises ab element code (*value* attribute), name (*label* attribute), IFC entity of the extended IFC-based bridge schema (*IfcEntityName* attribute), the IFC4 entity (*Ifc4EntityName* attribute), and the description (*description* attribute). Figure 8 shows a part of the PBS document developed in this study. In this module, the model object can be connected to not only the IFC schema entities developed for bridges, but also to the IFC4 entities, such that other commercial BIM authoring tools can utilize the IFC data generated through this module. Figure 9 shows a part of the iMapDoc containing essential information for the conversion to an extended IFC-based IPF, as well as for connecting a 3D object, IFC, and document fragments. We have designed the iMapDoc based on XML that can generate a relationship between bridge components, as well as connect attributes on the corresponding components. Various tested computer libraries related to the use of XML enable us to skip the verification process of the designed XML schema (iMapDoc).

The *Project* element in line 2 of Figure 9 stores the entire project-related information included in Revit, and is an element that is mapped with the *IfcProject* entity and its attributes of the IFC schema. The child elements of the *StructuralElement* in line 3 are mapped with the physical/spatial entities of the IFC. We used the *IfcSite* entity as the topmost element of all of the spatial elements. The spatial objects generated by the user, shown in Figure 7b, are arranged as *IfcSite*'s sub-entities. The spatial element representing the entire bridge itself is mapped to *IfcBridge* as an extended schema entity and *IfcBuildingStorey* as an IFC4 entity (line 7 of Figure 9). In this case, the *IfcBuilding* entity can be substituted for *IfcBuildingStorey*. Since no entities support bridge structures in IFC4, it is not essential to consider which entities of the IFC4 are used in this study. The entity just needs to be a spatial entity. The spatial objects constituting the entire bridge, such as the bridge section and the upper and lower structures, are mapped with each IFC entity, as shown in lines 11–17 of Figure 9. Line 19 of Figure 9 shows one of the physical elements of the bridge structure, i.e., *Concrete_Pier*, which includes not only the IFC entities (*IfcEntityName*, *Ifc4EntityName*), but also the object number (*OidInSWDB*) managed in Revit and the classification number (*PBSCode*) defined in the PBS. These data can serve as insights for linking information between the Revit objects and the IFC entity software independently, as well as between the IFC entities and engineering documents described in the following sections. The 3D geometries generated in the Revit are represented in the form of boundary representations (B-rep) in the IPF in this study. The vertex and edge information of the B-rep are composed of a separate file, as shown in Figure 10. The iMapDoc can manage this information through the identifier shown in line 24 of Figure 9. The *CompNo* attribute points to the ID of the vertex and triangle data, as shown in Figure 10. The iMapDoc can be converted to the IPF using subtypes of the *IfcRelationship* entity to connect "spatial object–spatial object," "spatial object–physical object," and "physical object–physical object" while sequentially reading elements of the iMapDoc. Property sets (lines 26–33 of Figure 9) in each element are connected to the parent object using the *IfcRelDefinesByProperties* entity in the IFC.

The structure of the document contents can be represented via the tree view, list view, or plain view features provided by the .NET framework, since the engineering document has been translated to the XML format (see Figure 7e). The document content query process is discussed in the next section.

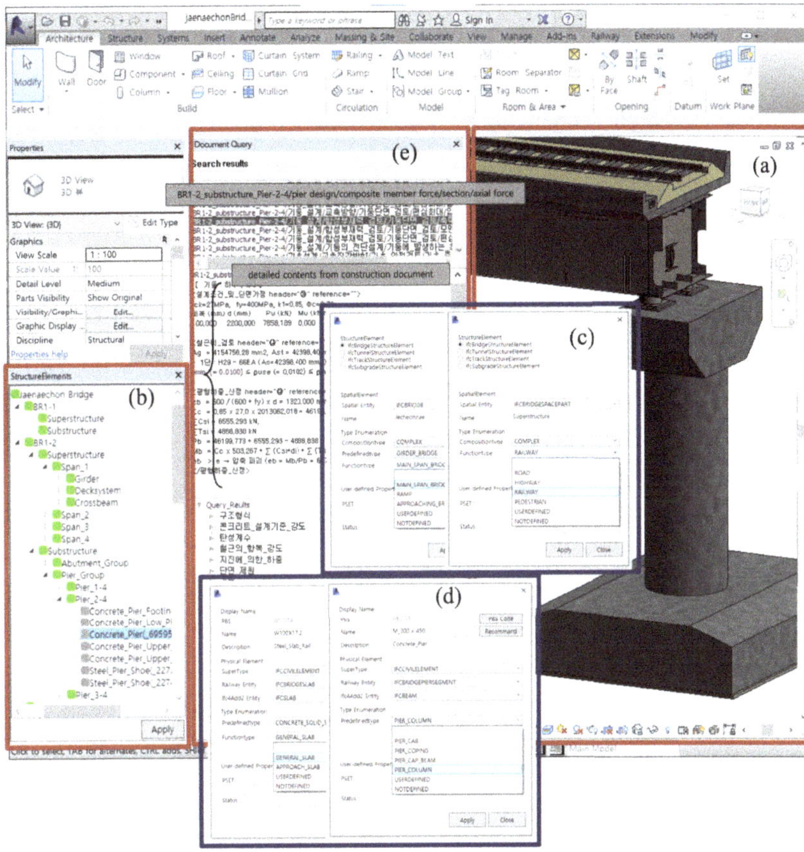

Figure 7. Module developed for connecting 3D model, IFC objects, and engineering document: (**a**) 3D bridge model, (**b**–**d**) user-interfaces for the IFC-based bridge modeling, and (**e**) user-interface for document fragment retrieved based on query.

```xml
<PBS>
  <StructureElements>
    ...
    <Element VALUE="AB" LABEL="Bridge" FILTER="A" IFCENTITYNAME="IfcCivilElementProxy" IFC4ENTITYNAME="IfcBuildingElementProxy"
    DESCRIPTION="It applies to each bridge with respect to roadbed bridges except temporary overpass bridges.">
      <Element VALUE="AB0" LABEL="General" FILTER="AB" IFCENTITYNAME="IfcCivilElementProxy"
      IFC4ENTITYNAME="IfcBuildingElementProxy" DESCRIPTION="It applies to common facilities of bridges.">
        <Element VALUE="AB000" LABEL="General" FILTER="AB0" IFCENTITYNAME="IfcCivilElementProxy"
        IFC4ENTITYNAME="IfcBuildingElementProxy" DESCRIPTION="It applies to common facilities of bridges." />
      </Element>
      <Element VALUE="AB1" LABEL="Superstructure" FILTER="AB" IFCENTITYNAME="IfcCivilElementProxy"
      IFC4ENTITYNAME="IfcBuildingElementProxy" DESCRIPTION="It applies to the upper structure including the plate and auxiliary
      equipment of the bridge installed above the substructure such as abutments or piers.">
        <Element VALUE="AB100" LABEL="General" FILTER="AB1" IFCENTITYNAME="IfcCivilElementProxy"
        IFC4ENTITYNAME="IfcBuildingElementProxy" DESCRIPTION="It applies to common facilities of the superstructure of the
        bridge." />
        <Element VALUE="AB101" LABEL="Plate" FILTER="AB1" IFCENTITYNAME="IfcBridgeSlab" IFC4ENTITYNAME="IfcSlab" DESCRIPTION="It
        applies to the plate of the superstructure of the bridge.">
          <Element VALUE="AB101A" LABEL="Slab" FILTER="AB101" IFCENTITYNAME="IfcBridgeSlab" IFC4ENTITYNAME="IfcSlab"
          DESCRIPTION="" />
          ...
        </Element>
      </Element>
    </Element>
  </StructureElements>
</PBS>
```

Figure 8. PBS document (shown partially) for mapping between 3D geometry object and IFC entity.

```xml
<MappingIFCRailway4Revit>
  <Project> <Atts Data="Globalid" ValueType="STRING" Value="JSW9naYtoUSHHv159LOLtQ" /> ... </Project>
  <StructureElements>
    <Element Globalid="LDyn4mJBakCAgUWlgS2zMQ" Data="Site" Description="" OidInSWDB="" IfcEntityName="IfcSite"
    isSpatialElement="True" Ifc4EntityName="IfcSite" PBSCode="">
      <Atts Data="Globalid" ValueType="STRING" Value="LDyn4mJBakCAgUWlgS2zMQ" isEnum="False" />
      ...
      <Element Globalid="0+6AkOSlBkGa2k0lqmE9SA" Data="Jaenaechon Bridge" Description="" OidInSWDB="" IfcEntityName="IFCBRIDGE"
      isSpatialElement="True" Ifc4EntityName="IfcBuildingStorey" PBSCode="">
        <Atts Data="Globalid" ValueType="STRING" Value="0+6AkOSlBkGa2k0lqmE9SA" isEnum="False" />
        ...
        <Atts Data="Functiontype" ValueType="IFCBRIDGEFUNCTIONTYPEENUM" Value="" isEnum="True" EnumData="MAIN_SPAN_BRIDGE" />
        <Element Globalid="OOyLl1P2y0GOtXVgrpyrrA" Data="BR1-2" Description="" OidInSWDB="" IfcEntityName="IFCBRIDGESPAN"
        isSpatialElement="True" Ifc4EntityName="IfcBuilding" PBSCode="">
          ...
          <Element Globalid="xmPXvzodgkK4b3qwdcgK/g" Data="Substructure" Description="" OidInSWDB=""
          IfcEntityName="IFCBRIDGESPACEPART" isSpatialElement="True" Ifc4EntityName="IfcBuildingStorey" PBSCode="">
            ...
            <Element Globalid="pC8mMNk8q0qI3DqQRVB1MQ" Data="Pier_Group" Description="" OidInSWDB=""
            IfcEntityName="IFCBRIDGESPACEPART" isSpatialElement="True" Ifc4EntityName="IfcBuildingStorey" PBSCode="">
              ...
              <Element Globalid="Ob/QaX0wv0Gnwtei0/RA6A" Data="Pier_2-4" Description="" OidInSWDB=""
              IfcEntityName="IFCBRIDGESPACEPART" isSpatialElement="True" Ifc4EntityName="IfcBuildingStorey" PBSCode="">
                ...
                <Element Globalid="pSMyBLhGB06HLAvQv7x8RA" Data="" Description="Concrete_Pier" OidInSWDB="695957"
                IfcEntityName="IFCBRIDGEPIER" isSpatialElement="False" Ifc4EntityName="IFCCOLUMN" PBSCode="AB202">
                  ...
                  <Atts Data="Predefinedtype" ValueType="IFCBRIDGEPIERTYPEENUM" Value="" isEnum="True" EnumData="T_TYPE" />
                  <Atts Data="Segments" ValueType="SETIFCBRIDGEPIERSEGMENT" Value="" isEnum="False" />
pier_design       <Doc xpathEle="/기둥_설계" xpathAtt="header" xpathAttVal="9." />
                  <Geometry> <Vertices Name="M_300 x 450" CompNo="327_0" /> <Triangles Name="M_300 x 450" CompNo="327_0" />
                  </Geometry>
                  <PSETList>
                    ...
                    <PSET Globalid="ppJgvWJtvUKlYQauq9XKxg" Name="Dimensions">
                      <Property Name="Area" Value="600.256173864946" />
                      <Property Name="Length" Value="39.3700787401575" />
                      ...
                    </PSET>
                  </PSETList>
                  ...
```

Figure 9. iMapDoc (shown partially) connecting 3D model object, IFC data, and engineering document fragment.

```xml
<Geometry>
    ...
    <Vertices CompNo="327_0">
        -98.735870406,3.922485127,0.000000000
        -98.608448124,3.916225271,0.000000000
        -98.482252987,3.897505991,0.000000000
        -98.358500326,3.866507563,0.000000000
        -98.238381944,3.823528519,0.000000000
        ...
    </Vertices>
    <Triangles CompNo="327_0">
        18,20,21
        22,23,21
        18,21,23
        15,16,17
        19,20,18
        ...
    </Triangles>
    ...
</Geometry>
```

Figure 10. Geometric information to represent the 3D object in IPF.

4. Experimental Verification via Retrieval of Document Fragments Related to the 3D Model Object

4.1. Process of the Experimental Verification for Connected Document Fragments and 3D Model Object

The contents in the XML-formatted structured engineering document can be retrieved using the information included in the selected object of the bridge components via the interface shown in Figure 7a or Figure 7b. XML element tags, attributes of the element, and text data (PCDATA of an XML element) are used as keywords to match a specific segment in the document fragments. The query, response, and management of the XML data process are performed by combining the document object model (DOM) to treat XML data for language independently adopted as standard by W3C [28], regular expression to search string patterns introduced by Kleene [29], and XPath defined by W3C as a standard XML data query language [30]. Figure 11 shows the conceptual process for retrieving document fragments from structured engineering documents. The entire document retrieval process occurs in the Revit environment, as shown in Figure 11. Information on the 3D model can use both the iMapDoc developed according to the process described in Section 3.2 and the data included in the IPF. Figure 12 illustrates the process that occurs during the *Query process* shown in Figure 11. Exact word matching is prioritized; however, if an exact matching element does not exist, then the module matches semantically similar words using the *Dictionary for synonyms* shown in Figure 11. The *Dictionary for synonyms* database was developed in this study to increase the accuracy of word matching; it comprises the attributes of *ID, Korean Word, English Word, Abbreviation, Symbol*, and *Synonym*.

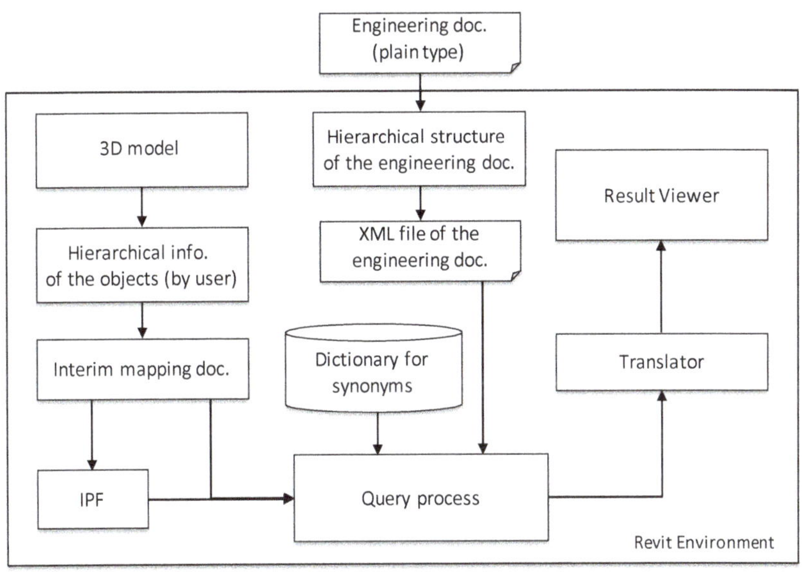

Figure 11. Process for retrieving document fragments from structured document in 3D model view.

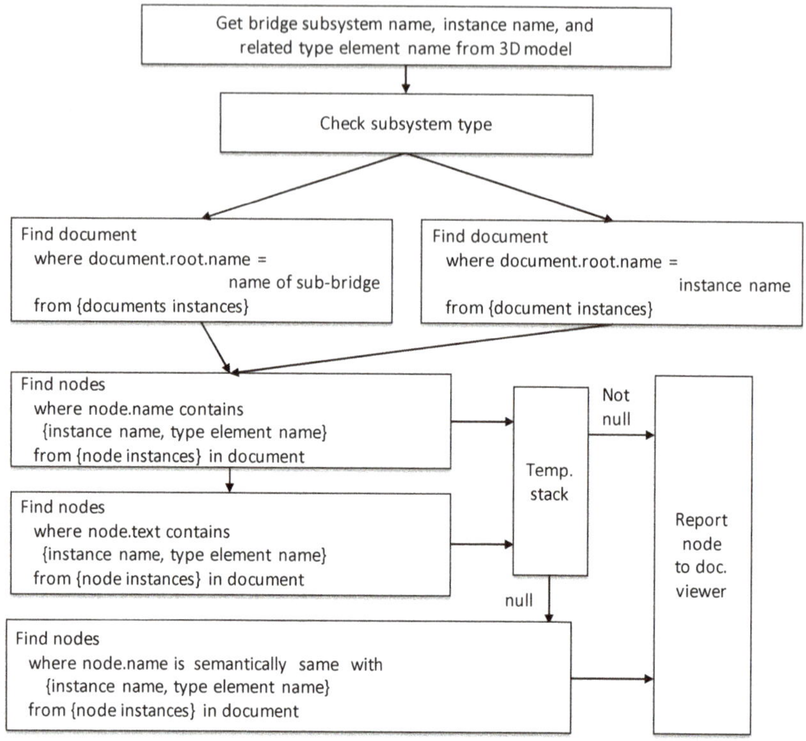

Figure 12. Query process for extracting specific nodes from structured engineering document in 3D model view.

4.2. Connection Review of the Retrieved Document Fragments and the 3D Model Objects

Figure 13 shows a part of the data of the retrieved structured engineering document related to the *Concrete_Pier* object under *Pier_2-4*. These data are presented in Figure 7e, where the results are shown in a list view (① of Figure 14). *Pier_2-4*, as well as its parent and children elements, can be used to retrieve *Concrete_Pier*-related elements and information because the object selected in Figure 7b is synchronized with iMapDoc (see line 19 of Figure 9 and ② of Figure 14). The *Dictionary for synonyms* database was also used to retrieve the related content. The results selected by the user among the retrieved document contents are integrated with the IPF. The connection information between the 3D object and document content is stored in the *Doc* element in iMapDoc, as shown in line 23 of Figure 9 (③ of Figure 14). The connection information indicates the location of the related document content translated into XML format according to the method described in Section 2 using XPath; line 23 of Figure 9 represents ./*pier_design*[@*header* = 9.]. The *Doc* element can be added to all children elements of *StructureElements* of iMapDoc. The selected object from the 3D model or iMapDoc can be connected to the IFC entity mutually through the element name and the *GlobalId* attribute of the iMapDoc. Figure 15 shows a part of the IPF generated using a developed converting module including the *Pier_2-4* object. The *Pier_2-4* object was implemented as the *IfcBridgeSpacePart* entity, and its attributes are consistent with the attributes of *GlobalId* and *Data* in line 17 of Figure 9. *Concrete_Pier* was implemented as the *IfcBridgePier* entity, as shown in *#2976952* in Figure 15, corresponding to line 19 of Figure 9 (④ of Figure 14). The elements of iMapDoc and the entities of IPF share the entity name and global ID; the IFC and Revit are connected through the information of *OidInSWDB* and *PBSCode* stored in iMapDoc. The *Representation* attribute representing the shape information of the *IfcBridgePier* entity uses a subtype of the *IfcProductRepresentation* entity, and the *IfcProductDefinitionShape* entity is used for the representation, as shown in *#1273678*. The actual data are implemented using the *IfcShapeRepresentation* entity (*#1274058*), which utilizes the information in line 24 of Figure 9 (⑤ of Figure 14). The *IfcBridgeSpacePart* (*#2971755*) and *IfcBridgePier* (*#2976952*) were connected through the *IfcRelContainedInSpatialStructure* entity (*#1268533*) since they are spatial and physical entities defined by the extended IFC, respectively. The *pier_design*-related information (line 23 of Figure 9) of the engineering document connected to the object of *Concrete_Pier* (line 19 of Figure 9) represents the *IfcDocumentInformation* entity (shown in #4852759 in Figure 15). Among the 17 attributes of the *IfcDocumentInformation* entity in the IFC4 schema, we generated values for attributes *Identification*, *Name*, *Description*, *Location*, *ElectronicFormat*, and *Status* automatically based on iMapDoc. The *Identification* attribute is used to identify documents, and the value data (*pier_design-pSMyBLhGB06HLAvQv7x8RA*) combined with the name (*pier_design*) of the connected document fragment and the *GlobalId* attribute data (*pSMyBLhGB06HLAvQv7x8RA*) of the parent element (*Concrete_Pier*) are used (⑥ of Figure 14). The document fragment name (*pier_design*) is used for the *Name* attribute, and the file name of the XML-formatted construction document (*Jaenaechon.xml*) is represented in the *Description* attribute. The *Location* attribute specifies the location of the document in the form of a uniform resource identifier (URI); it ($Jaenaechon/pier\_design[@header = 9.]$) is represented as using XPath, combining the path of the root element (*Jaenaechon*) of the construction document and the attribute data in the *Doc* element in Figure 9 to specify a concrete path for querying the document fragment. We specify *application/xml* according to the XML Multipurpose Internet Mail Extension (MIME-type) as the attribute data of *ElectronicFormat* for the type of media. The *Status* attribute indicates the document's status represented by the *IfcDocumentStatusEnum* type, and we designated the *.FINAL.* value by default.

This example verifies that 3D models, engineering documents, and IFC can be run together using the proposed integrated approach. In particular, the document files were not deliberately reconstructed and were used as they were created in the bridge design work. Furthermore, we can map information objects from different sets of information using the names of the sub-bridges and components. The same names for subsystems and bridge components are typically used in bridge design drawings and documents. Therefore, it is believed that this method can facilitate the establishment of an integrated information environment for 3D bridge models and engineering documents.

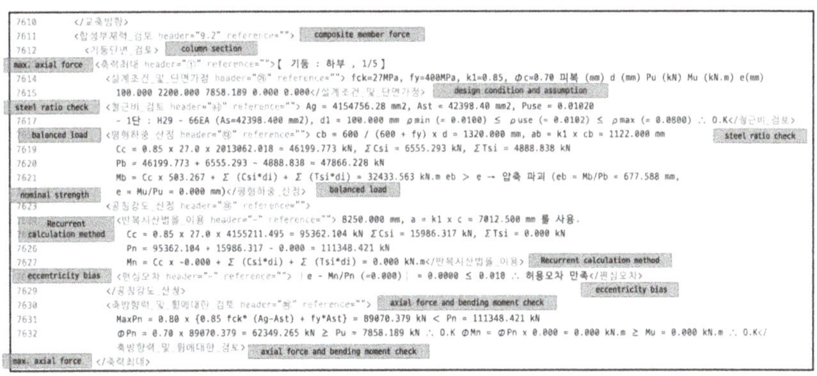

Figure 13. Data (shown partially) of retrieved structured engineering document related to *Concrete_Pier* object.

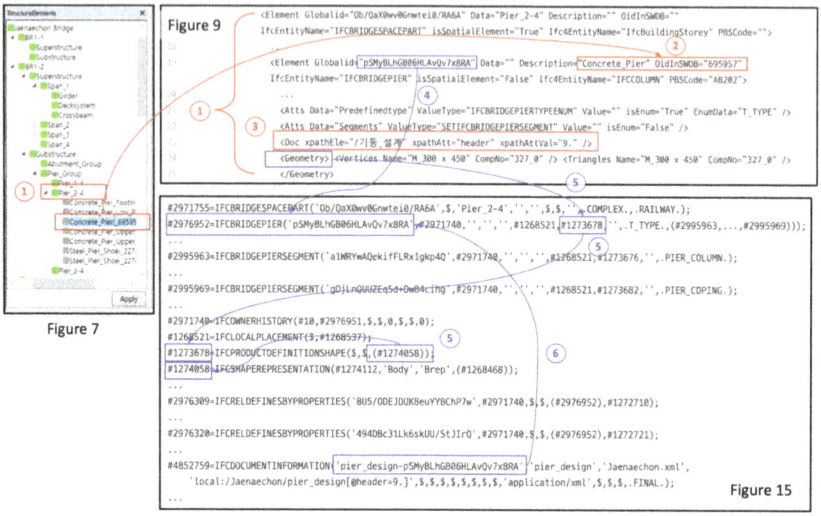

Figure 14. Information flow among BIM authoring tool–IFC–engineering document.

```
...
#1268533=IFCRELCONTAINEDINSPATIALSTRUCTURE('iOrfjRtaGEep7VgmZGFXSw',#2971740,'Pier_2-4',$,(...,#2976952,...),#2971755);
#1348174=IFCRELASSOCIATESDOCUMENT('Q/GRFDkNSGK5fo/gWpcRiw',$,$,$,(#2976952),#4852759);
...
#2971755=IFCBRIDGESPACEPART('Ob/QaX0wv0Gnwtei0/RA6A',$,'Pier_2-4','','',$,$,'',.COMPLEX.,.RAILWAY.);
#2976952=IFCBRIDGEPIER('pSMyBLhGB06HLAvQv7x8RA',#2971740,'','','',#1268521,#1273678,'',.T_TYPE.,(#2995963,...,#2995969)));
...
#2995963=IFCBRIDGEPIERSEGMENT('a1WRYwAQekifFLRx1gkp4Q',#2971740,'','','',#1268521,#1273676,'',.PIER_COLUMN.);
...
#2995969=IFCBRIDGEPIERSEGMENT('gDjLnQUUZEq5d+Dw84cihg',#2971740,'','','',#1268521,#1273682,'',.PIER_COPING.);
...
#2971740=IFCOWNERHISTORY(#10,#2976951,$,$,0,$,$,0);
#1268521=IFCLOCALPLACEMENT($,#1268537);
#1273678=IFCPRODUCTDEFINITIONSHAPE($,$,(#1274058));
#1274058=IFCSHAPEREPRESENTATION(#1274112,'Body','Brep',(#1268468));
...
#2976309=IFCRELDEFINESBYPROPERTIES('BU5/ODEJDUK8euYYBChP7w',#2971740,$,$,(#2976952),#1272710);
...
#2976320=IFCRELDEFINESBYPROPERTIES('494DBc31Lk6skUU/StJIrQ',#2971740,$,$,(#2976952),#1272721);
...
#4852759=IFCDOCUMENTINFORMATION('pier_design-pSMyBLhGB06HLAvQv7x8RA','pier_design','Jaenaechon.xml',
    'local:/Jaenaechon/pier_design[@header=9.]',$,$,$,$,$,$,$,$,'application/xml',$,$,$,.FINAL.);
...
```

Figure 15. IPF (shown partially) generated using integrated assistant module related to *Pier_2-4* object.

5. Conclusions

Providing appropriate information in a valid and accessible format to construction participants with various knowledge backgrounds is the most important consideration for successful construction and management activities. Research pertaining to the efficient interconnectivity between 3D models and engineering records stored in documents is still in its infancy, whereas research for providing pertinent information based on the visibility of 3D model applications has been actively conducted in recent decades. As one of the methods to effectively deliver information used in the construction and management of bridge structures to users, we focused on integrating the bridge information model and engineering documents, which serve as references throughout the lifecycle. To achieve this, we first proposed a structuralizing method to transform unstructured plain text-typed engineering documents to XML documents by classifying titles and contents, defining hierarchies, and reconstructing them using explicit and apparent semantic structures. Second, an extended IFC schema that can handle bridge structure information was adopted, and IFC entities that can connect IFC and document information were selected. Finally, a Revit-based add-in module was developed to assist in the integrated operation of bridge information models and engineering documents. The module includes the functions of generating IFC spatial objects, placing physical objects into spatial objects, and interconnecting XML document contents related to model objects. Furthermore, it generates an extended IFC-based IPF that retains hierarchical object relationships and document connection information.

The main contribution of this study was the proposal of a new approach toward achieving an interconnected and integrated operation of the BIM authoring tool–IFC–engineering document using an interim information document named iMapDoc beyond the IFC–engineering document linkage. Each specialized engineering process evolves continuously as new engineering techniques and design philosophies are developed, which can naturally support the process conducted in each independent engineering domain. We expect the proposed process in this study to serve as a reference for future studies as follows:

- Document content management using the IFC: Although the IFC contains most of the information that needs to be dealt with during the entire lifecycle of a facility, few studies suggest a technical process to control the document fragments. This study explained how IFC manages document fragments with examples;
- Smart infrastructure: The successful adaptation of BIM for buildings promotes the growth of BIM for infrastructure for operating smart infrastructure. Interoperability, as well as data mapping between physical and digital models, are considered to be some of the essential keys of successful BIM for infrastructure [31,32]. The integrated operation process of the BIM authoring tool–IFC–engineering document proposed in this study can be a good reference.

Several issues should be addressed to implement the proposed integrated approach more comprehensively in practical working environments. The proposed approach uses the names of instances and type objects (e.g., types of cross-section, material, and connection) to map different information sets. Therefore, a systematic naming rule should be established to identify bridge components and object types based on the consensus of the project participants. Another drawback is that the developed module cannot process contents in figures or tables. However, the text information contained in the tables can be handled using a specific element node in a structured document. This indicates that the figures and tables can be correctly provided with subtitles.

Author Contributions: Conceptualization, S.I.P., B.-G.K. and G.Z.; methodology, S.I.P.; software, S.I.P. and B.-G.K.; validation, S.I.P. and W.G.; formal analysis, S.I.P.; investigation, S.I.P. and B.-G.K.; resources, S.I.P. and B.-G.K.; data curation, S.I.P. and W.G.; writing—original draft preparation, S.I.P. and B.-G.K.; writing—review and editing, S.I.P. and G.Z.; visualization, S.I.P. and B.-G.K.; supervision, G.Z.; project administration, G.Z.; funding acquisition, G.Z. All authors have read and agreed to the published version of the manuscript.

Funding: This work was partially supported by the National Research Foundation of Korea (NRF) grant funded by the Korea government (MSIT; No. NRF-2021R1A5A1032433).

Conflicts of Interest: The authors declare no conflict of interest.

References

1. Tatum, C.B. Integration: Emerging management challenge. *J. Manag. Eng.* **1990**, *6*, 47–58. [CrossRef]
2. Park, S.I.; Park, J.; Kim, B.G.; Lee, S.H. Improving applicability for information model of an IFC-based steel bridge in the design phase using functional meanings of bridge components. *Appl. Sci.* **2018**, *8*, 2531. [CrossRef]
3. Preidel, C.; Borrmann, A. BIM-based code compliance checking. In *Building Information Modeling—Technology Foundations and Industry Practice*; Borrmann, A.; König, M., Koch, C., Beetz, J., Eds.; Springer International Publishing AG: Cham, Switzerland, 2018; pp. 367–381.
4. Lin, J.R.; Hu, Z.Z.; Zhang, J.P.; Yu, F.Q. A Natural-Language-Based Approach to Intelligent Data Retrieval and Representation for Cloud BIM. *Comput.-Aided Civ. Infrastruct. Eng.* **2016**, *31*, 18–33. [CrossRef]
5. ISO 10303-22:1998. *Industrial Automation Systems and Integration—Product Data Representation and Exchange—Part 22: Implementation Methods: Standard Data Access Interface*; International Organization for Standardization (ISO): Geneva, Switzerland, 1998.

6. Hjelseth, E.; Nisbet, N. Capturing normative constraints by use of the semantic mark-up rase methodology. In *CIB W78-W102 2011 28th International Conference—Applications of IT in the AEC Industry*; Blackwell Publishing Ltd.: Sophia Antipolis, Valbonne, France, 2011; pp. 241–250.
7. Beach, T.; Rezgui, Y.; Li, H.; Kasim, T. A rule-based semantic approach for automated regulatory compliance in the construction sector. *Expert Syst. Appl.* **2015**, *42*, 5219–5231. [CrossRef]
8. Sydora, C.; Stroulia, E. Rule-based compliance checking and generative design for building interiors using BIM. *Autom. Constr.* **2020**, *120*, 103368. [CrossRef]
9. Zhong, B.T.; Ding, L.Y.; Luo, H.B.; Zhou, Y.; Hu, Y.Z.; Hu, H.M. Ontology-based semantic modeling of regulation constraint for automated construction quality compliance checking. *Autom. Constr.* **2012**, *28*, 58–70. [CrossRef]
10. Zhou, P.; El-Gohary, N.M. Ontology-based automated information extraction from building energy conservation codes. *Autom. Constr.* **2017**, *74*, 103–117. [CrossRef]
11. Liu, K.; El-Gohary, N.M. Ontology-based semi-supervised conditional random fields for automated information extraction from bridge inspection reports. *Autom. Constr.* **2017**, *81*, 313–327. [CrossRef]
12. Al Qady, M.; Kandil, A. Concept relation extraction from construction documents using natural language processing. *J. Constr. Eng. Manag.* **2010**, *136*, 294–302. [CrossRef]
13. Zhang, J.; El-Gohary, N.M. Semantic NLP-based information extraction from construction regulatory documents for automated compliance checking. *J. Comput. Civ. Eng.* **2016**, *30*, 04015014. [CrossRef]
14. Song, J.; Lee, J.K.; Choi, J.; Kim, I. Deep learning-based extraction of predicate-argument structure (PAS) in building design rule sentences. *J. Comput. Des. Eng.* **2020**, *7*, 563–576. [CrossRef]
15. Wang, Z.; Wang, Y.; Gao, K. A new model of document structure analysis. In *Lecture Notes in Computer Science—Fuzzy Systems and Knowledge Discovery*; Wang, L., Jin, Y., Eds.; Springer: Berlin/Heidelberg, Germany, 2005; pp. 658–666.
16. Eastman, C.M.; Lee, J.m.; Jeong, Y.S.; Lee, J.K. Automatic rule-based checking of building designs. *Autom. Constr.* **2009**, *18*, 1011–1033. [CrossRef]
17. Jiang, S.; Wang, N.; Wu, J. Combining BIM and Ontology to Facilitate Intelligent Green Building Evaluation. *J. Comput. Civ. Eng.* **2018**, *32*, 04018039. [CrossRef]
18. Zhou, P.; El-Gohary, N.M. Semantic information alignment of BIMs to computer-interpretable regulations using ontologies and deep learning. *Adv. Eng. Inform.* **2021**, *48*, 101239. [CrossRef]
19. Ma, Z.; Li, H.; Shen, Q.P.; Yang, J. Using XML to support information exchange in construction projects. *Autom. Constr.* **2004**, *13*, 629–637. [CrossRef]
20. Kim, B.G.; Park, S.I.; Kim, H.J.; Lee, S.H. Automatic extraction of apparent semantic structure from text contents of a structural calculation document. *J. Comput. Civ. Eng.* **2010**, *24*, 313–324. [CrossRef]
21. Park, S.I.; Lee, S.H. Heuristic solution using decision tree model for enhanced XML schema matching of bridge structural calculation documents. *Front. Struct. Civ. Eng.* **2021**, *14*, 1403–1417. [CrossRef]
22. Choi, J.; Choi, J.; Kim, I. Development of BIM-based evacuation regulation checking system for high-rise and complex buildings. *Autom. Constr.* **2014**, *46*, 38–49. [CrossRef]
23. Opitz, F.; Windisch, R.; Scherer, R.J. Integration of document - and model-based information for project management support. *Procedia Eng.* **2014**, *85*, 403–411. [CrossRef]
24. Park, S.I.; Lee, S.H.; Almasi, A.; Song, J.H. Extended IFC-based strong form meshfree collocation analysis of a bridge structure. *Autom. Constr.* **2020**, *119*, 103364. [CrossRef]
25. buildingSMART International MSG. Industry Foundation Classes Release 4.0.2.1. 2017. Available online: https://standards.buildingsmart.org/IFC/RELEASE/IFC4/ADD2_TC1/HTML/ (accessed on 2 July 2021).
26. buildingSMART International MSG. Industry Foundation Classes Release 4.3RC4. 2021. Available online: https://github.com/bSI-InfraRoom/IFC-Documentation/tree/main/4_3_0_0/rc4 (accessed on 3 July 2021).
27. Autodesk. Revit—Multidisciplinary BIM Software for Higher-Quality, Coordinated Designs. 2021. Available online: https://www.autodesk.com/products/revit/overview (accessed on 2 July 2021).
28. World Wide Web Consortium (W3C). DOM. 2021. Available online: https://dom.spec.whatwg.org/#what (accessed on 18 July 2021).
29. Kleene, S.C. Representation of Events in Nerve Nets and Finite Automata. U.S. Air Force. 1951. Available online: https://www.rand.org/content/dam/rand/pubs/research_memoranda/2008/RM704.pdf (accessed on 20 December 2021).
30. World Wide Web Consortium (W3C). XML Path Language (XPath) 3.1. 2017. Available online: https://www.w3.org/TR/2017/REC-xpath-31-20170321/ (accessed on 10 July 2021).
31. Costin, A.; Adibfar, A.; Hu, H.; Chen, S.S. Building Information Modeling (BIM) for transportation infrastructure—Literature review, applications, challenges, and recommendations. *Autom. Constr.* **2018**, *94*, 257–281. [CrossRef]
32. Merenda, M.; Praticò, F.G.; Fedele, R.; Carotenuto, R.; Corte, F.G.D. A Real-Time Decision Platform for the Management of Structures and Infrastructures. *Electronics* **2019**, *8*, 1180. [CrossRef]

applied sciences

MDPI

Article

Slope Stability Classification under Seismic Conditions Using Several Tree-Based Intelligent Techniques

Panagiotis G. Asteris [1,*], Fariz Iskandar Mohd Rizal [2], Mohammadreza Koopialipoor [3], Panayiotis C. Roussis [4], Maria Ferentinou [5], Danial Jahed Armaghani [6,*] and Behrouz Gordan [7]

1 Computational Mechanics Laboratory, School of Pedagogical and Technological Equation, 15122 Maroussi, Greece
2 Department of Civil Engineering, Faculty of Engineering, University of Malaya, Kuala Lumpur 50603, Malaysia; fariziskandar@siswa.um.edu.my
3 Faculty of Civil and Environmental Engineering, Amirkabir University of Technology, Tehran 15914, Iran; mr.koopialipoor@aut.ac.ir
4 Department of Civil and Environmental Engineering, University of Cyprus, Nicosia 1678, Cyprus; roussis@ucy.ac.cy
5 School of Civil Engineering and Built Environment, Liverpool John Moores University, Liverpool L3 3AF, UK; M.Ferentinou@ljmu.ac.uk
6 Department of Urban Planning, Engineering Networks and Systems, Institute of Architecture and Construction, South Ural State University, 76, Lenin Prospect, 454080 Chelyabinsk, Russia
7 Department of Civil Engineering, Islamic Azad University, Varamin Pishva Branch, Tehran 15914, Iran; bh.gordan@iauh.ac.ir
* Correspondence: asteris@aspete.gr (P.G.A.); danialarmaghani@susu.ru (D.J.A.)

Abstract: Slope stability analysis allows engineers to pinpoint risky areas, study trigger mechanisms for slope failures, and design slopes with optimal safety and reliability. Before the widespread usage of computers, slope stability analysis was conducted through semi analytical methods, or stability charts. Presently, engineers have developed many computational tools to perform slope stability analysis more efficiently. The challenge associated with furthering slope stability methods is to create a reliable design solution to perform reliable estimations involving a number of geometric and mechanical variables. The objective of this study was to investigate the application of tree-based models, including decision tree (DT), random forest (RF), and AdaBoost, in slope stability classification under seismic loading conditions. The input variables used in the modelling were slope height, slope inclination, cohesion, friction angle, and peak ground acceleration to classify safe slopes and unsafe slopes. The training data for the developed computational intelligence models resulted from a series of slope stability analyses performed using a standard geotechnical engineering software commonly used in geotechnical engineering practice. Upon construction of the tree-based models, the model assessment was performed through the use and calculation of accuracy, F1-score, recall, and precision indices. All tree-based models could efficiently classify the slope stability status, with the AdaBoost model providing the highest performance for the classification of slope stability for both model development and model assessment parts. The proposed AdaBoost model can be used as a screening tool during the stage of feasibility studies of related infrastructure projects, to classify slopes according to their expected status of stability under seismic loading conditions.

Keywords: classification; slope stability; tree-based models; random forest; AdaBoost; decision tree

Citation: Asteris, P.G.; Rizal, F.I.M.; Koopialipoor, M.; Roussis, P.C.; Ferentinou, M.; Armaghani, D.J.; Gordan, B. Slope Stability Classification under Seismic Conditions Using Several Tree-Based Intelligent Techniques. *Appl. Sci.* **2022**, *12*, 1753. https://doi.org/10.3390/app12031753

Academic Editor: Chiara Bedon

Received: 16 December 2021
Accepted: 7 February 2022
Published: 8 February 2022

1. Introduction

Geotechnical engineers often employ analytical and empirical methods in order to estimate the safety factor, based on design parameters and engineering properties, of soil or rock material. It is a challenging task to develop an adequate model to efficiently simulate site specific engineering geological conditions and follow the appropriate design approach in order to eliminate the possibility of failure and propose the most cost-effective design.

Slope stability analysis is a standard practice in geotechnical engineering employed for the estimation of the stability of natural or man-made slopes such as embankments of highways, railways, earth dams, tailings, etc. The analysis of slope stability mainly involves the calculation of the factor of safety (FOS), which is defined as the ratio between shear strength and the acting shear stress. The key parameters that define the geometry of the slope (i.e., height and slope inclination) and the material properties (i.e., angle of internal friction, cohesion, and pore water pressure) influence the evaluation of stability of slopes [1–3]. Many sources of uncertainties, such as soil properties and loading, contribute to the stability of a slope [4–6]. The slopes can be classified as stable slopes (SS) or unstable slopes (US), depending on whether their FOS is greater or less than one [7]. The assessment of slope stability is usually performed using analytical techniques, such as the limit equilibrium method (LEM) and finite element methods.

The challenge associated with further development of slope stability analysis methods is to create a reliable generic design tool in order to perform precise evaluations of slope performance. Before the advent of computers, slope stability analysis was conducted using semi-graphical solutions, using manual calculations, or using stability charts [8]. Presently, engineers have developed many computational tools to perform slope stability analysis more efficiently. Geotechnical software based on analytical methods such as the limit equilibrium method (LEM) are widely used by engineers although this method is known to be inadequate when analysing complex slope conditions, requiring more efficient designs, where more sophisticated tools like finite element methods are used [9].

Statistical methods for slope stability classification are based on mathematical formulas that are used in the statistical analysis of research. Multiple regression is a statistical analysis method that can predict the nature of relationship among independent variables and dependent variables. Multiple regression is able to predict the relationship of multiple independent variables against an output variable. This technique is widely used in analysing slope stability problems [10]. For instance, Erzin and Cetin [11] used multiple regression to predict the FOS of homogeneous slopes. The cohesion of soil (c), angle of internal friction (ϕ), unit weight of soil (γ), and seismic coefficient (k) were used as input parameters, and the output parameter was FOS. It was concluded that the predictions made by the multiple regression model were acceptable. In a similar study, Chakraborty and Goswami [12] used the height of cut or slope height H, material properties, cohesion (c), friction (ϕ), slope inclination (β), unit weight (γ), and dimensionless parameter (m) as input parameters to predict the status of stability. They also reported a very similar conclusion to the study by Erzin and Cetin [11]. However, the analyses performed by statistical models are only statistical-based, and they are not able to provide a clear view to researchers and designers [13].

Artificial intelligence (AI) and machine learning (ML) techniques have been successfully implemented in the area of engineering and sciences [14–32] for the last 25 years. The same models were used to solve the slope stability problems [3,11,33–37]. Algorithms like ANFIS, (Adaptive Neuro-Fuzzy Inference System), were applied by Mohamed and Kasa [38] to predict the FOS of slopes and they compared their results from the LEM method. The predictions made by the ANFIS model were acceptable for applications in slope stability prediction. In another study, Kalatehjari et al. [39] utilized particle swarm optimization (PSO) to estimate the FOS of 3D slopes in comparison with a 3D finite element method (FEM) model using material properties (cohesion (c) and friction (ϕ) and unit weight (γ) as input variables. They confirmed a successful application of PSO for 3D slope stability conditions but lower performance for 2D slope stability analysis. Artificial neural network (ANN) as a basic and benchmark AI model was used by Sakellariou and Ferentinou [36], Ferentinou and Sakellariou [37], and Lu and Rosenbaum [40], and its performance was studied to estimate slope stability compared to the LEM slope stability analysis. The results produced by the ANN model were found to concur with the results obtained by the LEM and allowed for the classification of sample observations according to the anticipated failure mechanism. In another study, Samui [41] proposed a support vector machine (SVM) technique for the prediction of FOS and compared it with the ANN results. He found that the SVM was able to receive a slightly higher accuracy in comparison with the ANN technique. In addition, the same SVM model with different kernels, including polynomial, radial basis and spline, was proposed by Samui [35] to classify the FOS of slopes. The accuracy of the model was proven to be very high as it showed 100% similarity when compared to the expected slope stability classification results. It was concluded that the classifications made by the SVM model were acceptable for applications in slope stability predictions; however, when the size of the dataset and/or the dimension of the input vector were high, the performance of the developed models was poor. In a study carried out by Tien Bui et al. [42], decision tree (DT) was used to predict the FOS of slopes and was compared with the results obtained by some other ML/AI techniques such as SVM. The accuracy of the DT model was proven to be acceptable, but it was lower than the SVM model. It is clear that the AI/ML models have enough potential in classifying/predicting slope failure or FOS. Table 1 presents some of the classifications/prediction studies in the areas of slope stability using AI/ML models. In these studies, FOS was set as model output where the model performance was assessed using the coefficient of determination (R^2) and accuracy.

Table 1. Some of the classifications/prediction studies in the areas of slope stability using AI/ML models.

Reference	Model	Input	Data Size	R^2	Accuracy (%)
Amit and Geman [43]	DT	H, C, ϕ, β, rainfall and water level data	118	-	80
Sakellariou and Ferentinou [36]	ANN, SOM	H, c, φ, β, r_u and γ, k_{max}	45	0.94	
Ferentinou and Sakellariou [37]	ANN	H, c, φ, β, r_u and γ	46	0.95	
Lu and Rosenbaum [40]	ANN	H, c, ϕ,r_u and γ	30 datasets	-	99
Samui [41]	SVM	H, c, ϕ, r_u and γ	46 datasets	0.875	-
Hwang et al. [44]	DT	H, c, ϕ, β and γ	6828 datasets	-	72
Das et al. [7]	ANN	H, c, ϕ and γ	46 datasets	0.982	-
Samui [35]	SVM	H, c, ϕ, r_u and γ	32 datasets	1.0	-
Mohamed and Kasa [38]	ANFIS	H, c, ϕ and γ	300 datasets	0.980	-
Gelisli et al. [45]	ANN	H, c, ϕ and γ	100 datasets	0.99	-
Tao et al. [46]	SVM	H, c, ϕ, γ, rainfall data	20 datasets	-	88
Fattahi [47]	ANFIS	H, c, ϕ, β and γ	67 datasets	0.952	-
Qi and Tang [48]	ANN	H, β, γ,c	168 datasets	-	96
Hidayat et al. [49]	ANFIS	H, c, ϕ, γ, and γ	53 datasets	0.96	-
Ray et al. [10]	ANN	H, c, ϕ and γ	-	0.958	-
Sari et al. [50]	ANFIS	H, c, ϕ and γ	30 datasets	0.954	-

H: Height of cut, c: Cohesion of soil, ϕ: Angle of internal friction, β: Slope inclination, r_u: Pore water pressure ratio, k_{max}: seismic coefficient.

In the light of the above discussion, it is clear that ANN and ANN-based models are the main body for the previous investigations. On the other hand, some other techniques, namely, tree-based, performed well in the areas of geotechnics and civil engineering [51–54]. In this study, different classification systems are proposed for slope stability using decision trees (DT), random forest (RF), and AdaBoost tree-based techniques. As presented in Table 1, many researchers used key parameters (i.e., height (H), cohesion (c), friction (ϕ), and unit weight (γ)) for the classification of slope FOS under static conditions. According to our review, there is a limited number of studies aimed at FOS estimation or status of stability classification under dynamic conditions. In the current study, the horizontal component of peak ground acceleration (PGA) is included in the input parameters. Therefore, the contribution of this study concerns, firstly, the use of tree-based models in slope stability classification, and secondly, the inclusion of a component related to dynamic conditions in slope stability. This allows for a more reliable slope stability classification under dynamic loading conditions. The rest of this paper is outlined as follows:

Concepts of earthquake on soil slopes will be discussed in Section 2. Then, Section 3 describes the used models' concepts and fundamental facts. In addition, the same section will provide the needed information about data preparation used for modelling to the readers. Tree-based model developments for slope stability classifications will be provided in Section 4. The results of the study are evaluated and discussed in Section 5. In addition, the best tree-based model to classify slope stability will be discussed in the same section. Future work directions and the conclusion will offer some valuable input to the readers in Section 6.

2. Effect of Earthquake on Soil Slopes

If a slope is situated in a region subject to earthquakes, the design must satisfy these adverse conditions. The effect of the shaking depends on whether the shear strength of the soil material remains adequate during cyclic loading or shaking results in a significant loss of strength. Since deformation is the result of shearing or sliding movement, slope stability analysis is necessary to ensure that the factor of safety is adequate to satisfy dynamic loading and minimize the resulting deformation. In the case of loose, saturated, cohesionless material, the total lack of strength due to cyclic loading might induce liquefaction, which is when a cohesionless saturated or partially saturated soil loses structural strength as a result of an applied stress (such as trembling during an earthquake or another abrupt change in stress condition), and a material that is normally a solid acts as a liquid. Liquefaction assessment requires a more complex analysis and additional data, such as pore water pressure measurements, and is beyond the scope of this paper.

The susceptibility of a slope to failing due to a seismic event is also determined through the critical acceleration coefficient k_y. The coefficient of critical acceleration k_y is an appropriate measure of a soil or rock mass' resistance to earthquake induced sliding. The value of the coefficient depends on the slope inclination β. Essentially, k_y is as important for the sliding block model method [55], as the static safety factor is for the limit equilibrium method; these two variables are linearly related [56]. According to Sarma and Bhave [57], k_y is a measure of safety factor, and is the yield acceleration of the slope. Sarma and Bhave [57] proposed a method to relate these two coefficients which is independent of the assumed failure mechanism and the material properties. The coefficient of critical acceleration k_y is unique for each slope and is calculated when the safety factor is equal to one.

3. Material and Methods

3.1. Data Preparation

During the training process of developing a mathematical model to predict a parameter value as a function of a number of other variables, most researchers tend to focus on computational aspects, while at the same time paying less attention to the database being used for the training and development of the mathematical model.

However, we firmly believe that the main emphasis should be on the database to be used, as it is the database itself that describes the behaviour of the problem being modelled. The database, whether based on experimental or analytical data, is the available knowledge which must be properly utilized during the training process of the development of the mathematical model. In this regard, the database must be reliable with a sufficient amount of data to adequately describe the problem under study.

It should be noted that the phrase "sufficient amount of data" does not necessarily imply a high amount of data, but rather datasets that cover a wide range of combinations of input parameter values, thus assisting in the model's capability to simulate the problem. The demand for a reliable database is particularly crucial in the case of experimental databases, which are databases compiled using experimental results. In this case, significant deviations between experimental values are frequently noticed, not only between experiments conducted by different research teams and laboratories, but even between datasets derived from experiments conducted on specimens of the same synthesis, produced by the same technicians, cured under the same conditions, and tested implementing the same standards and testing instruments.

In light of the above discussion, in this study, in order to develop a comprehensive database for FOS classification under dynamic conditions, a series of models were constructed to calculate FOS using a standard geotechnical software. Figure 1 illustrates a generic limit equilibrium model for the simulated slope. In fact, many slope stability analysis tools use various versions of the methods of slices, such as Bishop simplified. The simplified Bishop method uses the method of slices to discretize the soil mass and determine the FOS. These methods were used in this research, the ordinary method of slices (Swedish circle method/Petterson/Fellenius), Spencer, Sarma, etc. Sarma and Spencer are called "rigorous methods" because they satisfy all three conditions of equilibrium: force equilibrium in both horizontal and vertical directions and moment equilibrium condition. Rigorous methods can provide more accurate results than non-rigorous methods. Bishop simplified or Fellenius are non-rigorous methods, satisfying only some of the equilibrium conditions and making some simplifying assumptions [58,59]. Some of these approaches are discussed below. Finally, slope stability analysis using Bishop simplified is a static or dynamic, analytical, or empirical method to evaluate the stability of earth and rock-fill dams, embankments, excavated slopes, and natural slopes in soil and rock. Slope stability refers to the ability of inclined soil or rock slopes to withstand or undergo movement.

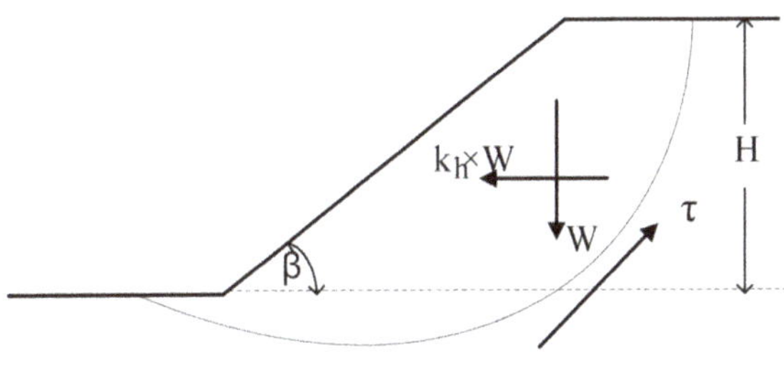

Figure 1. Limit equilibrium model for the stability analysis, (W: weight, τ: shear strength, k_h: seismic coefficient, g: acceleration due to gravity, β: is slope inclination, H: slope height).

The contribution of seismic loading is considered in the current slope stability analysis through the application of a horizontal force component of peak ground acceleration (PGA), that characterizes the amplitude of shaking within the sliding mass. Namely, the slope is assumed to be subjected to a force defined by

$$F_h = k_h W \tag{1}$$

where W is the weight of the sliding mass and k_h is a dimensionless coefficient defined by

$$k_h = PGA/g \tag{2}$$

The process was carried out in several phases to achieve a representative database. Boundary conditions, model dimensions, material properties, and seismic motion were the parameters considered in modelling. To do this, multiple homogeneous slopes with different conditions were modelled. Slopes with heights of 15, 20, 25, and 30 metres and inclinations of 20°, 25°, 30°, and 35° were produced. In terms of rigid behaviour, all of the models were placed on top of bedrock.

The failure criterion used in this method was the Mohr–Coulomb failure criterion

$$\tau = c + \sigma tan\varphi \quad (3)$$

where c: cohesion, φ: friction angle, σ: normal stress for slopes with soils with cohesion and internal friction, for a slope subjected to circular failure. The parametric values used were cohesion of 20, 30, 40, and 50 kPa and internal friction angle of 20°, 25°, 30°, 35°, and 40°. The effect of earthquake motion on slope behaviour was considered in the current analysis. For the purposes of this analysis, the soil unit weight was assumed to be 18 (kN/m^3). The amplitudes were defined as 0.1, 0.2, 0.3, and 0.4 g. On all of the slope models, thirty slices were used as slip surfaces. To achieve FOS values in this analysis, a grid and radius slip surface were used. The calculated FOS should be almost in the centre of the grid by using the grid and radius method. The FOS from the dataset was then separated manually into groups of safe slope or SS and unsafe slope or US in order to meet the objective of analysing and classifying all the slope stability cases in the dataset. Table 2 shows the input and output parameters used in the database development.

Table 2. Input and output variables for slope stability classification.

Property	Variable					
	Slope Height (m)	Angle of Inclination (°)	Cohesion (kPa)	Friction Angle (°)	Peak Ground Acceleration	Factor of Safety
Symbol	H	β	c	ϕ	PGA (m/s^2)	FOS
Category	Input	Input	Input	Input	Input	Output
Min	15	20	20	20	0	0.78
Max	30	35	50	40	3.92	2.46
Average	22.33	25.18	35.3	34.07	1.18	1.20
Std. Deviation	5.6	5	11.18	5.88	1.07	0.35
Variance	31.37	26	124.96	34.59	1.15	0.12

In this study, 700 homogeneous slopes were simulated using GeoStudio which utilizes the LEM method shown in Figure 1, along with the most critical FOS parameters. In these 700 slopes, different values of the mentioned parameters in Table 2 were used and their FOS values were recorded. Based on a literature review conducted, the parameters presented in Figure 1 are considered to be the most important. The best relationships between these input parameters and the output (i.e., FOS) were calculated. In this way, simple regression analysis (one to one relationship) was employed. The highest R^2 value was achieved by the PGA parameter through a polynomial trend-line (as the best trend-line among applied linear, exponential, logarithmic, and power) as follows:

$$\mathrm{FOS} = 0.0612(\mathrm{PGA})^2 - 0.3512(\mathrm{PGA}) + 1.4545 \quad (4)$$

A value of R^2 equal to 0.305 was reported for the above equation. Besides PGA, the parameter ϕ showed the best relationship with FOS values with $R^2 = 0.122$ through an exponential trend-line.

To determine the relative effect of each input parameter on the output parameter, a sensitivity analysis was performed. The following equation was used to perform the same analysis:

$$r_{ij} = \frac{\sum_{k=1}^{m} x_{ik} x_{jk}}{\sum_{k=1}^{m} x^2{}_{ik} \sum_{k=1}^{m} x^2{}_{jk}} \quad (5)$$

where, r_{ij} is the strength of relation between each input and output, x_{ik} is the ith sample of input k, j is the number of each sample in the output set, and m is the total number of data samples. Table 3 shows the strengths of the relations (r_{ij} values) between the inputs and output (FOS). The sensitivity analysis results showed that the input parameters have a great influence on the FOS. Parameter ϕ had the highest impact on FOS values followed by H, β, C, and PGA. The results obtained were in line with previous studies [60,61].

Table 3. Sensitivity analysis of input and output variables.

Input Parameter	H	β	C	ϕ	PGA
r_{ij}	0.930	0.924	0.915	0.962	0.616

3.2. Overview of Research Methodology

A review of past related studies that utilize AI in slope stability methods was first conducted in order to choose the parameters to be used in the dataset required for training and testing the DT, RF, and AdaBoost models. The review revealed an absence of studies considering the PGA as a parameter in the performance of slope stability analysis. Subsequently, the FOS values were estimated using intelligent techniques. For this purpose, DT, RF, and AdaBoost were utilized based on the most influential parameters for slope stability performance as mentioned before for the input parameters. The results of the DT, RF, and AdaBoost model were compared to the results from the GeoStudio software to observe the performance of the DT, RF, and AdaBoost methods. Results of both methods were evaluated using performance indicators and the best model was selected and introduced for the problem of this study. Figure 2 presents a flowchart of the research methodology followed in this study.

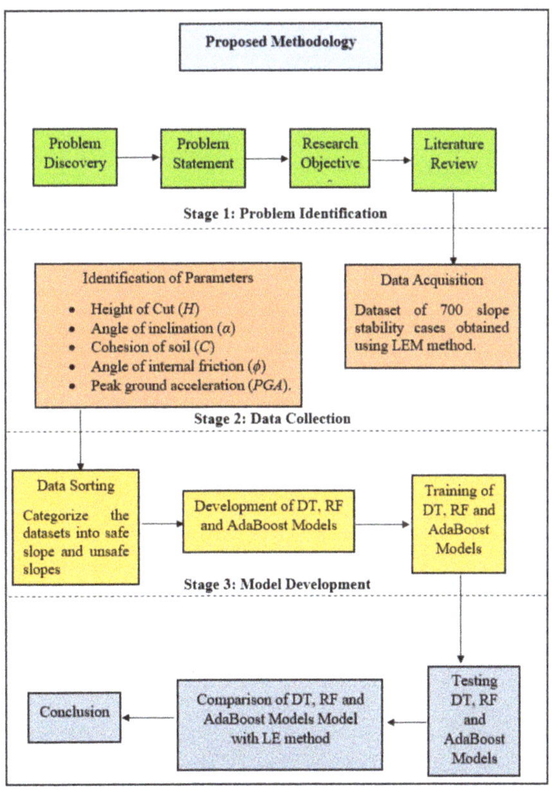

Figure 2. Procedure flowchart for FOS classification.

3.3. Decision Tree (DT)

DT is an AI technique that uses conditional judgement rules to divide predictor variables into homogeneous categories. The aim of DT specification is to find a set of decision rules for predicting an outcome from a set of input boundaries [62]. The DT is referred to as a predictive data mining tree depending on whether the target variables are objective or subjective [63]. Classifying the FOS of slopes from multiple input parameters is possible because modelling complex relationships between multiple input variables with an output variable is possible with a DT model as it will have both categorical and continuous variables without making any conclusions about the distribution of the provided data [64]. Furthermore, DT models are simple to implement, and the prediction results are simple to understand. The findings of the DT model revealed the relative significance of input parameters to the output parameter [65].

A root node, internal nodes, and leaf nodes make up a DT structure. All of the input variables are stored in the root node. A decision function is connected with an internal node, which may have two or three branches. The output of a given input vector is represented by a leaf node [42]. Figure 3 shows the flowchart of procedures conducted for the modelling of a DT model. The procedure of modelling a DT model is governed by two steps: tree building and pruning.

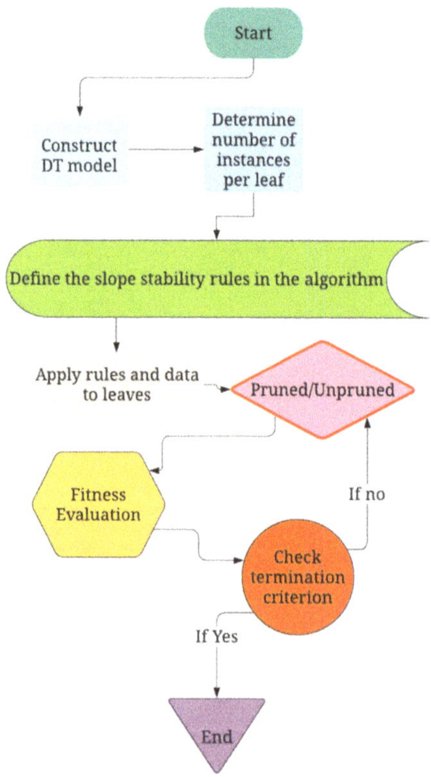

Figure 3. Methodology flowchart for DT modelling.

In the first step, the root node of the DT is defined by determining the input vector with the maximum gain ratio. The dataset is then divided into sub-nodes depending on the root values. For discrete input variables, each potential value is represented by a sub-node of the tree [66]. The gain ratio is then calculated for each of the sub-nodes separately in the second process, and the process is replicated until all of the instances in a node are classified the same way. Leaf nodes are such nodes, and their names are the class values. Since the tree produced during the design process will have a large number of branches, it will be vulnerable to over-fitting [67], it must be pruned in order to improve the prediction performance for new data. Tree pruning can be divided into two categories: pre-pruning and post-pruning. In the case of pre-pruning, the tree's development will be halted before another criterion is true; in the case of post-pruning, the whole tree will be grown first, and then the finished subtrees will be replaced by leaves based on the tree's flaw relation before and after eliminating sub-trees. More explanations regarding DT models can be found in [54].

3.4. Random Forest (RF)

RF, also known as random decision forest, is an ensemble modelling technique for grouping, regression, and other tasks that works by training a vast group of DTs and then outputting the category that is the average approximation (regression) of the individual trees [68]. The values of an independent random variable are used to develop the individual DTs. On the basis of voting, classification models estimate the value yielded by individual trees [69]. The basic RF algorithm utilizes the random subspace method. RFs are often used in industries as "black box" models because they provide accurate estimates over a broad variety of data with no configuration [70].

The DTs in the RF model recognize rules and patterns from the input data. The output parameter (FOS) can be easily measured using these rules and patterns for any new collection of results. The gain ratio formula can be used to rank the most important parameters of slope failures. To solve the issue of over-fitting, mathematical methods such as conservative pruning are used subsequently [71]. Figure 4 shows the flowchart of procedures for RF modelling.

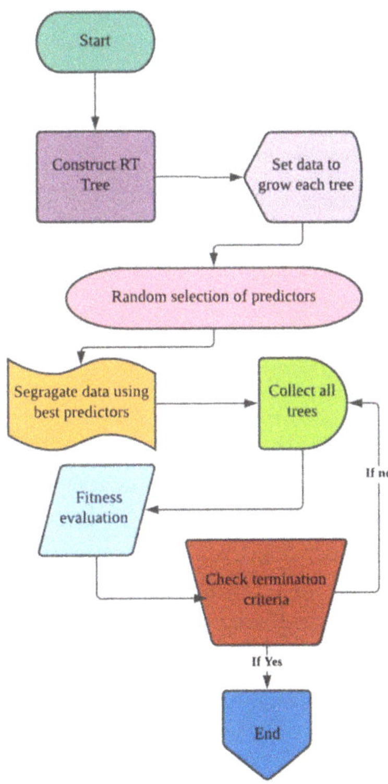

Figure 4. Methodology flowchart for RF modelling.

3.5. AdaBoost Algorithm

Adaptive Boosting, also known as AdaBoost, is a boosting algorithm that attempts to use weighted derivatives of the same testing dataset rather than sub-samples [72]. The benefit of this approach is that the algorithm does not need a large amount of data because it uses the same training dataset twice [73]. The algorithm is well-known for producing good results when constructing ensemble classifiers [74]. To get a classification model of the ensemble prediction function H:X → (−1, +1) shown in Equation (6), the AdaBoost machine learns using a series of weak learners or classifiers.

$$H(x) = sign\left(\sum_{m=1}^{M} a_m H_m(x)\right) \quad (6)$$

where $H(x)$ is the output of the developed ensemble classifier, a_1...., a_m, are a set of weights, and $Hm(x)$ is the performance of the weak learners m∈(1, ..., M) that are combined to get $H(x)$. In each round of the algorithm, the weights allocated to the training dataset are determined by how previous classifiers behaved. The algorithm then works on the specimens or data sets that have already been mistakenly classified in this case. Figure 5 shows the flowchart of procedures for AdaBoost modelling. More information on the AdaBoost concept can be found in the other studies ([75,76]).

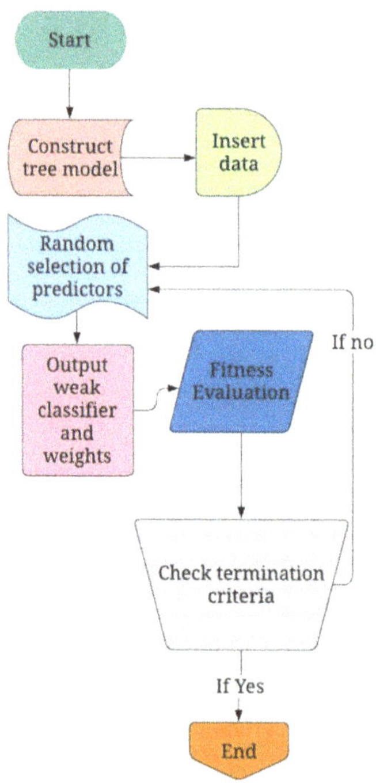

Figure 5. Methodology flowchart for AdaBoost modelling.

3.6. Performance Indicators

To measure the performance of the results obtained from the DT, RF, and AdaBoost models against each other and the expected results obtained from the GeoStudio software, a few performance indicators were used. These performance indicators were accuracy, precision, recall, F1-score, and ROC curve. All the models were subjected to the performance indicators to observe their effectiveness. Accuracy is the ratio of the number of correctly classified predictions divided by the total number of projections. It ranges from 0 to 1.

Equation (7) shows the calculation of accuracy where True Positive and True Negative are correct predictions made by the model.

$$\text{Accuracy} = \frac{\text{TruePositive} + \text{TrueNegative}}{\text{Total number of samples}} \tag{7}$$

Precision is the measurement of positive class predictions that actually belong to the positive class, which in turn calculates the accuracy of the minority class. This calculation is expressed in Equation (8) where the False Positive represents the false positive prediction made by the model.

$$\text{Precision} = \frac{\text{TruePositive}}{\text{TruePositive} + \text{FalsePositive}} \tag{8}$$

Recall is a statistic index that measures how many accurate positive assumptions were made out of all possible positive expectations. Unlike precision, which only considers true positive predictions out of all predictions, considering the positive predictions that were wrong. This calculation is expressed in Equation (9) where the False Negative represents the false negative prediction made by the model.

$$\text{Recall} = \frac{\text{TruePositive}}{\text{TruePositive} + \text{FalseNegative}} \tag{9}$$

F1-score is a method for combining precision and recall into a single measure that encompasses both. Neither precision nor recall can provide the full picture on their own. We may have excellent precision but poor recall, or vice versa, poor precision but good recall. With the F1-score, all issues with a single score can be expressed (Equation (10)).

$$F1 - score = 2 \times \frac{Precision \times Recall}{Precision + Recall} \tag{10}$$

ROC curve or receiver operating characteristic curve is a graph of the false positive rate (x-axis) vs. the precision (y-axis) with a variety of candidate thresholds ranging from 0.0 to 1.0. The false positive rate is determined by dividing the total number of false positives by the total number of false positives and true negatives. With all the performance indicators mentioned above, the area under the ROC curve could be obtained for each model. This value will represent the effectiveness of each model.

4. Development of Tree-Based Techniques

In order to develop the models implemented in this study, the hyperparameters of each model were optimized. A parametric analysis was performed on the parameters of each model because the models needed to be adjusted for each problem and dataset. Here, three types of DT, RF, and AdaBoost models were implemented, each of which had specific parameters related to its structure. In each section, these parameters were defined, and various values of their parameters were analysed in order to find the optimal structure. The details of each model are presented in the following.

4.1. DT Model

To obtain the most effective DT model, several models were developed using different numbers of parameters. Table 4 reports the parameters used for modelling in this study. Upon experimenting with the values of the number of instances in leaves, minimum limit of the split subset, and maximal tree depth, the most effective DT model with the optimal value of these parameters was obtained. In addition, Figure 6 shows the tree flowchart of the proposed DT model for classifying slope stability.

Table 4. The optimal parameters obtained by the DT model.

DT Parameter	Value
Minimum number of instances in leaves	7
Minimum limit of the split subset	5
Maximal tree depth	7

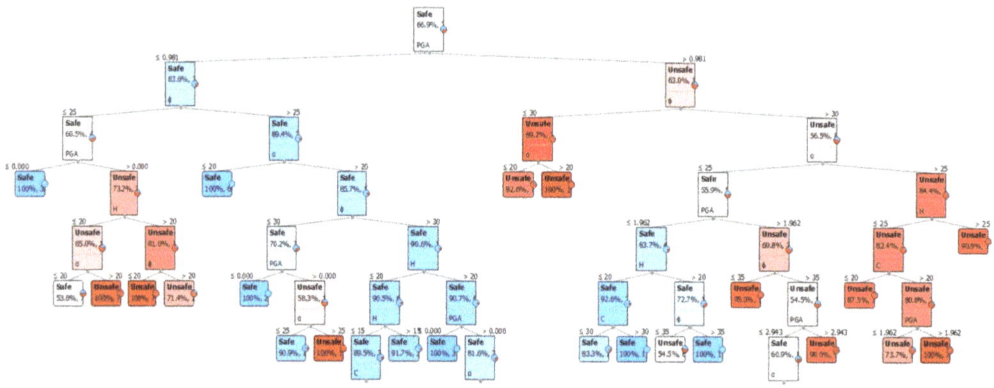

Figure 6. The optimal DT model for FOS classification.

In the training phase, 75% of the dataset was used (525 slope cases), which is similar to a study conducted by Piryonesi and El-Diraby [70]. The data was selected randomly, and the input parameters were inserted into the model. In the testing phase, 25% of the dataset was used, which corresponds to 175 slope cases. Figure 7 shows the results of the DT model in the classification of the FOS for training and testing sets. According to the training set, the DT model classified 300 safe slopes and 162 unsafe slopes accurately, while classifying 12 safe slopes and 21 unsafe slopes, wrongly. In addition, in the case of the testing set, the DT model classified 109 safe slopes and 47 unsafe slopes accurately, while classifying 3 safe slopes and 16 unsafe slopes, wrongly. Later, the results of the DT from both phases were observed using the performance indicators accuracy, precision, recall, F1-score, and ROC curve.

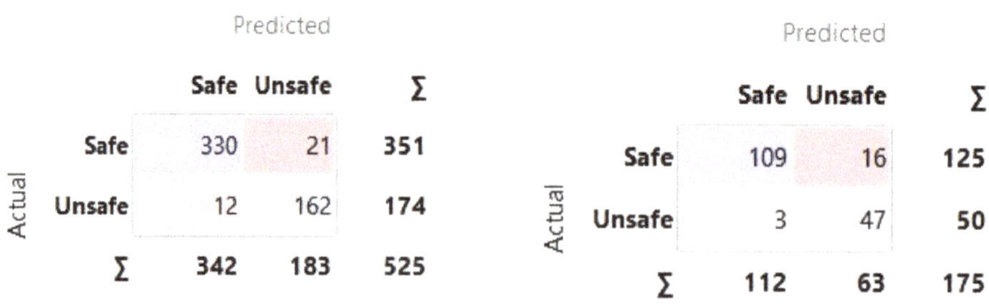

(A) (B)

Figure 7. The DT model results for FOS classification: (**A**) Training and (**B**) Testing.

4.2. RF Model

A similar modelling process was completed for the RF technique aiming at classification of slope stability considering FOS values of more than one as safe (SS) and less than one as unsafe (US). After experimenting with different numbers of trees and the minimum limit of split subsets, the most effective RF model with optimal values was obtained (Table 5). The same portions of DT model were used for the training and testing phases. Figure 8 displays the results obtained by the RF technique for the classification of slope stability for the training and testing phases. Considering the training phase, the RF technique classified 344 safe slopes and 169 unsafe slopes accurately, while classifying 5 safe slopes and 7 unsafe slopes, wrongly. In the case of the testing phase, the RF model was able to classify 116 safe slopes and 44 unsafe slopes accurately, while wrong classification of 9 safe slopes and 6 unsafe slopes, was reported. As with the DT model, the results obtained by the RF model are assessed and discussed later.

Table 5. The optimal parameters obtained by the RF model.

RF Parameter	Value
Number of trees	7
Minimum limit of the split subset	5

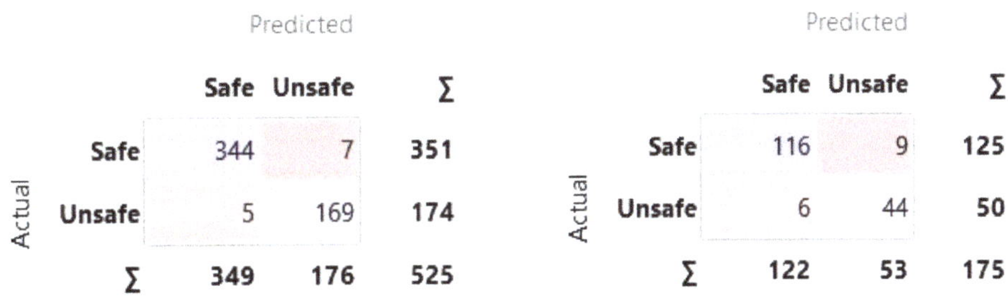

(A) (B)

Figure 8. The RF model results for FOS classification: (**A**) Training and (**B**) Testing.

4.3. *AdaBoost Model*

The same data with five input parameters under seismic condition was used to classify slopes as safe and unsafe. As with the previous parts, it was important to obtain the optimal parameters of the model, which was AdaBoost in this sub-section. Several parametric studies were conducted to get the most accurate AdaBoost model. The optimal AdaBoost parameters for the expressed aim are presented in Table 6. It should be mentioned that a different base model could be selected for the modelling of AdaBoost where DT was the best among them for solving the defined problem. As a result, the proposed AdaBoost model was able to classify 351 safe slopes and 174 unsafe slopes accurately, with no wrong classification results by AdaBoost in the training or model development phase (Figure 9). However, during the testing or model evaluation part, there were several wrong cases. An accurate value of 120 safe slopes and 43 unsafe slopes were reported for the testing part, while 7 safe cases and 5 unsafe cases were obtained wrongly (Figure 9). It seems that the classification results obtained by the AdaBoost model are slightly better than those obtained by the RF and DT techniques. It is important to mention that the evaluation of the proposed models was not the aim of this section and this will be reported in the following section.

Table 6. The optimal parameters obtained by the AdaBoost model.

AdaBoost Parameter	Value
Base Parameter	DT
Number of estimators	6

(**A**) (**B**)

Figure 9. The AdaBoost model results for FOS classification: (**A**) Training and (**B**) Testing.

5. Results and Discussion

This section presents the comparison of results obtained from the DT, RF, and AdaBoost models. The results obtained from these models were subjected to several performance indicators: namely, accuracy, precision, recall, F1-score, and the area under ROC curve or AUC (area under curve) to determine which method was the most accurate and effective for slope stability classification. Here, the testing phase of the datasets was considered for the validation of each tree-based model. This is a common method of evaluation or model assessment to understand the level of accuracy during training/model development. On the other hand, the training stage results showed that the proposed AdaBoost model could be considered as perfect, and therefore, there is no need to discuss further about this stage and have any comparison between models. Table 7 shows the comparison of the testing stage results obtained by the indicators: i.e., accuracy, precision, recall, F1-score, and AUC of ROC. In addition, the ranking procedure proposed by Zorlu et al. [77] was applied in this table. The ranking system is very easy to understand. In this system, the most accurate performance index receives the highest rank. According to Table 7, the model that showed the highest accuracy was AdaBoost as it obtained the highest rank value, which was 13. The second most accurate model was the RF, which obtained a total rank value of 10. The lowest accurate model was the DT model, with a total rank value of 7. Except for the AUC, AdaBoost achieved better accuracy and performance compared to the RF and DT models. It is important to note that the RF also received a high degree of accuracy, and it can be used for slope stability classification by the other researchers or engineers. For a better comparison, Figure 10 shows the classification results of the DT, RF, and AdaBoost models from the testing phase compared to the FOS results obtained with the GeoStudio software. As stated earlier, 175 data samples, which constituted 25% of the whole data, were used for each model in the testing phase. It is clear from Figure 10 that the AdaBoost technique was able to record an outstanding performance with the lowest number of unmatched answers (i.e., 11). The number of matched and unmatched for RF and DT were 160 and 15, and 156 and 19, respectively, confirming the RF model's superiority over the DT in slope stability classification. Overall, the error rate during the testing phase was very low, which reflected the high-performance level of the model development during the training phase. It was concluded that the best performing model for slope stability classification was the AdaBoost, and that it could be used in this field for the same purpose to minimize the associated risk.

Table 7. Modelling results for the testing datasets of DT, RF, and AdaBoost for slope stability classification.

Model	Performance Indicators					Rank					
	AUC	Accuracy	F1	Precision	Recall	AUC	Accuracy	F1	Precision	Recall	Total
DT	0.968	0.891	0.895	0.908	0.891	3	1	1	1	1	7
RF	0.961	0.914	0.915	0.916	0.914	2	2	2	2	2	10
AdaBoost	0.910	0.931	0.931	0.931	0.931	1	3	3	3	3	13

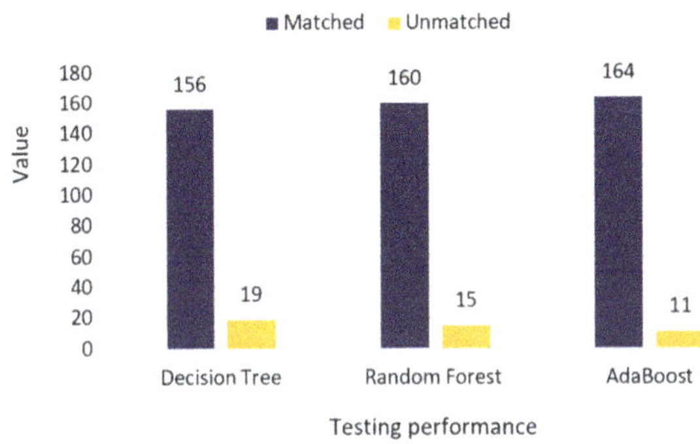

Figure 10. Chart of results obtained from the models compared to expected results.

6. Conclusions and Future Works

To achieve the aim of this study, tree-based models including DT, RF, and AdaBoost were developed to classify the stability of 700 slopes (464 safe slopes and 236 unsafe slopes) under seismic condition, which were modelled and analysed in GeoStudio software. The variables of H, β, C, ϕ, and PGA were set as model inputs for the classification of slopes where FOS ≥ 1 and FOS < 1 was considered for safe and unsafe slopes, respectively. To measure the performance of the DT, RF, and AdaBoost models, accuracy, precision, recall, F1-score, and AUC as performance indices were calculated for both stages of training and testing. After conducting modelling procedures of classification, the best technique was selected based on the performance indices' results. From the training part, it was found that the AdaBoost was a perfect technique capable of achieving the highest possible performance compared to the other employed models. Additionally, a higher degree of classification performance for the testing phase was reported for all calculated indices except AUC. Values of 0.910, 0.931, 0.931, 0.931, and 0.931; 0.961, 0.914, 0.915, 0.916 and 0.914; and 0.968, 0.891, 0.895, 0.908 and 0.891 were obtained for AUC, Accuracy, F1, Precision, and Recall of AdaBoost, RF, and DT models, respectively. These values confirmed the successful use of tree-based models in classifying slope stability. However, the better performance and higher capability for classification purpose goes to the proposed AdaBoost technique. Therefore, it can be introduced as a new technique for slope stability classification with the largest number of matched cases.

It is well established that to propose a new method for classifying slope stability cases using AI techniques, extensive investigation is required. Therefore, in order to develop a model for classifying slope stability, a comprehensive database comprising real cases must be gathered and utilized. Yet, collecting such database is very difficult and time consuming. By providing the mentioned data, slope stability classifications can be conducted using new (hybrid) AI techniques, such as RF or AdaBoost, combined with metaheuristic algorithms.

Moreover, the use of real slope stability data based on different types of soils considering other properties, such as unit weight, permeability, and ground water table, would be of interest and importance to geotechnical engineers. In this regard, model generalization as an important issue in classification and prediction problems can be considered, with the developed models covering a wider range of input parameters, as well as a larger number of effective problem variables.

Author Contributions: Conceptualization, D.J.A., B.G. and P.G.A.; methodology, D.J.A., F.I.M.R., M.K. and P.G.A.; software, D.J.A., F.I.M.R., M.K. and P.G.A.; formal analysis, D.J.A., F.I.M.R., M.K. and P.G.A.; writing—original draft preparation, D.J.A., F.I.M.R., M.K., P.G.A., B.G., P.C.R. and M.F.; writing—review and editing, D.J.A., F.I.M.R., M.K., P.G.A., B.G., P.C.R. and M.F.; supervision, D.J.A., B.G. and P.G.A.; Data curation, B.G. All authors have read and agreed to the published version of the manuscript.

Funding: This research received no external funding.

Institutional Review Board Statement: Not applicable.

Informed Consent Statement: Not applicable.

Data Availability Statement: The data are available upon request.

Acknowledgments: Authors of this study wish to express their appreciation to the University of Malaya for supporting this study and making it possible.

Conflicts of Interest: The authors declare no conflict of interest.

References

1. Gordan, B.; Raja, M.A.; Armaghani, D.J.; Adnan, A. Review on Dynamic Behaviour of Earth Dam and Embankment During an Earthquake. *Geotech. Geol. Eng.* **2021**, *40*, 3–33. [CrossRef]
2. Hajihassani, M.; Jahed Armaghani, D.; Kalatehjari, R. Applications of Particle Swarm Optimization in Geotechnical Engineering: A Comprehensive Review. *Geotech. Geol. Eng.* **2018**, *36*, 705–722. [CrossRef]
3. Bui, X.N.; Nguyen, H.; Choi, Y.; Nguyen-Thoi, T.; Zhou, J.; Dou, J. Prediction of slope failure in open-pit mines using a novel hybrid artificial intelligence model based on decision tree and evolution algorithm. *Sci. Rep.* **2020**, *10*, 9939. [CrossRef]
4. Koopialipoor, M.; Jahed Armaghani, D.; Hedayat, A.; Marto, A.; Gordan, B. Applying various hybrid intelligent systems to evaluate and predict slope stability under static and dynamic conditions. *Soft Comput.* **2019**, *23*, 5913–5929. [CrossRef]
5. Cai, M.; Koopialipoor, M.; Armaghani, D.J.; Thai Pham, B. Evaluating Slope Deformation of Earth Dams due to Earthquake Shaking using MARS and GMDH Techniques. *Appl. Sci.* **2020**, *10*, 1486. [CrossRef]
6. Gordan, B.; Armaghani, D.J.; Adnan, A.B.; Rashid, A.S.A. A New Model for Determining Slope Stability Based on Seismic Motion Performance. *Soil Mech. Found. Eng.* **2016**, *53*, 344–351. [CrossRef]
7. Das, S.K.; Biswal, R.K.; Sivakugan, N.; Das, B. Classification of slopes and prediction of factor of safety using differential evolution neural networks. *Environ. Earth Sci.* **2011**, *64*, 201–210. [CrossRef]
8. Hoek, E.; Bray, J.D. *Rock Slope Engineering*; CRC Press: Boca Raton, FL, USA, 1981; ISBN 1482267098.
9. Hammah, R. A Comparison of Finite Element Slope Stability Analysis With Conventional Limit-Equilibrium Investigation. *Proceedings of the 58th Canadian Geotechnical and 6th Joint IAH-CNC and CGS Groundwater Specialty Conferences–GeoSask 2005*, 2005.
10. Ray, A.; Kumar, V.; Kumar, A.; Rai, R.; Khandelwal, M.; Singh, T.N. Stability prediction of Himalayan residual soil slope using artificial neural network. *Nat. Hazards* **2020**, *103*, 3523–3540. [CrossRef]
11. Erzin, Y.; Cetin, T. The use of neural networks for the prediction of the critical factor of safety of an artificial slope subjected to earthquake forces. *Sci. Iran.* **2012**, *19*, 188–194. [CrossRef]
12. Chakraborty, A.; Goswami, D. Prediction of critical safety factor of slopes using multiple regression and neural network. *J. Geo-Eng. Sci.* **2018**, 1–10. [CrossRef]
13. Mahdiyar, A.; Hasanipanah, M.; Armaghani, D.J.; Gordan, B.; Abdullah, A.; Arab, H.; Majid, M.Z.A.; Danial, J.A.; Armaghani, J.; Gordan, B.; et al. A Monte Carlo technique in safety assessment of slope under seismic condition. *Eng. Comput.* **2017**, *33*, 807–817. [CrossRef]
14. Asteris, P.G.; Koopialipoor, M.; Armaghani, D.J.; Kotsonis, E.A.; Lourenço, P.B. Prediction of cement-based mortars compressive strength using machine learning techniques. *Neural Comput. Appl.* **2021**, *33*, 13089–13121. [CrossRef]
15. Harandizadeh, H.; Armaghani, D.; Asteris, P.G.; Gandomi, A. TBM performance prediction developing a hybrid ANFIS-PNN predictive model optimized by imperialism competitive algorithm. *Neural Comput. Appl.* **2021**, *33*, 16149–16179. [CrossRef]
16. Zhao, J.; Nguyen, H.; Nguyen-Thoi, T.; Asteris, P.G.; Zhou, J. Improved Levenberg–Marquardt backpropagation neural network by particle swarm and whale optimization algorithms to predict the deflection of RC beams. *Eng. Comput.* **2021**. [CrossRef]
17. Zhang, H.; Nguyen, H.; Bui, X.-N.; Pradhan, B.; Asteris, P.G.; Costache, R.; Aryal, J. A generalized artificial intelligence model for estimating the friction angle of clays in evaluating slope stability using a deep neural network and Harris Hawks optimization algorithm. *Eng. Comput.* **2021**, 1–14. [CrossRef]
18. Asteris, P.G.; Skentou, A.D.; Bardhan, A.; Samui, P.; Pilakoutas, K. Predicting concrete compressive strength using hybrid ensembling of surrogate machine learning models. *Cem. Concr. Res.* **2021**, *145*, 106449. [CrossRef]

19. Asteris, P.G.; Cavaleri, L.; Ly, H.-B.; Pham, B.T. Surrogate models for the compressive strength mapping of cement mortar materials. *Soft Comput.* **2021**, *25*, 6347–6372. [CrossRef]
20. Armaghani, D.J.; Mamou, A.; Maraveas, C.; Roussis, P.C.; Siorikis, V.G.; Skentou, A.D.; Asteris, P.G. Predicting the unconfined compressive strength of granite using only two non-destructive test indexes. *Geomech. Eng.* **2021**, *25*, 317–330.
21. Zhou, J.; Shen, X.; Qiu, Y.; Li, E.; Rao, D.; Shi, X. Improving the efficiency of microseismic source locating using a heuristic algorithm-based virtual field optimization method. *Geomech. Geophys. Geo-Energy Geo-Resour.* **2021**, *7*, 89. [CrossRef]
22. Zhou, J.; Qiu, Y.; Khandelwal, M.; Zhu, S.; Zhang, X. Developing a hybrid model of Jaya algorithm-based extreme gradient boosting machine to estimate blast-induced ground vibrations. *Int. J. Rock Mech. Min. Sci.* **2021**, *145*, 104856. [CrossRef]
23. Zhou, J.; Li, X.; Mitri, H.S. Classification of rockburst in underground projects: Comparison of ten supervised learning methods. *J. Comput. Civ. Eng.* **2016**, *30*, 4016003. [CrossRef]
24. Zhou, J.; Chen, C.; Wang, M.; Khandelwal, M. Proposing a novel comprehensive evaluation model for the coal burst liability in underground coal mines considering uncertainty factors. *Int. J. Min. Sci. Technol.* **2021**, *31*, 799–812. [CrossRef]
25. Yang, H.Q.; Li, Z.; Jie, T.Q.; Zhang, Z.Q. Effects of joints on the cutting behavior of disc cutter running on the jointed rock mass. *Tunn. Undergr. Sp. Technol.* **2018**, *81*, 112–120. [CrossRef]
26. Yang, H.; Wang, H.; Zhou, X. Analysis on the damage behavior of mixed ground during TBM cutting process. *Tunn. Undergr. Sp. Technol.* **2016**, *57*, 55–65. [CrossRef]
27. Liu, B.; Yang, H.; Karekal, S. Effect of Water Content on Argillization of Mudstone During the Tunnelling process. *Rock Mech. Rock Eng.* **2019**, *53*, 799–813. [CrossRef]
28. Yang, H.; Wang, Z.; Song, K. A new hybrid grey wolf optimizer-feature weighted-multiple kernel-support vector regression technique to predict TBM performance. *Eng. Comput.* **2020**, 1–17. [CrossRef]
29. Kardani, N.; Bardhan, A.; Samui, P.; Nazem, M.; Zhou, A.; Armaghani, D.J. A novel technique based on the improved firefly algorithm coupled with extreme learning machine (ELM-IFF) for predicting the thermal conductivity of soil. *Eng. Comput.* **2021**, 1–20. [CrossRef]
30. Parsajoo, M.; Armaghani, D.J.; Mohammed, A.S.; Khari, M.; Jahandari, S. Tensile strength prediction of rock material using non-destructive tests: A comparative intelligent study. *Transp. Geotech.* **2021**, *31*, 100652. [CrossRef]
31. Mahmood, W.; Mohammed, A.S.; Asteris, P.G.; Kurda, R.; Armaghani, D.J. Modeling Flexural and Compressive Strengths Behaviour of Cement-Grouted Sands Modified with Water Reducer Polymer. *Appl. Sci.* **2022**, *12*, 1016. [CrossRef]
32. Asteris, P.G.; Lourenço, P.B.; Roussis, P.C.; Adami, C.E.; Armaghani, D.J.; Cavaleri, L.; Chalioris, C.E.; Hajihassani, M.; Lemonis, M.E.; Mohammed, A.S. Revealing the nature of metakaolin-based concrete materials using artificial intelligence techniques. *Constr. Build. Mater.* **2022**, *322*, 126500. [CrossRef]
33. Abdalla, J.A.; Attom, M.F.; Hawileh, R. Prediction of minimum factor of safety against slope failure in clayey soils using artificial neural network. *Environ. Earth Sci.* **2015**, *73*, 5463–5477. [CrossRef]
34. Verma, A.K.; Singh, T.N.; Chauhan, N.K.; Sarkar, K. A hybrid FEM–ANN approach for slope instability prediction. *J. Inst. Eng. Ser. A* **2016**, *97*, 171–180. [CrossRef]
35. Samui, P. Support vector classifier analysis of slope. *Geomat. Nat. Hazards Risk* **2013**, *4*, 1–12. [CrossRef]
36. Sakellariou, M.G.; Ferentinou, M.D. A study of slope stability prediction using neural networks. *Geotech. Geol. Eng.* **2005**, *23*, 419. [CrossRef]
37. Ferentinou, M.D.; Sakellariou, M.G. Computational intelligence tools for the prediction of slope performance. *Comput. Geotech.* **2007**, *34*, 362–384. [CrossRef]
38. Mohamed, T.; Kasa, A. Application of fuzzy set theory to evaluate the stability of slopes. *Appl. Mech. Mater.* **2014**, *580–583*, 566–571. [CrossRef]
39. Kalatehjari, R.; Rashid, A.S.A.; Ali, N.; Hajihassani, M. The contribution of particle swarm optimization to three-dimensional slope stability analysis. *Sci. World J.* **2014**, *2014*, 973093. [CrossRef] [PubMed]
40. Lu, P.; Rosenbaum, M.S. Artificial neural networks and grey systems for the prediction of slope stability. *Nat. Hazards* **2003**, *30*, 383–398. [CrossRef]
41. Samui, P. Slope stability analysis: A support vector machine approach. *Environ. Geol.* **2008**, *56*, 255. [CrossRef]
42. Tien Bui, D.; Pradhan, B.; Lofman, O.; Revhaug, I. Landslide susceptibility assessment in vietnam using support vector machines, decision tree, and Naive Bayes Models. *Math. Probl. Eng.* **2012**, *2012*, 974638. [CrossRef]
43. Amit, Y.; Geman, D. Shape Quantization and Recognition with Randomized Trees. *Neural Comput.* **1997**, *9*, 1545–1588. [CrossRef]
44. Hwang, S.G.; Guevarra, I.F.; Yu, B.O. Slope failure prediction using a decision tree: A case of engineered slopes in South Korea. *Eng. Geol.* **2009**, *104*, 126–134. [CrossRef]
45. Gelisli, K.; Kaya, T.; Babacan, A.E. Assessing the factor of safety using an artificial neural network: Case studies on landslides in Giresun, Turkey. *Environ. Earth Sci.* **2015**, *73*, 8639–8646. [CrossRef]
46. Tao, G.L.; Yao, Z.S.; Tan, B.Z.; Gao, C.C.; Yao, Y.W. Application of Support Vector Machine for Prediction of Slope Stability Coefficient Considering the Influence of Rainfall and Water Level. *Appl. Mech. Mater.* **2016**, *851*, 840–845. [CrossRef]
47. Fattahi, H. Prediction of slope stability using adaptive neuro-fuzzy inference system based on clustering methods. *J. Min. Environ.* **2017**, *8*, 163–177.
48. Qi, C.; Tang, X. Slope stability prediction using integrated metaheuristic and machine learning approaches: A comparative study. *Comput. Ind. Eng.* **2018**, *118*, 112–122. [CrossRef]

49. Hidayat, S.; Alpiana; Rahmawati, D. Application of adaptive neuro-fuzzy inference system (ANFIS) for slope and pillar stability assessment. *IOP Conf. Ser. Earth Environ. Sci.* **2020**, *413*, 012003. [CrossRef]
50. Sari, P.A.; Suhatril, M.; Osman, N.; Mu'azu, M.A.; Katebi, J.; Abavisani, A.; Ghaffari, N.; Sadeghi Chahnasir, E.; Wakil, K.; Khorami, M.; et al. Developing a hybrid adoptive neuro-fuzzy inference system in predicting safety of factors of slopes subjected to surface eco-protection techniques. *Eng. Comput.* **2020**, *36*, 1347–1354. [CrossRef]
51. Han, H.; Armaghani, D.J.; Tarinejad, R.; Zhou, J.; Tahir, M.M. Random Forest and Bayesian Network Techniques for Probabilistic Prediction of Flyrock Induced by Blasting in Quarry Sites. *Nat. Resour. Res.* **2020**, *29*, 655–667. [CrossRef]
52. Zhou, J.; Asteris, P.G.; Armaghani, D.J.; Pham, B.T. Prediction of ground vibration induced by blasting operations through the use of the Bayesian Network and random forest models. *Soil Dyn. Earthq. Eng.* **2020**, *139*, 106390. [CrossRef]
53. Li, D.; Moghaddam, M.R.; Monjezi, M.; Jahed Armaghani, D.; Mehrdanesh, A. Development of a Group Method of Data Handling Technique to Forecast Iron Ore Price. *Appl. Sci.* **2020**, *10*, 2364. [CrossRef]
54. Liang, M.; Mohamad, E.T.; Faradonbeh, R.S.; Jahed Armaghani, D.; Ghoraba, S. Rock strength assessment based on regression tree technique. *Eng. Comput.* **2016**, *32*, 343–354. [CrossRef]
55. Newmark, N.M. Effects of earthquakes on dams and embankments. *Geotechnique* **1965**, *15*, 139–160. [CrossRef]
56. Ambraseys, N.N.; Menu, J.M. Earthquake-induced ground displacements. *Earthq. Eng. Struct. Dyn.* **1988**, *16*, 985–1006. [CrossRef]
57. Sarma, S.K.; Bhave, M.V. Critical acceleration versus static factor of safety in stability analysis of earth dams and embankments. *Geotechnique* **1974**, *24*, 661–665. [CrossRef]
58. Abramson, L.W.; Lee, T.S.; Sharma, S.; Boyce, G.M. *Slope Stability and Stabilization Methods*; John Wiley & Sons: Hoboken, NJ, USA, 2001; ISBN 0471384933.
59. Zhu, D.Y.; Lee, C.F.; Jiang, H.D. Generalised framework of limit equilibrium methods for slope stability analysis. *Geotechnique* **2003**, *53*, 377–395. [CrossRef]
60. Samui, P.; Kothari, D.P. Utilization of a least square support vector machine (LSSVM) for slope stability analysis. *Sci. Iran.* **2011**, *18*, 53–58. [CrossRef]
61. Erzin, Y.; Cetin, T. The prediction of the critical factor of safety of homogeneous finite slopes using neural networks and multiple regressions. *Comput. Geosci.* **2013**, *51*, 305–313. [CrossRef]
62. Myles, A.J.; Feudale, R.N.; Liu, Y.; Woody, N.A.; Brown, S.D. An introduction to decision tree modeling. *J. Chemom. A J. Chemom. Soc.* **2004**, *18*, 275–285. [CrossRef]
63. Jopp, F.; Reuter, H.; Breckling, B. *Modelling Complex Ecological Dynamics: An Introduction into Ecological Modelling for Students, Teachers & Scientists*; Springer: Berlin/Heidelberg, Germany, 2011; pp. 1–387.
64. Murthy, S.K. Automatic construction of decision trees from data: A multi-disciplinary survey. *Data Min. Knowl. Discov.* **1998**, *2*, 345–389. [CrossRef]
65. Kheir, R.B.; Greve, M.H.; Abdallah, C.; Dalgaard, T. Spatial soil zinc content distribution from terrain parameters: A GIS-based decision-tree model in Lebanon. *Environ. Pollut.* **2010**, *158*, 520–528. [CrossRef] [PubMed]
66. Tso, G.K.F.; Yau, K.K.W. Predicting electricity energy consumption: A comparison of regression analysis, decision tree and neural networks. *Energy* **2007**, *32*, 1761–1768. [CrossRef]
67. Zhao, Y.; Zhang, Y. Comparison of decision tree methods for finding active objects. *Adv. Sp. Res.* **2008**, *41*, 1955–1959. [CrossRef]
68. Ho, T.K. Random decision forests. *Proc. Int. Conf. Doc. Anal. Recognit. ICDAR* **1995**, *1*, 278–282.
69. Kainthura, P.; Sharma, N. Machine Learning Techniques to Predict Slope Failures in Uttarkashi, Uttarakhand (India). *J. Sci. Ind. Res.* **2021**, *80*, 66–74.
70. Piryonesi, S.M.; El-Diraby, T.E. Role of data analytics in infrastructure asset management: Overcoming data size and quality problems. *J. Transp. Eng. Part B Pavements* **2020**, *146*, 4020022. [CrossRef]
71. Boulesteix, A.L.; Janitza, S.; Kruppa, J.; König, I.R. Overview of random forest methodology and practical guidance with emphasis on computational biology and bioinformatics. *Wiley Interdiscip. Rev. Data Min. Knowl. Discov.* **2012**, *2*, 493–507. [CrossRef]
72. Kégl, B. The return of ADABOOST.MH: Multi-class Hamming trees. *arXiv* **2014**, arXiv:1312.6086.
73. Saadaari, F.; Olaleye, B.M. Development of a Stope Stability Prediction Model Using Ensemble Learning Techniques—A Case Study. *Ghana Min. J.* **2020**, *20*, 18–26. [CrossRef]
74. Wang, H.; Zhang, L.; Yin, K.; Luo, H.; Li, J. Landslide identification using machine learning. *Geosci. Front.* **2020**, *12*, 351–364. [CrossRef]
75. Pham, B.T.; Nguyen, M.D.; Nguyen-Thoi, T.; Ho, L.S.; Koopialipoor, M.; Quoc, N.K.; Armaghani, D.J.; Van Le, H. A novel approach for classification of soils based on laboratory tests using Adaboost, Tree and ANN modeling. *Transp. Geotech.* **2020**, *27*, 100508. [CrossRef]
76. Rojas, R. AdaBoost and the Super Bowl of Classifiers a Tutorial Introduction to Adaptive Boosting. Writing, Computer Science Department, Freie Universitat, Berlin, Germany. 2009. Available online: http://www.inf.fu-berlin.de/inst/ag-ki/adaboost4.pdf (accessed on 11 April 2016).
77. Zorlu, K.; Gokceoglu, C.; Ocakoglu, F.; Nefeslioglu, H.A.; Acikalin, S. Prediction of uniaxial compressive strength of sandstones using petrography-based models. *Eng. Geol.* **2008**, *96*, 141–158. [CrossRef]

applied sciences

MDPI

Article

Application of Soft Computing Techniques to Estimate Cutter Life Index Using Mechanical Properties of Rocks

Timur Massalov, Saffet Yagiz * and Amoussou Coffi Adoko

School of Mining and Geosciences, Nazarbayev University, Nur-Sultan 010000, Kazakhstan; timur.massalov@nu.edu.kz (T.M.); amoussou.adoko@nu.edu.kz (A.C.A.)
* Correspondence: saffet.yagiz@nu.edu.kz

Featured Application: The proposed models in this paper can be used to estimate the cutter life index for estimation of cutter wear and life where the actual index is not available.

Abstract: The wear of cutting tools is critical for any engineering applications dealing with mechanical rock excavations, as it directly affects the cost and time of project completion as well as the utilization rate of excavators in various rock masses. The cutting tool wear could be expressed in terms of the life of the tool used to excavate rocks in hours or cutter per unit volume of excavated materials. The aim of this study is to estimate disc cutter wear as a function of common mechanical rock properties including uniaxial compressive strength, Brazilian tensile strength, brittleness, and density. To achieve this goal, a database of cutter life was established by analyzing data from 80 tunneling projects. The data were then utilized for evaluating the relationship between rock properties and cutter consumption by means of cutter life index. The analysis was based on artificial intelligence techniques, namely artificial neural networks (ANN) and fuzzy logic (FL). Furthermore, linear and non-linear regression methods were also used to investigate the relationship between these parameters using a statistical software package. Several alternative models are introduced with different input variables for each model, to identify the best model with the highest accuracy. To develop these models, 70% of the dataset was used for training and the rest, for testing. The estimated cutter life by various models was compared with each other to identify the most reliable model. It appears that the ANN and FL techniques are superior to standard linear and non-linear multiple regression analysis, based on the higher correlation coefficient (R^2) and lower Mean square error (MSE).

Keywords: rock excavation; soft computing; cutter life index; rock strength; brittleness

Citation: Massalov, T.; Yagiz, S.; Adoko, A.C. Application of Soft Computing Techniques to Estimate Cutter Life Index Using Mechanical Properties of Rocks. *Appl. Sci.* **2022**, *12*, 1446. https://doi.org/10.3390/app12031446

Academic Editors: Arcady Dyskin, Danial Jahed Armaghani, Yixia Zhang, Pijush Samui, Ahmed Hussein Kamel Ahmed Elshafie and Aydin Azizi

Received: 28 November 2021
Accepted: 4 January 2022
Published: 28 January 2022

1. Introduction

The wear of rock drilling and cutting tools in mining, tunneling, and civil construction has always been a predominant factor for the costs of hard rock excavation. This fact is not only related to material and labor costs arising from cutting tools maintenance and replacement but also because of the direct and negative impact of wear on the drilling/cutting performance of worn cutters and bits [1]. Tool wear in hard rock drilling can be defined as a process of continuous loss of material from the surface of the cutting tool or drill bit due to mechanical contact and relative movement of the bit over the rock surface [2]. The potential of a rock or rock mass to cause wear on a rock-engaging tool can be described by abrasiveness. Disc cutter wear is the result of the rock–machine interaction in tunneling by tunnel boring machine (TBM). Indeed, the replacement of the disc cutter is a time-consuming and costly activity that can significantly reduce the TBM utilization (U) and advance rate (AR) and has a major effect on the total time and cost of TBM tunneling projects. Hence the importance of predicting the cutter life precisely can never be neglected. The abrasivity of rocks and the accompanied wear of cutting and drilling tools are commonly determined by simple and fast laboratory tests such as Cerchar abrasivity index (CAI) test; however, it has disadvantages in extreme cases such as soft or very hard rocks [3,4]. Over the years, several different indices have been developed to offer reasonable quantitative measures for rock abrasivity. Abrasivity value or "AV" and Abrasivity value of steel "AVS" were introduced by the Norwegian University of Science and Technology (NTNU) as a part of their suite of rock tests for quantifying the boreability of rocks [5]. These indices are used for estimating the life of rock cutting tools as bit wear index (BWI) as well as cutter life index (CLI). Plinninger et al. [2] developed a new composite index as rock abrasivity index (RAI) calculated by multiplying the uniaxial compressive strength with percentage of quartzite content. Schimazek and Knatz [6] introduced an index for rock abrasivity, especially for use in roadheader application. This index uses grain size and percent of quartzite to estimate the index (F) to represent the hardness of rock. The other index introduced in the mid-1980s [7] for abrasivity measurement of rock is CAI. The test is relatively simple and portable and useful to estimate the tool consumption in rock excavation. There are many publications about CAI test procedures and classifying the rock abrasivity. This test has also been standardized by ASTM and ISRM [8,9]. It is known that the cutter wear is not only related to CAI but also other rock properties and mineralogical features of the rocks. Meanwhile, several brittleness indices have been introduced in the past few decades [10–13] and they represent how well the rock will fracture and break. At present, the CAI is often used to estimate cutter consumption in the CSM model for TBM performance and cutter life estimation. Similarly, CLI is one of the main input parameters to estimate cutter wear in the NTNU TBM performance prediction model for hard rock [5,14–20].

Plinninger et al. [1] illustrated that "Abrasive wear" is the predominant wear process in excavation operation in most rock types. They stated that the abrasive wear leads to the removal of material from the tool surfaces while it is moving against the rock. Deketh [21] noted that according to the studies when the ratio of abrasiveness of two interacting materials exceeds 20% of their Vickers hardness, abrasive wear increases dramatically. Atkinson et al. [22] suggested that various factors affect the rock abrasiveness, and those factors can be evaluated and categorized as mineral composition, hardness of mineral constituents, grain shape and size, type of matrix material, and physical properties of rocks including strength, hardness, brittleness, and toughness. However, in the literature, most of the cutter consumption models were developed based on rock abrasivity and testing such as BWI that is part of the NTNU testing system, RAI, and CAI. The complex nature of the tool wear process leads to a vast number of factors that can dramatically influence tool wear. A summary of the field geology tools logistics and some of the main factors influencing the rate of penetration and type of tool wear are provided in Table 1.

Table 1. Summary of the main factors influencing cutting tool life and wear [1].

Geology	Tools	Logistics
Rock properties (mineral composition, rock strength, grain size, grain shape)	Tool characteristics (carbide composition, button shape, button number, steel composition)	Maintenance
Joint features (spacing, orientation, aperture, roughness)	Flushing (fluid, number and geometry of flushing holes and flutes, flushing pressure)	Tool handling
Weathering/alteration of rock	Feed and rotating velocity temperatures	Supporting methods
water situation composition of rock mass(homogenous/inhomogeneous) stress situation (stress direction, stress level)		

2. Background

Norwegian University of Science and Technology (NTNU) developed a model to evaluate the drillability of percussion drilling in the 1960s. This model has been used in major international mechanized underground construction projects and is considered as one of the most recognized and widely used methods for estimation of TBM performance and cutter life [5]. The NTNU rock drillability testing suite consists of a set of laboratory tests and different indices which are briefly introduced herein. A classification of the NTNU drillability indices Drilling Rate Index (DRI), CLI, and BWI has been available since 1998 [23]. In this study, CLI is examined in more detail, and some models are introduced to estimate CLI from more common mechanical rock properties.

2.1. Cutter Life Index

The CLI is computed based on the Sievers' J-value and the abrasion value of steel anvil or in short, AVS. The index can be used to estimate the lifetime of the TBM cutter discs, in the number of hours of cutter running on the face as the machine is excavating in the given rock type [5].

2.1.1. The Sievers' Miniature Drill Test

In order to evaluate the surface hardness of the rock, the Sj test was developed by H. Sievers in the 1950s. The Sievers' J-value is the depth of the drilled hole after 200 revolutions of the drill bit which is measured in 1/10 of mm. This test should be repeated 4 to 8 times and the mean value should be used as the final number [24]. The test showing a schematic view of the test is performed on a sawn sample (Figure 1).

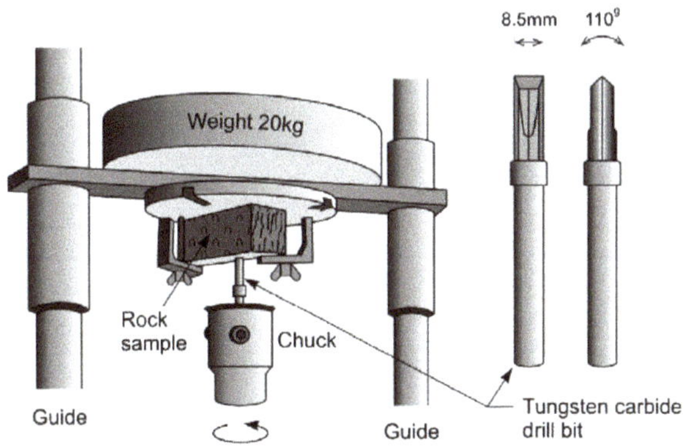

Figure 1. The Sievers' miniature drill test [24].

2.1.2. The Abrasion Value Steel

In abrasion value steel (AVS), rock powder in the size range of less than 1 mm is used to abrade the worn piece made of steel from a new cutter ring. The wear piece is under 10 kg dead load to increase the friction and contact pressure between rock grains and steel anvil. AVS is the weight loss of the worn piece after 20 rounds (1 min) of turn table rotation, which is measured in milligrams. Figure 2 shows the abrasion test and equipment [23,24].

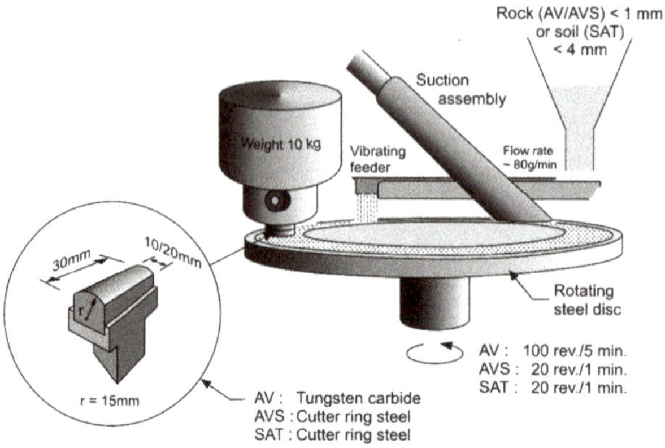

Figure 2. Outline of abrasion value (AV) and abrasion value cutter steel (AVS) [24].

2.1.3. Calculation of Cutter Life Index

After measurement of the AVS and SJ values, CLI can be calculated using the following formula. This formula is based on the real field data on actual cutter lifetime and related tested rock parameters.

$$\mathrm{CLI} = 13.84 \cdot \left[\frac{\mathrm{SJ}}{\mathrm{AVS}}\right]^{0.3847} \tag{1}$$

From CLI,

$$\mathrm{CL} = \mathrm{CLI} \cdot \mathrm{ROP} \cdot \mathrm{A} \tag{2}$$

Cutter life (CL) can be computed as a function of CLI, rate of penetration (ROP), and cross section area of the opening (A) within m^3/cutter. The cutter life index intervals are given in Table 2 as suggested in the literature [24].

Table 2. Category of intervals for cutter life index in NTNU Model [24].

Category	CLI
Extremely low	<5
Very low	5.0–5.9
Low	6.0–7.9
Medium	8.0–14.9
High	15.0–34
Very high	35–74
Extremely high	≥75

2.2. Cerchar Abrasivity Index (CAI)

Abrasivity is a good indicator of the wear life of cutting tools in any rock excavation operation. CAI is used for estimating the cutting tool life in the TBM performance prediction model of the Colorado School of Mines (CSM). Various rock abrasivity measurements have been introduced throughout the years to allow engineers to estimate cutting tool life. CAI is commonly utilized to characterize the abrasivity of rocks for estimation of cutting tool life and wear in various mining, civil and tunneling applications. The CAI [7] test has been introduced in the 1970s by the Centre d'Etudes et Recherches des Charbonages (CERCHAR) de France for abrasivity testing in coal-bearing rocks in mining industries while gradually being adopted for application in the tunneling industry [15,25–27]. Different generations of testing devices and the impact of various testing parameters on the test results have been discussed in the literature [15,18,28–33]. A typical CAI device is given in Figures 3 and 4. A typical rock sample and surface used for CAI is given in Figure 5.

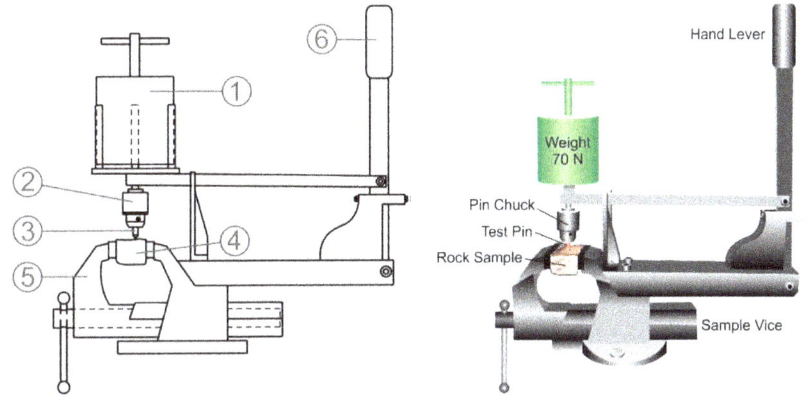

Figure 3. Setup of a first-generation Cerchar testing device [7]. (1) Weight, (2) pin chuck, (3) steel pin, (4) specimen, (5) vice, and (6) hand lever [1].

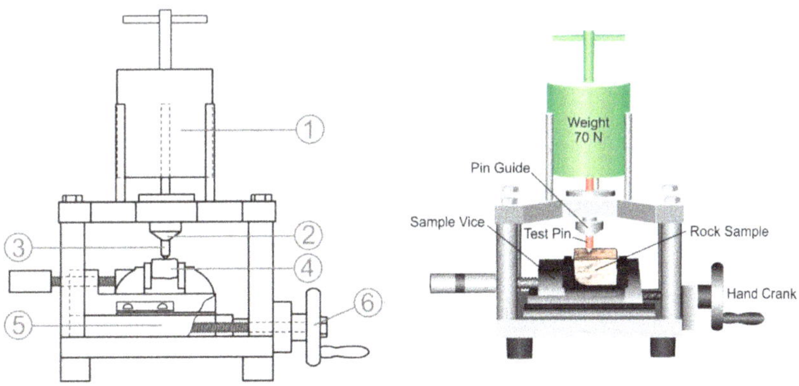

Figure 4. Setup of a second-generation Cerchar testing device [28].

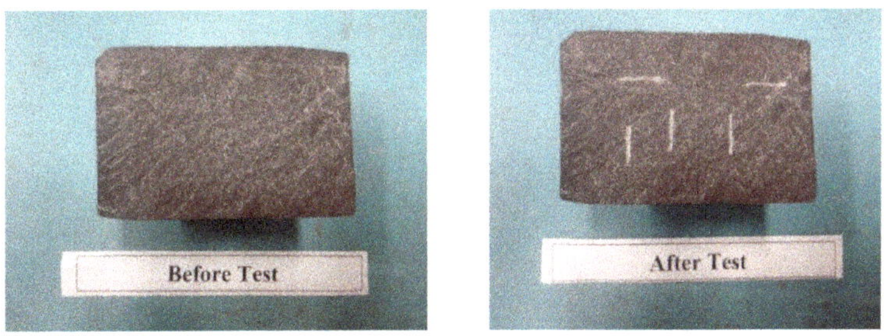

Figure 5. Example of rock surface conditions with rough surface (left over from BTS test) used for CAI test.

The CAI should be calculated as:

$$\mathrm{CAI} = \mathrm{d} \cdot 10 \quad (3)$$

where d is the wear tip surface measured to an accuracy of 0.01 mm. The dimensionless CAI value is reported as the arithmetic mean of five or more test replications together with standard deviation as suggested by [9]. While the original formal standard of the test was the French standard NF P 940-430-1 [34], recently an ASTM [8] and ISRM [9] standards for Cerchar testing were published. An effect of some Cerchar test parameters, including Cerchar pin properties, loading condition, and test length on the rock surface, on the values of CAI are given in Table 3.

Table 3. Effect of some Cerchar test parameters on the measured values of CAI [35].

Testing Factors	Effect on CAI Value
Test length	About 70% of wear occurs in the initial first millimeter of the scratch length, approximately 85% of after 2 mm of test slide, and only 15% of the wear flat is produced by the remaining 8 mm length [1,29,36,37]. In cases of harder and more abrasive rocks the CAI value length [38,39].
Static load	The CAI value increases linearly by changing static loads on the stylus [40,41].
Testing speed	The testing speed does not affect the CAI values significantly and commonly higher when conducted with 43 HRC pins at slow testing speeds [2,41]. The standardized testing speed is 10 mm/s for articulated hand lever type machine and 1 mm/s for hand-crank types [7,9].
Stylus hardness	Higher CAI values are obtained with soft CERCHAR test styli and vice versa [36,38,41–45].
Stylus metallurgy	No considerable effect on CAI value is observed by changes in the metallurgy of the stylus keeping regular hardness [43].

Rock abrasivity could be also examined based on the weighted average abrasivity of the constituent minerals. In this method, the percentage of each mineral in the rock is calculated and multiplied by its hardness or abrasivity, based on different available scales [21]. Abrasive mineral content (AMC), equivalent quartz content (EQC), and Vickers hardness number for rock (VHNR) are the most common methods to compute the abrasivity of rocks. AMC uses Mohs scratch hardness, while EQC uses Rosiwal [46] grinding hardness and VHNR benefits from Vickers indentation hardness (an indentation test in which the ratio of the force to the area of the indentation is considered as an index for abrasivity of the material) [21]. In the EQC method, constituent minerals of the rock would be identified either by microscopic or macroscopic mineral examination. Common methods utilized for abrasion measurement of rocks are given in Table 4.

Table 4. Common methods utilized for abrasion measurement and classification [47].

Method	Remarks	Advantage	Disadvantage	Ref.
Mohs scale	Mineral comparative scratch test	Simple to use	It is just a qualitative measure	[48]
Vickers hardness number rock (VHNR)	Based on indentation hardness (the ratio of force to the area of indentation) using a diamond tipped micro-indenter (Vickers)	Simple method to rate rock wear capacity based on available charted mineral VHNR values	Limited experience for TBM rock cutting	[40]
Rock abrasive index (RAI)	RAI = UCS × EQC	Simple method Presence of a chart for conical pick life	No chart for disc cutter life prediction	[30]
Abrasive mineral content	Uses Mohs scratch hardness	Simple to use	Limited experience for TBM rock cutting	[21]
Equivalent quartz content	Uses Rosiwal rating	Simple to use	Limited experience for TBM rock cutting	[21]
Wear index-F	F = Q × D.z.10 = equivalent quartz percentage, D—mean quartz grain size in mm, z—Brazilian tensile strength in MPa	Is developed for drag tool cutting	Specimen mean quartz grain size has high importance in the formula. In coarse grained metamorphic and igneous rocks, this index may lead to highly misleading results	[6]
Rosiwal mineral abrasivity rating	Rosiwal = 1000 × volume loss corundum/volume loss mineral specimen	Simple to use	Limited experience for TBM rock cutting	[46]
NTNU cutter life index (CLI)	CLI is obtained from AVS and Siever's J tests	Large database and presence of disc cutter life prediction charts	Correct tests can only be performed in SINTEF and the replicated testing equipment may show results with high discrepancy	[49]
Cerchar abrasivity index (CAI)	Steel pin tip diameter in 1/10th mm after 1 cm scratch test under 70N normal load	Widely used test in tunneling, simple, low cost, low sample requirement	Good only for rough surfaces, variability in the test results due to its sensitivity to method of tip reading, the rock surface condition, the non-constant cross-section of pin tip during the test	[50]

2.3. Mechanical Properties of Rocks

Both intact rock and rock mass properties are used as the main parameters for estimating the project time and cost for a tunneling project. Tool consumption and machine advance rate in any rock mass are very closely related to the rock parameters such as strength, brittleness, density, abrasivity, mineral content. In this paper, several common rock properties were examined to estimate cutter consumption based on rock properties.

2.3.1. Rock Strength

The uniaxial compressive strength (UCS) and Brazilian tensile strength (BTS) are two of the most important and commonly measured rock properties for rock excavation projects since those strengths are related to both, porosity, density, and brittleness behavior of rock under the indenter/cutter. Due to the importance of the strengths, both the BTS and UCS values of the rocks are measured according to the standards [8,9] and recorded in the database.

2.3.2. Density and Porosity

Density (D) and porosity of rocks are both crucial parameters and hence are commonly utilized for estimating the cutter consumption and machine performance. The density and porosity of rocks could be measured using standard procedures (either [8] or [9]). In this study, density is used as one of the input parameters to estimate the CLI.

2.3.3. Rock Brittleness

The brittleness (BI_i) value is another fundamental rock property that should be considered for the assessment of boreability and cutter consumption; however, there are no universally accepted standards to measure brittleness. Yagiz (2009) discussed various approaches to represent rock brittleness directly from the punch penetration test (PPT). However, the PPT test is not a commonly used and available test. Several alternative rock brittleness based on rock strength ratios introduced in the literature [10,11,13] are as follow:

$$B_1 = \frac{\sigma_c}{\sigma_t} \tag{4}$$

$$B_2 = \frac{\sigma_c - \sigma_t}{\sigma_c + \sigma_t} \tag{5}$$

$$B_3 = \sqrt{\frac{\sigma_c \cdot \sigma_t}{2}} \tag{6}$$

where σ_c is the compressive strength, σ_t is the Brazilian tensile strength and B_1, B_2, B_3 are brittleness indexes. As seen from the equations, all of them are different functions of the ratio of UCS to BTS. Although the BI is a function of the different combinations of rock properties rather than only ratios of the strength, there is no agreement on the measurement of the BI at the present time. Yagiz [13] has also introduced the BI value based on PPT and published equations that could be used for computing BI as follow:

$$BI_o = \frac{F_{max}}{P_{max}} \tag{7}$$

where, F_{max} is the max force and P_{max} is the corresponding penetration

$$BI_p = 0.19 \times \sigma_c - 2.174 \times \sigma_t + 0.913 \times \rho - 3.807 \tag{8}$$

He stated that the brittleness of a rock cannot be only identified based on the strengths of a rock, but also related to density or porosity as well. It should be mentioned that the BI used herein is directly measured from the laboratory PPT and computed as suggested by past studies [13].

3. Database Development

In this paper, 80 tunnel cases are used to examine the effect of rock properties including UCS, BTS, density (D), brittleness (BI_i) that is measured from the punch penetration tests (PPT), Cerchar abrasivity index of rocks on the disc cutter consumption. Unpublished data obtained from various sources [51] were utilized for this purpose. Rock types range from sedimentary, igneous, and metamorphic, to volcanic rocks. Rock strength, (UCS and BTS), density, and CAI was conducted in accordance with ISRM standards; however, punch penetration tests were performed according to the Colorado School of Mines testing procedure [13]. The PPT test is conducted, examined and the brittleness value was calculated directly from the tests based on Yagiz's Method [13,16]. To examine the effect of rock properties on the CLI, first CLI is computed as a function of CAI as follows [5]:

$$CLI = 2.87 \cdot CAI^2 - 35.62 \cdot CAI + 112.9 \quad (9)$$

In the dataset UCS of rock ranges from 9.5 to 317 MPa with averaged UCS values of rocks are 135 MPa. According to the ISRM classifications [52], the rock ranges from weak to very strong rock types. The BTS ranges from 2.5 to 17 with an average of 8.2 MPa. The density of rock ranges from 18 to 29.5 kN/m^3; however, most of the rock samples had a density of around 25–28. The BI values range from 46 (very high brittle rock) to ductile rock according to the classification recently published [32]. The qualitative statistical evaluation of the data is summarized in Table 5.

Table 5. Summary of rock properties and data ranges used for this paper [32].

Variables	N	Minimum	Maximum	Mean	Std. Dev.	Variance
D	80	17.69	29.53	25.70	2.03	4.14
UCS	80	9.50	327.00	131.24	54.91	3015.40
BTS	80	2.30	17.80	8.17	2.85	8.14
BI	80	9.68	46.00	27.78	8.55	73.09
CAI	80	0.66	6.40	3.52	1.27	1.62
CLI	80	2.49	90.64	27.68	22.32	498.02
Valid N	80					

CAI is dependent on rock strength and other properties such as quartz content, and brittleness. The CAI values in the database varied from 6.4 (extremely abrasive) to 0.66 (low abrasivity) according to the ISRM classification [9]. Furthermore, the cutter consumption rate is represented by CLI and estimated by examining different approaches [5,14,15,37]. CLI was computed as suggested by past studies [5,26].

4. Development of CLI Models

The established dataset is used to experiment with a series of models with different input parameters to obtain the best reliable model to estimate the CLI on the basis of the rock properties. The SPSS Statistics [53] program was used as a statistical tool for this research together with other artificial intelligence techniques including artificial neural network (ANN) and fuzzy logic (FL) developed in the Matbal program [54]. A simple regression analysis was first performed between a single variable and CLI in order to evaluate the influence of each individual parameter on that. Multiple linear (MLRM) and non-linear models (NLMRM) were subsequently run in order to obtain the most accurate equations from the datasets. In order to examine empirical correlations between variables methods, statistical software packages such as Excel or SPSS are commonly utilized.

4.1. Regression Analysis of Data

4.1.1. Simple Regression Analysis (Univariate)

In this study, non-linear ($y = ax^b$) simple regression analyses were performed among rock properties to find the best inputs to estimate the CLI. This means that CLI was used as a dependent, while other rock properties were used as independent variables to develop pertinent models. This resulted in several simple regression relations among the possible input variables and CLI. Plots of CLI as a function of individual parameters are shown in Figures 6–9.

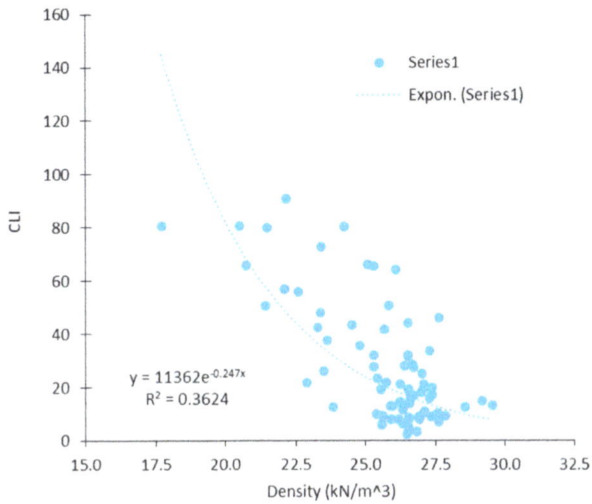

Figure 6. Relations between the density and CLI.

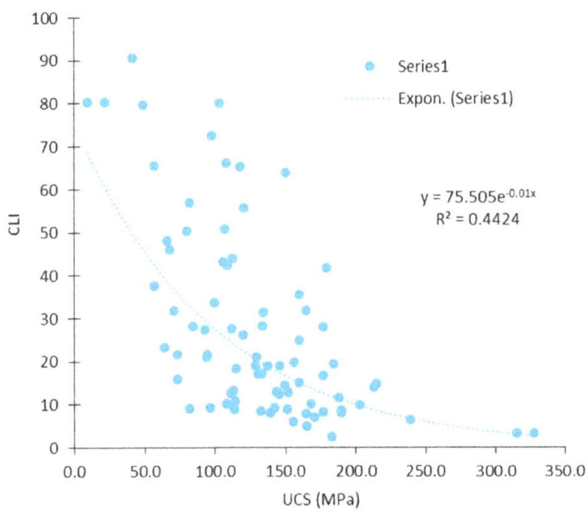

Figure 7. Relations between the UCS and CLI.

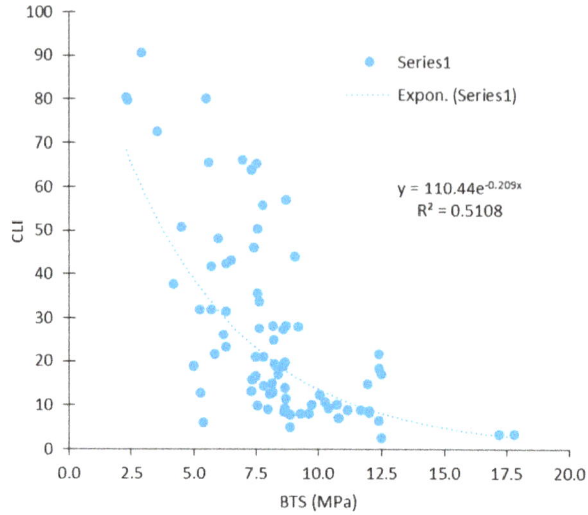

Figure 8. Relations between the BTS and CLI.

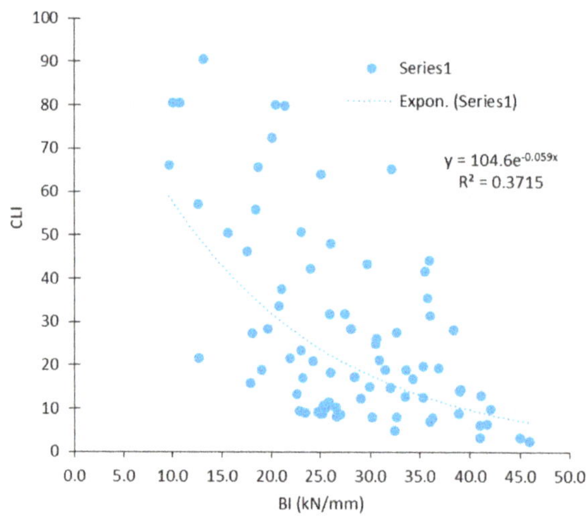

Figure 9. Relations between the BI and CLI.

4.1.2. Linear and Non-Linear Multi-Variable Regression Analysis

The prediction of CLI based on other rock properties can be carried out either using a linear or non-linear model. However, it is known that one of the common CLI estimation models from CAI [5] is polynomial a function. In this study, both linear and non-linear multivariable regression analyses were performed. While running the SPPS program, data were divided into two sets including: a training set (70%) and a testing set (30%). Table 6 is the summary of bi-linear multiple regression together with input variables of each model and statistical indices (#).

Table 6. Summary of multi-variable statistical analysis of database using linear functions. Statistical indices (#).

#	1	2	4	4	5	6	7
R	0.84	0.82	0.73	0.77	0.65	0.77	0.78
R^2	**0.71**	0.67	0.53	0.60	0.42	0.59	0.62
MSE	135.97	148.57	212.52	171.47	289.83	201.65	159.59
RMSE	11.66	12.19	14.58	13.09	17.02	14.20	12.63

Inputs: [1] UCS, BTS, D, BI; [2] UCS, D, BTS; [3] UCS, BTS; [4] UCS, D; [5] UCS, BI; [6] BTS, BI; [7] D, BI.

In this study, 7 different models were developed using both linear and non-linear regression as shown in Table 6. The results are very close to each other which means each rock property has some effect on CLI; hence, the models and related equations are listed in Table 7. The best model obtained via MLRM is given in Figures 10 and 11.

Similarly, multi-variable non-linear regression output equations and related performance indices are given in Tables 8 and 9, respectively, for each alternative model developed using the other rock properties. It could be stated that each rock property has some weight on CLI.

Table 7. Results of multi-variable linear regression of CLI as function of other rock properties.

#	Inputs	Equations for CLI
1	D, UCS, BTS, BI	$CLI = -4.425 \times D + 0.079 \times UCS - 3.408 \times BTS - 0.735BI + 180.855$
2	D, UCS, BTS	$CLI = -4.928 \times D - 0.005 \times UCS - 3.24 \times BTS + 183.232$
3	UCS, BTS	$CLI = -0.048 \times UCS - 4.601 \times BTS + 74.176$
4	D, UCS	$CLI = -6.055 \times D - 0.108 \times UCS + 198.304$
5	UCS, BI	$CLI = -0.106 \times UCS - 1.01 \times BI + 70.595$
6	BTS, BI	$CLI = -3.991 \times BTS - 0.761 \times BI + 83.729$
7	D, BI	$CLI = -5.858 \times D - 0.824 \times BI + 201.505$

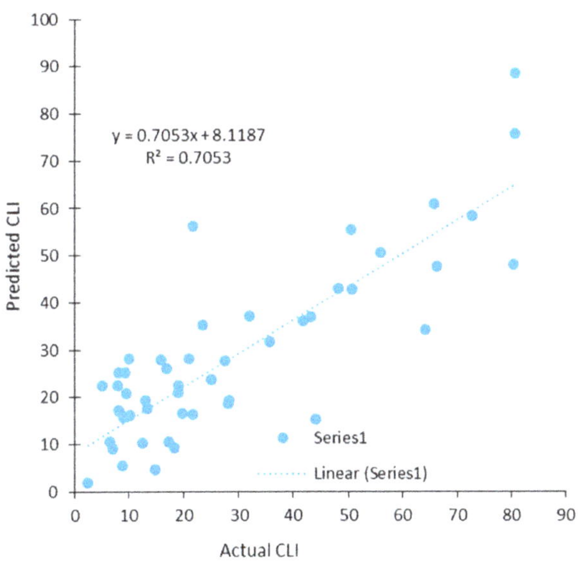

Figure 10. Relations between the measured CLI and predicted CLI for training (model 1-MLRM).

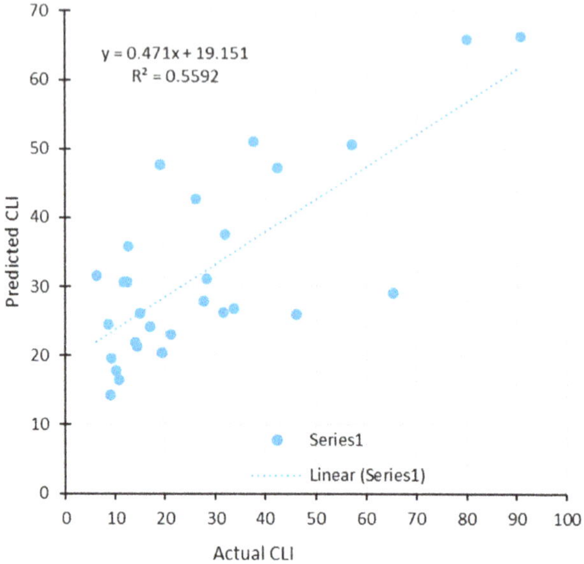

Figure 11. Relations between the measured CLI and predicted CLI for testing (model 1-MLRM).

Table 8. Summary of non-linear multi-variable statistical analysis of CLI.

Non-L	1	2	4	4	5	6	7
R	0.85	0.83	0.79	0.76	0.69	0.81	0.79
R^2	**0.72**	0.69	0.62	0.58	0.47	0.66	0.62
MSE	124.88	138.22	154.67	167.51	235.07	159.15	154.71
RMSE	11.18	11.76	12.44	12.94	15.33	12.62	12.44
VAF	72.33	68.86	62.39	57.58	47.30	66.01	62.07

Table 9. Results of multi-variable non-linear regression of CLI as function of other rock properties.

	Inputs	Equations for CLI
1	D, UCS, BTS, BI	$-4.099 \times D + 9.771 \times Ln(UCS) - 30.94 \times Ln(BTS) - 0.006 \times BI^2$ $-0.361 \times BI + 166.102$
2	D, UCS, BTS	$-3.938 \times D + 2.251 \times \ln(UCS) - 32.179 \times \ln(BTS) + 185.433$
3	UCS, BTS	$-3.38 \times \ln(UCS) - 40.63 \times \ln(BTS) + 128.817$
4	D, UCS	$-5.741 \times D - 10.418 \times \ln(UCS) + 225.352$
5	UCS, BI	$-13.933 \times \ln(UCS) + 0.025 \times BI^2 - 2.258 \times BI + 136.543$
6	BTS, BI	$-35.366 \times \ln(BTS) + 0.018 \times BI^2 - 1.604 \times BI + 131.049$
7	D, BI	$-6.344 \times D - 0.018 \times BI^2 + 0.256 \times BI + 199.683$

The results of the analysis for both linear and non-linear multi-variable regression show that models with three inputs can offer a reliable estimate of CLI with a coefficient of correlation of 0.82. This is very close to the correlation coefficient obtained from model 1 for training and testing data, respectively (Figures 12 and 13).

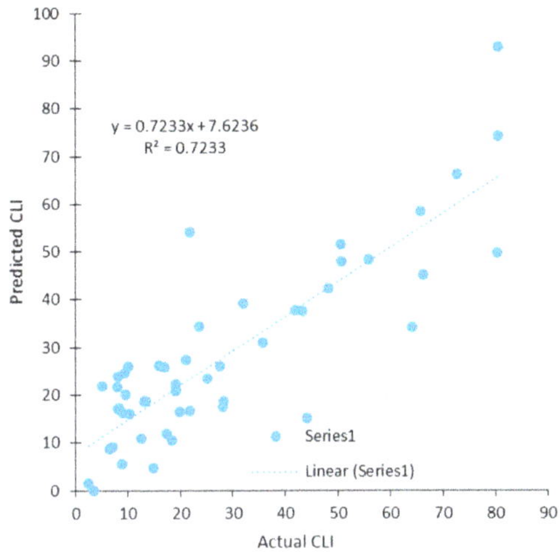

Figure 12. Relations between the measured CLI and predicted CLI for training (model 1-NLMRM).

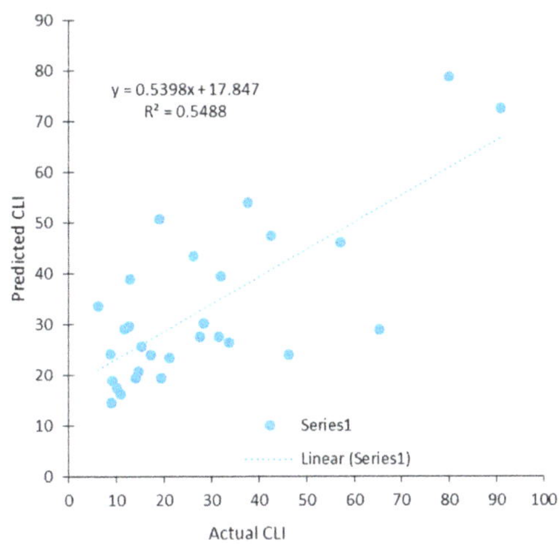

Figure 13. Relations between the measured CLI and predicted CLI for testing (model 1-NLMRM).

It should be noted that the linear regression analysis approach has clear advantages over the black box models provided by AI systems, simply by offering equations for use by everyone.

5. Soft Computing Techniques

Artificial intelligence (AI) is a common modeling technique for developing models to estimate an unknown parameter from knowing variables in rock engineering and tunneling as well as other engineering practices. In this research, two common artificial intelligence methods including artificial neural network (ANN) and fuzzy logic (FL) were utilized to estimate the CLI as a function of common rock properties including strength, density, and brittleness. Several alternative models with alternative input variables were developed and the best one among those models is highlighted herein.

5.1. Artificial Neural Networks (ANN)

The MATLAB environment [53] has a built-in application, neural net fitting tool, which is a computational tool that was used for developing an ANN model in this study. To set up input data for ANN analysis, it is required to set a portion of data for testing, validation, and training. In our case of 70% of input data was used for training purposes, another 15% was used for validation, and 15% was for testing [55]. ANN uses the Levenberg–Marquardt optimization algorithm due to its advantages, such as high speed of training on feed-forward networks of moderate size [56]. Moreover, the MATLAB environment has an efficient built-in function of the Levenberg–Marquardt algorithm with very efficient performance [57].

In this study, the ANN models were developed by using one hidden layer. Baheer and Hecht-Neilsen [58,59] claimed that neural networks with one hidden layer, in general, are enough for addressing the majority of issues. ANN models developed in the neural net fitting tool use a two-layer-forward network with linear output neurons, where one layer is the hidden layer and one is the output layer. This type of network is capable of solving multi-dimensional mapping issues with sufficient accuracy if ANN has reliable data for training and enough number of neurons for the hidden layer. The next step for the ANN model development is to define the number of neurons, which was considered to be the most crucial question in the process of determining the ANN structure [60]. If the number of hidden neurons is lower than needed, it will undergo the "under-fitting" problem in both generalization and training. Nevertheless, an excess number of hidden neurons can result in the problem of overfitting, which means the neural network overestimates the target problem's complexity [61]. This can cause a large variance in prediction results, and the generalization capacity drops considerably. Without performing a try-and-test during training and calculating the generalization error, determining the best number of hidden units is difficult. The number of hidden layers and hidden neurons, that is optimal to certain cases, is determined by the following factors: (a) network design complexity; (b) the number of input and output units; (c) the number of training samples; (d) the level of noise in the sample dataset; (e) the training algorithm [62]. In this regard, determining the correct number of hidden neurons to avoid overfitting or underfitting is crucial in the prediction process. The heuristic parameters suggested for this purpose are listed in Table 10.

Table 10. The heuristics proposed for optimal number of neurons.

Heuristic	References
$\leq 2 * N_i + 1$	[59]
$3N_i$	[63]
$(N_i + N_o)/2$	[64]
$\frac{2+N_o * N_i+0.5N_o * \left(N_o^2+N_i\right)-3}{N_i+N_o}$	[65]
$\frac{2N_i}{3}$	[66]
$\sqrt{N_i * N_o}$	[67]
$2N_i$	[68]

In Table 10, N_i refers to the number of input neurons and N_o refers to the number of output neurons, respectively. According to Table 10, the number of hidden neurons must be 1, 2, 5, or 6 for two inputs, while for a set of four input parameters, the number of hidden neurons was suggested to be between 2 and 12. However, the work of Ke and Liu [61] suggested that for 80 input samples and one hidden layer, the optimal number of neurons is 12. The trial-and-error approach is currently used to calculate the number of hidden neurons. This begins with a small number of neurons and progressively increasing the number till optimal performance is achieved. The downside is that it takes time and there is no assurance that the hidden neuron would be fixed [62].

To determine the number of hidden neurons, suggested heuristics in Table 10 were employed. Then, from 1 to 14 hidden neurons were used to establish the models, which were run at least 25 times to obtain the best output. This was followed by evaluating the performance of developed ANN models on the basis of average performance obtained, in terms of the R-value and MSE. The optimal number of hidden neurons was chosen as indicated in Table 11. An illustration of the ANN structure corresponding to four inputs with one hidden layer and 10 neurons is given in Figure 14.

Table 11. Number of hidden neurons for 80 sample inputs.

Input Parameters	# of Hidden Neurons
Density, UCS, BTS, BI	10
Density, UCS, BTS	8
UCS, BTS	9
Density, UCS	6
UCS, BI	8
BTS, BI	10
Density, BI	7

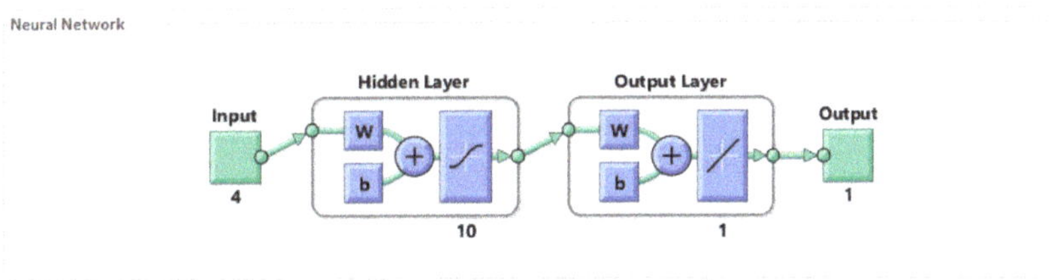

Figure 14. Generalized structure of ANN used in this study.

Developed ANN structures with a given number of inputs involve another component of the structure that is established and run to obtain average values of the model outputs. The results from each model are examined using several statistical indices such as coefficient of determination (r^2), correlation (r), and variance account for (VAF); root means square error (RMSE) and mean square error (MSE). A summary of the results for performance of various models is given in Table 12. It appears that each input rock property has some influence on the CLI and the output of model 7 is better than others. The best ANN model to estimate CLI is a function of the density and brittleness of rocks. The best results from various ANN models are given in Table 12, as a function of density and brittleness (BI) as a combination of rock properties (Figures 15 and 16).

Table 12. Statistical indices of the ANN models for calculation of CLI from other rock properties.

ANN	1	2	3	4	5	6	7
R	0.75	0.83	0.76	0.80	0.73	0.73	0.84
R^2	0.57	0.68	0.57	0.64	0.54	0.53	0.71
MSE	168.89	137.79	173.64	186.01	193.45	301.09	145.67
RMSE	13.00	11.74	13.18	13.64	13.91	17.35	12.07
VAF	56.78	68.12	57.31	64.12	53.75	51.36	69.99

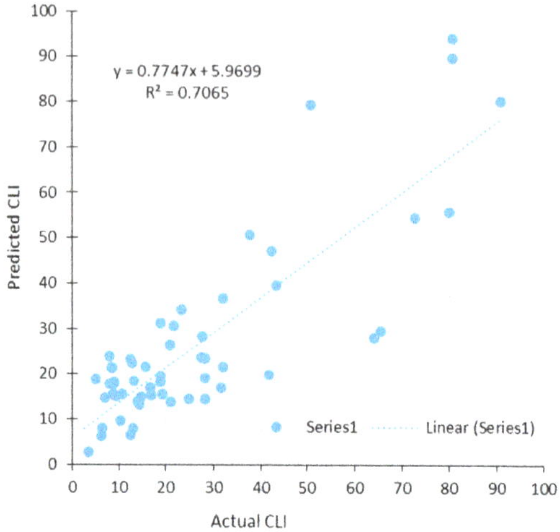

Figure 15. Relations between the actual and estimated CLI by ANN model 7 for training.

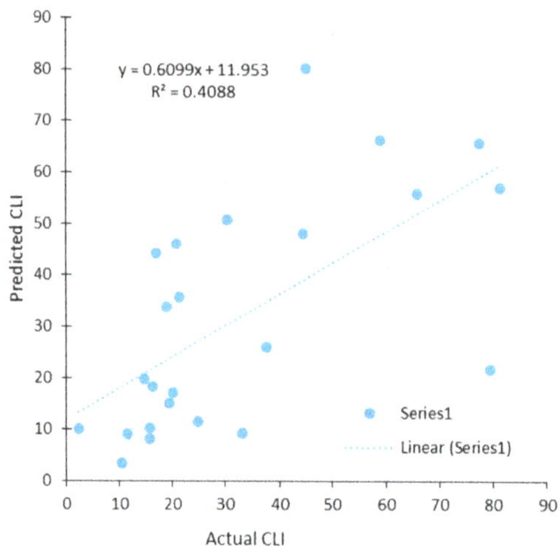

Figure 16. Relations between the actual and estimated CLI by ANN model 7 for testing.

5.2. Fuzzy Logic (FL)

In this research, another artificial intelligence model that was developed is based on Fuzzy Logic (FL). FL model requires special rules to control data behavior. To obtain these rules and membership functions, the Matlab built-in function "genfis2" was used. This function uses provided input and output data in separate matrices as input arguments. It also requires radii to be selected for input and output arguments, which specifies the center of cluster range of influence for each of the dimensions used in the data set. The data was then used to estimate the amount of membership functions and rules for the antecedents and consequents. The extracted rules were stored in a special FIS format variable, that later could be used to obtain predicted output for certain input variables. Similar to Section 5.1, different input rock characteristics were used as inputs for the fuzzy logic model and 70% of data was used for training, and the remaining 30% was used for testing to estimate the CLI. The FL model was also performed for seven different combinations of rock properties, and the output of the models is discussed in this paper. The generated FIS file contains membership functions that define to what degree a certain input belongs to a given set. The following figures are a representation of the fuzzy logic model created by the "genfis2" command, where input parameters are density and UCS with the output of CLI. These membership functions are based on gbellmf (generalized bell-shaped membership function), which is a function with a specific shape suitable for this analysis. This function was chosen automatically by the genfis2 command. The "anfis" function in MATLAB software uses the adaptive neuro-fuzzy inference system (ANFIS) to tune the FIS file obtained from "genfis2" and its use was recommended in the literature [69]. This approach is especially useful in engineering fields where traditional methods fail or are too difficult to use [70]. The ANFIS technique is effective for nonlinear system interpretation [71]. The "Anfis" command in the MATLAB environment provided some improvement in the modeling results as shown by higher correlation factors and MSE. The results of the FL analysis of CLI are presented in the next section of this paper. The final visualization of rules for the FIS file after the "anfis" tuning command is used can be seen in Figure 17 with two input cases for the dataset.

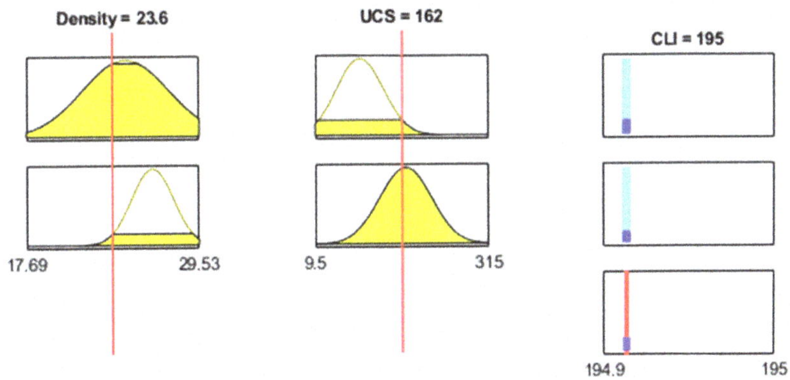

Figure 17. Fuzzy logic rules for 2 inputs and 1 output.

Given that the MATLAB [54] fuzzy logic designer tool is not capable of extracting rules that control the behavior of data, it had to be carried out with the built-in command "genfis2". To do so, input and output data were defined from the dataset, and then by using "dividerand" that sets what fraction of data can be used in training, testing, and validation; it randomly selects which samples of the dataset go to which part. The "genfis2" function is used on that randomly chosen 70% of data to be trained, then the "evalfis" command is used to check the model by predicting results on training and testing samples. It is performed inside of the loop where the variable "i" defines the radius of the cluster from 0.1 to 1 with 0.05 steps. Cluster radii is a scalar value, which is multiplied by the width of the data space. The "mse" code determines the MSE of models, while "mseALLtraining" collects these MSE measurements for finding optimal. After developing the rules and model structure for 80 tunneling cases to estimate the CLI as a function of rock properties, models run until they produce reliable and accurate results. To achieve the desired performance, seven different alternative input combinations were used and the results of the models were examined to find the best model among them. The initial steps involve evaluating the influence of each rock variable using the FL surface map as shown in Figures 18–21.

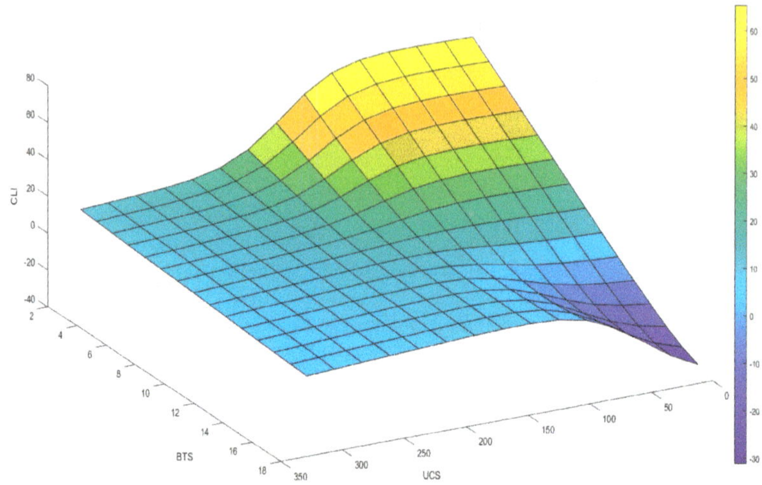

Figure 18. Fuzzy logic surface view for inputs (UCS and BTS) and CLI.

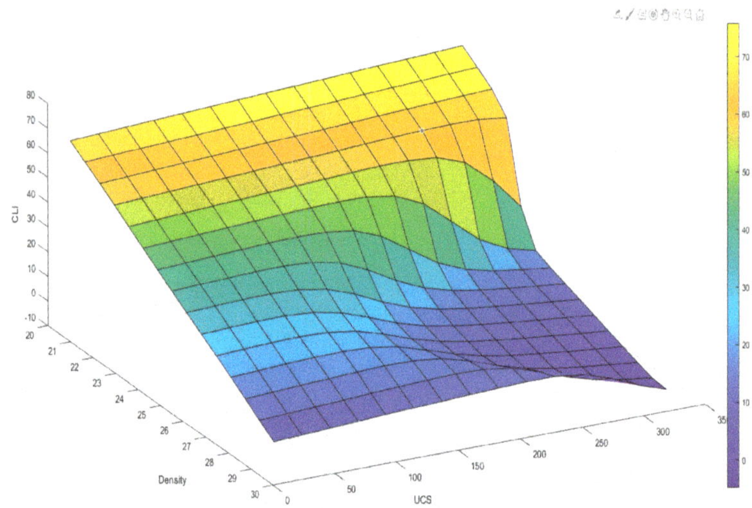

Figure 19. Fuzzy logic surface view for inputs (D, UCS) and CLI.

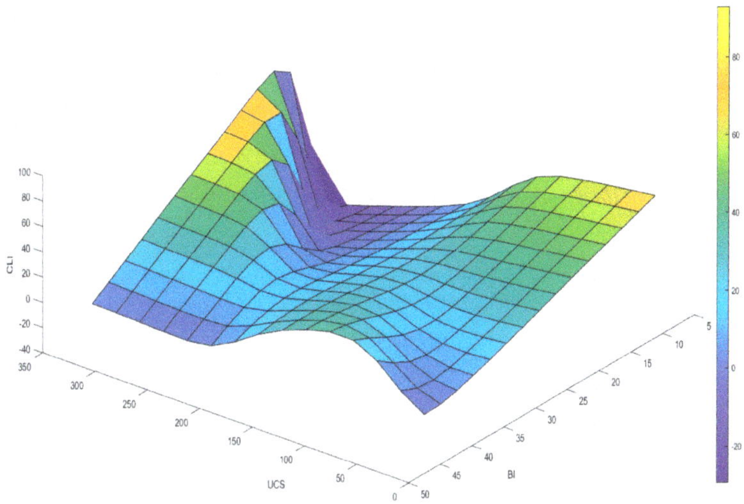

Figure 20. Fuzzy logic surface view for inputs (BI, UCS) and CLI.

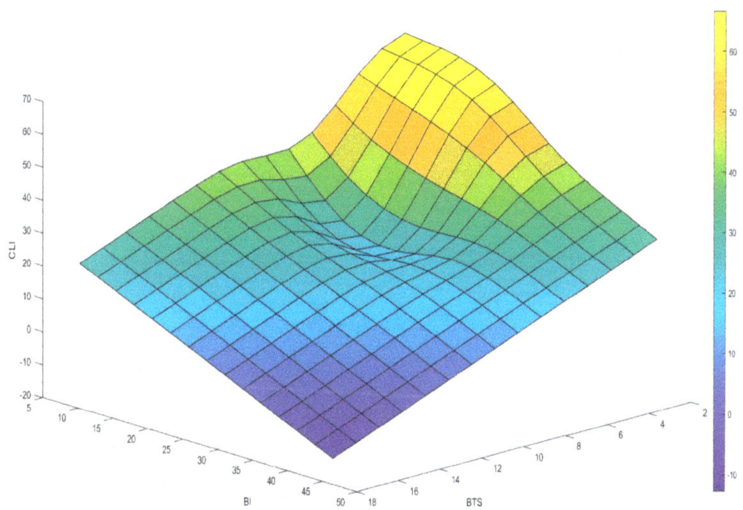

Figure 21. Fuzzy logic surface view for inputs (BTS, BI) and CLI.

As mentioned earlier, the FL models were run for various inputs as given in Table 13 and the best model was Model#1; model 7 was also very promising with two instead of four input parameters. Overall observation indicates that model 7 with input parameters including density and brittleness is also valid and offers acceptable results for practical applications.

Table 13. Statistical indices of the FL models obtained for each model.

FL	1	2	4	4	5	6	7
R	0.82	0.79	0.70	0.77	0.68	0.75	0.79
R^2	0.68	0.63	0.48	0.60	0.47	0.57	0.63
MSE	170.22	195.74	272.03	212.35	280.70	226.94	197.14
RMSE	13.05	13.99	16.49	14.57	16.75	15.06	14.04
VAF	67.71	62.87	48.39	59.72	46.75	56.95	62.60

It was also observed that the FL modeling technique is a powerful tool to estimate CLI from other rock parameters. The plots of predicted and measured CLI of model 1 for testing and training are shown in Figures 22 and 23, respectively.

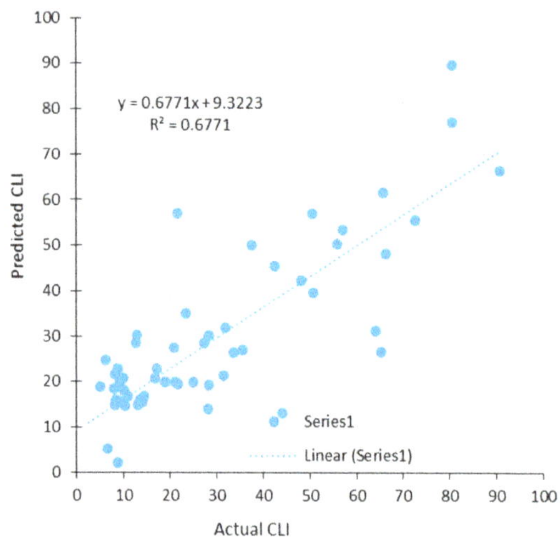

Figure 22. Relations between the actual CLI and estimated CLI by FL model 1 for training.

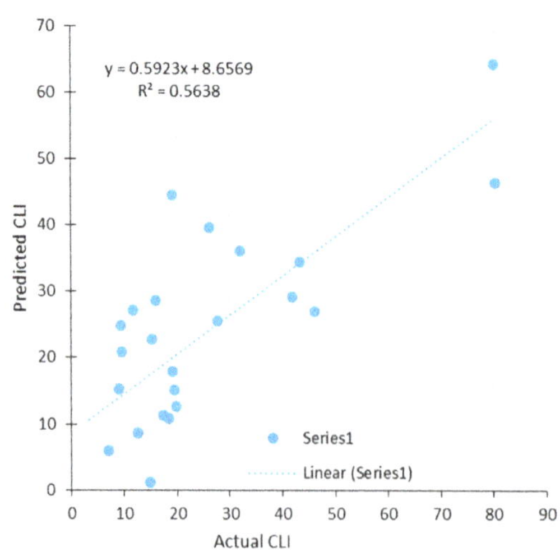

Figure 23. Relations between the actual CLI and estimated CLI by FL model 1 for testing.

In summary, various combinations of input parameters were used to examine the ability to predict CLI from other rock parameters and those models do not always produce reliable results, and the accuracy of the models depends on rock properties and the range of data in the original dataset used for modeling purposes.

6. Discussions

In this paper, four different methods were used to predict CLI from rock mechanical properties including UCS, BTS, density, and brittleness. For this purpose, a database of measured CLI and other rock mechanical properties was established and used in the analysis leading to the development of predictive models. These models were subsequently compared to each other via statistical indices. The analysis showed that the models offered a reasonable estimate of CLI, which can be used for the estimation of disc cutter consumption in the NTNU model. Table 14 presents the comparative statistical indices for the four methods used for developing the models.

Table 14. Comparison of accuracy of the models based on obtained statistical performance indices.

Inputs	BI						
	UCS	UCS					
	BTS	BTS	UCS	D	UCS	BTS	D
	D	D	BTS	UCS	BI	BI	BI
Linear	1	2	4	4	5	6	7
R	0.84	0.82	0.73	0.77	0.65	0.77	0.78
R^2	0.71	0.67	0.53	0.60	0.42	0.59	0.62
MSE	135.97	148.57	212.52	171.47	289.83	201.65	159.59
RMSE	11.66	12.19	14.58	13.09	17.02	14.20	12.63
VAF	70.53	67.48	53.26	59.74	42.03	58.56	61.62
Non-L	1	2	3	4	5	6	7
R	0.85	0.83	0.79	0.76	0.69	0.81	0.79
R^2	0.72	0.69	0.62	0.58	0.47	0.66	0.62
MSE	124.88	138.22	154.67	167.51	235.07	159.15	154.71
RMSE	11.18	11.76	12.44	12.94	15.33	12.62	12.44
VAF	72.33	68.86	62.39	57.58	47.30	66.01	62.07
ANN	1	2	3	4	5	6	7
R	0.75	0.83	0.76	0.80	0.73	0.73	0.84
R^2	0.57	0.68	0.57	0.64	0.54	0.53	0.71
MSE	168.89	137.79	173.64	186.01	193.45	301.09	145.67
RMSE	13.00	11.74	13.18	13.64	13.91	17.35	12.07
VAF	56.78	68.12	57.31	64.12	53.75	51.36	69.99
FL	1	2	4	4	5	6	7
R	0.82	0.79	0.70	0.77	0.68	0.75	0.79
R^2	0.68	0.63	0.48	0.60	0.47	0.57	0.63
MSE	170.22	195.74	272.03	212.35	280.70	226.94	197.14
RMSE	13.05	13.99	16.49	14.57	16.75	15.06	14.04
VAF	67.71	62.87	48.39	59.72	46.75	56.95	62.60

Since the output of the models, CLI, is sensitive to each rock property, obtained models give very similar results and it is not easy to choose the best among them. However, some of the models offer superior predictions compared to others depending on the number of input variables. For example, the models having two inputs including D and BI can offer reasonable results with only two parameters, however, BI requires punch penetration testing that is not very common. Yet, most of the models produced acceptable estimates of CLI.

7. Conclusions

In this paper, cutter consumption for TBMs as represented by CLI was examined and estimated as a function of several common rock properties to be used in practice. Datasets from several projects including uniaxial compressive strength, Brazilian tensile strength, density, and rock brittleness indices were compiled in a database and analyzed by various modeling techniques with different sets of input parameters. Statistical analysis (SPSS 26.0) software and MATLAB were used for the development of models based on regression and artificial intelligence models, respectively. The proposed models could estimate CLI, however, some of the models offered superior predictions.

Following are the the summary of findings in this study:

- Rock properties including strength, UCS, BTS, density, and brittleness indices have some influence on the CLI.
- Density and brittleness of rock are very important variables for estimating CLI and offer a better prediction of CLI compared to other variables. Moreover, while these two variables could be used for estimating the CLI when other parameters are not available, it is not recommended since density and BI do not reflect the abrasivity of the rock.
- Brazilian tensile strength of rock is the significant input when it is used with BI (Model 6, non-linear model).
- When comparing the variable for prediction of CLI on an individual basis, BTS shows a better correlation with CLI, perhaps since BTS is directly related to rock breakage and brittleness behavior of rock under the disc or indenter.

The dataset used in this study is based on real-world projects and proposed models could be valuable and applicable to practice; however, the output of the study should be improved by adding more rock data and inputs such as equivalent quartz content to the models.

Author Contributions: T.M and S.Y. wrote the manuscript; T.M. and S.Y. collected the data; A.C.A. edited and revised the manuscript; T.M. drew the figures; S.Y. designed research methods; T.M., S.Y. and A.C.A. analyzed the data. All authors have read and agreed to the published version of the manuscript.

Funding: This study was supported by the Faculty Development Competitive Research Grant program of Nazarbayev University, Grant No: 021220FD5151.

Institutional Review Board Statement: Not applicable.

Informed Consent Statement: Not applicable.

Data Availability Statement: Data are contained within the article. The data presented in this study can be requested from the corresponding author.

Conflicts of Interest: The authors declare no conflict of interest.

References

1. Plinninger, R.J. *Classification and Prediction of Tool Wear with Conventional, Mountain Solution Method in Solid Rock*; Munich Geological Books, Series B, Applied Geology; IAEG: Munich, Germany, 2002.
2. Plinninger, R.J.; Kasling, H.; Thuro, K. *Wear Prediction in Hardrock Excavation Using the CERCHAR Abrasiveness Index (CAI)*, EUROCK 2004 and 53rd Geomechanics Colloquium, ed.; Salzburg, Austria; pp. 599–604. Available online: http://www.plinninger.de/images/pdfs/2004_EUROCK_CAI.pdf (accessed on 1 December 2021).
3. Sun, Z.; Zhao, H.; Hong, K.; Chen, K.; Zhou, J.; Li, F.; Zhang, B.; Song, F.; Yang, Y.; He, R. A practical TBM cutter wear prediction model for disc cutter life and rock wear ability. *Tunn. Undergr. Space Technol.* **2019**, *85*, 92–99. [CrossRef]
4. Thuro, K.; Wilfing, L.; Wieser, C.; Ellecosta, P.; Käsling, H.; Schneider, E. Hard rock TBM Tunnelling—On the way to a better prognosis? *Goeomechanics Tunn.* **2015**, *8*, 191–199. [CrossRef]
5. Bruland, A. *Hard Rock Tunnel Boring—Geology and Site Investigation. Project Report 1D-98*; NTNU: Trondheim, Norway, 1998.
6. Schimazek, T.; Knatz, H. The influence of rock structures on the cutting speed and pick wear of heading machines. *Glücckauf March* **1970**, *106*, 275–278. (In German)
7. Cerchar. *Centre d ´ Etudes et Recherches de Charbonnages de France: The Cerchar Abrasiveness Index*; Cerchar: Verneuil, France, 1986.

8. ASTM. *Standard Test Method for Laboratory Determination of Abrasiveness of Rock Using the Cerchar Method*; ASTM Designation 2010, D7625-10 Vol. 04.08; American Society for Testing and Materials: West Conshohocken, PA, USA.
9. Alber, M.; Yaralı, O.; Dahl, F.; Bruland, A.; Käsling, H.; Michalakopoulos, T.N.; Cardu, M.; Hagan, P.; Aydın, H.; Özarslan, A. ISRM Suggested Method for Determining the Abrasivity of Rock by the CERCHAR Abrasivity Test. *Rock Mech. Rock Eng.* **2014**, *47*, 261–266. [CrossRef]
10. Hucka, V.; Das, B. Brittleness determination of rocks by different methods. *Int. J. Rock Mech. Min. Sci. Géoméch. Abstr.* **1974**, *11*, 389–392. [CrossRef]
11. Altindag, R. Assessment of some brittleness indexes in rock-drilling efficiency. *Rock Mech. Rock Eng.* **2009**, *43*, 361–370. [CrossRef]
12. Andreev, G.E. *Brittle Failure of Rock Materials: Test Results and Constitutive Models*; A. A. Balkema: Rotterdam, The Netherlands, 1995; p. 446.
13. Yagiz, S. Assessment of brittleness using rock strength and density with punch penetration test. *Tunn. Undergr. Space Technol.* **2009**, *24*, 66–74. [CrossRef]
14. Rostami, J. Development of a Force Estimation Model for Rock Fragmentation with Disccutters through Theoretical Modeling and Physical Measurement of Crushed Zone Pressure. Ph.D. Thesis, Colorado School of Mines, Golden, CO, USA, 1997; p. 249.
15. Rostami, J.; Huckami, A.; Gharahbagh, E.A.; Dogruoz, C.; Dahl, F. Study of dominant factors affecting Cerchar abrasivity index. *Rock Mech. Rock Eng.* **2014**, *47*, 1905–1919. [CrossRef]
16. Yagiz, S. Development of Rock Fracture and Brittleness Indices to Quantify the Effects of Rock Mass Features and Toughness in the CSM Model Basic Penetration for Hard Rock Tunneling Machines. Ph.D. Thesis, Department of Mining Engineering, Colorado School of Mines, Golden, CO, USA, 2002; 289p.
17. Yagiz, S. Utilizing rock mass properties for predicting TBM performance in hard rock condition. *Tunn. Undergr. Space Technol.* **2008**, *23*, 326–339. [CrossRef]
18. Yagiz, S.; Rostami, J.; Ozdemir, L. Recommended rock testing methods for predicting TBM performance: Focus on the CSM and NTNU Models. In Proceedings of the ISRM International Symposium-5th Asian Rock Mechanics Symposium, Tehran, Iran, 24–26 November 2008; pp. 1523–1530.
19. Frenzel, C. Modeling Uncertainty in Cutter Wear Prediction for Tunnel Boring Machines. *GeoCongress* **2012**, *2012*, 3239–3247. [CrossRef]
20. Parsajoo, M.; Mohammed, A.S.; Yagiz, S.; Armaghani, D.J.; Khandelwal, M. An evolutionary adaptive neuro-fuzzy inference system for estimating field penetration index of tunnel boring machine in rock mass. *J. Rock Mech. Geotech. Eng.* **2021**, *13*, 1290–1299. [CrossRef]
21. Deketh, H. Wear of Rock Cutting Tools. In *Wear of Rock Cutting Tools*; CRC Press: Boca Raton, FL, USA, 2020.
22. Atkinson, T.; Cassapi, V.B.; Singh, R.N. Assessment of abrasive wear resistance potential in rock excavation machinery. *Int. J. Min. Geol. Eng.* **1986**, *3*, 151–163. [CrossRef]
23. Dahl, F. *DRI, BWI, CLI Standards*; NTNU: Trondheim, Norway, 2003.
24. Dahl, F.; Bruland, A.; Drevland Jakobsen, P.; Nilsen, B.; Grøv, E. Classifications of properties influencing the drillability of rocks, based on the NTNU/SINTEF test method. *Tunnell. Undergr. Space Technol.* **2012**, *28*, 150–158. [CrossRef]
25. Alber, M. Stress dependency of the Cerchar abrasivity index (CAI) and its effects on wear of selected rock cutting tools. *Tunn. Undergr. Space Technol.* **2008**, *23*, 351–359. [CrossRef]
26. Massalov, T.; Yagiz, S.; Rostami, J. Relationship between key rock properties and Cerchar abrasivity index for estimation of disc cutter wear life in rock tunneling applications. In Proceedings of the ISRM International Symposium, Eurock 2020—Hard Rock Engineering, Trondheim, Norway, 14–19 June 2020. 7p.
27. Yagiz, S.; Frough, O.; Rostami, J. Evaluation of rock brittleness indices to estimate Cerchar Abrasivity Index for disc cutter weariness. In Proceedings of the 54th US Rock Mechanics and Geomechanics Symposium, Golden, CO, USA, 28 June–1 July 2020. 5p.
28. West, G. Rock abrasiveness testing for tunnelling. *Int. J. Rock Mech. Min. Sci. Géoméch. Abstr.* **1989**, *26*, 151–160. [CrossRef]
29. Plinninger, R.J. Abrasiveness Assessment for Hard Rock Drilling. *Géoméch. Tunn.* **2008**, *1*, 38–46. [CrossRef]
30. Plinninger, R.J.; Spaun, G.; Thuro, K. Prediction and classification of tool wear in drill and blast tunnelling. Engineering Geology for Developing Countries. In Proceedings of the 9th Congress of the International Association for Engineering Geology and the Environment, Durban, South Africa, 16–20 September 2002.
31. Plinninger, R.; Käsling, H.; Thuro, K.; Spaun, G. Testing conditions and geomechanical properties influencing the CERCHAR abrasiveness index (CAI) value. *Int. J. Rock Mech. Min. Sci.* **2003**, *40*, 259–263. [CrossRef]
32. Yagiz, S.; Yazitova, A.; Karahan, H. Application of differential evolution algorithm and comparing its performance with literature to predict rock brittleness for excavatability. *Int. J. Mining Reclam. Environ.* **2020**, *34*, 672–685. [CrossRef]
33. Yaralı, O.; Yaşar, E.; Bacak, G.; Ranjith, P. A study of rock abrasivity and tool wear in Coal Measures Rocks. *Int. J. Coal Geol.* **2008**, *74*, 53–66. [CrossRef]
34. AFNOR. Roches Determination du Pouvoir Abrasive d'uneroche Partie 1: Essai de Rayure Avec Une Pointe. Project Report NF P 94-430-1, Paris. 2000. Available online: https://www.boutique.afnor.org/fr-fr/norme/nf-p944301/roches-determination-du-pouvoir-abrasif-dune-roche-partie-1-essai-de-rayure/fa107119/17746 (accessed on 1 December 2021).
35. Majeed, Y.; Abu Bakar, M.Z. Statistical evaluation of CERCHAR Abrasivity Index (CAI) measurement methods and dependence on petrographic and mechanical properties of selected rocks of Pakistan. *Bull. Int. Assoc. Eng. Geol.* **2016**, *75*, 1341–1360. [CrossRef]

36. Al-Ameen, S.I.; Waller, M.D. The influence of rock strength and abrasive mineral content on the Cerchar Abrasive Index. *Eng. Geol.* **1994**, *36*, 293–301, ISSN 0013-7952.. [CrossRef]
37. Jamal, R.; Özdemir, L.; Amund, B.; Filip, D. Review of issues related to Cerchar abrasivity testing and their implications on geotechnical investigations and cutter cost estimates. In Proceedings of the Rapid Excavation and Tunnelling Conference, Society for Mining, Metallurgy, and Exploration Incorporated, Seattle, WA, USA, 13–15 June 2005; pp. 15–29.
38. Fowell, R.; Bakar, A.; Zubair, M. A Review of the Cerchar and LCPC Rock Abrasivity Measurement Methods. In Proceedings of the 11th Congress of the International Society for Rock Mechanics, Lisbon, Portugal, 9–13 July 2007.
39. Hamzaban, M.-T.; Memarian, H.; Rostami, J. Continuous Monitoring of Pin Tip Wear and Penetration into Rock Surface Using a New Cerchar Abrasivity Testing Device. *Rock Mech. Rock Eng.* **2013**, *47*, 689–701. [CrossRef]
40. Ghasemi, A. *Study of Cerchar Abrasivity Index and Potential Modifications for More Consistent Measurement of Rock Abrasion*; The Pennsylvania State University, Department of Energy and Mineral Engineering: State College, PA, USA, 2010; 88p.
41. Hamzaban, M.; Memarian, H.; Rostami, J.; Ghasemi-Monfared, H. Study of rock-pin interaction in Cerchar abrasivity test. *Int. J. Rock Mech. Min. Sci.* **2007**, *72*, 100–108. [CrossRef]
42. Michalakopoulos, T.; Anagnostou, V.; Bassanou, M.; Panagiotou, G. The influence of steel styli hardness on the Cerchar abrasiveness index value. *Int. J. Rock Mech. Min. Sci.* **2006**, *43*, 321–327. [CrossRef]
43. Stanford, J.; Hagan, P. An Assessment of the Impact of Stylus Metallurgy on Cerchar Abrasiveness Index. In Proceedings of the Coal Operators' Conference. Australia. 2009. Available online: https://ro.uow.edu.au/coal2009/ (accessed on 1 December 2021).
44. Kasling, H.; Thuro, K. Determining abrasivity of rock and soil in the laboratory. In Proceedings of the 11th IAEG Congress, Auckland, New Zealand, 5–10 September 2010; p. 235.
45. Gharahbagh, E.A.; Rostami, J.; Ghasemi, A.R.; Tonon, F. Review of rock abrasion testing. In Proceedings of the 45th US Rock Mechanics/Geomechanics Symposium, San Francisco, CA, USA, 26–29 October 2011; pp. 11–141.
46. Rosiwal, A. Neuere Ergebnisse der Härtebestimmung von Mineralien und Gesteinen.—Ein Absolutes Maß für die Härte spröder Körper.—Verhandlungen der k. k. Geologischen Reichsanstalt, 5+6: Pp. 117–147. (Recent Results of Hardness Determination of Minerals and Rocks. An Absolute Measure of the Hardness of Brittle Solids). Available online: https://www.zobodat.at/publikation_series.php?id=19695 (accessed on 1 December 2021).
47. Farrokh, E.; Kim, D.Y. A discussion on hard rock TBM cutter wear and cutterhead intervention interval length evaluation. *Tunn. Undergr. Space Technol.* **2018**, *81*, 336–357. [CrossRef]
48. Paez, C.V.G. Performance, Wear and Abrasion in Excavation Mechanized Tunneling in Heterogeneous Land. Ph.D. Thesis, Universitat Politècnica de Barcelona, Catalunya, Spain, 2014.
49. Johannessen, O. *NTH Hard Rock Tunnel Boring. Project Report 1–94*; NTH/NTNU: Trondheim, Norway, 1994.
50. West, G. A relation between abrasiveness and quartz content for some Coal Measures sediments. *Int. J. Min. Geol. Eng.* **1986**, *4*, 73–78. [CrossRef]
51. Yagiz, S. *Unpublished Database Obtained from Different Mechanical Tunnel Cases*; Earth Mechanics Institute of Colorado School of Mines: Golden, CO, USA, 2021.
52. Ulusay, R.; Hudson, J.A. (Eds.) ISRM Suggested Methods Published between 1974 and 2006 Are Compiled in The ISRM Blue Book: The Complete ISRM Suggested Methods for Rock Characterization, Testing and Monitoring: 2007–2014 Ankara, Turkey. Available online: https://link.springer.com/article/10.1007/s10064-009-0213-2 (accessed on 1 December 2021).
53. *Statistical Software Package*; SPSS 26.0; IBM: Armonk, NY, USA, 2020.
54. MATLab. 2021. Available online: https://www.mathworks.com/products/new_products/latest_features.html (accessed on 1 May 2021).
55. Yun, X.; Goodacre, R. On Splitting Training and Validation Set: A Comparative Study of Cross-Validation, Boot-strap and Systematic Sampling for Estimating the Generalization Performance of Supervised Learning. *J. Anal. Test.* **2018**, *2–3*, 249–262.
56. Mammadli, S. Financial time series prediction using artificial neural network based on Levenberg-Marquardt algorithm. *Procedia Comput. Sci.* **2017**, *120*, 602–607. [CrossRef]
57. Demuth, H.B.; Raele, M.H. *Neural Network Toolbox User's Guide for Use with Matlab*; MathWorks: Natick, MA, USA, 2009.
58. Baheer, I. Selection of methodology for modeling hysteresis behavior of soils using neural networks. *Comput. Aided Civ. Infrastruct. Eng.* **2000**, *5*, 445–463. [CrossRef]
59. Hecht-Nielsen, R. Kolmogorov's mapping neural network existence theorem. In Proceedings of the First IEEE International Conference on Neural Networks, San Diego, CA, USA, 16–18 October 1989; pp. 11–14.
60. Sonmez, H.; Gokceoglu, C.; Nefeslioglu, H.; Kayabasi, A. Estimation of rock modulus: For intact rocks with an artificial neural network and for rock masses with a new empirical equation. *Int. J. Rock Mech. Min. Sci.* **2006**, *43*, 224–235. [CrossRef]
61. Ke, J.; Liu, X. Empirical Analysis of Optimal Hidden Neurons in Neural Network Modeling for Stock Prediction. In Proceedings of the 2008 IEEE Pacific-Asia Workshop on Computational Intelligence and Industrial Application, Wuhan, China, 19–20 December 2008; Volume 2, pp. 828–832.
62. Sheela, K.G.; Deepa, S.N. Review on Methods to Fix Number of Hidden Neurons in Neural Networks. *Math. Probl. Eng.* **2013**, *2013*, 425740. [CrossRef]
63. Hush, D.R. Classification with neural networks: A performance analysis. In Proceedings of the IEEE International Conference on Systems Engineering, Dayton, OH, USA, 1–3 August 1989; pp. 277–280.

64. Ripley, B.D. Statistical aspects of neural networks. In *Networks and Chaos-Statistical and Probabilistic Aspects*; Barndoff Neilsen, O.E., Jensen, J.L., Kendall, W.S., Eds.; Chapman & Hall: London, UK, 1993; pp. 40–123.
65. Paola, J.D. Neural Network Classification of Multispectral Imagery. Master's Thesis, The University of Arizona, Tucson, AZ, USA, 1994.
66. Wang, C. A Theory of Generalization in Learning Machines with Neural Application. Ph.D. Thesis, The University of Pennsylvania, Philadelphia, PA, USA, 1994.
67. Masters, T. *Practical Neural Network Recipes in C++*; Academic Press: Boston, MA, USA, 1994.
68. Kaastra, I.; Boyd, M. Designing a neural network for forecasting financial and economic time series. *Neurocomputing* **1996**, *10*, 215–236. [CrossRef]
69. Naderloo, L.; Alimardani, R.; Omid, M.; Sarmadian, F.; Javadikia, P.; Torabi, M.Y.; Alimardani, F. Application of ANFIS to predict crop yield based on different energy inputs. *Measurement* **2012**, *45*, 1406–1413. [CrossRef]
70. Cheng, C.B.; Cheng, C.J.; Lee, E.S. Neuro-fuzzy and genetic algorithm in multiple response optimization. *Comput. Math. Appl.* **2002**, *44*, 1503–1514. [CrossRef]
71. Arkhipov, M.; Krueger, E.; Kurtener, D. *Evaluation of Ecological Conditions Using Bioindicators: Application of Fuzzy Modeling, Computational Science and Its Applications–ICCSA 2008*; Springer: Belrin, Germany, 2008; pp. 491–500.

applied sciences

MDPI

Article

Classification of Imbalanced Travel Mode Choice to Work Data Using Adjustable SVM Model

Yufeng Qian [1,*], Mahdi Aghaabbasi [2,*], Mujahid Ali [3,*], Muwaffaq Alqurashi [4], Bashir Salah [5], Rosilawati Zainol [2], Mehdi Moeinaddini [6] and Enas E. Hussein [7]

1 School of Science, Hubei University of Technology, Wuhan 430068, China
2 Centre for Sustainable Urban Planning and Real Estate (SUPRE), Department of Urban and Regional Planning, Faculty of Built Environment, University of Malaya, Kuala Lumpur 50603, Malaysia; rosilawatizai@um.edu.my
3 Department of Civil and Environmental Engineering, Universiti Teknologi PETRONAS, Seri Iskandar 32610, Perak, Malaysia
4 Department of Civil Engineering, College of Engineering, Taif University, Taif 21944, Saudi Arabia; m.gourashi@tu.edu.sa
5 Department of Industrial Engineering, College of Engineering, Kind Saud University, P.O. Box 800, Riyadh 11421, Saudi Arabia; bsalah@ksu.edu.sa
6 Centre for Public Health, Queen's University Belfast, Belfast BT12 6BA, UK; m.moeinaddini@qub.ac.uk
7 National Water Research Center, P.O. Box 74, Shubra El-Kheima 13411, Egypt; enas_el-sayed@nwrc.gov.eg
* Correspondence: yfqian@aliyun.com (Y.Q.); mahdi@um.edu.my (M.A.); mujahid_19001704@utp.edu.my (M.A.)

Abstract: The investigation of travel mode choice is an essential task in transport planning and policymaking for predicting travel demands. Typically, mode choice datasets are imbalanced and learning from such datasets is challenging. This study deals with imbalanced mode choice data by developing an algorithm (SVM_{AK}) based on a support vector machine model and the theory of adjusting kernel scaling. The kernel function's choice was evaluated by applying the likelihood-ratio chi-square and weighting measures. The empirical assessment was performed on the 2017 National Household Travel Survey–California dataset. The performance of the SVM_{AK} model was compared with several other models, including neural networks, XGBoost, Bayesian Network, standard support vector machine model, and some SVM-based models that were previously developed to handle the imbalanced datasets. The SVM_{AK} model outperformed these models, and in some cases improved the accuracy of the minority class classification. For the majority class, the accuracy improvement was substantial. This algorithm can be applied to other tasks in the transport planning domain that deal with uneven data distribution.

Keywords: imbalanced data; travel mode choice data; hybrid support vector machine-based model

Citation: Qian, Y.; Aghaabbasi, M.; Ali, M.; Alqurashi, M.; Salah, B.; Zainol, R.; Moeinaddini, M.; Hussein, E.E. Classification of Imbalanced Travel Mode Choice to Work Data Using Adjustable SVM Model. *Appl. Sci.* **2021**, *11*, 11916. https://doi.org/10.3390/app112411916

Academic Editor: Danial Jahed Armaghani

Received: 21 October 2021
Accepted: 3 December 2021
Published: 15 December 2021

1. Introduction

A considerable amount of people's daily trips is associated with their work. Transport planners and engineers attempt to discover work-related travel behaviors and establish strategies for reducing the adverse impacts of motorized transport on traffic, health, and the environment. One of these important behaviors is work mode choice which refers to the process where an individual chooses a certain mode for his/her trip to work. According to the literature, a variety of factors influence the work mode choice. Socioeconomic factors [1–3], household attributes e.g., [4], trip characteristics e.g., [5], job e.g., [6–9], and built environment [2,10,11] are some of these factors [3,11].

Mode choice data include a wide range of variables and samples. Typically, these data are complex and incomplete [12]. Furthermore, since motorized transport is dominant in most parts of the world, the travel surveys yield unbalanced mode choice classes; that is, there are more people who use cars than people who use other commute modes.

To date, many studies investigated the choice of travel mode to work (Table 1). These methods employed both traditional statistical methods and machine learning (ML) techniques. However, the former is criticized because of its linearity assumptions concerning mode choice data [13–15]. Thus, the employment of ML techniques has receieved more attention recently [16–23]. The classification of the new cases established concerning the existing samples is an essential task in ML models. If at least one of the categories comprises a smaller number of samples than other categories, the classification process becomes complex [24]. The class imbalance issue is simply an uneven data distribution amongst the different categories of the target. The precision of the classification algorithms will be unreliable when they are influenced by the majority class. In this case, the new samples are distributed to the majority category since the classification model tends to predict the minority category with less accuracy, which is an undesirable consequence [25].

Support vector machine (SVM) is a renowned ML technique for classification [26]. This algorithm also was used as a base to cope with imbalanced data. Batuwita and Palade [27] developed the fuzzy SVM model and dealt with imbalanced data in the presence of noises and outliers. Wang and Japkowicz [28] suggested boosting-SVMs with asymmetric cost. Their model runs by adjusting the classifier utilizing cost assignation, though it compensates the bias presented with adjustment through utilizing a combination system that is comparable, in effect, to adjust the distribution of data. Wu and Chang [29] suggested the class-boundary alignment algorithm to augment the SVM model to deal with imbalanced data. They modified the class boundary by converting the kernel function if data is represented in a vector space. This modification can also be performed by adjusting the kernel matrix if the data do not possess a vector-space representation. To enhance forecast performance, Liu, et al. [30] suggested consolidating an integrated sampling system, which mixes both over-sampling and under-sampling, with an ensemble of SVM. These studies investigated the binary classification problem based on the SVM model; however, less examinations have been done concerning multiclass imbalanced classification based on this model.

Many studies in other domains, including medicine, economy, crash severity, and so on, tried to reduce the issues of imbalanced data, e.g., [31–33]. However, to the best of the authors' knowledge, a very small number of studies have investigated the issues of imbalanced travel mode choice data and proposed a solution for it [33].

So far, many scholars have provided useful strategies to manage the issue of class imbalance. These strategies have been helpful and competent in explaining the issue partially through enhancing classifiers' performance. The majority of models developed for binary category imbalance issues are improper for the multiclass imbalanced datasets like work mode choice. In addition, rare studies have provided a solution for the imbalanced mode choice datasets. The shortcomings mentioned above prompted the authors to cope with the multiclass imbalance mode choice data issue and contribute to the body of research on this topic. Thus, this study developed the adjustable kernel-based SVM classification algorithm (SVM_{AK}) that is suitable for handling multiclass imbalanced data. Initially, the estimated hyperplane is obtained employing the regular SVM model. Subsequently, the parameter function and the weighting factor concerning each support vector in every iteration is determined. The likelihood-ratio chi-square test is utilized to estimate these parameters. Following this, the kernel transformation or the new kernel functions are determined. The unequal boundaries of class are enlarged, and data skewness is adjusted, thanks to this function of kernel conversion. Consequently, the estimated hyperplane is remedied through the developed model, and it also solves the problem of performance degradation.

The rest of this paper is structured as follows. In Section 2, the source of data, dataset characteristics, and methodology used for improving the performance of the SVM model for classification of imbalanced data are presented, and evaluation metrics are provided. Section 3 presents the results obtained with the model as well as a series of comparisons against other ML models and SVM-based models for classifying imbalanced datasets.

Section 4 describes the sensitivity analysis method and its outcomes. Finally, a conclusion of the paper is presented in Section 5.

Table 1. Some investigation on work mode choice.

Author	Main Factors Used	Modelling Method
Lu and Kawamura [34]	Mode preferences and responsiveness to level-of-service	Multinomial logit
Badoe [4]	Households (two-workers)	Multinomial logit
Xie et al. [35]	Sociodemographic and Level-of-service attributes	Multinomial logit, Decision trees (DT), and Neural Networks (NN)
Patterson et al. [36]	Gender	Multinomial logit
Al-Ahmadi [37]	Cultural, socioeconomic, safety, and religious parameters	Disaggregate models and utility maximization
Gang [1]	Socioeconomic	Multinomial logit
Vega and Reynolds-Feighan [5]	Travel time, travel cost, and employment destinations	Binary logit and GIS
Vega and Reynolds-Feighan [38]	Central and non-central and suburban employment patterns.	GIS and Cross-Nested Logit (CNL)
Day, Habib and Miller [6]	Commuter trip timing, occupation groups, labor rates, work hour rules, free parking availability, and the spatial distribution of work locations	Multinomial logit
Habib [7]	Work start time and work duration	Multinomial logit
Heinen, Maat and van Wee [8]	Office culture and colleagues' and employers' attitudes	Binary logit
Hamre and Buehler [9]	free car parking, public transportation benefits, showers/lockers, and bike parking at work	Multinomial logit
Heinen and Bohte [39]	Attitudes Toward Mode Choice	Multinomial logit
Tran, Zhang, Chikaraishi and Fujiwara [10]	Neighborhood and travel preferences, land use policy, land use diversity and population density	Multinomial logit
Kunhikrishnan and Srinivasan [40]	Contextual heterogeneity	Binary logit
Franco [41]	Downtown parking supply	Spatial general equilibrium model
Simons, De Bourdeaudhuij, Clarys, De Geus, Vandelanotte, Van Cauwenberg and Deforche [2]	Gender, socio-economic-status (SES) and living environment (urban vs. rural)	Zero-inflated negative binomial (ZINB) regression
Indriany et al. [42]	Risk and uncertainty	Binomial logit
Irfan et al. [43]	Econometric Modeling	Multinomial logit
Hatamzadeh et al. [44]	Gender	Binary logit

2. Methods and Data

2.1. Data

This study employed the 2017 National Household Travel Survey (NHTS)–California dataset. These data are provided by the US Federal Highway Administration and the California Department of Transportation and are freely available to all researchers and practitioners [45]. The NHTS is the definitive source on public travel behavior in the United States. It is the only national source of data that lets researchers and practitioners look at patterns in personal and household travel. This data comprises non-commercial travel information by all modes on a daily basis and the characteristics of the people who travel, their households, and their transport means. It appeared that 26,095 household samples of California were involved in this dataset. This research eliminated records that contained incomplete or inaccurate data. Additionally, the dataset included 458 variables. Thus, based on the literature, the authors selected those variables that linked to work mode choice. Finally, the dataset included 151,597 samples (based on the individuals' records), 26 inputs, and one target variable (mode choice to work). However, at the same time, it was found that the target variable included uneven distribution of classes, which is called imbalanced data. Table 2 shows the composition of the dataset used in this study. The work mode choice had nine classes, and as expected, "car" is the majority class. The imbalance ratio is large (777.5). A list of variables used in this study is provided in Table 3.

Table 2. Description data.

2017 National Household Travel Survey–California Dataset	
Total sample size	151,597
Number of inputs	26
Number of classes	9
Number of samples in each class	
Walk	345
Bicycle	2091
Car	108,885
SUV	12,582
Van	2998
Pickup truck	22,188
Motorcycle/moped	14
Public or commuter bus	2454
Private/charter/tour/shuttle bus	40
Imbalance ratio	777.5

Table 3. Variables included in the final dataset.

Variable	Description	Type	Mean/Mode
Sociodemographic			
age	Age of respondent	Continuous	36.947
educ	Highest education level	Categorial	3.00
gt1jblwk	More than one job	Flag	2.00
flextime	Flexibility of work start time	Flag	2.00
race	Race	Nominal	1.00
sex	Gender	Flag	1.00
wkftpt	Full-time or part-time worker	Flag	1.00

Table 3. *Cont.*

Variable	Description	Type	Mean/Mode
Household information			
drvrcnt	Number of drivers in household	Continuous	3.003
hh_ontd	Number of household members on trip including respondent	Continuous	1.884
hhfaminc	Household income	Categorial	7.00
hhsize	Count of household members	Continuous	4.576
hhvehcnt	Count of household vehicles	Continuous	3.647
numadlt	Count of adult household members at least 18 years old	Continuous	3.194
vehowned wrkcount	Owned vehicle longer than a year	Flag	1.00
youngchild	Number of workers in household	Continuous	0.25
Trip characteristics			
bikemore	Reason for not biking	Nominal	7.00
timetowk	Trip time to work in	Continuous	22.97
walkmore	Reason for not walking	Nominal	5.00
Health			
health	Opinion of health	Categorial	2.00
medcond	Any condition or handicap that makes it difficult to travel outside of the home?	Flag	2.00
medcond6	Medical condition, how long?	Categorial	2.00
Living environment			
urbansize	Urban area size where home address is located	Categorial	6.00
urbrur	Household in urban/rural area	Categorial	1.00
homeloc	Home location/reason to chose your current home location?	Nominal	1.00
wrktrans *	Mode to work	Nominal	3.00

* Target variable.

2.2. Proposed Approach

The algorithm proposed in this study aims at dealing with the imbalanced mode choice data effectively. The theory of adjusting the kernel scaling method [46] is behind the model developed in this research to manage the multi-category imbalanced data. This study combines the SVM classification algorithm with the adjusting kernel scaling technique, which is named SVM_{AK}.

2.2.1. Standard Support Vector Machine Model

Support Vector Machine (SVM) is a broadly employed and praised ML technique for classifying data [26]. The principal purpose of creating this model was to draw the input data into high dimensional space with the aid of the kernel function in such a way that the categories can be linearly divisible [47–49]. For the binary class issue, the greatest boundary that can divide the hyperplanes is as follows:

$$w \cdot a + l = 0 \quad (1)$$

The decision function for SVM based on the optimal pair (w_0, l_0) is expressed by:

$$f(a) = \sum_{i \in SV} \lambda_i y_i \langle a, a_i \rangle + l \quad (2)$$

where, λ_i stands for support vector, a_i denotes data sample and $i = 1, 2, \ldots, K$. Concerning greater dimensional feature space, the value of $\langle a.a_i \rangle$ is substituted by the kernel function Q $\langle a_i.a_j \rangle$, that is:

$$Q\langle a_i.a_j \rangle = \langle a_i.a_j \rangle \quad (3)$$

From the regular SVM, the kernel function was selected for estimating the boundaries' space. In the beginning, the dataset S is divided into different samples, which are S^1, S^2, S^3, ..., S^i, and subsequently, the kernel transformation function is implemented (Equation (4)).

$$f(a) = \begin{cases} e^{-z_1 h(a)^2}, \; if \; a \in S^1 \\ e^{-z_2 h(a)^2}, \; if \; a \in S^2 \\ \cdot \\ \cdot \\ \cdot \\ e^{-z_K h(a)^2}, \; if \; a \in S^K \end{cases} \tag{4}$$

where, $h_a = \sum_{i \in SV} \lambda_i y_i \langle a, a_i \rangle + l$ (where, λ_i signifies support vector), S^i refers to the ith sample of the training dataset, z_i is calculated using likelihood-ratio chi-square, which is described in the following sections.

2.2.2. Likelihood-Ratio Chi-Square

Likelihood-ratio chi-square (G^2) is a renowned non-parametric test which assesses the target-input independence and is suitable for categorial attributes. G^2 ascertains a frequency distribution-based relationship among the categorical attribute assortments. To put it another way, it can be said that this technique should be employed to assess the association between the groups. The importance of determining the G^2 is to ascertain the connection amongst the samples of every class and parameter z_i. Equation (5) presents the analytical formulation for estimating G^2.

$$G^2 = 2 \sum D_o log\left(\frac{D_o}{D_e}\right) \tag{5}$$

where, D_o and D_e signify observed and expected frequencies, respectively.

2.2.3. Calculating the Factor of Weighting

Ascertaining the factor of weighting is a challenging and vital task while handling an imbalanced category because finding a suitable weight is comparatively complicated. A practical technique to manage such issues is to assign smaller weight to the mainstream category and larger weight to the minority category through fulfilling the weight condition $z_i \in (0,1)$. For dealing with the multi-category imbalance issue in the SVM_{AK} algorithm, this study employed Equation (6).

$$w_i = \frac{N}{n_i \sum_{i=1}^{K} \frac{N}{n_i}} \tag{6}$$

where, C and N express class and training sample sizes, respectively. n_i symbolizes the size of each class when i = 1, 2, ..., K.

For computing the parameter z_i, let S denotes the dataset that comprises the N number of samples and K classes. The z_i parameter is estimated employing Equations (2) and (3). The G^2 value in optimal distribution can be calculated as follows:

$$G^2 = 2 \sum_{i=1}^{K} n_i log\left(\frac{n_i}{N/K}\right) \tag{7}$$

Let $X_i = n_i log\left(\frac{n_i}{N/K}\right)$.

Then,

$$G^2 = 2 \sum_{i=1}^{K} A_i \tag{8}$$

Hence, the parameter Z_i can be characterized as

$$Z_i = w_i \times \frac{A_i}{G^2} \tag{9}$$

In Equation (8), place the G^2 value

$$Z_i = w_i \times \frac{A_i}{2\sum_{i=1}^{K} A_i} \tag{10}$$

where, n_i is the size of the sample in the ith class and $I = 1, 2, \ldots, K$.

2.3. Model Development Steps

In the beginning, the NHTS data was prepared and cleansed. Later, these data were used for achieving the primary partition. Then, the authors determined the weighting factor (w_i) value as well as Z_i parameters for all support vectors in every iteration. The Z_i value was estimated using the likelihood-ratio chi-square test. The kernel conversion function was estimated in the next step. Eventually, utilizing the newly estimated kernel matrix K_{mt}, the model was retrained. Figure 1 indicates the flowchart of the suggested algorithm.

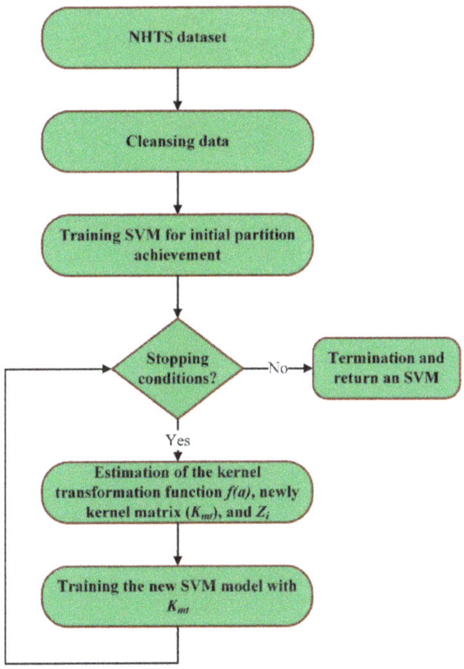

Figure 1. Developed algorithm flowchart.

2.4. Evaluation Metrics

This study employed four evaluation criteria to evaluate the performance of the models developed in this study. These criteria included accuracy, precision, recall, and F1 score. The formulas for calculating these criteria are shown in Equations (11)–(14). Accuracy refers to the ratio of the precisely forecasted class across the whole experiment class. Precision indicates the proportion of the true positive class over the whole number of an actual positive and false-positive category. Recall refers to the quantity of forecasted positive categories that fall out of whole positive cases in the data. F1 score shows the equilibrium between recall and precision.

$$Accuracy = \frac{TP + TN}{TP + TN + FP + FN} \times 100 \quad (11)$$

$$Precision = \frac{TP}{TP + FP} \quad (12)$$

$$Recall = \frac{TP}{TP + FN} \quad (13)$$

$$F1\ score = \frac{2 \times (precision \times recall)}{precision + recall} \quad (14)$$

where, TP, TN, FP, and FN denote the number of true positives, true negatives, false positives, and false negatives, respectively.

3. Models' Development and Evaluation

The authors of this study present the results of the SVM_{AK} model as well as other classification models, including the standard SVM model, BN, ANN model, and some SVM-based models in literature proposed for handling the imbalance data.

It is a challenging task to determine the most suitable classification model for handling imbalance data issues. The travel mode choice dataset was taken for the empirical investigation. Figure 2 shows the nine classes plotted on the x-axis, and the size of samples in each class plotted on the y-axis. As can be seen in this figure, it is obvious that the NHTS dataset includes uneven category distribution; practically, it is called imbalance. Hence, it grows to be more complicated to manage such a circumstance through regular classification techniques. The category-wise distributions of the dataset based on the sample size are: The car category contains 108,885 samples, the pickup truck category consists of 22,188 samples, the SUV class contains 12,582 samples, the van category consists of 2998 samples, the public or commuter bus class includes 2454 samples, the bicycle class category comprises 2091 samples, walk contains 345 samples, the private/charter/tour/shuttle bus category consists of 40 samples, and finally the motorcycle/moped category holds 14 samples. The class imbalance ratio of the dataset is 777.5.

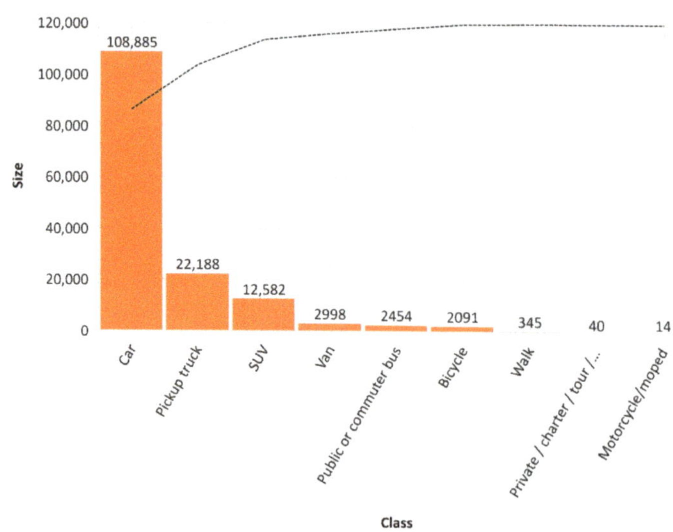

Figure 2. Category wise distribution of NHTS dataset.

The principal intention of this study is to discover the most competent classification technique that can examine the imbalanced class. Periodically, various scholars had proposed useful techniques to handle class imbalance issue. The majority of the proposed methods were for binary category imbalance issues, which were not fit for the multi-category imbalance issue. These shortcomings urged authors to adjust the algorithm which can effectively handle binary category and multi-category imbalance issues without jeopardizing the performance of algorithms. The classification mentioned above will further assist in attaining the desirable answer toward urban transport planning and prediction of travel mode choice. This study employed four renowned conventional classification techniques and SVM-based models proposed in the literature along with the SVM_{AK} model suggested in this study for the empirical assessment. The model applied in this study was compared with other models to ascertain effectiveness, fitness, and precision. This study evaluated the performance of the models developed using six criteria. Moreover, the authors employed a 10-fold cross-validation procedure as the validation scheme.

Four criteria, including accuracy, F1 score, precision, and recall were used to evaluate the outcomes of the classification algorithms applied in this study. The authors validated the classification techniques using the accuracy of classification. As is known, the NHTS dataset includes imbalanced category distribution, which may influence the performance of classification techniques. The overall performance of the models developed in this part of the study is shown in Table 4. All models achieved an overall accuracy above 80%. However, the SVM_{AK} outperformed other models. The worst model was BN. Regarding other evaluation criteria, SVM_{AK} again had the best values. It is worth mentioning that the SVM_{AK} improved the performance of the SVM_S model, which shows the capability of the proposed model of this study to handle the imbalanced mode choice data and enhance the performance of the typical SVM model for dealing with such data.

Table 4. Models' overall performance.

	Accuracy (%)	Precision	Recall	F1 Score
SVM_{AK}	99.81	0.99	1.00	0.99
SVM_S	93.18	0.87	0.89	0.88
XGBoost	85.4	0.20	0.28	0.21
NN	83.06	0.37	0.43	0.39
BN	80.54	0.71	0.86	0.77

An evaluation of the models developed by each class also is provided in Figure 3. For the class of car, which had the largest sample size, the SVM_{AK} improved the prediction accuracy from 82.33% (BN model) to 99.81%. Concerning the category of motorcycle/moped, which had the smallest sample size, the models developed yielded almost a similar accuracy. For other classes, the SVM_{AK} model almost achieved better accuracy.

As previously mentioned, the performance of the SVM_{AK} model was compared with some existing SVM models, which tried to alleviate the severe effects of using imbalanced data. In these methods the SVM model was hybridized with some techniques, including boosting [28], fuzzy [27], and class-boundary alignment [29], and ensemble [30]. The outcomes of the mentioned comparison are presented in Table 5. As can be seen, the SVM_{AK} obtained the highest overall accuracy among all models developed.

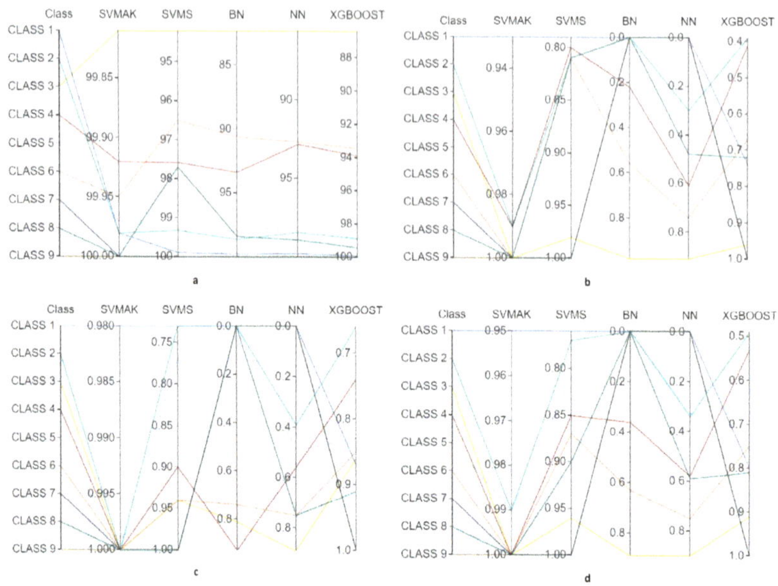

Figure 3. Models' performance by each class: (**a**) accuracy (%); (**b**) precision; (**c**) recall, (**d**) F1 score; CLASS 1 = Walk; CLASS 2 = Bicycle; CLASS 3 = Car; CLASS 4 = SUV; CLASS 5 = Van; CLASS 6 = Pickup truck; CLASS 7 = Motorcycle/moped; CLASS 8 = Public or commuter bus; CLASS 9 = Private/charter/tour/shuttle bus.

Table 5. The overall accuracy of SVM-based models and the SVM_{AK} model.

Model	Accuracy (%)
Wang and Japkowicz [28]	83.33
Batuwita and Palade [27]	92.99
Wu and Chang [29]	93.89
Liu, Yu, Huang and An [30]	90.16
SVM_{AK}	99.81

4. Sensitivity Analysis

Many factors impact the travel mode choice; however, their effects are not the same. Thus, it is necessary to ascertain the magnitude of these impacts and identify the most influential factors on travel mode choice. For this purpose, the authors employed the mutual information (MI) test method [50], which computes the importance of the inputs. MI means a filtering system that captures the random association between inputs and the target. MI examines the dependence among variables and confirms the strength of the connection among them. The MI size among inputs is measured employing the information gain:

$$Gain(C, D) = Ent(C) - \sum_{h-1}^{H} \frac{|C^h|}{|C|} Ent\left(C^h\right) \tag{15}$$

where, h denotes the number of all probable values of D, C^h is the set of C when D takes the value D_s, and $Ent(C)$ signifies the information entropy. The larger the value of Gain (C, D), the better the relationship between D and C.

Ultimately, the importance magnitude of each attribute for predicting travel mode choice was achieved based on the scores obtained in the MI test. The outcomes of this analysis are shown in Figure 4. The most important attributes were reason for not walking (walkmore), number of drivers in household (drvcnet), and count of adult household members at least 18 years old (numadlt). On the other hands, the lowest scores belonged to flexibility of work start time (flextime), owned vehicle longer than a year (vehowned), and gender (r_sex).

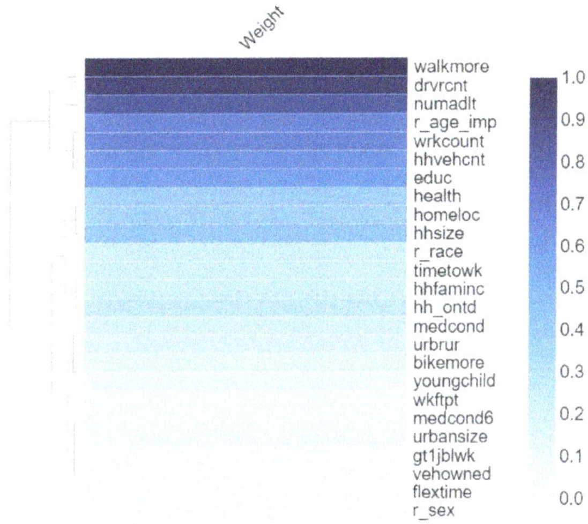

Figure 4. Outcomes of sensitivity analysis.

Reasons for not walking among respondents included unsafe street crossings, heavy traffic, and insufficient night lighting. It is clear that any improvement in these street conditions can encourage people to shift from motorized transport to walking. Thus, it makes sense that this factor is among the most influential travel mode choice factors to work [51–53]. The significance of the number of drivers in households can be attributed to its influence on the usage of vehicles and the generation of more trips. In practice, the likelihood of choosing active transportation options reduces as the number of drivers in a family grows [54]. As the number of adults in a family increases, the need for independent trips rises. Because of the different responsibilities that each adult in the family has, it is not easy to consolidate trips into one trip. This can be easily one of the principal sources of more trip generation and use of motorized transportation.

The flexibility of work start time was among the least important factors. This could be attributable to work culture of the respondents in the US. However, several previous studies showed that the flexibility of work start time influences the mode choice e.g., [54,55]. The possession of a vehicle for longer than a year was also an unimportant factor for predicting the choice of travel mode to work. A possible reason for this is that people usually look for flexible and convenient travel options to work. Usually, they are reluctant to replace their private cars with healthy travel modes unless they face new challenges. These challenges can be health problems, heavy traffic, so on. Thus, it is sensible that this factor does not influence the mode choice substantially.

5. Conclusions

In this research, the authors offered a novel method for learning from imbalanced mode choice data by the adjustable kernel based SVM classification model (SVM_{AK}). The likelihood-ratio chi-square test and weighting measures were used in this suggested method for selecting the kernel function. The aforementioned kernel transformation function makes it possible to increase the class limits and offset the irregular class limits. The authors also performed a sensitivity analysis which showed that the reason for not walking (walkmore), the number of drivers in the household (drvcnet), and the count of adult household members no less than 18 years old (numadlt) were the most influential factors. On the other hand, the lowest scores found for flexibility of work start time (flextime), owning a vehicle longer than a year (vehowned), and gender (r_sex) were the most influential factors on travel mode choice.

The outcomes of this model were compared with those of various SVM-based models and ML models. The authors employed four criteria, including accuracy, F1 score, precision, and recall to perform this comparison. The results of this study showed that the SVM_{AK} model achieved the best results and outperformed other models. The results also showed that this model improved the classification accuracy of most categories, especially the car class that had the largest samples.

Prediction of travel mode choice is an essential component of transport planning and traffic engineering. An accurate prediction is viable only if the data are precisely classified. Therefore, precise mode choice classification utilizing the algorithm suggested in this study would be efficacious for enhancing the current transport systems and further boosting the capacities for an efficient response to the worst traffic and transport scenarios.

The classes of choice of travel mode to work in the NHTS dataset are distinct from other mode choice datasets since this dataset considered the "SUV" category as different from the "car" category. However, future studies could combine these two classes and create a new class of travel mode to work. In the US, private motorized transport is dominant. Thus, the NHTS and, in turn, the result of this study is affected by this issue. Future studies can employ the method developed in this study to predict choice of travel mode to work in different environments, such as those in which walking, cycling, and public transport are dominant.

Author Contributions: Conceptualization, M.A. (Mahdi Aghaabbasi); methodology, M.A. (Mahdi Aghaabbasi); software, M.A. (Mahdi Aghaabbasi); formal analysis, M.A. (Mahdi Aghaabbasi); investigation, M.A. (Mahdi Aghaabbasi); resources, Y.Q., M.A. (Muwaffaq Alqurashi), M.A. (Mujahid Ali), B.S., E.E.H.; data curation, M.A. (Mahdi Aghaabbasi); writing—original draft preparation, M.A. (Mahdi Aghaabbasi) and M.M.; writing—review and editing, M.A. (Mahdi Aghaabbasi) and M.A. (Mujahid Ali); supervision, R.Z.; funding acquisition, Y.Q., M.A. (Muwaffaq Alqurashi), M.A. (Mujahid Ali), B.S., E.E.H. All authors have read and agreed to the published version of the manuscript.

Funding: This work was funded by the King Saud University, Saudi Arabia, through researchers supporting project number (RSP-2021/145), and the Taif University, Saudi Arabia, through Taif University Researchers Supporting Project, grant number [TURSP-2020/324].

Institutional Review Board Statement: Not applicable.

Informed Consent Statement: Not applicable.

Acknowledgments: The authors of this paper thank King Saud University, Saudi Arabia for funding this work through researchers supporting project numbers (RSP-2021/145). The authors also would like to acknowledge Taif University Researches Supporting Project number (TURSP-2020/324), Taif University, Taif, Saudi Arabia for supporting this work.

Conflicts of Interest: The authors declare no conflict of interest.

References

1. Gang, L. A behavioral model of work-trip mode choice in Shanghai. *China Econ. Rev.* **2007**, *18*, 456–476.
2. Simons, D.; De Bourdeaudhuij, I.; Clarys, P.; De Geus, B.; Vandelanotte, C.; Van Cauwenberg, J.; Deforche, B. Choice of transport mode in emerging adulthood: Differences between secondary school students, studying young adults and working young adults and relations with gender, SES and living environment. *Transp. Res. Part A Policy Pract.* **2017**, *103*, 172–184. [CrossRef]
3. Ali, M.; de Azevedo, A.R.G.; Marvila, M.T.; Khan, M.I.; Memon, A.M.; Masood, F.; Almahbashi, N.M.Y.; Shad, M.K.; Khan, M.A.; Fediuk, R.; et al. The Influence of COVID-19-Induced Daily Activities on Health Parameters—A Case Study in Malaysia. *Sustainability* **2021**, *13*, 7465. [CrossRef]
4. Badoe, D. Modelling work-trip mode choice decisions in two-worker households. *Transp. Plan. Technol.* **2002**, *25*, 49–73. [CrossRef]
5. Vega, A.; Reynolds-Feighan, A. Employment sub-centres and travel-to-work mode choice in the Dublin region. *Urban Stud.* **2008**, *45*, 1747–1768. [CrossRef]

6. Day, N.; Habib, K.N.; Miller, E.J. Analysis of work trip timing and mode choice in the Greater Toronto Area. *Can. J. Civ. Eng.* **2010**, *37*, 695–705. [CrossRef]
7. Habib, K.M.N. Modeling commuting mode choice jointly with work start time and work duration. *Transp. Res. Part A Policy Pract.* **2012**, *46*, 33–47. [CrossRef]
8. Heinen, E.; Maat, K.; van Wee, B. The effect of work-related factors on the bicycle commute mode choice in the Netherlands. *Transportation* **2013**, *40*, 23–43. [CrossRef]
9. Hamre, A.; Buehler, R. Commuter mode choice and free car parking, public transportation benefits, showers/lockers, and bike parking at work: Evidence from the Washington, DC region. *J. Public Transp.* **2014**, *17*, 4. [CrossRef]
10. Tran, M.T.; Zhang, J.; Chikaraishi, M.; Fujiwara, A. A joint analysis of residential location, work location and commuting mode choices in Hanoi, Vietnam. *J. Transp. Geogr.* **2016**, *54*, 181–193. [CrossRef]
11. Ali, M.; Dharmowijoyo, D.B.; Harahap, I.S.; Puri, A.; Tanjung, L.E. Travel behaviour and health: Interaction of Activity-Travel Pattern, Travel Parameter and Physical Intensity. *Solid State Technol.* **2020**, *63*, 4026–4039.
12. Aghaabbasi, M.; Shekari, Z.A.; Shah, M.Z.; Olakunle, O.; Armaghani, D.J.; Moeinaddini, M. Predicting the use frequency of ride-sourcing by off-campus university students through random forest and Bayesian network techniques. *Transp. Res. Part A Policy Pract.* **2020**, *136*, 262–281. [CrossRef]
13. Stylianou, K.; Dimitriou, L.; Abdel-Aty, M. Big data and road safety: A comprehensive review. *Mobil. Patterns Big Data Transp. Anal.* **2019**, 297–343. [CrossRef]
14. Rashidi, S.; Ranjitkar, P.; Hadas, Y. Modeling bus dwell time with decision tree-based methods. *Transp. Res. Rec.* **2014**, *2418*, 74–83. [CrossRef]
15. Ali, M.; Dharmowijoyo, D.B.E.; de Azevedo, A.R.G.; Fediuk, R.; Ahmad, H.; Salah, B. Time-Use and Spatio-Temporal Variables Influence on Physical Activity Intensity, Physical and Social Health of Travelers. *Sustainability* **2021**, *13*, 12226. [CrossRef]
16. Parsajoo, M.; Armaghani, D.J.; Mohammed, A.S.; Khari, M.; Jahandari, S. Tensile strength prediction of rock material using non-destructive tests: A comparative intelligent study. *Transp. Geotech.* **2021**, *31*, 100652. [CrossRef]
17. Harandizadeh, H.; Armaghani, D.J.; Asteris, P.G.; Gandomi, A.H. TBM performance prediction developing a hybrid ANFIS-PNN predictive model optimized by imperialism competitive algorithm. *Neural Comput. Appl.* **2021**, *33*, 16149–16179. [CrossRef]
18. Li, E.; Zhou, J.; Shi, X.; Armaghani, D.J.; Yu, Z.; Chen, X.; Huang, P. Developing a hybrid model of salp swarm algorithm-based support vector machine to predict the strength of fiber-reinforced cemented paste backfill. *Eng. Comput.* **2020**, *37*, 1–22. [CrossRef]
19. Jahed Armaghani, D.; Kumar, D.; Samui, P.; Hasanipanah, M.; Roy, B. A novel approach for forecasting of ground vibrations resulting from blasting: Modified particle swarm optimization coupled extreme learning machine. *Eng. Comput.* **2021**, *37*, 3221–3235. [CrossRef]
20. Armaghani, D.J.; Harandizadeh, H.; Momeni, E.; Maizir, H.; Zhou, J. An optimized system of GMDH-ANFIS predictive model by ICA for estimating pile bearing capacity. *Artif. Intell. Rev.* **2021**, *54*, 1–38. [CrossRef]
21. Li, Z.; Yazdani Bejarbaneh, B.; Asteris, P.G.; Koopialipoor, M.; Armaghani, D.J.; Tahir, M. A hybrid GEP and WOA approach to estimate the optimal penetration rate of TBM in granitic rock mass. *Soft Comput.* **2021**, *25*, 11877–11895. [CrossRef]
22. Yu, C.; Koopialipoor, M.; Murlidhar, B.R.; Mohammed, A.S.; Armaghani, D.J.; Mohamad, E.T.; Wang, Z. Optimal ELM–Harris Hawks optimization and ELM–Grasshopper optimization models to forecast peak particle velocity resulting from mine blasting. *Nat. Resour. Res.* **2021**, *30*, 2647–2662. [CrossRef]
23. Menardi, G.; Torelli, N. Training and assessing classification rules with imbalanced data. *Data Min. Knowl. Discov.* **2014**, *28*, 92–122. [CrossRef]
24. Daskalaki, S.; Kopanas, I.; Avouris, N. Evaluation of classifiers for an uneven class distribution problem. *Appl. Artif. Intell.* **2006**, *20*, 381–417. [CrossRef]
25. Vapnik, V. *The Nature of Support Vector Machine*; Springer: Berlin/Heidelberg, Germany, 1999.
26. Batuwita, R.; Palade, V. FSVM-CIL: Fuzzy Support Vector Machines for Class Imbalance Learning. *IEEE Trans. Fuzzy Syst.* **2010**, *18*, 558–571. [CrossRef]
27. Wang, B.X.; Japkowicz, N. Boosting support vector machines for imbalanced data sets. *Knowl. Inf. Syst.* **2009**, *25*, 1–20. [CrossRef]
28. Wu, G.; Chang, E.Y. Class-Boundary Alignment for Imbalanced Dataset Learning. In Proceedings of the Workshop Learningfrom Imbalanced Datasets II, Washington, DC, USA, 21 August 2003; pp. 49–56.
29. Liu, Y.; Yu, X.; Huang, J.X.; An, A. Combining integrated sampling with SVM ensembles for learning from imbalanced datasets. *Inf. Process. Manag.* **2011**, *47*, 617–631. [CrossRef]
30. Liu, Z.; Tang, D.; Cai, Y.; Wang, R.; Chen, F. A hybrid method based on ensemble WELM for handling multi class imbalance in cancer microarray data. *Neurocomputing* **2017**, *266*, 641–650. [CrossRef]
31. Hordri, N.F.; Yuhaniz, S.S.; Azmi, N.F.M.; Shamsuddin, S.M. Handling class imbalance in credit card fraud using resampling methods. *Int. J. Adv. Comput. Sci. Appl.* **2018**, *9*, 390–396. [CrossRef]
32. Kim, S.; Lym, Y.; Kim, K.-J. Developing crash severity model handling class imbalance and implementing ordered nature: Focusing on elderly drivers. *Int. J. Environ. Res. Public Health* **2021**, *18*, 1966. [CrossRef]
33. Rezaei, S.; Khojandi, A.; Haque, A.M.; Brakewood, C.; Jin, M.; Cherry, C. Performance evaluation of mode choice models under balanced and imbalanced data assumptions. *Transp. Lett.* **2021**, *13*, 1–13. [CrossRef]
34. Lu, Y.; Kawamura, K. Data-mining approach to work trip mode choice analysis in Chicago, Illinois, area. *Transp. Res. Rec.* **2010**, *2156*, 73–80. [CrossRef]

35. Xie, C.; Lu, J.; Parkany, E. Work travel mode choice modeling with data mining: Decision trees and neural networks. *Transp. Res. Rec.* **2003**, *1854*, 50–61. [CrossRef]
36. Patterson, Z.; Ewing, G.; Haider, M. Gender-based analysis of work trip mode choice of commuters in suburban Montreal, Canada, with stated preference data. *Transp. Res. Rec.* **2005**, *1924*, 85–93. [CrossRef]
37. Al-Ahmadi, H. Development of intercity work mode choice model for Saudi Arabia. In *WIT Transactions on The Built Environment*; WITT Press: Southampton, UK, 2007; Volume 96.
38. Vega, A.; Reynolds-Feighan, A. A methodological framework for the study of residential location and travel-to-work mode choice under central and suburban employment destination patterns. *Transp. Res. Part A Policy Pract.* **2009**, *43*, 401–419. [CrossRef]
39. Heinen, E.; Bohte, W. Multimodal commuting to work by public transport and bicycle: Attitudes toward mode choice. *Transp. Res. Rec.* **2014**, *2468*, 111–122. [CrossRef]
40. Kunhikrishnan, P.; Srinivasan, K.K. Choice set variability and contextual heterogeneity in work trip mode choice in Chennai city. *Transp. Lett.* **2019**, *11*, 174–189. [CrossRef]
41. Franco, S.F. Downtown parking supply, work-trip mode choice and urban spatial structure. *Transp. Res. Part B Methodol.* **2017**, *101*, 107–122. [CrossRef]
42. Indriany, S.; Sjafruddin, A.; Kusumawati, A.; Weningtyas, W. Mode choice model for working trip under risk and uncertainty. In Proceedings of the AIP Conference Proceedings, Maharashtra, India, 5–6 July 2018; p. 020041.
43. Irfan, M.; Khurshid, A.N.; Khurshid, M.B.; Ali, Y.; Khattak, A. Policy implications of work-trip mode choice using econometric modeling. *J. Transp. Eng. Part A Syst.* **2018**, *144*, 04018035. [CrossRef]
44. Hatamzadeh, Y.; Habibian, M.; Khodaii, A. Walking mode choice across genders for purposes of work and shopping: A case study of an Iranian city. *Int. J. Sustain. Transp.* **2020**, *14*, 389–402. [CrossRef]
45. Transportation Secure Data Center. *2017 National Household Travel Survey—California*; Transportation Secure Data Center: Golden, CO, USA, 2019.
46. Maratea, A.; Petrosino, A.; Manzo, M. Adjusted F-measure and kernel scaling for imbalanced data learning. *Inf. Sci.* **2014**, *257*, 331–341. [CrossRef]
47. Vapnik, V. *The Nature of Statistical Learning Theory*; Springer Science & Business Media: Berlin/Heidelberg, Germany, 2013.
48. Wang, L. *Support. Vector Machines: Theory and Applications*; Springer Science & Business Media: Berlin/Heidelberg, Germany, 2005; Volume 177.
49. Foody, G.M.; Mathur, A. Toward intelligent training of supervised image classifications: Directing training data acquisition for SVM classification. *Remote Sens. Environ.* **2004**, *93*, 107–117. [CrossRef]
50. Verron, S.; Tiplica, T.; Kobi, A. Fault detection and identification with a new feature selection based on mutual information. *J. Process. Control.* **2008**, *18*, 479–490. [CrossRef]
51. Aghaabbasi, M.; Shah, M.Z.; Zainol, R. Investigating the Use of Active Transportation Modes Among University Employees Through an Advanced Decision Tree Algorithm. *Civil. Sustain. Urban. Eng.* **2021**, *1*, 26–49. [CrossRef]
52. Aghaabbasi, M.; Moeinaddini, M.; Shah, M.Z.; Asadi-Shekari, Z. A new assessment model to evaluate the microscale sidewalk design factors at the neighbourhood level. *J. Transp. Health* **2017**, *5*, 97–112. [CrossRef]
53. Tabatabaee, S.; Aghaabbasi, M.; Mahdiyar, A.; Zainol, R.; Ismail, S. Measurement Quality Appraisal Instrument for Evaluation of Walkability Assessment Tools Based on Walking Needs. *Sustainability* **2021**, *13*, 11342. [CrossRef]
54. Sultana, S. *Factors Affecting Parents' Choice of Active Transport Modes for Children's Commute to School: Evidence from 2017 NHTS Data*; The University of Toledo: Toledo, OH, USA, 2019.
55. Thorhauge, M.; Cherchi, E.; Rich, J. How flexible is flexible? Accounting for the effect of rescheduling possibilities in choice of departure time for work trips. *Transp. Res. Part A Policy Pract.* **2016**, *86*, 177–193. [CrossRef]

applied sciences

MDPI

Article

Use of the Gene-Expression Programming Equation and FEM for the High-Strength CFST Columns

Huanjun Jiang [1,2], Ahmed Salih Mohammed [3], Reza Andasht Kazeroon [4] and Payam Sarir [1,2,*]

1 State Key Laboratory of Disaster Reduction in Civil Engineering, Tongji University, Shanghai 200092, China; jhj73@tongji.edu.cn
2 College of Civil Engineering, Tongji University, Shanghai 200092, China
3 Civil Engineering Department, College of Engineering, University of Sulaimani, Sulaymaniyah 46001, Iraq; ahmed.mohammed@univsul.edu.iq
4 Faculty of Civil Engineering, College of Engineering, Universiti Teknologi Mara (UiTM), Shah Alam 40450, Selangor, Malaysia; reza.andasht@gmail.com
* Correspondence: payamsarir@tongji.edu.cn; Tel.: +86-21-65985097; Fax: +86-21-65985099

Abstract: The ultimate strength of composite columns is a significant factor for engineers and, therefore, finding a trustworthy and quick method to predict it with a good accuracy is very important. In the previous studies, the gene expression programming (GEP), as a new methodology, was trained and tested for a number of concrete-filled steel tube (CFST) samples and a GEP-based equation was proposed to estimate the ultimate bearing capacity of the CFST columns. In this study, however, the equation is considered to be validated for its results, and to ensure it is clearly capable of predicting the ultimate bearing capacity of the columns with high-strength concrete. Therefore, 32 samples with high-strength concrete were considered and they were modelled using the finite element method (FEM). The ultimate bearing capacity was obtained by FEM, and was compared with the results achieved from the GEP equation, and both were compared to the respective experimental results. It was evident from the results that the majority of values obtained from GEP were closer to the real experimental data than those obtained from FEM. This demonstrates the accuracy of the predictive equation obtained from GEP for these types of CFST column.

Keywords: confinement of concrete; CFST composite column; artificial intelligence; gene-expression programming; hybrid techniques; finite element method (FEM)

Citation: Jiang, H.; Mohammed, A.S.; Kazeroon, R.A.; Sarir, P. Use of the Gene-Expression Programming Equation and FEM for the High-Strength CFST Columns. *Appl. Sci.* **2021**, *11*, 10468. https://doi.org/10.3390/app112110468

Academic Editor: Yixia Zhang

Received: 16 October 2021
Accepted: 4 November 2021
Published: 8 November 2021

1. Introduction

Analysis of previous experimental tests reveals that there are some key parameters impacting both performance and strength of concrete-filled steel tube (CFST) columns which are length to diameter ratio, wall thickness, and the interaction between steel tube and concrete [1–3]. In recent years, applying artificial intelligence (AI) techniques such as artificial neural network (ANN), genetic programming (GP), and gene expression programming (GEP) to predict and optimize engineering problems, has become very popular. Usually, this is because the AI approaches are faster and with less complexity in comparison with the finite element analysis (FEA). The external and local imperfections, boundary conditions, and residual stresses could be perfectly simulated by FEA. It provides load-deflection curves and determines the strain and stress at each stage; however, the accuracy of prediction by FE models is highly impacted by input elements, some of which cannot be modelled thoroughly [4]. Artificial intelligence (AI) techniques and particularly the gene expression programming (GEP) trump the FEA methods in terms of ease and speed of analysis, and better accuracy in predictions and optimizations specifically in obtaining the ultimate bearing capacity of CFST columns [5–7].

In CFST columns, concrete provides a radial lateral pressure to the steel tube. Besides, the steel tube supports concrete from spalling by means of confinement. In addition, the concrete core avoids inward buckling of the steel. Therefore, ductility and strength of these columns are usually high. Furthermore, the confinement of concrete in circular columns is more efficient compared to other types of sections which results in a better interaction and less slip between steel tube and concrete. It is consequently can lead to a higher axial strength and a better structural integrity [8,9].

There is a few research which focused on the analysis of the CFST composite columns using artificial intelligence (AI) techniques. In the previous study [10] an extensive data collection was performed and they were analyzed by GEP. A valuable equation was generated to estimate the ultimate strength of the CFST columns and it was optimized afterwards [1]. Most studies in civil engineering are based on experimental work and numerical analysis using the finite element method (FEM). Normally, using an equation-based analysis is less time consuming and less costly than the laboratory tests and computational modelling and analysis. Therefore, this study tried to ensure that the proposed GEP-based equation could be reliable to be used in prediction of the bearing capacity of CFST composite columns. Furthermore, in order to show fidelity of the results from the GEP equation, in addition to comparison of them with the finite element analysis (FEA), the outcomes were also compared with the experimental ones in the literature.

In previous research [1,10], 303 samples of CFST columns were collected from literature and they were trained and tested using GEP. The assessment was performed five times and each time led to an equation. The equations were classified based on the coefficient of determination (R^2) and the one with the closest R square to 1 was picked up for the further analysis. Then the GEP prediction equation has been optimized using AI optimization techniques. In this research, however, 32 different samples of those were collected in the previous research are considered with high-strength concrete but still within the range of the previous study. They are numerically modelled and analyzed to obtain their ultimate bearing capacity for the comparison purpose

FEM is a tool for numerical analysis which can model different types of composite columns. It could be a useful method in many of such cases specifically in the prediction of ultimate bearing capacity. However, after applying AI techniques for the composite columns [11–14], these novel methods were presented more promising results compared with the results of the FEA. Recently, AI techniques such as ANN and GEP are widely being used in the civil engineering to predict and optimize various criteria in this regard [15–21]. Following this, conventional mechanical calculations and also numerical analysis can be optimized by these intelligent methods following trends observed from recent reports [22–37]. Nevertheless, there is lack of research in the prediction of load bearing capacity or the compressive strength of composite columns specifically CFST columns with the AI approaches.

2. A Brief Background of Gene Expression Programming (GEP) and Finite Element Analysis (FEA)

Recently, GEP has gained recognition among civil engineers. In fact, it has shown a good capability in prediction of some engineering problems with an acceptable accuracy [38]. Table 1 indicates a list of recent studies using AI approaches together with the numbers of samples and the analyzed parameters. In solving structural engineering problems, AI has become quite popular recently, specifically in prediction of the ultimate bearing capacity of composite columns [39–41].

GEP benefits from two main chromosomes, and the expression tree (ET) provides solutions for removing the limitations of two older algorithms. The codifications are shown in the form of a string in GEP, which is in fact obtained from Karva programing language and can present a behavior like ETs. One of the interesting functions of GEP is that it can present its own models using mathematical equations. In fact, mathematical equations create relation between independent parameters. The GEP modeling process begins with the random creation of chromosomes for determined numbers, which follows Karva language (Karva is a symbolic language to introduce chromosomes). These symbolic chromosomes should be then defined as trees with different size and shapes (expression trees). These points are investigated by the functions that are responsible for controlling models and their adaptability. These functions have different types that can be defined by different criteria [10,19,38].

Table 1. Review studies for application of artificial intelligence (AI) for the composite columns.

Technique	Input	Output	Description	Reference
GEP	D, t, f_y, f_c, L, t_f,f_f, E_f	P	92 FRP-CFST columns	[39]
ANN	D, t, f_y, f_c, L	P	633 CCFST columns	[40]
GEP	D, t, f_y, f_c, L	P	314 CCFST columns	[21]
ANN	D, t, f_y, f_c, L	P	272 CCFST columns	[33]
ANN	D, t, f_y, f_c, L	P	205 CFST columns	[42]
SVM	B, t, f_y, f_c, L, E_c, E_s	P	180 SCFST columns	[34]
ANFIS-GA; ANFIS-PSO	D, t, t_p, f_y, f_c, L	P	57 steel Y-section columns	[35]
ANN	f_c, L, D, t_f, E_f, ε_{fu}, ε_{cc}	P	465 FRP-CCFST columns	[36]
GP	f_c, L, D, t_f, E_f	P	832 FRP-CCFST columns	[37]

D:diameter; t: thickness; f_y: yield stress of steel tube; f_c: concrete compressive strength; L: length; t_f: thickness of ply multiplied by FRP layers; f_f: tensile strength of FRP; E_f: modulus of elasticity for FRP, B: width; E_c: concrete modulus of elasticity; E_s: steel modulus of elasticity; t_p: thickness of the plate ε_{fu}: ultimate strain of FRP sheet; ε_{cc}: strain of confined column; GEP: gene expression programming; ANN: artificial neural network; SVM: support vector machine; ANFIS: adaptive neuro-fuzzy inference system; GA: genetic algorithm; PSO: particle swarm optimization.

On the other hand, a lot of scholars applied numerical methods to model and assess the mechanical behavior of CFST columns since experimental investigations are costly and time-consuming, and also have some limitations such as materials preparation [4,43–45]. In some of these studies, comparisons were made between the results of ultimate bearing capacity achieved from the FEA and the outcome reached by the experimental tests. They indicated a good agreement with the results obtained by FEA methods [46,47]. Nevertheless, it is believed that the GEP equation, generated from the previous study [10], is able to predict the maximum strength of CFST columns with a better accuracy compared to the FEA. Therefore, this study presents a valuable comparison between the methods which can be beneficial for the engineers to calculate the ultimate load bearing capacity faster and more accurately.

3. Prediction of Bearing Capacity of Concrete-Filled Steel Tube (CFST) Columns by Finite Element Analysis (FEA)

3.1. Brief Introduction

The accuracy of prediction from FEA is significantly impacted by input parameters, specifically selection of appropriate concrete model [4]. Moreover, there could be slip between two connected materials with different properties although a fully shear connection has to be assumed and simulated. In order to demonstrate the superior accuracy of predictions based on the GEP equation (Equation (1)) [10], 32 CFST columns were collected from recent literature and numerically modelled using FEA. Figure 1 shows a summary of the process of this modelling. As is evident, first of all, the data were collected for modelling and then, in order to achieve trustworthy results, convergence and mesh studies were performed for each column separately and after that the models were analyzed using ABAQUS [48].

$$P_{exp} = (\sqrt{d_0} \times (2 \times d_1) + (d_4 \times (\sqrt{d_1} - (6.219 - d_3))) + [(8.078 \times d_4) + (0.626 \times d_2)]/\tanh(-2.831) \quad (1)$$

where: $d(0) = f_C(\mathrm{MPa})$, $d(1) = D$ (mm), $d(2) = L$ (mm), $d(3) = t$ (mm),$d(4) = f_y$ (MPa).

Even though tremendous effort was previously made to model concrete-filled steel tubular column using FEA, this methodology may not be helpful in specific cases, particularly while assuming quick development and usage of thin-walled steel tube or the high-strength concrete [4,49,50]. FEA is performed according to the circular CFST (CCFST) specimens from the numbers of studies [51–56] which carry different geometries, material properties, boundary and support conditions, and loading functions. However, for the AI investigation, all of them were considered to be in the range of selected parameters. For the ultimate strength of the CFST columns, 1242.2 kN to 9187 kN was considered and, 36.2 MPa to 193.3 MPa was chosen for the concrete compressive strength. For the columns' length and diameter, 200 mm to 756 mm and 114.3 mm to 219.1 mm were considered respectively. In addition, the range of 2.5 mm to 10 mm was selected for the thickness, and the yield stress of steel tube is considered as 227 MPa to 428 MPa.

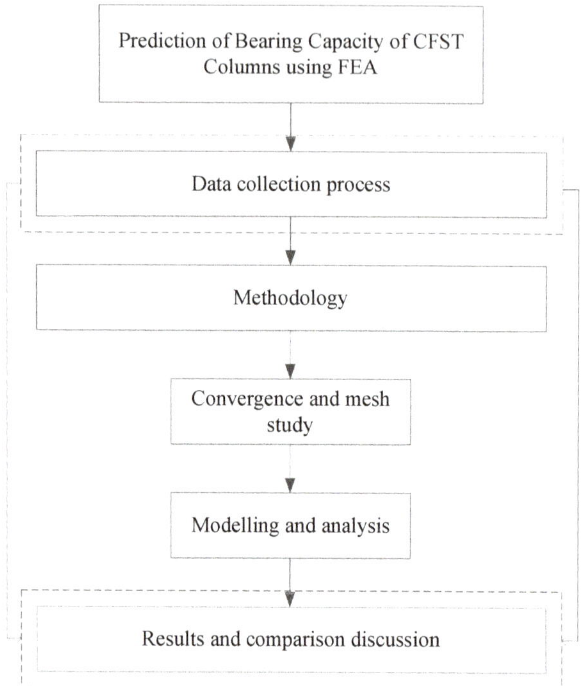

Figure 1. Diagram of finite element analysis (FEA) of the concrete-filled steel tube (CFST) columns.

3.2. FE Modelling and Analysis

Thirty-two CFST samples from literature [51–60] were collected and their specifications summarized in Table A1 (Appendix A). As is obvious from Table A1, different parameters such as diameter (D), length (L), thickness (t), length to diameter ratio (L/D), diameter to thickness ratio (D/t), yield stress of the steel tube (f_y), and the concrete compressive strength (f_c), the modulus of elasticity of concrete (E_c), the modulus of elasticity of the steel tube (E_s), and the Poisson's ratio (υ) were considered. Developing new techniques to analyze, assess, and predict the behaviour of composite columns is usually challenging for the engineers [51,61]. For the modelling purpose in FEA, different types of elements were used; in order to model concrete, a 3D deformable solid element and for the steel tube, the shell element, were applied. Furthermore, for the end plates, a solid element C3D8R was applied. For the plastic behaviour of concrete, concrete damage plasticity model is considered and the specifications are assumed according to the previous similar studies which led to acceptable results in analysis [43–47,56]. For simulation of the boundary conditions of models, the reference point was pinned which means the displacement assumed zero but rotation was free (U1 = U2 = U3 = 0). In addition, for the bottom end plate, it was assumed that there is no movement in x and y direction (U1 = U2 = 0). However, in the z direction the column was free to move upwards or downwards to simulate the loading condition.

In the FE modelling and in order to achieve better simulation of the interaction between the steel tube and the concrete, the gap element was applied to achieve more accurate results. This is an element in ABAQUS software which can be inserted and adjusted for a better simulation of interaction specifically between two different materials. The gap element allows the simulation of slip in the way that it prevents the penetration of one node into the node next to it. The shear stress can be simulated through a friction coefficient assigned to the gap element [58,59]. Therefore, in order to simulate more accurately, the interaction between materials was introduced to the system using the option of "allow separation after contact" and for this purpose, three different contacts were applied; (1) contact between steel tube and concrete with free normal and tangential separation; (2) contact between endplate and concrete, which is considered as normal interaction; and (3) contact between endplate and steel tube, which was applied as the tie connection to act like solid welding. Considering all processing data, boundary conditions, interaction between concrete and steel, and the modelling procedures, the final results were obtained and verified; afterwards they were compared with the artificial intelligence (AI) results generated from Equation (1).

3.3. The Results of FE Simulation

Some of the CFST samples which were modelled by the finite element method (FEM) are indicated in Figure 2a,b. In all of the cases, the column was subjected to axial loading in the center of the column along the Y axis. In this study, different columns with various L/D and D/t were modelled and their ultimate bearing capacity was obtained accordingly. The results were compared to the corresponding experimental results and the outcome is summarized in Table 2. The P_{exp} shows the experimental load-bearing capacity of the circular CFST columns while the P_{FE} indicates the result of the FE analysis for the ultimate bearing capacity of the column. The last column of Table 2 is a comparison between the experimental results and the finite element results which is reported as an error in percentage. In some cases, such as model Nos. 4, 5, 10–12, 14–18, 20–24, and 32, the bearing capacity predicted by FE was less than the measured experimental value of the bearing capacity. As is evident from Table 2, the ultimate bearing capacity from the experimental tests and those obtained from the numerical analysis are in good agreement and acceptable range with the maximum difference equal to or less than 15%.

In the model Nos. 3, 9, 19, 26, 27, the error is equal or even less than 5%; nevertheless, in the model Nos. 1, 2, 4–6, 8, 10, 12, 13, 15, 17, 21–24, 28, 30–32, the error is between 10–15% which are still in the acceptable range based on previous studies in the literature. In addition, Figures 3–8 show the axial strength of each finite element model versus the time which is 1 h for the model numbers 1–31 as they are normal columns and reacted to loading earlier, and 10 h for the Model No. 32 as it is considered as a huge, tall column which was taken more time to start reactions. As is evident from Figures 3–8, the maximum bearing capacity of each model is less than 10,000 kN; however, the running time to reach the ultimate bearing capacity is different for each model. The 32 FE models were divided into different groups based on the range of axial load bearing capacity of the column, the modes of failure, and the running time of the operation of the analysis. For this reason, model numbers 1–9 have been categorized as one group; model numbers 10–12 and 14 have been put as another group; model numbers 15–20, model numbers 21–25, and model numbers 26–28 have been categorized as three separated groups. This is due to a different running time and different range of ultimate axial strength which were resulted from each model. Nevertheless, the results of the above groups were taken into consideration on a 1-hour basis while the model number 13 and model numbers 29–32 have been indicated in a separated group and their performance was analyzed within 10 h of running as shown in Figure 8.

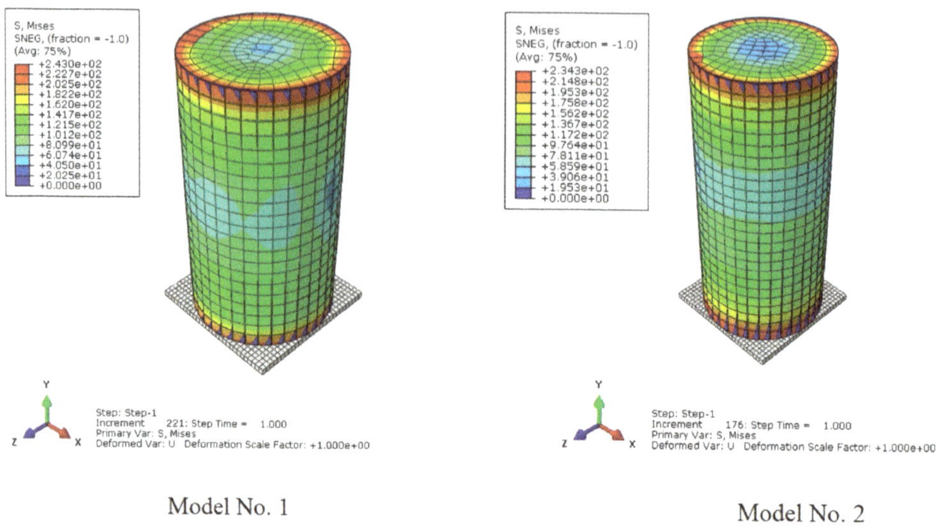

(**a**)

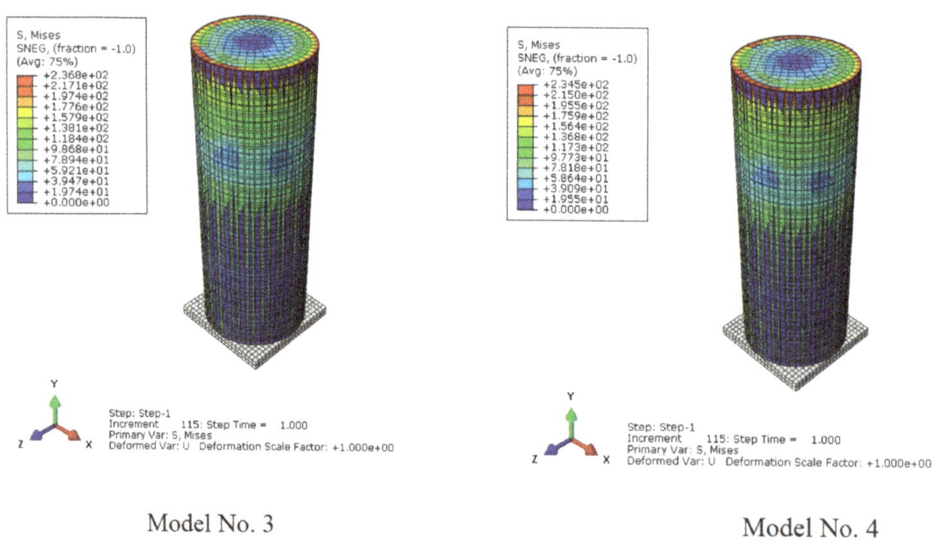

(**b**)

Figure 2. (**a**) FE modelling of Model No. 1 with PFE 3171 kN; Model No. 2 with PFE 2640 kN; (**b**) FE modelling of Model No. 3 with P_{FE} 7919 kN; Model No. 4 with P_{FE} 7928 kN.

Table 2. Comparison of the experimental results and FE results.

No.	P_{exp} (kN)	P_{FE} (kN)	Error $_{(FE)}$ (%)
1	2866	3171	11
2	2314	2640	14
3	7837	7919	1
4	9085	7928	13
5	9187	7994	13
6	6915	7946	15
7	7407	8032	8
8	6838	7791	14
9	7569	7890	4
10	1771	1554	12
11	3339	3037	9
12	3501	3050	13
13	4837	4320	11
14	4216	3850	9
15	4330	3812	12
16	4751	4313	9
17	4930	4379	11
18	5254	4738	10
19	2160	2184	1
20	2250	2092	7
21	1242.2	1100	11
22	1425.3	1225	14
23	1637.9	1432	13
24	1943.4	1672	14
25	2866	3088	8
26	2550	2642	4
27	3150	3195	1
28	3400	3926	15
29	3850	4073	6
30	5400	5993	11
31	3338	3854	15
32	8648	7694	11

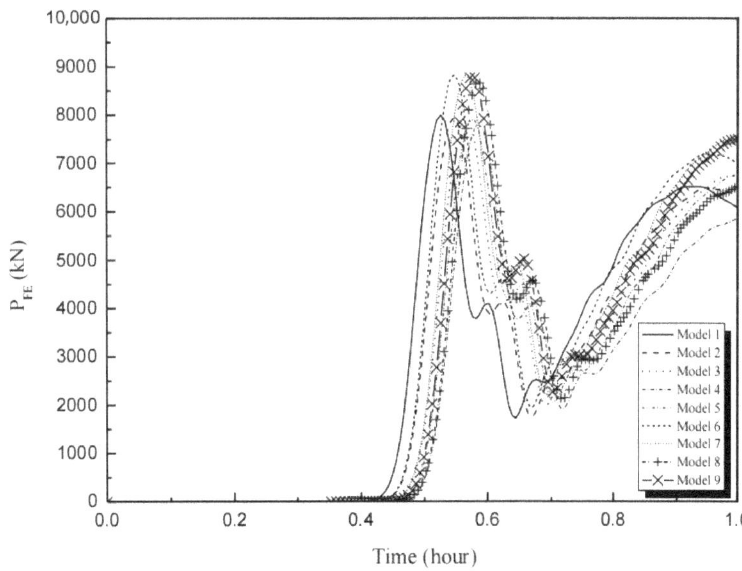

Figure 3. Axial strength versus time for model numbers 1–9.

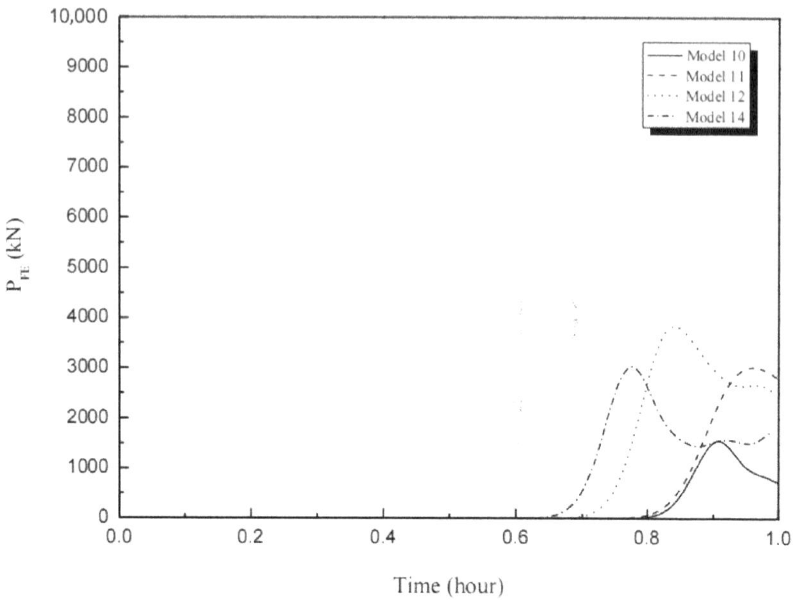

Figure 4. Axial strength versus time for model numbers 10–12 and 14.

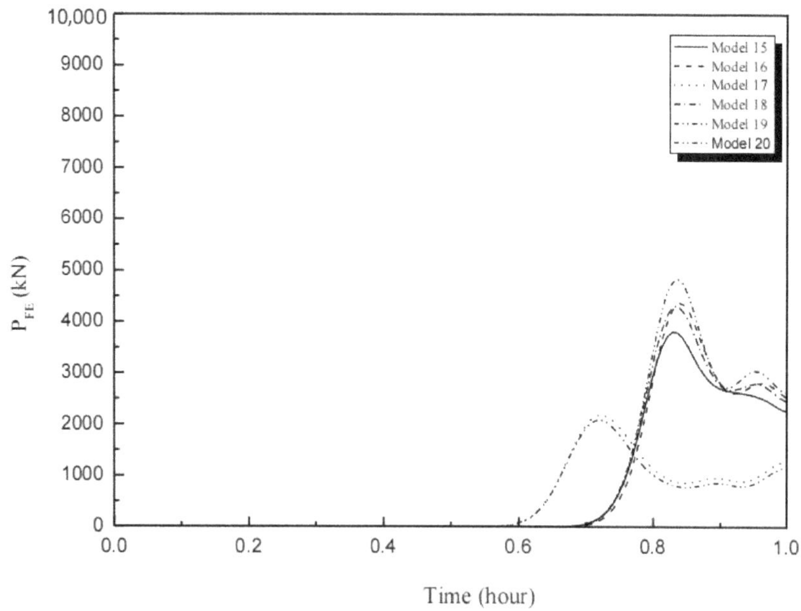

Figure 5. Axial strength versus time for model numbers 15–20.

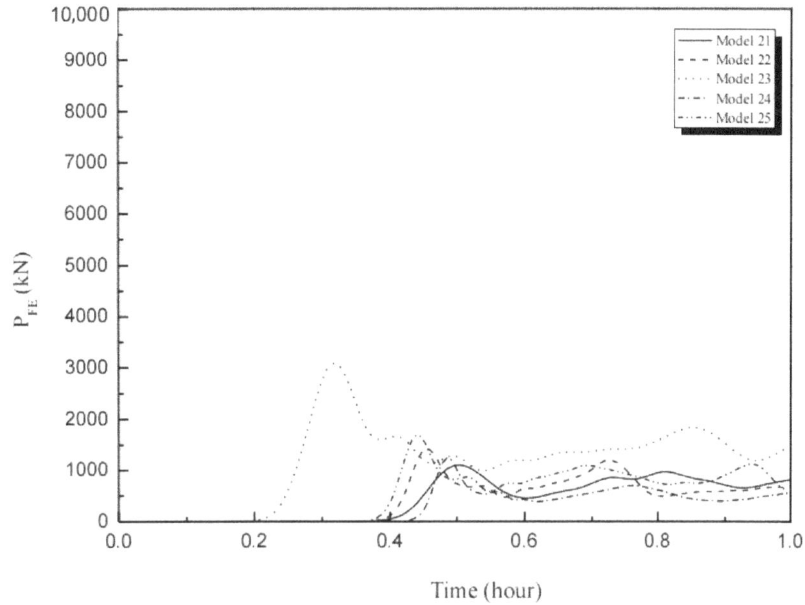

Figure 6. Axial strength versus time for model numbers 21–25.

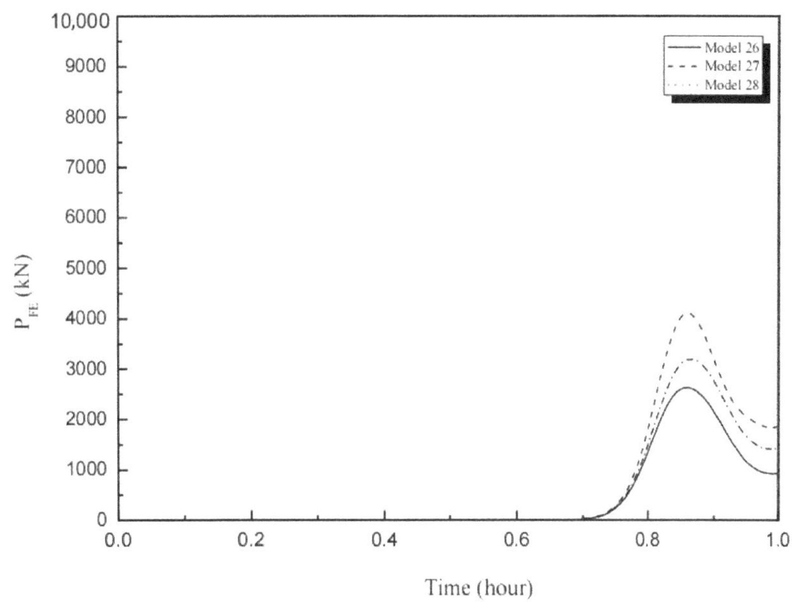

Figure 7. Axial strength versus time for model numbers 26–28.

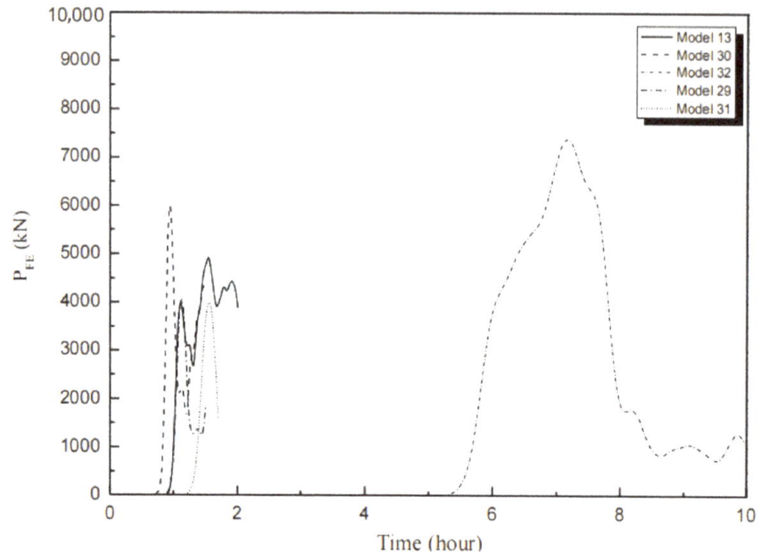

Figure 8. Axial strength versus time for model numbers 13, 29–32.

4. Results, Comparison, and Discussion of the GEP Equation Outcomes with the FEA Results

In this section, the data of those 32 samples were input into Equation (1) and the results were compared with those obtained from the FEA and the outcome summarized in Table 3. The last two columns in Table 3 present the error between the bearing capacity obtained from the FEA and the GEP equation with the experimental results. As is evident, in most of the cases, the ultimate bearing capacity achieved from the GEP equation is closer to the real data obtained from the experimental test. In five samples (3, 9, 23, 27, 30), FE results were manifest a better prediction of the maximum bearing capacity which could be due to specific conditions of the composite columns such as lower L/D ratio, slenderness rate, and the strength of the materials. Therefore, in these special cases, the prediction from the GEP equation may not be good enough due to wide ranges of input datasets. Nevertheless, more than 85% of the predictions by Equation (1) showed very good agreement with the experimental results. This proves the feasibility of this equation and the GEP approach for estimation of the maximum bearing capacity of such composite columns. If the maximum and minimum strength of the CFST columns resulting from the experimental work and their corresponding values from FEA and GEP are taken into consideration, it can be seen that the column sample No. 5 with the maximum experimental bearing capacity of 9187 kN has the closest predicted value of 8929.05 kN from the GEP equation. On the other hand, the column No. 21 with the minimum experimental bearing capacity of 1242.2 kN also has the most accurate value of 1242.01 kN from its corresponding GEP equation result. Therefore, the GEP-based equation yields very good results compared to those from the FEA.

Furthermore, according to the Table 3, model numbers 3, 9, 23, 27, and 30 presented a closer value of ultimate bearing capacity to the corresponding experimental outcome in the FEA instead of AI analysis using the GEP equation. In fact, this is because these models are short columns, stub columns, or composite column with a low or a very high strength materials, or they are the columns subjected to a kind of impact loading; Therefore, the prediction by the GEP equation led to less accuracy compared to the FEA. Nevertheless, as can be clearly inferred, the GEP equation results for 27 models out of 32 models indicated more accurate prediction and this shows that the generated prediction equation from GEP is sufficiently capable to predict the ultimate bearing capacity of such columns.

Table 3. Comparison of the GEP equation results and the FEA results with the experimental test results.

No.	P_{exp} (kN)	P_{FE} (kN)	P_{GEP} (kN)	Error $_{(FE)}$ (%)	Error $_{(GEP)}$ (%)
1	2866	3171	3151.69	11	9.9
2	2314	2640	2113.89	14	8.6
3	7837	7919	6894.05	1	12.0
4	9085	7928	8798.49	13	3.2
5	9187	7994	8929.05	13	2.8
6	6915	7946	6637.38	15	4.0
7	7407	8032	6849.84	8	7.5
8	6838	7791	6391.7	14	6.5
9	7529	7890	6834.93	4	9.7
10	1771	1554	1815.07	12	2.5
11	3339	3037	3086.68	9	7.6
12	3501	3050	3331.11	13	4.9
13	4837	4320	4438.7	11	8.2
14	4216	3850	4098.37	9	2.8
15	4330	3812	4549.91	12	5.1
16	4751	4313	4587.73	9	3.4
17	4930	4379	4808.45	11	2.5
18	5254	4738	5756.6	10	9.6
19	2160	2184	2160.52	1	0.0
20	2250	2092	2143.02	7	4.8
21	1242.2	1100	1242.01	11	0.0
22	1425.3	1225	1625.42	14	14
23	1637.9	1432	2022.22	13	23
24	1943.4	1672	2213.29	14	13.9
25	2866	3088	3077.52	8	7.4
26	2550	2642	2465.14	4	3.3
27	3150	3195	2852.55	1	9.4
28	3400	3926	3031.56	15	10.8
29	3850	4073	3672.81	6	4.6
30	5400	5993	4703.31	11	12.9
31	3338	3854	3569.71	15	6.9
32	8648	7694	8870	11	2.6

5. Concluding Remarks

Prediction of the bearing capacity of CFST columns is always important for the engineers. Due to some limitations in analysis by means of FEA, much efforts have been expended in attempts to find some other feasible and less time-consuming methods for prediction of the maximum capacity that CFST columns can withstand, and among them using AI techniques are widely being considered. This is because AI approaches such as GEP in most of the cases have shown better results in comparison with the FEA. In this study, 32 CFST columns, in the acceptable specification range of the AI-GEP analysis, were modeled and analyzed using the FEM. The results obtained from ABAQUS software were compared with the outcomes achieved from the GEP-based equation (Equation (1)). According to the results:

- A very good agreement is evident between the experimental outcomes and the GEP equation results with a less than 14% difference of the estimated bearing capacity for the majority of the cases;
- More than 85% of the results from Equation (1) were in accordance with the experimental results which proves the suitability and workability of this GEP-based equation for the prediction of the ultimate bearing capacity of the CFST columns;
- Only five models showed considerable differences in predicted values by GEP compared to the experimental data which could be because of the specific conditions of the composite columns such as low L/D ratio, slenderness rate, use of very high or very low strength for materials and the application of heavy loads to the columns.

It is possible to further study the CFST columns with the special conditions as explained above and find a more suitable GEP-based equation for those columns by training and testing more data. For this reason, it is recommended to set limitations on the specifications of the CFST columns to provide an equation with better prediction in comparison with the experimental results.

Author Contributions: This paper represents a result of collaborative teamwork. The conceptualization of the article is proposed by H.J. who is a professional in structural engineering. The methodology was outlined by A.S.M. who is an expert in civil engineering and the software analysis and validation were performed by P.S. and checked and investigated by R.A.K. All authors have read and agreed to the published version of the manuscript.

Funding: This research was funded by National Natural Science Foundation of China, grant number 51978526.

Acknowledgments: The authors' great gratitude goes to Shanghai Jiao Tong University and Tongji University as we have benefited from the facilities and the library access of these universities during this study. In addition, the financial support from National Natural Science Foundation of China, grant number 51978526, is highly appreciated.

Conflicts of Interest: The authors declare no conflict of interest.

Appendix A

Table A1. Specification of the samples for the FEA.

No.	Ref. No	D (mm)	L (mm)	t (mm)	L/D	D/t	f_y (MPa)	E_s (GPa)	f'_c (MPa)	E_c (GPa)	υ
1	[25]	114.3	210	6.3	1.84	18.14	428	209	173.5	63	0.3
2	[25]	114.3	250	3.6	2.19	31.75	403	213	184.2	63	0.3
3	[25]	219.1	600	5	2.74	43.82	380	205	185.1	66	0.3
4	[25]	219.1	600	10	2.74	21.91	381	212	185.1	66	0.3
5	[25]	219.1	600	10	2.74	21.91	381	212	193.3	67	0.3
6	[25]	219.1	600	6.3	2.74	34.78	300	202	163	62	0.3
7	[25]	219.1	600	6.3	2.74	34.78	300	202	175.4	58	0.3
8	[25]	219.1	600	6.3	2.74	34.78	300	202	148.8	54	0.3
9	[25]	219.1	600	6.3	2.74	34.78	300	202	174.5	56	0.3
10	[6]	168.6	645	3.9	3.83	43.23	363	206	36.2	33	0.3
11	[6]	168.6	645	3.9	3.83	43.23	363	206	95.8	33	0.3
12	[6]	164.2	652	2.5	3.97	65.68	377	206	158.46	33	0.3
13	[6]	189	756	3	4.00	63.00	398	206	158.46	33	0.3
14	[6]	168.6	648	3.9	3.84	43.23	363	206	165.49	33	0.3
15	[6]	169	645	4.8	3.82	35.21	399	206	167.87	33	0.3
16	[6]	168.7	645	5.2	3.82	32.44	405	206	158.75	33	0.3
17	[6]	168.8	650	5.7	3.85	29.61	452	206	151.91	33	0.3
18	[6]	168.1	645	8.1	3.84	20.75	409	206	158.75	33	0.3
19	[6]	165	500	2.81	3.03	58.72	350	212	67.94	67	0.3
20	[6]	165	500	2.76	3.03	59.78	350	212	67.94	67	0.3
21	[6]	114.3	342.9	3.35	3.00	34.12	287.33	212	86.21	67	0.3
22	[6]	114.3	342.9	6	3.00	19.05	342.95	212	56.99	67	0.3
23	[6]	114.3	342.9	6	3.00	19.05	342.95	212	86.21	67	0.3
24	[6]	114.3	342.9	6	3.00	19.05	342.95	212	102.43	67	0.3
25	[6]	114.3	200	6.3	1.75	18.14	428	212	164.35	67	0.3
26	[7]	200	600	1.945	3.00	102.83	227	212	52.7	67	0.3
27	[7]	200	600	1.945	3.00	102.83	227	212	67.7	67	0.3
28	[7]	200	600	1.945	3.00	102.83	227	205	74.4	58	0.3
29	[7]	260	780	1.945	3.00	133.68	227	205	52.7	58	0.3
30	[7]	260	780	1.945	3.00	133.68	227	205	85.4	58	0.3
31	[14]	299	848	1.68	2.84	177.98	267.5	205	47.2	58	0.3
32	[26]	273	4195	10	15.37	27.30	412	205	180	58	0.3

References

1. Sarir, P.; Shen, S.L.; Wang, Z.F.; Chen, J.; Horpibulsuk, S.; Pham, B.T. Optimum model for bearing capacity of concrete-steel columns with AI technology via incorporating the algorithms of IWO and ABC. *Eng. Comput.* **2021**, *37*, 797–807. [CrossRef]
2. Han, L.H.; An, Y.-F. Performance of concrete-encased CFST stub columns under axial compression. *J. Constr. Steel Res.* **2014**, *93*, 62–76. [CrossRef]
3. Han, L.H.; Li, W.; Bjorhovde, R. Developments and advanced applications of concrete-filled steel tubular (CFST) structures: Members. *J. Constr. Steel Res.* **2014**, *100*, 211–228. [CrossRef]
4. Tao, Z.; Wang, Z.B.; Yu, Q. Finite element modelling of concrete-filled steel stub columns under axial compression. *J. Constr. Steel Res.* **2013**, *89*, 121–131. [CrossRef]
5. Keshavarz, A.; Mehramiri, M. New gene expression programming models for normalized shear modulus and damping ratio of sands. *Eng. Appl. Artif. Intell.* **2015**, *45*, 464–472. [CrossRef]
6. Khandelwal, M.; Faradonbeh, R.S.; Monjezi, M.; Armaghani, D.J.; Majid, M.Z.A.; Yagiz, S. Function development for appraising brittleness of intact rocks using genetic programming and non-linear multiple regression models. *Eng. Comput.* **2017**, *33*, 13–21. [CrossRef]
7. Koopialipoor, M.; Armaghani, D.J.; Haghighi, M.; Ghaleini, E.N. A neuro-genetic predictive model to approximate overbreak induced by drilling and blasting operation in tunnels. *Bull. Eng. Geol. Environ.* **2019**, *78*, 981–990. [CrossRef]
8. Uddin, M.N.; Li, L.Z.; Khan, R.K.M.; Shahriar, F.; Sob, L.W.T. Axial capacity prediction of concrete-filled steel tubular short members using multiple linear regression and artificial neural network. *Mater. Sci. Forum* **2021**, *1047*, 220–226. [CrossRef]
9. Chen, W.; Sarir, P.; Bui, X.N.; Nguyen, H.; Tahir, M.M.; Armaghani, D.J. Neuro-genetic, Neuro-imperialism and genetic programing models in predicting ultimate bearing capacity of pile. *Eng. Comput.* **2020**, *36*, 1101–1115. [CrossRef]
10. Sarir, P.; Chen, J.; Asteris, P.G.; Armaghani, D.J.; Tahir, M.M. Developing GEP tree-based, Neuro-swarm, and whale optimization models for evaluation of bearing capacity of concrete-filled steel tube columns. *Eng. Comput.* **2021**, *37*, 1–19. [CrossRef]
11. Javadi, A.; Mehravar, M.; Faramarzi, A.; Ahangar-Asr, A. An artificial intelligence based finite element method. *ISAST Trans. Comput. Intell. Syst.* **2009**, *6*, 11–17.
12. Alani, Y.R.A.; Agarwal, V. Nonlinear finite element study on the circular concrete filled steel tubular columns. *Int. J. Innov. Technol. Explor. Eng.* **2013**, *3*, 52–55.
13. Armaghani, D.J.; Hajihassani, M.; Bejarbaneh, B.Y.; Marto, A.; Mohamad, E.T. Indirect measure of shale shear strength parameters by means of rock index tests through an optimized artificial neural network. *Meas. J.* **2014**, *55*, 487–498. [CrossRef]
14. Armaghani, D.J.; Safari, V.; Fahimifar, A.; Amin, M.F.M.; Monjezi, M.; Mohammadi, M.A. Uniaxial compressive strength prediction through a new technique based on gene expression programming. *Neural Comput. Appl.* **2018**, *30*, 3523–3532. [CrossRef]
15. Elshekh, E.; Shafiq, N.; Nuruddin, I.; Fathi, A. High strength concrete. *J. Appl. Sci.* **2014**, *14*, 1073–1077. [CrossRef]
16. Evirgen, B.; Tuncan, A.; Taskin, K. Structural behavior of concrete filled steel tubular sections (CFT/CFSt) under axial compression. *Thin-Walled Struct.* **2014**, *80*, 46–56. [CrossRef]
17. Faradonbeh, R.S.; Armaghani, D.J.; Monjezi, M. Development of a new model for predicting flyrock distance in quarry blasting: A genetic programming technique. *Bull. Eng. Geol. Environ.* **2016**, *75*, 993–1006. [CrossRef]
18. Faradonbeh, R.S.; Armaghani, D.J.; Monjezi, M.; Mohamad, E.T. Genetic programming and gene expression programming for flyrock assessment due to mine blasting. *Int. J. Rock Mech. Min. Sci.* **2016**, *88*, 254–264. [CrossRef]
19. Faradonbeh, R.S.; Hasanipanah, M.; Amnieh, H.B.; Armaghani, D.J.; Monjezi, M. Development of GP and GEP models to estimate an environmental issue induced by blasting operation. *Environ. Monit. Assess.* **2018**, *190*, 351–363. [CrossRef] [PubMed]
20. Gordan, B.; Armaghani, D.J.; Hajihassani, M.; Monjezi, M. Prediction of seismic slope stability through combination of particle swarm optimization and neural network. *Eng. Comput.* **2016**, *32*, 85–97. [CrossRef]
21. Gordan, B.; Koopialipoor, M.; Clementking, A.; Tootoonchi, H.; Mohamad, E.T. Estimating and optimizing safety factors of retaining wall through neural network and bee colony techniques. *Eng. Comput.* **2019**, *35*, 945–954. [CrossRef]
22. Güneyisi, E.M.; Gültekin, A.; Mermerdaş, K. Ultimate capacity prediction of axially loaded CFST short columns. *International J. Steel Struct.* **2016**, *16*, 99–114. [CrossRef]
23. Hoang, A.L.; Fehling, E.; Thai, D.; Van Nguyen, C. Evaluation of axial strength in circular STCC columns using UHPC and UHPFRC. *J. Constr. Steel Res.* **2019**, *153*, 533–549. [CrossRef]
24. Khandelwal, M.; Armaghani, D.J.; Faradonbeh, R.S.; Ranjith, P.G.; Ghoraba, S. A new model based on gene expression programming to estimate air flow in a single rock joint. *Environ. Earth Sci.* **2016**, *75*, 739–745. [CrossRef]
25. Koopialipoor, M.; Armaghani, D.J.; Hedayat, A.; Marto, A.; Gordan, B. Applying various hybrid intelligent systems to evaluate and predict slope stability under static and dynamic conditions. *Soft Comput.* **2018**, *23*, 5913–5929. [CrossRef]
26. Koopialipoor, M.; Noorbakhsh, A.; Ghaleini, E.N. A new approach for estimation of rock brittleness based on non-destructive tests. *Nondestruct. Test. Eval.* **2019**, *16*, 214–223. [CrossRef]
27. Koopialipoor, M.; Fallah, A.; Armaghani, D.J.; Azizi, A.; Mohamad, E.T. Three hybrid intelligent models in estimating flyrock distance resulting from blasting. *Eng. Comput.* **2019**, *35*, 243–256. [CrossRef]
28. Koopialipoor, M.; Ghaleini, E.N.; Haghighi, M.; Kanagarajan, S.; Maarefvand, P.; Mohamad, E.T. Overbreak prediction and optimization in tunnel using neural network and bee colony techniques. *Eng. Comput.* **2018**, *5*, 4–12. [CrossRef]

29. Koopialipoor, M.; Ghaleini, E.N.; Tootoonchi, H.; Armaghani, D.J.; Haghighi, M.; Hedayat, A. Developing a new intelligent technique to predict overbreak in tunnels using an artificial bee colony-based ANN. *Environ. Earth Sci.* **2019**, *78*, 165–178. [CrossRef]
30. Kvočák, V.; Varga, G.; Vargová, R. Composite steel concrete filled tubes. *Procedia Eng.* **2012**, *40*, 469–474. [CrossRef]
31. Rabbat, B.G.; Russell, H.G. Friction coefficient of steel on concrete or grout. *J. Struct. Eng.* **1985**, *111*, 505–515. [CrossRef]
32. Shen, S.L.; Wang, Z.F.; Sun, W.J.; Wang, L.B.; Horpibulsuk, S. A field trial of horizontal jet grouting using the composite-pipe method in the soft deposits of Shanghai. *Tunn. Undergr. Space Technol.* **2013**, *35*, 142–151. [CrossRef]
33. Ahmadi, M.; Naderpour, H.; Kheyroddin, A. Utilization of artificial neural networks to prediction of the capacity of CCFT short columns subject to short term axial load. *Arch. Civ. Mech. Eng.* **2014**, *14*, 510–517. [CrossRef]
34. Ren, Q.; Li, M.; Zhang, M.; Shen, Y.; Si, W. Prediction of ultimate axial capacity of square concrete-filled steel tubular short columns using a hybrid intelligent algorithm. *J. Appl. Sci.* **2019**, *28*, 2802. [CrossRef]
35. Le, L.M.; Ly, H.B.; Pham, B.T.; Le, V.M.; Pham, T.A.; Nguyen, D.H.; Tran, X.T.; Le, T.T. Hybrid artificial intelligence approaches for prediction buckling damage of steel columns under axial compression. *J. Mater.* **2019**, *16*, 12–30.
36. Cascardi, A.; Mircelli, F.; Aiello, M.A. An artificial neural networks model for the prediction of the compressive strength of FRP-confined concrete circular columns. *J. Eng. Struct.* **2017**, *140*, 199–208. [CrossRef]
37. Lim, J.C.; Karakus, M.; Ozbakkaloglu, T. Evaluation of ultimate conditions of FRP-confined concrete columns using genetic programming. *J. Comput. Struct.* **2016**, *162*, 28–37. [CrossRef]
38. Ahmadi, M.; Naderpour, H.; Kheyroddin, A. ANN model for predicting the compressive strength of circular steel-confined concrete. *Int. J. Civ. Eng.* **2017**, *15*, 213–221. [CrossRef]
39. Güneyisi, E.M.; Abdikarim, I.N. Axial compression capacity of circular CFST columns transversely strengthened by FRP. *Eng. Struct.* **2019**, *191*, 417–431. [CrossRef]
40. Jayalekshmi, S.; Jegadesh, J.S.S.; Goel, A. Empirical approach for determining axial strength of circular concrete filled steel tubular columns. *J. Inst. Eng.* **2018**, *99*, 257–268. [CrossRef]
41. Ansari, Y.; Hashemi, A.H. Neural Network Approach in Assessment of Fiber Concrete Impact Strength. *J. Civ. Eng. Mater. Appl.* **2017**, *1*, 88–97. [CrossRef]
42. Yagoubi, M.H. Prediction of ultimate strength for concrete-filled steel tube columns using artificial neural network. Ph.D. Thesis, University of Zabol, Zabol, Iran, 2018.
43. Wang, Z.B.; Liu, L.Y. Finite element modelling of concrete-filled steel tube reinforced concrete stub columns under axial compression. *Appl. Mech. Mater.* **2013**, *142*, 351–359. [CrossRef]
44. Xiao, J.; Huang, Y.; Yang, J.; Zhang, C. Mechanical properties of confined recycled aggregate concrete under axial compression. *Constr. Build. Mater.* **2012**, *26*, 591–603. [CrossRef]
45. Kanishchev, R.; Kvocak, V. Stability and carrying capacity of the steel tubes. *Procedia Eng.* **2017**, *190*, 447–451. [CrossRef]
46. Singh, H.; Gupta, P. Numerical modeling of rectangular concrete-filled steel tubular short columns. *Contrib. Pap.* **2013**, 167–179.
47. Davoudi, S.A.M.; Naghipour, M. Studying the Buckling Behavior of Composite Columns (CFST) by Cyclic Loading. *J. Civ. Eng. Mater. Appl.* **2019**, *3*, 203–213. [CrossRef]
48. David, H. *ABAQUS Standard User's Manual*, version 6.12.1, USA, 2012.
49. Sarir, P.; Shen, S.-L.; Arulrajah, A.; Horpibulsuk, S. Concrete wedge and coarse sand coating shear connection system in GFRP concrete composite deck. *Constr. Build. Mater.* **2016**, *114*, 650–655. [CrossRef]
50. Ding, H.T.; Francis, A.; Kaczmarek, O.; Karsch, F.; Laermann, E.; Soeldner, W. Thermal dilepton rate and electrical conductivity: An analysis of vector current correlation functions in quenched lattice QCD. *Phys. Rev. D* **2011**, *83*, 34–50. [CrossRef]
51. Liang, Q.Q.; Fragomeni, S. Nonlinear analysis of circular concrete-filled steel tubular short columns under axial loading. *J. Constr. Steel Res.* **2009**, *65*, 2186–2196. [CrossRef]
52. Hoang, A.L.; Fehling, E. Numerical study of circular steel tube confined concrete (STCC) stub columns. *J. Constr. Steel Res.* **2017**, *136*, 238–255. [CrossRef]
53. Liu, J.; Teng, Y.; Zhang, Y.; Wang, X.; Chen, Y.F. Axial stress—Strain behavior of high-strength concrete confined by circular thin-walled steel tubes. *Constr. Build. Mater.* **2018**, *177*, 366–377. [CrossRef]
54. Xiong, M.; Xiong, D.; Liew, J.Y.R. Axial performance of short concrete filled steel tubes with high- and ultra-high- strength materials. *Eng. Struct.* **2017**, *136*, 494–510. [CrossRef]
55. Xiong, M.; Xiong, D.; Liew, J.Y.R. Behaviour of steel tubular members infilled with ultra-high strength concrete. *J. Constr. Steel Res.* **2017**, *138*, 168–183. [CrossRef]
56. Wang, Y.; Yang, L.; Yang, H.; Liu, C. Behaviour of concrete- filled corrugated steel tubes under axial compression. *Eng. Struct.* **2019**, *183*, 475–495. [CrossRef]
57. Stoffel, M.; Bamer, F.; Markert, B. Artificial neural networks and intelligent finite elements in non-linear structural mechanics. *Thin-Walled Struct.* **2018**, *131*, 102–106. [CrossRef]
58. Dai, X.H.; Lam, D.; Jamaluddin, N.; Ye, J. Numerical analysis of slender elliptical concrete filled columns under axial compression. *Thin-Walled Struct.* **2014**, *77*, 26–35. [CrossRef]
59. Roure, F.; Pastor, M.M.; Casafont, M.; Somalo, M.R. Stub column tests for racking design: Experimental testing, FE analysis and EC3. *Thin-Walled Struct.* **2011**, *49*, 167–184. [CrossRef]

60. Code, P. *Eurocode 4: Design of Structures for Earthquake Resistance-Part 1: General Rules, Seismic Actions and Rules for Buildings;* European Committee for Standardization: Brussels, Belgium, 2005.
61. Giakoumelis, G.; Lam, D. Axial capacity of circular concrete-filled tube columns. *J. Constr. Steel Res.* **2004**, *60*, 1049–1068. [CrossRef]

applied sciences

MDPI

Review

Designing for Hybrid Intelligence: A Taxonomy and Survey of Crowd-Machine Interaction

António Correia [1,2,*], Andrea Grover [2], Daniel Schneider [3,4], Ana Paula Pimentel [3], Ramon Chaves [5], Marcos Antonio de Almeida [5] and Benjamim Fonseca [1]

1 INESC TEC, University of Trás-os-Montes e Alto Douro, 5000-801 Vila Real, Portugal
2 College of Information Science & Technology, University of Nebraska at Omaha, Omaha, NE 68182, USA
3 Postgraduate Program in Informatics, PPGI/UFRJ, Rio de Janeiro 21941-916, Brazil
4 Tércio Pacitti Institute of Computer Applications and Research (NCE), Federal University of Rio de Janeiro, Rio de Janeiro 21941-916, Brazil
5 Systems Engineering and Computer Science Program (PESC/COPPE/UFRJ), Rio de Janeiro 21941-972, Brazil
* Correspondence: antonio.g.correia@inesctec.pt

Abstract: With the widespread availability and pervasiveness of artificial intelligence (AI) in many application areas across the globe, the role of crowdsourcing has seen an upsurge in terms of importance for scaling up data-driven algorithms in rapid cycles through a relatively low-cost distributed workforce or even on a volunteer basis. However, there is a lack of systematic and empirical examination of the interplay among the processes and activities combining crowd-machine hybrid interaction. To uncover the enduring aspects characterizing the human-centered AI design space when involving ensembles of crowds and algorithms and their symbiotic relations and requirements, a Computer-Supported Cooperative Work (CSCW) lens strongly rooted in the taxonomic tradition of conceptual scheme development is taken with the aim of aggregating and characterizing some of the main component entities in the burgeoning domain of hybrid crowd-AI centered systems. The goal of this article is thus to propose a theoretically grounded and empirically validated analytical framework for the study of crowd-machine interaction and its environment. Based on a scoping review and several cross-sectional analyses of research studies comprising hybrid forms of human interaction with AI systems and applications at a crowd scale, the available literature was distilled and incorporated into a unifying framework comprised of taxonomic units distributed across integration dimensions that range from the original time and space axes in which every collaborative activity take place to the main attributes that constitute a hybrid intelligence architecture. The upshot is that when turning to the challenges that are inherent in tasks requiring massive participation, novel properties can be obtained for a set of potential scenarios that go beyond the single experience of a human interacting with the technology to comprise a vast set of massive machine-crowd interactions.

Keywords: conceptual framework; crowd-machine hybrid interaction; design implications; hybrid intelligence; survey; taxonomy

Citation: Correia, A.; Grover, A.; Schneider, D.; Pimentel, A.P.; Chaves, R.; de Almeida, M.A.; Fonseca, B. Designing for Hybrid Intelligence: A Taxonomy and Survey of Crowd-Machine Interaction. *Appl. Sci.* **2023**, *13*, 2198. https://doi.org/10.3390/app13042198

Academic Editor: Danial Jahed Armaghani

Received: 30 December 2022
Revised: 29 January 2023
Accepted: 2 February 2023
Published: 8 February 2023

1. Introduction and Context

Crowd-centered design is far from a trivial undertaking, and this is even more challenging when trying to implement hybrid intelligence models incorporating human cognition into algorithmic-crowdsourcing workflows [1]. In fact, crowd-algorithm interaction has recently reached a certain level of maturity, and a vast range of crowd-powered algorithms have successfully been applied in areas like medical image segmentation [2] and games with a purpose (GWAP) [3]. In these instances, crowds of untrained (non-expert) online workers have proved to provide similar results in terms of detection accuracy when compared to other groups such as domain knowledge experts, medical students, and experienced crowd workers. Further investigations in this burgeoning domain have also shown that the use of crowd-algorithm hybrids can outperform crowd-only techniques in accomplishing tasks like examining protein interactions and chemical reactions that are very common in the field of network biology [4]. Nonetheless, the taxonomic rationale behind the mass interaction efforts between crowds and machines as an integrated and complex socio-technical system is not completely understood, and there is a need to find novel ways of characterizing this body of work in its whole range. To mitigate this brittleness, a review of the main activities and contexts in which such crowd-AI ensembles have been investigated was carried out to develop a taxonomic scheme as comprehensive as possible to capture the nuances that are unique in comparison with other types of interactions between humans and computational systems.

For more than three decades, taxonomy development has been seen as a crucial part of socio-technical research within the field of CSCW [5]. To some extent, taxonomies provide a useful guide and theoretical foundation for assessing technological developments due to their capability to organize complex concepts and knowledge structures into understandable formats [6]. By going back in the course of time, one may find several taxonomic approaches that formed the basis for the understanding of the task types that are currently present in many crowdsourcing systems. For a review of prior taxonomic proposals, the reader is referred to Harris and co-authors [7]. In retrospect, McGrath [8] proposed a circumplex model of group tasks intended to characterize their nature (e.g., decision-making) into four quadrants that reflect the processes involved in their execution (i.e., generate, choose, negotiate, and execute). When moving even further back in history, Shaw [9] asserted the importance of aspects like task difficulty and intrinsic interest which are seen as foundational in several conceptual frameworks proposed to characterize the broader crowdsourcing phenomena (e.g., [10,11]). According to some authors, Johansen's [12] time-space matrix is a landmark in the field of CSCW and inspired the development of descriptive models such as the Model of Coordinated Action (MoCA) [13], which frames each collaborative work arrangement on a continuum of synchronicity (synchronous vs. asynchronous), physical distribution, scale (i.e., number of participants), number of communities of practice involved, nascence and planned permanence of coordinated actions, and turnover. More recently, Renyi and colleagues [14] executed a set of data collection and processing procedures involving structured interviews in order to create a taxonomic scheme covering the components related to the collaboration technology support in home care work, while other authors have devoted most of their efforts to the design of innovative taxonomic interfaces [15]. In addition, there is now an emerging body of research documenting the different levels of hybrid intelligence in human-algorithm interactions.

From a more generic view, the concept of hybrid intelligence has been defined as the "combination of human and machine intelligence, augmenting human intellect and capabilities instead of replacing them and achieving goals that were unreachable by either humans or machines" [16]. Stemming from this definition, experiments have shown that the time is now appropriate to develop a new taxonomic proposal that can be used for planning and assessing activities among humans (crowds) and algorithms in a hybrid mode. To the best of the authors' knowledge, no other previous work has specifically focused on crowd-AI interaction, although there are some research works addressing the particularities of hybrid human-AI intelligence at a taxonomic level. For example, Pescetelli [17] stressed the role of algorithms as assistants, peers, facilitators, and system-level operators. On the other hand, Dellermann and associates [18] characterized the design space of hybrid intelligence systems and recalled the importance of the task itself and its characteristics as a central aspect of collaboration among humans and machines. In the same vein, Dubey et al. [19] proposed a taxonomy of human-AI teaming comprised of task properties, trust-related aspects, teaming characteristics (e.g., shared awareness), and the learning paradigm involved. However, these taxonomies have hitherto not yet fully explored the particularities of hybrid crowd-AI systems and their use cases in real-world applications. Through a qualitative inspection of conceptual frameworks, artifacts, case studies, and empirical results comprising some type of human-AI hybrid interaction at a massive scale, this article's contribution lies in systematically structuring a set of attributes and characteristics into an integrated taxonomy that arises as a continuum of co-evolving crowd-algorithmic partnerships intended to solve complex problems that neither humans nor machines can solve separately.

The article is set out as follows. After a discussion of background work in Section 2, a description of the methodological steps follows until the development of a taxonomy for hybrid crowd-AI systems is provided in Section 3. The resulting taxonomic framework is then presented and discussed in detail in Section 4, while Section 5 is concerned with the validation of the taxonomy proposed. Finally, possible extensions of this work are suggested in the Section 6 by looking toward the future of hybrid systems from a socio-technical view of human-centered systems design.

2. Background and Scope

The point of departure for building the taxonomy presented in this article was the existing work found on the intersectional space between human-computer interaction (HCI) and AI from a crowdsourcing perspective. Although the coining of the term 'crowdsourcing' took place in the mid-2000s, some may argue that its origin is rooted in the seminal work of the physicist and astronomer Denison Olmsted, who used news media as a crowdsourcing strategy for obtaining accurate observations on the Leonid meteor shower that was witnessed across the United States in 1833 [20]. What is interesting to note is that the sequential steps and general techniques used by Olmsted about nineteen decades ago constitute the basis for most of the current crowdsourcing applications. Aligned with this goal, a variety of taxonomies and conceptual frameworks have been developed to better characterize the way as information technology (IT)-enabled crowdsourcing operates. Among the known classifications of crowdsourcing activities, Corney and co-authors [21] were some of the first to frame this phenomenon from a taxonomic point of view by incorporating the nature of the crowd, the payment mechanisms or lack thereof, and the type of task into an integrated framework. In line with this, Rouse [22] proposed a taxonomy that comprises the different forms of intrinsic and extrinsic motivation that can lead to a successful crowdsourcing experience (e.g., social status, altruistic behavior, and personal achievement). This taxonomic proposal also addresses a set of aspects that are specific to the nature of the crowdsourcing task being undertaken by encompassing the expertise and complexity that are directly or indirectly involved in such initiatives. On the basis of insights from the history of group support systems, one would notice similar points to McGrath's [8] task circumplex taxons taking into consideration the different task types that can be executed by individuals in a group structure, which may include decision-making, idea generation and information gathering to name just a few examples.

To an extent, this research strand led to the proliferation of several taxonomies incorporating task-related elements (e.g., [23–30]). Consistent with the task properties discussed in most of these studies, a cursory look at the literature reveals certain commonalities related to crowd attributes (e.g., reputation), requester features (e.g., incentivization), and platform facilities such as aggregation and payment mechanisms [29]. Other research works have focused specifically on internal forms of crowdsourcing [31] or even on the use of crowdsourcing as a taxonomy development strategy by itself [32]. On a more generic level, Modaresnezhad and colleagues [10] made a clear distinction between the IT-enabled crowdsourcing requirements in business and non-business contexts by basing their proposal on the four collective intelligence "genes" proposed by Malone et al. [23]. However, these taxonomies fail to fully account for the hybrid nature of crowd-AI interaction and thus are unable to capture the variety of interactions and relations that occur when using a hybrid intelligence system.

During the last few years, the advances in the development of AI technologies have been silently leveraging the capacity of a large pool of crowd workers worldwide who provide data on a daily basis and thus contribute to the improvement of several models on a scale that had never been seen before. In fact, this intertwinement of algorithms with crowdsourcing workflows brought important advantages in a multiplicity of settings. Prior work has employed these principles and proved to be effective in detecting accessibility problems on public surfaces (e.g., sidewalks) through the use of street-level imagery [33]. In the same vein, Zhang and associates [34] proposed a system for identifying urban infrastructure damages, such as fallen street signs, when AI-based solutions fail to recognize them. These architectures have also been applied in the context of video object segmentation [35], cultural heritage damage identification [36], endoscopic image annotation [37], and historical portrait identification [38]. In addition, weaving together crowd- and AI-powered techniques has also resulted in positive outcomes in real-time and remote on-demand assistance [39]. In the literature, there are also examples of sensing systems embedded in real-world environments (e.g., domestic spaces) that resort to built-in cameras and crowdsourcing interfaces for dynamic image labeling [40]. That is, crowd-AI hybrid systems are now able to engage humans and machines through a massively collaborative joint action that spans research fields and temporal and geographical boundaries [41]. Drawing from previous studies on the characterization of hybrid intelligence systems from a taxonomic viewpoint [18], the work conducted herein expands upon what has been previously investigated by examining the many facets of crowd-machine hybrid systems and thus identifying key thematic elements derived from the literature.

3. Methodological Approach

Drawing on a literature review of extant studies on human-AI interaction with a crowd-in-the-loop, this article outlines a particular set of arrangements in which the research on this burgeoning area can inform the development of future hybrid intelligence systems while contributing to understanding the socio-technical practices that require humans and machines working together towards a common goal. To this end, this work takes a human-centered AI approach [42] guided by the evidence-based taxonomy development method proposed by Nickerson and colleagues [6], as depicted in Figure 1. Synoptically, the practice of taxonomic classification can be described as a full-fledged endeavor in fields like astrophysics [43] and genetics [44] that usually consists of a formal semantic model with empirically or conceptually derived dimensions and characteristics that are exhaustive and mutually exclusive by nature [6]. At their structural level, taxonomies may have hierarchical or non-hierarchical configurations [45] and be constantly subjected to updating revisions [15]. Building on these methodological elements, the present study draws on the HCI body of literature to create a taxonomy of crowd-AI hybrids and thus aid researchers, practitioners, and anyone concerned with the understanding and development of these technologies. With this in mind, a step forward is made by distilling a variety and breadth of conceptual units from studies that seek to address the complementary way in which human crowds interact with AI systems. Essentially, this study sheds light on the socio-technical dimensions of crowd-AI integration by acknowledging that both social and technical aspects must be taken into account to understand the functioning of a hybrid system as a whole.

In this study, a novel set of heuristics and theoretical aspects are proposed as a foundational structure for future research based on a scoping review that follows the guidelines of evidence-based practice [46]. From a methodological perspective, this approach seeks to systematically categorize research into a classification scheme that is then used as a foundation for taxonomy construction and validation. To operationalize the taxonomic process, a phenetic approach [47] was used throughout a set of iterative cycles until the ending conditions were met. To this end, this article explores the vast space covered by the literature on hybrid crowd-AI systems grounded in case studies, ethnographic fieldwork, conceptual frameworks, surveys, semi-structured interviews, experimental work, mixed methods, and technical artifacts (e.g., algorithms). The taxonomy-building process followed the formal definition of Nickerson et al. [6] to create a taxonomy T with "a set of n dimensions Di ($i = 1, \ldots, n$), each consisting of ki ($ki \geq 2$) mutually exclusive and collectively exhaustive characteristics Cij ($j = 1, \ldots, ki$) such that each object under consideration has one and only one Cij for each Di, or $\mathrm{T} = \{\mathrm{D}i, i = 1, \ldots, n \mid \mathrm{D}i = \{\mathrm{C}ij, j = 1, \ldots, ki; ki \geq 2\}\}$".

It is worth noting that the guidelines provided by Nickerson and associates [6] represent one of the most well-established methodological approaches for taxonomy development in the field of information systems (IS), as reported in a recent literature review [48]. In this vein, these guidelines were systematically applied in an effort to make the proposed taxonomy clear, concise, robust, comprehensive, explanatory, and extendible as nearly as possible to attend to the conditions advocated by Gerber [49] when addressing the creation of classification artifacts.

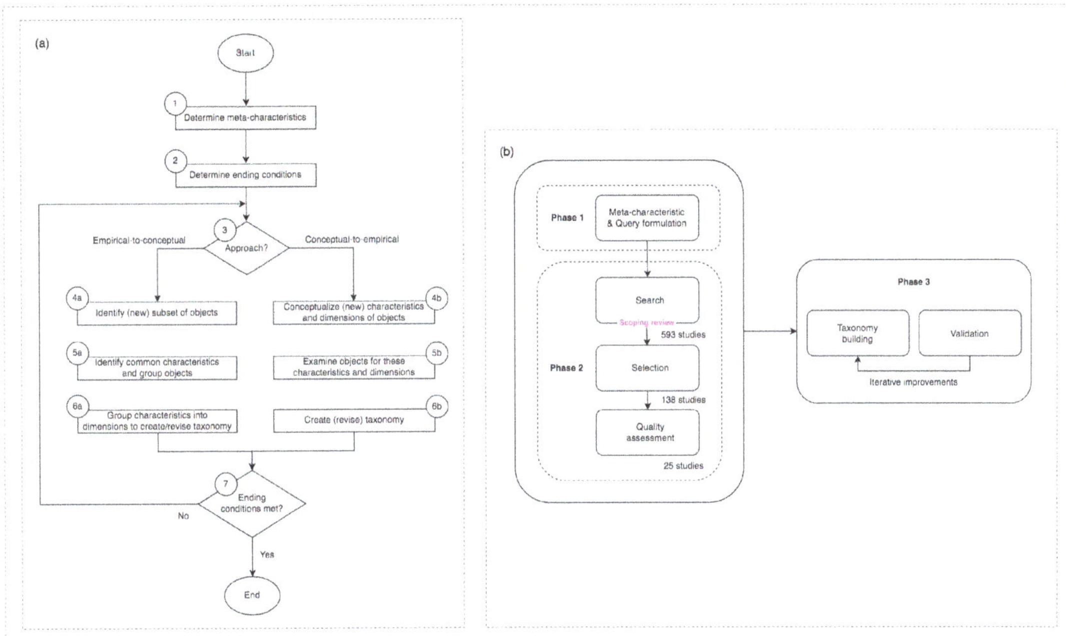

Figure 1. Iterative taxonomy development process flow (**a**) and methodological details underlying the work undertaken in this study (**b**). Adapted from Nickerson and co-authors [6].

The first phase of taxonomy development consisted of a descriptive literature analysis [50] to identify rationales for the use of crowd-AI hybrids. This was followed by a systematic examination of the insights extracted and further categorized into a literature classification scheme. In fact, this empirical-to-conceptual methodological approach has been a common procedure for data collection in the taxonomy-building activity (e.g., [51–53]), involving a set of systematic processes that range from a literature search to data filtering and classification. For taxonomic validation, a conventional approach for corpus construction was used as previously described in [54]. Essentially, the sample used in this study is an expanded version of that used in [41]. This was achieved by following a living systematic review protocol [55], where the search strategy is maintained and updated in a continuous manner as new studies become available. For the purpose of this review, a simplistic Boolean query formulation was applied using the following sequence of terms:

((*"crowd*-AI"* OR *"AI-crowd*"* OR *"crowd*-machine"* OR *"machine-crowd*"* OR *"crowd*-computing"*) AND (*"interact*"*))

This study expanded upon a previous corpus to accommodate a new set of possible settings in which crowd-AI interaction occurs. This was done due to two main reasons. First, a more recent picture of the state-of-the-art in this domain was needed. To this end, only papers published in the last five years (2018–2022) as of 17 December 2022 were inspected. Second, most of the studies considered for taxonomy validation in [41] comprised human-AI interaction at an individual level, while here, the focus is on evaluating arrangements involving crowds mixed with AI. The present work is also more restrictive in terms of peer-reviewed studies since this contribution only considered journal articles and conference papers. From a systematic search for publications indexed by the Dimensions database, which contains records from diverse digital libraries such as ACM Digital Library, IEEE Xplore, SpringerLink, and Science Direct with large coverage when compared to Web of Science and Scopus [56], content types such as adjunct/companion proceedings, panels, tutorials, book reviews, correspondence articles, introductions to special issues, doctoral colloquiums and student research competitions, keynote talks, commentaries, and course summaries were disregarded to ensure high-quality results. The search returned 593 publication records. After initial scrutiny of the titles and abstracts, along with the removal of papers that did not meet the inclusion criteria, a total of 138 studies were selected for further appraisal. To be eligible for inclusion, studies had to describe original research from primary or secondary literature addressing the broader domain of human-centered AI with a focus on crowd-AI interaction. As can be perceived from Figure 1, this selection resulted in 25 research studies published in English-written, peer-reviewed manuscripts (see Appendix A for details). The final set of papers chosen provided a reliable source of information for testing the taxonomic proposal since they presented a diverse set of scenarios.

As an integral part of the iterative taxonomy development process proposed in [6], the meta-characteristic of the taxonomy was determined to be its focus on *functional properties and attributes of hybrid crowd-AI systems*. Through a socio-technical lens grounded on the foundational aspects of crowd computation [57] and its embodiment in hybrid human-AI systems [58], the definition of this meta-characteristic allowed to frame and guide the taxonomy development process until the *subjective ending conditions* previously mentioned at the level of robustness, comprehensiveness, conciseness, extendibility, and explanatory nature of the taxonomy were fulfilled. Following the taxonomic work of Landolt and co-authors [59] on the use of deep neural networks in natural language processing (NLP) applications, this contribution also tried to meet *objective ending conditions* to ensure that each dimension and characteristic within the dimension were exclusive and no new characteristics or dimensions were added in the final iteration. Therefore, the original dimensions of the taxonomic proposal were validated within a literature matrix in order to verify whether these dimensions and characteristics are present in the final sample of studies addressing crowd-machine hybrid interaction. To some degree, the empirical validation of the taxonomy proposed here is inspired by the work of Straus [60], who took McGrath's [8] group task circumplex as the object of evaluation.

4. 'Inside the Matrix': In Pursuit of a Taxonomy for Hybrid Crowd-AI Interaction

The availability of crowdsourcing platforms has led many organizations to adopt them as continuous and highly available sources of data upon which the paradigm of open innovation [61] is founded and continues to develop. On its most generic level, these solutions are leveraged by a 24/7 digital workforce and represent a problem-solving and innovation-driven approach able to shorten the entire product lifecycle [62]. As novel AI-infused products and features become more and more prevalent and integral to many everyday life pursuits, the need to incorporate hybrid intelligence in highly complex and volatile scenarios (e.g., early warning and prompt response) become even more evident since the complementarity [63] and adaptivity [64] of human and AI-based systems co-evolving over time "as coequal partners" [65] can be of particular value to suppress each other's failures. In this vein, crowdsourcing has been applied to executing tasks such as obtaining ground-truth human labels [66], gathering ratings for data to be used in supervised machine learning [67], or even managing portfolio information [68]. In general terms, Kittur and associates [57] reported that crowd intelligence could be particularly useful in supervising, training, or even supplementing automation, while AI techniques can make the crowd more accurate while augmenting human capabilities and interactions through machine intelligence. This constitutes the point of departure for the proposal of a taxonomic framework for crowd-AI interaction, whose dimensions are shown in Figure 2 and briefly described in the following subsections.

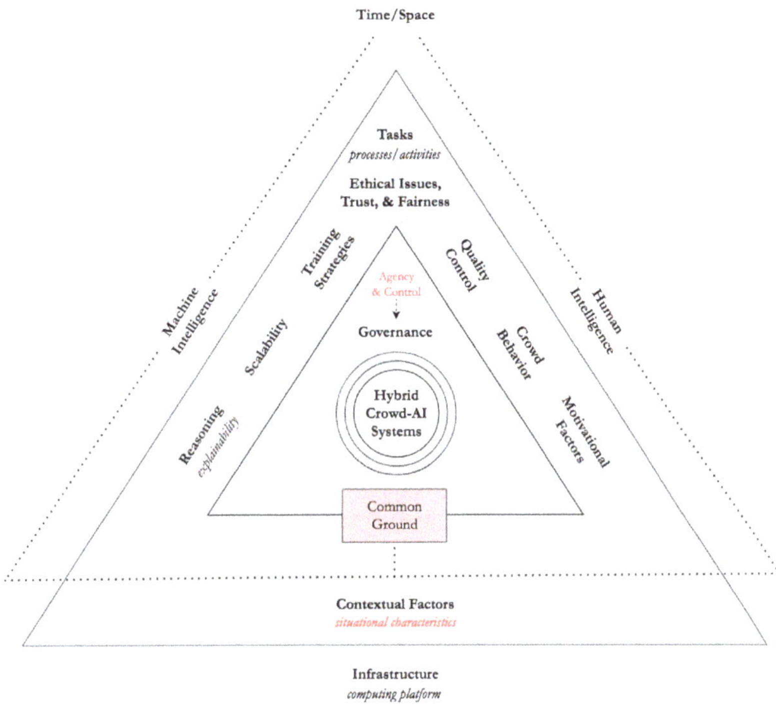

Figure 2. Taxonomy of hybrid crowd-AI systems. This taxonomic proposal integrates key conceptual dimensions of the human-centered AI framework introduced in [41] to characterize the configurations in which crowd-AI interaction occurs within the interplay between human and machine intelligence.

From a taxonomy-building methodological standpoint, the taxonomic design approach was largely inspired by the Work System Theory as depicted by Alter [69] and further explored by Venumuddala and Kamath [70], who conducted an ethnographic work grounded on a set of observations retrieved in an AI research laboratory. In addition, some elements from the Activity Theory [71] inspired model for assessing CSCW in distributed settings [72] were also introduced. As a result, a previous human-centered AI framework [41] was revised and extended to highlight the importance of agency and control, explainability, fairness, common ground, and situational awareness in the design space of hybrid crowd-AI systems.

4.1. Temporal and Spatial Axes of Crowd-AI Systems

Crowdsourcing can be seen as a gateway to obtain reliable solutions to problems of varying levels of difficulty when there is an urgent need for quick and prompt action or even when the development of a game, big-scale application, software module, sketch, etc., is required without the strict rigidity to be situated physically close [73]. At the interaction level, hybrid crowd-AI systems can be able to support real-time crowdsourcing activities involving chatting and live tracking services, and also those occurring asynchronously, such as post-match soccer video analysis. In framing this discussion within the time-space matrix originally described in the context of groupware applications [12], this article concentrates on the spatio-temporal patterns of human-AI partnerships at a crowd scale. Thus, one can argue that the notion of space has been reshaped to incorporate the provision of localization and navigation information into crowdsourcing settings as a way of exploring the full potential of local-and-remote on-demand real-time response in tasks like road data acquisition [74] and local news reporting [75]. That is, crowd workers can be physically or virtually distributed in a dispersed or co-located manner or even "synchronize in both time and physical space" [76]. As some scholars noted, the level of engagement in both paid and non-profit crowdsourcing communities can also be evaluated, taking into account the daily-devoted time of participants, periodicity of interactions, and activity duration [77]. In this regard, the contribution time and availability of the crowd constitute key information sources in crowd-AI hybrid settings.

4.2. Crowd-Machine Hybrid Task Execution and Delegation

The rapid progress of AI-based technology has led to novel ways of motivating humans to delegate tasks to AI for further fulfillment. Bouwer [78] proposed a four-quadrant taxonomic model for AI-based task delegation and stressed the importance of emotional/affective states as key deterministic factors for task delegation. In line with this, Lubars and Tan [79] mentioned the relevance of trust, motivation, difficulty, and risk as influential determinants of human-AI delegation decisions. In particular, trust and reliance assume a special significance in terms of delegation preferences. The strategic line behind most of the tasks that are commonly crowdsourced in current digital labor platforms is still grounded in microtask design settings [80], although some recent attention has been given to macrotasking activities (e.g., creative work) which involve crowd-powered tools designed to support computer-hard tasks that need specialized expertise and thus cannot be executed by AI algorithms in an effective manner [81]. By focusing on the task properties and attributes in crowdsourcing, Nakatsu and co-workers [27] introduced a taxonomy that classifies the structure (well-structured vs. unstructured) and level of interdependence (independent vs. interdependent) together with a third binary dimension involving the degree of commitment (low vs. high) required to accomplish a task.

Going back to the levels of complexity that may be present in crowdsourcing tasks, Hosseini et al. [29] briefly divided them into two main categories: simple and complex. Using this rationale, microtasks have been largely described as being simple for crowd workers to perform well and easily in the sense that they involve a lesser degree of context dependence [82]. Furthermore, these self-contained tasks are usually short by nature and take little time to finish. Zulfiqar and co-authors [83] go even further by underlining that microtasks do not require specialized skills, which enable any worker to contribute in a rapid and cognitive effortless manner. Extrapolating to more complex crowdsourcing processes, many forms of advanced crowd work have emerged throughout the years, and there is now a renewed focus on task assignment optimization involving algorithmically-supported teams of crowd workers acting collaboratively [84,85]. While the possibilities for optimization are manifold across a number of different task scenarios, robust forms of hybrid crowd-machine task allocation and delegation are needed to yield accurate results and reliable outcomes not only for crowd workers acting at the individual level but also in terms of team composition and related performance.

4.3. Contextual Factors and Situational Characteristics in Crowd-Computing Arrangements

Any crowd-machine hybrid interaction has its own contextual characteristics and specificities. Dwelling on this issue, one may wish to claim that crowdsourcing settings are highly context-dependent and situational information is particularly critical to achieving successful interactions in a crowd-AI working environment since a crowd can be affected by contextual factors such as geo-location, temporal availability, and surrounding devices [86]. Considering the context from which a crowd worker is interacting with an intelligent system can help to personalize the way the actions are developed and thus improve processes, such as task assignment [87] while providing resources and contextually relevant information tailored to the needs of each individual based on content usage behaviors [42] and other forms of context extraction. This involves a set of environmental, social, and cultural contexts [88] that come with fundamental challenges for hybrid algorithmic-crowdsourcing applications in terms of infrastructural support for achieving efficient and accurate context detection and interpretation. When designing a crowd-AI hybrid system, user-generated inputs must be handled adequately in order to filter the relevant information and better adapt the interaction elements and styles to each particular case [89]. In hindsight, this is also somewhat related to the notions of explainability and trust in AI systems [90] since the trustworthy nature of these interactions will be affected by the quality of the contextual information provided and the degree to which a user perceives the AI system they are interacting with as useful for aiding their activities. In such scenarios, aspects like satisfaction shape the internal states of the actors [72] and can constrain the general performance of the crowd-AI partnerships if the system does not meet the expectations of the users.

4.4. Deconstructing the Crowd Behavior Continuum in Hybrid Crowd-Machine Supported Environments

To some extent, both paid and non-paid forms of crowdsourcing have served as "Petri dishes" for many behavioral studies involving experimental work [91]. A crowd can differ in terms of attention level, size, emotional state, motivation and preferences, and expertise/skills, among many other characteristics [86]. In this vein, Robert and Romero [92] found a considerable impact of diversity and crowd size on performance outcomes while testing the registered users of a WikiProject Film community. As such, online crowd behaviors are volatile by nature and vary given the contextual factors and situational complexity of the work, along with the surrounding environment of its members. Neale and co-authors [72] briefly explained the importance of context for creating a common ground which can be understood as the shared awareness among actors in their joint activities, including their mutual knowledge. That is, sustaining an appropriate shared understanding can constitute a critical success factor for achieving a successful interaction when designing intelligent systems [93]. This also applies to the range of crowd work activities that involve self-organized behaviors and transient identities [94], which imply a reinforced need for effective quality control mechanisms (e.g., gold standard questions) in crowd-AI settings [40]. Furthermore, some crowds are arbitrary, while others are socially networked or organized into teams that coalesce and dissolve in response to an open call for solutions where the nature of the task being crowdsourced is largely dependent on collective actions instead of individual effort only. In some specific cases, these tasks are non-decomposable and involve a shared context, mutual dependencies, changing requirements, and expert skills [95,96]. In this vein, some prior research has revealed the presence of "a rich network of collaboration" [97] through which the crowd constituents are connected and interact in a social manner, although there are many concerns about the bias introduced by these social ties. Seen from a human-machine teaming perspective, imbalanced crowd engagement [98], conflict management [99], and lack of common ground [100] are also key aspects that must be taken into account in such arrangements.

4.5. Hybrid Intelligence Systems at a Crowd Scale: An Infrastructural Viewpoint

As AI-infused systems thrive and expand, crowdsourcing platforms continue to play an active role in aggregating inputs that are used by companies and other requesters around the globe toward the ultimate goal of enabling algorithms with the ability to cope with complex problems that neither humans nor machines can solve alone [101]. However, designing for AI with a crowd-in-the-loop includes a set of infrastructure-level elements such as data objects, software elements, and functions that together must provide effective support for actions like assigning tasks, stating rewards, setting time periods, providing feedback, evaluating crowd workers, selecting the best submissions, and aggregating results [102]. To realize the full potential of these systems, online algorithms can be incorporated into task assignment optimization processes for different types of problems involving simple (decomposable), complex (non-decomposable), and well-structured tasks [85]. By showing reasonable results in terms of effectiveness, some algorithms have been proposed to organize teams of crowd workers as cooperative units able to perform joint activities and accomplish tasks of varying complexity [95,96,103]. From an infrastructural perspective fitted into the taxonomy proposed in this article, the contribution of this study builds on Kamar's [104] work to stress the importance of combining both human and machine capabilities in a co-evolving synergistic way.

Taken together, crowd and machine intelligence can offer a lot of opportunities for predicting future events while improving large-scale decision-making since online algorithms can learn from crowd behavior using different integration and coupling levels. In many settings, hybrid intelligence systems can help to draw novel conclusions by interpreting complex patterns in highly dynamic scenarios. In line with this, many have studied novel forms of incorporating explainable AI approaches, such as gamification [105], for enhancing human perceptions and interpretations of algorithmic decisions in a more transparent and understandable manner. Due to their scalability, crowd-AI architectures can constitute an effective instrument for handling complexity, and thus more research is needed to explore how to best develop hybrid crowd-AI-centered systems taking into account the requirements and personal needs of each crowd worker. In particular, this domain raises some questions about the use of AI to enhance the quality of crowdsourcing outputs through high-quality training data [67] and related interaction experiences, as seen from a human-centered design perspective [106]. To summarize, crowd-powered systems can present a wide variety of opportunities to train algorithms "in situ" [107] while providing learning mechanisms and configuration features for customizing the levels of automation over time.

4.6. *'Rebuilding from the Ruins': Hybrid Crowd-Artificial Intelligence and Its Social-Ethica Caveats*

There is a clarion call for an investigation on the ethical, privacy, and trust aspects of human-AI interaction from several causes. For instance, Amershi and colleagues [88] raised a set of concerns related to the need to avoid social biases and detrimental behaviors. To tackle those issues, it is necessary to dive deep into the harms provided by AI decisions in a contextualized way to ensure fairness, transparency, and accountability in such interactions [108]. This can be realized by materializing human agency and other strategies that can provide more control over machine behaviors [109–111]. From diversity to inclusiveness—and subsequently justice—there is still a long way until these goals are accomplished within the dynamic frame of human-AI interaction and hybrid intelligence augmentation. To address these shortcomings, system developers can play a critical role by considering the potential effects of AI-infused tools on user experiences.

Extrapolating to the crowdsourcing settings, Daniel and co-workers [112] reported a concern with the ethical conditions, terms, and standards aligned with the compliance towards regulations and laws that are sometimes overlooked in such arrangements. When considering crowd work regulation, aspects of intellectual property, privacy, and confidentiality in terms of participant identities constitute pivotal points [113]. A look into previous works (e.g., [114]) shows multiple concerns regarding worker rights, ambiguous task descriptions, acknowledgment of crowd contributions, licensing and consent, low wages, and unjustified rejected work. Such ethical and legal issues are even more expressive in the context of hybrid crowd-AI systems where there are not only online experiments and other human intelligence tasks (HITs) running on crowdsourcing platforms but also machine-in-the-loop processes within the entire hybrid workflow. In a particular setting, strategies like shared decision-making and informed consent can be particularly helpful to mitigate the threats of bad conduct and malicious work if based on a governance strategy where the guidelines, rules, actions, and policies are socially organized by the crowd itself [115]. In this vein, the potential impacts of the aforementioned socio-ethical concerns surrounding crowd-powered hybrid intelligence systems must be further elucidated and investigated from several lenses to draw a realistic picture of the current situation.

5. Validation and Assessment of the Proposed Taxonomy

This study proposes a taxonomic framework aimed at accommodating a diverse set of infrastructurally supported crowd-algorithm interactions that occur in a certain time and space within two separate orders of intelligence, which, therefore, can be combined in a hybrid model architecture. The interactions occurring in this hybrid space have a set of unique contextual and situational aspects and must be guided by ethical guidelines, rules, and principles in order to combine crowd and machine workflows effectively and transparently. To validate the proposed taxonomy and demonstrate its utility, this contribution examined the applicability of the taxonomy in a total of twenty-five studies presenting some type of crowd-machine interaction. This is in line with the need for a methodologically rigorous inspection of the possible effects of hybrid intelligence in practical settings. For instance, substantial literature on human-AI interaction has developed quickly across different areas [116], but few attempts have been made to gather evidence about this intersectional space at a crowd scale and thus understand the uses and limitations of hybrid crowd-AI systems from a socio-technical design viewpoint. The results of the taxonomy-based review are provided in Figure 3, accompanied by an example of a scheme used to explain the rationale behind the taxonomic classification (Figure 4). Further details regarding the 14 journal articles and 11 conference papers selected for taxonomy-based literature analysis are given in Tables A1 and A2 in the Appendix A. In order to determine whether each category of the taxonomy was either present or absent, the following levels were considered:

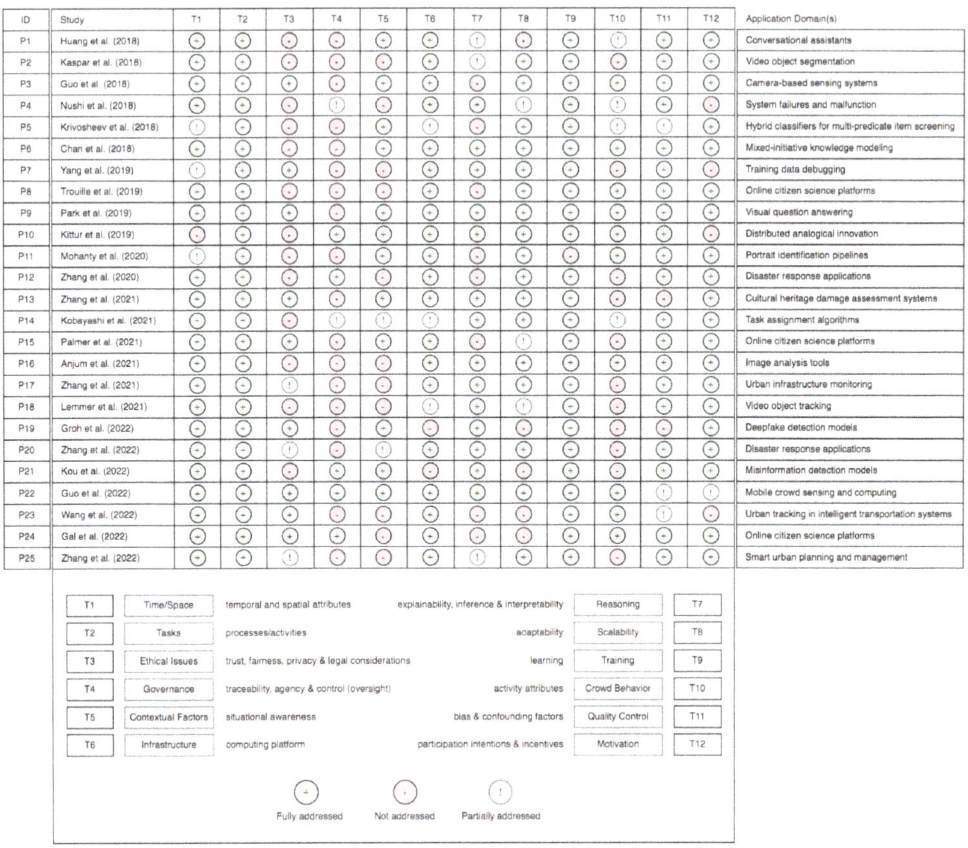

Figure 3. Synthesis of the literature analysis based on the taxonomy proposed.

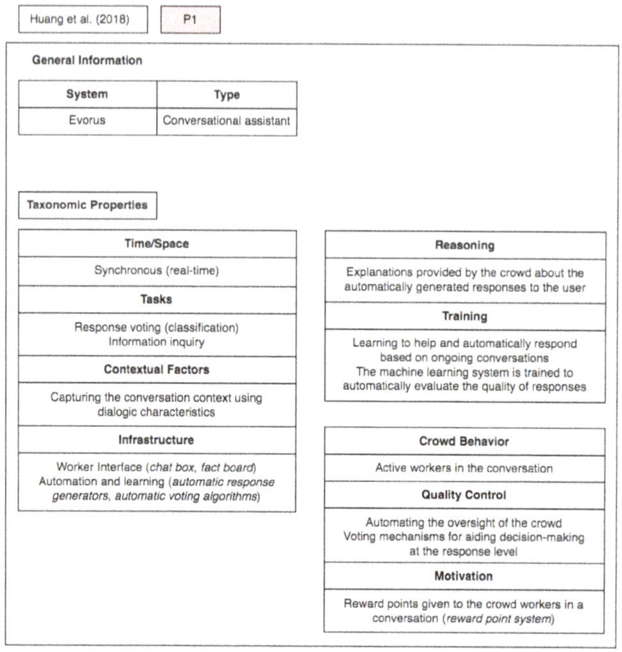

Figure 4. Example of a taxonomic scheme used to classify a crowd-AI interaction scenario [39].

Fully addressed: The manuscript clearly emphasizes the specific elements underlying the taxonomic category by addressing one or more of its unique attributes, with a potential experiment, solution, or case study demonstrating applicability. For instance, Mohanty and co-authors [38] make explicit reference to the contextual information (e.g., biographical details) provided to the user about each portrait in Photo Sleuth, a crowd-AI-enabled face recognition platform where a crowd of both expert and non-expert volunteers can tag a picture using this supplementary piece of contextual data to aid the decision process.

Not addressed: The work does not directly address any of the aspects that are inherent to the category under consideration.

Partially addressed: The study provides details that can be used to address the particular taxonomic category, even if not explicitly mentioned in the manuscript. By way of example, Kobayashi et al. [117] do not directly provide details about the contextual information required in the natural disaster response setting used for demonstrating the proposed method, but the situational awareness and subsequent timely information required to manage the rapidly evolving scenarios toward well-informed and up-to-date decision-making are implicitly stated.

On the basis of insights from previous analytical work, this taxonomically grounded literature review process has been adopted in areas like business intelligence and analytics [118] as a way of iteratively developing and refining taxonomic dimensions and characteristics while pinpointing areas requiring further investigation.

As can be seen from Figure 3, the taxonomy presented in this article is far from comprehensive enough to accommodate all types of possible scenarios involving crowd-AI interaction. Instead, the goal is to facilitate a cohesive understanding as a basis for further scrutiny of crowd-computing hybrids in real-world applicative contexts. Note that there are some categories that can co-exist, taking into account the specificity of each situation or use case. As such, the first taxonomic unit contains the spatio-temporal elements (T1) that frame crowd-AI interaction in relation to the original time-space matrix proposed by Johansen [12]. In brief terms, this classification model categorizes interactions as follows: same place/time, different places/same time, same place/different times, or different places/different times. To a broad extent, crowd-AI interactions can occur in asynchronous or real-time settings where the individuals that constitute the crowd can be physically and virtually co-located or geographically dispersed (remote). In addition, the worker location and task duration time [11] were also considered, as the latter is intimately connected to the time frame or limit that is set to complete a task. In the example provided in Figure 4, a nearly real-time on-demand crowd-powered system is proposed to collect responses from crowd workers that can be at any location but need to be available to provide contributions in real-time due to the quickly changing contextual requirements underlying the type of tasks performed. Looking at the results of the taxonomy-based literature review in detail, a total of 84% (n = 21) of included papers have reported temporal and/or spatial aspects of crowd activity. As a brief example, Chan and colleagues [119] introduced a mixed-initiative system with an annotation time of 1 min per paper on average in analogy matching tasks. In terms of real-time crowd-AI settings, some primary studies (e.g., [36,39,40,98,120]) presented synchronous interactions between crowd members, although most of the crowdsourcing systems relied on an asynchronous model.

Consistent with the previous literature, the most addressed taxonomic unit is related to task design, assignment, and execution (T2), with a total of 25 primary studies. In crowdsourcing experiments, task design is seen as a cornerstone to achieving the goals of a project or campaign since the characteristics and configuration of crowdsourced tasks influence the general outcomes obtained from the crowd [91]. In general, different types of tasks were found in the selected sample. As mentioned before, tasks differ both in terms of attributes, complexity, and granularity [11]. For instance, Scalpel-CD [121] generates label inspection microtasks in a dynamic way, while Evorus [39] focuses on classification tasks in the form of voting. A slightly different task specification is employed in Photo Sleuth [38], where crowd workers are invited to perform person identification/recognition tasks that are therefore augmented with visual tags to allow portrait seeking. Moreover, CollabLearn [36] is based on crowd query tasks where human processing is needed to highlight damaged areas from cultural heritage imagery. A somewhat related body of work (e.g., [34]) has sought to support the execution of crowd-in-the-loop interactive image labeling tasks with the ultimate goal of enhancing AI-powered damage scene assessment algorithms. All in all, the task-related aspects discussed in the growing literature on the interplay between crowdsourcing and AI systems have been playing an indispensable role in explaining complex relationships among crowd inputs and further integration into hybrid workflows.

Extrapolating to the ethical principles and standards in crowd-AI settings (T3), the review only identified nine papers (36%) that explicitly discuss ethical behaviors from a requester-, crowd- or even AI-centered standpoint. Despite the recognized need for fair payment and long-term career building in online crowd work platforms [122], this study shows that the ethical concerns underlying the interaction-centric crowd-AI activity are often overlooked from a practical perspective, despite some examples of strategies presented in the crowdsourcing literature such as ensuring fair compensation by paying crowd workers in conformity with the complexity of the task being performed [123]. Based on the findings from the chosen sample, Palmer and co-authors [124] provide one of the few examples of studies calling attention to possible unethical actions associated with the disclosure of sensitive information from images and videos. In a similar way, only 20% of primary studies (n = 5) fully describe machine and human (crowd) agency, governance practices, or control (oversight) (T4), although extensive research has been conducted about the potential risks and unintentional harms associated with the lack of an effective governance strategy able to regulate algorithmic actions [125]. In this regard, trust building [126,127] appears among the most critical factors affecting technology acceptance when considering human-AI interaction at a massive scale.

One enduring taxonomic unit that has been largely addressed since the very beginning of the field of CSCW is concerned with the contextual and situational information (T5) that is then used to support awareness about the environment in which the interaction takes place. This includes what goes on in the environment, who is available, who leaves, and how individuals "remain sensitive to the conduct of others so that an event or action, which may have some passing significance, can be displayed to each other without it necessarily gaining interactional or sequential import" [128]. If the entire sample is considered, 48% of studies (n = 12) mentioned some kind of contextual or situational issues. For instance, Huang et al. [39] proposed a crowd-machine hybrid system where the conversation context is used to provide response candidates using recorded facets and previous chat conversation logs. In particular, the task-specific contextual data is captured with the help of the crowd (by using chat logs) to improve the quality of responses based on current and past conversations. Moreover, Park and associates [129] used self-adapting mechanisms based on reinforcement learning (RL) and contextual features extracted to increase crowdsourcing participation over time, while Guo and co-workers [40] considered the lack of context as a determining factor for failure in smart environments.

Turning to the role of infrastructural support (T6) in interactive human-AI practices at a crowd level, the review disclosed a total of 20 studies (80%) where infrastructure or the characteristics of a crowd-computing platform are reported. In CSCW, the concept of 'infrastructure' and its ecological nature [130] has developed over the years to characterize socio-technical assemblages "that underpins and enables action, engagement, and awareness" [131]. On the basis of their research review, Hosseini and colleagues [29] gave a detailed description of the features that are commonly found in crowdsourcing platforms. In line with this, Santos and co-authors [102] stressed that a crowdsourcing system must provide functions and components able to support workflows involving actions such as task assignment, pre-selecting crowd workers, stating rewards, and selecting contributions. From payment mechanisms to result aggregation, a crowd-computing platform must combine crowd-, requester-, task- and platform-related information and facilities (i.e., infrastructural elements) that act in unison to carry out tasks in accordance with the different requirements. From an infrastructural perspective, Huang and associates [39] described the conversational worker interface used for chatting and real-time response modeling along with the automatic response voting and generating algorithms deployed to operate in a continuous manner as the conversation continues. Using a crowd-AI hybrid intelligence lens, the results showed a total of 14 studies addressing algorithmic reasoning, inference, explainability, and interpretability (T7). For instance, human-AI decision-making processes are complex by nature, and AI-infused systems require a certain level of explainability [132] and interpretability [133] to provide insights about the algorithmic actions taken during the AI-enabled experience. However, several studies agree that these explanations must manifestly be comprehensible, transparent, and actionable (i.e., how humans use or find the explanations useful) to ensure traceability and trust in AI-advised crowd decision-making [134]. Moreover, incorporating reasoning capabilities into hybrid intelligence systems at a massive scale can provide support for better decisions since RL and related algorithms can learn from crowd behavior [104] while offering a lot of possibilities to improve decision-making at a large scale.

This points to the notions of scalability and adaptability (T8) and their importance in highly dynamic and unpredictable environments. Due to their flexibility, hybrid crowd-algorithm methods represent a means of handling complexity and gathering high-quality training data. From the entire sample, 17 studies (68%) addressed scalability and/or crowd-AI adaptability. As an example, Anjum et al. [135] stressed the value of scalable image annotation, while Trouille and co-authors [136] have drawn attention to scalable application programming interfaces with the ability to quickly configure a citizen science campaign. A further focus of the taxonomic-based review presented here is on the learning and training processes (T9) behind the current AI models. In crowd-machine settings, humans may "feed" the algorithm to act in situ in an automatic fashion based on data inputs that can work as training samples [137]. On this point, 96% of included studies (n = 24) addressed aspects related to this taxonomic unit. For instance, Kaspar and colleagues [35] proposed a crowd-

AI hybrid workflow in which the training data is generated through video segmentation. Further expanding the scope, a related important question is how to train the crowd itself when an AI output is used [117]. Accordingly, Zhang and associates [36,120] call for more research into aspects like AI bias mitigation and the detection of imperfect or biased inputs from the crowd as factors that may compromise the system's reliability. A look at the work conducted by Huang et al. [39] denotes that the machine learning model that works behind the conversational assistant proposed is fed with training data from past up/down votes given by crowd workers. This continuous learning approach allows optimization of the entire automatic voting process based on the assessment of the quality of the human responses.

Stemming from the literature of social and behavioral sciences, the extraction of behavior features from crowd activity (T10) has been particularly relevant to unravel the complexities of crowdsourcing practice and improving the synergistic interaction between humans (crowds) and algorithms. However, the results from this scoping review show that only 40 percent of the literature sample (n = 10) focused on aspects of crowd activity from a behavioral standpoint. Building on the collective intelligence genome [23], the understanding of what, why, who, how, and the circumstances under which such interaction takes place can be enhanced through the behavioral analysis of traces of past activity [138,139]. In hybrid crowd-algorithm interactive settings, user activity tracking involving keystroke, eye tracking, time duration, and mouse click recording (e.g., window resizing) can contribute to the cognitive, physical, and perceptual augmentation of the crowd with practical implications for improving task assignment, performance estimation, and worker pre-selection and/or recommendation based on reliability measures [140–143]. From a behavioral point of view, identifying active workers can play a critical role in systems such as Evorus [39] since the model strongly depends on human inputs, while capturing crowd members' meta-information is important to personalize the experience to the user in more intelligent ways. Although the development of AI systems supported by online interfaces able to log user actions has a great capacity to conduct behavior analysis [144], recent research works (e.g., [145]) have shown that there are a lot of resources required to realize the effective capture of these behavioral traces from an infrastructural lens.

A closely related line of investigation involves the quality control mechanisms (T11) that are used in crowdsourcing systems to reduce the occurrence of inaccuracies and biased inputs provided by malicious (or poorly motivated) crowd workers. Empirically, this work shows that there were only five papers (20%) that did not explicitly report strategies for ensuring quality control and modeling crowd bias. In general terms, quality control strategies for detecting low-quality work can vary from input and output agreement to majority voting/consensus, ground truth (e.g., gold standard questions), contributor evaluation, expert review, real-time support, or even fine-grained behavioral traces [146]. Yet, as pointed out by Daniel and co-authors [112] and further developed by Jin et al. [86], a quality assessment process can be performed computationally (e.g., task execution log analysis), collaboratively (e.g., peer review), or even individually (e.g., qualification test). Regarding the latter, worker pre-selection has been used by requesters as a common approach to filter unqualified workers by taking into consideration factors like reputation and credentials. In the example of the scenario shown in Figure 4, the system has a high error tolerance for imperfect automated actions from voting algorithms and chatbots since the oversight is done by the (human) crowd.

Throughout the last decades, several scholars have stressed the importance of motivational factors (T12) as a quality assurance determinant and also a catalyst for sustained participation in crowdsourcing [147]. Briefly, the taxonomy-based review identified 20 primary studies (80%) addressing motivation and incentive mechanisms regarding the use of algorithmic systems powered by crowdsourcing techniques. This includes extrinsic incentives (e.g., immediate payoffs) and also intrinsic (hedonic) motives like inherent satisfaction and entertainment [112]. For example, Evorus [39] provides a continuously updated scoreboard that displays the reward points given to each crowd worker according to his/her performance on a particular task, where the value is automatically converted into a monetary bonus. As Truong et al. [148] have noted, crowdsourcing contests are also considered intuitive ways for incentivizing crowd workers and are frequently used in macrotask crowdsourcing for solving problems with an elevated degree of complexity [81,149]. In general terms, the incentives reported in the literature range from monetary rewards to gifts and gamification strategies [112]. Concerning the former, the review presented here also provides a summary of the primary studies from the sample that presented experimental work based on monetary rewards. As Table 1 depicts, 60% of the papers included in the taxonomy-based literature review (n = 15) have reported paid experiments in remote settings. For paid crowdsourcing experiments where the crowd had to execute the whole experiment remotely, this part of the analysis considered the time allotted, pre-selection mechanism(s), crowd size, platform(s) used, and reward in terms of cost per HIT in US Dollars ($). This is in line with previous studies (e.g., [91]) reporting aspects related to the several stages of experimental design in crowdsourcing settings.

Table 1. Methodological remarks extracted from primary studies reporting paid crowdsourcing experiments conducted remotely.

ID	Experimental Settings	Pre-Selection Mechanism(s)	Cost per HIT and Platform(s)	Time Allotted
P1	5-month-long deployment and testing with real users (n = 80 crowd workers)	-	$0.142 (Phase-1 deployment); $0.211 (Control Phase); MTurk; Hangoutsbot	~10 min (per conversation)
P2	Ensemble method combining multiple results on individual frame segmentations and crowd-based propagated segmentation results (n = 70 crowd workers)	-	$0.90 (Segmentation); $0.15 (Scribble); MTurk	142.6 s (per frame segmentation); 2.5 s (per method scribbles)
P3	4-week testing (n = 17 participants), with an unspecified number of crowd workers	>95% assignment approval rate; Gold standard question sensor instances	~$10/hour ($0.02 for each task performed on MTurk)	~3 s (per labeled question sensor instance)
P5	Classification of potential studies for a systematic literature review (n = 147 crowd workers)	>70% overall accuracy; Worker screening based on two test questions	$10/hour; MTurk	-
P6	Purpose-mechanism annotation analogical search (n = 3 crowd workers per document), with an unspecified number of crowd workers	>=95% acceptance rate; Training step based on a gold standard example before the task execution	$30/hour (Upwork-worker 1); $20/hour (Upwork-worker 2); $10/hour ($0.70 for each task performed on MTurk)	1.3 min (per document annotation); 4 min (overall task completion)
P9	Contextual bandit algorithm and agent deployment powered by AI-based request strategies for visual question answering, with an unspecified number of crowd workers	Training step using examples and a qualifying task	$12/hour; MTurk	-
P13	Performance evaluation of a crowd-AI hybrid framework through real-world datasets (n = 3 crowd workers per image in a crowd query) with an unspecified number of crowd workers	>95% overall task approval rate; >=1000 HITs completed	$0.20 for each worker per-image annotation; Labelme; MTurk	-
P14	A method for AI worker evaluation that uses a "divide-and-conquer" strategy for dynamic task assignment with an unspecified number of crowd workers	No strategies were deployed to target malicious workers	240$ for 2 h of labor; MTurk	-
P16	Evaluation of hybrid crowd-algorithmic workflows for image annotation based on time completion and quality, with an unspecified number of crowd workers	>92% approval rate; >500 HITs completed	$9/hour ($0.20 for each task performed on MTurk)	80 s (per HIT completion)

Table 1. *Cont.*

ID	Experimental Settings	Pre-Selection Mechanism(s)	Cost per HIT and Platform(s)	Time Allotted
P17	Evaluation of crowd responses and computational performance in identifying damages from urban infrastructure imagery data (n = 2 to 5 crowd workers per query), with an unspecified number of crowd workers	>95% overall task approval rate; >=1000 HITs completed	$0.05 for each worker per image classification; MTurk	0.0227 (average time taken to accomplish each streaming urban monitoring task using a hybrid crowd-AI model)
P18	Evaluation of model performance to re-query or not crowdsourced initializations for bounding-box annotations (n = 26 crowd workers located in the United States)	A gold standard for identifying inattentive workers; Annotators with more than 15% incorrect annotations were disregarded	~$12/hour ($0.06 for each bounding-box annotation); MTurk	-
P19	Randomized online experiments comparing the performance of a computer vision model and a crowd of 15,016 individuals in tasks related to the detection of authentic vs. deepfake videos (n = 5524 participants: Experiment 1; n = 9492 participants: Experiment 2)	-	$7.28/hour plus bonus payments of 20% to the top participants; Experiment hosted on an external website (i.e., Detect Fakes); 304 participants recruited from Prolific	15 min (per task completion)
P20	Performance evaluation of a dynamic optimal neural architecture searching framework that leverages crowdsourcing for handling disaster damage assessment problems with an unspecified number of crowd workers	>95% overall task approval rate; >=1000 HITs completed	$0.20 for each crowd worker per-image labeling; MTurk	0.0198 s (average time with varying crowd query frequency); 0.0201 s (average time with varying numbers of crowd workers)
P21	Evaluation of a hybrid framework combining expert and crowd intelligence with explainable AI for misinformation detection (n = 3 crowd workers per HIT plus 5 experts), with an unspecified number of crowd workers	>=95% task acceptance rate	Unspecified amount above the minimum requirement on MTurk ($0.01 per assignment)	61 s (average time of task completion); 21.4 h (total waiting time to collect and aggregate contributions from crowd workers)
P25	Development of a crowd-AI system for optimizing smart urban sensing applications (n = 3 to 7 crowd workers per task), with an unspecified number of crowd workers	>95% overall task approval rate; >=1000 HITs completed	$0.05 for each crowd worker per image classification; MTurk	-

Regarding the filtering mechanisms used for early pre-selection of crowd workers, the review of the literature showed five studies where the HIT acceptance rate was set to more than 95%. Moreover, this contribution also identified four studies where the number of tasks completed by a potential crowd worker had to be at least 1000. From this scoping review, a total of five experiments involved some type of ground truth in the form of a gold standard or test question. The selected sample also contained cases in which no pre-selection strategies were applied, while one of the experiments disregarded crowd workers with more than 15 percent of incorrect answers. It is also worth noting that one of the primary studies contained workers located in the United States only. Taken all together, the utilization of these pre-selection techniques can be useful to specify the characteristics of potential contributors improve the likelihood that only skilled, high performing, and/or trustworthy crowd workers are allowed to participate. When considering the platforms used to recruit participants, the results show a clear preference for the use of MTurk (n = 14). Although some tasks were paid up to $0.20, some workers only received $0.05 per task performed. Going back to the payment imbalances and unfair compensation that challenge ethical norms in crowdsourcing marketplaces [150,151], a lens into the literature has revealed that there is an increasing awareness of the crowd worker's conditions and that the monetary compensation must be set in a fair manner when adopting crowdsourcing for tasks such as data collection and analysis. Overall, this study also revealed different average times of HIT completion in accordance with the complexity and requirements of each task, while a remarkable number of primary studies (n = 10) did not mention the total number of crowd workers involved in the experiment. Nonetheless, some studies involve both crowd workers and experts in their experimental settings, with a crowd size ranging from 2 to 7 crowd workers per task and a maximum size of 147 paid online workers in a single experiment.

6. Concluding Discussion and Challenges Ahead

Owing to the difficulty in handling problems of increasing complexity involving noisy and complex data streams, hybrid crowd-machine interactive workflows have been implemented to efficiently scale training data and parameter models in order to produce insights and support decision-making processes in a way that was not possible using conventional methods. In various problem domains, new patterns can be identified from complex decision rules for further verification in a human-in-the-loop basis encapsulated in crowd-AI systems and architectures able to support tasks like content regulation and medical diagnosis. Considering the latter, machine learning skills are now increasingly crowdsourced in the form of contests or competitions running on predictive modeling and analytics services where both monetary and non-monetary incentives are used to aggregate crowd knowledge and thus help to better streamline the early detection and treatment processes that are critical in healthcare settings. However, building trust in crowd-machine interaction while making AI more efficient and adaptable are among the prevalent challenges in crowdsourcing and are usually seen as hindering factors for the successful adoption and use of these systems in practice.

In this study, an initial taxonomy of crowd-AI hybrid interaction was proposed as a guiding framework for system developers, public and private health professionals, scientists, and other stakeholders worldwide interested in this emerging area. Despite the contribution towards a comprehensive scheme to explain how crowd-machine hybrid interaction has been addressed in various scenarios presented in the literature, this article constitutes only one piece of a much larger puzzle. In other words, the information obtained from work presented here is considered a basis for further expansions and testing scenarios in real-world contexts in the form of continuous observation of the co-evolving relations between humans and algorithms with the goal of informing the design of intelligent systems adequately and cohesively. Framing a territory in constant expansion like crowd-AI hybrids is a challenging task. Overall, the taxonomy-based review found a gap in terms of understanding, both empirically and conceptually, the role of ethical principles and perceived fairness in building and deploying AI responsibly and with adequate governance strategies. This study also shows that more experimentation and additional investigative steps will be needed to cope with inconsistent records from crowd workers. Moreover, there are also a number of directions for future work that should be beneficial to extend in the near future for new types of research practices involving crowd-computing hybrids so that scientific institutions, companies, and the general public can all benefit from the knowledge generated from this convergence and therefore better respond to the volatile nature and changing demands of the current environments.

Author Contributions: Conceptualization, methodology, formal analysis & writing–original draft, A.C.; supervision, A.G.; writing–review and editing, D.S.; investigation, A.P.P.; investigation, R.C.; investigation, M.A.d.A.; supervision & validation, B.F. All authors have read and agreed to the published version of the manuscript.

Funding: This research has been sponsored by National Funds through FLAD—Luso-American Development Foundation and FCT—Portuguese Foundation for Science and Technology. The work of António Correia is supported by FCT grant SFRH/BD/136211/2018.

Institutional Review Board Statement: Not applicable.

Informed Consent Statement: Not applicable.

Data Availability Statement: Not applicable.

Conflicts of Interest: The authors declare that there are no conflict of interest or competing financial interests related to this manuscript.

Appendix A

Table A1. List of primary studies included in the taxonomic validation process.

ID	Author(s)	Year	Title
P1	Huang et al.	2018	Evorus: A crowd-powered conversational assistant built to automate itself over time
P2	Kaspar et al.	2018	Crowd-guided ensembles: How can we choreograph crowd workers for video segmentation?
P3	Guo et al.	2018	Crowd-AI camera sensing in the real world
P4	Nushi et al.	2018	Towards accountable AI: Hybrid human-machine analyses for characterizing system failure
P5	Krivosheev et al.	2018	Combining crowd and machines for multi-predicate item screening
P6	Chan et al.	2018	SOLVENT: A mixed initiative system for finding analogies between research papers
P7	Yang et al.	2019	Scalpel-CD: Leveraging crowdsourcing and deep probabilistic modeling for debugging noisy training data
P8	Trouille et al.	2019	Citizen science frontiers: Efficiency, engagement, and serendipitous discovery with human-machine systems
P9	Park et al.	2019	AI-based request augmentation to increase crowdsourcing participation
P10	Kittur et al.	2019	Scaling up analogical innovation with crowds and AI
P11	Mohanty et al.	2020	Photo Sleuth: Identifying historical portraits with face recognition and crowdsourced human expertise
P12	Zhang et al.	2020	Crowd-assisted disaster scene assessment with human-AI interactive attention
P13	Zhang et al.	2021	CollabLearn: An uncertainty-aware crowd-AI collaboration system for cultural heritage damage assessment
P14	Kobayashi et al.	2021	Human+AI crowd task assignment considering result quality requirements
P15	Palmer et al.	2021	Citizen science, computing, and conservation: How can "Crowd AI" change the way we tackle large-scale ecological challenges?
P16	Anjum et al.	2021	Exploring the use of deep learning with crowdsourcing to annotate images
P17	Zhang et al.	2021	StreamCollab: A streaming crowd-AI collaborative system to smart urban infrastructure monitoring in social sensing
P18	Lemmer et al.	2021	Crowdsourcing more effective initializations for single-target trackers through automatic re-querying
P19	Groh et al.	2022	Deepfake detection by human crowds, machines, and machine-informed crowds
P20	Zhang et al.	2022	On streaming disaster damage assessment in social sensing: A crowd-driven dynamic neural architecture searching approach
P21	Kou et al.	2022	Crowd, expert & AI: A human-AI interactive approach towards natural language explanation based COVID-19 misinformation detection
P22	Guo et al.	2022	CrowdHMT: Crowd intelligence with the deep fusion of human, machine, and IoT
P23	Wang et al.	2022	Graph optimized data offloading for crowd-AI hybrid urban tracking in intelligent transportation systems
P24	Gal et al.	2022	A new workflow for human-AI collaboration in citizen science
P25	Zhang et al.	2022	CrowdOptim: A crowd-driven neural network hyperparameter optimization approach to AI-based smart urban sensing

Table A2. Distribution of publications per venue.

Conference Proceedings	AAAI Conference on Artificial Intelligence AAAI Conference on Human Computation and Crowdsourcing (4) ACM Conference on Human Factors in Computing Systems (3) ACM Conference on Information Technology for Social Good ACM Web Conference International Joint Conference on Artificial Intelligence
Journal/Transactions	ACM Transactions on Interactive Intelligent Systems Human Computation (2) IEEE Internet of Things Journal IEEE Transactions on Computational Social Systems IEEE Transactions on Intelligent Transportation Systems Knowledge-Based Systems Proceedings of the ACM on Human-Computer Interaction (3) Proceedings of the ACM on Interactive, Mobile, Wearable, and Ubiquitous Technologies Proceedings of the National Academy of Sciences (3)

References

1. Lofi, C.; El Maarry, K. Design patterns for hybrid algorithmic-crowdsourcing workflows. In Proceedings of the IEEE 16th Conference on Business Informatics, Geneva, Switzerland, 14–17 July 2014; pp. 1–8.
2. Heim, E.; Roß, T.; Seitel, A.; März, K.; Stieltjes, B.; Eisenmann, M.; Lebert, J.; Metzger, J.; Sommer, G.; Sauter, A.W.; et al. Large-scale medical image annotation with crowd-powered algorithms. *J. Med. Imaging* **2018**, *5*, 034002. [CrossRef] [PubMed]
3. Vargas-Santiago, M.; Monroy, R.; Ramirez-Marquez, J.E.; Zhang, C.; Leon-Velasco, D.A.; Zhu, H. Complementing solutions to optimization problems via crowdsourcing on video game plays. *Appl. Sci.* **2020**, *10*, 8410. [CrossRef]
4. Bharadwaj, A.; Gwizdala, D.; Kim, Y.; Luther, K.; Murali, T.M. Flud: A hybrid crowd–algorithm approach for visualizing biological networks. *ACM Trans. Comput. Interact.* **2022**, *29*, 1–53. [CrossRef]
5. Grudin, J.; Poltrock, S. Taxonomy and theory in computer supported cooperative work. *Oxf. Handb. Organ. Psychol.* **2012**, *2*, 1323–1348. [CrossRef]
6. Nickerson, R.C.; Varshney, U.; Muntermann, J. A method for taxonomy development and its application in information systems. *Eur. J. Inf. Syst.* **2013**, *22*, 336–359. [CrossRef]
7. Harris, A.M.; Gómez-Zará, D.; DeChurch, L.A.; Contractor, N.S. Joining together online: The trajectory of CSCW scholarship on group formation. *Proc. ACM Hum.-Comput. Interact.* **2019**, *3*, 1–27. [CrossRef]
8. McGrath, J.E. *Groups: Interaction and Performance*; Prentice-Hall: Englewood Cliffs, NJ, USA, 1984.
9. Shaw, M.E. Scaling group tasks: A method for dimensional analysis. *JSAS Cat. Sel. Doc. Psychol.* **1973**, *3*, 8.
10. Modaresnezhad, M.; Iyer, L.; Palvia, P.; Taras, V. Information technology (IT) enabled crowdsourcing: A conceptual framework. *Inf. Process. Manag.* **2020**, *57*, 102135. [CrossRef]
11. Bhatti, S.S.; Gao, X.; Chen, G. General framework, opportunities and challenges for crowdsourcing techniques: A comprehensive survey. *J. Syst. Softw.* **2020**, *167*, 110611. [CrossRef]
12. Johansen, R. *Groupware: Computer Support for Business Teams*; The Free Press: New York, NY, USA, 1988.
13. Lee, C.P.; Paine, D. From the matrix to a model of coordinated action (MoCA): A conceptual framework of and for CSCW. In Proceedings of the 18th ACM Conference on Computer Supported Cooperative Work & Social Computing, Vancouver, BC, Canada, 14–18 March 2015; pp. 179–194.
14. Renyi, M.; Gaugisch, P.; Hunck, A.; Strunck, S.; Kunze, C.; Teuteberg, F. Uncovering the complexity of care networks—Towards a taxonomy of collaboration complexity in homecare. *Comput. Support. Cooperative Work. (CSCW)* **2022**, *31*, 517–554. [CrossRef]
15. Thomer, A.K.; Twidale, M.B.; Yoder, M.J. Transforming taxonomic interfaces: "Arm's length" cooperative work and the maintenance of a long-lived classification system. *Proc. ACM Hum.-Comput. Interact.* **2018**, *2*, 1–23. [CrossRef]
16. Akata, Z.; Balliet, D.; de Rijke, M.; Dignum, F.; Dignum, V.; Eiben, G.; Fokkens, A.; Grossi, D.; Hindriks, K.V.; Hoos, H.H.; et al. A research agenda for hybrid intelligence: Augmenting human intellect with collaborative, adaptive, responsible, and explainable artificial intelligence. *Computer* **2020**, *53*, 18–28. [CrossRef]
17. Pescetelli, N. A brief taxonomy of hybrid intelligence. *Forecasting* **2021**, *3*, 633–643. [CrossRef]
18. Dellermann, D.; Calma, A.; Lipusch, N.; Weber, T.; Weigel, S.; Ebel, P. The future of human-AI collaboration: A taxonomy of design knowledge for hybrid intelligence systems. In Proceedings of the 52nd Hawaii International Conference on System Sciences, Maui, HI, USA, 8–11 January 2019; pp. 274–283.
19. Dubey, A.; Abhinav, K.; Jain, S.; Arora, V.; Puttaveerana, A. HACO: A framework for developing human-AI teaming. In Proceedings of the 13th Innovations in Software Engineering Conference, Jabalpur, India, 27–29 February 2020; pp. 1–9.
20. Littmann, M.; Suomela, T. Crowdsourcing, the great meteor storm of 1833, and the founding of meteor science. *Endeavour* **2014**, *38*, 130–138. [CrossRef]

21. Corney, J.R.; Torres-Sánchez, C.; Jagadeesan, A.P.; Regli, W.C. Outsourcing labour to the cloud. *Int. J. Innovation Sustain. Dev.* **2009**, *4*, 294–313. [CrossRef]
22. Rouse, A.C. A preliminary taxonomy of crowdsourcing. In Proceedings of the Australasian Conference on Information Systems, Brisbane, Australia, 1–3 December 2010; Volume 76.
23. Malone, T.W.; Laubacher, R.; Dellarocas, C. The collective intelligence genome. *IEEE Eng. Manag. Rev.* **2010**, *38*, 38–52. [CrossRef]
24. Zwass, V. Co-creation: Toward a taxonomy and an integrated research perspective. *Int. J. Electron. Commer.* **2010**, *15*, 11–48. [CrossRef]
25. Doan, A.; Ramakrishnan, R.; Halevy, A.Y. Crowdsourcing systems on the world-wide web. *Commun. ACM* **2011**, *54*, 86–96. [CrossRef]
26. Saxton, G.D.; Oh, O.; Kishore, R. Rules of crowdsourcing: Models, issues, and systems of control. *Inf. Syst. Management* **2013**, *30*, 2–20. [CrossRef]
27. Nakatsu, R.T.; Grossman, E.B.; Iacovou, C.L. A taxonomy of crowdsourcing based on task complexity. *J. Inf. Sci.* **2014**, *40*, 823–834. [CrossRef]
28. Gadiraju, U.; Kawase, R.; Dietze, S. A taxonomy of microtasks on the web. In Proceedings of the 25th ACM Conference on Hypertext and Social Media, Santiago, Chile, 1–4 September 2014; pp. 218–223.
29. Hosseini, M.; Shahri, A.; Phalp, K.; Taylor, J.; Ali, R. Crowdsourcing: A taxonomy and systematic mapping study. *Comput. Sci. Rev.* **2015**, *17*, 43–69. [CrossRef]
30. Alabduljabbar, R.; Al-Dossari, H. Towards a classification model for tasks in crowdsourcing. In Proceedings of the Second International Conference on Internet of Things and Cloud Computing, Cambridge, UK, 22–23 March 2017; pp. 1–7.
31. Chen, Q.; Magnusson, M.; Björk, J. Exploring the effects of problem- and solution-related knowledge sharing in internal crowdsourcing. *J. Knowl. Manag.* **2022**, *26*, 324–347. [CrossRef]
32. Chilton, L.B.; Little, G.; Edge, D.; Weld, D.S.; Landay, J.A. Cascade: Crowdsourcing taxonomy creation. In Proceedings of the CHI Conference on Human Factors in Computing Systems, Paris, France, 27 April–2 May 2013; pp. 1999–2008.
33. Sharif, A.; Gopal, P.; Saugstad, M.; Bhatt, S.; Fok, R.; Weld, G.; Dey, K.A.M.; Froehlich, J.E. Experimental crowd+AI approaches to track accessibility features in sidewalk intersections over time. In Proceedings of the 23rd International ACM SIGACCESS Conference on Computers and Accessibility, Virtual Event, 18–22 October 2021; pp. 1–5.
34. Zhang, D.Y.; Huang, Y.; Zhang, Y.; Wang, D. Crowd-assisted disaster scene assessment with human-AI interactive attention. In Proceedings of the Thirty-Fourth AAAI Conference on Artificial Intelligence, New York, NY, USA, 7–12 February 2020; pp. 2717–2724.
35. Kaspar, A.; Patterson, G.; Kim, C.; Aksoy, Y.; Matusik, W.; Elgharib, M. Crowd-guided ensembles: How can we choreograph crowd workers for video segmentation? In Proceedings of the CHI Conference on Human Factors in Computing Systems, Montreal, QC, Canada, 21–26 April 2018.
36. Zhang, Y.; Zong, R.; Kou, Z.; Shang, L.; Wang, D. CollabLearn: An uncertainty-aware crowd-AI collaboration system for cultural heritage damage assessment. *IEEE Trans. Comput. Soc. Syst.* **2021**, *9*, 1515–1529. [CrossRef]
37. Maier-Hein, L.; Ross, T.; Gröhl, J.; Glocker, B.; Bodenstedt, S.; Stock, C.; Heim, E.; Götz, M.; Wirkert, S.J.; Kenngott, H.; et al. Crowd-algorithm collaboration for large-scale endoscopic image annotation with confidence. In Proceedings of the 19th International Conference on Medical Image Computing and Computer-Assisted Intervention, Athens, Greece, 17–21 October 2016; pp. 616–623.
38. Mohanty, V.; Thames, D.; Mehta, S.; Luther, K. Photo Sleuth: Combining human expertise and face recognition to identify historical portraits. In Proceedings of the 24th International Conference on Intelligent User Interfaces, Marina del Ray, CA, USA, 17–20 March 2019; pp. 547–557.
39. Huang, T.H.; Chang, J.C.; Bigham, J.P. Evorus: A crowd-powered conversational assistant built to automate itself over time. In Proceedings of the CHI Conference on Human Factors in Computing Systems, Montreal, QC, Canada, 21–26 April 2018; p. 295.
40. Guo, A.; Jain, A.; Ghose, S.; Laput, G.; Harrison, C.; Bigham, J.P. Crowd-AI camera sensing in the real world. *Proc. ACM Interactive, Mobile, Wearable Ubiquitous Technol.* **2018**, *2*, 1–20. [CrossRef]
41. Correia, A.; Paredes, H.; Schneider, D.; Jameel, S.; Fonseca, B. Towards hybrid crowd-AI centered systems: Developing an integrated framework from an empirical perspective. In Proceedings of the IEEE International Conference on Systems, Man and Cybernetics, Bari, Italy, 6–9 October 2019; pp. 4013–4018.
42. Xu, W.; Dainoff, M.J.; Ge, L.; Gao, Z. Transitioning to human interaction with AI systems: New challenges and opportunities for HCI professionals to enable human-centered AI. *Int. J. Human–Computer Interact.* **2022**, *39*, 494–518. [CrossRef]
43. Colazo, M.; Alvarez-Candal, A.; Duffard, R. Zero-phase angle asteroid taxonomy classification using unsupervised machine learning algorithms. *Astron. Astrophys.* **2022**, *666*, A77. [CrossRef]
44. Mock, F.; Kretschmer, F.; Kriese, A.; Böcker, S.; Marz, M. Taxonomic classification of DNA sequences beyond sequence similarity using deep neural networks. *Proc. Natl. Acad. Sci. USA* **2022**, *119*, e2122636119. [CrossRef]
45. Rasch, R.F. The nature of taxonomy. *Image J. Nurs. Scholarsh.* **1987**, *19*, 147–149. [CrossRef]
46. Tricco, A.C.; Lillie, E.; Zarin, W.; O'Brien, K.; Colquhoun, H.; Kastner, M.; Levac, D.; Ng, C.; Sharpe, J.P.; Wilson, K.; et al. A scoping review on the conduct and reporting of scoping reviews. *BMC Med. Res. Methodol.* **2016**, *16*, 15. [CrossRef]
47. Sokal, R.R. Phenetic taxonomy: Theory and methods. *Annu. Rev. Ecol. Syst.* **1986**, *17*, 423–442. [CrossRef]
48. Oberländer, A.M.; Lösser, B.; Rau, D. Taxonomy research in information systems: A systematic assessment. In Proceedings of the 27th European Conference on Information Systems, Stockholm and Uppsala, Sweden, 8–14 June 2019.

49. Gerber, A. Computational ontologies as classification artifacts in IS research. In Proceedings of the 24th Americas Conference on Information Systems, New Orleans, LA, USA, 16–18 August 2018.
50. Webster, J.; Watson, R.T. Analyzing the past to prepare for the future: Writing a literature review. *MIS Q.* **2002**, *26*, 2.
51. Schmidt-Kraepelin, M.; Thiebes, S.; Tran, M.C.; Sunyaev, A. What's in the game? Developing a taxonomy of gamification concepts for health apps. In Proceedings of the 51st Hawaii International Conference on System Sciences, Hilton Waikoloa Village, HI, USA, 3–6 January 2018; pp. 1–10.
52. Sai, A.R.; Buckley, J.; Fitzgerald, B.; Le Gear, A. Taxonomy of centralization in public blockchain systems: A systematic literature review. *Inf. Process. Manag.* **2021**, *58*, 102584. [CrossRef]
53. Andraschko, L.; Wunderlich, P.; Veit, D.; Sarker, S. Towards a taxonomy of smart home technology: A preliminary understanding. In Proceedings of the 42nd International Conference on Information Systems, Austin, TX, USA, 12–15 December 2021.
54. Larsen, K.R.; Hovorka, D.; Dennis, A.; West, J.D. Understanding the elephant: The discourse approach to boundary identification and corpus construction for theory review articles. *J. Assoc. Inf. Syst.* **2019**, *20*, 15. [CrossRef]
55. Elliott, J.H.; Turner, T.; Clavisi, O.; Thomas, J.; Higgins, J.P.T.; Mavergames, C.; Gruen, R.L. Living systematic reviews: An emerging opportunity to narrow the evidence-practice gap. *PLoS Med.* **2014**, *11*, e1001603. [CrossRef]
56. Singh, V.K.; Singh, P.; Karmakar, M.; Leta, J.; Mayr, P. The journal coverage of Web of Science, Scopus and Dimensions: A comparative analysis. *Scientometrics* **2021**, *126*, 5113–5142. [CrossRef]
57. Kittur, A.; Nickerson, J.V.; Bernstein, M.; Gerber, E.; Shaw, A.; Zimmerman, J.; Lease, M.; Horton, J.J. The future of crowd work. In Proceedings of the ACM Conference on Computer Supported Cooperative Work, San Antonio, TX, USA, 23–27 February 2013; pp. 1301–1318.
58. Zhang, D.; Zhang, Y.; Li, Q.; Plummer, T.; Wang, D. CrowdLearn: A crowd-AI hybrid system for deep learning-based damage assessment applications. In Proceedings of the 39th IEEE International Conference on Distributed Computing Systems, Dallas, TX, USA, 7–10 July 2019; pp. 1221–1232.
59. Landolt, S.; Wambsganss, T.; Söllner, M. A taxonomy for deep learning in natural language processing. In Proceedings of the 54th Hawaii International Conference on System Sciences, Kauai, HI, USA, 5 January 2021; pp. 1061–1070.
60. Straus, S.G. Testing a typology of tasks: An empirical validation of McGrath's (1984) group task circumplex. *Small Group Research* **1999**, *30*, 166–187. [CrossRef]
61. Chesbrough, H.W. *Open Innovation: The New Imperative for Creating and Profiting from Technology*; Harvard Business Press: Boston, MA, USA, 2003.
62. Karachiwalla, R.; Pinkow, F. Understanding crowdsourcing projects: A review on the key design elements of a crowdsourcing initiative. *Creativity Innov. Manag.* **2021**, *30*, 563–584. [CrossRef]
63. Hemmer, P.; Schemmer, M.; Vössing, M.; Kühl, N. Human-AI complementarity in hybrid intelligence systems: A structured literature review. In Proceedings of the 25th Pacific Asia Conference on Information Systems, Virtual Event, Dubai, United Arab Emirates, 12–14 July 2021; p. 78.
64. Holstein, K.; Aleven, V.; Rummel, N. A conceptual framework for human-AI hybrid adaptivity in education. In Proceedings of the 21st International Conference on Artificial Intelligence in Education, Ifrane, Morocco, 6–10 July 2020; pp. 240–254.
65. Siemon, D. Elaborating team roles for artificial intelligence-based teammates in human-AI collaboration. *Group Decis. Negot.* **2022**, *31*, 871–912. [CrossRef]
66. Weber, E.; Marzo, N.; Papadopoulos, D.P.; Biswas, A.; Lapedriza, A.; Ofli, F.; Imran, M.; Torralba, A. Detecting natural disasters, damage, and incidents in the wild. In Proceedings of the 16th European Conference on Computer Vision, Glasgow, UK, 23–28 August 2020; pp. 331–350.
67. Vaughan, J.W. Making better use of the crowd: How crowdsourcing can advance machine learning research. *J. Mach. Learn. Res.* **2017**, *18*, 7026–7071.
68. Hamadi, R.; Ghazzai, H.; Massoud, Y. A generative adversarial network for financial advisor recruitment in smart crowdsourcing platforms. *Appl. Sci.* **2022**, *12*, 9830. [CrossRef]
69. Alter, S. Work system theory: Overview of core concepts, extensions, and challenges for the future. *J. Assoc. Inf. Syst.* **2013**, *14*, 2. [CrossRef]
70. Venumuddala, V.R.; Kamath, R. Work systems in the Indian information technology (IT) industry delivering artificial intelligence (AI) solutions and the challenges of work from home. *Inf. Syst. Front.* **2022**, 1–25. [CrossRef] [PubMed]
71. Nardi, B. *Context and Consciousness: Activity Theory and Human-Computer Interaction*; MIT Press: Cambridge, MA, USA, 1996.
72. Neale, D.C.; Carroll, J.M.; Rosson, M.B. Evaluating computer-supported cooperative work: Models and frameworks. In Proceedings of the ACM Conference on Computer Supported Cooperative Work, Chicago, IL, USA, 6–10 November 2004; pp. 112–121.
73. Lee, S.W.; Krosnick, R.; Park, S.Y.; Keelean, B.; Vaidya, S.; O'Keefe, S.D.; Lasecki, W.S. Exploring real-time collaboration in crowd-powered systems through a UI design tool. *Proc. ACM Human-Computer Interact.* **2018**, *2*, 1–23. [CrossRef]
74. Wang, X.; Ding, L.; Wang, Q.; Xie, J.; Wang, T.; Tian, X.; Guan, Y.; Wang, X. A picture is worth a thousand words: Share your real-time view on the road. *IEEE Trans. Veh. Technol.* **2016**, *66*, 2902–2914. [CrossRef]
75. Agapie, E.; Teevan, J.; Monroy-Hernández, A. Crowdsourcing in the field: A case study using local crowds for event reporting. In Proceedings of the Third AAAI Conference on Human Computation and Crowdsourcing, San Diego, CA, USA, 8–11 November 2015; pp. 2–11.

76. Lafreniere, B.J.; Grossman, T.; Anderson, F.; Matejka, J.; Kerrick, H.; Nagy, D.; Vasey, L.; Atherton, E.; Beirne, N.; Coelho, M.H.; et al. Crowdsourced fabrication. In Proceedings of the 29th Annual Symposium on User Interface Software and Technology, Tokyo, Japan, 16–19 October 2016; pp. 15–28.
77. Aristeidou, M.; Scanlon, E.; Sharples, M. Profiles of engagement in online communities of citizen science participation. *Comput. Hum. Behav.* **2017**, *74*, 246–256. [CrossRef]
78. Bouwer, A. Under which conditions are humans motivated to delegate tasks to AI? A taxonomy on the human emotional state driving the motivation for AI delegation. In *Marketing and Smart Technologies*; Springer: Singapore, 2022; pp. 37–53.
79. Lubars, B.; Tan, C. Ask not what AI can do, but what AI should do: Towards a framework of task delegability. In Proceedings of the Annual Conference on Neural Information Processing Systems, Vancouver, BC, Canada, 8–14 December 2019; pp. 57–67.
80. Sun, Y.; Ma, X.; Ye, K.; He, L. Investigating crowdworkers' identify, perception and practices in micro-task crowdsourcing. *Proc. ACM Hum.-Comput. Interact.* **2022**, *6*, 1–20. [CrossRef]
81. Khan, V.J.; Papangelis, K.; Lykourentzou, I.; Markopoulos, P. *Macrotask Crowdsourcing—Engaging the Crowds to Address Complex Problems*; Human-Computer Interaction Series; Springer: Cham, Switzerland, 2019.
82. Teevan, J. The future of microwork. *XRDS Crossroads ACM Mag. Stud.* **2016**, *23*, 26–29. [CrossRef]
83. Zulfiqar, M.; Malik, M.N.; Khan, H.H. Microtasking activities in crowdsourced software development: A systematic literature review. *IEEE Access* **2022**, *10*, 24721–24737. [CrossRef]
84. Rahman, H.; Roy, S.B.; Thirumuruganathan, S.; Amer-Yahia, S.; Das, G. Optimized group formation for solving collaborative tasks. *VLDB J.* **2018**, *28*, 1–23. [CrossRef]
85. Schmitz, H.; Lykourentzou, I. Online sequencing of non-decomposable macrotasks in expert crowdsourcing. *ACM Trans. Soc. Comput.* **2018**, *1*, 1–33. [CrossRef]
86. Jin, Y.; Carman, M.; Zhu, Y.; Xiang, Y. A technical survey on statistical modelling and design methods for crowdsourcing quality control. *Artif. Intell.* **2020**, *287*, 103351. [CrossRef]
87. Moayedikia, A.; Ghaderi, H.; Yeoh, W. Optimizing microtask assignment on crowdsourcing platforms using Markov chain Monte Carlo. *Decis. Support Syst.* **2020**, *139*, 113404. [CrossRef]
88. Amershi, S.; Weld, D.; Vorvoreanu, M.; Fourney, A.; Nushi, B.; Collisson, P.; Suh, J.; Iqbal, S.T.; Bennett, P.N.; Inkpen, K.; et al. Guidelines for human-AI interaction. In Proceedings of the CHI Conference on Human Factors in Computing Systems, Glasgow, Scotland, UK, 4–9 May 2019.
89. Rafner, J.; Gajdacz, M.; Kragh, G.; Hjorth, A.; Gander, A.; Palfi, B.; Berditchevskiaia, A.; Grey, F.; Gal, K.; Segal, A.; et al. Mapping citizen science through the lens of human-centered AI. *Hum. Comput.* **2022**, *9*, 66–95. [CrossRef]
90. Shneiderman, B. Bridging the gap between ethics and practice: Guidelines for reliable, safe, and trustworthy human-centered AI systems. *ACM Trans. Interact. Intell. Syst.* **2020**, *10*, 1–31. [CrossRef]
91. Ramírez, J.; Sayin, B.; Baez, M.; Casati, F.; Cernuzzi, L.; Benatallah, B.; Demartini, G. On the state of reporting in crowdsourcing experiments and a checklist to aid current practices. *Proc. ACM Hum.-Comput. Interact.* **2021**, *5*, 1–34. [CrossRef]
92. Robert, L.; Romero, D.M. Crowd size, diversity and performance. In Proceedings of the 33rd Annual ACM Conference on Human Factors in Computing Systems, Seoul, Republic of Korea, 18–23 April 2015; pp. 1379–1382.
93. Blandford, A. Intelligent interaction design: The role of human-computer interaction research in the design of intelligent systems. *Expert Syst.* **2001**, *18*, 3–18. [CrossRef]
94. Huang, K.; Zhou, J.; Chen, S. Being a solo endeavor or team worker in crowdsourcing contests? It is a long-term decision you need to make. *Proc. ACM Hum.-Comput. Interact.* **2022**, *6*, 1–32. [CrossRef]
95. Venkatagiri, S.; Thebault-Spieker, J.; Kohler, R.; Purviance, J.; Mansur, R.S.; Luther, K. GroundTruth: Augmenting expert image geolocation with crowdsourcing and shared representations. *Proc. ACM Hum.-Comput. Interact.* **2019**, *3*, 1–30. [CrossRef]
96. Zhou, S.; Valentine, M.; Bernstein, M.S. In search of the dream team: Temporally constrained multi-armed bandits for identifying effective team structures. In Proceedings of the CHI Conference on Human Factors in Computing Systems, Montreal, QC, Canada, 21–26 April 2018.
97. Gray, M.L.; Suri, S.; Ali, S.S.; Kulkarni, D. The crowd is a collaborative network. In Proceedings of the 19th ACM Conference on Computer-Supported Cooperative Work & Social Computing, San Francisco, CA, USA, 27 February–2 March 2016; pp. 134–147.
98. Zhang, X.; Zhang, W.; Zhao, Y.; Zhu, Q. Imbalanced volunteer engagement in cultural heritage crowdsourcing: A task-related exploration based on causal inference. *Inf. Process. Manag.* **2022**, *59*, 103027. [CrossRef]
99. McNeese, N.J.; Demir, M.; Cooke, N.J.; She, M. Team situation awareness and conflict: A study of human–machine teaming. *J. Cogn. Eng. Decis. Mak.* **2021**, *15*, 83–96. [CrossRef]
100. Dafoe, A.; Bachrach, Y.; Hadfield, G.; Horvitz, E.; Larson, K.; Graepel, T. Cooperative AI: Machines must learn to find common ground. *Nature* **2021**, *593*, 33–36. [CrossRef]
101. Alorwu, A.; Savage, S.; van Berkel, N.; Ustalov, D.; Drutsa, A.; Oppenlaender, J.; Bates, O.; Hettiachchi, D.; Gadiraju, U.; Gonçalves, J.; et al. REGROW: Reimagining global crowdsourcing for better human-AI collaboration. In Proceedings of the CHI Conference on Human Factors in Computing Systems, Extended Abstracts, New Orleans, LA, USA, 29 April–5 May 2022; pp. 1–7.
102. Santos, C.A.; Baldi, A.M.; de Assis Neto, F.R.; Barcellos, M.P. Essential elements, conceptual foundations and workflow design in crowd-powered projects. *J. Inf. Sci.* **2022**. [CrossRef]

103. Valentine, M.A.; Retelny, D.; To, A.; Rahmati, N.; Doshi, T.; Bernstein, M.S. Flash organizations: Crowdsourcing complex work by structuring crowds as organizations. In Proceedings of the CHI Conference on Human Factors in Computing Systems, Denver, CO, USA, 6–11 May 2017; pp. 3523–3537.
104. Kamar, E. Directions in hybrid intelligence: Complementing AI systems with human intelligence. In Proceedings of the Twenty-Fifth International Joint Conference on Artificial Intelligence, New York, NY, USA, 9–15 July 2016; pp. 4070–4073.
105. Tocchetti, A.; Corti, L.; Brambilla, M.; Celino, I. EXP-Crowd: A gamified crowdsourcing framework for explainability. *Front. Artif. Intell.* **2022**, *5*, 826499. [CrossRef] [PubMed]
106. Barbosa, N.M.; Chen, M. Rehumanized crowdsourcing: A labeling framework addressing bias and ethics in machine learning. In Proceedings of the CHI Conference on Human Factors in Computing Systems, Glasgow, Scotland, UK, 4–9 May 2019; pp. 1–12.
107. Basker, T.; Tottler, D.; Sanguet, R.; Muffbur, J. Artificial intelligence and human learning: Improving analytic reasoning via crowdsourcing and structured analytic techniques. *Comput. Educ.* **2022**, *3*, 1003056.
108. Mirbabaie, M.; Brendel, A.B.; Hofeditz, L. Ethics and AI in information systems research. *Commun. Assoc. Inf. Syst.* **2022**, *50*, 38. [CrossRef]
109. Sundar, S.S. Rise of machine agency: A framework for studying the psychology of human–AI interaction (HAII). *J. Comput. Commun.* **2020**, *25*, 74–88. [CrossRef]
110. Liu, B. In AI we trust? Effects of agency locus and transparency on uncertainty reduction in human–AI interaction. *J. Comput. Commun.* **2021**, *26*, 384–402. [CrossRef]
111. Kang, H.; Lou, C. AI agency vs. human agency: Understanding human–AI interactions on TikTok and their implications for user engagement. *J. Comput. Commun.* **2022**, *27*, zmac014. [CrossRef]
112. Daniel, F.; Kucherbaev, P.; Cappiello, C.; Benatallah, B.; Allahbakhsh, M. Quality control in crowdsourcing: A survey of quality attributes, assessment techniques, and assurance actions. *ACM Comput. Surv.* **2018**, *51*, 1–40. [CrossRef]
113. Pedersen, J.; Kocsis, D.; Tripathi, A.; Tarrell, A.; Weerakoon, A.; Tahmasbi, N.; Xiong, J.; Deng, W.; Oh, O.; de Vreede, G.-J. Conceptual foundations of crowdsourcing: A review of IS research. In Proceedings of the 46th Hawaii International Conference on System Sciences, Wailea, HI, USA, 7–10 January 2013; pp. 579–588.
114. Hansson, K.; Ludwig, T. Crowd dynamics: Conflicts, contradictions, and community in crowdsourcing. *Comput. Support. Coop. Work.* **2019**, *28*, 791–794. [CrossRef]
115. Gimpel, H.; Graf-Seyfried, V.; Laubacher, R.; Meindl, O. Towards artificial intelligence augmenting facilitation: AI affordances in macro-task crowdsourcing. *Group Decis. Negot.* **2023**, 1–50. [CrossRef]
116. Wu, T.; Terry, M.; Cai, C.J. AI chains: Transparent and controllable human-AI interaction by chaining large language model prompts. In Proceedings of the CHI Conference on Human Factors in Computing Systems, New Orleans, LA, USA, 29 April–5 May 2022.
117. Kobayashi, M.; Wakabayashi, K.; Morishima, A. Human+AI crowd task assignment considering result quality requirements. In Proceedings of the Ninth AAAI Conference on Human Computation and Crowdsourcing, Virtual, 14–18 November 2021; pp. 97–107.
118. Eggert, M.; Alberts, J. Frontiers of business intelligence and analytics 3.0: A taxonomy-based literature review and research agenda. *Bus. Res.* **2020**, *13*, 685–739. [CrossRef]
119. Chan, J.; Chang, J.C.; Hope, T.; Shahaf, D.; Kittur, A. SOLVENT: A mixed initiative system for finding analogies between research papers. *Proc. ACM Hum.-Comput. Interact.* **2018**, *2*, 1–21. [CrossRef]
120. Zhang, Y.; Shang, L.; Zong, R.; Wang, Z.; Kou, Z.; Wang, D. StreamCollab: A streaming crowd-AI collaborative system to smart urban infrastructure monitoring in social sensing. In Proceedings of the Ninth AAAI Conference on Human Computation and Crowdsourcing, Virtual, 14–18 November 2021; pp. 179–190.
121. Yang, J.; Smirnova, A.; Yang, D.; Demartini, G.; Lu, Y.; Cudré-Mauroux, P. Scalpel-CD: Leveraging crowdsourcing and deep probabilistic modeling for debugging noisy training data. In Proceedings of the World Wide Web Conference, San Francisco, CA, USA, 13–17 May 2019; pp. 2158–2168.
122. Schlagwein, D.; Cecez-Kecmanovic, D.; Hanckel, B. Ethical norms and issues in crowdsourcing practices: A Habermasian analysis. *Inf. Syst. J.* **2018**, *29*, 811–837. [CrossRef]
123. Gadiraju, U.; Demartini, G.; Kawase, R.; Dietze, S. Crowd anatomy beyond the good and bad: Behavioral traces for crowd worker modeling and pre-selection. *Comput. Support. Cooperative Work.* **2018**, *28*, 815–841. [CrossRef]
124. Palmer, M.S.; Huebner, S.E.; Willi, M.; Fortson, L.; Packer, C. Citizen science, computing, and conservation: How can "crowd AI" change the way we tackle large-scale ecological challenges? *Hum. Comput.* **2021**, *8*, 54–75. [CrossRef]
125. Mannes, A. Governance, risk, and artificial intelligence. *AI Mag.* **2020**, *41*, 61–69. [CrossRef]
126. Choung, H.; David, P.; Ross, A. Trust and ethics in AI. *AI Soc.* **2022**, 1–13. [CrossRef]
127. Zheng, Q.; Tang, Y.; Liu, Y.; Liu, W.; Huang, Y. UX research on conversational human-AI interaction: A literature review of the ACM Digital Library. In Proceedings of the CHI Conference on Human Factors in Computing Systems, New Orleans, LA, USA, 29 April–5 May 2022.
128. Heath, C.; Svensson, M.S.; Hindmarsh, J.; Luff, P.; Vom Lehn, D. Configuring awareness. *Comput. Support. Coop. Work.* **2002**, *11*, 317–347. [CrossRef]

129. Park, J.; Krishna, R.; Khadpe, P.; Fei-Fei, L.; Bernstein, M. AI-based request augmentation to increase crowdsourcing participation. In Proceedings of the Seventh AAAI Conference on Human Computation and Crowdsourcing, Stevenson, WA, USA, 28–30 October 2019; pp. 115–124.
130. Star, S.L.; Ruhleder, K. Steps towards an ecology of infrastructure: Complex problems in design and access for large-scale collaborative systems. In Proceedings of the ACM Conference on Computer Supported Cooperative Work, Chapel Hill, NC, USA, 22–26 October 1994; pp. 253–264.
131. Mosconi, G.; Korn, M.; Reuter, C.; Tolmie, P.; Teli, M.; Pipek, V. From Facebook to the neighbourhood: Infrastructuring of hybrid community engagement. *Comput. Support. Coop. Work (CSCW)* **2017**, *26*, 959–1003. [CrossRef]
132. Ehsan, U.; Liao, Q.V.; Muller, M.; Riedl, M.O.; Weisz, J.D. Expanding explainability: Towards social transparency in AI systems. In Proceedings of the CHI Conference on Human Factors in Computing Systems, Yokohama, Japan, 8–13 May 2021; pp. 1–19.
133. Thieme, A.; Cutrell, E.; Morrison, C.; Taylor, A.; Sellen, A. Interpretability as a dynamic of human-AI interaction. *Interactions* **2020**, *27*, 40–45. [CrossRef]
134. Walzner, D.D.; Fuegener, A.; Gupta, A. Managing AI advice in crowd decision-making. In Proceedings of the International Conference on Information Systems, Copenhagen, Denmark, 9–14 December 2022; p. 1315.
135. Anjum, S.; Verma, A.; Dang, B.; Gurari, D. Exploring the use of deep learning with crowdsourcing to annotate images. *Hum. Comput.* **2021**, *8*, 76–106. [CrossRef]
136. Trouille, L.; Lintott, C.J.; Fortson, L.F. Citizen science frontiers: Efficiency, engagement, and serendipitous discovery with human-machine systems. *Proc. Natl. Acad. Sci. USA* **2019**, *116*, 1902–1909. [CrossRef] [PubMed]
137. Zhou, Z.; Yatani, K. Gesture-aware interactive machine teaching with in-situ object annotations. In Proceedings of the 35th Annual ACM Symposium on User Interface Software and Technology, Bend, OR, USA, 29 October–2 November 2022; pp. 1–14.
138. Avdic, M.; Bødker, S.; Larsen-Ledet, I. Two cases for traces: A theoretical framing of mediated joint activity. *Proc. ACM Hum.-Comput. Interact.* **2021**, *5*, 1–28. [CrossRef]
139. Tchernavskij, P.; Bødker, S. Entangled artifacts: The meeting between a volunteer-run citizen science project and a biodiversity data platform. In Proceedings of the Nordic Human-Computer Interaction Conference, Aarhus, Denmark, 8–12 October 2022; pp. 1–13.
140. Rzeszotarski, J.M.; Kittur, A. Instrumenting the crowd: Using implicit behavioral measures to predict task performance. In Proceedings of the 24th Annual ACM Symposium on User Interface Software and Technology, Santa Barbara, CA, USA, 16–19 October 2011; pp. 13–22.
141. Newman, A.; McNamara, B.; Fosco, C.; Zhang, Y.B.; Sukhum, P.; Tancik, M.; Kim, N.W.; Bylinskii, Z. TurkEyes: A web-based toolbox for crowdsourcing attention data. In Proceedings of the CHI Conference on Human Factors in Computing Systems, Honolulu, HI, USA, 25–30 April 2020; pp. 1–13.
142. Goyal, T.; McDonnell, T.; Kutlu, M.; Elsayed, T.; Lease, M. Your behavior signals your reliability: Modeling crowd behavioral traces to ensure quality relevance annotations. In Proceedings of the Sixth AAAI Conference on Human Computation and Crowdsourcing, Zürich, Switzerland, 5–8 July 2018; pp. 41–49.
143. Hettiachchi, D.; Van Berkel, N.; Kostakos, V.; Goncalves, J. CrowdCog: A cognitive skill based system for heterogeneous task assignment and recommendation in crowdsourcing. *Proc. ACM Hum.-Comput. Interact.* **2020**, *4*, 1–22. [CrossRef]
144. Zimmerman, J.; Oh, C.; Yildirim, N.; Kass, A.; Tung, T.; Forlizzi, J. UX designers pushing AI in the enterprise: A case for adaptive UIs. *Interactions* **2020**, *28*, 72–77. [CrossRef]
145. Hettiachchi, D.; Kostakos, V.; Goncalves, J. A survey on task assignment in crowdsourcing. *ACM Comput. Surv.* **2022**, *55*, 1–35. [CrossRef]
146. Pei, W.; Yang, Z.; Chen, M.; Yue, C. Quality control in crowdsourcing based on fine-grained behavioral features. *Proc. ACM Hum.-Comput. Interact.* **2021**, *5*, 1–28. [CrossRef]
147. Bakici, T. Comparison of crowdsourcing platforms from social-psychological and motivational perspectives. *Int. J. Inf. Manag.* **2020**, *54*, 102121. [CrossRef]
148. Truong, N.V.-Q.; Dinh, L.C.; Stein, S.; Tran-Thanh, L.; Jennings, N.R. Efficient and adaptive incentive selection for crowdsourcing contests. *Appl. Intell.* **2022**, 1–31. [CrossRef]
149. Correia, A.; Jameel, S.; Paredes, H.; Fonseca, B.; Schneider, D. Hybrid machine-crowd interaction for handling complexity: Steps toward a scaffolding design framework. In *Macrotask Crowdsourcing—Engaging the Crowds to Address Complex Problems*; Human-Computer Interaction Series; Springer: Cham, Switzerland, 2019; pp. 149–161.
150. Sutherland, W.; Jarrahi, M.H.; Dunn, M.; Nelson, S.B. Work precarity and gig literacies in online freelancing. *Work Employ. Soc.* **2019**, *34*, 457–475. [CrossRef]
151. Salminen, J.; Kamel, A.M.S.; Jung, S.-G.; Mustak, M.; Jansen, B.J. Fair compensation of crowdsourcing work: The problem of flat rates. *Behav. Inf. Technol.* **2022**, 1–22. [CrossRef]

MDPI
St. Alban-Anlage 66
4052 Basel
Switzerland
Tel. +41 61 683 77 34
Fax +41 61 302 89 18
www.mdpi.com

Applied Sciences Editorial Office
E-mail: applsci@mdpi.com
www.mdpi.com/journal/applsci

www.ingramcontent.com/pod-product-compliance
Lightning Source LLC
Chambersburg PA
CBHW061509220926
51850CB00004B/208
9783036571065